# Psychoneuroendocrine Dysfunction

# Psychoneuroendocrine Dysfunction

Edited by
### NANDKUMAR S. SHAH, Ph.D.[†]
*Director, Ensor Foundation Research Laboratory*
*William S. Hall Psychiatric Institute*
*Research Professor, Department of Neuropsychiatry and Behavioral Science*
*University of South Carolina School of Medicine*
*Columbia, South Carolina*

and
### ALEXANDER G. DONALD, M.D.
*Director, William S. Hall Psychiatric Institute*
*Professor and Chairman, Department of Neuropsychiatry and Behavioral Science*
*University of South Carolina School of Medicine*
*Columbia, South Carolina*

PLENUM MEDICAL BOOK COMPANY
New York and London

Library of Congress Cataloging in Publication Data

Main entry under title:

Psychoneuroendocrine dysfunction.

Includes bibliographical references and index.
1. Mental illness—Physiological aspects—Addresses, essays, lectures. 2. Neuroendocrinology—Addresses, essays, lectures. 3. Neuropsychiatry—Addresses, essays, lectures. I. Shah, Nandkumar S. II. Donald, Alexander G., 1928-     . [DNLM: 1. Mental disorders—Physiopathology. 2. Neuroregulators—Physiology. 3. Psychopharmacology. 4. Mental disorders—Chemically induced. 5. Hypothalamic hormones—Physiology. 6. Pituitary hormones—Physiology. WM 100 P98925]
RC455.4.B5P78   1983                     616.89'07                          83-19202
ISBN-13:978-1-4684-4531-2         e-ISBN-13:978-1-4684-4529-9
DOI:10.1007/978-1-4684-4529-9

© 1984 Plenum Publishing Corporation
Softcover reprint of the hardcover 1st edition 1984

233 Spring Street, New York, N.Y. 10013

Plenum Medical Book Company is an imprint of Plenum Publishing Corporation

All rights reserved

No part of this book may be reproduced, stored in a retrieval system, or transmitted in any form or by any means, electronic, mechanical, photocopying, microfilming, recording, or otherwise, without written permission from the Publisher

# In Memoriam

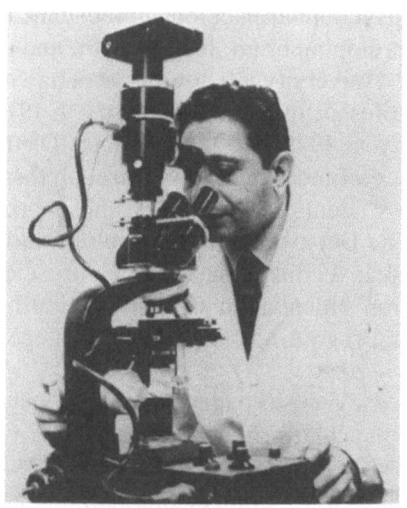

**Nandkumar S. Shah**
**(1928–1983)**

Nandkumar S. Shah died on May 23, 1983, at the age of 55. Dr. Shah was Chief of Research Services and Director of the Ensor Research Laboratory at the William S. Hall Psychiatric Institute and Research Professor, Department of Neuropsychiatry and Behavioral Science, at the University of South Carolina School of Medicine. He had completed the majority of the work involved in the publication of this volume at the time of his death.

Dr. Shah, a son of the late Shankarbhai and Parvati Shah, was born in Nandurbar, India. He received his B.S. and M.S. in biochemistry from Poona University, India. He completed his Ph.D. in pharmacology at the University of Florida, Gainesville, Florida, in 1965.

Dr. Shah was the epitome of a scholar and an excellent teacher. In addition, he was a superb researcher. He made significant contributions to the medical community of Columbia, South Carolina, through his research and teaching.

Through his many publications and presentations at national and international symposia, his work became known worldwide.

Dr. Shah's professional career began in 1955 at Poona University and at the Gandhi Memorial Hospital, where he conducted studies on lathyrism. He was later selected by the Atomic Energy Commission to continue his research in clinical biochemistry. He came to the United States in 1961 and studied at the College of Medicine at the University of Florida. On completion of his studies, he joined the staff at the Thudichum Psychiatric Research Laboratory in Galesburg, Illinois, working with the late Dr. Harold Himwich. He subsequently became Chief of the Psychopharmacology Radioisotope Laboratory, conducting studies on biogenic amines and hallucinogens: uptake, distribution, metabolism, and excretions under various conditions.

From 1970 to the time of his death, Dr. Shah directed research programs including projects in psychopharmacology, biochemical mechanisms, neurochemical aspects of developing brain, fetal growth, and drug metabolism at the Hall Institute. At the University of South Carolina School of Medicine he became a Research Professor in Neuropsychiatry in 1975 and an Adjunct Professor in Pharmacology in 1978. He was Adjunct Distinguished Professor in the College of Pharmacy of the University. Through formal lectures, he taught residents and medical students in psychiatry and pharmacology. He directed graduate programs in the Department of Psychology and in the School of Pharmacy. Dr. Shah served as a consultant to the Neuroscience Laboratory at the Veterans Administration Hospital in Columbia, South Carolina, and to the Biomedical Research Program at Voorhees College in Denmark, South Carolina.

Dr. Shah published over 140 papers related to psychopharmacology and the biological basis of psychiatry in national and international journals. In addition, he co-edited two major books entitled *Endorphins and Opiate Antagonists in Psychiatric Research: Clinical Implications* and *GABA Neurotransmission: Current Developments in Physiology and Neurochemistry*.

For the last 13 years, Dr. Shah conducted an annual research symposium of international significance on topics of pharmacological interest, under the auspices of the William S. Hall Psychiatric Institute. These symposia, over the years, have brought scientists of international stature to Columbia, providing the local scientific, medical, and psychiatric communities with a unique opportunity to personally exchange ideas with these outstanding men.

Dr. Shah earned several awards from national and international scientific and pharmaceutical associations: (1) from the American Society for Pharmacology and Experimental Therapeutics (1975–1976); (2) from the National Institute for General Medical Sciences (1980); (3) from the Deutscher Akademischer Austauschdienst (German Academic Exchange Service) to work at Max Planck Institute for Psychiatry, Munich, West Germany, in psychoneuroendocrinology (1983); and (4) the Director's Award, from the William S. Hall Psychiatric Institute. He was listed in The Marquis *Who's Who, Who's Who in the Biomedical Sciences,* and the *International Directory of Investigators in*

*Psychopharmacology* published by the World Health Organization and the National Institute of Mental Health. He was also a member of the Ad Hoc Committee of the NASA Biochemical Research Program in space and an editor/advisor for several professional journals.

Dr. Shah was strongly attached to his family. He had affection for his brothers, nephews, and nieces, helping to guide and shape their careers and financing their education when needed. Neeta, his wife of 22 years, is a remarkable person who shared his experiences, successes, and disappointments. His daughter shares her father's love of science and is pursuing a career in medicine.

Dr. Shah was an excellent teacher, an internationally known researcher, a humane colleague, and a gentleman who is missed by all. In June, 1983, when he was selected as the first recipient (posthumously) of the Director's Award of the William S. Hall Psychiatric Institute for the most significant contribution to its mission and purpose during the preceding year, Dr. Shah was described as "always interested in and willing to help students . . . the epitome of the scholar and the teacher—an extremely diligent and hard-working faculty member who set a high standard of excellence for students in all disciplines."

This volume is respectfully dedicated to the memory of Dr. Nandkumar S. Shah in appreciation of his accomplishments and his compassion for his fellow man.

Alexander G. Donald

# Contributors

ALESSANDRO AGNOLI • I Clinica Neurologica, University of Rome, Rome, Italy

SYED I. ALI • Department of Research, Illinois State Psychiatric Institute, Chicago, Illinois

ARNOLD E. ANDERSEN • Department of Psychiatry, Johns Hopkins University Medical Center, Baltimore, Maryland

MASSIMO BALDASSARRE • Clinica Neurologica, University of L'Aquila, L'Aquila, Italy

TOMMASO BARRECA • Istituto Scientifico di Medicina Interna, Cattedra di Patologia Medica, University of Genoa, Genoa, Italy

MATHIAS BERGER • Max-Planck-Institut für Psychiatrie, Munich, FRG

LLOYD A. BING • Adult Psychiatry Branch, Division of Special Mental Health Research, Intramural Research Program, National Institute of Mental Health, Saint Elizabeths Hospital, Washington, D.C.

FRANCESCA BRAMBILLA • Ospedale Psichiatrico Paolo Pini, Milano Affori, Italy

GREGORY M. BROWN • Department of Neurosciences, McMaster University, Hamilton, Ontario, Canada

WALTER ARMIN BROWN • Neuroendocrine Research Program, Providence Veterans Administration Medical Center, and Department of Psychiatry, Brown University, Providence, Rhode Island

SHIRLEY L. BUCHANAN • Neuroscience Laboratory, Wm. Jennings Bryan Dorn Veterans' Hospital, and University of South Carolina, Columbia, South Carolina

EARL A. BURCH • William S. Hall Psychiatric Institute, and Department of Neuropsychiatry and Behavioral Sciences, University of South Carolina School of Medicine, Columbia, South Carolina

REGINA CASPER • Department of Research, Illinois State Psychiatric Institute, Chicago, Illinois, and Department of Psychiatry, University of Illinois College of Medicine, Chicago, Illnois

DANIELA COCCHI • Department of Pharmacology, School of Medicine, University of Milan, Milan, Italy

DAVID H. COY • Tulane University School of Medicine, New Orleans, Louisiana

PIER GIORGIO CROSIGNANI • II Department of Medicine, Fatebenefretelli Hospital, and IV Department of Obstetrics and Gynecology, University of Milan, Milan, Italy

JOHN M. DAVIS • Department of Research, Illinois State Psychiatric Institute, Chicago, Illinois, and Department of Psychiatry, University of Illinois College of Medicine, Chicago, Illinois

GIACOMO D'ELIA • Department of Psychiatry, University of Linköping, Linköping, Sweden

ALESSANDRO DENARO • Department of Neurology, University of Rome, Rome, Italy

PETER DOERR • Max-Planck-Institut für Psychiatrie, Munich, FRG

ALEXANDER G. DONALD • William S. Hall Psychiatric Institute and Department of Neuropsychiatry and Behavioral Sciences, University of South Carolina School of Medicine, Columbia, South Carolina

RAYMON DURSO • Department of Neurology, Boston University School of Medicine, Boston, Massachusetts

ROSARIA D'URSO • V Clinica Medica, University of Rome, Rome, Italy

H. M. EMRICH • Max-Planck-Institut für Psychiatrie, Munich, FRG

IRL EXTEIN • Falkirk Hospital, Central Valley, New York

FABIO FACCHINETTI • Cattedra di Patologia Ostetrica-Ginecologica, Università di Cagliari, Cagliari, Italy

PAOLO FALASCHI • V Clinica Medica, University of Rome, Rome, Italy

CARLO FERRARI • II Department of Medicine, Fatebenefretelli Hospital, and IV Department of Obstetrics and Gynecology, University of Milan, Milan, Italy

SVEND-OTTO FREDERIKSEN • Department of Psychiatry, University of Linköping, Linköping, Sweden

WILLIAM J. FREED • Adult Psychiatry Branch, Division of Special Mental Health Research, Intramural Research Program, National Institute of Mental Health, Saint Elizabeths Hospital, Washington, D.C.

ANDREA GENAZZANI • Cattedra di Patologia Ostetrica-Ginecologica, Università di Cagliari, Cagliari, Italy

ROBERT H. GERNER • University of California, Irvine, California, and Long Beach Veterans Administration Hospital, Long Beach, California

G. L. GESSA • Institute of Pharmacology, University of Cagliari, Cagliari, Italy

# CONTRIBUTORS

MARK S. GOLD • Research Facilities, and Psychiatric Diagnostic Laboratories of America, Fair Oaks Hospital, Summit, New Jersey, and Department of Psychiatry, Yale University, New Haven, Connecticut

EVA GROF • Department of Psychiatry, McMaster University, Hamilton, Ontario, Canada

PAUL GROF • Department of Psychiatry, McMaster University, and Hamilton Psychiatric Hospital, Hamilton, Ontario, Canada

HARRY E. GWIRTSMAN • National Institutes of Health, Bethesda, Maryland

NOBORU HATOTANI • Department of Psychiatry, Mie University School of Medicine, Tsu, Mie, Japan

ABBA J. KASTIN • Veterans Administration Medical Center and Tulane University School of Medicine, New Orleans, Louisiana

STANLEY R. KAY • Bronx Psychiatric Center and Albert Einstein College of Medicine, New York, New York

ISAO KITAYAMA • Department of Psychiatry, Mie University School of Medicine, Tsu, Mie, Japan

IRWIN J. KOPIN • Laboratory of Clinical Science, National Institute of Mental Health, Bethesda, Maryland

N. M. KURTZ • Clinical Investigation Unit, Sepulveda Veterans Administration Hospital, and Department of Psychiatry, University of California at Los Angeles, Sepulveda, California

S. LAL • Douglas Hospital Research Centre, Department of Psychiatry, Montreal General Hospital, and Department of Psychiatry, McGill University, Montreal, Quebec, Canada

IOANA LANCRANJAN • Department of Clinical Research, Sandoz Ltd., Basel, Switzerland

J. J. LEGROS • Neuroendocrinology Section, CHU University of Liège, Liège, Belgium

VITTORIO LOCATELLI • Department of Pharmacology, School of Medicine, University of Milan, Milan, Italy

PETER T. LOOSEN • Clinical Research Unit, Dorothea Dix Hospital, Raleigh, North Carolina

JOSEPH B. MARTIN • Department of Neurology, Massachusetts General Hospital, Harvard Medical School, Boston, Massachusetts

NICOLA MARTUCCI • Clinica Neurologica, University of L'Aquila, L'Aquila, Italy

ANNA MARIA MATTEI • II Department of Medicine, Fatebenefretelli Hospital, and IV Department of Obstetrics and Gynecology, University of Milan, Milan, Italy

JULIEN MENDLEWICZ • Department of Psychiatry, University Clinics of Brussels, Erasme Hospital, Free University of Brussels, Brussels, Belgium

DONALD W. MORGAN • William S. Hall Psychiatric Institute and Department of Neuropsychiatry and Behavioral Sciences, University of South Carolina School of Medicine, Columbia, South Carolina

EUGENIO E. MÜLLER • Department of Pharmacology, School of Medicine, University of Milan, Milan, Italy

FRANZ MÜLLER • Psychiatric Hospital of the University of Munich, Munich, FRG

ETHAN V. MUNSON • Department of Psychiatry, University of California at San Diego, La Jolla, California

GIOVANNI MURIALDO • Semeiotica Medica, I.S.M.I., University of Genoa, Genoa, Italy

DIETER NABER • Psychiatric Hospital of the University of Munich, Munich, FRG

N. P. V. NAIR • Douglas Hospital Research Centre, and Department of Psychiatry, McGill University, Montreal, Quebec, Canada

NORBERT NEDOPIL • Psychiatric Hospital of the University of Munich, Munich, FRG

JUNICHI NOMURA • Department of Psychiatry, Mie University School of Medicine, Tsu, Mie, Japan

GAYLE A. OLSON • Department of Psychology, University of New Orleans, New Orleans, Louisiana

RICHARD D. OLSON • Department of Psychology, University of New Orleans, New Orleans, Louisiana

GHANSHYAM N. PANDEY • Department of Research, Illinois State Psychiatric Institute, Chicago, Illinois, and Department of Psychiatry, University of Illinois College of Medicine, Chicago, Illinois

A. CARTER POTTASH • Research Facilities, and Psychiatric Diagnostic Laboratories of America, Fair Oaks Hospital, Summit, New Jersey, and Department of Psychiatry, Yale University, New Haven, Connecticut

D. A. POWELL • Neuroscience Laboratory, Wm. Jennings Bryan Dorn Veterans' Hospital, and University of South Carolina, Columbia, South Carolina

ARTHUR J. PRANGE, JR. • Department of Psychiatry, Biological Sciences Research Center, University of North Carolina at Chapel Hill, School of Medicine, Chapel Hill, North Carolina

JEFFREY L. RAUSCH • Ensor Foundation Research Laboratory, William S. Hall Psychiatric Institute, Columbia, South Carolina, and Department of Psychiatry, University of California at San Diego, La Jolla, California

S. RICHARDSON • Department of Medicine, McGill University, Royal Victoria Hospital, Montreal, Quebec, Canada

ALAN D. ROGOL • University of Virginia Medical Center, Charlottesville, Virginia

ERMANNO ROLANDI • Istituto Scientifico di Medicina Interna, Cattedra di Patologia Medica, University of Genoa, Genoa, Italy

STEFANO A. RUGGIERI • Department of Neurology, University of Rome, Rome, Italy

ECKART RÜTHER • Psychiatric Hospital of the University of Munich, Munich, FRG

S. CHARLES SCHULZ • Western Psychiatric Institute and Clinic, University of Pittsburgh, Pittsburgh, Pennsylvania

G. SERRA • Institute of Pharmacology, University of Cagliari, Cagliari, Italy

ARUNKUMAR B. SHAH • Ensor Foundation Research Laboratory, William S. Hall Psychiatric Institute, and Department of Neuropsychiatry and Behavioral Sciences, University of South Carolina School of Medicine, Columbia, South Carolina. *Present address:* Comprehensive Stroke Center, Department of Neurology, School of Medicine, The Oregon Health Sciences University, Portland, Oregon

NANDKUMAR S. SHAH • Ensor Foundation Research Laboratory, William S. Hall Psychiatric Institute, and Department of Neuropsychiatry and Behavioral Sciences, University of South Carolina School of Medicine, Columbia, South Carolina

MAN MOHAN SINGH • Creedmoor Psychiatric Center, Queens Village, New York

HARVEY A. STERNBACH • Department of Psychiatry, University of California at Los Angeles, Neuropsychiatric Institute, Los Angeles, California, and Department of Psychiatry, Brentwood Veterans Administration Hospital, Los Angeles, California

FABRIZIO STOCCHI • Clinica Neurologica, Universtity of L'Aquila, L'Aquila, Italy

CAROL A. TAMMINGA • Department of Psychiatry, University of Maryland, Baltimore, Maryland

G. TOLIS • Departments of Medicine, Obstetrics and Gynecology, McGill University, Royal Victoria Hospital, Montreal, Quebec, Canada

DANIEL P. VAN KAMMEN • Western Psychiatric Institute and Clinic, University of Pittsburgh, Pittsburgh, Pennsylvania

DETLEV VON ZERSSEN • Max-Planck-Institut für Psychiatrie, Munich, FRG

JOHANNES WEISS-BRUMMER • Psychiatric Hospital of the University of Munich, Munich, FRG

RICHARD JED WYATT • Adult Psychiatry Branch, Division of Special Mental Health Research, Intramural Research Program, National Institute of Mental Health, Saint Elizabeths Hospital, Washington, D.C.

GEORGE G. YARBROUGH • Merck Institute for Therapeutic Research, West Point, Pennsylvania

CLAUDIO ZAULI • Semeiotica Medica, I.S.M.I., University of Genoa, Genoa, Italy

# Foreword

There is no area in medicine that has affected biological psychiatry more profoundly than the developments that have occurred in the last 15 years in endocrinology and more specifically in neuroendocrinology. In the 1960s, the regulation of endocrine function was considered to rest primarily in the feedback system between the pituitary and the secretions of various target organs. In R. H. Williams' Fourth Edition of the *Textbook of Endocrinology* published in 1968, the chapter on neuroendocrinology did refer to the median eminence gland with a relatively brief mention of various releasing factors that were the subject of ongoing studies. Only six years later, in the Fifth Edition published in 1974, Seymour Reichlin's chapter on neuroendocrinology listed nine specific hypothalamic releasing factors of which three had already been isolated and purified and thus were referred to as hormones. Most recently in the current Sixth Edition, published in 1981, the chapter on neuroendocrinology contains a detailed description of the physiology of the various hypothalamic releasing factors and hormones, but also significant emphasis is given to the various neurotransmitters that have been shown to regulate the synthesis and release of these important hypothalamic hormones. In addition, there appeared for the first time in this classic textbook a chapter on psychoendocrinology.

One may wonder why there is so much interest not only in endocrinology but more recently in psychology and psychiatry about psychoneuroendocrine function. Several reasons may be suggested. It has been known for some time that peripheral hormones strongly affect brain function. Thyroid hormones are necessary for normal brain development; sex hormones such as testosterone, estrogen, and progesterone significantly influence both sexual and aggressisve behavior in animals and in man. The catecholamines—epinephrine and norepinephrine—also significantly impact upon behavior as seen in the alarm response of Cannon. But something new over this last decade and a half has been added.

The reader will recall the tremendous contribution made by Ungerstedt, Fuxe, Hokfelt, and others in demonstrating new neurotransmitter pathways in brain. It was as recently as the late 1960s that the existence of these norepinephrine, serotonin, and dopamine pathways in the brain was established. In a distinct, but later, related field, Schally and Guillemin isolated the first hypothalamic hormones and clarified their peptide structure, for which they received the Nobel Prize. It soon became clear that the hypothalamic hormones were in great part under the regulation of the recently discovered neurotransmitters, i.e., norepinephrine, serotonin, and dopamine. The brain was thus clearly established as the master endocrine organ, limbic brain → hypothalamus → pituitary.

The realization that the brain was significantly involved in endocrine regulation was an extremely important and exciting conclusion. Indeed, much of the work in stress physiology already suggested that the brain had significant influences on endocrine function. But other developments in psychiatry were to make the link with endocrinology even closer. In the last ten years, an enormous amount of research has provided very impressive evidence that these same neurotransmitter systems discovered earlier to regulate the secretion of the hypothalamic hypophyseal hormones (releasing factors/hormones) are also potentially implicated in depressive illness, schizophrenia, anorexia nervosa, and perhaps other major psychiatric syndromes. The dopamine hypothesis of schizophrenia remains alive and well today, whether it involves an excess of dopamine or an alteration in dopamine receptors. Possible deficiencies in norepinephrine and serotonin in depressive illness are also being actively investigated. It became apparent to many investigators that there might be parallels between the neurotransmitter regulation of hypothalamic hormones and these same transmitters having some important role in the development or expression of major psychiatric illness. Such a potential relationship was captured in Ed Sachar's concept that hypothalamic-pituitary-peripheral endocrime function could serve as a "window into the brain" for the study of psychiatric illness. As pointed out in this volume by Brown *et al.* (Chapter 18, "Psychoendocrinology of Depression"), the endocrine abnormalities that are seen in this illness suggest two important strategies. One is that these may become clinical tests for the diagnosis and course of the illness; the second is that they promise to provide very important research tools for defining the underlying abnormalities in the brain in depressive illness.

In the past few years, the interaction between psychiatry and endocrinology has grown even more complex as a result of the discovery of many new hormones. These fall primarily into two classes with many members in each class. There are the endogenous opiates, endorphins and enkephalins, and the gut hormones, such as cholecystokinin, somatostatin, substance p, and so forth, with both groups often referred to as neuropeptides. It is clear that many of these hormones not only occur in relatively large concentrations in the gut but also in the brain. They appear to function as neurotransmitters or neuromodulators. However, as most of these hormones recently discovered to exist in the brain are still difficult to measure with significant sensitivity and specificity, their role in the brain is for the most part still unknown. However, it has already been established that some are important regulators of peripheral endocrine function. For example, as outlined in several chapters of this volume, somatostatin is present both in the GI tract and in the pancreas as well as functioning as a hypothalamic releasing factor which inhibits growth hormone secretion. Somatostatin's function in other parts of the brain, where it is also found in substantial quantities, is still unclear.

There are several strategies that the reader will find in reviewing the many chapters of this most interesting volume. It presents work at the forefront. Most all the material presented here is truly new in that the research has been conducted only in the last few years. Several chapters report on whether or not there are altered levels of these newer hormones in psychiatric illness compared to controls; others provide very new information on the interaction between established transmitters and their putative role in regulating various hypothalamic hormones that are also distributed throughout the brain. Two such examples are the chapter on the cholinergic control of TRH and the chapter reviewing norepinephrine control over various neuropeptides. Perhaps the most extensive and richest contribution of this volume is provided by a number of reports utilizing various pharmacological probes to study potentially altered endocrine responsivity in various psychiatric illnesses.

Much research in psychiatry is currently focused on potential abnormalities in receptor physiology in depressive illness and schizophrenia. Much of this research utilizes specific pharmacological agonists and antagonists of these receptors and this research has been most useful in elucidating the potential mechanism of various psychoactive drugs. These same agonists and antagonists, of course, can be used to assess receptor physiology by measuring changes in endocrine secretion, in addition to altering behavior. Thus apomorphine, a dopamine agonist, not only intensifies schizophrenic symptomatology but also stimulates growth hormone secretion as dopamine receptors stimulate growth hormone-releasing factor. Studying changes in hormones as a method for elucidating receptor physiology parallels the assessment of these pharmacological probes of behavior and often with clearer results.

In summary, this is a volume depicting the state of the art. Most of the research reported is too new to yield definitive conclusions and consensus. Many of these studies utilize radioimmunoassay or radioreceptor assays which are being pushed to the limit of their sensitivity. Of course, at the same time, we are all aware of the heterogeneous nature of various psychiatric syndromes. Thus, it is too early to expect there will be homogeneity of endocrine findings among our patients. Perhaps the endocrine abnormalities may define the clinical subtypes. Psychoneuroendocrinology is a brash newcomer, raising more questions to date than it has provided answers. Research in this area not only promises to be some of the most exciting in psychiatry but also, it is hoped, will provide some insight into mechanism and possibly even future treatment specificity.

Robert M. Rose

*Department of Psychiatry and Behavioral Sciences*
*The University of Texas Medical Branch*
*Galveston, Texas*

# Contents

FOREWORD .................................................................. xv
    Robert M. Rose

INTRODUCTION. PSYCHONEUROPHARMACOLOGY AND NEUROENDOCRINOLOGY: IMPLICATIONS IN NEUROPSYCHIATRIC RESEARCH............................... 1
    Arunkumar B. Shah, Nandkumar S. Shah, Donald W. Morgan, and Alexander G. Donald

1. NEUROENDOCRINOLOGY AND BRAIN PEPTIDES: AN EMERGING NEW FRONTIER IN NEUROBIOLOGY............... 15
    Joseph B. Martin

2. NEUROPHARMACOLOGICAL INFLUENCES ON HYPOTHALAMIC–PITUITARY SECRETION....................... 41
    Ermanno Rolandi and Tommaso Barreca

3. TRH INTERACTIONS WITH CHOLINERGIC MECHANISMS AND CONSEQUENT THERAPEUTIC IMPLICATIONS............ 73
    George G. Yarbrough

4. CALCITONIN AS AN ANORECTIC AGENT ........................ 83
    William J. Freed, Lloyd A. Bing, Arnold E. Andersen, and Richard Jed Wyatt

5. GONADAL DYSFUNCTION IN ANOREXIA NERVOSA: A MINIREVIEW ...................................................... 111
    S. Richardson and G. Tolis

6. ABNORMALITIES IN ANOREXIA NERVOSA OF DEXAMETHASONE SUPPRESSION TEST AND URINARY MHPG SUGGEST A NOREPINEPHRINE HYPOTHESIS ......... 129

   Harry E. Gwirtsman, Robert H. Gerner, and Harvey Sternbach

7. ROLE OF BRAIN MONOAMINES AND PEPTIDES IN THE REGULATION OF MALE SEXUAL BEHAVIOR .................... 141

   G. Serra and G. L. Gessa

8. PLASMA CATECHOLAMINES AS AN INDEX OF A NEUROENDOCRINE RESPONSE ................................. 157

   Irwin J. Kopin

9. NONSPECIFIC PITUITARY RESPONSES TO HYPOTHALAMIC HORMONES IN BASIC AND CLINICAL RESEARCH ............. 173

   Daniela Cocchi, Vittorio Locatelli, and Eugenio E. Müller

10. NEUROENDOCRINE STUDIES IN HUNTINGTON'S DISEASE .. 209

    Raymon Durso, Stefano A. Ruggieri, Alessandro Denaro, and Carol A. Tamminga

11. $ACTH_{4-10}$: EFFECTS ON PAVLOVIAN CONDITIONING .......... 231

    D. A. Powell and Shirley L. Buchanan

12. $ACTH_{4-10}$ AND MEMORY IN PSYCHIATRIC PATIENTS .......... 243

    Giacomo d'Elia and Svend-Otto Frederiksen

13. VASOPRESSIN IN NEUROPSYCHIATRIC DISORDERS ......... 255

    J. J. Legros and Ioana Lancranjan

14. ENDOGENOUS OPIATE SYSTEMS MAY MODULATE LEARNING AND MEMORY ........................................ 279

    Gayle A. Olson, Richard D. Olson, Abba J. Kastin, and David H. Coy

15. POSSIBLE ROLE OF OPIOIDS IN MENTAL DISORDERS: PRESENT STATE OF KNOWLEDGE ............................ 293

    H. M. Emrich

| | | |
|---|---|---|
| 16. | ENDOGENOUS OPIOID PEPTIDES IN SCHIZOPHRENIA AND AFFECTIVE DISORDERS............................................. | 309 |
| | Fräncesca Brambilla, Andrea Genazzani, and Fabio Facchinetti | |
| 17. | CHANGES OF BRAIN MONOAMINES IN THE ANIMAL MODEL FOR DEPRESSION......................................... | 331 |
| | Noboru Hatotani, Junichi Nomura, and Isao Kitayama | |
| 18. | PSYCHOENDOCRINOLOGY OF DEPRESSION.................... | 343 |
| | Gregory M. Brown, Paul Grof, and Eva Grof | |
| 19. | NEUROENDOCRINE DYSFUNCTION IN SUBTYPES OF DEPRESSION......................................................... | 357 |
| | Detlev von Zerssen, Mathias Berger, and Peter Doerr | |
| 20. | BIOLOGICAL TESTS IN THE DIAGNOSIS AND TREATMENT OF AFFECTIVE DISORDERS ........................................ | 383 |
| | Harvey A. Sternbach, Harry E. Gwirtsman, and Robert H. Gerner | |
| 21. | THE DEXAMETHASONE SUPPRESSION TEST IN CLINICAL PSYCHIATRY ......................................................... | 399 |
| | N. M. Kurtz and Jeffrey L. Rausch | |
| 22. | THE TRH TEST IN THE DIAGNOSIS OF AFFECTIVE DISORDERS AND SCHIZOPHRENIA.............................. | 413 |
| | Mark S. Gold, A. Carter Pottash, and Irl Extein | |
| 23. | ASPECTS OF THYROID AXIS FUNCTION IN DEPRESSION: A REVIEW............................................................... | 431 |
| | Arthur J. Prange, Jr., and Peter T. Loosen | |
| 24. | CHRONONEUROENDOCRINOLOGY OF DEPRESSION: AN INTERPRETATION OF NEUROENDOCRINE RHYTHMS ON THE BASIS OF PHARMACOLOGICAL STUDIES................. | 443 |
| | Alessandro Agnoli, Massimo Baldassarre, Fabrizio Stocchi, Nicola Martucci, Giovanni Murialdo, Claudio Zauli, Rosaria D'Urso, and Paolo Falaschi | |

25. ALTERATIONS IN CIRCADIAN SECRETION OF PITUITARY AND PINEAL HORMONES IN AFFECTIVE DISORDERS ........ 465

    Julien Mendlewicz

26. A STUDY OF CIRCADIAN VARIATION OF PLATELET SEROTONIN UPTAKE AND SERUM CORTISOL IN PATIENTS WITH MAJOR DEPRESSION ...................................... 473

    Jeffrey L. Rausch, Nandkumar S. Shah, Earl A. Burch, Alexander G. Donald, and Ethan V. Munson

27. NEUROENDOCRINE EVALUATION OF CATECHOLAMINERGIC FUNCTION IN MAN: APPLICATION TO RESEARCH IN PSYCHIATRY AND NEUROLOGY ............ 485

    S. Lal and N. P. V. Nair

28. NEUROENDOCRINE STUDIES IN SCHIZOPHRENIA ............ 503

    Ghanshyam N. Pandey, Syed I. Ali, Regina Casper, and John M. Davis

29. EXOGENOUS PEPTIDES AND SCHIZOPHRENIA ................ 517

    Man Mohan Singh and Stanley R. Kay

30. HORMONAL RESPONSES TO D-AMPHETAMINE IN SCHIZOPHENIA ..................................................... 549

    Daniel P. van Kammen, S. Charles Schulz, and Alan D. Rogol

31. NEUROLEPTICS AND PROLACTIN: A SECOND LOOK ......... 569

    Walter Armin Brown

32. NEUROENDOCRINE CHANGES DURING THE COURSE OF NEUROLEPTIC TREATMENT OF SCHIZOPHRENIC PATIENTS 583

    Norbert Nedopil, Johannes Weiss-Brummer, and Eckart Rüther

33. EFFECTS OF LONG-TERM NEUROLEPTIC TREATMENT ON SERUM LEVELS OF PROLACTIN, TSH, LH, AND NOREPINEPHRINE AND ON α-ADRENERGIC AND DOPAMINERGIC RECEPTOR SENSITIVITY: RELATIONS TO TARDIVE DYSKINESIA ............................................. 599

    Dieter Naber and Franz Müller

34. CENTRAL NERVOUS SYSTEM AND PITUITARY
DOPAMINERGIC DEFECTS IN HYPERPROLACTINEMIC
STATES.................................................................. 613

    Carlo Ferrari, Anna Maria Mattei, and
    Pier Giorgio Crosignani

INDEX ........................................................................ 631

INTRODUCTION

# Psychoneuropharmacology and Neuroendocrinology
## Implications in Neuropsychiatric Research

ARUNKUMAR B. SHAH, NANDKUMAR S. SHAH,
DONALD W. MORGAN, and ALEXANDER G. DONALD

## 1. INTRODUCTION

Psychoneuroendocrinology is a fascinating field of study among several disciplines including the neurosciences, neurology, psychiatry, and endocrinology. The last decade has witnessed a phenomenal growth in our understanding of the role of neuroendocrine hormones and neuropeptides in relation to a wide range of psychiatric and neurological disorders. A study of interaction between endocrine hormones and the brain has thus become a rapidly expanding area which has been spurred on by an avalanche of new knowledge. The overlap of the endocrine and nervous systems was well elucidated by classical studies during the first half of this century. The concepts of "neurosecretion," where the nerve cell could have a secretory function (Scharrer and Scharrer, 1940; Bargmann and Scharrer, 1951), and hypothalamic regulation of anterior pituitary secretion were well documented during that time period (Hinsey and Markee, 1933; Harris, 1952). During the last two decades, advances in immunoassay techniques, immunocytochemistry and biobehavioral assays, have immensely contributed to our knowledge of neuroendocrine relationships with higher cortical functions. A number of hypothalamic peptides involved in the regulation of anterior pituitary secretion have been found in extrahypothalamic areas, and biobehavioral assays have postulated the importance of these peptides in mam-

---

ARUNKUMAR B. SHAH, NANDKUMAR S. SHAH, DONALD W. MORGAN, and ALEXANDER G. DONALD • Ensor Foundation Research Laboratory, William S. Hall Psychiatric Institute, and Department of Neuropsychiatry and Behavioral Sciences, University of South Carolina School of Medicine, Columbia, South Carolina 29208. *Present address of A.B.S.:* Comprehensive Stroke Center, Department of Neurology, School of Medicine, The Oregon Health Sciences University, Portland, Oregon 94201.

malian behavior. The posterior pituitary peptides, vasopressin and oxytocin, have also been found in extrahypothalamic projections (Sterba, 1974; Buijs *et al.*, 1978; Rossor *et al.*, 1981). The classical gastrointestinal peptide hormones have been found to occur in both endocrine and nerve cells in the gut and the brain (Dockray, 1976; Dockray *et al.*, 1978). Psychoneuropharmacological and behavioral studies indicate that the peptides mentioned above do modulate neural functions; thus, the brain appears to be an important "target organ for hormones."

Classically, the clinical significance of brain–endocrine interactions was suggested by Fröhlich in 1901, well before the existence of such interactions was known. The neuropsychiatric manifestations of primary endocrine disorders are now well recognized. The role of the hypothalamic pathology in the genesis of endocrine dysfunction is equally well recognized. In recent years, the main emphasis has been on the study of possible neuroendocrine dysfunctions in various disorders which are primarily considered to be neuropsychiatric in nature.

Martin, in his review of anatomical, biochemical, and physiological aspects (Chapter 1), observes that several peptides have multiple roles in the regulation of brain functions and may in turn be involved in a wide variety of psychiatric and neurological disorders.

Psychoneuroendocrine studies have evolved from following presumptions:

1. Neuroendocrine abnormalities may serve as biological markers for changes in the central monoamine neurotransmitters which are implicated in the etiology and/or pharmacotherapy of psychopathological states.
2. The novel peptides recently discovered in extrahypothalamic areas of the brain have profound behavioral effects; hence, their dysfunction may possibly be involved in human psychopathology.
3. Utilization of neuroendocrine strategies may serve useful purpose to clarify neurotransmitter mechanisms involved in the actions of neuropharmacological agents.

With these postulations, a host of studies in both basic sciences and clinical areas have been carried out during the last few years, some of which have been reviewed in this volume. In the present article, an overview of some of the recent developments in these areas is briefly presented to evaluate the implications of such studies in neuropsychiatric research.

## 2. NEUROPHARMACOLOGY OF HYPOTHALAMIC PEPTIDES

The hypothalamic and pituitary oligopeptides exert important influences on the integrative functions of the CNS and consequently they seem to be an important class of chemical substances involved in the regulation of mammalian behavior. Several recent monographs and reviews, detailing various aspects

## INTRODUCTION

of their distribution in extrahypothalamic areas, CNS effects, and clinical applications, have appeared (Plotnikoff and Kastin, 1977; Gotto et al., 1979; de Wied and van Keep, 1980). Despite the widespread interest in these peptides, their physiological role and precise neurochemical mode of actions in extrahypothalamic areas are as yet far from clear.

### 2.1. Thyrotropin-Releasing Hormone

Thyrotropin-releasing hormone (TRH) was the first hypothalamic tripeptide (pGlu-His-Pro-NH$_2$) to be isolated from ovine and porcine hypothalamic tissues (Bøler et al., 1969; Burgus et al., 1969). Shortly thereafter, evidence for a possible action of this tripeptide in the CNS was first provided by Plotnikoff et al. (1972), who demonstrated that TRH enhanced the stimulant properties of L-dopa in pargyline-treated mice. The latter effect was shown to be independent of the pituitary–thyroid axis (Plotnikoff et al., 1974a,b). The ongoing pharmacobehavioral studies have further characterized the CNS effects of TRH (Table 1), and have led to the hypotheses that TRH forms an endogenous ergotropic or analeptic system in the CNS (Metcalf and Dettmar, 1981).

The pharmacological dissection of these actions has been the most useful exercise in elucidating the mechanisms underlying the CNS effects of this peptide. The arousal effects of TRH are antagonized by muscarcinic cholinergic antagonists, suggesting the participation of cholinergic mechanisms in the former effects of TRH. However, the thermoregulatory effects are not antagonized and have obviously different mechanisms (Yarbrough, 1979). Similarly, cholinergic mechanisms are not involved in many of the behavioral effects. This diversity and dissociation of actions coupled with the presence of this peptide in nerve terminals (Hökfelt et al., 1975), its release induced by depolarization (Maeda and Frohman, 1980), and presence of high-affinity receptors for this peptide in extrahypothalamic regions of the brain (Ogawa et al., 1981), tends

TABLE 1
Effects of TRH on the CNS

---

Potentiation of behavioral effects of L-dopa + pargyline (Plotnikoff et al., 1972)
Active in mouse serotonin protentiation test (Green and Grahame-Smith, 1974)
Antagonism of sedation and hypothermia induced by bartiturates, ethanol, diazepam (Prange et al., 1979b)
Tremorogenic activity in mice (Kruse, 1976)
Hyperthermia in rabbits and reversal of reserpine-induced hypothermia (not antagonized by muscarinic cholinergic antagonists) (Horita and Carino, 1975; Brewster et al., 1981)
Potentiation of motor activity (Smith et al., 1976)
Alteration in neuronal excitability on microiontophoresis (Winokur and Beckman, 1978)
Modulatory role on synaptic processes (evidenced by alteration in the neurotransmitter-induced neuronal effects) (Yarbrough, 1979)
Antagonism of many effects of opioid peptides except analgesia (Holaday et al., 1978, 1981)
Increased turnover of NE and Ach in the brain (Yarbrough, 1979; Malthe-Sorensen et al., 1978)

to suggest that TRH functions as a neurotransmitter and/or neuromodulator with a specific physiological role.

The release of hypothalamic TRH is stimulated by norepinephrine (NE) and dopamine (DA), and inhibited by serotonin (Reichlin *et al.*, 1974; Kirkegaard *et al.*, 1977; Martin *et al.*, 1977). This led to the postulation that the alteration in the central monoaminergic transmission may produce changes in the hypothalamic–pituitary–thyroid (HPT) axis. This presumption has been used in the study of affective disorders and forms the basis of the TRH test for these disorders (Kastin *et al.*, 1972; Kirkegaard *et al.*, 1978; Extein *et al.*, 1980).

Therapeutic usefulness of TRH or its analogs remains to be evaluated in future clinical trials. Besides its partial effectiveness in alleviating depressive symptoms in some patients, the tripeptide has shown some promise in the treatment of alcoholism, hyperkinetic syndrome, childhood autism, and schizophrenia (Prange *et al.*, 1979a). In Chapter 23 of this volume, Prange and Loosen have emphasized that a defect in the HPT axis exists in some patients with mental depression and that the hormones of HPT axis such as TSH, TRH, triiodothyronine, and thyroxine may find some therapeutic value in depressed patients. Other prospective uses, listed in Chapter 3, include diseases such as Huntington's chorea, Alzheimer's disease, and other senile dementias with suspected hypocholinergic functions; TRH and its long-acting analogs—MK-771, CG 3703, DN-1417, and RX 77368—influence cholinergic function probably through TRH receptors postulated to be present on cholinergic neurons (Metcalf, 1982).

## 2.2. Somatostatin

Similar advances have occurred in our knowledge of neuropharmacological effects of yet another hypothalamic peptide, somatostatin—a tetradecapeptide. This peptide also seems to have a variety of behavioral and neuronal effects (Table 2) which seem to be the opposite of TRH (Table 1). Somatostatin has

TABLE 2
Effects of Somatostatin on the CNS

Depressant effect on CNS:
    Tranquilizing effect in monkeys (Siler *et al.*, 1973)
    Potentiation of barbiturate-induced sedation and hypothermia (Prange *et al.*, 1975)
    Interference with strychnine-induced seizures (Brown and Vale, 1975)
Suppression of locomotor activity—catalepsy (Cohn and Cohn, 1975)
Rotational behavior—antagonized by atropine in rats (Cohn and Cohn, 1975)
Partial agonist–antagonist of opiate receptor in CNS (Terenius, 1973)
Marked behavioral effects on central and hippocampal administration (Rezek *et al.*, 1976)
Elevation in cAMP content in CNS (Havlicek *et al.*, 1976)
Depression of neuronal excitability on microiontophoresis (Renaud *et al.*, 1975)
Increased turnover of DA, 5HT, and NE (Garcia-Sevilla *et al.*, 1978)

been demonstrated in specific neuronal systems with particularly high concentrations in the nerve endings where it is stored in storage vesicles; its release from the latter is $K^+$ stimulated and $Ca^{2+}$ dependent (Wakabayashi et al., 1977; Iversen et al., 1978). This coupled with the direct neuronal effects suggests that somatostatin could function as a neurotransmitter or modulator with predominant depressant effects as opposed to the stimulant effects of TRH. Clinical studies with respect to somatostatin in neuropsychiatric dysfunction are limited. Reduced levels of somatostatin have been observed in brain from patients with senile dementia of Alzheimer's type (Davies et al., 1980; Rossor et al., 1980). The implications of this finding are as yet not clear. An attempt also has been made to study the CSF levels of somatostatin in various neurological disorders; however, this has failed to yield any disease-specific pattern (Patel et al., 1977). Some aspects of somatostatin are discussed in Chapter 2.

### 2.3. Neurotropic Activity of Several Peptides

The significance of the pituitary as an important source of peptides with neurotropic activity has been well elucidated during the last decade. The initial evidence for the role of hypophyseal factors in the maintenance of normal behavioral patterns was clearly demonstrated by the observation of impaired learning behavior in animals after removal of the pituitary (de Wied, 1969). Since then, this area has been well investigated by several workers. Melanocyte-stimulating hormone/adrenocorticotropic hormone (MSH/ACTH) like peptides have been shown to modulate attentional/perceptual function (Sandman et al., 1981). $ACTH_{4-10}$ has been reported to play an important role in memory (Chapter 12). Vasopressin appears to facilitate the consolidation of learned experience and also promotes retrieval of stored information (de Wied, 1980). Its possible therapeutic applications in some neuropsychiatric disorders are reviewed in Chapter 13. While many peptides such as substance P and cholecystokinin have been identified in the brain, their physiological functions have yet to be established. Finally, the discoveries in the 1970s of opioid receptors and their endogenous ligands—enkephalins and endorphins—seem to represent only the first step in a fascinating and relatively new field, the clinical importance of which cannot be underestimated (for details see Shah and Donald, 1982). Disturbances in the regulation and secretion of these peptides may lead to alterations in brain functions and may underlie psychopathological states. Aspects concerning the possible involvement of these peptides in learning, memory, schizophrenia, and affective disorders are discussed in Chapters 14, 15, and 16.

## 3. NEUROENDOCRINE STRATEGIES: BIOLOGICAL MARKERS

Evidence supporting the involvement of monoamine neurotransmitters in the regulation of secretion of hypothalamic releasing hormones (Martin and

Ganong, 1976) and the proposed abnormalities in monoamine functions in several psychiatric and neurological disorders have led to the suggestion that the study of regulation of neuroendocrine hormone secretion may serve as useful diagnostic tests in probing these diseases.

One of the interesting phenomena in psychoneuroendocrine studies is the circadian rhythms which many endocrine hormones have unquestionably shown. Observations of disturbed circadian rhythms, for example, the cortisol secretion, provided a useful guideline in the diagnosis and perhaps in the treatment of affective disorders. Details on disturbed circadian rhythms and their possible significance in neuropsychiatric disorders are discussed in Chapters 13, 21, 25, and 26.

Despite voluminous reports compiled over the last decade, the controversies surrounding the usefulness and specificities of these tests for diagnostic purposes still exist. Although the utility of these tests for diagnostic purposes is still an open issue, the readers of this book will find several chapters that furnish detailed information and critical reviews on this subject. Briefly, most studies involved are centered around the investigation of three major neuroendocrine axis: (1) hypothalamic–pituitary–adrenal (HPA); (2) hypothalamic–pituitary–thyroid (HPT); (3) hypothalamic–pituitary–gonadal (HPG). In addition, the central dopaminergic and noradrenergic functions have been assessed by examining the serum prolactin (PRL) and growth hormone (GH) levels and by measuring urinary MHPG, respectively; PRL and GH are discussed in Sections 4.1 and 4.3.

### 3.1. Dexamethasone Suppression Test and HPA Axis

So far the most thoroughly investigated neuroendocrine test is the overnight dexamethasone suppression test (DST) for the study of the HPA axis in depressive disorders. Several aspects of this test are reported in a number of chapters of this book. More specific problems in the methods and interpretation are dealt with in Chapters 6, 18, 19, 20, and 21.

### 3.2. TSH Response to TRH and HPT Axis

Another neuroendocrine test more frequently used is the TSH response to TRH administration for the diagnosis of the abnormal functioning of the HPT axis and focuses mainly for its diagnostic value in affective disorders. Using the TSH response to TRH, Prange and Loosen (see Chapter 23) produced evidence demonstrating a defective function of the HPT axis in some patients with depression. Authors of other chapters (2, 9, 20, 22, 24) report several aspects of this test including the diagnostic value in mental illness.

### 3.3. LH and FSH Responses to LH-RH and HPG Axis

Details concerning the physiological regulation and functional role of the HPG axis hormones are focused in Chapter 2. Neuroendocrine test of the HPG axis involves measurement of altered plasma levels of luteinizing hormone (LH) or follicle-stimulating hormone (FSH) in response to administration of LH-releasing hormone (LH-RH). The significance of the HPG axis as a diagnostic test for psychiatric illnesses is far from clear and needs further assessment. One study (Ettigi *et al.*, 1979) reported in Chapter 18 showed an increase of the LH response to LH-RH in secondary depression. Some studies claiming blunted LH and FSH responses to LH-RH in patients with anorexia nervosa have been reported in Chapter 5. Although no definitive value of this test has been established, it is hoped that additional work may eventually determine its utility for the diagnosis of this disease which afflicts females mostly during adolescent age. Another diagnostic test proposed for anorexia nervosa involves the measurement of plasma LH response to estrogen or the estrogen antagonist clomiphene (see Chapter 5).

## 4. NEUROENDOCRINE TECHNIQUES IN THE STUDY OF PSYCHOACTIVE DRUGS

Neuropharmacological studies have helped in clarifying the neurotransmitter regulation of neuroendocrine function. The knowledge gained in this area has made it possible to use neuroendocrine strategies to clarify the neurotransmitter effects of psychoactive drugs. These techniques have obvious limitations in view of the fact that several neuropharmacological agents exert their action on more than one neurotransmitter. Furthermore, the hormonal levels measured in plasma are influenced by many other factors in addition to monoamine neurotransmitters. Despite these limitations, this approach has been useful in clarifying some neuropharmacological issues.

### 4.1. Use of PRL Secretion as an Index of Brain Dopaminergic Function

Galactorrhea as a side effect of neuroleptic therapy was reported soon after their introduction in the therapy of psychiatric disorders (Gäde and Heinrich, 1955; Winnik and Tennenbaum, 1955). Since then, it has been established that neuroleptics block DA receptors and increase PRL secretion.

The predominant hypothalamic influence on the release of PRL from the anterior pituitary is mediated largely by dopaminergic inhibition. DA released from nerve terminals of tuberoinfundibular DA neurons (TIDA) is transported to the lactotrope cells of the anterior pituitary through portal capillaries. DA binds to plasma membrane DA receptors on the lactotrope. This results in activation of an as yet unidentified mechanism which results in tonic inhibition of PRL release. Thus, serum PRL levels inversely reflect both the tone of DA pathways and the sensitivity of DA receptors of the lactotrope cells. However, whether the tone of the TIDA pathway reflects similar alterations in other central pathways (nigrostriatal, mesolimbic, and mesocortical) is questionable. The regulation of DA synthesis and release from the TIDA pathway in fact differs from the other dopaminergic pathways. The TIDA pathway lacks autoreceptors as well as classical neuronal "feedback" mechanisms which control DA synthesis and release in other central DA pathways (Moore et al., 1980). Finally, the TIDA pathway is highly sensitive to hormonal feedback mechanisms, and the latter do not affect DA turnover in extrapyramidal or limbic structures (Moore et al., 1980). In addition to the differences in regulating mechanisms, it should also be emphasized that lactotrope DA receptors and TIDA nerve terminals lie outside the blood–brain barrier. This may make them more prone to pharmacological manipulation than the other central pathways described above which are within the blood–brain barrier. Neuroleptics with diverse chemical structures increase serum PRL in man. The latter action of neuroleptics is antagonized by dopaminergic agents and also by DA infusion, suggesting that neuroleptics act at the level of the pituitary. Initially, it was believed that for several neuroleptics, clinical antipsychotic potency correlates with their potency in increasing PRL secretion (Gruen et al., 1978). However, it has now been shown that clinically equipotent antipsychotic doses of different medications can have variable stimulatory effects on PRL (Meltzer et al., 1979), and some compounds that are very potent in elevating PRL have only poor antipsychotic effects (Mielke et al., 1977).

With the development of radioreceptor assay (RRA), it has been possible to determine the DA receptor-blocking activity of neuroleptics by measuring the drug levels in the plasma of patients (Creese and Snyder, 1977). In *in vitro* studies, clinically effective neuroleptic drugs block the stereospecific binding of [$^3$H]haloperidol to striatal DA receptors at concentrations which correlate directly with their clinical potency (Seeman et al., 1976). However, *in vivo* monitoring of plasma neuroleptic activity (DA receptor-binding activity) by RRA has revealed that the neuroleptic activity in plasma differs for different medications when given in clinically equivalent (CPZ equivalents) antipsychotic doses (Javaid et al., 1980; A. B. Shah, N. S. Shah, and R. Wiscovitch, unpublished observations). The RRA measures the total neuroleptic activity of the drug, including the drug bound to plasma proteins as well as all the active metabolites present in plasma (Creese and Snyder, 1977). Some of the factors which may contribute to the observed discrepancy in the plasma levels of DA-blocking activity of different neuroleptics administered in clinically equivalent doses include variations in plasma protein binding and variable transport of the drug or active metabolites across the blood–brain barrier. If the latter explanation is true, it should be possible to correlate the DA receptor-blocking activity in plasma following administration of a neuroleptic with its stimulatory effects on PRL secretion. Significant positive correlations between the serum neuroleptic levels determined by RRA and serum PRL levels have been reported by Brown et al. (1981) for the entire group of patients treated with different neuroleptic agents.

Examples stressing the diagnostic value of the assay of serum PRL levels in determining the clinical response to neuroleptics, bioavailability of neuroleptics, extrapyramidal side effects, CNS and pituitary DA dysfunction, and controversies surrounding these issues are detailed in Chapters 10, 24, 30, 31,

32, 33, and 34. Other aspects of PRL are covered in Chapters 1, 2, 7, 18, and 25.

## 4.2. Calcium Entry Blockers: PRL Secretion and Psychotropic Effects

Despite these limitations, the potential utility of this approach can be well demonstrated by considering the neuroendocrine effects of a new class of drugs, "calcium entry blockers," which are used widely in cardiovascular disorders. This group of drugs has the potential ability to alter $Ca^{2+}$ fluxes at the neuronal membrane. This has been demonstrated by *in vitro* effects of verapamil on amphetamine-induced DA synthesis (Uretsky and Snodgrass, 1979), and cocaine-induced NE release (Gothert *et al.*, 1979). However, the possibility of *in vivo* effect of this drug on neuronal/synaptosomal function has been raised by the observation of hyperprolactinemia resulting from verapamil therapy (Gluskin *et al.*, 1981). This may be due to verapamil-induced inhibition of TIDA "tone" resulting from inhibition of DA release from the nerve terminals of TIDA neurons which lie outside the blood–brain barrier. Our preliminary studies have indicated that chronic treatment with nitrendipine and nifedipine causes hyperprolactinemia in adult male rats (unpublished data). Furthermore, it was recently observed that nisoldipine, a dihydropyridine group of calcium antagonist, effectively blocked phencyclidine (PCP)-induced stereotypy in neonatal rats (3 and 4 weeks old), indicating that PCP-induced increase in the central catecholaminergic activity is antagonized by nisoldipine (Shah *et al.*, 1982, 1983). Finally, in a recent clinical report, verapamil was successfully used in the treatment of manic symptoms (Dubovsky *et al.*, 1982). Even though all these findings are preliminary and a number of questions remain to be answered, these series of developments are good indicators of the contributions of neuroendocrinology to the progress in neuropsychopharmacology.

## 4.3. GH

DA, NE, and serotonin exert a positive control on GH release (Frohman and Stachura, 1975; Brown *et al.*, 1979; Meites and Sonntag, 1981). Several pharmacological agents affect GH secretion through their actions on putative monoamine neurotransmitters. DA agonists enhance and antagonists inhibit GH secretion; for example, amphetamine, levodopa, apomorphine, piribedil—all DA-stimulating agents—cause an increase in GH secretion, whereas reserpine and most antipsychotic drugs—DA-blocking agents—cause suppression. Noradrenergic drugs, e.g., clonidine ($\alpha$-adrenergic agonist), stimulate and phentolamine ($\alpha$-adrenergic antagonist) inhibits GH release.

Changes in monoamine neurotransmitter functions in the CNS are expected to alter GH secretion. Indeed, many studies have shown abnormalities in GH secretion in various psychiatric and neurological disorders associated with disturbances in brain monoaminergic functions.

Measurement of plasma levels of GH as an index of central monoaminergic functions has provided a useful tool and this strategy has been utilized by several investigators to determine neuroendocrine abnormalities associated with Huntington's disease (Chapters 10, 27), affective disorders (Chapters 18, 24), anorexia nervosa (Chapters 9, 27), Parkinson's disease (Chapter 27), schizophrenia (Chapters 9, 27, 28, 30, 32, 33), tardive dyskinesia (Chapters 27, 33), and Metabolic diseases (Chapter 9). Altered responses to GH secretion by psychopharmacological agents, hypothalamic neuropeptides as well as hormones of pituitary and target organ glands have been examined in detail and clarified in Chapters 2, 9, 27, 30, 33.

## 5. PSYCHONEUROENDOCRINE DYSFUNCTION

Changes in psychoneuroendocrine functions have been associated with a wide range of disorders. Several hypothalamic peptides and the anterior pituitary hormones have been shown to interact with the brain to elicit a variety of behavioral effects. One of the concerns of this volume is to focus on the role of hypothalamic peptides, pituitary hormones, and brain putative neurotransmitters and their mechanisms of action in elucidating the etiologies of neuroendocrine-related disorders. Although concerned primarily with the neuroendocrine aspects of psychiatric and neurological disorders such as schizophrenia, affective disorders, anorexia nervosa, Huntington's chorea, senile dementia of Alzheimer's type, and many more, this volume also provides a valuable source of information on a wide range of endocrine and metabolic abnormalities, PRL-, ACTH-, and FSH-secreting tumors, endogenous opioid peptides and opiate addiction, primary hypogonadism, male sexual behavior, Cushing's disease, acromegaly, and several others.

## 6. CONCLUSIONS

Several hypothalamic and pituitary hormones have been shown to exert a wide range of pharmacological and behavioral effects. The present chapter briefly illustrates this view by giving examples of a few selected neuropeptides. Abnormalities in neuroendocrine hormones have been linked to several psychiatric and neurological illnesses; a somewhat limited discussion is presented in this chapter since several other chapters in this book have covered this area more explicitly presenting detailed accounts on this subject matter. Neuroendocrine abnormalities may also serve as biological markers for altered functions of brain putative neurotransmitters which are also involved in the regulation of hypothalamic releasing peptide hormones. Finally, while the readers may find psychoneuroendocrine approach and hormonal markers as controversial areas, it is our strong hope that continued research efforts in this exciting area will eventually help to offer future directions in resolving the conflicting views currently clouding the question of usefulness of neuroendocrine strategies in psychiatry and neurology.

## 7. REFERENCES

Bargmann, W., and Scharrer, E., 1951, The site of origin of the hormones of the posterior pituitary, *Am. Sci.* **39**:255.

Bøler, J., Enzmann, F., Folkers, K., Bowers, C. Y., and Schally, A. V., 1969, The identity of chemical and hormonal properties to the thyrotropin releasing hormone and pyroglutamyl-histidyl-proline-amide, *Biochem. Biophys. Res. Commun.* **37**:705.

Brewster, D., Dettmar, P. W., and Metcalf, G., 1981, Biologically stable analogues of TRH with increased neuropharmacological potency, *Neuropharmacology* **20**:497.

Brown, G. M., Friend, W. C., and Chambers, J. W., 1979, Neuropharmacology of hypothalamic-pituitary regulation, in: *Clinical Neuroendocrinology: A Pathophysiological Approach* (G. Tolis, F. Labrie, J. B. Martin, and F. Naftolin, eds.), pp. 47–81, Raven Press, New York.

Brown, M., and Vale, W., 1975, Central nervous system effects of hypothalamic peptides, *Endocrinology* **96**:1333.

Brown, W. A., Laughren, T. O., and Williams, B., 1981, Differential effects of neuroleptic agents of the pituitary–gonadal axis in men, *Arch. Gen. Psychiatry* **38**:1270.

Buijs, R. M., Swaab, D. F., Dogterom, J., and van Leeuwen, F. W., 1978, Intra and extrahypothalamic vasopressin and oxytocin pathways in the rat, *Cell Tissue Res.* **186**:423.

Burgus, R., Dunn, T. F., Desiderio, D., and Guillemin, R., 1969, Structure moleculaire de facteur hypothalamique hypophysiotrope TRF d. origine ovine: Mise en evidence par spectrometric de masse de la sequence PCA-His-Pro-NH$_2$, *C.R. Acad. Sci. Paris* **269**:1870.

Cohn, M. L., and Cohn, M., 1975, Barrel rotation: Induced by somatostatin in the non-lesioned rat, *Brain Res.* **96**:138.

Creese, I., and Snyder, S. H., 1977, A simple and sensitive radioreceptor assay for antischizophrenic drugs in blood, *Nature (London)* **270**:180.

Davies, P., Katzman, R., and Terry, R. D., 1980, Reduced somatostatin-like immunoreactivity in cerebral cortex from cases of Alzheimer disease and Alzheimer senile dementia, *Nature (London)* **288**:279.

de Wied, D., 1969, Effects of peptide hormones on behavior, in: *Frontiers in Neuroendocrinology* (W. F. Ganong and L. Martini, eds.), pp. 97–133, Oxford University Press, London.

de Wied, D., 1980, Hormonal influences on motivation, learning, memory, and psychosis, in: *Neuroendocrinology* (D. T. Krieger and J. C. Hughes, eds.), pp. 194–204, Sinauer, Sunderland, Mass.

de Wied, D., and van Keep, P. A., 1980, *Hormones and the Brain*, University Park Press, Baltimore.

Dockray, G. J., 1976, Immunochemical evidence of cholecystokinin-like peptides in brain, *Nature (London)* **264**:568.

Dockray, G. J., Gregory, R. A., Hutchison, J. B., Harris, J. I., and Runswick, M. J., 1978, Isolation, structure and biological activity of two cholecystokinin octapeptides from sheep brain, *Nature (London)* **274**:711.

Dubovsky, S. L., Franks, R. D., Lifshitz, M., and Cohen, P., 1982, Effectiveness of verapamil in the treatment of a manic patient, *Am. J. Psychiatry* **139**:502.

Ettigi, P. G., Brown, G. M., and Seggie, J. A., 1979, TSH and LH responses in subtypes of depression, *Psychosom. Med.* **41**:203.

Extein, I., Pottash, A. L. C., Gold, M. S., Cadet, J., Sweeney, D. R., Davies, R. K., and Martin, D. M., 1980, The thyroid-stimulating hormone response to thyrotropin-releasing hormone in mania and bipolar depression, *Psychiatr. Res.* **2**:199.

Fröhlich, A., 1901, Ein Fall von Tumor der Hypophysis Cerebri ohne Akromegalie, *Wien. Klm. Rundsch.* **15**:883, 906.

Frohman, L. A., and Stachura, M. E., 1975, Neuropharmacologic control of neuroendocrine function in man, *Metabolism* **24**:211.

Gäde, E. B., and Heinrich, K., 1955, Klinische Beobachtungen bei Megaphenbehandlung in der Psychiatrie, *Nervenarzt* **26**:49.

Garcia-Sevilla, J. A., Magnusson, T., and Carlsson, A., 1978, Effect of intracerebroventricularly administered somatostatin on brain monoamine turnover, *Brain Res.* **155**:159.

Gluskin, L. E., Strasberg, B., and Shah, J. H., 1981, Verapamil-induced hyperprolactinemia and galactorrhea, *Ann. Intern. Med.* **95:**66.

Gothert, M., Nawroth, P., and Pohl, I.-M., 1979, Effects of verapamil and presynaptic modulators on calcium-induced noradrenaline release, in: *Catecholamines: Basic and Clinical Frontiers* (E. Usdin, I. J. Kopin, and J. Barchas, eds.), pp. 289–291, Pergamon Press, Elmsford, N.Y.

Gotto, A. M., Jr., Peck, E. J., Jr., and Boyd, A. E., III, 1979, *Brain Peptides: A New Endocrinology*, Elsevier/North-Holland, Amsterdam.

Green, A. R., and Grahame-Smith, D. G., 1974, TRH potentiates behavioral changes following increased brain 5-hydroxytryptamine accumulation in rats, *Nature (London)* **251:**524.

Gruen, P. H., Sachar, E. J., Langer, G., Altman, N., Leifer, M., Frantz, A., and Halpern, F. S., 1978, Prolactin responses to neuroleptics in normal and schizophrenic subjects, *Arch. Gen. Psychiatry* **35:**108.

Harris, G. W., 1952, Hypothalamic control of the pituitary gland, *Ciba Found. Colloq. Endocrinol. Proc.* **4:**106.

Havlicek, V., Herchl, R., Rezek, M., Kroeger, E., Hughes, K. R., Lesiuk, H., and Friesen, H., 1976, Somatostatin (SRIF) action on cyclic AMP in rat brain: Role of central adrenergic mechanisms, *Fed. Proc.* **35:**782.

Hinsey, J. C., and Markee, J. E., 1933, Pregnancy following bilateral section of cervical sympathetic trunks in rabbit, *Proc. Soc. Exp. Biol. Med.* **31:**270.

Hökfelt, T., Fuxe, K., Johansson, O., Jeffcoate, S., and White, N., 1975, Thyrotropin releasing hormone (TRH)-containing nerve terminals in certain brain stem nuclei and in the spinal cord, *Neurosci. Lett.* **1:**133.

Holaday, J. W., Tseng, L. F., Loh, H. H., and Li, C. H., 1978, Thyrotropin releasing hormone antagonizes β-endorphin hypothermia and catalepsy, *Life Sci.* **22:**1537.

Holaday, J. W., D'Amato, R. J., and Faden, A. I., 1981, Thyrotropin-releasing hormone improves cardiovascular function in experimental endotoxic and hemorrhagic shock, *Science* **213:**216.

Horita, A., and Carino, M. A., 1975, Thyrotropin-releasing hormone-induced hyperthermia and behavioral excitation in rabbits, *Psychopharmacol. Commun.* **1:**403.

Iversen, L. L., Iversen, S. D., Bloom, F., Douglas, C., Brown, M., and Vale, W., 1978, Calcium-dependent release of somatostatin and neurotensin from rat brain *in vitro*, *Nature (London)* **273:**161.

Javaid, J. I., Pandey, G. N., Duslak, B., Hu, H.-Y., and Davis, J. M., 1980, Measurement of neuroleptic concentrations by GLC and radioreceptor assay, *Commun. Psychopharmacol.* **4:**467.

Kastin, A. J., Schalach, D. S., Enrensing, R. H., and Anderson, M. S., 1972, Improvement in mental depression with decreased thyrotropin response after administration of thyrotropin-releasing hormone, *Lancet* **II:**740.

Kirkegaard, C., Bjørum, N., Cohn, D., Faber, J., Lauridsen, U. B., and Nerup, J., 1977, Studies on the influence of biogenic amines and psychoactive drugs on the prognostic value of the TRH stimulation test in endogenous depression, *Psychoneuroendocrinology* **2:**131.

Kirkegaard, C., Bjørum, N., Cohn, D., and Lauridsen, U. B., 1978, Thyrotropin-releasing hormone stimulation test in manic depressive illness. *Arch. Gen. Psychiatry* **35:**1017.

Kruse, L., 1976, Thyrotropin-releasing hormone: Restoration of oxotremorine tremor in mice, *Naunyn-Schmiedebergs Arch. Exp. Pathol. Pharmakol.* **294:**39.

Maeda, K., and Frohman, L. A., 1980, Release of somatostatin and thyrotropin-releasing hormone from rat hypothalamic fragments *in vitro*, *Endocrinology* **106:**1837.

Malthe-Sorensen, D., Wood, P. L., Cheney, D. L., and Costa, E., 1978, Modulation of the turnover rate of acetylcholine in rat brain by intraventricular injections of thyrotropin-releasing hormone, somatostatin, neurotensin and angiotensin II, *J. Neurochem.* **31:**685.

Martin, J. B., Reichlin, S., and Brown, G. M. (eds.), 1977, *Clinical Neuroendocrinology*, Davis, Philadelphia.

Martin, L., and Ganong, W. (eds.), 1976, *Frontiers in Neuroendocrinology*, Vol. 4, Raven Press, New York.

Meites, J., and Sonntag, W. E., 1981, Hypothalamic hypophysiotropic hormones and neurotransmitter regulation: Current views, *Annu. Rev. Pharmacol. Toxicol.* **21:**295.

Meltzer, H. Y., Goode, D. J., Schyve, P. M., Young, M., and Fang, V. S., 1979, Effect of clozapine on human serum prolactin levels, *Am. J. Psychiatry* **136**:1550.

Metcalf, G., 1982, Regulatory peptides as a source of new drugs—The clinical prospectus for analogues of TRH which are resistant to metabolic degradation, *Brain Res. Rev.* **4**:389.

Metcalf, G., and Dettmar, P. W., 1981, Is thyrotropin releasing hormone an endogenous ergotropic substance in the brain?, *Lancet* **I**:586.

Mielke, D. H., Gallant, D. M., and Kessler, C., 1977, An evaluation of a unique new antipsychotic agent, sulpiride: Effects on serum prolactin and growth hormone levels, *Am. J. Psychiatry* **134**:1371.

Moore, K. E., Demarest, K. T., Johnston, C. A., and Alper, R. H., 1980, Pharmacological and endocrinological manipulations of tuberoinfundibular and tuberohypophyseal dopaminergic neurons, in: *Neuroactive Drugs in Endocrinology* (E. E. Müller, ed.), pp. 109–122, Elsevier/North-Holland, Amsterdam.

Ogawa, N., Yamawaki, Y., Kuroda, H., Ofuji, T., Itoga, E., and Kito, S., 1981, Discrete regional distributions of thyrotropin-releasing hormone (TRH) receptor binding in monkey central nervous system, *Brain Res.* **205**:169.

Patel, Y. C., Rao, H., and Reichlin, S., 1977, Somatostatin in human cerebrospinal fluid, *N. Engl. J. Med.* **296**:529.

Plotnikoff, N. P., and Kastin, A. J., 1977, Neuropharmacological review of hypothalamic releasing factors, in: *Advances in Biochemical Psychopharmacology* (L. H. Miller, C. A. Sandman, and A. J. Kastin, eds.), pp. 88–108, Raven Press, New York.

Plotnikoff, N. P., Prange, A. J., Breese, G. R., Anderson, M. S., and Wilson, I. C., 1972, Thyrotropin-releasing hormone: Enhancement of DOPA activity by a hypothalamic hormone, *Science* **178**:417.

Plotnikoff, N. P., Prange, A. J., Breese, G. R., Anderson, M. S., and Wilson, I. C., 1974A, The effects of thyrotropin-releasing hormone on DOPA response in normal, hypophysectomized and thyroidectomized animals, in: *The Thyroid Axis, Drugs and Behavior* (A. J. Prange, ed.), pp. 103–113, Raven Press, New York.

Plotnikoff, N. P., Prange, A. J., Breese, G. R., and Wilson, I. C., 1974b, Thyrotropin-releasing hormone: Enhancement of DOPA activity in thyroidectomized rats, *Life Sci.* **14**:1271.

Prange, A. J., Breese, G. R., Jahnke, G. D., Martin, B. R., Cooper, B. R., Cott, J. M., Wilson, I. C., Alltop, L. B., and Lipton, M. A., 1975, Modification of pentobarbital effects by natural and synthetic polypeptides: Dissociation of brain and pituitary effects, *Life Sci.* **16**:1907.

Prange, A. J., Loosen, P. T., and Nemeroff, C. B., 1979a, Peptides: Application to research in nervous and mental disorders, in: *New Frontiers in Psychotropic Drug Research* (S. Fielding and R. C. Ettland, eds.), pp. 117–189, Futura, Mount Kisco, N.Y.

Prange, A. J., Jr., Nemeroff, C. B., Loosen, P. T., Bissette, G., Osbahr, A. J., III, Wilson, I. C., and Lipton, M. A., 1979b, Behavioral effects of thyrotropin-releasing hormone in animals and man: A review, in: *Central Nervous System Effects of Hypothalamic Hormones and Other Peptides* (R. Collu, A. Barbeau, J. R. Ducharme, and J.-G. Rochefort, eds.), pp. 75–96, Raven Press, New York.

Reichlin, S., Jackson, I., Seyler, L. E., and Grimm-Jorgenson, Y., 1974, Regulation of the secretion of TRH and LH-RH, in: *Frontiers in Neurology and Neuroscience Research* (P. Seeman and G. M. Brown, eds.), pp. 48–59, University of Toronto Press, Toronto.

Renaud, L. P., Martin, J. B., and Brazeau, P., 1975, Depressant action of TRH, LH-RH and somatostatin on activity of central neurons, *Nature (London)* **255**:233.

Rezek, M., Havlicek, V., Hughes, K. R., and Friesen, H., 1976, Central site of action of somatostatin (SRIF): Role of hippocampus, *Neuropharmacology* **15**:499.

Rossor, M. N., Emson, P. C., Mountjoy, C. Q., Roth, M., and Iversen, L. L., 1980, Reduced amounts of immunoreactive somatostatin in the temporal cortex in senile dementia of Alzheimer type, *Neurosci. Lett.* **20**:373.

Rossor, M. N., Iversen, L. L., Hawthorn, J., Ang, V. T. Y., and Jenkins, J. S., 1981, Extrahypothalamic vasopressin in human brain, *Brain Res.* **214**:349.

Sandman, C. A., Kastin, A. J., and Schally, A. V., 1981, Neuropeptide influences on the central nervous system: A psychobiological perspective, in: *Neuroendocrine Regulation and Altered Behavior* (P. D. Hrdina and R. L. Singhal, eds.), pp. 5–25, Plenum Press, New York.

Scharrer, E., and Scharrer, B., 1940, Secretory cells within the hypothalamus, in: *The Hypothalamus and Central Levels of Autonomic Function* (J. F. Fulton, ed.), Res. Publ. Assoc. Nerv. Ment. Dis. Vol. 20, pp. 170–174, Hafner, New York.

Seeman, P., Lee, T., Chau-Wong, M., and Wong, K., 1976, Antipsychotic drug doses and neuroleptic/dopamine receptors, *Nature (London)* **261**:717.

Shah, A. B., Poiletman, R., and Shah, N. S., 1982, Influence of nisoldipine—a "calcium channel blocking agent" on drug induced stereotyped behavior in young rats: A new approach, *Pharmacologist* **24**:160.

Shah, A. B., Poiletman, R. M. and Shah, N. S., 1983, The influence of nisoldipine—a "calcium entry blocker" on drug induced stereotyped behavior in rats, *Prog. Neuropsychopharmacol. Biol. Psychiatry* in press.

Shan, N. S., and Donald, A. G. (eds.), 1982, *Endorphins and Opiate Antagonists in Psychiatric Research: Clinical Implications*, Plenum Press, New York.

Siler, T. M., Vandenberg, G., Yen, S. S. C., Brazeau, P., Vale, W., and Guillemin, R., 1973, Inhibition of growth hormone release in humans by somatostatin, *J. Clin. Endocrinol. Metab.* **37**:632.

Smith, J. R., Latham, T. R., Chestnut, R. M., Carino, M. A., and Horita, A., 1976, Thyrotropin-releasing hormone: Stimulation of colonic activity followig intracerebroventricular administration, *Science* **196**:660.

Sterba, G., 1974, Ascending neurosecretory pathways of the peptidergic type, in: *Neurosecretion—The Final Neuroendocrine Pathway* (F. Knowles and L. Vollrath, eds.), pp. 38–47, Springer-Verlag, Berlin.

Terenius, L., 1973, Somatostatin and ACTH are peptides with partial antagonist-like selectivity for opiate receptors, *Eur. J. Pharmacol.* **38**:211.

Uretsky, N. J., and Snodgrass, S. R., 1979, The involvement of calcium in the amphetamine-induced stimulation of dopamine synthesis in striatal slices, in: *Catecholamines: Basic and Clinical Frontiers* (E. Usdin, I. J. Kopin, and J. Barchas, eds.), pp. 82–84, Pergamon Press, Elmsford, N.Y.

Wakabayashi, I., Miyazawa, Y., Kanda, M., Miki, N., Demura, R., Demure, H., and Shizume, K., 1977, Stimulation of immunoreactive somatostatin release from hypothalamic synaptosomes by high [$K^+$] and dopamine, *Endocrinol. Jpn.* **24**:601.

Winnik, H. Z., and Tennenbaum, L., 1955, Apparition de galactorrhee au conours du traitement de largactil, *Presse Med.* **63**:1092.

Winokur, A., and Beckman, A. L., 1978, Effects of thyrotropin-releasing hormone, norepinephrine and acetylcholine on the activity of neurons in the hypothalamus, septum and cerebral cortex of the rat, *Brain Res.* **150**:205.

Yarbrough, G. G., 1979, On the neuropharmacology of tryrotropin releasing hormone (TRH), *Prog. Neurobiol.* **12**:291.

CHAPTER 1

# Neuroendocrinology and Brain Peptides
## An Emerging New Frontier in Neurobiology

JOSEPH B. MARTIN

## 1. INTRODUCTION

This chapter will describe recent advances in the field of neuroendocrinology and brain peptides. First, principles of neurosecretion, including the processes of peptide hormone biosynthesis, are considered. Then, two examples of hypothalamic–pituitary regulatory systems, those for prolactin and ACTH, are examined because they illustrate principles of neuroendocrine control. Last, the emerging field of brain peptides is reviewed. It is apparent that peptide substances, which include the hormones of the hypothalamus and pituitary, are of great importance for an understanding of brain function.

## 2. NEUROSECRETION

The importance of the hypothalamus in the regulation of various autonomic functions such as temperature regulation, water balance, food intake, and emotional behavior has long been recognized. These effects are mediated primarily by the autonomic nervous system. The hypothalamus also serves as the "final common pathway" for neuroendocrine regulation. This function is achieved by special neurons called *neurosecretory cells*.

Neurosecretory neurons can be defined as specialized cells that provide a link between the nervous and endocrine systems. As such they provide the final common pathway for neuroendocrine integration. Several types of communication have been described (Fig. 1).

---

JOSEPH B. MARTIN • Department of Neurology, Massachusetts General Hospital, Harvard Medical School, Boston, Massachusetts 02114.

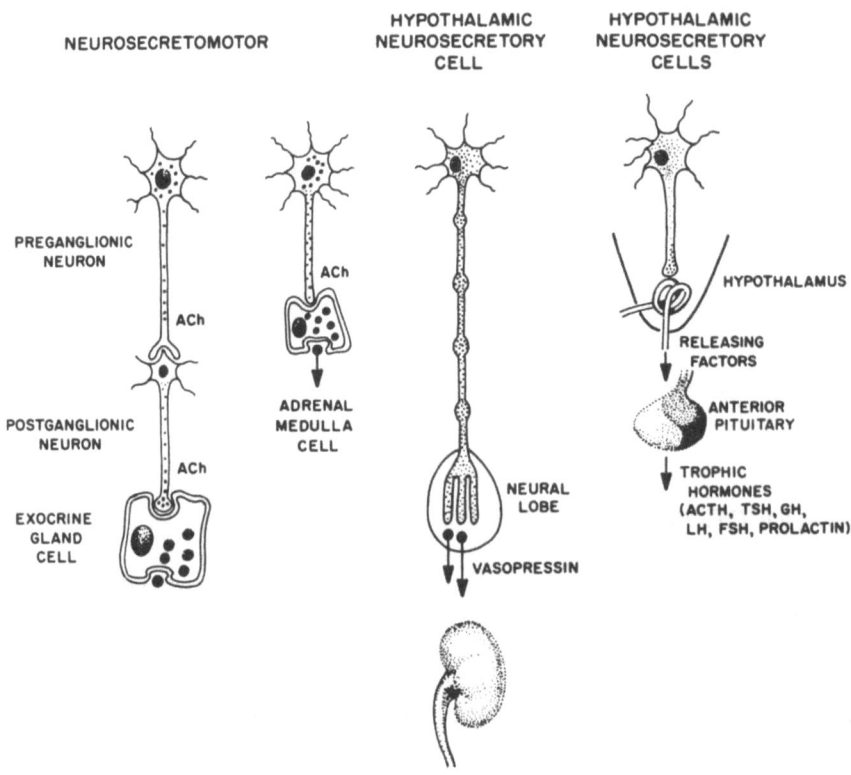

FIGURE 1. Examples of neurosecretory functions attained by peptide-producing cells.

The neurosecretory system that has been most intensively investigated is the hypothalamic–neurohypophysial system. This system consists of neurons, the cell bodies of which are located in the supraoptic and paraventricular nuclei of the hypothalamus. The unmyelinated axons arising from these cells traverse the basal hypothalamus to form the infundibulum, ending in the posterior lobe of the pituitary (neurohypophysis). Some fibers from the paraventricular nucleus also end in the median eminence. Axons terminate directly upon capillaries of the posterior lobe and median eminence and hormones released from the terminals have immediate access to the general circulation. The cells of the supraoptic and paraventricular neurons and their processes contain "neurosecretory material" (NSM). Electron microscopic studies have demonstrated the morphological counterpart of NSM as membrane-bound, central electron-dense neurosecretory granules (NSG) of diameter 1000–3000 Å. The polypeptide hormones, vasopressin and oxytocin, are contained within NSG in association with a precursor molecule, neurophysin, which has a molecular weight of approximately 10,000. It is now known that vasopressin is derived from a large-molecular-weight precursor, the structure of which has now been determined. The precursor in the bovine hypothalamus contains vasopressin, neurophysin, and an additional glycoprotein of unknown function (Fig. 2) (Land et al., 1982).

In order to establish the concept of a neurosecretory neuron, it is necessary to show that such cells are, in fact, neurons, and thus capable of those functions classically attributed to neurons. Morphologically, neurosecretory cells show the characteristics of neurons; axons and dendrites, well-developed endoplasmic reticulum, Golgi complexes, and neurofibrillae. Electrophysiologic studies have documented the electrical characteristics of these cells. Action potentials can be recorded from the neurohypophysial stalk in association with release of posterior pituitary hormones, and retrograde cell discharges have been demonstrated by microelectrodes in the region of the supraoptic nuclei following stimulation of the neurohypophysis or infundibulum. Microelectrode studies have shown alterations in firing rates following physiological stimuli known to produce hormonal release.

## Synthesis and Transport of NSG

There is convincing evidence that vasopressin and oxytocin are synthesized within the perikaryon of the neuron in a manner similar to that occurring

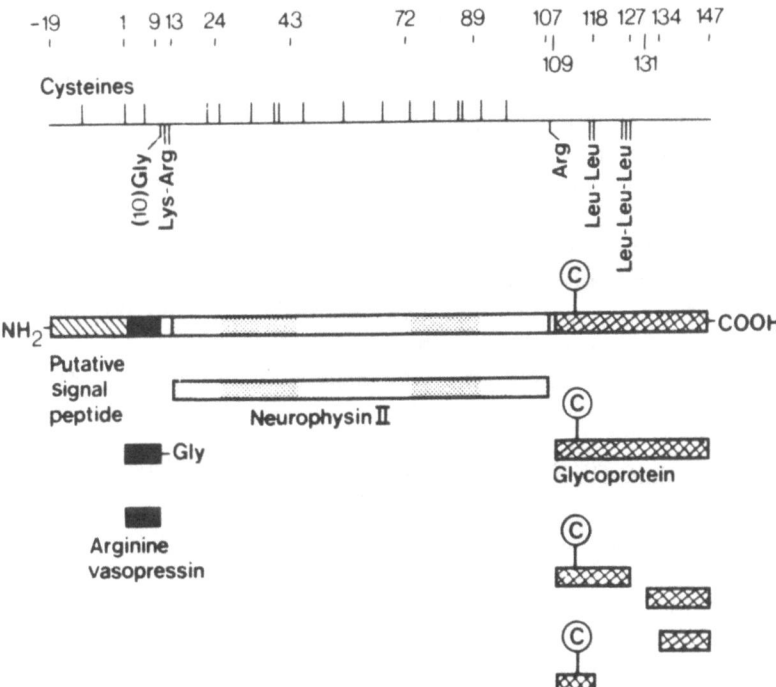

FIGURE 2. Structure of the precursor for vasopressin. The nucleotide sequence was obtained from a cloned cDNA that encodes for the bovine vasopressin—neurophysin II—precursor. [From Land et al., 1982, Nature (London) **295**:299–303, with permission.]

in other hormone-secreting cells. Thus, the polypeptides are synthesized in the endoplasmic reticulum and carried to the Golgi apparatus where they are packaged into NSG.

The evidence for transport of NSG down the axons to the neurohypophysis is also convincing. Section of the hypothalamic–hypophysial tract results in a distal depletion of stainable NSM with an accumulation proximal to the cut. The incorporation of labeled amino acids into vasopressin occurs first in the hypothalamus following administration of the label into the third ventricle, and only later appears in the posterior pituitary; section of the pituitary stalk prevents the appearance of labeled vasopressin in the neurohypophysis. *In vitro* incubation studies using hypothalamic tissue have shown that such tissue is capable of *de novo* vasopressin synthesis, whereas isolated neurohypophysial tissue is not. The process of production and storage within NSG of vasopressin, neurophysin, and the larger glycoprotein appears to be a common mechanism shared by other neurons that produce brain peptides. Thus, large precursors for the endorphins, enkephalins, and somatostatin have also been described. Recent anatomic observations indicate that the hypothalamic magnocellular neurons that produce vasopressin and neurophysin are distributed in several additional anatomic pathways (Swanson and Sawchenko, 1980; Swanson et al., 1980; Zimmerman, 1981). Axons with identified secretory granules can be traced between ependymal cells that line the third ventricle to end blindly in the cerebrospinal fluid (CSF). Arginine vasopressin can be detected in the CSF by radioimmunoassay, and it has been speculated that CSF distribution of the peptide may have functional importance. Another anatomic pattern of distribution of the hypothalamic neurosecretory system occurs, arising primarily from the paraventricular neurons, with extrahypothalamic terminations in the septum and thalamus, the locus coeruleus and nucleus tractus solitarius of the brain stem, and the dorsal and intermediate gray matter of the spinal cord. Retrograde tracing studies with horseradish peroxidase suggest that the cells in the paraventricular nucleus that give rise to these extrahypothalamic projections are distinct from those with neural-lobe or median-eminence terminations, although they arise in the same magnocellular cluster of cells.

## 3. HYPOTHALAMIC REGULATION OF ANTERIOR PITUITARY

The embryological origin of the anterior and posterior divisions of the pituitary is believed to be separate. The neurohypophysis arises as a downward evagination of the base of the hypothalamus and, as such, forms a true extension of the brain. The neural regulation of this structure occurs by direct nerve fiber connections. The adenohypophysis is derived from the primitive mouth cavity (stomatodeum).

The importance of the hypothalamus in the control of certain functions of the anterior pituitary was first recognized at the turn of the century when clinicians associated hypothalamic disease with anterior pituitary insufficiency.

## TABLE 1
### Anterior Pituitary and Hypophysiotrophic Hormones

| Pituitary hormone | Hypophysiotrophic hormones | |
|---|---|---|
| | Name | Structure |
| Thyrotropin (TSH) | Thyrotropin-releasing hormone (TRH) | Tripeptide |
| Adrenocorticotropin (ACTH)[a] | Corticotropin-releasing factor (CRF) | 41 amino acids |
| β-Lipotropin (β-LPH)[a] | CRF | 41 amino acids |
| Luteinizing hormone (LH) | Luteinizing hormone-releasing hormone (LHRH) OR | Decapeptide |
| Follicle-stimulating hormone (FSH) | Gonadotropin-releasing hormone (GnRH) | |
| Growth hormone (GH) | Growth hormone-releasing factor (GRF) | 44 amino acids[d] |
| | Growth hormone release-inhibiting hormone[b] (somatostatin, GIH) | 14-amino-acid peptide |
| Prolactin | Prolactin release-inhibiting factor (PIF) | Dopamine |
| | Prolactin-releasing factor (PRF)[c] | Unknown |

[a] Share a common precursor.
[b] Somatostatin also blocks TRH-stimulated TSH release.
[c] TRH stimulates prolactin release.
[d] For recent review see Martin, J. B., 1983, *Trends Neurosci.* 6:1–3.

With the recognition of the dependency of the anterior pituitary on intact hypothalamic function, the search began for a mechanism to explain such regulation.

The innervation of the anterior pituitary is remarkably sparse and the nerve fibers present are fine and probably exclusively of postganglionic sympathetic origin. Anterior pituitary regulation is achieved by secretion of peptides and biogenic amines into the pituitary portal circulation. During the past decade, several hypothalamic hormones responsible for hypothalamic regulation have been identified (Table 1). The structures of thyrotropin-releasing hormone (TRH), luteinizing hormone-releasing hormone (LHRH), somatostatin, and corticotropin-releasing factor (CRF) are now known (Fig. 3) (Vale *et al.*, 1981).

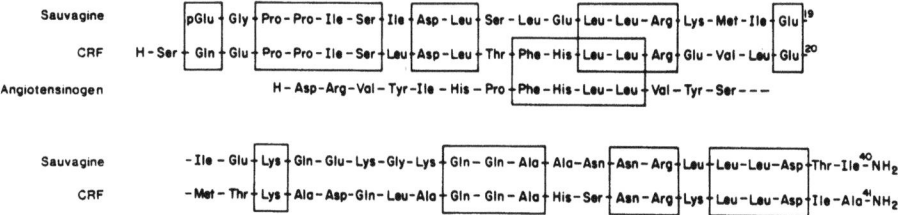

FIGURE 3. Primary sequence of ovine CRF, also showing homologies with sauvagine and angiotensinogen (bovine) (From Vale *et al.*, 1981, *Science* 213:1394–1397, with permission.)

## 3.1. Structure of the Median Eminence Gland

The cells of origin of the axons which end on the portal capillaries appear to be located almost exclusively along the periventricular margin of the third ventricle (for review see Martin et al., 1977). These nuclear groups are composed of small neurons. The axons of these cells can be traced in an inferior direction to end in the median eminence forming the tuberohypophysial tract. As the axons descend through the median eminence, they crisscross with the axons of the supraoptic and paraventricular–hypophysial tracts which are found in the deep portion of the median eminence clustered on either side of the midline. The median eminence anatomically can be divided into three zones; (1) an inner ependymal zone which comprises the lining of the inferior portion of the third ventricle, (2) an inner palisade layer which contains the hypothalamic–neurohypophysial tract, and (3) an outer palisade layer where terminals of the tuberohypophysial tract end on the capillary loops of the portal plexus. The capillary loops arise from the branches of the superior hypophysial arteries (which originate from the internal carotids) and penetrate the surface of the median eminence to the region of the inner palisade layer.

## 3.2. Regulation of Prolactin Secretion

Pituitary prolactin secretion is regulated by a tonic hypothalamic inhibitory mechanism, which is primarily dependent upon dopamine release from the median eminence (Riskind and Martin, 1982). Current evidence indicates that dopamine is the principal prolactin inhibitory factor. Basal prolactin levels in the adult, as measured by radioimmunoassay, range up to 15 ng/ml during the day. In one report of a large series of women, mean levels averaged $7.9 \pm 0.4$ (S.E.M.) ng/ml; those in men measured $5.2 \pm 0.6$ ng/ml. The 24-hr secretion of prolactin, assessed by frequent blood sampling, is characterized by an episodic release pattern throughout the day and night with highest levels (up to 20–40 ng/ml) found during the early morning hours from 3 to 6 a.m. There is an abrupt fall to basal morning values upon awakening. The nightime rise in prolactin can be inhibited by dopaminergic drugs and by methysergide, an antiserotonergic agent that also possesses antidopaminergic activity. The rise is not affected by opioid antagonists such as naloxone and therefore does not appear to depend upon opioid peptides.

Prolactin levels rise during pregnancy, with maximal levels attained in the third trimester. Levels remain elevated after parturition and increase with each episode of nursing. Data from experimental animals indicate that this increase is mediated by serotonin. During lactation ovulatory cycles are inhibited. The prolonged postpartum amenorrhea that accompanies lactation is believed to be due, in part, to the inhibitory feedback effect of prolactin on hypothalamic LHRH secretion. The elucidation of this mechanism has aided in our understanding of the causative role of hyperprolactinemia in anovulatory amenorrhea.

In the human, agents that act as dopamine receptor antagonists cause a rise in serum prolactin levels. Thus, antidopaminergic drugs such as haloperidol, chlorpromazine, or pimozide, which block dopaminergic receptors, interfere with normal hypothalamic inhibition. Other drugs that deplete biogenic amines, such as reserpine and α-methyldopa, also may increase prolactin levels. In constrast, dopaminergic drugs such as L-dopa, apomorphine, or bromocriptine suppress prolactin secretion. These effects are mediated directly on the pituitary lactotrope, which has been shown to contain dopamine receptors on its cellular membrane. *In vitro* studies indicate that dopamine, in concentrations known to be present in the pituitary portal blood, binds to pituitary receptors to suppress prolactin secretion. Attachment of dopamine to these receptors is hypothesized to be followed by internalization of the dopamine–receptor complex. The net result of this receptor activation is inhibition of both prolactin synthesis and release. In addition, animal studies suggest that dopamine may exert a direct antiproliferative effect upon anterior pituitary cells. Dopamine or bromocriptine decreases the mitotic activity of cultured anterior pituitary cells, while antidopaminergic drugs enhance mitosis. Furthermore, bromocriptine facilitates degradation of prolactin *in vitro*. Thus, there is theoretical evidence to support the idea that dopamine agonists may decrease the size of lactotropes, as well as enhance degradation of prolactin, inhibit the synthesis of prolactin, and block the release of synthesized prolactin.

Several additional factors appear to function in the regulation of prolactin secretion. Prolactin is released by administration of TRH, a tripeptide known to be of physiologic significance in the regulation of thyrotropin (TSH). The stimulation of prolactin secretion by TRH occurs with concentrations of TRH equivalent to those required for stimulation of TSH secretion. However, the importance of TRH in the physiologic regulation of prolactin secretion is uncertain. Physiologic and biochemical evidence exists for the presence of a separate prolactin-releasing factor that is presumably responsible for increased secretion of prolactin following stress or suckling, stimuli that do not lead to increases in TSH levels. Opioid peptides also stimulate prolactin release; the significance of this effect is also unclear. Additional neurotransmitters influence prolactin secretion. Serotonin precursors result in an increase in prolactin secretion in man and this effect is mediated at a hypothalamic level by altering release of dopamine or of the prolactin-releasing factor. Histamine may also be an important mediator of hypothalamic control of prolactin secretion, as blockade of $H_2$ receptors causes a rapid increase in prolactin levels. This effect is reported to be absent in patients with prolactin-secreting pituitary tumors. More importantly, the defective response to $H_2$-receptor blockade is reported to persist in such patients after surgical removal of their adenomas, even when pituitary responses to TRH have normalized. If substantiated, this defect would suggest that a hypothalamic disorder may be responsible for the subsequent development of some prolactinomas. Supporting this idea are observations that a dysfunction of hypothalamic dopaminergic transmission may be the underlying impetus for formation of certain prolactinomas.

Pituitary lactotrope size and secretory responsivity are also regulated by estrogen effects. Increased estrogen secretion in pregnancy normally stimulates enlargement of the pituitary gland due to prolactin cell hyperplasia and may also markedly stimulate enlargement of prolactin-secreting pituitary adenomas. Estrogens may interact directly with the dopamine receptor on the pituitary lactotrope.

### 3.3. Regulation of ACTH Secretion

Human ACTH is a 39-amino-acid peptide. The half-life of ACTH in blood is short, about 10 min, and the effects on adrenocortical function are rapid and transient. The normal pituitary secretion of ACTH consists of a series of five to eight episodic bursts during a 24-hr period. Cortisol secretion is also episodic and follows the secretory patterns of ACTH. Increases in the frequency and amplitude of these secretory bursts begin during the late night and reach maximum levels shortly after awakening in the morning. The typical circadian or diurnal secretion of ACTH and cortisol represents the summation of these individual episodic surges. ACTH secretion can be further increased by a variety of stresses and by insulin-induced hypoglycemia.

Pituitary corticotropes also secrete several other peptide products that are synthesized from a common proopiocortin precursor of approximately 31,000 molecular weight. These peptides include β-lipotropin and β-endorphin, as well as a 16,000-molecular-weight fragment. The functional significance of the release of these substances is unclear but the same factors that regulate ACTH secretion also regulate their release.

The regulation of pituitary ACTH secretion is dependent on corticotropin-releasing factor (CRF), the structure of which has recently been characterized (Fig. 3) (Vale *et al.*, 1981). CRF is produced in neurons of the medial basal hypothalamus and is released into the portal circulation to stimulate ACTH secretion, which in turn activates the adrenal cortex to increase synthesis and release of cortisol. Cortisol acts on both the hypothalamus and the pituitary to exert feedback inhibition on ACTH secretion.

#### Pharmacology of ACTH

ACTH secretion is also regulated by other substances that appear to act directly on the pituitary. *In vitro* studies have shown that norepinephrine stimulates secretion of both ACTH and β-lipotropin, an effect mediated by α-adrenergic receptors, since the response is blocked by phenoxybenzamine and phentolamine. Vasopressin is also effective in eliciting ACTH secretion. Inhibition of CRF-stimulated ACTH secretion occurs after administration of progesterone, prostaglandins ($PGE_1$, $PGE_2$, and $PGF_2$), and glucocorticoids. The physiologic importance of each of these substances is still uncertain, although glucocorticoids clearly function to exert negative feedback inhibition.

Several hypothalamic biogenic amine systems are known to be important for regulation of CRF secretion. Most studies in experimental animals indicate that norepinephrine is inhibitory to CRF secretion by actions mediated on α-adrenergic receptors and that serotonin is stimulatory. Serotonin antagonists such as cyproheptadine and metergoline have been advocated in the therapy of Cushing's disease. A cholinergic stimulatory component of CRF regulation has also been suggested.

## 4. BRAIN PEPTIDES

The demonstration that a number of peptides are localized in the brain has resulted in major new developments in neuroendocrinology and the neurosciences. The surprise and disbelief occasioned by early reports of hypothalamic releasing hormones located in extrahypothalamic sites, and of the presence of pituitary and gastrointestinal hormones in brain, have given way to attempts to determine the role of each peptide in the central nervous system (CNS) (for more detailed review see Krieger and Martin, 1981).

The concept of a substance acting as both a neurotransmitter and a hormone is not unique, having been accepted for several decades in the cases of norepinephrine and epinephrine. Although the presence of the same peptide in the brain and in another tissue would suggest concurrent regulation and interaction in various physiological responses, to date there is little compelling evidence to support such a suggestion. Rather it appears that nature is both parsimonious and imaginative, using the same substance for multiple functions.

The various categories of peptides described in brain are indicated in Table 2. The hypothalamic releasing and release-inhibitory hormones, which serve a neuroendocrine function to regulate anterior pituitary hormone secretion, have a wide distribution throughout the CNS that is compatible with their additional postulated behavioral roles. The same is true for the posterior pituitary peptides vasopressin and oxytocin, which are found in extrahypothalamic fiber projections in addition to the hypothalamic–neurohypophysial system. A number of peptides, originally found in secretory elements of the gastrointestinal tract including substance P, cholecystokinin (CCK), and vasoactive intestinal polypeptide (VIP), are now described as occurring in both endocrine and nerve cells in the gut and also in brain.

It is now recognized that ACTH and β-LPH, which were initially isolated separately from anterior pituitary extracts, are derived from a common precursor molecule (Mains *et al.*, 1977; Roberts and Herbert, 1977; Nakaniski *et al.*, 1979) and that both are found in the brain. This precursor also contains several recurrent sequences identified as α-, β-, and γ-melanocyte-stimulating-hormone (α-, β-, γ-MSH). The peptide β-endorphin corresponds to residues 61–91 of β-LPH.

Thus, it is evident that two groups of cells synthesize this precursor and produce a "family" of peptides: (1) the corticotrophs in the anterior pituitary,

TABLE 2
Categories of Brain Peptides

Hypothalamic releasing hormones
    Thyrotropin-releasing hormone (TRH)
    Gonadotropin-releasing hormone (GNRH)
    Somatostatin
    Corticotropin-releasing factor (CRF)
Neurohypophysial hormones
    Vasopressin
    Oxytocin
    Neurophysin(s)
Gastrointestinal peptides
    Vasoactive intestinal polypeptide (VIP)
    Cholecystokinin (CCK-8)
    Gastrin
    Substance P
    Neurotensin
    Methionine enkephalin
    Leucine enkephalin
    Insulin
    Glucagon
    Bombesin
    Secretin
    Somatostatin
    TRH
Pituitary peptides
    Adrenocorticotropic hormone (ACTH)
    β-Endorphin
    Melanocyte-stimulating hormone (α-MSH)
    Prolactin
    Growth hormone
    Luteinizing hormone
    Thyrotropin
Others
    Angiotensin II
    Bradykinin
    Carnosine
    Sleep peptide(s)
    Calcitonin

which secrete the peptides into the blood as hormones, and (2) neurons of the arcuate nucleus of the hypothalamus, which send fibers both to the median eminence and to other extrahypothalamic brain regions.

In general, hormones are present in brain in much lower concentrations ($10^{-12}$ to $10^{-15}$ M/mg protein) than those of the classical biogenic amine neurotransmitters, such as acetylcholine, norepinephrine, and dopamine, which are found in concentrations of $10^{-9}$ to $10^{-10}$ M/mg protein. The amino acid neurotransmitters γ-aminobutyric acid (GABA), aminoacetic acid (glycine), and glutamate are found in even higher concentrations ($10^{-6}$ to $10^{-8}$ M/mg protein). Hypothalamic and pancreatic concentrations of somatostatin are approximately equivalent, whereas the concentration of brain ACTH is approximately 1/1000 that found in the anterior pituitary, and that of brain insulin is approximately 1/1000 that present in pancreas. CCK seems thus far to be the only peptide with a CNS concentration greater than that described in the gastrointestinal tract. In general, however, *CNS concentrations of peptides derived from homogenized tissue studies do not reliably indicate their functional importance.* Concentration does not indicate turnover; additionally, regional, cellular, and subcellular distribution of peptides are likely of greater biologic importance than overall determinations of concentrations in gross areas of the brain and spinal cord.

Because it is estimated that nonpeptide neurotransmitters may account for only 40% of the synapses in the CNS, it is apparent that the elucidation of the role of peptides both discovered and unknown will provide major insights into the functions of the CNS.

As additional peptides are described in brain because of advances in modern peptide purification, sequencing, and synthesis and because of the development of antibodies for radioimmunoassays and immunocytochemical studies, it is essential to validate the methods used to detect them and to ascertain whether the forms identified in the CNS are identical to those present in the originally described site of origin. In attempting to assign a functional role for each peptide, it is also necessary to determine whether they are synthesized in the brain or occur there as a result of vascular transport—either from the general circulation or, in the case of the pituitary hormones, through retrograde circulation in the pituitary portal venous system.

Table 3 shows the types of questions that arise when a peptide that has previously been known to be present in a given tissue is discovered in another, along with the methods available to answer such questions (Krieger and Martin, 1981). Similarity of structure and cellular localization have been investigated mainly by immunoassay and immunocytochemistry, with only occasional confirmation by bioassay or peptide sequencing. Neither immunocytochemistry nor immunoassay alone provides absolute identification. Questions of specificity arise with these techniques both with respect to antibody cross-reactivity with similar amino acid sequences in known but unrelated peptides, and potentially with similar sequences in presently unknown peptides. Furthermore, at the low dilutions of multivalent antisera used in immunocytochemistry, cross-reactivity may occur with substances that exhibit no cross-reactivity at the higher antibody dilutions used in radioimmunoassay. Immunocytochemistry is also less sensitive than radioimmunoassay. Thus, reports of negative results in cell-body localization cannot be taken as firm evidence. Indeed, the use of additional maneuvers such as colchicine treatment and nerve ligation, which block axonal transport, has shown cell bodies to be present where they were previously declared absent. Problems in the identification of peptide receptors in the brain and issues regarding the biosynthesis of peptides in neural tissues are discussed below.

## TABLE 3
### Characterization of a Peptide Occurring in Different Tissues[a]

| Question | Methods |
| --- | --- |
| 1. Are the forms detected in the tissue similar? | Bioassay<br>Immunoassay (multiple antibodies)<br>Physicochemical characterization<br>Sequence determination |
| 2. What is its cellular localization? | Immunocytochemistry<br>Regional immunoassay<br>Subcellular localization |
| 3. Are peptide receptors present? Are they similar in the two tissues, and what is their distribution? | Membrane binding studies<br>Radioautography |
| 4. Is it produced locally or transported from another site? | Effects of lesions of one postulated site of production<br>Determination of blood–brain transport<br>Demonstration of synthesis:<br>  Accumulation in cells and media during culture (indirect)<br>  Incorporation of labeled amino acids into peptide products<br>  Use of cDNA probe derived from one tissue to identify mRNA in other<br>  Extraction of mRNA and translation in a cell-free system |
| 5. If synthesis is demonstrated, are the precursor form (if present) and its processing similar in the tissues? | Physicochemical characterization of synthesized forms (size, charge, tryptic digests)<br>Pulse-labeling experiments |

[a] From the *New England Journal of Medicine* (Krieger and Martin, 1981).

### 4.1. Evolutionary and Embryological Aspects

The presence of similar peptides in different tissues has raised important evolutionary issues that may provide insight into the structure, function, mutation, and duplication of genes. Neurotransmission and neurosecretion of peptides occur in coelenterates and annelids that have no recognizable endocrine tissue (Dockray, 1979). The presence of CCK and somatostatin in the gastrointestinal tract and brain has been noted as early in invertebrate evolution as the lamprey (Vale *et al.*, 1976). Even more striking is the demonstration of substances similar to mammalian insulin in unicellular eukaryotes (*Tetrahymena pyriformis, Neurospora crassa, Aspergillis fumigatis, E. coli*) (Le Roith *et al.*, 1980). Immunoreactive and bioreactive ACTH and endorphin have also been described in *Tetrahymena* (Le Roith *et al.*, 1981). These observations raise the possibility that most cells have a constant low level of expression of many peptides. In simpler organisms, peptides may have functions different from those that have evolved in higher organisms; with increasing evolutionary complexity, there may be selective enhanced expression of peptide biosynthesis in specific tissues. There is also evidence for evolution of specific peptide receptors (Mains et al., 1977).

Reports of similar peptides in different tissues also raise questions concerning the embryological origins of peptides. Observations that peptide expression may be a common feature of many cells mitigate against the necessity for attempting, as has been done previously, to establish a common embryological origin for tissues that express the same peptides. Pearse (1969) postulated that all tissues containing similar peptides were of neural crest origin. It is now realized that the pineal gland, anterior pituitary, and hypothalamus arise not from the neural crest but from neuroectoderm or specialized ectodermal placodes. The origin of the peptide-producing cells of the gastrointestinal tract remains uncertain.

## 4.2. Anatomical Pathways of Peptidergic Neurons

The generation of antibodies specific for each of the neuropeptides has provided a powerful tool to investigate their neuronal localization and fiber pathways in the CNS (Hökfelt et al., 1978, 1980). To date, immunocytochemical studies have shown that perikarya containing certain neuropeptides are confined to the hypothalamus and preoptic area. These substances include vasopressin and oxytocin and their respective neurophysins; the proopiocortin-ACTH, β-lipotropin, and β-endorphin; LHRH; and bradykinin. In contrast, somatostatin, TRH, neurotensin, substance P, CCK, VIP, insulin, and the enkephalins are found in perikarya distributed widely throughout the brain and spinal cord.

Subcellular studies of extrahypothalamic brain tissue have shown that neuropeptide concentration is greatest in the synaptosomal nerve terminal fraction (Epelbaum et al., 1977). These observations are similar to those obtained in studies of conventional neurotransmitters, including norepinephrine, acetylcholine, and dopamine, and they indicate that the principal storage site for the neuropeptides in brain is in the synaptic region. *In vitro* incubation studies of hypothalamus, of other brain regions, and of synaptosomes have shown that neuropeptide release occurs with neuronal depolarization induced either by excess potassium or by electrical stimulation (Iversen et al., 1978; Terry et al., 1980). Release is dependent on cellular uptake of calcium and is facilitated by calcium ionophores. Release can also be stimulated or inhibited by application of conventional neurotransmitters, indicating interaction between the neurotransmitters and peptides, although the data are not sufficient to permit definite conclusions about such interactions.

Light microscopic immunocytochemical studies of serial sections of brain have also provided morphologic evidence for coexistence of neuropeptides and conventional neurotransmitters within the same cell. Coexistence of peptides and biogenic amines is known to occur in endocrine cells of the gastrointestinal tract. As described by Hökfelt et al. (1980), serotonin has been found together

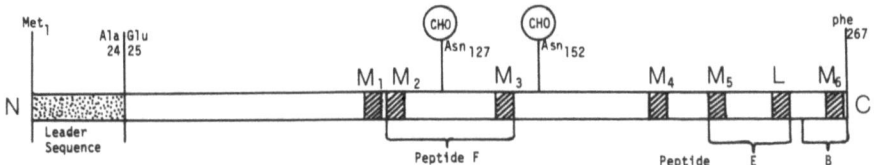

FIGURE 4. Low-resolution model of enkephalin precursor showing distribution of Met-enkephalin ($M_1$–$M_6$) and Leu-enkephalin (L). (From Comb et al., 1982, Nature **295**:663–666, with permission.)

with substance P and with TRH in neurons of the medulla oblongata. CCK and dopamine are postulated to occur in cells of the mesencephalon ($A_{10}$ region) adjacent to the substantia nigra. Autonomic ganglia sympathetic nerves, and the adrenal medulla contain several different peptides and amines, some of which undoubtedly reside in the same cells. In many nerve endings, both "synaptic" vesicles of small diameter (30–50 nm) and larger granular vesicles (100–150 nm) can be identified. Immunocytochemical analysis with the electron microscope has identified reaction products for neuropeptides in synaptic terminals, usually in relation to the larger dense-core granules.

Of great importance is the discovery that enkephalins, first isolated from porcine brain (Pert and Synder, 1973; Hughes et al., 1975), are derived from a separate precursor (Fig. 4) (Comb et al., 1982) from that giving rise to β-endorphin. This precursor, fully characterized by cDNA probes in bovine adrenal tissue and in human pheochromocytoma, contains recurring sequences of Met-enkephalin. Thus, one precursor can serve as the source of more than one peptide, including several molecules of the same structure.

At the present time there are at least three separate families of endogenous opioid peptides in the mammalian brain, β-endorphin, the enkephalins, and dynorphin (Goldstein et al., 1979).

## 4.3. Peptide Receptors in Brain

The presence of peptides in the CNS has stimulated intense interest in uncovering cellular receptors for them. Current studies are directed at determining whether such receptors exist, the nature of their localization (i.e., presynaptic, postsynaptic, blood vessel), whether their distribution coincides with that described for the peptide, and whether more than one receptor type is present for a given peptide form or forms. At present, there is convincing evidence for several varieties of opioid receptors (Pert and Snyder, 1973; Hughes et al., 1975; Goldstein et al., 1979; Goodman et al., 1980; Comb et al., 1982).

There has been only preliminary characterization of receptor sites for other peptides. It is of interest that such characterization has been for those peptides such as VIP (Taylor and Pert, 1979), CCK (Saito et al., 1980), and insulin (Havrankova et al., 1978) which have a diffuse rather than a localized CNS representation. CCK receptor sites, labeled with [$^{125}$I]-CCK-33, have been shown in pancreas and brain, with different peptide specificities of the receptors in the two tissues. This observation may represent another example in which the presence of receptors for a given peptide in two different tissues, coupled with differential processing of the same peptide in each tissue, could explain the diverse physiological functions of the same peptide. A correlation has been observed between the location of immunoreactive CCK and its specific binding sites in the brain. There is also an intriguing report that CCK 27-33 binds to opiate receptors, albeit at markedly greater molar concentrations than that of opiates (Schiller et al., 1978).

Insulin receptors were found in the brain before there was detection of insulin itself. Studies using autoradiography by light and electron microscopy have indicated the presence of insulin-specific binding sites in association with rat brain blood vessels (van Houten and Posner, 1979), as well as on axons and axon terminals in the medial basal hypothalamus (van Houten et al., 1979). No difference in content of insulin receptors has been detected in brains obtained from normal and from diabetic rats (Pacold and Blackard, 1979). There are also limited reports of specific receptors in brain for substance P, TRH, somatostatin, and neurotensin.

### 4.4. Electrophysiological Effects

Neurophysiologic techniques, widely used to investigate the neuronal effects of conventional neurotransmitters, have also been applied to the neuropeptides. Neuropeptides applied by micropipettes to the extracellular environment of single neurons *in vivo* have given variable effects. Application of TRH, LHRH, and somatostatin to cells in the hypothalamus, spinal cord, cerebellum, and cerebral cortex in the rat resulted in rapid depression of ongoing electrical activity in some but not all neurons tested (Renaud et al., 1975; Phillis and Kirkpatrick, 1980). The proportion of cells that responded varied from region to region. Subsequent studies of neurons in cerebral cortex in rabbits have shown that somatostatin is predominantly excitatory (Ioffe et al., 1978). Application of Leu-enkephalin and Met-enkephalin to CNS neurons causes inhibition (Frenk et al., 1978). With an important exception noted on hippocampal neurons where excitation occurred, an effect that may correlate with the observed seizure-inducing effects of opioid substances.

Application of substance P to dorsal horn sensory neurons results in excitation, an effect that often occurs only after an initial latent period (Henry, 1976). Both somatostatin and Met-enkephalin are reported to inhibit neuronal discharges in the dorsal horn of the spinal cord in the cat (Randic and Miletic, 1978); both peptides are postulated to function in relay of pain impulses (see below).

Such *in vivo* studies have shown great variation in responses; the same peptide may be found to have either excitatory or inhibitory effects in a given brain area. Furthermore, the percentage of cells that respond is variable and the quantity of neuropeptide released is difficult to assess.

Although a great amount of work remains to be done to establish the effects of peptides on neuronal excitability, current observations indicate that neuropeptides act, at least in part, by rapid effects on neuronal membranes. It remains to be shown whether these effects are direct or are mediated by presynaptic influences on the release of other neurotransmitters.

## 4.5. Peptide Functions in the CNS

### 4.5.1. Brain Peptides as Neuronal Markers

Most degenerative neurological disorders are characterized by premature cell death in discrete neuronal populations. In some cases, the cellular components affected can be characterized partially by their functional roles in neuronal systems. Further delineation of the cellular basis of such neurological disorders may contribute important new information about their pathogenesis. Immunocytochemical characterization of the cell types commonly affected in degenerative diseases, together with quantitative measurement of regional peptide concentrations, may provide clues to the neuronal specificity (or lack of it) involved in a given disorder. It is conceivable that one or more of the metabolic processes subserving synthesis, processing, or degradation of specific peptides might be the locus of a pathological process resulting in or contributing to cell death. Neuronal dysfunction might also result in defects in peptide release leading to loss of transsynaptic or trophic effects normally exerted on other neurons.

The idea that a specific neurotransmitter deficiency is important in the pathophysiology and clinical manifestations of a neuronal system degeneration has been developed most completely in Parkinson's disease and Huntington's chorea. Both disorders have specific clinical and neuropathologic features. Their pathogenesis is different, since Parkinson's disease is usually sporadic, whereas Huntington's disease is an autosomal dominant disorder with a high degree of penetrance.

Huntington's disease affects other neuronal populations: cellular loss is most extensive in the putamen and caudate nucleus and less widespread in the lower laminae of the cerebral cortex. Other areas of the basal ganglia are less affected. There is no explanation for this pattern of cell death. Neurons in these regions originate from different areas of the brain during development, differ markedly in cytological features and connectivity, have dissimilar functions, and contain several neurotransmitters. The cellular loss in the putamen and caudate is accompanied by a loss of the neurotransmitter GABA, and of glutamic acid decarboxylase, its synthesizing enzyme (Bird, 1980). However, it is clear that the deficiency of this neurotransmitter is not in itself the cause of cell death because GABA-containing neurons elsewhere in the brain survive.

The neuronal loss in Huntington's disease is also accompanied by changes in the regional content of acetylcholine and certain peptides. The concentration of substance P is decreased in the globus pallidus and the putamen (Bird, 1980). Levels of CCK (Rehfeld, 1980) and Met-enkephalin (Emson *et al.*, 1980), are reported to be decreased in the basal ganglia in Huntington's disease, whereas VIP levels are normal (Emson *et al.*, 1979). We have recently found that concentrations of somatostatin are significantly increased in the caudate, putamen, and globus pallidus (Aronin *et al.*, 1983). Spindel *et al.* (1980) have also found an increase in TRH in postmortem tissue in cases of Huntington's disease. These findings point to a degree of neuronal selectivity in the degenerative process.

Alzheimer's disease, the most common cause of dementia in the United States, is characterized by loss of neurons in specific areas of the cerebral cortex, including the association areas of the orbital frontal, parietal, and temporal lobes. Changes are particularly extensive in the hippocampus. The disorder is usually sporadic, but occasionally familial, and occurs with increasing incidence in the seventh decade and beyond. The nature of the degenerative process is entirely unknown. The earliest cellular lesion, the neurofibrillary tangle, consists of an ultrastructurally identifiable alteration in the neurofilaments and neurotubules. Several studies have documented a decrease in acetylcholine content and choline acetyltransferase activity in the cortical regions affected. Recent reports provide evidence that a decrease in somatostatin concentration also occurs in the hippocampus and other cortical regions in this disease (Davies *et al.*, 1980).

### 4.5.2. Peptides in the CSF

Various peptides have been detected by radioimmunoassay in the CSF (Table 4) (Jackson, 1980; Martin and Landis, 1981). The mechanism by which peptides find access to the CSF is unknown. The CSF is generally considered to be an important route for the removal of substances from the extracellular space of the brain, which is contiguous with the CSF, but it is also possible that such substances may penetrate from the CSF space into the brain extracellular space. Thus, secretion of a peptide into the CSF could serve a function for dissemination of the peptide to the brain parenchyma at a remote site. Such a mechanism has not as yet been demonstrated.

It is of interest to note that the behavioral effects of peptides appear often to be rather specific for an individual peptide, after administration into the cerebroventricular or cisternal route. Although the concentrations given generally exceed those normally found in CSF under physiological conditions, the specificity of such effects implies that the peptide circulates via the CSF to reach target receptor sites that evoke selective responses. Immunoreactivity of peptides in the CSF appears to be quite stable, even when fluid is kept at room temperature for several hours, suggesting that degradation is minimal.

TABLE 4
Peptides Detected in Human CSF

TRH
LHRH
Somatostatin
Prolactin
Growth hormone
ACTH
β-Endorphin
Enkephalins
Neurohypophysial hormones (vasopressin, oxytocin, neurophysin)
Gastrin
CCK
VIP
Angiotensin II
Substance P
Calcitonin

However, full characterization of the molecules that demonstrate peptide immunoreactivity in the CSF has not been accomplished for most peptides.

Several reports have documented changes in the CSF concentrations of peptides in neurological and psychiatric disorders. Levels of somatostatin are increased in the lumbar CSF in a variety of destructive and degenerative disorders of the CNS (Patel et al., 1977). CSF levels of substance P are reported to be lowered in certain peripheral neuropathies and in idiopathic postural hypotension and to be elevated in spinal arachnoiditis (Nutt et al., 1980). Concentrations of immunoreactive β-endorphin and ACTH in the CSF are reported to be normal in schizophrenia and acromegaly; a significant reduction in levels of both occurs in Cushing's disease (Nakao et al., 1980). It is possible that assessment of peptide concentrations in chronic pain conditions, narcolepsy, dystonia musculorum deformans, schizophrenia, and manic-depressive illness may contribute useful clinical information. A limiting factor, as in measurements of the CSF concentrations of the metabolic products of conventional neurotransmitters, is the fact that lumbar CSF samples may reflect principally the release of peptides from the spinal cord rather than the brain.

### 4.5.3. Pain

The relay of painful sensations from peripheral tissues to the CNS occurs via unmyelinated or thinly myelinated nerve fibers which terminate in the outer laminae of the dorsal horn of the spinal cord. Fibers from the face end in the medulla oblongata in the spinal nucleus of the fifth cranial nerve. Recent experiments have implicated the importance of substance P as a potential neurotransmitter for the relay of nociception (Jessell, 1981). In addition to the initial processing of nociceptive inputs that occurs at the spinal cord level, there are also descending inputs from brain stem centers to spinal cord that exert effects on local circuits.

The evidence that substance P is a primary sensory neurotransmitter is convincing (Nicoll *et al.*, 1980). It is released from spinal cord after depolarization by potassium or electrical stimulation. Peripheral nerve stimulation of nociceptive fibers elicits release of the peptide into the CSF. Microiontophoretic application of substance P results in prolonged excitation of dorsal horn neurons which can also be shown to be activated by noxious peripheral stimuli.

Analgesia can be induced by treatment with agents that cause substance P depletion. Capsaicin, a substance obtained from red pepper and shown to be structurally related to homovanillic acid, causes release of substance P from the spinal cord, which is accompanied by elevated peripheral thresholds to painful stimuli (Yaksh *et al.*, 1979). An interaction of substance P with enkephalins at the spinal cord level is suggested by evidence that enkephalins inhibit release of substance P from sensory neurons in culture (Mudge *et al.*, 1979).

Peptides also function in pain pathways at other sites in the CNS. In the early 1970s it was shown that electrical stimulation of the ventrolateral periaqueductal gray (PAG) of the midbrain produces analgesia in animals and man (Basbaum and Fields, 1978). The presence of opioid receptors has been documented in similar brain regions. Considerable overlap is found between the sites of stimulation-produced analgesia and analgesia induced by intracerebral opiate injection, the PAG being the most effective site for both. These data suggest that endogenous opioids might be active in this system. In addition to high receptor concentration in the PAG, immunohistochemical studies show a rich field of enkephalin terminations. Further evidence that opioid peptides are involved in the pain relay at this level is the finding that analgesia produced by PAG stimulation in humans is partially blocked by naloxone and is associated with a rise in β-endorphin immunoreactivity in the CSF (Akil *et al.*, 1978). Intraventricular administration of β-endorphin is effective in causing analgesia in experimental animals, whereas the intravenous route is not, presumably because of failure to cross the blood–brain barrier. Intrathecal injection of β-endorphin in man is effective in causing profound analgesia.

### 4.5.4. Effects of Peptides on Behavior

Evidence has accumulated, over the past two decades, to indicate that impaired conditioned avoidance behavior, present in hypophysectomized animals, is corrected by systemic or cerebroventricular administration of ACTH, α-melanotropin (α-MSH), and vasopressin (Dogterom *et al.*, 1977; de Kloet and de Wied, 1980). This effect can also be induced by fragments or analogs of these hormones that are devoid of any endocrine effects. ACTH fragments 4–7 and 4–10, which are sequences common to ACTH, α-MSH, the "β-MSH-like" part of β-LPH and of γ-MSH, are effective in this regard. Similar behavioral effects of these peptides can also be demonstrated in intact animals. The data suggest that these peptides can enhance the rate of acquisition of avoidance behavior, interpreted as memory formation, and inhibit the extinction of such avoidance behavior—an effect interpreted as memory formation—and can inhibit the extinction of such avoidance behavior on learned response. Further studies suggest that ACTH fragments are involved in short-term memory processes, while vasopressin is involved in long-term memory. It has subsequently been shown that vasopressin and oxytocin have opposite effects on avoidance behavior (Kovacs et al., 1979). Some of the effects of vasopressin in correcting passive avoidance behavior deficits in Brattleboro rats with hereditary vasopressin deficiency may be due to enhanced central availability of oxytocin, since oxytocin levels in the CSF of these rats are "elevated."

The effects of vasopressin on learning may be mediated by modulation of neurotransmission in distinct catecholamine systems distributed via the dorsal noradrenergic bundle. Evidence from electrophysiological studies suggests that the mechanism responsible for ACTH effects on behavior occurs by facilitation of a selective arousal state of limbic–midbrain structures, resulting in increased attention and perception and enhancement of stimulus-specific behavioral responses. The biochemical basis for these effects, whether through alterations in cellular second-messenger systems such as cAMP, protein synthesis, membrane phosphorylation, or neurotransmitter turnover, remains to be established. When behavioral effects of exogenously administered peptides were initially reported, it was concluded that they might be derived from endogenous pituitary hormones secreted into plasma, with degradation fragments entering the brain. However, to date, there has been no demonstration of generation of such fragments in plasma; it is also doubtful, in view of the short half-life of such fragments in blood after exogenous administration, that such a mechanism could be important. The demonstration of the synthesis of the ACTH precursor molecule in brain suggests the possibility that these fragments are derived locally; anatomical pathways from the site of synthesis in the hypothalamic arcuate nucleus to the limbic system have been reported.

Initially, the physiological relevance of these animal studies was questioned since the major conditioned avoidance response studied was that motivated by fear. Subsequent studies have indicated the effectiveness of these peptides in operant conditioning situations motivated by hunger or sex. To date, very few studies have been performed in a double-blind manner; there is still limited information with regard to dose–response characteristics, relation of dose administered to physiological concentrations of peptides present in the animal, sex differences in response, and circadian variability of response. Questions also arise as to whether the active $ACTH_{4-7}$ fragment (Met-Glu-His-Phe) is derived from the ACTH/MSH molecule or whether this sequence occurs in another as yet uncharacterized peptide. Since the dextro 7 isomer of $ACTH_{4-7}$ has effects similar to the levo isomer in avoidance acquisition but opposite effects on extinction, this raises questions as to the nature of the receptor involved. There are also reports indicating an affinity of ACTH and its fragments for opiate receptors. It is apparent that although a role for these peptides in cognitive function appears to have been demonstrated, the specificity of these responses and their biochemical and physiological basis requires extended investigation.

The appearance of these studies has occasioned questions as to their applicability to human learning and memory. There are reports that systemic administration of $ACTH_{4-10}$ improved the attention of mentally retarded subjects (Sandman et al., 1976) and increased visual discrimination in normal males, while augmenting verbal skills in women (Sandman and Kastin, 1978). Negative findings have also been reported and the number of clinical observations thus far is insufficient to yield any definite conclusions. Most of the studies to date have been performed in young, healthy volunteers; studies in individuals in whom cognitive functions are disturbed, such as the elderly, may provide more consequential information.

Studies using vasopressin or its long-acting analog 1-desamino-D-arginine vasopressin have reported improvements of ability to learn new information, and beneficial effects in patients suffering from long-term amnesia as the result of car accidents (Legros et al., 1978; Oliveros et al., 1978; Weingartner et al., 1981). Beneficial effects of vasopressin have also been noted in several attentional and memory tests in elderly people.

### 4.5.5. Feeding Behavior

The relationship of food intake to caloric needs represents one of the body's major homeostatic mechanisms. Previous studies have described satiety centers located in the ventromedial nucleus, which receives stimulatory serotoninergic inputs from the raphe nuclei and inhibitory inputs from the ventral adrenergic bundles, and feeding centers in the lateral hypothalamus, which receives a prominent dopaminergic input. There is controversy as to whether dopamine plays a stimulatory or inhibitory role. There appears to be a reciprocal interaction between the ventromedial hypothalamus and the lateral hypothalamus, characterized by mutual inhibition—that is, activation of one area results in inhibition of the other, while suppression of one results in activation of the other.

Recent studies in animals have implicated several peptides in the regulation of feeding behavior. CCK (Gibbs et al., 1973), TRH (Vijayan and McCann, 1979), and insulin (Woods et al., 1979) are reported to be satiety factors, decreasing food intake, whereas β-endorphin has been implicated in states of increased food ingestion (Margules et al., 1978). The site of action of these peptides in affecting feeding behavior has been presumed to be within the CNS, with controversy as to whether or not this is mediated via the ventromedial hypothalamus; interaction with the noradrenergic system has been proposed. Suppression of feeding by CCK has been reported to also be mediated via a parenteral abdominal site (Smith and Gibbs, 1981).

Peptidergic effects on glucoregulation have also been demonstrated. Systemic administration of neurotensin and bombesin and intraventricular administration of CCK have been shown to produce hyperglycemia. Studies in human subjects, however, using physiologically relevant doses of neurotensin, which were fivefold less than those employed in animal studies, have failed to demonstrate a hyperglycemic effect.

It has been suggested that the low levels of CCK reported to be present in brains of genetically obese (*ob/ob*) mice may be causally related to the hyperphagia present in these animals. Other studies, however, have failed to show significant alteration in brain CCK in obese animals. It has also been suggested that the elevated pituitary β-endorphin concentrations present in genetically obese animals are causally related to such obesity. This has been questioned, since such elevations follow but do not precede the development of obesity. In addition, there are reports of reversal of hyperphagia by naloxone in several experimental models of obesity, not all of which were characterized by increased plasma β-endorphin concentrations. It would be expected, however, that central rather than peripheral β-endorphin concentrations would provide a more meaningful insight into the role of β-endorphin in obese states, and there are recent reports that brain β-endorphin concentrations are increased in genetically obese rats and that starvation decreases such concentrations (Straus and Yalow, 1979; Schneider *et al.*, 1979; Oku *et al.*, 1980). In view of the known interaction of β-endorphin with hypothalamic serotonin and dopamine turnover, it is indeed possible that interactions between the peptidergic and monoaminergic systems are involved in feeding regulation. There is an intriguing report that the body weight of the genetically obese rat is greatly reduced following chronic ingestion of L-dopa.

### 4.5.6. Temperature Regulation

Temperature regulation represents a complex interplay between peripheral and central receptors (Martin *et al.*, 1977). There have been numerous studies of the effects of changes of temperature on neurotransmitter release (centrally and peripherally), and on hormone release. Conversely, the effect of neurotransmitters (usually administered intracisternally) on temperature regulation has been studied. With regard to central mechanisms involved in temperature regulation research studies have emphasized the role of noradrenergic and serotonergic involvement, although controversy exists as to their specific effect on body temperature among various species. The effects of central cooling on peripheral release of norepinephrine and thyroid hormone also appear to be well substantiated.

Recent studies have implicated several brain peptides in the regulation of body temperature (Brown *et al.*, 1978, 1981). These effects have yet to be fully integrated into any of the above schemes. Intracisternal injection of TRH causes hyperthermia, whereas small amounts ($10^{-10}$ to $10^{-15}$ M) of β-endorphin, neurotensin, and bombesin result in hypothermia. There is some evidence of an interaction of neurotensin with dopaminergic systems and with the hypothalamic–pituitary–thyroid axis. Bombesin effects show species differences, and it is still uncertain whether the hypothermic effect occurs at ambient temperatures or only in animals exposed to the cold. The effects of bombesin are reversed by administration of either TRH, $PGE_1$ and $PGE_2$, or naloxone. TRH also reverses the hypothermic effect of endorphin, but not the antinociceptive effect of morphine. The effect of TRH is observed in hypophysectomized animals, implying a central rather than a peripheral effect. A recent report has described a function of vasopressin in suppression of the febrile response in the newborn sheep (Cooper *et al.*, 1979). A determination of the physiological significance of these effects will require further study.

## 5. SUMMARY

Recent discoveries of brain peptides have important implications for our understanding of neurologic and psychiatric disorders. Hormones in the hypothalamus regulate anterior pituitary functions serving as a crucial link between brain and soma. Hormones in the gut are also found in nerve cells of the brain. The current level of understanding of the functions of neuropeptides is primitive but important advances can be anticipated.

ACKNOWLEDGMENTS. Supported by USPHS Grants AM 26252 and NS 16367 (Huntington's Disease Center Without Walls). The excellent secretarial assistance of Kathy Sullivan is acknowledged.

## 6. REFERENCES

Akil, H., Richardson, D. E., Barchas, J. D., and Li, C. H., 1978, Appearance of β-endorphin-like immunoreactivity in human ventricular cerebrospinal fluid upon analgesic electrical stimulation, *Proc. Natl. Acad. Sci. USA* **75**:5170.

Aronin, N., Cooper, P E., Lorenz, L. J., Bird, E. D., Sagar, S. M., Leeman, S. E., and Martin, J. B., 1983, Somatostatin is increased in the basal ganglia in Huntington's disease, *Ann. Neurol.* **13**:519.

Basbaum, A. I., and Fields, H. L., 1978, Endogenous pain control mechanisms: Review and hypothesis, *Ann. Neurol.* **4**:451.

Bird, E. D., 1980, Chemical pathology of Huntington's disease, *Annu. Rev. Pharmacol. Toxicol.* **20**:533.

Brown, M. R., Rivier, J. E., Kobayashi, R., and Vale, W. W., 1978, Neurotensin-like and bombesin-like peptides: CNS distribution and actions, in: *Gut Hormones* (S. R. Bloom, ed.), pp. 550–558, Churchill Livingston, Edinburgh.

Brown, M. R., Tache, Y., Rivier, J., and Pittman, Q., 1981, Peptides and regulation of body temperature in: *Neurosecretion and Brain Peptides: Implication for Brain Function and Neurological Disease* (J. B. Martin, S. Reichlin, and K. L. Bick, eds.), pp. 397–408, Raven Press, New York.

Comb, M., Seeburg, P. H., Adelman, J., Eiden, L., and Herbert, E., 1982, Primary structure of the human Met- and Leu-enkephalin precursor and its mRNA, *Nature (London)* **295**:663.

Cooper, K. E., Kasting, N. W., Lederis, K., and Veale, W. L., 1979, Evidence supporting a role for endogenous vasopressin in natural suppression of fever in the sheep, *J. Physiol.* **295**:33.

Davies, P., Katzman, R., and Terry, R. D., 1980, Reduced somatostatin-like immunoreactivity in cerebral cortex from cases of Alzheimer disease and Alzheimer senile dementia, *Nature (London)* **288**:279.

de Kloet, R., and de Weid, D., 1980, The brain as target tissue for hormones of pituitary origin: Behavioral and biochemical studies, in: *Frontiers in Neuroendocrinology*, Vol. 6 (L. Martini and W. F. Ganong, eds.), pp. 157–201, Raven Press, New York.

Dockray, G. J., 1979, Evolutionary relationships of the gut hormones, *Fed. Proc.* **38**:2295.
Dogterom, J., van Wimersma Griedanus, T. B., and Swaab, D. F., 1977, Evidence for the release of vasopressin and oxytocin into cerebrospinal fluid: Measurements in plasma and CSF of intact and hypophysectomized rats, *Neuroendocrinology* **24**:108.
Emson, P. C., Fahrenkrug, J., and Spokes, E. G. S., 1979, Vasoactive intestinal polypeptide (VIP): Distribution in normal human brain and in Huntington's disease, *Brain Res.* **173**:174.
Emson, P. C., Arregui, A., Clement-Jones, V., Sandberg, B. E. B., and Rossor, M., 1980, Regional distribution of methionine-enkephalin and substance P-like immunoreactivity in normal human brain and in Huntington's disease, *Brain Res.* **199**:147.
Epelbaum, J., Brazeau, P., Tsang, D., Brawer, J., and Martin, J. B., 1977, Subcellular distribution of radioimmunoassayable somatostatin in rat brain, *Brain Res.* **126**:309.
Frenk, H., Urca, G., and Liebeskind, J. C., 1978, Epileptic properties of leucine- and methionine-enkephalin: Comparison with morphine and reversibility by naloxone, *Brain Res.* **147**:327.
Gibbs, J., Young, R. C., and Smith, G. P., 1973, Cholecystokinin elecits satiety in rats with open gastric fistulas, *Nature (London)* **245**:323.
Goldstein, A., Tachibana, S., Lowney, L. L., Hunkapiller, M., and Hood, L., 1979, Dynorphin-(1–13), an extraordinarily potent opioid peptide, *Proc. Natl. Acad. Sci. USA* **76**:6666.
Goodman, R. R., Snyder, S. H., Kuhar, M. J., and Young, W. S., III, 1980, Differentiation of delta and mu opiate receptor localizations by light microscopic autoradiography, *Proc. Natl. Acad. Sci. USA* **77**:6239.
Havrankova, J., Roth, J., and Brownstein, M., 1978, Insulin receptors are widely distributed in the central nervous system of the rat, *Nature (London)* **272**:827.
Henry, J. L., 1976, Effects of substance P on functionally identified units in cat spinal cord, *Brain Res.* **114**:439.
Hökfelt, T., Elde, R. P., Johansson, O., Ljungdahl, A., Schultzberg, M., Fuxe, K., Goldstein, M., Nilsson, G., Pernow, B., Terenius, L., Ganten, D., Jeffcoate, S. L., Rehfeld, J., and Said, S., 1978, Distribution of peptide-containing neurons, in: *Psychopharmacology: A Generation of Progress* (M. A. Lipton, A. DiMascio, and K. F. Killam, eds.), pp. 39–66, Raven Press, New York.
Hökfelt, T., Johansson, O., Ljungdahl, A., Lundberg, J. M., and Schultzberg, M., 1980, Peptidergic neurons, *Nature (London)* **284**:515.
Hughes, J., Smith, T. W., Kosterlitz, H. W., Fotergill, L. A., Morgan, B. A., and Morris, H. R., 1975, Identification of two related pentapeptides from the brain with potent opiate agonist activity, *Nature (London)* **258**:577.
Ioffe, S., Havlicek, V., Friesen, H., and Chernick, V., 1978, Effect of somatostatin (SRIF) and L-glutamate on neurons of the sensorimotor cortex in awake habituated rabbits, *Brain Res.* **153**:414.
Iversen, L. L., Iversen, S. D., Bloom, F., Douglas, C., Brown, M., and Vale, W., 1978, Calcium-dependent release of somatostatin and neurotensin from rat brain in vitro, *Nature (London)* **273**:161.
Jackson, I. M. D., 1980, Significance and function of neuropeptides in cerebrospinal fluid, in: *Neurobiology of Cerebrospinal Fluid*, Vol. 1 (J. H. Wood, ed.), pp. 625–650, Plenum Press, New York.
Jessell, T. M., 1981, The role of substance P in sensory transmission and pain perception, in: *Neurosecretion and Brain Peptides: Implication for Brain Function and Neurological Disease* (J. B. Martin, S. Reichlin, and K. L. Bick, eds.), pp. 189–197, Raven Press, New York.
Kovacs, G. I., Bohus, B., and Versteeg, D. H. G., 1979, The effects of vasopressin on memory processes: The role of noradrenergic neurotransmission, *Neuroscience* **4**:1529.
Krieger, D. M., and Martin, J. B., 1981, Brain peptides, *N. Engl. J. Med.* **304**:876, 944.
Land, H., Shutz, G., Schmale, H., and Richter, D., 1982, Nucleotide sequence of cloned cDNA encoding bovine arginine vasopressin-neurophysin II precursor, *Nature (London)* **295**:299.
Legros, J. J., Gilot, P., Seron, X., Claessens, J., Adam, A., Moeglen, J. M., Audibert, A., and Berchier, P., 1978, Influence of vasopressin on learning and memory, *Lancet* **I**:41.

LeRoith, D., Shiloach, J., Roth, J., and Lesniak, M. A., 1980, Evolutionary origins of vertebrate hormones: Substances similar to mammalian insulins are native to unicellular eukaryotes (*Tetrahymena/Neurospora*), *Proc. Natl. Acad. Sci. USA* **77**:6184.

LeRoith, D., Shiloach, J., Liotta, A. S., Krieger, D. T., Lewis, M., and Pert, C. B., 1981, Evolutionary origins of vertebrate hormones: Material very similar to adrenocorticotropic hormone, β-endorphin and dynorphin in protozoa, *Trans. Assoc. Am. Physicians* **94**:52.

Mains, R. E., Eipper, B. A., and Ling, N., 1977, Common precursor to corticotropins and endorphins, *Proc. Natl. Acad. Sci. USA* **74**:3014.

Margules, D. L., Moisset, B., Lewis, M. J., Shibuya, H., and Pert, C. B., 1978, Beta-endorphin is associated with overeating in genetically obese mice (*ob/ob*) and rats (*fa/fa*), *Science* **202**:988.

Martin, J. B., and Landis, D. M. D., 1981, Potential implications of brain peptides in neurological disorders, in: *Neurosecretion and Brain Peptides: Implications for Brain Function and Neurological Disease* (J. B. Martin, S. Reichlin, and K. Bick, eds.), pp. 673–690, Raven Press, New York.

Martin, J. B., Reichlin, S., and Brown, G. M., 1977, *Clinical Neuroendocrinology*, Davis, Philadelphia.

Mudge, A. W., Leeman, S. E., and Fischbach, G. D., 1979, Enkephalin inhibits release of substance P from sensory neurons in culture and decreases action potential duration, *Proc. Natl. Acad. Sci. USA* **76**:526.

Nakanishi, S., Inoue, A., Kita, T., Nakamura, M., Chang, A. C. Y., Cohen, S. N., and Numa, S., 1979, Nucleotide sequence of cloned cDNA for bovine corticotropin-β-lipotropin precursor, *Nature (London)* **278**:423.

Nakao, K., Oki, S., Tanaka, I., Horii, K., Nakai, Y., Furui, T., Fukushima, M., Kuwayama, A., Kageyama, N., and Imura, H., 1980, Immunoreactive β-endorphin and adrenocorticotropin in human cerebrospinal fluid, *J. Clin. Invest.* **66**:1383.

Nicoll, R. A., Schenker, C., and Leeman, S. E., 1980, Substance P as a transmitter candidate, *Annu. Rev. Neurosci.* **3**:227.

Nutt, J. G., Mroz, E. A., Leeman, S. E., Williams, A. C., Engel, W. K., and Chase, T. N., 1980, Substance P in human cerebrospinal fluid: Reductions in peripheral neuropathy and autonomic dysfunction, *Neurology* **30**:1280.

Oku, J., Glick, Z., Shimomura, Y., Inoue, S., Bray, G. A., and Walsh, J., 1980, Cholecystokinin and obesity, *Clin. Res.* **28**:281 (abstract).

Oliveros, J. C., Jandali, M. K., Timsit-Berthier, M., Remy, R., Benghezal, A., Audibert, A., and Moeglen, J. M., 1978, Vasopressin in amnesia, *Lancet* **I**:42.

Pacold, S. T., and Blackard, W. G., 1979, Central nervous system insulin receptors in normal and diabetic rats, *Endocrinology* **105**:1452.

Patel, Y., Rao, K., and Reichlin, S., 1977, Somatostatin in human cerebrospinal fluid, *N. Engl. J. Med.* **296**:529.

Pearse, A. G. E., 1969, The cytochemistry and ultrastructure of polypeptide hormone-producing cells of the APUD series and the embryologic, physiologic, and pathologic implications of the concept, *J. Histochem. Cytochem.* **17**:303.

Pert, C. B., and Snyder, S. H., 1973, Opiate receptor: Demonstration in nervous tissue, *Science* **179**:1011.

Phillis, J. W., and Kirkpatrick, J. R., 1980, The actions of motilin, luteinizing hormone releasing hormone, cholecystokinin, somatostatin, vasoactive intestinal peptide and other peptides on rat cerebral cortical neurons, *Can. J. Physiol. Pharmacol.* **58**:612.

Randic, M., and Miletic, V., 1978, Depressant actions of methionine-enkephalin and somatostatin in cat dorsal horn neurones activated by noxious stimuli, *Brain Res.* **152**:196.

Rehfeld, J., 1980, Cholecystokinin, *Trends Neurosci.* **3**:65.

Renaud, L. P., Martin, J. B., and Brazeau, P., 1975, Depressant action of TRH, LH-RH and somatostatin on activity of central neurons, *Nature (London)* **255**:233.

Riskind, P., and Martin, J. B., 1982, Management of pituitary secretory adenomas, in: *Harrison's Principles of Internal Medicine*, Vol. 3 (K. Isselbacher, R. D. Adams, E. Braunwald, J. B. Martin, R. G. Petersdorf, and J. Wilson, eds.), pp. 235–252, McGraw-Hill, New York.

Roberts, J. L., and Herbert, E., 1977, Characterization of a common precursor to corticotropin and β-lipotropin: Identification of β-lipotropin peptides and their arrangement relative to corticotropin in the precursor synthesized in a cell-free system, *Proc. Natl. Acad. Sci. USA* **74**:5300.

Saito, A., Sankaran, H., Goldfine, I. D., and Williams, J. A., 1980, Cholecystokinin receptors in the brain: Characterization and distribution, *Science* **208**:1155.

Sandman, C. A., and Kastin, A. J., 1978, A behavioral strategy for the CNS actions of the neuropeptides, in: *Current Studies of Hypothalamic Function*, Vol. 2 (W. L. Veale and K. Lederis, eds.), pp. 163–174, Karger, Basel.

Sandman, C. A., George, J., Walker, B. B., Nolan, J. D., and Kastin, A. J., 1976, Neuropeptide MSH/-ACTH 4–10 enhances attention in the mentally retarded, *Pharmacol. Biochem. Behav.* **5**(Suppl. 1):23.

Schiller, P. W., Lipton, A., Horrobin, D. F., and Bodanszky, M., 1978, Unsulfated C-terminal 7-peptide of cholecystokinin: A new ligand of the opiate receptor, *Biochem. Biophys. Res. Commun.* **85**:1332.

Schneider, B. S., Monahan, J. W., and Hirsch, J., 1979, Brain cholecystokinin and nutritional status in rats and mice, *J. Clin. Invest.* **64**:1348.

Smith, G. P., and Gibbs, J., 1981, Brain–gut peptides and the control of food intake, in: *Neurosecretion and Brain Peptides: Implication for Brain Function and Neurological Disease* (J. B. Martin, S. Reichlin, and K. L. Bick, eds.), pp. 389–395, Raven Press, New York.

Spindel, E. R., Wurtman, R. J., and Bird, E. D., 1980, Increased TRH content of the basal ganglia in Huntington's disease, *N. Engl. J. Med.* **303**:1235.

Straus, E., and Yalow, R. S., 1979, Cholecystokinin in the brains of obese and nonobese mice, *Science* **203**:68.

Swanson, L. W., and Sawchenko, P. E., 1980, Paraventricular nucleus: A site for the integration of neuroendocrine and autonomic mechanisms, *Neuroendocrinology* **31**:410.

Swanson, L. W., Sawchenko, P. E., Wiegand, S. J., and Price, J. L., 1980, Separate neurons in the paraventricular nucleus project to the median eminence and to the medulla or spinal cord, *Brain Res.* **198**:190.

Taylor, D. P., and Pert, C. B., 1979, Vasoactive intestinal polypeptide: Specific binding to rat brain membranes, *Proc. Natl. Acad. Sci. USA* **76**:660.

Terry, L. C., Rorstad, O. P., and Martin, J. B., 1980, The release of biologically and immunologically reactive somatostatin from perifused hypothalamic fragments, *Endocrinology* **107**:794.

Vale, W., Ling, N., Rivier, J., Villarreal, J., Rivier, C., Douglas, C., and Brown, M., 1976, Anatomic and phylogenetic distribution of somatostatin, *Metabolism* **25**:1491.

Vale, W., Spiess, J., Rivier, C., and Rivier, J., 1981, Characterization of a 41-residue ovine hypothalamic peptide that stimulates secretion of corticotropin and β-endorphin, *Science* **213**:1394.

van Houten, M., and Posner, B. I., 1979, Insulin binds to brain blood vessels *in vivo, Nature (London)* **282**:623.

van Houten, M., Posner, B. I., Kopriwa, B. M., and Brawer, J. R., 1979, Insulin binding sites localized to nerve terminals in rat median eminence and arcuate nucleus, *Science* **207**:1081.

Vijayan, E., and McCann, S. M., 1979, Suppression of feeding and drinking activity in rats following intraventricular injection of thyrotropin releasing hormone (TRH), *Endocrinology* **100**:1727.

Weingartner, H., Gold, P., Ballenger, J. C., Smallberg, S. A., Summers, R., Rubinow, D. R., Post, R. M., and Goodwin, F. K., 1981, Effects of vasopressin on human memory functions, *Science* **211**:601.

Woods, S. C., Lotter, E. C., McKay, L. D., and Porte, D., Jr., 1979, Chronic intracerebroventricular infusion of insulin reduces food intake and body weight of baboons, *Nature (London)* **282**:503.

Yaksh, T. L., Farb, D. H., Leeman, S. E., and Jessell, T. M., 1979, Intrathecal capsaicin depletes substance P in the rat spinal cord and produces prolonged thermal analgesia, *Science* **206**:481.

Zimmerman, E. A., 1981, The organization of oxytocin and vasopressin pathways, in: *Neurosecretion and Brain Peptides: Implications for Brain Functions and Neurological Diseases* (J. B. Martin, S. Reichlin, and K. L. Bick, eds.), pp. 63–75, Raven Press, New York.

CHAPTER 2

# Neuropharmacological Influences on Hypothalamic–Pituitary Secretion

ERMANNO ROLANDI and TOMMASO BARRECA

## 1. INTRODUCTION

The past 10 years have seen an increasing number of important studies on the hypothalamic–pituitary axis, focusing mainly on the biochemical characterization of the hypothalamic peptide hormones and the neuroendocrine mechanisms regulating the secretory patterns of pituitary hormones. Major advances in this field have included the isolation, characterization, and synthesis of three hypothalamic peptides, which have indisputable hypophysiotropic activities: the gonadotropin-releasing hormone (GnRH), the thyrotropin-releasing hormone (TRH), and the growth hormone-release-inhibiting hormone (GIH or somatostatin). Another peptide isolated from brain tissues is the putative inhibiting hormone of the melanocyte-stimulating hormone (MIF). The amino acid sequences of the remaining hypophysiotropic hormones are still unknown. Furthermore, it is well established that hypothalamic secretion is subject to a double control: the first through peripheral hormones (feedback mechanism) and the second through the cerebral neurotransmitters. These monoamine neurotransmitters include dopamine (DA), norepinephrine (NE), histamine, serotonin, acetylcholine, and γ-aminobutyric acid (GABA). Recently, it has been proposed that various substances, such as neurotensin, substance P, vasoactive intestinal polypeptide (VIP), glucagon, bombesin, gastrin, and particularly endorphins, can act as neurotransmitters or neuromodulators and also affect hypothalamic–pituitary secretions. Finally, experimental and clinical studies have shown significant changes in hypothalamic–pituitary function induced by an ever-increasing number of drugs.

---

ERMANNO ROLANDI and TOMMASO BARRECA • Istituto Scientifico di Medicina Interna, Cattedra di Patologia Medica, University of Genoa, 16132 Genoa, Italy.

We believe that these neuropharmacological studies are especially promising and could provide important information to help clarify the mechanisms regulating hypothalamic–pituitary functions in physiological and pathological processes.

## 2. PROLACTIN SECRETION

### 2.1. Hypothalamic Control

It has been well documented that the control of prolactin (PRL) secretion by the hypothalamus is mainly inhibitory (Fig. 1). Hypothalamic lesions (Arimura *et al.*, 1972; Krulich *et al.*, 1975) or disruption of hypothalamic–pituitary connections by stalk section (Kanematsu and Sawyer, 1973; Langer *et al.*, 1978) result in increased PRL secretion. This hypothalamic control should be exerted by a prolactin-inhibiting factor (PIF), whose chemical nature is not entirely known. Van Maanen and Smelik (1968) first proposed that DA is the physiological PIF and actually many findings suggest this possibility. In fact, *in vitro* experiments showed that physiological amounts of DA in pituitary tissue cultures significantly inhibit PFL release (Shaar and Clemens, 1974). Furthermore, it has been demonstrated that tuberoinfundibular DA neurons (TIDA) are involved in the control of PRL release from the pituitary. DA released from terminals of these neurons is directly transported by the blood in the portal vessels to the anterior pituitary, where it activates the specific receptors located

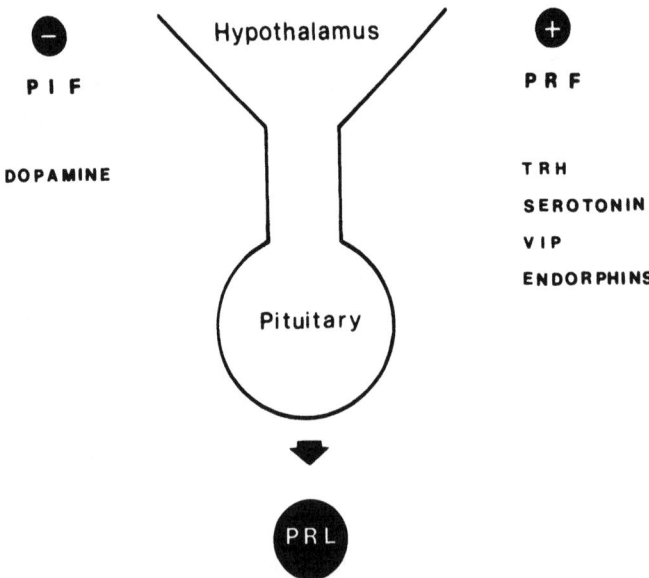

FIGURE 1. Hypothalamic control of PRL secretion in man.

on PRL-secreting cells (Weiner and Ganong, 1978). The activity of TIDA is conditioned by PRL levels; it has been reported that TIDA activity is stimulated when high PRL levels occur (e.g., pregnancy, lactation, or after PRL administration) (Moore et al., 1980). However, it is also possible that other PRL-inhibiting factors (PIFs), different from DA, exist. In fact, when DA is eliminated from hypothalamic extracts by adsorption on alumina, PIF activity persists (Enjalbert et al., 1977) and this activity is not blocked by DA receptor antagonists (Kordon and Enjalbert, 1980). Recently, a naturally occurring metabolite of TRH has been proposed as another putative PIF (Kordon and Enjalbert, 1980). This substance, histidyl-proline diketopiperazine, present in the median eminence and pituitary, inhibits *in vitro* PRL release. Schally et al. (1977) demonstrated that GABA, present in hypothalamic extracts, can also inhibit PRL release *in vitro* and *in vivo*, but this effect is not confirmed by studies employing a GABA agonist in man (Tamminga et al., 1978).

More uncertain is the existence and chemical characterization of a hypothalamic prolactin-releasing factor (PRF). TRH is capable of inducing PRL release and it has been considered as a possible PRF. Nevertheless, PRL may also be released without thyrotropin, e.g., during suckling (Gautvik et al., 1974), and PRL appears unchanged after TRH neurtralization by specific antibodies (Harris et al., 1978). Serotonin also promotes PRL release (Kato et al., 1974) and appears to participate in the suckling-induced rise in serum PRL, but serotonin is not able to stimulate hormonal release *in vitro* (Birge et al., 1970) and no definitive evidence exists that this neurotransmitter is the PRF. Many other peptides found in significant amounts in hypothalamic–pituitary portal vessels elicit a stimulation of PRL release: bombesin, alytensin, ranatensin, litorin (Rivier et al., 1978), vasotocin (Johnson, 1978), substance P, neurotensin (Rivier et al., 1977), and VIP (Kato et al., 1978; Ruberg et al., 1978). Among these peptides, only VIP is able to release PRL from rat pituitary halves incubated *in vitro*. For this reason, a possible role of VIP as the physiological PRF has recently been proposed (Clemens and Shaar, 1980).

## 2.2. Drugs Causing Hyperprolactinemia

The above considerations clearly indicate that PRL secretion is subject to a multifactorial control. Therefore, many pharmacological agents can induce changes in blood levels.

Hyperprolactinemia occurs after the administration of several drugs (Table 1). The PRL-releasing activity of TRH, previously recorded among putative PRF peptides, acts directly on lactotroph cells (MacLeod, 1976). The majority of the substances mentioned act as DA synthesis inhibitor (α-methylparatyrosine) (Meltzer, 1980), DA receptor blockers (phenothiazines, butyrophenones, benzamides), or DA-depleting drugs (reserpine, methyldopa) and induce hyperprolactinemia after acute or chronic administration (Thorner, 1977; Rolandi et al., 1979a; L'Hermite et al., 1979). The stimulation of PRL secretion caused by estrogens is due to their potent antidopaminergic activity at the

TABLE 1
Pharmacological Agents Increasing PRL Release in Man

TRH
Insulin-induced hypoglycemia
Estrogens and oral contraceptives
Amino acids: arginine, tryptophan, 5HT
Antipsychotic drugs: phenothiazines, butyrophenones
Antidepressant drugs: tricyclics, MAOI
Opiates: morphine, methadone, pentazocine, nalorphine, DAMME
Benzamides: metoclopramide, sulpiride, sultopride, tiapride
Hypotensive drugs: methyldopa, reserpine
$H_2$-histamine receptor blockers: cimetidine, ranitidine
Antiemetic drugs: domperidone, thiethylperazine
Aromatic L-amino acid decarboxylase inhibitors: carbidopa, benserazide
Miscellaneous: monoiodotyrosine, GHB, PTH, muscimol, bombesin, α-methylparatyrosine

pituitary level (Raymond et al., 1978). Estrogens alone are more effective in elevating PRL levels than combined estrogen/progesterone oral contraceptives (Abu-Fadil et al., 1976). PRL release induced by tryptophan, 5-OH-tryptophan, and insulin-induced hypoglycemia seems to be due to stimulation of the serotonin receptor (Del Pozo and Lancranjan, 1978). Among the antidepressant compounds, some tricyclic drugs (e.g., Clomipramine) increase serum PRL probably acting through a serotoninergic mechanism (Francis et al., 1976; Asnis et al., 1980) and the same effect has been observed after monoamine oxidase inhibitors (MAOI), chlorgyline, and pargyline are administered (Slater et al., 1977). A hyperprolactinemizing effect of other tricyclic antidepressants (imipramine and amitriptyline) has not been adequately demonstrated (Meltzer, 1980). Conversely, PRL release after administration of two aromatic L-amino acid decarboxylase inhibitors (carbidopa and benserazide) is markedly increased (Brown et al., 1976; Pontiroli et al., 1977), but their mode of action is unclear. Other hyperprolactinemizing drugs worth noting are the $H_2$-histamine receptor blockers, cimetidine (Rolandi et al., 1979b) and possibly ranitidine (Delitala et al., 1980a), some GABA derivatives such as γ-hydroxybutyric acid (GHB) (Takahara et al., 1977), baclofen (Cavagnini et al., 1977), γ-amino-β-hydroxybutyric acid (GABOB) (Fioretti et al., 1979), muscimol (Tamminga et al., 1978), and two potent antiemetic drugs, thiethylperazine (Sachar, 1978) and domperidone (Pourmand et al., 1980). Recently, the effect of opiate peptides and related drugs on PRL secretion has been emphasized. In fact, after the report of Tolis et al. (1975a) attesting to a significant PRL release after morphine administration, many studies on the endocrine effects of opioid drugs and endorphins have been performed. In particular, a marked increase in serum PRL values after therapeutic doses of methadone and pentazocine (Rolandi and Barreca, 1978), nalorphine (Rolandi et al., 1980a), β-endorphin (Catlin et al., 1980), and an enkephalin analog (DAMME) (Stubbs et al., 1978) were found. However, we have not been able to demonstrate hyperprolactinemia after Met-enkephalin administration (Rolandi *et al.*, 1980b). The mechanism of action of opiates is not well established, but experimental studies suggest the possibility that the pituitary effects of these substances can be mediated by an activation of the serotoninergic pathway (Spampinato *et al.*, 1979; Koenig *et al.*, 1979) or by a decrease of DA activity in the median eminence (Van Vugt *et al.*, 1979). Finally, increased serum PRL values have been observed after the administration of monoiodotyrosine (Smythe *et al.*, 1975), parathyroid hormone (Isaac *et al.*, 1978), and bombesin (Pontiroli *et al.*, 1980a), but these observations need to be further investigated.

### 2.3. Drugs Causing Hypoprolactinemia

Twelve years ago, Malarkey *et al.* (1971) noted that oral administration of 500 mg of L-dopa induced a marked decrease in serum PRL levels, but this inhibition appeared to be of rather short duration (Table 2). Later, it was well established that DA suppresses PRL release both in animals and in man by directly affecting PRL-secreting cells (Crosignani *et al.*, 1976; Del Pozo and Lancranjan, 1978; Tuomisto, 1978). Among DA agonist substances which lower PRL secretion, the most clinically employed are the ergot derivatives, which produce prolonged PRL suppression (Del Pozo *et al.*, 1972). In particular, 2-Br-α-ergocriptine (bromocriptine), metergoline, lisuride, lergotrile, methysergide and ergonovine are able to suppress PRL secretion in basal conditions and in the course of physiological, pharmacological, and pathological hyperprolactinemias (Del Pozo *et al.*, 1972; Shane and Naftolin, 1974; Mendelson *et al.*, 1975; Delitala *et al.*, 1976a; Horowski *et al.*, 1978; Ferrari *et al.*, 1978). In addition to a DA-stimulating action it is probable that some of these substances such as metergoline and methysergide can also act as serotonin antagonists (Ferrari *et al.*, 1978; Lotti *et al.*, 1979). Recently, PRL suppression by a long-acting DA agonist, pergolide mesylate, has been reported (Kleinberg *et al.*, 1980), whereas PRL effects of methylergometrine are still contradictory (Del Pozo and Lancranjan, 1978). Other DA agonists like apomorphine (Martin *et al.*, 1974) and piribedil (Slatopolsky-Cantis *et al.*, 1977) can also induce PRL inhibition both in normal and in hyperprolactinemic conditions. A possible

TABLE 2
Pharmacological Agents Decreasing PRL Release in Man

L-Dopa and dopamine
Ergot derivatives: bromocriptine, lergotrile, lisuride, metergoline, methysergide, pergolide, ergonovine
Apomorphine
Piribedil
Pyridoxine
Antiestrogens: clomiphene
Antidepressant drugs: nomifensine, trazodone, bupropion
Calcitonin

stimulation of the hypothalamic dopaminergic pathway has been proposed, also to explain the PRL inhibitory effect of pyridoxine (Delitala *et al.*, 1976b). Furthermore, a moderate decrease in serum PRL levels has been observed after administration of the antiestrogen drugs clomiphene and tamoxifen (Willis *et al.*, 1976; Masala *et al.*, 1977), and after calcitonin (Rolandi *et al.*, 1979c; Isaac *et al.*, 1980). Among psychoactive drugs, nomifensine, trazodone, and bupropion (Müller *et al.*, 1978; Rolandi *et al.*, 1981; Stern *et al.*, 1979) are able to reduce blood PRL values. Finally, naloxone, a pure opiate antagonist, has also been proposed as a PRL-inhibiting drug (Rubin *et al.*, 1979), but this action has not been confirmed by others (Morley *et al.*, 1980; Barreca *et al.*, 1980).

## 3. GROWTH HORMONE SECRETION

### 3.1. Hypothalamic Control

The secretion of GH, like that of PRL, is subject to a double hypothalamic control (Fig. 2). This regulation is due to the combined effects of a releasing factor (GRF) and an inhibiting factor (GIH or somatostatin). Various experimental studies suggest that the hypothalamic areas involved in this control include the ventromedial (VMN) and arcuate nuclei regions as GH-stimulating areas and the anterior hypothalamic or parachiasmatic region as a GH-inhibiting area (see Martin, 1976). This hypothalamic control consists of an predominantly stimulatory influence. In fact, a sharp drop in GH secretion is seen in patients with hypothalamic lesions or after hypothalamic–pituitary disconnections. The

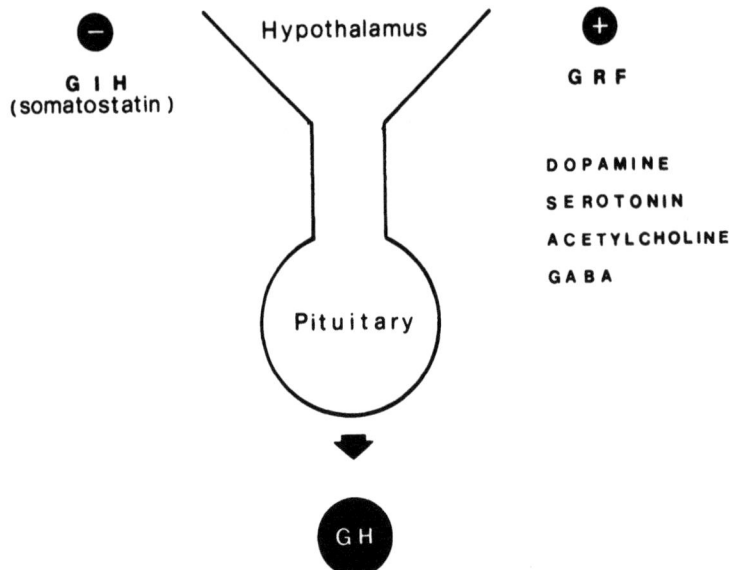

FIGURE 2. Hypothalamic control of GH secretion in man.

chemical structure of GRF is still unknown and the decapeptide proposed by Schally et al. (1971b) has not been shown to have a true stimulatory effect on GH release. On the other hand, the chemical characteristics of the GH-inhibiting factor, somatostatin, are well known (Brazeau et al., 1974). It is a tetradecapeptide (H·Ala-Gly-Cys-Lys-Asn-Phe-Phe-Trp-Lys-Thr-Phe-Thr-Ser-Cys·OH), present both in the hypothalamus and in extrahypothalamic sites, and is able to inhibit basal GH secretion as well as the hormonal release observed during sleep (Parker et al., 1974), subsequent to insulin-induced hypoglycemia or to L-dopa administration (Siler et al., 1973). Under certain conditions the peptide also inhibits the secretion of other pituitary hormones, such as TSH (Siler et al., 1974), ACTH (Tyrell et al., 1975), and PRL (Barreca et al., 1979), and of several extrapituitary hormones (gastrin, glucagon, insulin, VIP, and renin) (see Besser, 1976).

Other unrelated peptides, such as TRH, ADH (Giordano et al., 1971), α-MSH (Strauch et al., 1973), and glucagon (Wieland et al., 1973), have been shown to stimulate GH release. Among these peptides the effect of TRH has been well studied and has been shown to affect GH secretion in acromegalic patients (Faglia et al., 1973).

GH secretion appears to be strongly affected by cerebral monoamines, which probably act via GRF or somatostatin. With regard to these possibilities, both catecholaminergic and serotoninergic hypotheses have been proposed (Smythe, 1977). The first is supported by the observation that DA increases serum GH levels in normal subjects (Boyd et al., 1970) and chlorpromazine, a DA-blocking drug, inhibits GH release (Sherman et al., 1971). The second hypothesis is prompted by the observed increase in serum GH levels after 5HT administration (Imura et al., 1973a). However, the possibility that DA and serotonin act on the same hypothalamic receptors cannot be ruled out. Furthermore, a possible role of other neurotransmitters such as histamine, GABA, and acetylcholine has been proposed (Krulich, 1979).

## 3.2. Substances Stimulating GH Secretion in Man

The classic GH-stimulating effect is seen in insulin-induced hypoglycemia, which probably occurs through serotoninergic stimulation (Bivens et al., 1973; Smythe and Lazarus, 1974) (Table 3). The same stimulatory effect follows administration of L-dopa (Boyd et al., 1970), DA (Leebaw et al., 1978), or DA agonists such as ergot derivatives (in particular bromocriptine) (Tolis et al., 1975b; Camanni et al., 1975), apomorphine (Lal et al., 1975), and piribedil (Martin, 1978). Methylphenidate and amphetamines also produce an increase in GH secretion (Brown and Williams, 1976). This effect is due to activation of the catecholaminergic pathway. The increased GH secretion after clonidine (Lal et al., 1975) and NE (Blackard and Hubbell, 1970) administration is promoted by stimulation of α-adrenergic receptors. Analogously, desimipramine and chlorimipramine, a more potent 5HT reuptake blocker, promote GH release (Laakmann et al., 1977) but this effect, according to Meltzer (1980), re-

### TABLE 3
#### Pharmacological Agents Increasing GH Release in Man

Insulin-induced hypoglycemia
L-Dopa and dopamine
Dopamine agonists: ergot derivatives, apomorphine, piribedil
$\alpha$-Adrenergic agonists: NE, clonidine
5HT reuptake blockers: chlorimipramine
Amphetamines
Serotonin precursors: tryptophan, 5HT
GABA and GHB
Cholinergic agonists: $\beta$-methycholine, edrophonium
Individual amino acids: arginine, glycine, proline, leucine, etc.
Benzamide derivatives
Diazepam
DAMME

quires further verification. β-Adrenoceptor blockade appears to enhance the GH response to most physiological stimuli (Checkley, 1980).

An enhancement of GH secretion after 5HT administration has been reported (Imura et al., 1973a; Lancranjan et al., 1977); tryptophan, on the other hand, produced a slight effect (Müller et al., 1974). A good GH secretory response is observed after the administration of other amino acids (arginine, lysine, leucine, etc.), probably due to activation of adrenergic or serotoninergic pathways. In fact, phenotalamine or cyproheptadine pretreatment blocks this response (Buckler et al., 1969; Nakai et al., 1974). In recent studies, GABA has been shown to be capable of inducing a significant elevation of blood GH levels (Takahara et al., 1977; Cavagnini et al., 1980). However, contradictory findings have been observed in previous investigations employing baclofen, a GABA derivative (Cavagnini et al., 1977), and muscimol, a GABA agonist (Tamminga et al., 1978); the first substance blunted GH response to insulin-induced hypoglycemia, while the second significantly raised GH secretion.

Cholinergic stimulation also induces an increase in GH secretion and the administration of cholinergic agonists and antagonists promotes and inhibits GH release, respectively (Mendelson et al., 1978; Leveston and Cryer, 1980).

Among others, diazepam (Sylvälahti and Kando, 1975) and benzamide derivatives such as metoclopramide (Cohen et al., 1979) enhanced GH secretion, but their mechanism of action is still not clear. Finally, DAMME, an above-mentioned long-acting analog of enkephalin, simulates GH release (Stubbs et al., 1978), whereas Met-enkephalin (Rolandi et al., 1980b; Golstein et al., 1981) and β-endorphin do not affect GH secretion in man (Catlin et al., 1980).

### 3.3. Substances Inhibiting GH Secretion in Man

It is well known that serum GH values are very low under basal conditions in healthy men (0–5 ng/ml in our laboratory) and for this reason it is difficult

TABLE 4
Pharmacological Agents Inhibiting GH Release in Man

Somatostatin
Hyperglycemia
Serotonin antagonists: methysergide, cyproheptadine, $p$-chlorophenylalanine
α-Adrenergic blockers (phentolamine)
Cholinergic antagonists (methoscopolamine)
Imipramine, chlorpromazine, pimozide, pizotifen
(In pathological conditions: L-dopa and DA agonists)

to prove the suppressibility of GH release by pharmacological agents. Furthermore, many drugs seem to be able to lower hormonal secretion only under pathological conditions such as acromegaly.

In normal subjects, some drugs can be considered GH release inhibitors for their capacity to blunt or block hormonal increases triggered by various stimuli (Table 4). Among these, we mention the serotonin antagonists methysergide and cyproheptadine, which inhibit GH discharge caused by insulin-induced hypoglycemia, exercise, or sleep (Bivens et al., 1973; Smythe and Lazarus, 1974; Chihara et al., 1976). Similar effects are shown by phentolamine, which also blocks GH response to arginine (Buckler et al., 1969), vasopressin (Heidingsfelder and Blackard, 1968), and L-dopa (Kansal et al., 1972), and by cholinergic antagonists such as methoscopolamine, which blocks both sleep-associated with insulin-induced GH secretion (Mendelson et al., 1978). Some psychoactive drugs can also exert an inhibitory influence on this hormone; they are imipramine (Takahashi et al., 1968), chlorpromazine (Sherman et al., 1971), pimozide (Schwinn et al., 1976), and pizotifen (Del Pozo and Lancranjan, 1978). Hyperglycemia has been shown to prevent sleep-related and L-dopa-induced GH releases (Lucke and Glick, 1971; Mims et al., 1973).

In pathological conditions, GH secretion can be depressed by drugs usually stimulating GH release in healthy subjects. These inhibitory effects were observed in many acromegalic patients after L-dopa, apomorphine, and bromocriptine administration (Liuzzi et al., 1972; Chiodini et al., 1974), but the mechanism of this paradoxical effect is still unknown, even if an altered receptor sensitivity can offer a partial explanation.

## 4. TSH SECRETION

TSH secretion is controlled by hypothalamic secretions and by thyroid hormones, the latter exerting an inhibitory effect at hypothalamic and pituitary levels.

### 4.1. Hypothalamic Control

Hypothalamic control of TSH secretion (Fig. 3) is mainly performed by TRH, the first of the hypothalamic hormones to be identified and simultane-

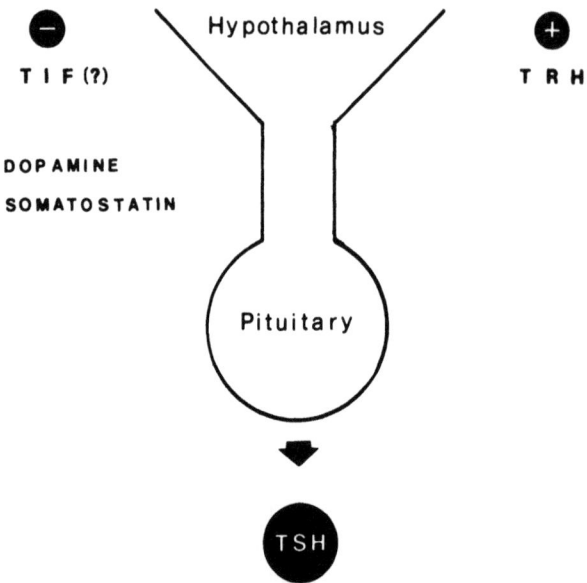

FIGURE 3. Hypothalamic control of TSH secretion in man.

ously synthesized by the groups of Guillemin and Schally (Burgus et al., 1969; Bøler et al., 1969). This hormone (pyro·Glu-His-Pro·NH$_2$) shows a potent TSH-releasing activity even at very low doses (Hershman, 1978). It is synthesized within peptidergic neurons in the "thyrotropic" area of the hypothalamus and is also present in extrahypothalamic sites which contain 75–80% of the total TRH present in the brain (Martin et al., 1977; Reichlin, 1978a). The wide distribution of TRH outside the hypothalamus lends credence to the possibility that this hormone may be a primitive neurotransmitter which has been "co-opted" for pituitary regulation as suggested by Reichlin (1978b). Its extrapituitary effects are numerous, involving the CNS, gastrointestinal tract, and pancreas (Morley, 1979). Its action at the pituitary level results in activation of membrane-bound adenylate cyclase and an increase in intracellular cAMP (Wilber and Utiger, 1968). Increased serum values not only of TSH but also of PRL follow TRH administration. For this reason the peptide has been considered a physiological PRF (Noel et al., 1974), but, as mentioned previously, further studies suggest that other PRFs, different from TRH and not yet chemically identified, exist. Further pituitary TRH effects, such as stimulation of LH (Franchimont, 1972) and FSH (Mortimer et al., 1973) secretions, have been reported. Finally, under certain pathological conditions, TRH administration also induces GH release, as has been observed in acromegaly (Faglia et al., 1973), renal failure (Gomez-Pan et al., 1975), depressive patients (Maeda et al., 1975), subjects suffering from anorexia nervosa, Nelson's syndrome, and severe liver disease (Maeda et al., 1975; Krieger and Luria, 1977; Panerai et al., 1977), and in hypothyroid patients (Collu et al., 1977). Recently, some experimental evidence seems to indicate that the hypothalamus can also exert an inhibitory control on TSH secretion through a thyrotropin-inhibiting factor (TIF) (Lamberg and Gordin, 1978). This factor could be somatostatin, whose above-mentioned inhibitory action may also influence TSH secretion. In fact, it has been demonstrated that somatostatin suppresses both the TSH response to TRH (Siler et al., 1974) and the nocturnal increase of TSH (Weeke and Laurberg, 1976).

## 4.2. Pharmacological Agents Stimulating TSH Secretion in Man

Serum TSH increase (Table 5) following TRH administration in euthyroid subjects has previously been mentioned. This effect is blocked in hyperthyroid patients and amplified in hypothyroid subjects (Besser and Mortimer, 1976). Furthermore, $T_3$ or $T_4$ administration suppresses this TSH response (Shenkman et al., 1973; Sawin et al., 1977). Estrogen administration enhances the TRH-induced thyrotropin release in healthy men (Faglia et al., 1973) but not in women (Reymond and Lemarchand-Béraud, 1976). In women, TSH response to TRH is higher than in men (Ormston et al., 1971) and is much greater during the follicular phase of the menstrual cycle (Sanchez-Franco et al., 1973). These observations suggest that estrogenic hormones sensitize the pituitary to TRH effects, probably by increasing the binding of TRH to thyrotrophs (De Lean et al., 1977).

Antidopaminergic drugs, such as metoclopramide (Scanlon et al., 1978), sulpiride (Massara et al., 1978), and domperidone (Pourmand et al., 1980), have been shown to cause a significant rise in blood TSH levels in normal subjects, though these effects have not been observed after administration of other DA antagonists such as chlorpromazine, pimozide, and in some studies after the administration of sulpiride or metoclopramide themselves (L'Hermite et al., 1978; Faglia et al., 1979; Spitz et al., 1979). However, a stimulatory action on TSH of some of these drugs is evident in hypothyroid patients (Scanlon et al., 1977; Faglia et al., 1979). TSH secretion is also positively affected by theophylline, which inhibits phosphodiesterase, an enzyme involved in cyclic nucleotide degradation (Faglia et al., 1972), and by benserazide, an inhibitor of peripheral L-aromatic amino acid decarboxylase (Delitala et al., 1980b), and

TABLE 5
Pharmacological Agents Increasing TSH Release in Man

| |
|---|
| TRH |
| Estrogens |
| DA antagonists (benzamine derivatives, domperidone) |
| Benserazide |
| Monoiodotyrosine |
| Theophylline |
| Naloxone |

by monoiodotyrosine (Scanlon *et al.*, 1980), a competitive inhibitor of the enzyme tyrosine hydroxylase. Finally, TSH secretion seems to be enhanced by naloxone (Zanoboni *et al.*, 1980) and spironolactone (Smals *et al.*, 1979) administration.

### 4.3. Pharmacological Agents Inhibiting TSH Secretion in Man

In addition to somatostatin, whose effects have been reported, many drugs decrease TSH release (Table 6). It has been shown that L-dopa reduces TSH levels in hypothyroid patients (Rapoport *et al.*, 1973; Minozzi *et al.*, 1975), whereas this effect is not always seen in normal subjects (Burrow *et al.*, 1974). DA infusion is followed both in normal and in hypothyroid subjects by a significant fall in serum TSH levels (Delitala, 1977; Kaptein *et al.*, 1980) and similar inhibiting effect is obtained by ergot derivatives (Miyai *et al.*, 1974; Felt and Nedvikova, 1977; Yap *et al.*, 1978) though the decrease in TSH values is much less evident in normal subjects (Faglia *et al.*, 1979). A depression in TSH release was also observed in normal subjects after lisuride administration (Delitala *et al.*, 1979) and after fusaric acid in hypothyroid patients (Yoshimura *et al.*, 1977). Serotoninergic blockade by cyproheptadine, metergoline, and methysergide can affect TSH secretion, but their effect has not been thoroughly investigated (Ferrari *et al.*, 1976; Delitala *et al.*, 1978; Golstein *et al.*, 1979). Among α-adrenergic blockers, phentolamine alone seems to significantly lower TSH secretion both under basal conditions and after TRH stimulation (Zgliczynski and Kaniewski, 1980). Both acetylsalicylic acid and glucocorticoids inhibit TSH response to TRH through two different mechanisms: the former probably increases serum free $T_3$ levels (Larsen, 1972), while the latter seems to act directly at the pituitary level (Wilber and Utiger, 1969; Lamberg and Gordin, 1978). A blunted TSH response to TRH has been observed in patients in barbiturate coma due to attempted suicide, but whether this effect is due to the drug appears unclear, since this diminished response is also seen in depressed patients (Naeije *et al.*, 1978). Salmon calcitonin is able to modify pituitary secretion and also inhibits TSH release induced by TRH (Leicht *et al.*, 1974). Finally, TSH secretion is paradoxically depressed after administration of some DA

TABLE 6
Pharmacological Agents Decreasing TSH Release in Man

Somatostatin
L-Dopa and DA agonists (ergot derivatives, fusaric acid)
Serotonin antagonists (cyproheptadine, metergoline, methysergide)
α-Adrenergic blockers (phentolamine)
Acetylsalicylic acid
Glucocorticoids
Barbiturates
Calcitonin
Others: pimozide, chlorpromazine, thioridazine

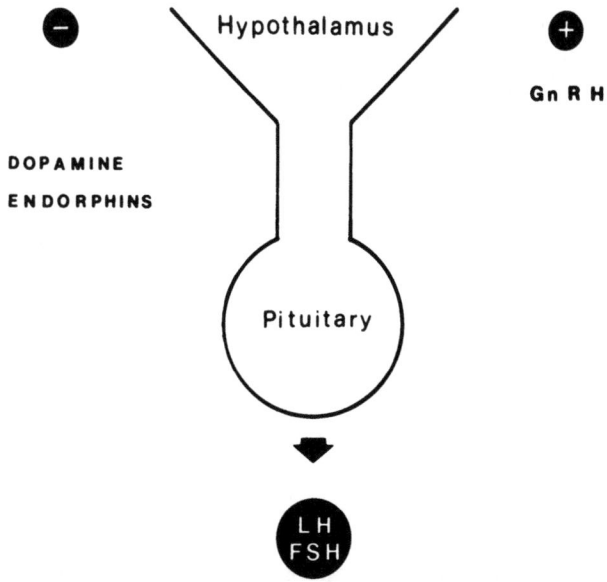

FIGURE 4. Hypothalamic control of LH and FSH secretion in man.

antagonists such as pimozide (Collu *et al.*, 1975; Faglia *et al.*, 1979), chlorpromazine, and thioridazine (Lamberg *et al.*, 1977).

## 5. GONADOTROPIN SECRETION

Gonadotropin secretion (Fig. 4) is subject to two main controlling mechanisms exerted by the hypothalamus and gonads.

### 5.1. Hypothalamic Control

In 1971 Schally's and Guillemin's groups contemporaneously isolated, characterized, and synthesized a decapeptide (pyro·Glu-His-Trp-Ser-Tyr-Gly-Leu-Arg-Pro-Gly·NH$_2$) capable of promoting gonadotropin release (Burgus *et al.*, 1971; Matsuo *et al.*, 1971; Schally *et al.*, 1971a). This peptide, initially named luteinizing hormone-releasing factor (LRF) or hormone (LHRH) and later gonadotropin-releasing hormone (GnRH), is produced by peptidergic neurons in the hypothalamic areas. Using immunohistochemical methods it has been possible to demonstrate that the highest GnRH amounts are present in the median eminence and arcuate nucleus; however, small amounts of this peptide have also been found in the preoptic area and other regions (Sétaló *et al.*, 1975; Zimmerman, 1976; Elde and Hökfelt, 1978). Hypothalamic secretion of GnRH is of two types: the first, "basal" (male pattern), is regulated by the arcuate and ventromedial nuclei; the second, "cyclic" (female pattern), is stimulated by the preoptic suprachiasmatic nuclei (Reichlin et al., 1976). Decapeptide administration induces a sharp increase of both gonadotropins, but the increase in serum LH after GnRH is greater than that of FSH. For this reason, it has been proposed that an FSH-releasing factor (FRF) distinct from LRF may exist (Shin and Kraicer, 1974). However, the chemical structure of the proposed hormone (FRF) is not known and the existence of this hormone is controversial. Nevertheless, it is well established that in humans decapeptide administration, at doses ranging from 25 to 500 µg, causes a prompt increase in serum values of LH and FSH, comparable to the LH and FSH ovulatory surges.

The secretion of GnRH and gonadotropins is also regulated by cerebral neurotransmitters, but their role in man is not well elucidated. However, experimental evidence in rats suggests a major regulatory role of NE, which promotes LH secretion (Vijayan and McCann, 1978a). This neurotransmitter also causes the release of GnRH from hypothalamic fragments (Negro-Vilar et al., 1978). On the other hand, a possible DA influence on GnRH and gonadotropin secretion is still under debate. In fact, it has been reported that injection of DA into the third cerebral ventricle of rats triggers the release of GnRH and LH (Schneider and McCann, 1969), but this effect has not been seen by others (Kreig and Sawyer, 1976). Furthermore, intravenous injection of DA or apomorphine elevated serum LH in ovariectomized rats, whereas intraperitoneal administration reduced LH values in the same animals (Vijayan and McCann, 1978b). In man an inhibitory influence of DA on LH and FSH secretion seems certain (Kaptein et al., 1980; Huseman et al., 1980).

The role of serotonin on gonadotropin secretion remains to be determined (Smythe, 1977). In rats, the inhibition of serotonin biosynthesis does not affect gonadotropin release (Donoso et al., 1971); in man the administration of anti-5HT drugs does not affect LH release (Pontiroli et al., 1980b). Information on the function of other neurotransmitters with respect to gonadotropin secretion is very scant, even though some experimental data on possible release effect of acetylcholine, histamine (Libertun and McCann, 1976), and GABA (Pass and Ondo, 1977) have been reported.

A possible inhibitory influence of PRL on gonadotropin secretion is suggested by several clinical observations; in particular, hypogonadism associated with hyperprolactinemia has frequently been observed (Thorner, 1977). The mechanisms of PRL intervention seems to consist of a hypothalamic inhibition of GnRH release, probably mediated by an increase of DA turnover in the median eminence and consequently an inhibition of gonadotropin secretion (Porter et al., 1980). A possible role of the pineal gland in gonadotropin secretion has also been suggested. The clinical observations of association of either precocious or delayed puberty with pinealoma seem to support this possibility, and some experimental evidence of gonadotropin inhibition by melatonin and another pineal factor has been reported (Motta et al., 1967; Benson et al., 1971; Frohman and Stachura, 1975).

The possible influence of glucocorticoids on gonadotropin secretion still appears uncertain. In fact, inhibition of gonadotropin secretion after adrenal steroid administration in normal subjects or in patients with Cushing's disease has not always been observed (Rolandi *et al.*, 1974; McKenna *et al.*, 1979). It has been suggested that prostaglandins can release gonadotropins by an activation of GnRH neurons. This possibility is supported by an observed inhibition of LH release after blockade of prostaglandin synthesis with indomethacin (McCann, 1980). Finally, an inhibition of GnRH-induced LH secretion by calcitonin (Leicht *et al.*, 1974) and by α-MSH (Reid *et al.*, 1981) has been reported.

### 5.2. Pharmacological Agents Affecting Gonadotropin Secretion in Man

Gonadotropin secretion is poorly influenced by pharmaceutics (Table 7) with the exception of some nonsteroidal compounds such as the antiestrogen agents. These substances, e.g., clomiphene, classically induce a discharge of GnRH and gonadotropins, preventing the normal feedback inhibition exerted by estrogens at the hypothalamic and pituitary levels. The reports regarding the effects of L-dopa (500 mg) administration on gonadotropin secretion in man are not in good agreement (Eddy *et al.*, 1971; Boden *et al.*, 1972), whereas DA infusion inhibits LH secretion (Leblanc *et al.*, 1976; Huseman *et al.*, 1980; Kaptein *et al.*, 1980) and similar effects were obtained employing DA agonists (Lachelin *et al.*, 1977). Inhibitory effects on this secretion were also observed in man after administration of a DA receptor blocker, pimozide (Collu *et al.*, 1975), but not after chlorpromazine (Santen and Barden, 1973). Reduced concentrations of both gonadotropins in basal conditions and after GnRH stimulation were observed in children treated with phenobarbitone (Masala *et al.*, 1980). Other anticonvulsant drugs, such as phenytoin alone, or phenytoin associated with either primidone or phenobarbitone, induced an increase in serum LH levels in male epileptic patients. This effect is mediated by an increase in sex steroid binding globulin and consequently in free testosterone values (Toone *et al.*, 1980). A drop in both gonadotropin levels in man was seen after administration of DAMME, an enkephalin analog (Stubbs *et al.*, 1978), and

TABLE 7
Pharmacological Agents Affecting Gonadotropin Release in Man

Stimulating agents
    GnRH
    Antiestrogens
    Naloxone and naltrexone
    Metoclopramide
Inhibiting agents
    Dopaminergic agents
    Pimozide
    Barbiturates (phenobarbitone)
    Met-enkephalin and DAMME
    Diacetylmorphine and tetrahydrocannabinol

after acute i.v. heroin administration (Mirin *et al.*, 1976), in male heroin users (Brambilla *et al.*, 1977) and marijuana users (Kolodny *et al.*, 1974), whereas we have not seen any significant changes in serum gonadotropins after acute administration of 250 µg of Met-enkephalin (Rolandi *et al.*, 1980b). Recently, Golstein *et al.* (1981) noted some inhibitory influence following slow infusion of Met-enkephalin in healthy subjects. Conversely, a mild increase in serum LH values was observed in healthy males after high doses of naloxone or naltrexone (Mandelson *et al.*, 1978; Morley *et al.*, 1980; Delitala *et al.*, 1980c). This effect was not seen after a single bolus of low doses of naloxone (Barreca *et al.*, 1980; Quigley *et al.*, 1980). Lastly, metoclopramide administration results in immediate dose-related increments in circulating levels of gonadotropin in hyperprolactinemic patients, but not in normal women (Quigley *et al.*, 1979).

## 6. ACTH SECRETION

The secretion of corticotropin is controlled by both the hypothalamus (Fig. 5) and adrenal steroids. The hormones, particularly cortisol, act at the hypothalamic and pituitary levels by a negative feedback mechanism.

### 6.1. Hypothalamic Control

The hypothalamus exerts a stimulatory control on ACTH release via a peptide or peptides called corticotropin-releasing factor(s) (CRF or CRFs). The

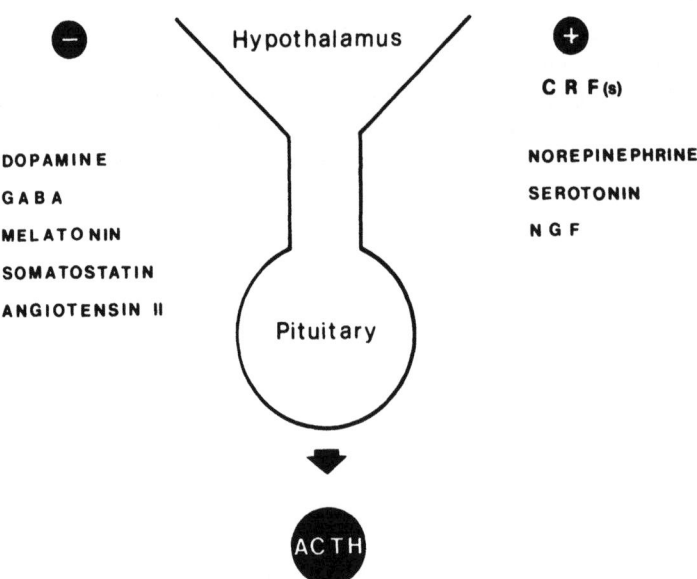

FIGURE 5. Hypothalamic control of ACTH secretion in man.

chemical nature of this factor(s) is not yet established, despite two decades of intensive investigations. However, ACTH-releasing activity in hypothalamic extracts, distinct from GnRH and TRH activity, has been definitively demonstrated (Pearlmutter et al., 1975; Takebe et al., 1975) and CRF-like substances have also been isolated from extrahypothalamic brain tissue (Witorsch and Brodish, 1972). In addition, bioassay studies showed that CRF activity is the highest in the hypothalamus and especially in the arcuate and suprachiasmatic nuclei (Palkovits, 1980). A large body of research concerns the possible influences exerted on CRF-ACTH secretions by neurotransmitters, but the reports existing in the literature are still confusing. In general, basal ACTH secretion is enhanced via the adrenergic, serotoninergic, and cholinergic pathways and inhibited via the dopaminergic pathway (Frohman and Stachura, 1975). In particular NE may promote ACTH secretion by the stimulation of $\alpha$-adrenoreceptors (Rees et al., 1970; Nakai et al., 1973) and similar effects were observed after administration of a serotonin precursor, 5HT (Imura et al., 1973a), and after a cholinomimetic agent, acetyl-$\beta$-methycholine (Soulairac et al., 1968). Conversely, L-dopa administration does not seem to affect basal ACTH secretion (Eddy et al., 1971), but inhibits the hypoglycemia-induced cortisol release (Balestreri et al., 1979). Finally, some evidence indicates that angiotensin II may inhibit ACTH secretion by an action on the brain–hypothalamus–pituitary complex (Semple et al., 1979).

### 6.2. Pharmacological Agents Increasing ACTH Secretion in Man

ACTH–cortisol release determined by insulin-induced hypoglycemia and by lysine-vasopressin (LVP) has been well known for many years and is due to a nonspecific stress effect. Increased ACTH secretion (Table 8) is found after D-amphetamine, and methylamphetamine administration is entirely inhibited by thymoxamine, an $\alpha$-adrenergic blocker (Besser et al., 1969; Rees et al., 1970). The stimulatory role of the adrenergic system is confirmed by observed pituitary–adrenal responses to two -adrenergic agonists, clonidine and

TABLE 8
Pharmacological Agents Stimulating ACTH Release in Man

| |
|---|
| Insulin-induced hypoglycemia |
| Lysine-vasopressin |
| Amphetamines (D-amphetamine, methylamphetamine) |
| $\alpha$-Adrenergic agonists (clonidine, methoxamine) |
| $\beta$-Adrenergic blockers (propranolol) |
| D blockers (haloperidol) |
| 5HTP |
| Naloxone |
| Cholinergic agents (acetyl-$\beta$-methycholine) |
| Nerve growth factor |

methoxamine (Balestreri *et al.*, 1979), and to a β-adrenergic blocker, propranolol (Imura *et al.*, 1973b). Similar effects on corticotropin secretion have been observed after the administration of tryptophan (Modlinger *et al.*, 1980), 5HT (Imura *et al.*, 1973a), and acetyl-β-methycholine (Soulairac *et al.*, 1968). Conversely, a possible inhibitory role of the DA system seems to be supported by an observed increase in cortisol secretion after DA receptor blockade by haloperidol (Balestreri *et al.*, 1979).

The activation of the pituitary–adrenal axis after naloxone, a selective opiate antagonist, suggests that ACTH secretion may be subject to a tonic inhibition via an opioid pathway in man (Blankstein *et al.*, 1980). Recently, stimulation of the pituitary–adrenocortical axis has been observed after nerve growth factor (NGF) administration (Otten *et al.*, 1980).

### 6.3. Pharmacological Agents Decreasing ACTH Secretion in Man

Inhibition of ACTH–cortisol secretion is obtained after various neuropharmacological manipulations (Table 9). It has been reported that a reduction of serotonin control by cyproheptadine or metergoline induces a depression of plasma cortisol levels and blunts the ACTH response to insulin hypoglycemia (Delitala *et al.*, 1975; Chihara *et al.*, 1976; Cavagnini *et al.*, 1976). These effects are not observed after methysergide, which is also an antiserotonin drug, and in some studies after cyproheptadine itself (Kletzky *et al.*, 1980). However, acute administration of melatonin, a serotonin derivative, causes a significant drop in circulating cortisol values and also inhibits the cortisol response to insulin-induced hypoglycemia (Smythe and Lazarus, 1974). It is possible that the changes in ACTH–cortisol secretion promoted by the above antiserotonin agents may not be mediated by a selective serotonin receptor blockade, but cholinergic or dopaminergic pathways may be involved (Smythe, 1977). The reports concerning the influence of L-dopa administration on ACTH–cortisol secretion are controversial; in fact, both an inhibitory and a stimulatory effect of this drug have been reported (Balestreri *et al.*, 1979). Some DA agonist agents, such as bromocriptine, significantly reduce ACTH secretion in Cush-

TABLE 9
Pharmacological Agents Inhibiting ACTH Release in Man

Serotonin antagonists (cyproheptadine, metergoline)
Melatonin
Dopaminergic agonists (L-dopa, bromocriptine)
α-Adrenergic antagonists (phentolamine)
β-Adrenergic agonists (isoxsuprine)
$H_1$-receptor antagonist (meclastine)
Benzodiazepines
GABA
Somatostatin

ing's and Nelson's diseases (Benker et al., 1976; Lamberts and Birkenhäger, 1976). However, DA does not appear to play a major role in the control of ACTH secretory mechanisms (Johnstone and Ferrier, 1981). The administration of α-adrenergic antagonists and β-adrenergic agonists, such as phentolamine and isoxsuprine, respectively, decrease ACTH secretion (Balestreri et al., 1979) and the same effect after the administration of a GABA derivative, baclofen, has been reported (Invitti et al., 1976). However, a role of GABA in the regulation of CRF–ACTH secretion has not yet been established. Recently, a decrease in serum ACTH values has been observed after administration of meclastine, an $H_1$-receptor antagonist (Allolio et al., 1980), and diazepam (Johnstone and Ferrier, 1980). Finally, somatostatin can decrease ACTH secretion (or release) only in pathological conditions such as pituitary adenoma (Tyrell et al., 1975).

## 7. MSH SECRETION

It is well known that MSH is subject to hypothalamic control, probably exerted by two MSH-release-inhibiting hormones, MIF-I (H·Pro-Leu-Gly·NH$_2$) and MIF-II (H·Pro-His-Phe-Arg-Gly·NH$_2$) (Burgus et al., 1976; Kastin et al., 1974). The first of these peptides can be produced from oxytocin by an enzyme present in hypothalamic tissue. The origin of MIF-II has not been determined. The existence of an MSH-releasing factor (MRF) has also been proposed (Martin et al., 1977).

Several experimental investigations regarding the secretion of MSH have been reported, but studies in humans are not available.

## 8. CONCLUSION

Pharmacological influences on hypothalamic–pituitary secretions are numerous and only partially elucidated. A great number of studies have been reported regarding the effects of DA agonists and antagonists on PRL secretion, whereas little data are available regarding the possible drug-induced changes on other pituitary hormones. In particular, secretion of glycoproteic hormones (LH, FSH, TSH) and ACTH are barely modified by various pharmacological agents. Moreover, no major progress in the isolation and chemical characterization of other hypothalamic releasing hormones has been made in the last decade, though the discovery of endorphins aroused great interest, especially in the possible endocrine effects of these peptides. Recently, some peptides have been isolated at the cerebral level which had previously been considered gastrointestinal hormones (e.g., gastrin, cholecystokinin, bombesin, VIP, and others); these could have a role as neurotransmitters and have been "co-opted" to act in the regulation of hypothalamic–pituitary secretions.

From what has been described in this chapter it is apparent that the data in this field are enormous and at first glance somewhat confusing; however, many interesting directions for research have been indicated which could help to clarify the effects of pharmacological agents on hypothalamic–pituitary function.

ACKNOWLEDGMENTS. The authors wish to thank Mr. Raffanti for his help in preparation of the manuscript.

This work was partially supported by the National Research Council (CNR) (CT 80.01862.04).

## 9. REFERENCES

Abu-Fadil, S., de Vane, G., Siler, T. M., and Yen, S. S. C., 1976, Effects of oral contraceptive steroids on pituitary prolactin secretion, *Contraception* **13**:79.

Allolio, B., Winkelmann, W., Deuss, U., Heesen, D., Hipp, F. X., and Mies, R., 1980, Effect of the antihistaminic agent meclastine on plasma ACTH, *Acta Endocrinol. (Kbh).* **94**(Suppl. 234):143.

Arimura, A., Dunn, J. D., and Schally, A. V., 1972, Effect of infusion of hypothalamic extracts on serum prolactin levels in rats treated with nembutal, CNS depressants, or bearing hypothalamic lesions, *Endocrinology* **90**:378.

Asnis, G. M., Nathan, R. S., Halbreich, U., Halpern, F. S., Sachar, M. A., and Sachar, E. J., 1980, Prolactin changes in major depressive disorders, *Am. J. Psychiatry* **137**:1117.

Balestreri, R., Bertolini, S., and Castello, C., 1979, The neural regulation of ACTH secretion in man, in: *Neuroendocrinology: Biological and Clinical Aspects* (A. Polleri and R. M. MacLeod, eds.), pp. 155–185, Academic Press, New York.

Barreca, T., Cicchetti, V., Perria, C., Masturzo, P., and Rolandi, E., 1979, Effets de la somatostatine sur la sécrétion de prolactine: Etude chez des sujets normaux et chez des patients ayant un adénome hypophysaire, *Nouv. Presse Med.* **8**:331.

Barreca, T., Marabini, A., Magnani, G., Sannia, A., and Rolandi, E., 1980, Naloxone does not affect pituitary secretion in healthy man, *IRCS Med. Sci.* **8**:908.

Benker, G., Hackenberg, K., Hamburger, B., and Reinwein, D., 1976, Effects of growth hormone release-inhibiting hormone and bromocryptin (CB 154) in states of abnormal pituitary–adrenal function, *Clin. Endocrinol.* **5**:187.

Benson, B., Matthews, M. J., and Rodin, A. E., 1971, A melatonin-free extract of bovine pineal with antigonadotropic activity, *Life Sci.* **10**:607.

Besser, G. M., 1976, Clinical implications of growth hormone release inhibiting hormone (GH-RIH), in: *Hypothalamus and Endocrine Functions* (F. Labrie and G. Pelletier, eds.), pp. 115–125, Plenum Press, New York.

Besser, G. M., and Mortimer, C. H., 1976, Clinical neuroendocrinology, in: *Frontiers in Neuroendocrinology* (L. Martini and W. F. Ganong, eds.), pp. 227–254, Raven Press, New York.

Besser, G. M., Butler, P. W., Landon, J., and Rees, L., 1969, Influence of amphetamines on plasma corticosteroid and growth hormone levels in man, *Br. Med. J.* **4**:528.

Birge, C. A., Jacobs, L. S., Hammer, C. T., and Daughaday, W. H., 1970, Catecholamine inhibition of prolactin secretion by isolated rat adenohypophyses, *Endocrinology* **86**:120.

Bivens, C. H., Lebovitz, H. E., and Feldman, J. M., 1973, Inhibition of hypoglycemia-induced growth hormone secretion by the serotonin antagonists cyproheptadine and methysergide, *N. Engl. J. Med.* **289**:236.

Blackard, W. G., and Hubbell, G. J., 1970, Stimulatory effect of exogenous catecholamines on plasma HGH concentrations in presence of beta adrenergic blockade, *Metabolism* **19**:547.

Blankstein, J., Reyes, F. I., Winter, J. S., and Faiman, G., 1980, Effects of naloxone upon prolactin and cortisol in normal women, *Proc. Soc. Exp. Biol. Med.* **164**:363.

Boden, G., Lundy, L. E., and Owen, O. E., 1972, Influence of levodopa on serum levels of anterior pituitary hormones in man, *Neuroendocrinology* **10**:309.

Bøler, J., Enzmann, F., Folkers, K., Bowers, C. Y., and Schally, A. V., 1969, The identity of chemical and hormonal properties of the thyrotropin releasing hormone and pyroglutamyl-histidyl-proline-amide, *Biochem. Biophys. Res. Commun.* **37**:705.

Boyd, A. E., III, Lebovitz, H. E., and Pfeiffer, J. B., 1970, Stimulation of human-growth-hormone secretion by L-Dopa, *N. Engl. J. Med.* **283**:1425.

Brambilla, F., Sacchetti, E., and Brunetta, M., 1977, Pituitary–gonadal function in heroin addicts, *Neuropsychobiology* **3**:160.

Brazeau, P., Vale, W., Burgus, R., and Guillemin, R., 1974, Isolation of somatostatin (a somatotropin release inhibiting factor) of ovine hypothalamic origin, *Can. J. Biochem.* **52**:1067.

Brown, G. M., Garfinkel, P. E., Warsh, J. J., and Stancer, H. G., 1976, Effect of carbidopa on prolactin, growth hormone and cortisol secretion in man, *J. Clin. Endocrinol. Metab.* **43**:236.

Brown, W. A., and Williams, B. W., 1976, Methylphenidate increases serum growth hormone concentrations, *J. Clin. Endocrinol. Metab.* **43**:937.

Buckler, J. M. H., Bold, A. M., Taberner, M., and London, D. R., 1969, Modification of hormonal response to arginine by α-adrenergic blockade, *Br. Med. J.* **3**:153.

Burgus, R., Dunn, T., Desiderio, D., and Guillemin, R., 1969, Structure moléculaire du facteur hypothalamique hypophysiotrope TRF d'origine ovine: Mise en evidence par spectrometre de masse de la séquence PGA-His-Pro-NH$_2$, *C.R. Acad. Sci.* **269**:1870.

Burgus, R., Butcher, M., Ling, N., Monahan, M., Rivier, J., Fellows, R., Amoss, M., Blackwell, R., Vale, W., and Guillemin, R., 1971, Structure moléculaire du facteur hypothalamique (LRF) d'origine ovine contrôlant la sécrétion de l'hormone gonadotrope hypophysaire de lutéinisation, *C.R. Acad. Sci.* **273**:1611.

Burgus, R., Amoss, M., Brazeau, P., Brown, M., Ling, N., Rivier, C., Rivier, J., Vale, W., and Villareal, J., 1976, Isolation and characterization of hypothalamic peptide hormones, in: *Hypothalamus and Endocrine Functions* (F. Labrie, J. Meites, and G. Pelletier, eds.), pp. 355–372, Plenum Press, New York.

Burrow, G. N., Spaulding, S. W., Donabedian, R., Van Woert, M., and Ambani, L., 1974, The effect of L-DOPA on the hypothalamic–pituitary–thyroid axis, *Adv. Neurol.* **5**:489.

Camanni, F., Massara, F., Belforte, L., and Molinatti, G. M., 1975, Changes in plasma growth hormone levels in normal and acromegalic subjects following administration of 2-bromo-ergocryptine, *J. Clin. Endocrinol. Metab.* **40**:363.

Catlin, D. H., Poland, R. E., Gorelick, D. A., Gerner, R. H., Hui, K. K., Rubin, R. T., and Li, C. H., 1980, Intravenous infusion of beta-endorphin increases serum prolactin, but not growth hormone or cortisol, in depressed subjects and withdrawing methadone addicts, *J. Clin. Endocrinol. Metab.* **50**:1021.

Cavagnini, F., Raggi, U., Micossi, P., Di Landro, A., and Invitti, C., 1976, Effect of an antiserotoninergic drug, metergoline, on the ACTH and cortisol response to insulin hypoglycemia and lysine-vasopressin in man, *J. Clin. Endocrinol. Metab.* **43**:306.

Cavagnini, F., Invitti, C., Di Landro, A., Tenconi, L., Maraschini, C., and Girotti, G., 1977, Effects of gamma aminobutyric acid (GABA) derivative, baclofen, on growth hormone and prolactin secretion in man, *J. Clin. Endocrinol. Metab.* **45**:579.

Cavagnini, F., Invitti, C., Pinto, M., Maraschini, G., Di Landro, A., Dubini, A., and Marelli, A., 1980, Effect of acute and repeated administration of gamma aminobutyric acid (GABA) on growth hormone and prolactin secretion in man, *Acta Endocrinol. (Kbh.)* **93**:149.

Checkley, S. A., 1980, Neuroendocrine tests of monoamine function in man: A review of basic theory and its application to the study of depressive illness, *Psychol. Med.* **10**:35.

Chihara, K., Kato, Y., Maeda, K., Matsukura, S., and Imura, H., 1976, Suppression by cyproheptadine of human growth hormone and cortisol secretion during sleep, *J. Clin. Invest.* **57**:1393.

Chiodini, P. G., Liuzzi, A., Botalla, L., Cremascoli, G., and Silvestrini, F., 1974, Inhibitory effect of dopaminergic stimulation on GH release in acromegaly, *J. Clin. Endocrinol. Metab.* **38**:200.

Clemens, J. A., and Shaar, C. J., 1980, Control of prolactin secretion in mammals, *Fed. Proc.* **39**:2588.

Cohen, H. N., Hay, I. D., Thomson, J. A., Logue, F., Ratcliff, W. A., and Beastall, G. H., 1979, Metoclopramide stimulation: A test of growth hormone reserve in adolescent males, *Clin. Endocrinol.* **11**:89.

Collu, R., Jequier, J. C., Leboeuf, G., Letarte, J., and Ducharme, J. R., 1975, Endocrine effects of pimozide, a specific dopaminergic blocker, *J. Clin. Endocrinol. Metab.* **41**:981.

Collu, R., Leboeuf, G., Letarte, J., and Ducharme, J. R., 1977, Increase in plasma growth hormone levels following thyrotropin-releasing hormone injection in children with primary hypothyroidism, *J. Clin. Endocrinol. Metab.* **44**:743.

Crosignani, P. G., D'Alberton, A., Peracchi, M., and Reschini, E., 1976, Dopamine-induced inhibition of prolactin secretion in amenorrhoea-galactorrhoea, *Lancet* **II**:975.

De Lean, A., Ferland, L., Drouin, J., Kelly, P. A., and Labrie, F., 1977, Modulation of pituitary thyrotropin releasing hormone receptor levels by estrogens and thyroid hormones, *Endocrinology* **100**:1496.

Delitala, G., 1977, Dopamine and T.S.H. secretion in man, *Lancet* **II**:760.

Delitala, G., Masala, A., Alagna, S., and Devilla, L., 1975, Effect of cyproheptadine on the spontaneous diurnal variations of plasma ACTH-cortisol and ACTH-GH secretion induced by L-dopa, *Biomedicine* **23**:406.

Delitala, G., Masala, A., Alagna, S., Devilla, L., and Lotti, G., 1976a, Growth hormone and prolactin release in acromegalic patients following metergoline administration, *J. Clin. Endocrinol. Metab.* **43**:1382.

Delitala, G., Masala, A., Alagna, S., and Devilla, L., 1976b, Effect of pyridoxine on human hypophyseal trophic hormone release: A possible stimulation of hypothalamic dopaminergic pathway, *J. Clin. Endocrinol. Metab.* **42**:603.

Delitala, G., Rovasio, P. P., Masala, A., Alagna, S., and Devilla, L., 1978, Metergoline inhibition of thyrotrophin and prolactin secretions in primary hypothyroidism, *Clin. Endocrinol. (Oxford)* **8**:69.

Delitala, G., Wass, J. A., Stubbs, W. A., Jones, A., Williams, S., and Besser, G. M., 1979, The effect of lisuride hydrogen maleate, an ergot derivative on anterior pituitary hormone secretion in man, *Clin. Endocrinol.* **11**:1.

Delitala, G., Devilla, L., Pende, A., and Lotti, G., 1980a, Stimolazione prolattinica indotta dalla ranitidina, farmaco antagonista dei recettori istaminici $H_2$ nell'uomo, *J. Endocrinol. Invest.* **3**(Suppl. 1):12.

Delitala, G., Devilla, L., and Pende, A., 1980b, Pituitary hormone secretion after inhibition of L-aromatic amino acid decarboxylase activity in man, *Acta Endocrinol. (Kbh).* **95**:438.

Delitala, G., Devilla L., and Di Biaso, D., 1980c, Dopamine inhibits the naloxone-induced gonadotropin rise in man, *Clin. Endocrinol.* **13**:515.

Del Pozo, E., and Lancranjan, I., 1978, Clinical use of drugs modifying the release of anterior pituitary hormones, in: *Frontiers in Neuroendocrinology* (W. F. Ganong and L. Martini, eds.), pp. 207–247, Raven Press, New York.

Del Pozo, E., Del Re, R. B., Varga, L., and Friesen, H., 1972, The inhibition of prolactin secretion in man by CB-154 (2-Br-α-ergocryptine), *J. Clin. Endocrinol. Metab.* **35**:768.

Donoso, A. O., Bishop, W., Fawcett, C. P., Kvalich, L., and McCann, S. M., 1971, Effects of drugs that modify brain monoamine concentrations on plasma gonadotropin and prolactin levels in the rat, *Endocrinology* **89**:774.

Eddy, R. L., Jones, A. L., Chakmakjian, Z. H., and Silverthorne, M. C., 1971, Effect of levodopa (L-DOPA) on human hypophyseal tropic hormone release, *J. Clin. Endocrinol. Metab.* **33**:709.

Elde, R., and Hökfelt, T., 1978, Distribution of hypothalamic hormones and other peptides in the brain, in: *Frontiers in Neuroendocrinology* (W. F. Ganong and L. Martini, eds.), pp. 1–33, Raven Press, New York.

Enjalbert, A., Priam, M., and Kordon, C., 1977, Evidence in favour of the existence of a dopamine-free prolactin-inhibiting factor (PIF) in rat hypothalamic extracts, *Eur. J. Pharmacol.* **41**:243.

Faglia, G., Ambrosi, B., Beck-Peccoz, P., Travaglini, P., and Ferrari, C., 1972, The effect of theophylline on plasma thyrotropin (HTSH) response to thyrotropin releasing factor (TRF) in man, *J. Clin. Endocrinol. Metab.* **34**:906.

Faglia, G., Beck-Peccoz, P., Ferrari, C., Travaglini, P., Ambrosi, B., and Spada, A., 1973, Plasma growth hormones response to thyrotropin-releasing hormone in patients with active acromegaly, *J. Clin. Endocrinol. Metab.* **36**:1259.

Faglia, G., Ferrari, C., Paracchi, A., and Beck-Peccoz, P., 1979, Monoaminergic regulation of thyrotropin secretion in humans, in: *Neuroendocrinology: Biological and Clinical Aspects* (A. Polleri and R. M. MacLeod, eds.), pp. 187–196, Academic Press, New York.

Felt, V., and Nedviková, J., 1977, Effect of bromocryptine on the secretion of thyrotropic hormone (TSH), prolactin (Pr), human growth hormone (HGH), thyroxine ($T_4$) and triiodothyronine ($T_3$) in hypothyroidism, *Horm. Metab. Res.* **9**:274.

Ferrari, C., Paracchi, A., Rondena, M., Beck-Peccoz, P., and Faglia, G., 1976, Effect of two serotonin antagonists on prolactin and thyrotrophin secretion in man, *Clin. Endocrinol.* **5**:575.

Ferrari, C., Caldara, R., Rampini, P., Telloli, P., Romussi, M., Bertazzoni, A., Polloni, G., Mattei, A., and Crosignani, P. G., 1978, Inhibition of prolactin release by serotonin antagonists in hyperprolactinemic subjects, *Metabolism* **27**:1499.

Fioretti, P., Melis, G. B., Paoletti, A. M., Murru, M. S., Nasi, A., Parodo, G., Corsini, U., and Martini, L., 1979, Gabaergic drugs and pituitary function in: *Neuroendocrinology: Biological and Clinical Aspects* (A. Polleri and R. M. MacLeod, eds.), pp. 359—364, Academic Press, New York.

Franchimont, P., 1972, Thyrotropin releasing hormone, in: *Frontiers in Hormone Research* (R. Hall, I. Werner, and H. Holgate, eds.), pp. 139–140, Karger, Basel.

Francis, A. F., Williams, P., William, R., and Cole, E. N., 1976, The effect of clomipramine on prolactin levels—Pilot studies, *Postgrad. Med. J.* **52**(Suppl. 3):87.

Frohman, L. A., and Stachura, M. E., 1975, Neuropharmacologic control of neuroendocrine function in man, *Metabolism* **24**:211.

Gautvik, K. M., Tashjian, A. H., Jr., Kourides, I. A., Weintraub, B. D., Graeber, C. T., Maloof, F., Suzuki, K., and Zuckerman, J. E., 1974, Thyrotropin-releasing hormone is not the sole physiologic mediator of prolactin release during suckling, *N. Engl. J. Med.* **290**:1162.

Giordano, G., Marugo, M., Barreca, T., and Minuto, F., 1971, Vasopressina ed increzione dell'ormone somatotropo, *Folia Endocrinol.* **24**:268.

Golstein, J., Vanhaelst, L., Bruno, O. D., and L'Hermite, M., 1979, Effect of cyproheptadine on thyrotrophin and prolactin secretion in normal man, *Acta Endocrinol. (Kbh.)* **92**:205.

Golstein, J., Cantrane, F., Copinschi, G., L'Hermite, M., Pipeleers, D., Robyn, C., Velkeniers, B., and Vanhaelst, L., 1981, Effects of enkephalins infusion on hormonal levels in man: Inhibition of the episodic secretion of LH by metenkephalin, *IRCS Med. Sci.* **9**:218.

Gomez-Pan, A., Alvarez-Ude, F., Evered, D. C., Duns, A., Hall, R., and Kerr, D. N. S., 1975, Pituitary responses to thyrotrophin-releasing hormone in chronic renal failure, *Clin. Sci. Mol. Med.* **49**:23P.

Harris, A. R., Christianson, D., Smith, M. S., Fang, S. L., Braverman, L. E., and Vagenaki, A. G., 1978, The physiological role of thyrotropin-releasing hormone in the regulation of thyroid-stimulating hormone and prolactin secretion in the rat, *J. Clin. Invest.* **61**:441.

Heidingsfelder, S. A., and Blackard, W. G., 1968, Adrenergic control mechanism for vasopressin-induced plasma growth hormone response, *Metabolism* **17**:1019.

Hershman, J. M., 1978, Use of thyrotropin-releasing hormone in clinical medicine, *Med. Clin. North Am.* **62**:313.

Horowski, R., Wendt, H., and Gräf, K.-J., 1978, Prolactin-lowering effect of low doses of lisuride in man, *Acta Endocrinol. (Kbh).* **87**:234.

Huseman, C. A., Kugler, J. A., and Schneider, I. G., 1980, Mechanism of dopaminergic suppression of gonadotropin secretion in men, *J. Clin. Endocrinol. Metab.* **51**:209.

Imura, H., Nakai, Y., and Yoshimi, T., 1973a, Effect of 5-hydroxytryptophan (5-HPT) on growth hormone and ACTH release in man, *J. Clin. Endocrinol. Metab.* **36**:204.

Imura, H., Nakai, Y., Kato, Y., Yoshimoto, Y., and Moridera, K., 1973b, Effect of adrenergic agents on growth hormone and ACTH secretion, in: *Endocrinology* (R. O. Scow, ed.), pp. 156–165, Excerpta Medica, Amsterdam.

Invitti, C., Di Landro, A., Pinto, M., and Cavagnini, F., 1976, Effectto inibente di un derivato del GABA, il baclofen, sulla risposta dell'ormone somatotropo e del cortisolo, Proc. XVI Congr. Soc. It. Endocrinol., Serono Symp., Bari, abstract C-33.

Isaac, R., Merceron, R. E., Caillens, G., Raymond, J. P., and Ardaillou, R., 1978, Effect of parathyroid hormone on plasma prolactin in man, *J. Clin. Endocrinol. Metab.* **47**:18.

Isaac, R., Merceron, R., Caillens, G., Raymond, J. P., and Ardaillou, R., 1980, Effects of calcitonin on basal and thyrotropin-releasing hormone-stimulated prolactin secretion in man, *J. Clin. Endocrinol. Metab.* **50**:1011.

Johnson, L. Y., 1978, The effect of arginine vasotocin, a pineal peptide, on prolactin secretion in the female rat, *Anat. Rec.* **190**:433.

Johnstone, E. C., and Ferrier, I. N., 1980, Neuroendocrine markers of CNS drug effects, *Br. J. Clin. Pharmacol.* **10**:5.

Johnstone, E. C., and Ferrier, I. N., 1981, Neural mechanisms controlling the secretion of corticotropin releasing factor, *Br. J. Clin. Pharmacol.* **11**:219.

Kanematsu, S., and Sawyer, C. H., 1973, Elevation of plasma prolactin after hypophysial stalk section in the rat, *Endocrinology* **93**:238.

Kansal, P. C., Buse, J., Talbert, O. R., and Buse, M., 1972, The effect of L-Dopa on plasma growth hormone, insulin, and thyroxine, *J. Clin. Endocrinol. Metab.* **34**:99.

Kaptein, E. M., Kletzky, O. A., Spencer, C. A., and Nicoloff, J. T., 1980, Effects of prolonged dopamine infusion on anterior pituitary function in normal males, *J. Clin. Endocrinol. Metab.* **51**:488.

Kastin, A. J., Schally, A. V., Ehrensing, R. H., and Barbeau, A., 1974, Endocrine and extraendocrine studies of hypothalamic hormones in man, in: *Recent Studies of Hypothalamic Function* (K. Lederis and K. E. Cooper, eds.), pp. 196–206, Karger, Basel.

Kato, Y., Nakai, Y., Imura, H., Chihara, K., and Ahgoi, S., 1974, Effect of 5-hydroxytryptophan (5-HTP) on plasma prolactin levels in man, *J. Clin. Endocrinol. Metab.* **38**:695.

Kato, Y., Iwasaki, Y., Iwasaki, J., Abe, H., Yanaihara, N., and Imura, H., 1978, Prolactin release by vasoactive intestinal polypeptides in rats, *Endocrinology* **103**:554.

Kleinberg, D. L., Lieberman, A., Todd, J., Greising, J., Neophytides, A., and Kupersmith, M., 1980, Pergolide mesylate: A potent day-long inhibitor of prolactin in rhesus monkeys and patients with Parkinson's disease, *J. Clin. Endocrinol. Metab.* **51**:152.

Kletzky, O. A., Marrs, R. P., and Nicoloff, J. T., 1980, Effects of cyproheptadine on insulin-induced hypoglycaemia secretion of prolactin, growth hormone, and cortisol, *Clin. Endocrinol.* **13**:231.

Koenig, J. I., Mayfield, M. A., McCann, S. M., and Krulich, L., 1979, Stimulation of prolactin secretion by morphine: Role of the central serotonergic system, *Life Sci.* **25**:853.

Kolodny, R. C., Masters, W. H., Kolodner, R. M., and Toron, G., 1974, Depression of plasma testosterone levels after chronic intensive marihuana use, *N. Engl. J. Med.* **290**:872.

Kordon, C., and Enjalbert, A., 1980, Prolactin inhibiting and stimulating factors, in: *Central and Peripheral Regulation of Prolactin* (R. M. MacLeod and U. Scapagnini, eds.), pp. 69–77, Raven Press, New York.

Kreig, R. J., and Sawyer, C. H., 1976, Effects of intraventricular catecholamines on luteinizing hormome release in ovariectomized–steroid primed rats, *Endocrinology* **99**:411.

Krieger, D. T., and Luria, M., 1977, Plasma ACTH and cortisol responses to TRF, vasopressin or hypoglycemia in Cushing's disease and Nelson's syndrome, *J. Clin. Endocrinol. Metab.* **41**:361.

Krulich, L., 1979, Central neurotransmitters and the secretion of prolactin, GH, LH, and TSH, *Annu. Rev. Physiol.* **41**:603.

Krulich, L., Hefco, E., and Aschenbrenner, J. E., 1975, Mechanism of the effects of hypothalamic deafferentation on prolactin secretion in the rat, *Endocrinology* **96**:107.

Laakmann, G., Schumacher, G., Benkert, O., and Werder, K. R., 1977, Stimulation of growth hormone secretion by desimipramine and chlorimipramine in man, *J. Clin. Endocrinol. Metab.* **44**:1010.

Lachelin, G. C., Leblanc, H., and Yen, S. S., 1977, The inhibitory effect of dopamine agonists on LH release in women, *J. Clin. Endocrinol. Metab.* **44**:728.

Lal, S., Martin, J. B., De la Vega, C. E., and Friesen, H. G., 1975, Comparison of the effect of apomorphine and L-Dopa on serum growth hormone levels in normal men, *Clin. Endocrinol.* **4**:227.

Lamberg, B. A., and Gordin, A., 1978, Abnormalities of thyrotrophin secretion and clinical implications of the thyrotrophin releasing hormone stimulation test, *Ann. Clin. Res.* **10**:171.

Lamberg, B. A., Linnoila, M., Fogelholm, R., Olkinuora, M., Kotilainen, P., and Saarinen, P., 1977, The effect of psychotropic drugs on the TSH-response to thyroliberin (TRH), *Neuroendocrinology* **24**:90.

Lamberts, S. W. J., and Birkenhäger, J. C., 1976, Bromocriptine in Nelson's syndrome and Cushing's disease, *Lancet* **II**:811.

Lancranjan, I., Wirz-Justice, A., Pühringer, W., and Del Pozo, E., 1977, Effect of 1-5 hydroxytryptophan infusion on growth hormone and prolactin secretion in man, *J. Clin. Endocrinol. Metab.* **45**:588.

Langer, G., Ferin, M., and Sacher, E. J., 1978, Effect of haloperidol and L-dopa on plasma prolactin in stalk-sectioned and intact monkeys, *Endocrinology* **102**:367.

Larsen, P. R., 1972, Salicylate-induced increases in free triiodothyronine in human serum: Evidence of inhibition of triiodothyronine binding to thyroxine-binding globulin and thyroxine-binding prealbumin, *J. Clin. Invest.* **51**:1125.

Leblanc, H., Lachelin, G. C., Abu-Fadil, S., and Yen, S. S. C., 1976, Effects of dopamine infusion on pituitary hormone secretion in humans, *J. Clin. Endocrinol. Metab.* **43**:668.

Leebaw, W. F., Lee, L. A., and Woolf, P. P., 1978, Dopamine affects basal and augmented pituitary hormone secretion, *J. Clin. Endocrinol. Metab.* **47**:480.

Leicht, E., Birò, G., and Weinges, K. F., 1974, Inhibition of releasing hormone-induced secretion of TSH and LH by calcitonin, *Horm. Metab. Res.* **6**:410.

Leveston, S. A., and Cryer, P. E., 1980, Endogenous cholinergic modulation of growth-hormone secretion in normal and acromegalic humans, *Metabolism* **29**:703.

L'Hermite, M., Denayer, P., Golstein, J., Virasoro, E., Vanhaelst, L., Copinschi, G., and Robyn, C., 1978, Acute endocrine profile of sulpiride in the human, *Clin. Endocrinol.* **9**:195.

L'Hermite, M., Michaux-Duchéne, A., and Robyn, C., 1979, Tiapride-induced chronic hyperprolactinaemia: Interference with the human menstrual cycle, *Acta Endocrinol. (Kbh.)* **92**:214.

Libertun, C., and McCann, S. M., 1976, The possible role of histamine in the control of prolactin and gonadotropin release, *Neuroendocrinology* **20**:110.

Liuzzi, A., Chiodini, P. G., Botalla, L., Cremascoli, G., Müller, E. E., and Silvestrini, F., 1972, Inhibitory effect of L-Dopa on GH release in acromegalic patients, *J. Clin. Endocrinol. Metab.* **35**:941.

Lotti, G., Delitala, G., Masala, A., Alagna, S., and Devilla, L., 1979, Hypoprolactinaemizing drugs in clinical practice, in: *Neuroendocrinology: Biological and Clinical Aspects* (A. Polleri and R. M. MacLeod, eds.), pp. 233–256, Academic Press, New York.

Lucke, C., and Glick, S. M., 1971, Experimental modification of the sleep-induced peak of growth hormone secretion, *J. Clin. Endocrinol. Metab.* **32**:729.

McCann, S. M., 1980, Fifth Geoffrey Harris Memorial Lecture, Control of anterior pituitary hormone release by brain peptides, *Neuroendocrinology* **31**:355.

McKenna, T. J., Lorber, D., Lacroix, A., and Rabin, D., 1979, Testicular activity in Cushing's disease, *Acta Endocrinol. (Kbh)* **91**:501.

MacLeod, R. M., 1976, Regulation of prolactin secretion, in: *Frontiers in Neuroendocrinology* (L. Martini and W. F. Ganong, eds.), pp. 169–194, Raven Press, New York.

Maeda, K., Kato, Y., Ohgo, S., Chihara, K., Yoshimoto, Y., Yamaguchi, N., Kuromaru, S., and Imura, H., 1975, Growth hormone and prolactin release after injection of thyrotropin-releasing hormone in patients with depression, *J. Clin. Endocrinol. Metab.* **40:**501.

Malarkey, W. B., Jacobs, L. S., and Daughaday, W. H., 1971, Levodopa suppression of prolactin in nonpuerperal galactorrhea, *N. Engl. J. Med.* **285:**1160.

Mandelson, J. H., Ellingboe, J., Keuhnle, J. C., and Mello, N. K., 1978, Effects of naltrexone on mood and neuroendocrine function in normal adult males, *Psychoneuroendocrinology* **3:**231.

Martin, J. B., 1976, Brain regulation of growth hormone secretion, in: *Frontiers in Neuroendocrinology* (L. Martini and W. F. Ganong, eds.), pp. 129–168, Raven Press, New York.

Martin, J. B., 1978, Neural regulation of growth hormone secretion, *Med. Clin. North Am.* **62:**327.

Martin, J. B., Lal, S., Tolis, G., and Friesen, H. G., 1974, Inhibition by apomorphine of prolactin secretion in patients with elevated serum prolactin, *J. Clin. Endocrinol. Metab.* **39:**180.

Martin, J. B., Reichlin, S., and Brown, G. M., 1977, *Clinical Neuroendocrinology*, Davis, Philadelphia.

Masala, A., Delitala, G., LoDico, G., Alagna, S., and Devilla, L., 1977, Effect of clomiphene on release of prolactin induced by mechanical breast emptying in women post partum, *J. Endocrinol.* **74:**501.

Masala, A., Meloni, T., Alagna, S., Rovasio, P. P., Mele, G., and Franca, V., 1980, Pituitary responsiveness to gonadotropin-releasing and thyrotrophin-releasing hormones in children receiving phenobarbitone, *Br. Med. J.* **281:**1175.

Massara, F., Camanni, F., Belforte, L., Vergano, V., and Molinatti, G. M., 1978, Increased thyrotrophin secretion induced by sulpiride in man, *Clin. Endocrinol.* **9:**419.

Matsuo, H., Baba, Y., Nair, R. M., Arimura, A., and Schally, A. V., 1971, Structure of the porcine LH and FSH releasing hormone. I. The proposed amino acid sequence, *Biochem. Biophys. Res. Commun.* **43:**1334.

Meltzer, H. Y., 1980, Effect of psychotropic drugs on neuroendocrine function, *Psychiatr. Clin. North Am.* **3:**277.

Mendelson, W. B., Jacob, L. S., Reichman, J. D., Othmer, E., Cryer, P. E., Trivedi, B., and Daughaday, W. H., 1975, Methysergide: Suppression of sleep-related prolactin secretion and enhancement of sleep-related growth hormone secretion, *J. Clin. Invest.* **56:**690.

Mendelson, W. B., Sitaram, N., Wyatt, R. J., Gillan, J. C., and Jacob, L. S., 1978, Methoscopolamine inhibition of sleep-related growth hormone secretion: Evidence for a cholinergic secretory mechanism, *J. Clin. Invest.* **61:**1683.

Mims, R. B., Scott, C. L., Modebe, O. M., and Bethune, J. E., 1973, Prevention of L-dopa induced growth hormone stimulation by hyperglycemia, *J. Clin. Endocrinol. Metab.* **37:**660.

Minozzi, M., Faggiano, M., Lombardi, G., Garella, G., Criscuolo, T., and Scapagnini, U., 1975, Effects of L-Dopa on plasma TSH levels in primary hypothyroidism, *Neuroendocrinology* **17:**147.

Mirin, S. M., Mendelson, J. H., Ellingboe, J., and Meyer, R. E., 1976, Acute effects of heroin and naltrexone on testosterone and gonadotrophin secretion, a pilot study, *Psychoneuroendocrinology* **1:**359.

Miyai, K., Onishi, T., Hosokawa, M., Ishibashi, K., and Kumahara, Y., 1974, Inhibition of thyrotropin and prolactin secretions in primary hypothyroidism by 2-Br-alpha-ergocryptin, *J. Clin. Endocrinol. Metab.* **39:**391.

Modlinger, R. S., Schonmuller, J. M., and Arora, S. P., 1980, Adrenocorticotropin release by tryptophan in man, *J. Clin. Endocrinol. Metab.* **50:**360.

Moore, K. E., Demaerst, K. T., and Johnston, C. A., 1980, Influence of prolactin on dopaminergic neuronal systems in the hypothalamus, *Fed. Proc.* **39:**2912.

Morley, J. E., 1979, Extrahypothalamic thyrotropin releasing hormone (TRH), its distribution and its functions, *Life Sci.* **25:**1539.

Morley, J. E., Baranetsky, N. G., Wingert, T. D., Carlson, H. E., Hershman, J. M., Melmed, S., Levin, S. R., Jamison, K. R., Weitzman, R., Chang, R. J., and Varner, A. A., 1980, Endocrine effects of naloxone-induced opiate receptor blockade, *J. Clin. Endocrinol. Metab.* **50:**251.

Mortimer, C. H., Besser, G. M., McNeilly, A. T., Tunbridge, W. M. G., Gomez-Pan, A., and Hall, R., 1973, Interaction between secretion of the gonadotropins, prolactin, growth hormone, thyrotropin and corticosteroid in man: The effects of LH/FSH-RH, TRH and hypoglycemia along and in combination, *Clin. Endocrinol. (Oxford)* **2**:317.

Motta, M., Fraschini, F., and Martini, L., 1967, Endocrine effects of pineal gland and of melatonin, *Proc. Soc. Exp. Biol. Med.* **126**:431.

Müller, E. E., Brambilla, F., Cavagnini, F., Peracchi, M., and Panerai, A. E., 1974, Slight effect of L-tryptophan on growth hormone release in normal human subjects, *J. Clin. Endocrinol. Metab.* **39**:1.

Müller, E. E., Genazzani, A. R., and Murru, S., 1978, Nomifensine: Diagnostic test in hyperprolactinemic states, *J. Clin. Endocrinol. Metab.* **47**:1352.

Naeije, R., Golstein, J., Zegers de Beyl, D., Linkowski, P., Mendlewicz, J., Copinschi, G., Badawi, M., Leclercq, R., L'Hermite, M., and Vanhaelst, L., 1978, Thyrotrophin, prolactin and growth hormone responses to TRH in barbiturate coma and in depression, *Clin. Endocrinol.* **9**:49.

Nakai, Y., Imura, H., Yoshimi, T., and Matsukura, S., 1973, Adrenergic control mechanisms for ACTH secretion in man, *Acta Endocrinol. (Kbh)* **74**:263.

Nakai, Y., Imura, H., Sakurai, H., Kurahachi, H., and Yoshimi, T., 1974, Effect of cyproheptadine on human growth hormone secretion, *J. Clin. Endocrinol. Metab.* **38**:446.

Negro-Vilar, A., Ojeda, S. R., and McCann, S. M., 1978, In vitro release of somatostatin and LRH by MBH and ME fragments: Effect of catecholamines (CAs), *Fed. Proc.* **37**:445 (abstr.).

Noel, G. L., Dimond, R. C., Wartofsky, L., Earll, J. M., and Frantz, A. G., 1974, Studies of prolactin and TSH secretion by continuous infusion of small amounts of thyrotrophin-releasing hormone (TRH), *J. Clin. Endocrinol. Metab.* **39**:6.

Ormston, B. J., Garry, R., Cryer, R. J., Besser, G. M., and Hall, R., 1971, Thyrotrophin-releasing hormone as a thyroid function test, *Lancet* **II**:10.

Otten, U., Baumann, J. B., and Girard, J., 1980, Stimulation of the pituitary–adrenocortical axis by the nerve growth factor, *Acta Endocrinol.* **94**(Suppl. 234):138.

Palkovits, M., 1980, Topography of chemically identified neurons in the central nervous system: Progress in 1977–1979, *Med. Biol.* **58**:188.

Panerai, A. E., Salerno, F., Manneschi, M., Cocchi, D., and Müller, E. E., 1977, Growth hormone and prolactin responses to thyrotropin-releasing hormone in patients with severe liver disease, *J. Clin. Endocrinol. Metab.* **45**:134.

Parker, D. C., Rossman, L. G., Siler, T. M., Rivier, J., Yen, S. S., and Guillemin, R., 1974, Inhibition of the sleep-related peak in physiologic human growth hormone release by somatostatin, *J. Clin. Endocrinol. Metab.* **38**:496.

Pass, K. A., and Ondo, J. G., 1977, The effects of gamma-aminobutyric acid on prolactin and gonadotropin secretion in the unanesthetized rat, *Endocrinology* **100**:1437.

Pearlmutter, A. F., Rapino, E., and Saffran, M., 1975, The ACTH-releasing hormone of the hypothalamus requires a co-factor, *Endocrinology* **977**:1336.

Pontiroli, A. E., Castegnaro, E., Vettaro, M. P., Viberti, G., and Pozza, G., 1977, Stimulatory effect of the dopa-decarboxylase inhibitor Ro 4-4602 on prolactin release; inhibition by L-dopa, metergoline, methysergide and 2-Br-alpha-ergocryptine, *Acta Endocrinol. (Kbh.)* **84**:36.

Pontiroli, A. E., Alberetto, M., Restelli, L., and Facchinetti, A., 1980a, Effect of bombesin and ceruletide on prolactin, growth hormone, luteinizing hormone, and parathyroid hormone release in normal human males, *J. Clin. Endocrinol. Metab.* **51**:1303.

Pontiroli, A. E., Alberetto, M., Pelliciotta, G., De Castro e Silva, E., De Pasqua, A., Girardi, A. M., and Pozza, G., 1980b, Interaction of dopaminergic and antiserotoninergic drugs in the control of prolactin and LH release in normal women, *Acta Endocrinol. (Kbh).* **93**:271.

Porter, J. C., Nansel, D. D., Gudelsky, G. A., Foreman, M. M., Pilotte, N. S., Parker, C. R., Jr., Burrows, G. H., Bates, G. W., and Madden, J. D., 1980, Neuroendocrine control of gonadotropin secretion, *Fed. Proc.* **39**:2896.

Pourmand, M., Rodriguez-Arnao, M. D., Weightman, D. R., Hall, R., Cook, D. B., Lewis, M., and Scanlon, M. F., 1980, Domperidone: A novel agent for the investigation of anterior pituitary function and control in man, *Clin. Endocrinol.* **12**:211.

Quigley, M. E., Judd, S. J., Gilliland, G. B., and Yen, S. S., 1979, Effect of a dopamine antagonist on the release of gonadotropin and prolactin in normal women and women with hyperprolactinemic anovulation, *J. Clin. Endocrinol. Metab.* **48**:718.

Quigley, M. E., Sheehan, K. L., Casper, R. F., and Yen, S. S., 1980, Evidence for an increased opioid inhibition of luteinizing hormone secretion in hyperprolactinemic patients with pituitary microadenoma, *J. Clin. Endocrinol. Metab.* **50**:427.

Rapoport, B., Refetoff, S., Fang, V. S., and Friesen, H. G., 1973, Suppression of serum thyrotropin (TSH) by L-Dopa in chronic hypothyroidism: Interrelationships in the regulation of TSH and prolactin secretion, *J. Clin. Endocrinol. Metab.* **36**:256.

Raymond, V., Beaulieu, M., Labrier, F., and Boissier, J., 1978, Potent antidopaminergic activity of estradiol at the pituitary level on prolactin release, *Science* **200**:1173.

Rees, L., Butler, P. W., Gosling, C., and Besser, G. M., 1970, Adrenergic blockade and the corticosteroid and growth hormone responses to methylamphetamine, *Nature (London)* **288**:565.

Reichlin, S., 1978a, Neural control of the pituitary gland: Normal physiology and pathophysiologic implications. Current concepts, Upjohn Co., New York.

Reichlin, S., 1978b, Regulation of the hypothalamic–pituitary–thyroid axis, *Med. Clin. North Am.* **62**:305.

Reichlin, S., Saperstein, R., Jackson, I. M., Boyd, A. E., III, and Patel, Y., 1976, Hypothalamic hormones, *Annu. Rev. Physiol.* **38**:389.

Reid, R. L., Ling, N., and Yen, S. S., 1981, Alpha-melanocyte stimulating hormone induces gonadotropin release, *J. Clin. Endocrinol. Metab.* **52**:159.

Reymond, M., and Lemarchand-Béraud, T., 1976, Effects of oestrogens on prolactin and thyrotrophin responses to TRH in women during the menstrual cycle and under oral contraceptive treatment, *Clin. Endocrinol.* **5**:429.

Rivier, C., Brown, M., and Vale, W., 1977, Effects of neurotensin, substance P and morphine sulphate on the secretion of prolactin and growth hormone in the rat, *Endocrinology* **100**:751.

Rivier, C., Rivier, J., and Vale, W., 1978, The effect of bombesin and related peptides on prolactin and growth hormone secretion in the rat, *Endocrinology* **102**:519.

Rolandi, E., and Barreca, T., 1978, Effects of two analgesic opiates (methadone and pentazocine) on the serum prolactin levels in breast cancer, *Acta Endocrinol. (Kbh.)* **88**:452.

Rolandi, E., Masala, A., and Delitala, G., 1974, Effetto del desametasone sui valori ematici di LH ed HGH e su quelli urinarî di FSH e testosterone in due maschi adulti sani, *Studi Sassar.* **52** (Sez. 1-2-3):105.

Rolandi, E., Barreca, T., Gallamini, A., Gianrossi, R., Masturzo, P., Murialdo, G., and Nizzo, M. C., 1979a, Physiological, pharmacological and pathological hyperprolactinemias, in: *Neuroendocrinology: Biological and Clinical Aspects* (A. Polleri and R. M. MacLeod, eds.), pp. 257–286, Academic Press, New York.

Rolandi, E., Masturzo, P., and Barreca, T., 1979b, Inhibition of cimetidine-induced hyperprolactinaemia by pretreatment with levodopa or bromocriptine, *Clin. Endocrinol.* **10**:93.

Rolandi, E., Sannia, A., Milesi, G. M., and Barreca, T., 1979c, Effect of salmon calcitonin administration on serum prolactin levels in man, *IRCS Med. Sci.* **8**:570.

Rolandi, E., Magnani, G., Sannia, A., and Barreca, T., 1980a, Prolactin changes induced by a partial opiate antagonist, nalorphine, *IRCS Med. Sci.* **8**:570.

Rolandi, E., Cicchetti, V., Sannia, A., Magnani, G., and Barreca, T., 1980b, Failure of intravenous metenkephalin administration to affect pituitary secretion in man, *IRCS Med. Sci.* **8**:235.

Rolandi, E., Magnani, G., Milesi, G. M., and Barreca, T., 1981, Effect of psychoactive drug, trazodone, on prolactin secretion in man, *Neuropsychobiology* **7**:17.

Ruberg, M., Rotsztejn, W. H., Arancibia, S., Besson, J., and Enjalbert, A., 1978, Stimulation of prolactin release by vasoactive intestinal polypeptide (VIP), *Eur. J. Pharmacol.* **51**:319.

Rubin, P., Swezey, S., and Blaschke, T., 1979, Naloxone lowers plasma-prolactin in man, *Lancet* **I**:1293.

Sachar, E. J., 1978, Neuroendocrine responses to psychotropic drugs, in: *Psychopharmacology:*

*A Generation of Progress* (M. A. Lipton, A. DiMascio, and K. F. Killam, eds.), pp. 499–507, Raven Press, New York.

Sanchez-Franco, F., Garcia, M. D., Cacicedo, L., Martin-Zurro, A., and Escobar de Rey, F., 1973, Influence of sex phase of the menstrual cycle on thyrotropin (TSH) response to thyrotropin-releasing hormone (TRH), *J. Clin. Endocrinol. Metab.* **37**:736.

Santen, R. J., and Barden, C. W., 1973, Episodic luteinizing hormone secretion in man: Pulse analysis, clinical interpretation, physiologic mechanisms, *J. Clin. Invest.* **52**:2617.

Sawin, C. T., Hershman, J. M., and Chopra, I. J., 1977, The comparative effect of $T_4$ and $T_3$ on the TSH response to TRH in young adult men, *J. Clin. Endocrinol. Metab.* **44**:273.

Scanlon, M. F., Weightman, D. R., Mora, B., Heath, M., Shale, D. J., Snow, M. H., and Hall, R., 1977, Evidence for dopaminergic control of thyrotrophin secretion in man, *Lancet* **II**:421.

Scanlon, M. F., Ree, S., Smith, B., and Hall, R., 1978, Thyroid-stimulating hormone: Neuroregulation and clinical applications, Part 1, *Clin. Sci. Mol. Med.* **55**:1.

Scanlon, M. F., Rodriguez-Arnao, M. D., Pourmand, M., Shale, D. J., Weightman, D. R., Lewis, M., and Hall, R., 1980, Catecholaminergic interactions in the regulation of thyrotropin (TSH) secretion in man, *J. Endocrinol. Invest.* **3**:125.

Schally, A. V., Arimura, A., Baba, Y., Nair, R. M., Matsuo, H., Redding, T. W., Debeljuk, L., and White, W. F., 1971a, Isolation and properties of the FSH and LH releasing hormone, *Biochem. Biophys. Res. Commun.* **43**:393.

Schally, A. V., Baba, Y., Nair, R. M., and Bennett, C. D., 1971b, The amino acid sequence of a peptide with growth hormone releasing activity isolated from porcine hypothalamus, *J. Biol. Chem.* **246**:6647.

Schally, A. V., Redding, T. W., Arimura, A., Dupont, A., and Linthicum, G. L., 1977, Isolation of gamma-amino butyric acid from pig hypothalami and demonstration of its prolactin release-inhibiting (PIF) in vivo and in vitro, *Endocrinology* **100**:681.

Schneider, H. P., and McCann, S. M., 1969, Possible role of dopamine as transmitter to promote discharge of LH-releasing factor, *Endocrinology* **85**:121.

Schwinn, G., Schwarck, H., McIntosh, C., Milstrey, H. R., Williams, B., and Köbberling, J., 1976, Effect of the dopamine receptor blocking agent pimozide on the growth hormone response to arginine and exercise and on the spontaneous growth hormone fluctuations, *J. Clin. Endocrinol. Metab.* **43**:1183.

Semple, P. F., Buckingham, J. C., Mason, P. A., and Fraser, R., 1979, Suppression of plasma ACTH concentration by angiotensin II infusion in normal humans and in a subject with a steroid 17 alpha-hydroxylase defect, *Clin. Endocrinol.* **10**:137.

Sétaló, G., Vigh, S., Schally, A. V., Arimura, A., and Flerkó, B., 1975, LH-RH containing neural elements in the rat hypothalamus, *Endocrinology* **96**:135.

Shaar, C. J., and Clemens, J. A., 1974, The role of catecholamines in the release of anterior pituitary prolactin in vitro, *Endocrinology* **95**:1202.

Shane, J. M., and Naftolin, F., 1974, Effect of ergonovine maleate on purerperal prolactin, *Am. J. Obstet. Gynecol.* **120**:129.

Shenkman, L., Mitsuma, T., and Hollander, C. S., 1973, Modulation of pituitary responsiveness to thyrotropin-releasing hormone by iodothyronine, *J. Clin. Invest.* **52**:205.

Sherman, L., Kim, S., Benjamin, F., and Kolodny, H. D., 1971, Effect of chlorpromazine on serum growth hormone concentration in man, *N. Engl. J. Med.* **284**:72.

Shin, S. H., and Kraicer, J., 1974, LH-RH radioimmunoassay and its applications: Evidence of antigenically distinct FSH-RH and a diurnal study of LH-RH and gonadotrophins, *Life Sci.* **14**:281.

Siler, T. M., Van den Berg, G., and Yen, S. S., 1973, Inhibition of growth hormone release in humans by somatostatin, *J. Clin. Endocrinol. Metab.* **37**:632.

Siler, T. M., Yen, S. S., Vale, W., and Guillemin, R., 1974, Inhibition by somatostatin on the release of TSH induced in man by thyrotropin-releasing factor, *J. Clin. Endocrinol. Metab.* **38**:742.

Slater, S. L., Lipper, S., Shiling, D. J., and Murphy, D. L., 1977, Elevation of plasma prolactin by monoamine-oxidase inhibitors, *Lancet* **II**:275.

Slatopolsky-Cantis, M., Guitelman, A., Razumny, J., Baudini, R., and Rubel, H., 1977, Le piribedil et son emploi dans la suppression de la lactation, *Pharmatherapeutica* **1**:610.

Smals, A. G., Kloppenborg, P. W., Hoefnagels, W. H., and Drayer, J. I., 1979, Pituitary–thyroid function in spironolactone treated hypertensive women, *Acta Endocrinol. (Kbh.)* **90**:577.

Smythe, G. A., 1977, The role of serotonin and dopamine in hypothalamic–pituitary function, *Clin. Endocrinol.* **7**:325.

Smythe, G. A., and Lazarus, L., 1974, Growth hormone responses to melatonin in man, *Science* **184**:1373.

Smythe, G. A., Compton, P. J., and Lazarus, L., 1975, The stimulation of human prolactin secretion by 3-iodo-L-tyrosine, *J. Clin. Endocrinol. Metab.* **40**:714.

Soulairac, A., Schaub, C., Franchimont, P., Aymard, N., and Van Cawenberg, H., 1968, Etude de l'activation pharmacologique du pôle central de l'axe hypothalamo-hypophysaire, *Ann. Endocrinol.* **29**:45.

Spampinato, S., Locatelli, V., Cocchi, D., Vicentini, L., Bajusz, S., Ferri, S., and Müller, E. E., 1979, Involvement of brain serotonin in the prolactin-releasing effect of opioid peptides, *Endocrinology* **105**:163.

Spitz, I. M., Trestian, S., Cohen, H., Arnon, N., and Le Roith, D., 1979, Failure of metoclopramide to influence LH, FSH, and TSH secretion or their responses to releasing hormones, *Acta Endocrinol. (Kbh.)* **92**:640.

Stern, W. C., Rogers, J., Fang, V., and Meltzer, H. Y., 1979, Influence of bupropion HCl (Wellbatrin), a novel antidepressant, on plasma levels of prolactin and growth hormone in man and rat, *Life Sci.* **25**:1717.

Strauch, G., Girault, D., Rifai, M., and Bricaire, H., 1973, Alpha-MSH stimulation of growth hormone release, *J. Clin. Endocrinol. Metab.* **37**:990.

Stubbs, W. A., Delitala, G., Jones, A., Jeffcoate, W. J., Edwards, C. R. W., Ratter, S. J., Besser, G. M., Bloom, S. R., and Alberti, K. G. M. M., 1978, Hormonal and metabolic responses to an enkephalin analogue in normal man, *Lancet* **II**:1225.

Sylvälahti, E. K., and Kando, J. H., 1975, Serum growth hormone, serum immunoreactive insulin and blood glucose response to oral and intravenous diazepam in man, *Int. J. Clin. Pharmacol. Biopharm.* **12**:74.

Takahara, J., Yunoki, S., Yakushiji, W., Yamuchi, J., Yamane Y., and Ofuji, N., 1977, Stimulatory effects of gamma-hydroxybutyric acid on growth hormone and prolactin release in humans, *J. Clin. Endocrinol. Metab.* **44**:1014.

Takahashi, Y., Kipnis, D. M., and Daughaday, W. H., 1968, Growth hormone secretion during sleep, *J. Clin. Invest.* **47**:2079.

Takebe, K., Yasuda, N., and Greer, M. A., 1975, A sensitive and simple in vitro assay for corticotropin-releasing substances utilizing ACTH release from cultured anterior pituitary cells, *Endocrinology* **97**:1248.

Tamminga, C. A., Neophytides, A., Chase, T. N., and Frohman, L. A., 1978, Stimulation of prolactin and growth hormone secretion by muscimol, a α-aminobutyric acid agonist, *J. Clin. Endocrinol. Metab.* **47**:1348.

Thorner, M. O., 1977, Prolactin, *Clin. Endocrinol. Metab.* **6**:201.

Tolis, G., Hickey, J., and Guyda, H., 1975a, Effect of morphine on serum growth hormone, cortisol, prolactin and thyroid stimulating hormone in man, *J. Clin. Endocrinol. Metab.* **41**:797.

Tolis, G., Pinter, E. J., and Friesen, H. G., 1975b, The acute effect of 2-bromo-α-ergocyptine (CB 154) on anterior pituitary hormones and free fatty acids in man, *Int. J. Clin. Pharmacol. Biopharm.* **12**:281.

Toone, B. K., Wheeler, M., and Fenwick, P. B., 1980, Sex hormone changes in male epileptics, *Clin. Endocrinol.* **12**:391.

Tuomisto, J., 1978, Neuropharmacological intervention on the pituitary–hypothalamic relationship, *Ann. Clin. Res.* **10**:120.

Tyrell, J. B., Lorenzi, M., Forsham, P. H., and Gerich, J. E., 1975, The effect of somatostatin on secretion of adrenocorticotropin in normal subjects and in patients with Nelson's syndrome and Cushing's disease, *Endocrinology* **96**:A350.

Van Maanen, J. H., and Smelik, P. G., 1968, Induction of pseudopregnancy in rats following local depletion of monoamines in the median eminence of the hypothalamus, *Neuroendocrinology* 3:177.

Van Vugt, D. A., Bruni, J. F., Sylvester, P. W., Chen, H. T., Ieiri, T., and Meites, J., 1979, Interaction between opiates and hypothalamic dopamine on prolactin release, *Life Sci.* 24:2361.

Vijayan, E., and McCann, S. M., 1978a, Reevaluation of the role of catecholamines in control of gonadotropin and prolactin release, *Neuroendocrinology* 25:150.

Vijayan, E., and McCann, S. M., 1978b, The effect of systemic administration of dopamine and apomorphine on plasma LH and prolactin concentrations in conscious rats, *Neuroendocrinology* 25:221.

Weeke, J., and Laurberg, P., 1976, Diurnal TSH variations in hypothyroidism, *J. Clin. Endocrinol. Metab.* 43:32.

Weiner, R. I., and Ganong, W. F., 1978, Role of brain monoamines and histamine in regulation of anterior pituitary secretion, *Physiol. Rev.* 58:905.

Wieland, R. G., Hallberg, M. C., and Zorn, E. M., 1973, Growth hormone response to intramuscular glucagon, *J. Clin. Endocrinol. Metab.* 37:329.

Wilber, J. F., and Utiger, R. D., 1968, In vitro studies on mechanism of action of thyrotropin releasing factor, *Proc. Soc. Exp. Biol. Med.* 127:488.

Wilber, J. F., and Utiger, R. D., 1969, The effect of glucocorticoids on thyrotropin secretion, *J. Clin. Invest.* 48:2096.

Willis, K. J., London, D. R., and Butt, W. R., 1976, Proceedings: Hormonal effects of tamoxifen in women with carcinoma of the breast, *J. Endocrinol.* 69:51P.

Witorsch, R. J., and Brodish, A., 1972, Evidence for acute ACTH release by extrahypothalamic mechanisms, *Endocrinology* 90:1160.

Yap, P. L., Davidson, N. M., Lidgard, G. P., and Fyffe, J. A., 1978, Bromocriptine suppression of the thyrotrophin response to thyrotrophin releasing hormone, *Clin. Endocrinol.* 9:179.

Yoshimura, M., Hachiya, T., Ochi, Y., Nagasaka, A., Takeda, A., Hidaka, H., Refetoff, S., and Fang, V. S., 1977, Suppression of elevated serum TSH levels in hypothyroidism by fusaric acid, *J. Clin. Endocrinol. Metab.* 45:95.

Zanoboni, A., Zanoboni-Muciaccia, W., and Zanussi, C., 1980, Influenza del naloxone sul test di stimolazione ipofisaria con TRH: Potenziamento della risposta del TSH, *J. Clin. Endocrinol. Invest.* 3(Suppl.):10.

Zgliczynski, S., and Kaniewski, M., 1980, Evidence for alpha-adrenergic receptors mediated TSH release in men, *Acta Endocrinol. (Kbh.)* 95:172.

Zimmerman, E. A., 1976, Localization of hypothalamic hormones by immunocytochemical techniques, in: *Frontiers in Neuroendocrinology* (L. Martini and W. F. Ganong, eds.), pp. 25–62, Raven Press, New York.

CHAPTER 3

# TRH Interactions with Cholinergic Mechanisms and Consequent Therapeutic Implications

GEORGE G. YARBROUGH

## 1. INTRODUCTION

Subsequent to the claim that thyrotropin-releasing hormone (TRH; pyroglutamyl-histidyl-proline amide) was effective in the treatment of depression (Prange and Wilson, 1972) and the demonstration that TRH enhanced the stimulant properties of L-dopa in mice (Plotnikoff et al., 1972), a great deal of experimental work has been devoted to elucidating both the basic pharmacological properties of this small peptide and its underlying mode of action. To date, a unifying hypothesis that would satisfactorily account for the multiple and unique actions of exogenously administered TRH is not apparent. Similarly, the functional significance of endogenous, extrahypothalamic TRH, which is found in both mammalian and nonmammalian species, is largely a matter of conjecture. However, with continued research in a variety of experimental preparations an unusual property, which appears to underlie many of the pharmacological effects of TRH, has emerged; namely that TRH, at all levels of the neuraxis and in a unique fashion, appears to facilitate cholinergic transmission via a unique action on central cholinergic neurons. The evidence for these phenomena and consequent therapeutic indications are summarized below.

## 2. TRH EFFECTS ON ACETYLCHOLINE METABOLISM, TURNOVER, AND RELEASE

High-affinity, $Na^+$-dependent choline uptake is thought to be an index of *in vivo* cholinergic neuronal activity (Atweh et al., 1975). While not affecting

---

GEORGE G. YARBROUGH • Merck Institute for Therapeutic Research, West Point, Pennsylvania 19486.

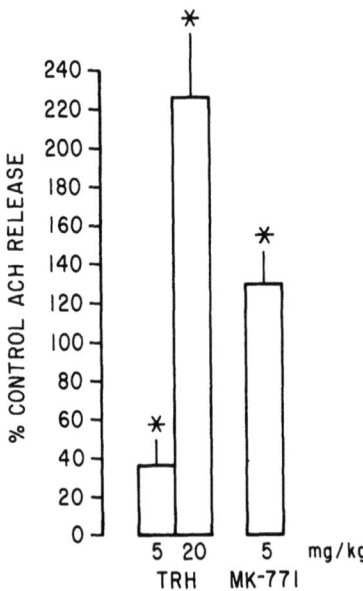

FIGURE 1. ThH and MK-771 effects on ACh release from the cerebral cortical surface of anesthetized rabbits. ACh was collected and assayed essentially as described by Phillis et al. (1973). Adult, male rabbits (2.5–3.8 kg) were anesthetized with 50 mg/kg i.v. of vinbarbital. The parietal cortices were exposed bilaterally and small Perspex cups placed on each hemisphere. Each cup contained 300 µl of artificial CSF containing 50 µg/ml of neostigmine which was left in contact with the cortical surface for 10 min. At the end of that time the samples were removed (replaced with fresh CSF), combined, and frozen on dry ice for subsequent bioassay on the isolated hearts of *Mercenaria mercenaria*. After an equilibration period of 30 min, samples were collected every 10 min for the next 60 min. At the end of the third collection interval, the animals received an i.v. injection of saline, TRH, or MK-771 and three more 10-min samples obtained. The average amounts of released ACh for the entire 30-min interval postinjection were calculated and expressed as a percentage ($\pm$ S.E.M.) of that obtained from the cortices of saline-injected animals. Four animals were used per treatment. It can be seen that TRH caused an apparently dose-related increase in ACh release with MK-771 at the one dose tested being more effective than TRH. $p < 0.05$ (determined by ANOVA and Duncan multiple range test) indicated by asterisks.

regional brain levels of acetylcholine (ACh) or choline uptake in normal rats, intraperitoneally administered TRH was found to antagonize barbiturate-induced decreases in high-affinity choline uptake in the cerebral cortex, hippocampus, and midbrain (Schmidt, 1977; Yarbrough et al., 1978). Similarly, MK-771 (L-pyro-2-aminoadipyl-histidyl-thiazolidine-4-carboxamide), which is a potent analog of TRH (Veber et al., 1976), was found after intraventricular administration to antagonize pentobarbital-induced decreases in choline uptake in the same brain areas (Santori and Schmidt, 1980). The direct inference from these data is that these peptides can antagonize barbiturate-induced decreases in cholinergic neuronal activity. Employing more direct biochemical assessments of cholinergic dynamics, Malthe-Sørenssen et al. (1978) found that TRH stimulated the turnover rate of ACh in the parietal cortex of unanesthetized rats and Yarbrough et al. (1978) observed that MK-771 enhanced the incorporation of [$^3$H]choline into [$^3$H]-ACh in the brains of pentobarbital-treated mice. *In vitro*, TRH does not influence acetylcholinesterase activity or the uptake and K$^+$-evoked release of choline (Renaud et al., 1979) nor does MK-771 alter the binding of [$^3$H]quinuclidinyl benzilate ([$^3$H]-QNB) to brain synaptosomes (M. Williams, personal communication). Finally, it has recently been demonstrated that TRH and MK-771 enhance the *in vivo* release of endogenous ACh from the cerebral cortical surface of anesthetized rabbits (Fig. 1). Thus, the available biochemical data indicate that these peptides can stimulate central cholinergic neurons and that this action is more readily apparent when neuronal activity has been depressed by barbiturates.

## 3. ACTIONS ON CENTRAL CHOLINERGIC PATHWAYS AND NEURONS

In an extensive series of microinjection experiments, Kalivas and Horita (1980) determined that the most likely anatomical substrate mediating, at least in part, the analeptic actions of TRH is the septohippocampal system. Since it is well established that there is an efferent cholinergic projection from the septum to the hippocampus, it was concluded that the biochemical changes in ACh metabolism in the hippocampus (alluded to above) resulted from an activation of cholinergic neurons in the septum. Further support for a septohippocampal locus of action of TRH is derived from experiments recording hippocampal electroencephalograms (EEG) and multiunit activity. Thus, McNaughton et al. (1977) found that TRH reversed the effects of amylobarbitone on septal-stimulated theta activity in the hippocampus. More recently, Kalivas et al. (1980) demonstrated that TRH induced hippocampal EEG synchrony in pentobarbital-pretreated rats. This and previous findings demonstrating that TRH increased hippocampal neuronal interspike intervals (Koranyi et al., 1977) clearly suggest a facilitory interaction of TRH with this classical cholinergic pathway in the brain.

Another indisputably cholinergic pathway where such an effect is equally evident is in the spinal cord. TRH and MK-771 exert pronounced excitatory effects on spinal motoneurons (Nicoll, 1977; Yarbrough and Singh, 1979; Phillis and Kirkpatrick, 1980) and it is conceivable that this provides the basis for the pronounced tremorogenic effects of these peptides. It is known that these peptides induce a graded, observable tremor in rats (Schenkel-Hulliger et al., 1974) and mice (Yarbrough, 1979) and cause a pronounced, centrally mediated activation of electromyographic activity in cats (Cooper and Boyer, 1978) and rats (Yarbrough and McGuffin-Clineschmidt, 1979). Additionally, although there is no direct information available, it appears likely that the well-known respiratory stimulant effects of TRH are mediated finally through an excitatory effect of the peptide on the montoneuron pool of the phrenic nerve.

The intimate association of the so-called "diffuse reticular activating system" with cholinergic mechanisms is well known (cf. Phillis, 1970) and it is likely that the final mediator in the cerebral cortex that underlies the behavioral and EEG arousal subsequent to reticular formation activation is in fact ACh. Bearing this in mind, and the observations that TRH has an excitatory effect on reticular formation neurons (Koranyi et al., 1977; Briggs, 1979), it becomes entirely conceivable that some of the arousal effects of the peptide are secondary to an activation of rostrally projecting reticular mechanisms which are ultimately expressed through a release of ACh. This notion would be compatible with the documented effects of TRH on ACh release and the atropine-sensitive effects of the peptide to activate the cortical EEG (Beale et al., 1977) and induce analepsis (Breese et al., 1975; Horita et al., 1976; Nagai et al., 1980).

Several laboratories have now reported on the presence (Yarbrough, 1978; Braitman et al., 1980) or absence (Winokur and Beckman, 1978; Renaud et al., 1979; Phillis and Kirkpatrick, 1979) of a facilitory effect of TRH on ACh-induced excitations of cerebral cortical neurons. Similarly, discrepancies exist even in reports from the same laboratory (cf. Renaud and Martin, 1975; Renaud *et al.*, 1979) as to the direct effects of TRH on cerebral cortical neuronal excitability. Suffice it to say that further speculation on the possible sources of these discrepancies, in the absence of new data, will not prove very edifying. Unfortunately, it is apparent that at least in the cerebral cortex, where cholinergic projections and possibly interneurons are known to have important functions, utilization of the microiontophoretic technique has not yet led to a consensus as to an interaction of TRH and MK-771 with cholinergic receptors. Nonetheless, on the more positive side, a facilitory interaction has been observed and as discussed by Yarbrough (1979) such an interaction is entirely compatible with the known antianesthetic effects of these peptides.

## 4. EFFECTS ON PARASYMPATHETIC OUTFLOW TO THE PUPIL

As assessed by both indirect and direct means, TRH and MK-771 have been convincingly shown to activate the efferent cholinergic projection from the Edinger–Westphal nucleus to the pupil of the eye in cats (Koss, 1980; Koss and Stone, 1980). Thus, these investigators found that these peptides (1) antagonized clonidine-induced pupillary dilatation in both normal and phenoxybenzamine-pretreated preparations, (2) antagonized the pupillary dilatation but not the contracture of the nictitating membrane caused by stimulation of the superior cervical nerve, and (3) in direct recordings of ciliary nerve activity reversed the depressant effects of clonidine and by themselves stimulated nerve activity both in normal cats and in preparations with transections of the sympathetic nerves to the eye, spinal cord, or spinal cord plus optic tracts. These findings clearly indicate that these peptides exert a direct facilitory effect on the cranial optic oculomotor nucleus to increase parasympathetic tone to the iris.

## 5. ACTIONS ON THE GASTROINTESTINAL TRACT

In a manner quite analogous to the central effects of TRH on the parasympathetic tone of the eye, the peptide appears to elicit centrally mediated, vagus-dependent stimulations of gut motility and gastric acid secretion. Thus, it has been found that intracerebroventricular administration of TRH causes an atropine-sensitive stimulation of muscular activity of the colon of rabbits (Smith *et al.*, 1977) and an increase in the electroenteromyographic activity of the rat duodenum (Tonoue and Nomoto, 1979). In the latter study the effects of TRH were abolished by vagotomy or atropine pretreatment. While TRH may also exert peripheral neurogenic and direct myogenic effects on intestinal muscle (cf. Furukawa *et al.*, 1980), these findings indicate that TRH has a central facilitory effect on vagal efferents regulating the excitability of at least portions of the intestinal tract. Similarly, intracisternal administration of TRH to rats has recently been shown to elicit a scopolamine-sensitive secretion of gastric acid (Tache *et al.*, 1980). Interestingly, intravenous TRH in cats did not affect basal acid secretion but reduced insulin-induced elevations in gastric acid (Gascoigne *et al.*, 1980). As discussed by these authors, this finding suggests that TRH is exerting an influence on brain centers that regulate the activity of vagal nerves since the expression of insulin-induced elevations of gastric acid are mediated centrally through the vagus. However, these data would also be consonant with an inhibitory effect of TRH on vagal efferents regulating gastric acid secretion in this species.

## 6. TRH RECEPTORS, CHOLINERGIC NEURONS, AND ANTIMUSCARINIC DRUGS

Neither TRH nor MK-771 appears to possess any intrinsic cholinomimetic activity. It therefore seems likely that central cholinergic neuronal cell bodies possess specific receptors for TRH that when activated result in the release of ACh postsynaptically onto muscarinic or nicotinic receptors. When the behavioral, neuronal, or muscular response to TRH is mediated by muscarinic cholinergic receptors, the final measurable effects of TRH agonism are susceptible to antagonism by antimuscarinic agents. Thus, as previously discussed, the analeptic, cortical EEG activation and gastrointestinal stimulant effects of TRH are reduced by scopolamine or atropine.

There does seem to be a tendency in the literature (on the part of the present author as well; cf. Yarbrough, 1979) to assume that if an effect of TRH is not antagonized by antimuscarinic agents, then there is not likely to be a cholinergic involvement in the phenomenon. This is obviously incomplete reasoning ignoring the apparent involvement of nonmuscarinic cholinergic receptors in, for instance, the respiratory and EMG stimulant effects of TRH. Furthermore, there are no documented examples of any direct effects of TRH or MK-771 being antagonized by antimuscarinic agents. In fact, the excitatory effects of TRH on reticular neurons (Briggs, 1979) and of MK-771 on ciliary nerve activity (Koss and Stone, 1980) were not affected by antimuscarinic agents.

## 7. CONCLUSIONS AND THERAPEUTIC INDICATIONS

A schematic of the documented sites of action of TRH on central cholinergic pathways is shown in Fig. 2. It should be first stated that it is unlikely that the depicted interactions can account for all of the effects of exogenously administered TRH or its agonists such as MK-771. However, many of the pharmacological actions of these peptides appear to be initiated through activation of central cholinergic systems possibly through a stimulation of TRH

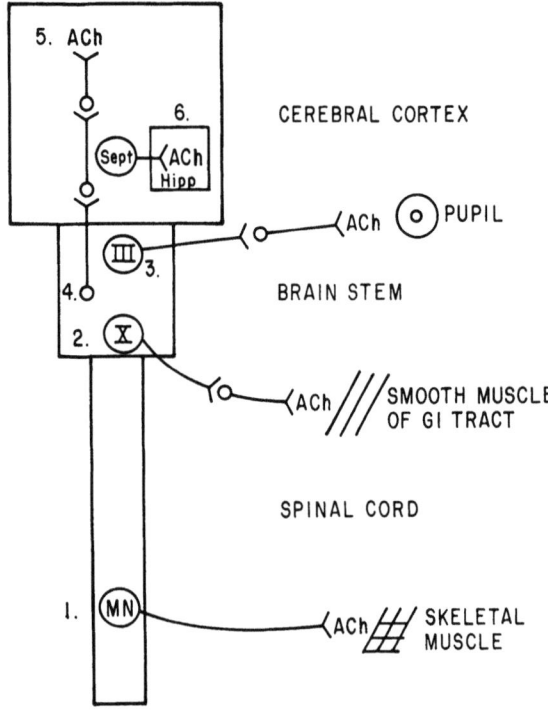

FIGURE 2. TRH excitatory effects on cholinergic systems. As documented in the text, TRH has been shown to (1) excite spinal motoneurons, (2) activate vagal efferents regulating parasympathetic tone of the gut, (3) increase via a central action parasympathetic tone of the pupil, (4) excite reticular formation neurons which might result in an activation of the reticular activating system leading to an enhanced release of ACh from the cerebral cortex, (5) perhaps also cause a sensitization of cortical cholinergic receptors to ACh, and (6) activate the septohippocampal cholinergic projection.

receptors located on cholinergic neurons. Furthermore, it is conceivable that one of the functions of endogenous, extrahypothalamic TRH, distributed throughout the mammalian neuraxis, is to participate in the regulation of the level of excitability of these cholinergic systems. From the above speculation, it follows that some of the clinical syndromes generally thought to be the result of cholinergic hypofunction may actually be reflections of deficits in TRH dynamics in the CNS. In any event, based on the known interactions of these peptides with cholinergic systems in experimental animals, some rational predictions concerning the therapeutic usefulness of TRH, or an improved agonist such as MK-771, can be made.

Broadly states, these peptides might be therapeutic in disease states where a deficit in cholinergic tone is suspected. Thus, Huntington's chorea, Alzheimer's disease and other senile dementias might all be expected to benefit from TRH agonist therapy. In view of the arousal properties of TRH, which are undoubtedly due in part to a diffuse cholinergic activation, and the suspected involvement of cholinergic mechanisms in memory and learning processes, a TRH agonist might offset the lethargy and memory loss often attendant with the normal aging process. Deficits in somatic and visceral efferent motor unit activity such as those occurring in a variety of CNS degenerative disorders could conceivably be ameliorated by these peptides. Finally, the effects of TRH on gut motility might prove of benefit to patients experiencing paralytic ileum following surgical anesthesia. Furthermore, the analeptic, arousing, and respiratory stimulant properties of these peptides might prove globally useful to facilitate postsurgical recovery.

Clearly, the groundwork has been laid for some rational and creative clinical pharmacology with TRH and TRH agonists such as MK-771.

ACKNOWLEDGMENTS. The author is most grateful to Professor J. W. Phillis (University of Saskatchewan) for performing the ACh bioassays of the CSF samples obtained from the rabbit cerebral cortex and to Dr. N. Bohidar (Merck Institute for Therapeutic Research) for statistical analysis of the data.

## 8. REFERENCES

Atweh, S., Simon, J. R., and Kuhar, M. J., 1975, Utilization of sodium-dependent high affinity choline uptake in vitro as a measure of the activity of cholinergic neurons in vivo, *Life Sci.* **17**:1535.

Beale, J. S., White, R. P., and Huang, S. P., 1977, EGG and blood pressure effects of TRH in rabbits, *Neuropharmacology* **16**:499.

Braitman, D. J., Auker, C. R., and Carpenter, D. O., 1980, Thyrotropin-releasing hormone has multiple actions in cortex, *Brain Res.* **194**:244.

Breese, G. R., Cott, J. M., Cooper, B. R., Prange, A. J., Lipton, M. A., and Plotnikoff, N. P., 1975, Effects of thyrotropin-releasing hormone on the actions of pentobarbital and other centrally active drugs, *J. Pharmacol. Exp. Ther.* **193**:11.

Briggs, I., 1979, Excitatory effects of TRH on medullary reticular neurones, *Neurosci. Lett.* **15**:33.

Cooper, B. R., and Boyer, C. E., 1978, Stimulant action of thyrotropin releasing hormone on cat spinal cord, *Neuropharmacology* **17**:153.

Furukawa, K., Nomoto, T., and Tonoue, T., 1980, Effects of thyrotropin-releasing hormone (TRH) on the isolated small intestine and taenia coli of the guinea pig, *Eur. J. Pharmacol.* **64**:279.

Gascoigne, A. D., Hirst, B. H., Reed, J. D., and Shaw, B., 1980, Effects of thyrotrophin-releasing hormone, and methionine-enkephalin on gastric acid and pepsin secretion in the cat, *Br. J. Pharmacol.* **69**:527.

Horita, A., Carino, M. A., and Chesnut, R. M., 1976, Influence of thyrotropin-releasing hormone (TRH) on drug induced narcosis and hypothermia in rabbits, *Psychopharmacol. Bull.* **49**:57.

Kalivas, P. W., and Horita, A., 1980, Thyrotropin-releasing hormone: Neurogenesis of actions in the pentobarbital narcotized rat, *J. Pharmacol. Exp. Ther.* **212**:203.

Kalivas, P. W., Halpern, L. M., and Horita, A., 1980, Synchronization of hippocampal and cortical electroencephalogram by thyrotropin-releasing hormone, *Exp. Neurol.* **69**:627.

Koranyi, L., Sawyer, C., and Whitmoyer, D., 1977, Effect of thyrotropin-releasing hormone, luteinizing hormone-releasing hormone, and somatostatin on neuronal activity of brain stem reticular formation and hippocampus in the female rat, *Exp. Neurol.* **57**:807.

Koss, M. C., 1980, Stimulant action of thyrotropin-releasing hormone on ciliary nerve activity, *Eur. J. Pharmacol.* **65**:105.

Koss, M., and Stone, C., 1980, Increase of tonic parasympathetic outflow to the pupil produced by an analog of TRH (MK-771), *Regul. Peptides* **1**:31.

McNaughton, N., Jamus, D. T. D., Stewart, J., Gray, J. A., Valevo, I., and Drewnowski, A., 1977, Septal driving of hippocampal theta rhythm as a function of frequency in the male rat: Effects of drugs, *Neuroscience* **2**:1019.

Malthe-Sørenssen, D., Wood, P. L., Cheney, D. L., and Costa, E., 1978, Modulation of the turnover rate of acetylcholine in rat brain by intraventricular injections of thyrotropin-releasing hormone, somatostatin, neurotensin and angiotensin II, *J. Neurochem.* **31**:685.

Nagai, Y., Narumi, S., Nagawa, Y., Sakurada, O., Ueno, H., and Ishii, S., 1980, Effect of thyrotropin-releasing hormone (TRH) on local cerebral glucose utilization, by the autoradiographic 2-deoxy[$^{14}$C]glucose method, in conscious and phenobarbitalized rats, *J. Neurochem.* **35**:963.

Nicoll, R. A., 1977, Excitatory action of TRH on spinal motoneurones, *Nature (London)* **265**:242.

Phillis, J. W. (ed.), 1970, *The Pharmacology of Synapses*, Pergamon Press, Elmsford, N.Y.

Phillis, J. W., and Kirpatrick, J. R., 1979, Actions of various gastrointestinal peptides on the isolated amphibian spinal cord, *Can. J. Physiol. Pharmacol.* **57**:887.

Phillis, J. W., and Kirkpatrick, J., 1980, The actions of motilin, luteinizing hormone releasing hormone, cholecystokinin, somatostatin, vasoactive intestinal peptide, and other peptides on rat cerebral cortical neurons, *Can. J. Physiol. Pharmacol.* **58**:612.

Phillis, J. W., Mullin, W. J., and Pinsky, C., 1973, Morphine enhancement of acetylcholine release into the lateral ventricle and from the cerebral cortex of unanaesthetized cats, *Comp. Gen. Pharmacol.* **4**:189.

Plotnikoff, N. P., Prange, A. J., Jr., Breese, G. R., Anderson, M. S., and Wilson, I. C., 1972, Thyrotropin-releasing hormone: Enhancement of DOPA activity by a hypothalamic hormone, *Science* **178**:417.

Prange, A. J., and Wilson, I. C., 1972, Thyrotropin-releasing hormone (TRH) for the immediate relief of depression: A preliminary report, *Psychopharmacologia* **26**:82.

Renaud, L. P., and Martin, J. B., 1975, Thyrotropin-releasing hormone: Depressant action on central neuronal activity, *Brain Res.* **86**:150.

Renaud, L. P., Blume, H. W., Pittman, Q. J., Lamour, Y., and Tan, A. T., 1979, Thyrotropin-releasing hormone selectively depresses glutamate excitation of cerebral cortical neurons, *Science* **205**:1275.

Santori, E., and Schmidt, D., 1980, Effects of MK-771, a TRH analog, on pentobarbital-induced alterations of cholinergic parameters in discrete regions of rat brain, *Regul. Peptides* **1**:69.

Schenkel-Hulliger, L., Koella, W. P., Hartmann, A., and Maitre, L., 1974, Tremorogenic effect of thyrotropin-releasing hormone in rats, *Experientia* **30**:1168.

Schmidt, D. E., 1977, Effects of thyrotropin-releasing hormone (TRH) on pentobarbital-induced decrease in cholinergic neuronal activity, *Commun. Psychopharmacol.* **1**:469.

Smith, J. R., La Hann, T. R., Chestnut, R. M., Carino, M. A., and Horita, A., 1977, Thyrotropin-releasing hormone: Stimulation of colonic activity following intracerebroventricular administration, *Science* **196**:660.

Tache, Y., Brown, M., and Vale, W., 1980, Thyrotropin-releasing hormone—CNS action to stimulate gastric acid secretion, *Nature (London)* **287**:149.

Tonoue, T., and Nomoto, T., 1979, Effect of intracerebroventricular administration of thyrotropin-releasing hormone upon the electroenteromyogram of rat duodenum. *Eur. J. Pharmacol.* **58**:369.

Veber, D. F., Holly, F. W., Varga, S. L., Hirschmann, R., Nutt, R. F., Lotti, V. J., and Porter, C. C., 1976, The dissociation of hormonal and CNS effects in analogs of TRH, Proceedings 14th European Peptide Symposium (A. Loffet, ed.), pp. 167–172, University of Brunds Press, Belgium.

Winokur, A., and Beckman, A. L., 1978, Effects of thyrotropin releasing hormone, norepinephrine and acetylcholine on the activity of neurons in the hypothalamus, septum and cerebral cortex of the rat, *Brain Res.* **150**:205.

Yarbrough, G. G., 1978, Studies on the neuropharmacology of thyrotropin releasing hormone (TRH) and a new TRH analog, *Eur. J. Pharmacol.* **48**:19.

Yarbrough, G. G., 1979, On the neuropharmacology of thyrotropin releasing hormone (TRH), *Prog. Neurobiol.* **12**:291.

Yarbrough, G. G., and McGuffin-Clineschmidt, J. C., 1979, MK-771-induced electromyographic (EMG) activity in the rat: Comparison with thyrotropin releasing hormone (TRH) and antagonism by neurotensin, *Eur. J. Pharmacol.* **60**:41.

Yarbrough, G. G., and Singh, D. K., 1979, Effects of MK-771 on the isolated amphibian spinal cord: Comparison with thyrotropin-releasing hormone, *Can. J. Physiol. Pharmacol.* **57**(8):920.

Yarbrough, G. G., Haubrich, D. R., and Schmidt, D. E., 1978, Thyrotropin releasing hormone (TRH) and MK-771 interactions with CNS cholinergic mechanisms, in: *Iontophoresis and Transmitter Mechanisms in the Mammalian Central Nervous System* (R. Ryall and J. S. Kelly, eds.), pp. 136–138, Elsevier/North-Holland, Amsterdam.

CHAPTER 4

# Calcitonin as an Anorectic Agent

WILLIAM J. FREED, LLOYD A. BING,
ARNOLD E. ANDERSEN, and
RICHARD JED WYATT

## 1. INTRODUCTION

Calcitonin is a peptide hormone secreted by the C cells of the thyroid gland. Its primary physiological effect is to decrease plasma calcium and phosphorus concentrations. This action of calcitonin is particularly pronounced whenever plasma calcium becomes elevated (cf. Anast and Conaway, 1972; Copp *et al.*, 1962; Copp, 1969; Hirsch *et al.*, 1963; Munson and Gray, 1970). Calcitonin was discovered in 1962, after Copp and his colleagues (1962) found that perfusion of the thyroid–parathyroid complex with hypercalcemic solutions caused a lowering of plasma calcium concentrations. These results could not be explained in terms of decreased secretion of parathyroid hormone, which increases plasma calcium. Thus, it eventually became established that the thyroid secretes calcitonin, a hormone that decreases plasma calcium (Anast and Conaway, 1972; Copp, 1969; Talmage *et al.*, 1980).

The two primary sites of action of calcitonin are bone and kidney. In bone, calcitonin inhibits bone resorption and calcium release (Hirsch, 1967; Holtrop *et al.*, 1974). In kidney, calcitonin decreases tubular reabsorption of calcium, phosphorus, and other electrolytes (Ardaillou, 1975; Clifton-Bligh *et al.*, 1980). Calcitonin, however, may have additional sites of action: Calcitonin increases protein synthesis in the intestine, and decreases protein synthesis in the pancreas (Nakhla, 1979). Systemically administered calcitonin accumulates in muscle and blood vessel walls, as well as in kidney and bone (Forslund *et al.*, 1980). Recently, a substantial literature has begun to accumulate suggesting that the brain is another major site of action of calcitonin.

---

WILLIAM J. FREED, LLOYD A. BING, and RICHARD JED WYATT • Adult Psychiatry Branch, Division of Special Mental Health Research, Intramural Research Program, National Institute of Mental Health, Saint Elizabeths Hospital, Washington, D.C. 20032.    ARNOLD E. ANDERSEN • Department of Psychiatry, Johns Hopkins University Medical Center, Baltimore, Maryland 21218.

## 2. IS THE BRAIN A TARGET ORGAN FOR CALCITONIN?

### 2.1. Calcitonin in the Brain and Pituitary

There have been reports that calcitonin is present in human cerebrospinal fluid (CSF) (Becker et al., 1980; Pavlinac et al., 1980),as well as in the brain parenchyma (Becker et al., 1979). It has been established that calcitonin is present in the pituitary gland (Cooper et al., 1980a; Deftos et al., 1978, 1980). Whether brain and CSF calcitonin is thyroid-derived is unknown; however, pituitary calcitonin is not decreased by thyroidectomy (Cooper et al.,1980a; Watkins et al., 1980), and is, therefore, thought to be produced by the pituitary. Secretion of pituitary calcitonin has not been demonstrated (Cooper et al., 1980a).

### 2.2. Binding Sites

The rat brain contains calcitonin binding sites, which are similar to binding sites for calcitonin that have been studied in other tissues (Fischer et al., 1981). As in some other tissues, binding of salmon calcitonin to rat brain was only partially reversible, in that about one-third of bound labeled calcitonin could not be dissociated by addition of unlabeled calcitonin. This result could not be explained by metabolic degradation (also cf. Findlay et al., 1980; Goltzman, 1980; Queener et al., 1975). Porcine calcitonin was about 50 times as potent as human calcitonin, while salmon calcitonin was 7000 times as potent as human calcitonin, which is similar to results obtained for binding of calcitonin to other tissues (cf. Goltzman, 1980; Marx et al., 1972, 1974).The binding to rat brain also shows a regional distribution, with the greatest amounts of binding occurring in the hypothalamus (Rizzo and Goltzman, 1981; Koida et al., 1980; Fischer et al., 1981).

### 2.3. Analgesia

Several groups have reported that calcitonin produces analgesia when injected intracerebrally (Pecile et al., 1975; Braga et al., 1978; Satoh et al., 1979; Yamamoto et al., 1979).But the dosages of calcitonin that have been used to produce this effect have been large, ranging from 8 to 100 U/kg body wt, injected intracerebrally. Opiate antagonists do not block the analgesic effects of calcitonin (Braga et al., 1978; Yamamoto et al., 1979). The analgesic effect of porcine calcitonin, however, was blocked by concomitant administration of 0.1 μmole of calcium (Satoh et al., 1979).

There is some evidence that the analgesic effect is mediated by the brain rather than by leakage to peripheral sites (Yamamoto et al., 1981). Nevertheless, the fact that such large dosages are required casts considerable doubt as to the physiological significance of this effect.

## 2.4. Prolactin Secretion

Calcitonin also has been reported to decrease plasma prolactin concentrations in human patients (Carman and Wyatt, 1977; Isaac et al., 1980). This effect may also be mediated by the brain. Olgiati et al. (1981) reported that plasma prolactin in rats could be decreased by calcitonin administered either intravenously or intraventricularly, with intraventricular administration producing substantial decreases in plasma prolactin with dosages as small as 25 ng, or less than 0.2 U, per rat. Lesions of the median eminence blocked this effect, suggesting that calcitonin was acting via the hypothalamus, rather than directly on the pituitary. The decreases in plasma prolactin that were induced by calcitonin were also blocked by haloperidol, a dopamine antagonist (Olgiati et al., 1981). One study, however, has reported that calcitonin *increased* plasma prolactin when injected intraventricularly, whether salmon, porcine, human, or eel calcitonins were used (Iwasaki et al., 1979). The reasons for the difference between studies in terms of the direction of the effect are unclear. However, this area of study is particularly interesting in light of evidence that prolactin is involved in the regulation of calcium metabolism (Mattheij et al., 1980; Pahuja and DeLuca, 1981).

## 2.5. Gastric Acid Secretion

A final action of calcitonin that may be mediated by the brain is inhibition of gastric acid secretion. A number of studies have shown that calcitonin inhibits secretion of gastric acid when administered peripherally (Becker et al., 1973; Hesch et al., 1971; Bieberdorf et al., 1974). Recently, Morley and Levine (1981) have reported that intracerebral administration of salmon calcitonin in dosages as small as 0.0002 U decreased gastric acid secretion. In contrast, approximately 1000-fold larger dosages were required to inhibit gastric secretion by subcutaneous administration. Therefore, inhibition of gastric acid secretion may be another effect of calcitonin that is mediated by the CNS.

Tepperman and Evered (1980) have recently demonstrated that gastric acid secretion can be stimulated by intrahypothalamic injections of small amounts of gastrin. This finding suggests that the CNS control of gastric acid secretion is localized in the hypothalamus. If so, the hypothalamus may also be the site at which calcitonin acts to inhibit gastric acid secretion.

## 3. CALCITONIN AND GASTROINTESTINAL STIMULI

Numerous physiological and chemical stimuli increase calcitonin secretion. These include feeding, calcium, gut hormones such as cholecystokinin and gastrin, glucagon, triglycerides, AMP, and amines (Anast and Conaway, 1972; Avioli et al., 1969; Bell, 1970; Care et al., 1970, 1971; Cooper et al.,

1971; Metz et al., 1978; Roos and Deftos, 1976; Sethi et al., 1981; Swaminathan et al., 1973). Many of these stimuli are related to feeding and to subsequent digestive processes, and it has been argued that a major function of calcitonin may be to regulate the disposition of calcium absorbed from foods (Cooper et al., 1977; Talmage et al., 1980).

Ingestion of food has been shown to stimulate calcitonin secretion in several species, and under a variety of conditions including during lactation and infancy (Cooper et al., 1977; Garel and Jullienne, 1977; Roos et al., 1977; Swaminathan et al., 1973; Talmage et al., 1975). In humans, stimulation of calcitonin secretion by eating is controversial (cf. Grubb et al., 1979; Parthemore and Deftos, 1978). In newborn rats allowed to suckle normally, plasma calcitonin increases from about 800 pg/ml 2 hr after birth to about 2000 pg/ml 3 days later (Garel and Jullienne, 1977). These authors also reported that rats delivered by cesarean section and not allowed to suckle for 16 hr had little or no detectable plasma calcitonin until they were allowed to suckle (Garel and Jullienne, 1977). After 8 to 12 hr of fasting, 1 hr of suckling increased serum calcitonin of infant rats from below 200 pg/ml to above 400 pg/ml (Cooper et al., 1977). Thus, the presence of high concentrations of calcitonin in infant rats appears to be dependent upon normal feeding.

There is evidence that food constituents other than calcium play some role in stimulating calcitonin secretion. Garel and Besnard (1979) have studied the relative role of the milk constituents lactose, casein, triglycerides, and calcium in stimulating calcitonin secretion. Casein hydrolysate and normal saline did not stimulate calcitonin secretion, and lactose induced only a small increase in plasma calcitonin. Calcium produced a fourfold increase in plasma calcitonin. Triglycerides, however, were even more effective than calcium, producing an eightfold increase in plasma calcitonin concentrations. The increased secretion of calcitonin caused by triglycerides could not be explained as an effect secondary to increased plasma calcium, glucose, phosphorus, or insulin. These results, therefore, suggest that calcitonin may have a role in some metabolic process in addition to the regulation of calcium disposition.

Several hormones that stimulate calcitonin secretion, such as cholecystokinin, gastrin, and glucagon, also inhibit eating behavior (Martin and Novin, 1977; Smith and Gibbs, 1975; Smith et al., 1974). The anorectic effects of gastrin in the rat are weak (Smith et al., 1974; Lorenz et al., 1979); however, gastrin is a weak calcitonin secretagogue in the rat (Cooper et al., 1978). There is conflicting evidence suggesting that cholecystokinin and gastrin are not anorexigenic when injected intracerebrally (Manaker et al., 1979; Maddison, 1977; Nemeroff et al., 1978).

Feeding increases plasma concentrations of several calcitonin secretagogues, such as triglycerides, calcium, glucagon, cholecystokinin, and gastrin. These stimuli in turn stimulate secretion of calcitonin. Our hypothesis is that this postprandial secretion of calcitonin is utilized by the brain as a negative feedback signal to inhibit subsequent feeding behavior. We therefore suspected that calcitonin might have anorectic effects.

## 4. EFFECTS OF PARENTERAL CALCITONIN ON FEEDING BEHAVIOR IN RATS

### 4.1. Effects of Subcutaneous Calcitonin on 24-hr Food Intake

To test the possibility that calcitonin is involved in the regulation of feeding behavior, we investigated the effects of calcitonin injections on eating in rats. Initially, we studied the effects of subcutaneous salmon calcitonin on 24-hr food and water intake. Rats were given single subcutaneous injections of synthetic salmon calcitonin, and their eating, drinking, urine and fecal excretion, and weights were measured for the following 24 hr. Calcitonin produced a dose-dependent inhibition of eating, resulting in a 40% decrease in food intake with a dose of 50 medical research council units/kg (Fig. 1). There was a concomitant decrease in fecal excretion and body weight, all of these effects recovering over the succeeding 24 hr.

Calcitonin did not significantly influence drinking, but did increase urine output. This increased urine output was followed, from 24 to 48 hr after calcitonin injections, by an increase in drinking which was apparently compensatory to the increased urine excretion. It has previously been reported that calcitonin increases urine volume and excretion of sodium and other electrolytes (Ardaillou, 1975; Clark and Wideman, 1980; Paillard et al., 1972). Calcitonin did not produce diuresis in a study of marsh mice (Bryson and Bischoff, 1980), but did decrease food intake in these animals.

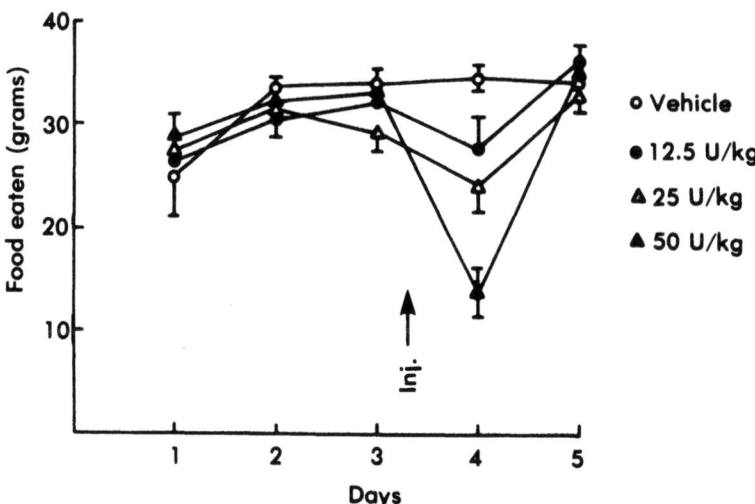

FIGURE 1. Inhibition of eating in rats by subcutaneous injections of synthetic salmon calcitonin. Twenty-four-hour food intake in grams (means ± S.E.M.) is shown for the 3 days preceding and 2 days following injections of calcitonin in dosages of 12.5, 25, or 50 U/kg, or vehicle.

## 4.2. Time Course

The fact that calcitonin was effective over an entire 24-hr period suggested that calcitonin is capable of inhibiting eating for at least several hours after injection. We therefore studied the time course of the inhibition of feeding by calcitonin. Rats were adapted to a feeding schedule, by being given an opportunity to eat a moist mash (7 parts ground rat food to 10 parts water by weight) for 30 min each day for several days. Animals were then given salmon calcitonin in a dose of 12.5 U/kg at various time intervals from 1 to 22 hr prior to feeding. The maximum inhibitory effect of calcitonin occurred between 4 and 8 hr after injection. Injections given 1, 3, or 22 hr prior to feeding were ineffective (Fig. 2).

Thus, a single subcutaneous injection of salmon calcitonin is effective in decreasing eating for as long as 8 hr, which is not surprising in view of the fact that salmon calcitonin is degraded very slowly in comparison to mammalian calcitonin, having an *in vitro* plasma half-life in excess of 24 hr (Bennett and McMartin, 1979; Habener *et al.*, 1971; Huwyler *et al.*, 1979). The fact that calcitonin was relatively ineffective when administered 1 or 3 hr prior to feeding was unexpected. We therefore determined whether larger dosages of salmon calcitonin would also be ineffective in inhibiting eating 1 hr after administration.

Additional groups of rats were adapted to eating moistened mash for 30 min every 24 hr, and then given 50 U/kg of synthetic salmon calcitonin or vehicle either 1 or 6 hr prior to feeding. Despite the larger dosage, subcutaneous calcitonin did not significantly inhibit eating when administered 1 hr prior to testing, while eating was significantly inhibited by injections given 6 hr before feeding (Fig. 3). Thus, the peak effect of subcutaneously administered salmon calcitonin on eating occurs several hours after administration. This delayed

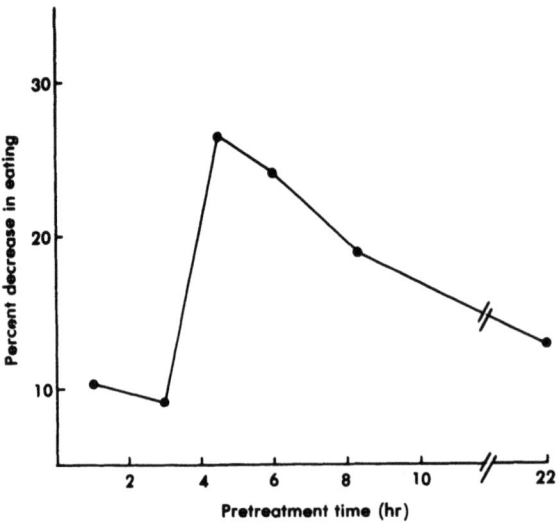

FIGURE 2. Inhibition of eating in rats by subcutaneous calcitonin as a function of time from injection to food access. The vertical axis shows the percentage decrease in the amount of food eaten by animals injected with synthetic salmon calcitonin (12.5 U/kg) as compared to controls injected with vehicle. Each point represents an independent experiment for which 6 to 12 animals received calcitonin and at least 6 animals received vehicle. At various times after injection, animals that had been deprived of food for 23½ hr were given moistened rat food for 30 min.

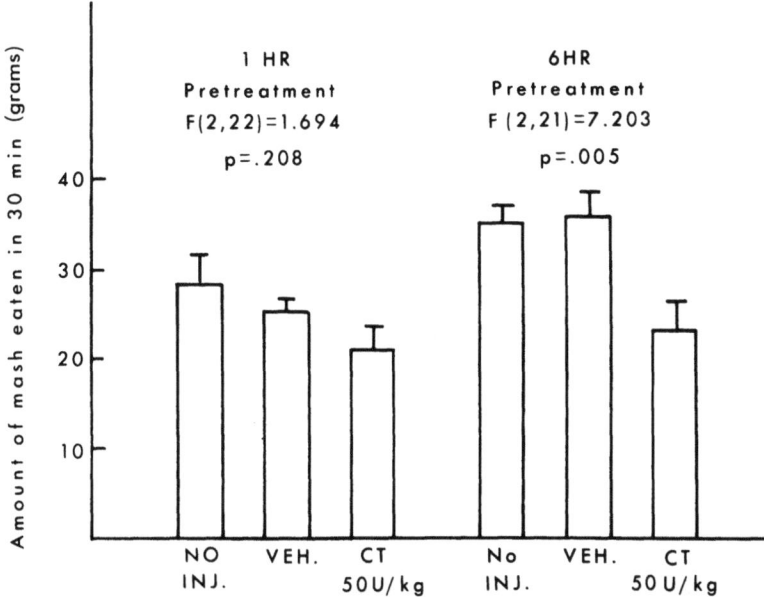

FIGURE 3. Inhibition of eating by subcutaneous injections of synthetic salmon calcitonin (50 U/kg) given 1 or 6 hr prior to feeding. Animals received either no injection ($N = 12$), vehicle ($N = 6$), or calcitonin ($N = 6$). The rats that received no injection in the 1-hr experiment subsequently received injections in the 6-hr experiment, and vice versa. Feeding was measured during a 30-min period of access to moistened rat food after $23\frac{1}{2}$ hr of food deprivation.

effect might be due to slow absorption and distribution of calcitonin after subcutaneous administration, combined with the slow degradation of salmon calcitonin in mammals (Bennett and McMartin, 1979; Habener *et al.*, 1971; Huwyler *et al.*, 1979). An alternative possibility is that the behavioral effects of salmon calcitonin are due to a smaller peptide which is a fragment of the intact molecule.

### 4.3. Effects of Calcitonin in Food-Deprived Rats

Under normal conditions, some of the water intake of rats is associated with eating (Hsiao and Pertsulakes, 1970; Jacobs, 1964). Therefore, because calcitonin inhibits eating, it would also be expected to decrease drinking. On the other hand, the diuretic properties of calcitonin would be expected to tend to increase drinking. Any direct effects of calcitonin on drinking would be obscured by these other effects under normal testing conditions. To determine whether calcitonin influences drinking directly, we examined the effects of calcitonin on drinking and urine excretion in food-deprived animals.

Animals were deprived of food, and 1 day later were given injections of calcitonin (50 U/kg, subcutaneous, $N = 9$) or vehicle ($N = 8$). Drinking and urine excretion were measured for the following 3 days. The results of this

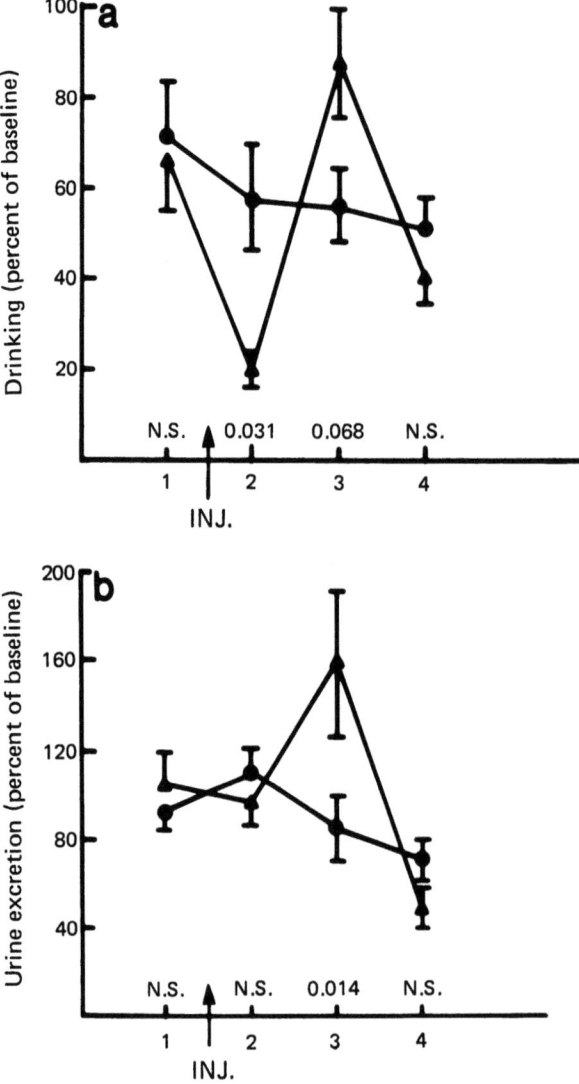

FIGURE 4. Effects of synthetic salmon calcitonin (triangles; dosage of 50 U/kg, subcutaneous, $N = 9$) or vehicle (circles; $N = 8$) on drinking (a) and urine excretion (b) of food-deprived rats. Data are shown for 1 day prior to and for 3 days following injections. Values shown are expressed as a percentage of baseline values obtained on the preceding day before food deprivation (means ± S.E.M.). The main effect of treatment was not significant $[F(1, 15) = 0.41, p = 0.54]$ but the main effect of days was significant $[F(3, 45) = 8.41, p = 0.0003]$. Interaction effects were also significant $[F(3, 45) = 5.71, p = 0.0025]$. Levels of significance shown on the graph are from Scheffe' tests (N.S. = not significant, $p > 0.20$). There was also a significant decrease in drinking in the calcitonin-treated group from day 1 to day 2 ($T = 3.89, p = 0.0005$, Scheffe').

experiment are shown in Fig. 4. Calcitonin had a diuretic effect under these conditions, but only on the second day after injection. Drinking was substantially decreased for 24 hr. On the second day after injections, drinking was increased, presumably as a compensatory response to the diuresis and the prior decrease in drinking.

Therefore, calcitonin decreases drinking under conditions of food deprivation, and appears to have a direct inhibitory effect on drinking in rats.

## 5. CALCITONIN AND EATING IN PRIMATES AND HUMANS

### 5.1. Calcitonin and Eating and Drinking in Monkeys

Salmon calcitonin has been shown to decrease both eating and drinking in rhesus monkeys (Perlow *et al.*, 1980). Monkeys were given a simple subcutaneous injection of synthetic salmon calcitonin (either 30 U/kg or 2 U/kg) and their 24-hr food and water intake was measured over the course of the

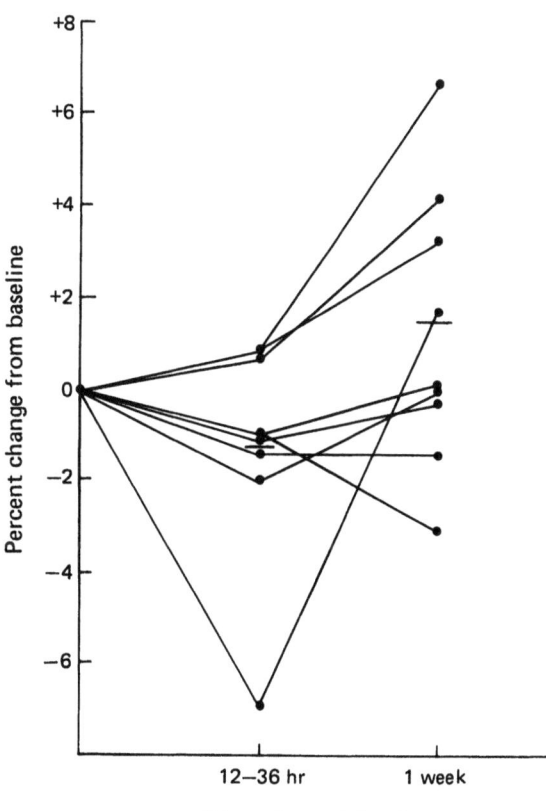

FIGURE 5. Changes in body weight of nine psychiatric patients 12 to 36 hr, and 1 week, after subcutaneous administration of synthetic salmon calcitonin (2 U/kg). The data are expressed as a percentage change from baseline weights, obtained the previous week. Subjects received placebo injections on each week that they did not receive calcitonin. Horizontal bars indicate means. (From Carman and Wyatt, 1979a.)

succeeding several days. The larger dosage of calcitonin (30 U/kg) decreased eating by about 80% for 2 days, and food intake did not return to baseline levels until 5 days after the injections. Water intake was also decreased by about 50% for 2 days. The smaller dosage of calcitonin (2 U/kg) had no effect. Thus, salmon calcitonin is a long-lasting and potent suppressor of eating in rhesus monkeys.

## 5.2. Calcitonin and Body Weight in Man

In light of the finding that calcitonin decreases eating in monkeys, the weights of nine patients who had received calcitonin (2 U/kg, subcutaneous) during the course of unrelated experiments (Carman and Wyatt, 1977, 1979a) were obtained retrospectively from their records 1 week before, 12–36 hr after, and 1 week after calcitonin. These patients had been given placebo injections the weeks before and after calcitonin injections. In general their weights were slightly decreased when measured 12–36 hr after calcitonin administration (Car-

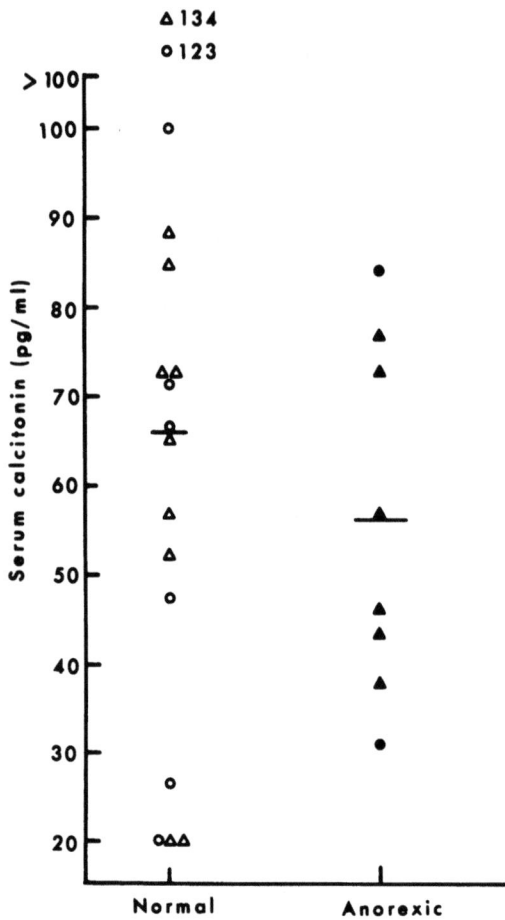

FIGURE 6. Serum calcitonin concentrations in anorexic patients (filled symbols) and normal controls (open symbols). Calcitonin was measured by a homologous double-antibody radioimmunoassay (Immunonuclear Corp.). Triangles indicate females, and circles indicate males. Horizontal bars indicate means.

man and Wyatt, 1979b; Perlow et al., 1980). This weight loss was recovered within 1 week (Fig. 5). We are not able to determine whether this loss of weight was due to diuresis, as seen in rats, or decreased eating.

### 5.3. Serum Calcitonin in Patients with Anorexia Nervosa

The possibility that calcitonin plays a role in the regulation of feeding suggests that a disorder of calcitonin secretion might play a role in some feeding disorders. For example, some cases of anorexia nervosa, a disorder that includes weight loss as a result of decreased food intake, might conceivably be due to an excess secretion of calcitonin.

To test this hypothesis, serum samples from male and female anorexia nervosa patients and from normal controls of similar ages were obtained in the morning prior to eating. The protocol for the study was approved by the Johns Hopkins University Institutional Review Board, and all patients gave informed consent for the study. Serum calcitonin concentrations were measured by a homologous double-antibody radioimmunoassay kit (Immunonuclear Corporation, Stillwater, Minn.) using human calcitonin as a standard. Recovery of a 125-pg internal standard in human serum was $97.5 \pm 3.5\%$ (mean ± range). Calcitonin concentrations did not differ between normal and anorexic subjects (Fig. 6). There was also no difference between the anorexic patients that had binge patterns of eating and those that did not. These data, therefore, do not support the hypothesis that increases in serum calcitonin concentrations play a role in anorexia nervosa.

## 6. INTRAVENTRICULAR CALCITONIN AND EATING IN RATS

### 6.1. Effects of Intraventricular Calcitonin on 24-hr Food Intake

A series of experiments involving intracerebral injections of calcitonin were carried out in order to test the hypothesis that calcitonin inhibits eating by a direct effect on the brain. Rats were given intraventricular injections of synthetic salmon calcitonin and their 24-hr food and water intake, urine and fecal excretion, and body weight were measured for the following 2 days. Food intake was decreased in a manner similar to the inhibition that is produced by subcutaneous injections (Fig. 7). Significant decreases in eating were produced by dosages as small as 43 ng, or 0.2 U. Fecal excretion was also decreased paralleling the decreased food intake.

The diuresis that was produced by subcutaneous injections was not evident; however, a delayed increase in drinking did occur. Figure 8 illustrates the daily water intake of rats before and after subcutaneous or intraventricular administration of calcitonin. The intraventricular injections produced a decreased water intake for 24 hr, which might be explainable in terms of decreased feeding-associated drinking (cf. Hsiao and Pertsulakes, 1970; Jacobs, 1964).

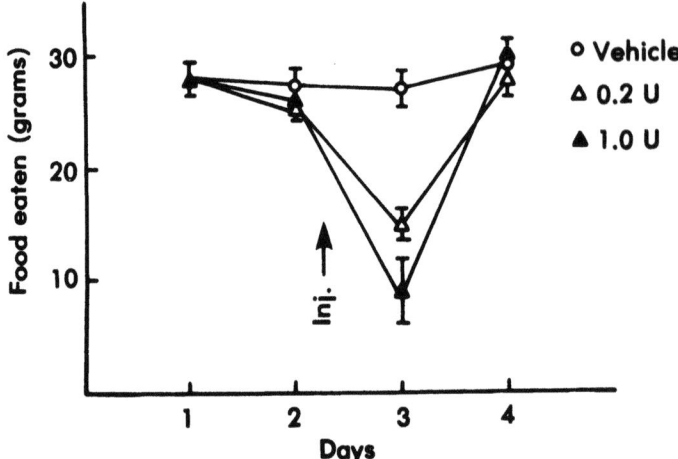

FIGURE 7. Twenty-four-hour food intake (means ± S.E.M.) of rats for 2 days before and 2 days after intraventricular administration of synthetic salmon calcitonin ($N = 6$ per group). One unit = 213 ng.

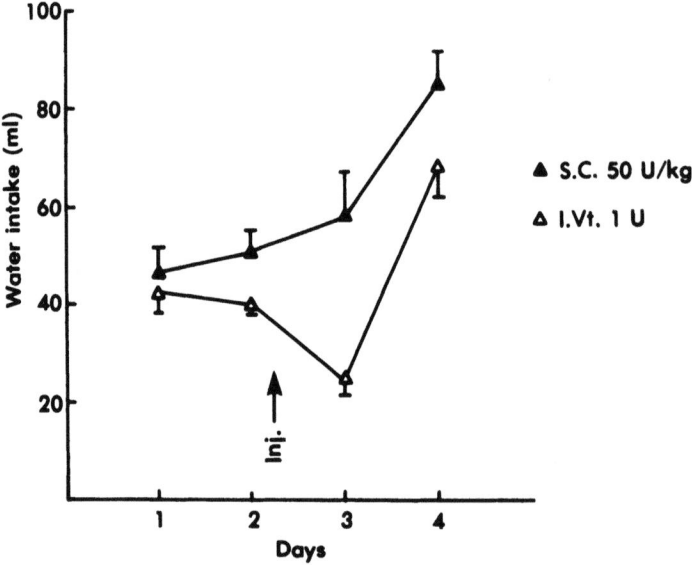

FIGURE 8. Twenty-four-hour water intake (means ± S.E.M.) of rats for 2 days before and 2 days after administration of synthetic salmon calcitonin by subcutaneous (50 U/kg) or intraventricular (1 U) administration ($N = 6$ per group).

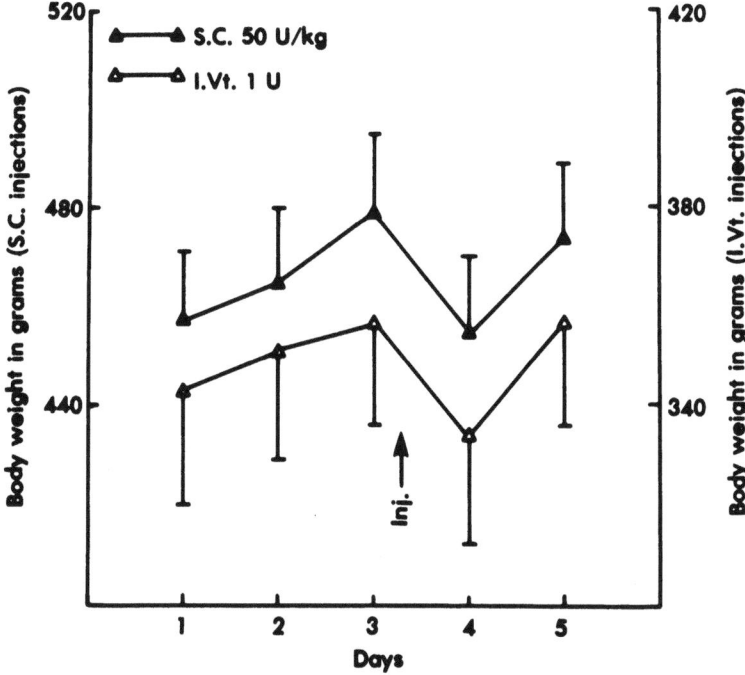

FIGURE 9. Body weight (means ± S.E.M.) of rats for 3 days before and 2 days after administration of synthetic salmon calcitonin by subcutaneous (50 U/kg) or intraventricular (1 U) administration ($N = 6$ per group).

The diuresis produced by subcutaneous injections apparently obscured any tendency for drinking to be decreased and resulted in elevated fluid consumption.

Figure 9 illustrates the weight loss induced by intraventricular injections as compared to subcutaneous administration of calcitonin. The two curves are nearly parallel. Intraventricular injections of calcitonin did not produce diuresis, but intraventricular and subcutaneous injections had similar effects on food intake. Therefore, the weight loss induced by subcutaneous calcitonin in rats is probably due to decreased food intake, rather than to uncompensated diuresis.

## 6.2. Denaturation of Calcitonin

To verify that the anorexic effects of intraventricular calcitonin were due to the structural properties of the calcitonin molecule rather than to an impurity or to nonspecific effects, we studied the anorexic properties of calcitonin that had been degraded by several treatments that are known to eliminate its hypocalcemic effect (Tashjian and Warnock, 1967). Animals were given either vehicle, 1.0 U of synthetic salmon calcitonin intraventricularly, or 1.0 U of

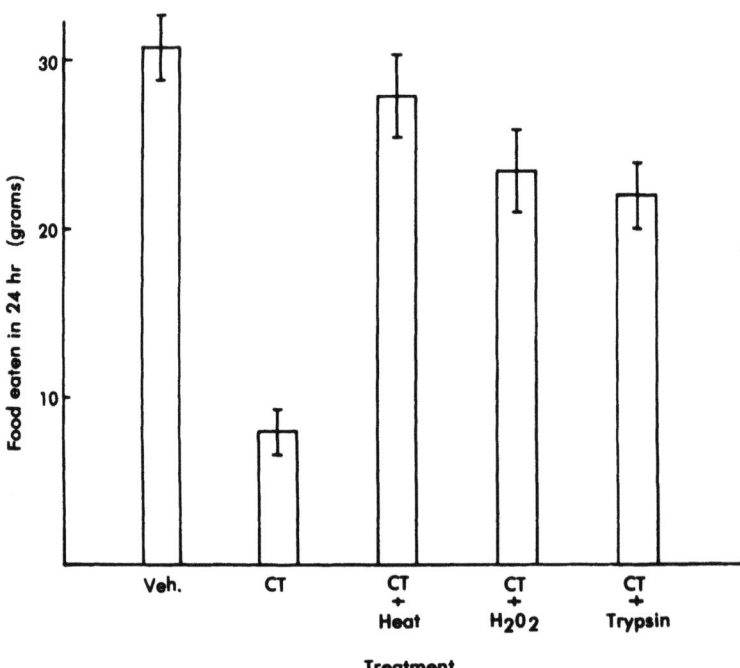

FIGURE 10. Food intake (means ± S.E.M.) during the 24 hr following intraventricular administration of vehicle ($N = 6$), calcitonin (1 U, $N = 6$), or 1 U of calcitonin denatured by heat ($N = 5$), hydrogen peroxide ($N = 5$), or trypsin digestion ($N = 5$). Heat denaturation consisted of incubation at 90°C for 18 hr, hydrogen peroxide treatment involved incubation in 0.2 M $H_2O_2$ at 37°C for 30 min, and trypsin digestion (1 part trypsin to 500 parts calcitonin) consisted of incubation at 25°C for 1 hr.

calcitonin treated with heat (90°C for 18 hr), with hydrogen peroxide (0.2 M $H_2O_2$ at 37°C for 30 min) or digested with trypsin (1 part trypsin to 500 parts calcitonin, by weight, incubated at 25°C for 1 hr). Each of these treatments essentially eliminated the anorexic effects of the peptide (Fig. 10). Therefore, several treatments that are known to eliminate the hypocalcemic effect of calcitonin also eliminate its anorexic effect.

### 6.3. Inhibition of Eating in Rats by Human Calcitonin

Mammalian calcitonins in general, and human calcitonin in particular, are much less potent than salmon calcitonin (see Section 2.2). Also, human calcitonin is much less stable than salmon calcitonin, its *in vitro* plasma half-life in rats being on the order of 2 hr as compared with about 24 hr for salmon calcitonin (Bennett and McMartin, 1979). The anorexic effect of salmon calcitonin might, therefore, be peculiar to its great potency and long half-life. Cooper and colleagues (1980b) have reported that human calcitonin does not significantly inhibit 24-hr food intake in rats when injected intraventricularly.

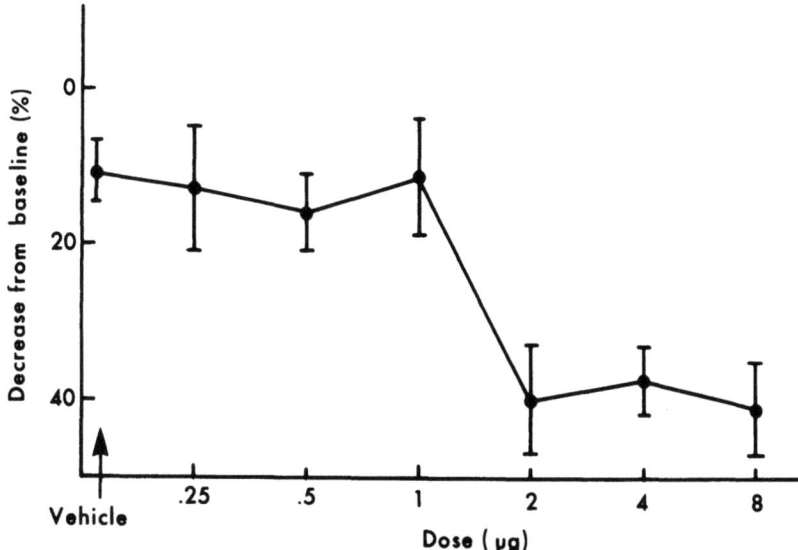

FIGURE 11. Food intake of rats deprived of food for 23½ hr, 1 hr after intraventricular administration of synthetic human calcitonin in dosages from 0.25 to 8.0 μg (0.025 to 0.8 U). Animals were given moistened rat food for 30 min starting 1 hr after injections ($N$ = 8 to 10 per group). The main effect of treatment was significant by a one-way analysis of variance [$F(6, 62) = 5.77, p = 0.0002$].

The fact that salmon calcitonin is effective over an entire 24 hr period is probably due to its long half-life. Human calcitonin, which has an *in vitro* plasma half-life of about 2 hr in the rat, would be expected to be effective for no more than a few hours. We therefore have studied the effects of human calcitonin on food intake in rats, using a paradigm modified so as to be appropriate for the short half-life of human calcitonin.

Rats were adapted to a restricted feeding schedule by being allowed access to dry rat food for 1 hr every 24 hr for several days. Water was continuously available. They were then given synthetic human calcitonin by freehand intraventricular injection (Noble *et al.*, 1967) under light ether anesthesia 60 ± 15 min prior to the start of the 1-hr feeding period. It was found that calcitonin in dosages of 2 μg or above, or 0.2 U, caused a 40% reduction in food intake (Fig. 11). This effect was not dose-related, with a similar inhibition produced by all dosages from 2 to 8 μg with smaller dosages having no effect. Therefore, human calcitonin, which is the least potent calcitonin molecule, is nonetheless capable of inhibiting eating behavior in rats on a short-term basis. Whether human calcitonin is able to inhibit eating in man is unknown.

### 6.4. Stress-Induced Eating

Levine and Morley (1981) have reported that eating induced by stress (pinching of the tail) in rats can be inhibited by synthetic salmon calcitonin.

These investigators found that subcutaneous dosages of calcitonin as small as 10 U/kg and intraventricular dosages of calcitonin as small as 0.002 U produced significant decreases in tail-pinch-induced eating. Calcitonin (2 U, intraventricular) was also able to block eating induced by intracerebral $CaCl_2$, as well as spontaneous eating. Spontaneous eating was decreased for the first 10 hr after injection (during the rats normal nocturnal feeding cycle) and had returned to normal by 32 hr after injection. These authors (Morley *et al.*, 1981) also reported that intraventricular administration of 2 U of calcitonin inhibited muscimol-induced eating in rats.

## 6.5. Conclusions

These results suggest that calcitonin can act directly on the CNS to inhibit eating. A substantial decrease in food intake is produced by small dosages of calcitonin and this effect depends upon the structural integrity of the calcitonin molecule. Stress-induced eating, as well as spontaneous eating, is inhibited by calcitonin. Human calcitonin also inhibits eating, in a dosage range that is consistent with its potency, when appropriate testing conditions are employed. Therefore, calcitonin has a general inhibitory effect on eating through a direct action on the CNS.

## 7. DOES CALCITONIN PRODUCE ILLNESS?

The question arises as to whether the inhibition of eating by calcitonin is a specific, direct effect on eating, or is secondary to some other process such as sedation or production of illness or general malaise.

Calcitonin has been reported to cause a decrease in the activity of rats (Carman and Wyatt, 1979b). This is probably not an explanation for the anorectic effects of calcitonin, because sedative drugs generally stimulate eating, while eating is inhibited by stimulants such as amphetamine (cf. Feldman and Smith, 1978; Hoebel, 1977; Seoane and Baile, 1973a).

A question that is more difficult to answer is whether calcitonin produces nausea or illness which, in turn, inhibits eating. We are convinced that this is not the case because: (1) Food-deprived rats that had received calcitonin injections which inhibit eating invariably began to eat rapidly and avidly, and did not exhibit any grossly observable signs of illness or nausea while eating. (2) Psychiatric patients receiving calcitonin seldom reported nausea (Carman and Wyatt, 1979a),and the literature on administration of calcitonin to humans does not reveal a significant incidence of nausea and illness (Pak *et al.*, 1968; Haddad *et al.*, 1970; Shai *et al.*, 1971). (3) Calcitonin did not act as an aversive stimulus in rats, in terms of producing a conditioned taste aversion (Freed *et al.*, 1979, 1981).

The "conditioned aversion" paradigm (Garcia *et al.*, 1974; Rozin, 1967) has frequently been used to test the possibility that various drugs and other agents cause illness or malaise in rats. This paradigm, in essence, measures reductions in the consumption of preferred novel fluids (such as saccharin solutions) when the first exposure to the solution has been followed by illness, such as the illness produced by radiation or lithium chloride. Recently, however, the utility of this paradigm for this particular purpose has become questionable for two reasons. First, it has become apparent that a very wide range of drugs, including many drugs that do not cause significant nausea clinically, are capable of producing conditioned taste aversions when sufficiently sensitive tests are used (Berger, 1972; Carey, 1973; Corcoran *et al.*, 1974; Goudie and Dickins, 1978; Vogel and Nathan, 1975). Rats will even self-administer drugs in dosages equivalent to those that produce conditioned taste aversions (Wise *et al.*, 1976). Second, certain agents that are extremely toxic, such as gallamine and cyanide, do not induce taste aversions (Ionescu and Buresová, 1977; Nachman and Hartley, 1975).

We have conducted a number of experiments in order to determine whether calcitonin can produce conditioned taste aversions in rats (Freed *et al.*, 1981), aware of the limitations of such experiments. Despite the use of a variety of experimental designs that employed refinements such as two-stimulus tests (Deutsch and Hardy, 1977; Grote and Brown, 1971) and backward conditioning (Barker *et al.*, 1977; Domjan and Gregg, 1977), no clear-cut statistically significant conditioned taste aversion could be demonstrated. In one experiment, which is illustrated in Fig. 12, rats were allowed access to a saccharin solution

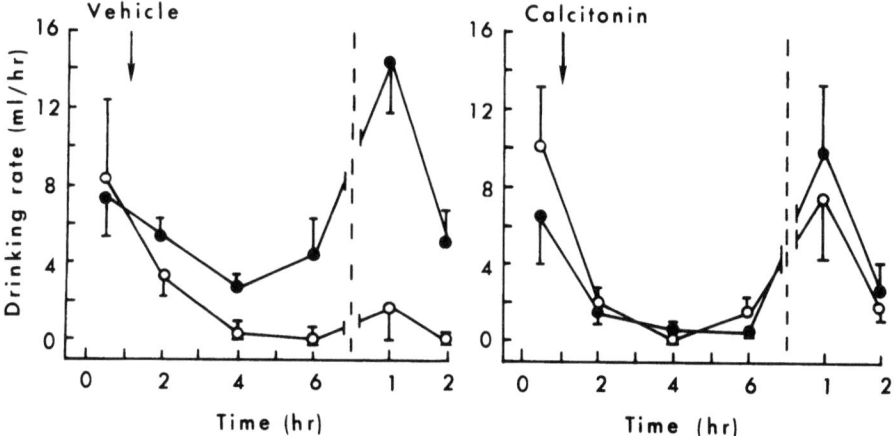

FIGURE 12. Intake of 0.2% sodium saccharin solution (closed circles) and water (open circles) following subcutaneous administration of synthetic salmon calcitonin (50 U/kg) or vehicle. Fluid intake is shown for 6 hr on the day of injections and for 2 hr on the following day. Animals were deprived of water 18 hr prior to testing ($N = 6$ per group). A three-way analysis of variance for two repeated measures revealed no significant main effects of treatment [$F(1, 10) = 1.59$, $p = 0.235$], and no significant interactions [$F(1, 10) = 3.01$, $p = 0.111$ for treatment × fluid, $F(5, 50) = 0.58$, $p = 0.281$ for treatment × time, and $F(5, 50) = 0.59$, $p = 0.291$ for treatment × fluid × time]. However there was a significant overall preference for saccharin as compared to water in the vehicle-treated group [$F(1, 5) = 7.73$, $p = 0.038$, two-way analysis of variance] but not in the calcitonin-treated group [$F(1, 5) = 0.02$, $p = 0.896$].

and water for 6 hr following administration of synthetic salmon calcitonin (50 U/kg) or vehicle. The following day, the animals were again allowed access to water and saccharin for 2 hr. Calcitonin directly inhibited saccharin consumption, but this inhibition was not conditioned, as it occurred immediately as well as 24 hr after the injections. More importantly, the animals that had received calcitonin increased their consumption of saccharin on the second day, even though these animals drank very little saccharin during the last 4 hr of the first 6-hr testing session. The animals treated with vehicle did not show a relatively greater increase in saccharin consumption. In this experiment, the animals always had water available as an alternative beverage, so that if the saccharin had become aversive through conditioning, it could easily have been avoided. Other experiments gave similar results: Conditioned aversions to saccharin could not be produced by injections of calcitonin (50 U/kg) given either 3 hr before exposure to saccharin (Freed et al., 1981) or immediately after exposure to saccharin (Freed et al., 1979). In a two-bottle preference test, calcitonin tended to decrease consumption of one particular fluid, but this effect was not statistically significant, and conditioned taste aversions were not produced to other combinations of fluids (Freed et al., 1981).

Therefore, calcitonin does not appear to be capable of producing a conditioned taste aversion. Together with the fact that calcitonin does not produce other obvious signs of illness in animals and does not produce a high incidence of illness in man, the data suggest that calcitonin produces its inhibition of eating through mechanisms other than the production of illness and nausea.

## 8. CALCITONIN IN OBESE RATS

Margules and his colleagues (1979) have reported that obese Zucker rats have increased concentrations of calcitonin in their pituitary and thyroid glands. Calcitonin concentrations in these glands were found to be increased two- to threefold. Serum calcitonin concentrations did not show a statistically significant increase, although they tended to be higher. The mean serum concentration of calcitonin in the obese animals was 10.1 ng/ml as compared with 3.6 ng/ml in the lean rats. This difference probably did not reach statistical significance due to the large variability of calcitonin concentrations in serum. Margules and his colleagues suggested the possibility that obese Zucker rats do not release calcitonin normally, and that this inability to release calcitonin contributes to their obesity. However, it would be expected that if calcitonin is involved in the regulation of eating, then calcitonin would be decreased in obese animals.

It is possible, however, that the obese animals are not responding to normal or increased calcitonin concentrations. Increased insulin secretion occurs in some cases of diabetes, and diabetes accompanied by obesity (Berger et al., 1978; Reaven and Olefsky, 1978). Insulin secretion is controlled by some of the same factors as calcitonin (Allan and Tepperman, 1969). There could be an interesting parallel between increased insulin secretion in diabetes and altered calcitonin secretion in obesity. The findings of Margules and his colleagues (1979) raise interesting possibilities regarding the relationship between obesity and calcitonin secretion, and suggest that investigations of calcitonin in other models of obesity in animals and man would be worthwhile.

## 9. EFFECTS OF PARATHYROID HORMONE ON FOOD AND WATER INTAKE

Parathyroid hormone (PTH) has effects on calcium disposition that are generally opposite to those of calcitonin. PTH mobilizes calcium from bone in response to decreased plasma calcium concentrations, and promotes excretion of phosphate by the kidney, thus slowing reprecipitation of calcium phoshate (Arnaud and Tenenhouse, 1970; Martin, 1976). PTH is not known to have a CNS effect. Nevertheless, because its physiological effects are generally opposite to those of calcitonin, we have studied the effects of intraventricular PTH on 24-hr food and water intake in rats.

Groups of adult male and female rats (6 rats per group, 48 rats total) were adapted to individual cages, and given injections of parathyroid substance (Sigma Chemical Co.) in dosages of 0.018, 0.18, and 1.8 U (0.1, 1.0, and 10 µg) or vehicle by freehand intraventricular injection under ether anesthesia (Noble et al., 1967) in lactated Ringer's in a volume of 10 µl.

There was no effect of any dosage of PTH on food or water intake, urine or feces excretion, or body weight in either male or female rats (data not shown). Thus, PTH and calcitonin do not appear to have reciprocal effects on eating behavior. Whether PTH can exert some more specific or short-term influence on eating or drinking is a possibility that we have not yet examined.

## 10. DISCUSSION

The primary function of calcitonin is currently thought to be in the regulation of plasma calcium. Increases in plasma calcium result in large increases in calcitonin secretion. However, a variety of foodstuffs, such as triglycerides, also increase calcitonin secretion. In rats, triglycerides actually stimulate calcitonin secretion to a greater degree than does an increase in plasma calcium. In addition, several hormones (glucagon, cholecystokinin, and gastrin) known to be released in response to food intake also stimulate the release of calcitonin. Recent evidence suggests that some of these hormones may exert an anorectic effect in addition to their well-established roles in metabolic processes (see Section 3). On the basis of these observations, it was hypothesized that the anorectic effects of these hormones and of certain foods are mediated by calcitonin. In other words, increased plasma calcitonin concentrations resulting from food intake and subsequent increases in hormone secretion might exert a negative feedback effect on subsequent food intake.

Experiments performed to test this hypothesis have indeed shown that calcitonin has anorectic properties. Salmon calcitonin is a potent and long-lasting anorectic agent in rats and in monkeys when injected subcutaneously. When injected intracerebrally, calcitonin decreases spontaneous eating in rats in dosages as small as 0.2 U. Others have reported that tail-pinch-induced eating in rats can be decreased by calcitonin in dosages as small as 0.002 U (Levine and Morley, 1981). Subcutaneous injections of calcitonin also caused a transient loss of body weight and a pronounced diuresis. Intracerebral injections of calcitonin caused a similar loss of body weight but no diuresis. This suggests that the body-weight loss that is produced by subcutaneous calcitonin is caused by decreased food intake rather than uncompensated diuresis.

Human calcitonin, the least-potent form of calcitonin known, was found to be capable of inhibiting eating in rats on a short-term basis. Human patients who had received synthetic salmon calcitonin in the course of other experiments (Carman and Wyatt, 1977, 1979a) had a slight decrease in body weight the day after calcitonin injections. But effects of calcitonin on eating behavior have not been directly studied in man.

A study by Margules and his colleagues (1979) has raised the interesting possibility that altered calcitonin production or secretion may be associated with some disorder of eating behavior or appetite. They found that obese rats had increased concentrations of calcitonin in their thyroid and pituitary glands, but not in serum. Because serum calcitonin tended to be higher, not decreased, in the obese animals, the role of calcitonin in the development of this obesity is unclear. Calcitonin concentrations in serum from anorexia nervosa patients were measured, but were found to be normal. The possibility that calcitonin is associated with these and other disorders of eating or appetite, such as cancer cachexia, still exists and may warrant further investigation.

At the present time, the mode of action of calcitonin on the CNS is unknown. On the basis of studies in mouse Ehrlich ascites tumor cells, it has been suggested that the primary effect of calcitonin is to promote redistribution of calcium ions between cellular and extracellular compartments (Rasmussen and Pechet, 1970). It has recently been reported that calcitonin reduces *in vitro* uptake of calcium into hypothalamic explants (Levine and Morley, 1981). Thus, calcitonin may also alter cellular distributions of calcium in the brain.

Administration of calcium into the cerebral ventricles stimulates feeding in satiated rats (Myers *et al.*, 1972; Myers and Bender, 1973) and sheep (Seoane and Baile, 1973b). This effect appears to be mediated by the hypothalamus (Myers *et al.*, 1976; Seoane *et al.*, 1975). Other effects of intracerebral calcium administration have also been described, including hypothermia (Myers *et al.*, 1976), decreased paradoxical sleep, slow-wave cortical EEG patterns, and wet-dog shakes (Toszeghi *et al.*, 1978). It is, therefore, plausible that the effects of calcitonin on eating could be mediated by alterations of brain calcium distributions. The fact that calcium-induced feeding is mediated by the hypothalamus (Myers *et al.*, 1976; Seoane *et al.*, 1975) suggests that the hypothalamus might also be the site of action for inhibition of eating by calcitonin. Several other findings are consistent with this hypothesis: (1) The greatest amounts of calcitonin binding within the brain are found in the hypothalamus (see Section 2.2). (2) Calcitonin inhibits calcium uptake in hypothalamic explants (Levine and Morley, 1981). (3) Effects of calcitonin on prolactin secretion and gastric acid secretion may be mediated by the hypothalamus (see Sections 2.4 and 2.5). (4) The blood–brain barrier is more permeable in the hypothalamus than in the rest of the brain (Bradbury, 1979; Rapaport, 1976). This final point would explain why peripheral injections of calcitonin are effective if the site of action is within the CNS. Therefore, one possible mechanism through which calcitonin might inhibit feeding is by entering the brain in or near the hypothalamus, and causing a redistribution of hypothalamic calcium. It is also possible, however, that calcitonin binds to specific receptors in the hypothalamus or elsewhere in the brain that influence feeding behavior but have no direct relationship to calcium.

Therefore, calcitonin is a potent inhibitor of eating behavior and appears to produce this effect through an action on the CNS. A number of other recently discovered effects of calcitonin, including suppression of prolactin secretion, inhibition of gastric acid secretion, and possibly the induction of analgesia, also seem to be mediated by the CNS. Undoubtedly, calcitonin has other as yet undiscovered effects on the CNS as well. The parameters of the effects of calcitonin on eating, prolactin secretion, and other functions suggest that calcitonin will prove to be an interesting and useful psychopharmacological agent.

## 11. SUMMARY

Calcitonin, a peptide hormone secreted by the C cells of the thyroid gland, plays an important role in calcium homeostasis. Calcitonin is secreted postprandially, and it has been suggested that one function of calcitonin is to promote mineralization of the skeleton via calcium absorbed from foods. Ingestion of food also, however, inhibits subsequent eating, which led us to speculate that calcitonin might be a hormonal signal involved in the regulation of appetite.

To test this hypothesis, we administered synthetic salmon calcitonin to rats subcutaneously, and found eating to be substantially decreased for the following 24 hr. Calcitonin in both small and large doses was maximally effective when administered several hours prior to feeding; injections given either 1 or 22 hr before food access were ineffective. Calcitonin was also found to inhibit drinking in rats when food was not made available.

Calcitonin has been reported to inhibit feeding in rhesus monkeys, and to cause a slight decrease in body weight in man. Calcitonin concentrations in serum from patients with anorexia nervosa were, however, found to be normal. This finding does not support the hypothesis that altered calcitonin secretion is involved in anorexia nervosa.

To determine whether calcitonin can inhibit eating by acting directly on the brain, small dosages of salmon calcitonin were injected into the lateral cerebral ventricles of rats. Calcitonin in dosages as small as 43 ng significantly decreased food intake for 24 hr. Denaturation of calcitonin by heat, hydrogen peroxidase, or trypsin eliminated is anorectic effect. Human calcitonin, a weak form of calcitonin, was also found to have a transient inhibitory effect on feeding in amounts consistent with its relative potency.

It is concluded that calcitonin is a potent anorectic agent. This effect appears to be mediated by the brain, possibly by the hypothalamus. Calcitonin may play a role in the regulation of feeding and other homeostatic functions, either through an interaction with specific receptors in the brain or by causing a redistribution of hypothalamic calcium.

ACKNOWLEDGMENTS. The authors thank J. P. Aldred of the Armour Pharmaceutical Company, Kankakee, Illinois, for providing synthetic salmon calcitonin and for advice and suggestions. We also thank the DSMR support staff for their invaluable assistance, especially Theresa Hoffman for preparing the manuscript.

## 12. REFERENCES

Allan, W., and Tepperman, H. M., 1969, Stimulation of insulin secretion in the rat by glucagon, secretin, and pancreozymin: Effect of aminophylline, *Life Sci.* **8**:307.

Anast, C. S., and Conaway, H. H., 1972, Calcitonin, *Clin. Orthop. Relat. Res.* **84**:207.

Ardaillou, R., 1975, Kidney and calcitonin, *Nephron* **15**:250.

Arnaud, C. D., and Tenenhouse, A., 1970, Parathyroid hormone, in: *International Encyclopedia of Pharmacology and Therapeutics*, Section 51, Vol. 1, *Parathyroid Hormone, Thyrocalcitonin, and Related Drugs* (H. Rasmussen, ed.), pp. 197–235, Pergamon Press, Elmsford, N.Y.

Avioli, L. V., Birge, S. J., Scott, S., and Shieber, W., 1969, Role of the thyroid gland during glucagon-induced hypocalcemia in the dog, *Am. J. Physiol.* **216**:939.

Barker, L. M., Smith, J. C., and Suarez, E. M., 1977, "Sickness" and the backward conditioning of taste aversions, in: *Learning Mechanisms in Food Selection* (L. M. Barker, M. R. Best, and M. Doonjan, eds.), pp. 533–553, Baylor University Press, Waco, Tex.

Becker, H. D., Konturek, S. J., Reeder, D. D., and Thompson, J. C., 1973, Effect of calcium and calcitonin on gastrin and gastric secretion in cats, *Am. J. Physiol.* **225**:277.

Becker, K. L., Snider, R. H., Moore, C. F., Monoghan, K. G., and Silva, O., 1979, Calcitonin in extrathyroidal tissues of man, *Acta Endocrinologica* **92**:746.

Becker, K. L., Silva, O. L., Post, R. M., Ballenger, J. C., Carmen, J. S., Snider, R. H., and Moore, C. F., 1980, Immunoreactive calcitonin in cerebrospinal fluid of man, *Brain Res.* **194**:598.

Bell, N. H., 1970, Effects of glucagon, dibutyryl cyclic 3',5'-adenosine monophosphate, and theophylline on calcitonin secretion in vitro, *J. Clin. Invest*, **49**:1368.

Bennett, H. P. J., and McMartin, C., 1979, Peptide hormones and their analogues: Distribution, clearance from the circulation, and inactivation in vivo, *Pharmacol. Rev.* **30**:247.

Berger, B. D., 1972, Conditioning of food aversions by injections of psychoactive drugs, *J. Comp. Physiol. Psychol.* **81**:21.

Berger, M., Muller, W. A., and Renold, A. E., 1978, Relationship of obesity to diabetes: Some facts, many questions, in: *Advances in Modern Nutrition*, Vol. 2, *Diabetes, Obesity, and Vascular Disease* (H. M. Katzen and R. J. Mahler, eds.), pp. 211–228, Wiley, New York.

Bieberdorf, F. A., Gray, T. K., Walsh, J. H., and Fordtran, J. S., 1974, Effect of calcitonin on meal-stimulated gastric acid secretion and serum gastrin concentration, *Gastroenterology* **66**:343.

Bradbury, M., 1979, *The Concept of a Blood–Brain Barrier*, Wiley, New York.
Braga, P., Ferri, S., Santagostino, A., Olgiati, V. R., and Pecile, A., 1978, Lack of opiate receptor involvement in centrally induced calcitonin analgesia, *Life Sci.* **22**:971.
Bryson, G., and Bischoff, F., 1980, Long-term estrogenization in mammals. III. Vasopressin and calcitonin in the regulation of water intake in estrogenized Marsh mice, *Res. Commun. Chem. Pathol. Pharmacol.* **29**:499.
Care, A. D., Bates, R. F., and Gitelman, H. J., 1970, A possible role for the adenyl cyclase system in calcitonin release, *J. Endocrinol.* **48**:1.
Care, A. D., Bruce, J. B., Boelkins, J., Kenny, A. D., Conaway, H., and Anast, C. S., 1971, Role of pancreozymin-cholecystokinin and structurally related compounds as calcitonin secretogogues, *Endocrinology* **89**:262.
Carey, R. J., 1973, Long-term aversion to a saccharin solution induced by repeated amphetamine injections, *Pharmacol. Biochem. Behav.* **1**:265.
Carman, J. S., and Wyatt, R. J., 1977, Reduction of serum-prolactin after subcutaneous salmon calcitonin, *Lancet* **1**:1267.
Carman, J. S., and Wyatt, R. J., 1979a, Use of calcitonin in psychotic agitation or mania, *Arch. Gen. Psychiatry* **36**:72.
Carman, J. S., and Wyatt, R. J., 1979b, Calcium: Bivalent cation in the bivalent psychoses, *Bio. Psychiatry* **14**:295.
Clark, N. B., and Wideman, R. F., Jr., 1980, Calcitonin stimulation of urine flow and sodium excretion in the starling, *Am. J. Physiol.* **238**:406.
Clifton-Bligh, P., Robinson, B., Poon, T., and Posen, S., 1980 Calcitonin and the kidney, *Prog. Biochem. Pharmacol.* **17**:204.
Cooper, C. W., Schwesinger, W. H., Mahgoub, A. M., and Ontijes, D. A., 1971, Thyrocalcitonin: Stimulation of secretion by pentagastrin, *Science* **172**:1238.
Cooper, C. W., Obie, J. F., Toverud, S. U., and Munson, P. L., 1977, Elevated serum calcitonin and serum calcium during suckling in the baby rat, *Endocrinology* **101**:1657.
Cooper, C. W., Bolman, R. M., III, Linehan, W. M., and Wells, S. A., Jr., 1978, Interrelationships between calcium, calcemic hormones and gastrointestinal hormones, *Recent Prog. Horm. Res.* **34**:259.
Cooper, C. W., Peng, T. C., Obie, J. F., and Garner, S. C., 1980a, Calcitonin-like immunoreactivity in rat and human pituitary glands: Histochemical, in vitro and in vivo studies, *Endocrinology* **107**:98.
Cooper, C. W., Obie, J. F., Margules, D. L., Flynn, J. J., and Walker, J., 1980b, Differences in the ability of calcitonins to inhibit food consumption in the rat, in: Abstracts of the 2nd Annual Meeting of the American Society for Bone and Mineral Research, p. 7A.
Copp, D. H., 1969, Endocrine control of calcium homeostasis, *J. Endocrinol.* **43**:137.
Copp, D. H., Cameron, E. C., Cheney, B. A., Davidson, A. G., and Henze, K. G., 1962, Evidence for calcitonin—A new hormone from the parathyroid that lowers blood calcium, *Endocrinology* **70**:638.
Corcoran, M. E., Bolotow, I., Amit, Z., and McCaughran, J. A., 1974, Conditioned taste aversions produced by active and inactive cannabinoids, *Pharmacol. Biochem. Behav.* **2**:725.
Deftos, L. J., Burton, D., Bone, H. G., Catherwood, B. D., Parthemore, J. G., Moore, R. Y., Minick, S., and Guillemin, R., 1978, Immunoreactive calcitonin in the intermediate lobe of the pituitary gland, *Life Sci.* **23**:743.
Deftos, L. J., Burton, D. W., Watkins, W. B., and Catherwood, B. D., 1980, Immunohistological studies of ariodactyl and teleost pituitaries with antisera to calcitonin, *Gen. Comp. Endocrinol.* **42**:9.
Deutsch, J. A., and Hardy, W. T., 1977, Cholecystokinin produces bait shyness in rats, *Nature (London)* **266**:196.
Domjan, M., and Gregg, B., 1977, Long-delay backward taste-aversion conditioning with lithium, *Physiol. Behav.* **18**:59.
Feldman, R. S., and Smith, W. C., 1978, Chlordiazepoxide–fluoxetine interactions on food intake in free-feeding rats, *Pharmacol. Biochem. Behav.* **8**:749.
Findlay, D. M., deLuise, M., Michelangeli, V. P., Ellison, M., and Martin, T. J., 1980, Properties

of a calcitonin receptor and adenylate cyclase in BEN cells, a human cancer cell line, *Cancer Res.* **40**:1311.

Fischer, J. A., Sagar, S. M., and Martin, J. B., 1981, Characterization and regional distribution of calcitonin binding sites in the rat brain, *Life Sci.* **29**:663.

Forslund, K., Slanina, P., Stridsberg, M., and Appelgren, L. E., 1980, Whole body autoradiography of $^3$H- and $^{125}$I-labelled calcitonin in young rats, *Acta Pharmacol. Toxicol.* **46**:398.

Freed, W. J., Perlow, M. J., and Wyatt, R. J., 1979, Calcitonin: Inhibitory effect on eating in rats, *Science* **206**:850.

Freed, W. J., Bing, L. A., and Wyatt, R. J., 1981, Calcitonin: Aversive effects in rats? *Science* **211**:733.

Garcia, J., Hankins, W. G., and Rusiniak, K. W., 1974, Behavioral regulation of the milieu interne in man and rat: Food preferences set by delayed visceral effects facilitate memory research and predator control, *Science* **185**:824.

Garel, J. M., and Besnard, P., 1979, Milk factors controlling the plasma calcitonin level in the newborn rat, *Endocrinology* **104**:1617.

Garel, J. M., and Jullienne, A., 1977, Plasma calcitonin levels in pregnant and newborn rats, *J. Endocrinol.* **75**:373.

Goltzman, D., 1980, Examination of interspecies differences in renal and skeletal receptor binding and adenylate cyclase stimulation with human calcitonin, *Endocrinology* **106**:510.

Goudie, A. J., and Dickins, D. W., 1978, Nitrous oxide-induced conditioned taste aversions in rats: The role of duration of drug exposure and its relation to the taste aversion–self-administration "paradox," *Pharmacol. Biochem. Behav.* **9**:587.

Grote, F. W., and Brown, R. T., 1971, Conditioned taste aversions: Two-stimulus tests are more sensitive than one-stimulus tests, *Behav. Res. Methods Instrum.* **3**:311.

Grubb, S. A., Decker, S. A., Taft, T. N., and Talmage, R. V., 1979, Effect of long-term thyroidectomy on post-prandial plasma calcium changes in man, *J. Endocrinol. Invest.* **2**:75.

Habener, J. F., Singer, F. R., Deftos, L. J., Neer, R. J., and Potts, J. T., Jr., 1971, Explanation for unusual potency of salmon calcitonin, *Nature New Biol.* **232**:91.

Haddad, J. G., Jr., Birge, S. J., and Avioli, L. V., 1970, Effects of prolonged thyrocalcitonin administration on Paget's disease of bone, *N. Engl. J. Med.* **283**:549.

Hesch, R. D., Hüfner, M., Hausenhager, B., and Creutzfield, W., 1971, Inhibition of gastric secretion by calcitonin in man, *Horm. Metab. Res.* **3**:140.

Hirsch, P. F., 1967, Thyrocalcitonin inhibition of bone resorption induced by parathyroid extract in thryoparathyroidectomized rats, *Endocrinology* **80**:539.

Hirsch, P. F., Gauthier, G. F., and Munson, P. L., 1963, Thyroid hypocalcemic principle and recurrent laryngeal nerve injury as factors affecting the response to parathyroidectomy in rats, *Endocrinology* **73**:244.

Hoebel, B. G., 1977, Pharmacologic control of feeding, *Annu. Rev. Pharmacol. Toxicol.* **17**:605.

Holtrop, M. E., Raisz, L. G., and Simmons, H. A., 1974, The effects of parathyroid hormone, colchicine, and calcitonin on the ultrastructure and the activity of osteoclasts in organ culture, *J. Cell Biol.* **60**:346.

Hsiao, S., and Pertsulakes, W., 1970, Feeding–drinking interaction: Food rationing and intake of liquids varying in taste, caloric and osmotic properties, *Physiol. Behav.* **5**:1495.

Huwyler, R., Born, W., Ohnhaus, E. E., and Fischer, J. A., 1979, Plamsa kinetics and urinary excretion of exogenous human and salmon calcitonin in man, *Am. J. Physiol.* **236**:E15.

Ionescu, E., and Buresová, O., 1977, Failure to elicit conditioned taste aversion by severe poisoning, *Pharmacol. Biochem. Behav.* **6**:251.

Isaac, R., Merceron, R., Caillens, G., Raymond, J. P., and Ardaillou, R., 1980, Effects of calcitonin on basal and thyrotropin-releasing hormone-stimulated prolactin secretion in man, *J. Clin. Endocrinol. Metab.* **50**:1011.

Iwasaki, Y., Chihara, K., Iwasaki, J., Abe, H., and Fujita, T., 1979, Effect of calcitonin on prolactin release in rats, *Life Sci.* **25**:1243.

Jacobs, H. L., 1964, The interaction of hunger and thirst: Experimental separation of osmotic and oral-gastric factors in the regulation of caloric intake, in: *Thirst First International Symposium on Thirst in the Regulation of Body Water* (M. J. Wagner, ed.), pp. 117–137, Pergamon Press, Elmsford, N.Y.

Koida, M., Nakamuta, H., Furukawa, S., and Orlowski, R. C., 1980, Abundance and location of $^{125}$I-salmon calcitonin binding site in rat brain, *Jpn. J. Pharmacol.* **30**:575.

Levine, A. S., and Morley, J. E., 1981, Reduction of feeding in rats by calcitonin, *Brian Res.* **22**:187.

Lorenz, D. N., Kreielsheimer, G., and Smith, G. P., 1979, Effect of cholecystokinin, gastrin, secretin, and GIP on sham feeding in the rat, *Physiol. Behav.* **23**:1065.

Maddison, S., 1977, Intraperitoneal and intracranial cholecystokinin depress operant responding for food, *Physiol. Behav.* **19**:819.

Manaker, S., Ackerman, S. H., and Weiner, H., 1979, Intracerebroventricular pentagastrin fails to affect feeding and acid secretion in the rat, *Physiol. Behav.* **23**:395.

Margules, D. L., Flynn, J. J., Walker, J., and Cooper, C. W., 1979, Elevation of calcitonin immunoreactivity in the pituitary and thyroid glands of genetically obese rats (fa/fa)*Brain Res. Bull.* **4**:589.

Martin, C. R., 1976, *Textbook of Endocrine Physiology*, Oxford University Press, London.

Martin, J. R., and Novin, D., 1977, Decreased feeding in rats following hepatic-portal infusion of glucagon, *Physiol. Behav.* **19**:461.

Marx, S. J., Woodard, C. J., and Aurbach, G. D., 1972, Calcitonin receptors of kidney and bone, *Science* **178**:999.

Marx, S. J., Aurbach, G. D., Gavin, J. R., III, and Buell, D. W., 1974, Calcitonin receptors on cultured human lymphocytes, *J. Biol. Chem.* **249**:6812.

Mattheij, J. A. M., Sterrenberg, L., and Swarts, H. J., 1980, Current concepts. III. Prolactin and calcium metabolism in the rat, *Life Sci.* **27**:2031.

Metz, S. A., Deftos, L. J., Baylink, D. J., and Robertson, R. P., 1978, Neuroendocrine modulation of calcitonin and parathyroid hormone in man, *J. Clin. Endocrinol. Metab.* **47**:151.

Morley, J. E., and Levine, A. S., 1981, Intraventricular calcitonin inhibits gastric acid secretion, *Science* **214**:671.

Morley, J. E., Levine, A. S., and Kneip, J., 1981, Muscimol induced feeding: A model to study the hypothalamic regulation of appetite, *Life Sci.* **29**:1213.

Munson, P. L., and Gray, T. K., 1970, Function of thyrocalcitonin in normal physiology, *Fed. Proc.* **29**:1206.

Myers, R. D., and Bender, S. A., 1973, Action of excess calcium ions in the brain on motivated feeding in the rat: Attenuation by pharmacological antagonists, *Pharmacol. Biochem. Behav.* **1**:569.

Myers, R. D., Bender, S. A., Krstíc, M. K., and Brophy, P. D., 1972, Feeding produced in the satiated rat by elevating the concentration of calcium in the brain, *Science* **176**:1124.

Myers, R. D., Melchior, C. L., and Gisolfi, C. V., 1976, Feeding and body temperature in the rat: Diencephalic localization of changes produced by excess calcium ions, *Brain Res. Bull.* **1**:33.

Nachman, M., and Hartley, P. L., 1975, Role of illness in producing learned taste aversions in rats: A comparison of several rodenticides, *J. Comp. Physiol. Psychol.* **89**:1010.

Nakhla, A. M., 1979, Calcitonin: Effect on protein synthesis in different rat tissues, *Experientia* **35**:1525.

Nemeroff, C. B., Osbahr, A. J., III, Bissette, G., Jahnke, G., Lipton, M. A., and Prange, A. J., Jr., 1978, Cholecystokinin inhibits tail pinch-induced eating in rats, *Science* **200**:793.

Nobel, E. P., Wurtman, R. J., and Axelrod, J., 1967, A simple and rapid method for injecting H$^3$-norepinephrine into the lateral ventricle of the rat brain, *Life Sci.* **6**:281.

Olgiati, U. R., Guidobono, F., Luisetto, G., Hetti, C., Bianchi, C., and Pecile, A., 1981, Calcitonin inhibition of physiological and stimulated prolactin secretion in rats, *Life Sci.* **29**:585.

Pahuja, D. N., and DeLuca, H. F., 1981, Stimulation of intestinal calcium transport and bone calcium mobilization by prolactin in vitamin D-deficient rats, *Science* **214**:1038.

Paillard, F., Ardaillou, R., Malendin, H., Fillastre, J.-P., and Prier, S., 1972, Renal effects of salmon calcitonin in man, *J. Lab. Clin. Med.* **80**:200.

Pak, C. Y., Wills, M. R., Smith, G. W., III, and Bartter, F. C., 1968, Tretament with thyrocalcitonin of the hypercalcemia of parathyroid carcinoma, *J. Clin. Endocrinol.* **28:**1657.

Parthemore, J. G., and Deftos, L. J., 1978, Calcitonin secretion in normal human subjects, *J. Clin. Endocrinol. Metab.* **47:**184.

Pavlinac, D. M., Lenhard, L. W., Parthemore, J. G., and Deftos, L. J., 1980, Immunoreactive calcitonin in human cerebrospinal fluid, *J. Clin. Endocrinol. Metab.* **50:**717.

Pecile, A., Ferri, S., Braga, P. C., and Olgiati, V. R., 1975, Effects of intracerebroventricular calcitonin in the conscious rabbit, *Experientia* **31:**332.

Perlow, M. J., Freed, W. J., Carman, J. S., and Wyatt, R. J., 1980, Calcitonin reduces feeding in man, monkey, and rat, *Pharmacol. Biochem. Behav.* **12:**609.

Queener, S. F., Fleming, J. W., and Bell, N. H., 1975, Solubilization of calcitonin-responsive renal cortical adenylate cyclase, *J. Biol. Chem.* **250:**7586.

Rapaport, S. I., 1976, *Blood–Brain Barrier in Physiology and Medicine*, Raven Press, New York.

Rasmussen, H., and Pechet, M., 1970, Calcitonin and thyrocalcitonin, in: *International Encyclopedia of Pharmacology and Therapeutics*, Section 51, Vol. 1, *Parathyroid Hormone, Thyrocalcitonin, and Related Drugs* (H. Rasmussen, ed.), pp. 237–260, Pergamon Press, Elmsford, N.Y.

Reaven, G. M., and Olefsky, J. M., 1978, Role of insulin resistance in the pathogenesis of hyperglycemia, in: *Advances in Modern Nutrition*, Vol. 2, *Diabetes, Obesity, and Vascular Disease* (H. M. Katzen and R. J. Mahler, eds.), pp. 229–266, Wiley, New York.

Rizzo, A. J., and Goltzman, D., 1981, Calcitonin receptors in the central nervous system of the rat, *Endocrinology* **108:**1672.

Roos, B. A., and Deftos, L. J., 1976, Calcitonin secretion *in vitro*. II. Regulatory effects of enteric mammalian polypeptide hormones on trout C-cell cultures, *Endocrinology* **98:**1284.

Roos, B. A., Bergeron, G., Guggenheim, K., and Deftos, L. J., 1977, Maturational increase in plasma calcitonin related to gastrointestinal function, *Clin. Res.* **25:**398a.

Rozin, P., 1967, Specific aversions as a component of specific hungers, *J. Comp. Physiol. Psychol.* **64:**237.

Satoh, M., Amano, H., Nakazawa, T., and Takagi, H., 1979, Inhibition by calcium of analgesia induced by intracisternal injection of porcine calcitonin in mice, *Res. Commun. Chem. Pathol. Parmacol.* **26:**213.

Seoane, J. R., and Baile, C. A., 1973a, Feeding behavior in sheep as related to the hypnotic activities of barbiturates injected into the third ventricle, *Pharmacol. Biochem. Behav.* **1:**47.

Seoane, J. R., and Baile, C. A., 1973b, Feeding elicited by injections of $Ca^{++}$ and $Mg^{++}$ into the third ventricle of sheep, *Experientia* **29:**61.

Seoane, J. R., McLaughlin, C. L., and Baile, C. A., 1975, Feeding following intrahypothalamic injections of calcium and magnesium ions in sheep, *J. Dairy Sci.* **58:**349.

Sethi, R., Kukreja, S. C., Bowser, E. N., Hareis, G. K., and Williams, G. A., 1981, Effects of secretin on parathyroid hormone and calcitonin secretion, *J. Clin. Endocrinol. Metab.* **53:**153.

Shai, F., Baker, R. K., and Wallach, S., 1971, The clinical and metabolic effects of porcine calcitonin on Paget's disease of bone, *J. Clin. Invest.* **50:**1927.

Smith, G. P., and Gibbs, J., 1975, Cholecystokinin: A putative satiety signal, *Pharmacol. Biochem. Behav.* **3:**135.

Smith, G. P., Gibbs, J., and Young, R. C., 1974, Cholecystokinin and intestinal satiety in the rat, *Fed. Proc.* **33:**1146.

Swaminathan, R., Bates, R. F., Bloom, S. R., Ganguli, R. C., and Care, A. D., 1973, The relationship between food, gastro-intestinal hormones and calcitonin secretion, *J. Endocrinol.* **59:**217.

Talmage, R. V., Doppelt, S. H., and Cooper, C. W., 1975, Relationship of blood concentrations of calcium, phosphate, gastrin and calcitonin to the onset of feeding in the rat, *Proc. Soc. Exp. Biol. Med.* **149:**855.

Talmage, R. V., Grubb, S. A., Norimatsu, H., and Vanderwiel, C. J., 1980, Evidence for an immportant physiological role for càlcitonin, *Proc. Natl. Acad. Sci. USA* **77:**609.

Tashjian, A. H., Jr., and Warnock, D. R., 1967, Stability of the hypocalcemic activity of porcine thyrocalcitonin, *Endocrinology* **81**:306.

Tepperman, B. L., and Evered, M. D., 1980, Gastrin injected into the lateral hypothalamus stimulates secretion of gastric acid in rats, *Science* **209**:1142.

Toszeghi, P., Tobler, I., and Borbély, A., 1978, Cerebral ventricular infusion of excess calcium in the rat: Effects on sleep states, behavior and cortical EEG, *Eur. J. Pharmacol.* **51**:407.

Vogel, J. R., and Nathan, B. A., 1975, Learned taste aversions induced by hypnotic drugs, *Pharmacol. Biochem. Behav.* **3**:189.

Watkins, W. B., Moore, R. Y., Burton, D., Bone, H. G., III, Catherwood, B. D., and Deftos, L. J., 1980, Distribution of immunoreactive calcitonin in the rat pituitary gland, *Endocrinology* **106**:1966.

Wise, R. A., Yokel, R. A., and deWit, H., 1976, Both positive reinforcement and conditioned aversion from amphetamine and from apomorphine in rats, *Science* **191**:1273.

Yamamoto, M., Kumagai, F., Tachikawa, S., and Maeno, H., 1979, Lack of effect of levallorphan on analgesia induced by intraventricular application of porcine calcitonin in mice, *Eur. J. Pharmacol.* **55**:211.

Yamamoto, M., Tachikawa, S., and Maeno, H., 1981, Evoked potential studies of porcine calcitonin in rabbits, *Neuropharmacology* **20**:83.

CHAPTER 5

# Gonadal Dysfunction in Anorexia Nervosa
## A Minireview

### S. RICHARDSON and G. TOLIS

## 1. INTRODUCTION

Anorexia nervosa (AN) is a disease of self-imposed starvation which occurs most frequently in adolescent girls. All females with the disease suffer from amenorrhea which may begin prior to, concurrent with, or following the onset of weight loss. The associated endocrine changes in the hypothalamic–pituitary–ovarian axis have been studied in detail.

## 2. THE HYPOTHALAMIC–PITUITARY–OVARIAN AXIS IN ANOREXIA NERVOSA

### 2.1. Estrogens

During normal childhood, the plasma estradiol ($E_2$) level is low. It rises at puberty to adult levels, which are maintained until menopause.

Studies done during the acute stage of AN, i.e, the period of maximum weight loss, have found plasma $E_2$ levels to be below those of controls (Beumont *et al.*, 1976; Mecklenburg *et al.*, 1974; Sherman *et al.*, 1975; Travaglini *et al.*, 1976; Vigersky *et al.*, 1976; Wakeling *et al.*, 1977); similarly, urinary estrogens (Kanis *et al.*, 1974) and the effect of estrogens upon vaginal cytology (Rudolf *et al.*, 1979) have been described as subnormal.

The metabolism of estrogen is also altered at the time of weight loss with an increase in the estrone:estradiol ($E_1:E_2$) ratio (Bell *et al.*, 1966) and an increased excretion of catechol estrogens (Fishman *et al.*, 1975).

---

S. RICHARDSON • Department of Medicine   G. TOLIS • Departments of Medicine, Obstetrics and Gynecology, McGill University, Royal Victoria Hospital, Montreal, Quebec H3A 1A1, Canada.

## 2.2. Gonadotropins

Basal plasma gonadotropins and the pattern of their secretion also vary according to age in normals. Prepubertal children have low plasma LH and FSH with little fluctuation around the mean (see Fig. 1). In pubertal girls, the mean FSH levels rise early, then plateau. Mean LH rises at a later age. Early in puberty a circadian rhythm develops. LH secretion increases two- to three-fold at the onset of sleep and continues to be secreted in large bursts throughout the night (see Fig. 1). The adult pattern is that of bursts of equal magnitude approximately every 90 min during sleep and waking with a mean value greater than that during prepuberty (see Fig. 2). At different ages, the magnitude of gonadotropin response to LRH stimulation varies in proportion to basal LH

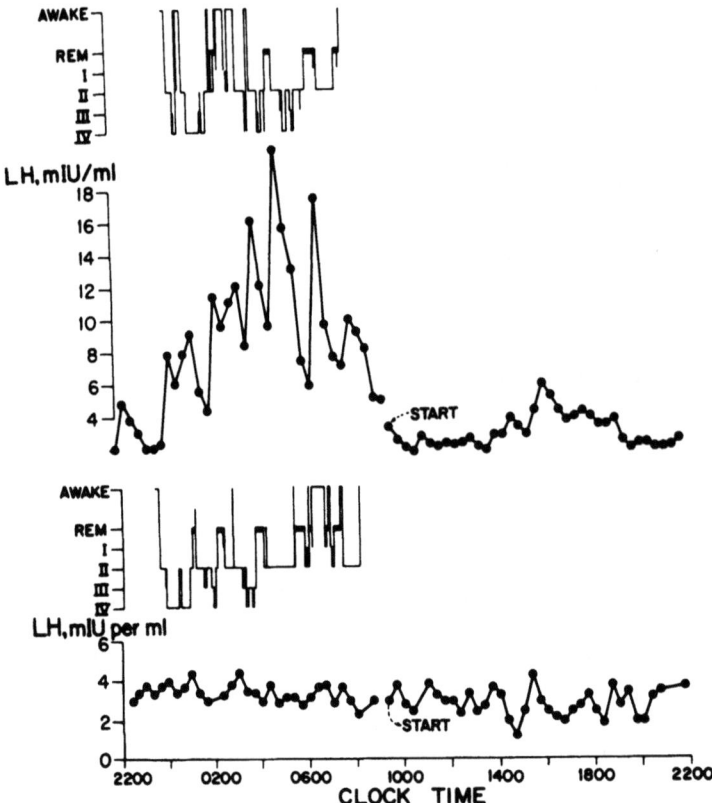

FIGURE 1. Plasma LH concentration every 20 min for 24 hr in a normal prepubertal girl (lower panel) and early pubertal girl (upper panel). The sleep histogram is shown above the period of nocturnal sleep. Sleep stages are awake, rapid eye movement (REM, —; stages I–IV by depth of line graph. Plasma LH concentrations are expressed as milli-international units per milliliter of Second International Reference Preparation of Human Menopausal Gonadotropin. (From Boyar et al., 1974.)

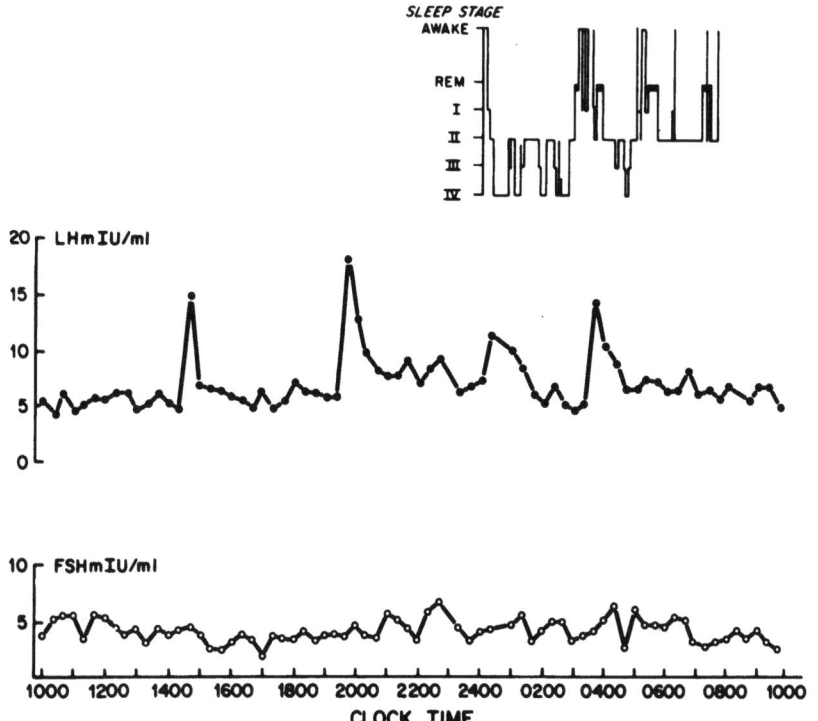

FIGURE 2. The 24-hr LH and FSH secretory patterns in a normal 14-year-old postmenarchal girl. Note the random LH secretory activity of equal magnitude during sleep and waking periods. (From Boyar and Katz, 1977.)

and FSH levels. Thus, during prepuberty, the LH response to LRH is minimal. It increases markedly during the pubertal period and is even greater in the adult (see Fig. 3). There is also a change in the FSH:LH ratios in response to LRH at puberty; i.e., in prepuberty $> 1$; in postpuberty $< 1$.

At the time of maximal weight loss in AN, basal plasma LH levels have been found to be low (Beumont *et al.*, 1976; Hurd *et al.*, 1977; Isaacs *et al.*, 1980; Kanis *et al.*, 1974; Marshall and Fraser, 1971; Mecklenburg *et al.*, 1974; Rudolf *et al.*, 1979; Sherman *et al.*, 1975; Travaglini *et al.*, 1976; Vigersky *et al.*, 1976; Wakeling *et al.*, 1977) and to correlate with both weight and percent body fat (Jeuniewic *et al.*, 1978).

Basal plasma FSH levels have been reported as low (Beumont *et al.*, 1976; Isaacs *et al.*, 1980; Palmer *et al.*, 1975; Sherman *et al.*, 1975; Vigersky *et al.*, 1976), normal (Jeuniewic *et al.*, 1978; Travaglini *et al.*, 1976), or even high (Hurd *et al.*, 1977). However, in the last study, the range was very wide; urinary gonadotropins were undetectable in 68%.

The circadian rhythm of LH in AN was studied by Boyar *et al.* (1974) in nine women suffering from the disease at the time of weight loss. Eight of the

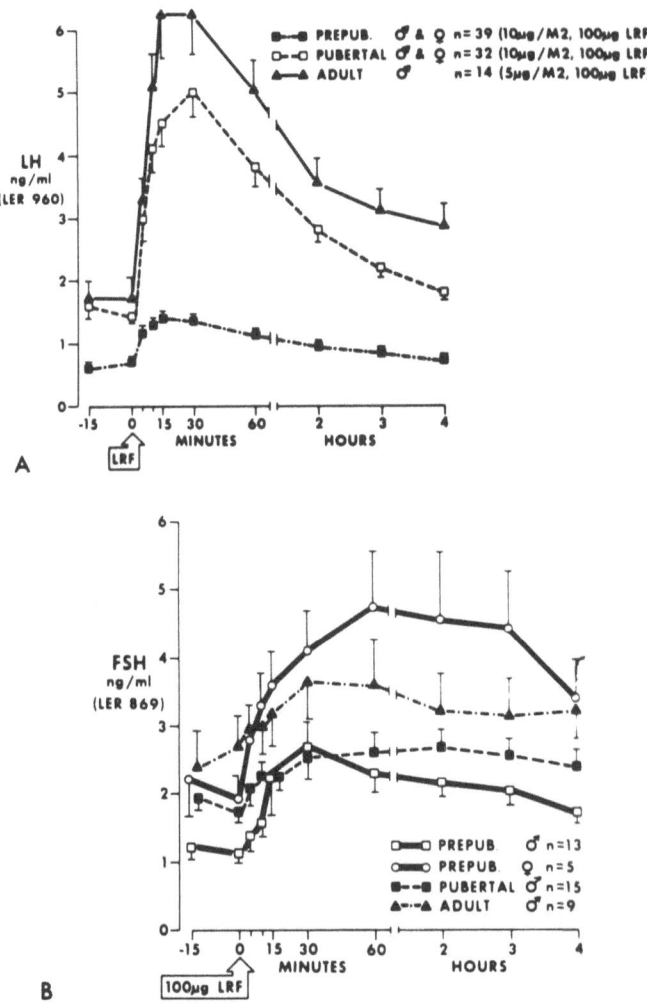

FIGURE 3. The change in plasma LH (A) and FSH (B) in prepubertal, pubertal, and adult subjects. Note the limited LH response in prepubertal children when compared with that of pubertal and adult subjects. The FSH response to LRF is similar in prepubertal, pubertal, and adult males. In females, the FSH response is significantly greater than that of prepubertal, pubertal, or adult males. (From Grumbach et al., 1974.)

nine had LH patterns which were inappropriate for their age, i.e., prepubertal or pubertal. This was confirmed by Pirke et al. (1979) who also found the LH secretory pattern to be "immature" during the acute stage of illness. After weight gain, many progressed to a more mature pattern (see below).

Most authors have found a blunted LH response to LRH in AN patients while they are at a low body weight (Beumont et al., 1976; Isaacs et al., 1980; Sherman et al., 1975). There is a positive correlation between body weight and magnitude of LH response (Isaacs et al., 1980; Jeuniewic et al., 1978; Kanis

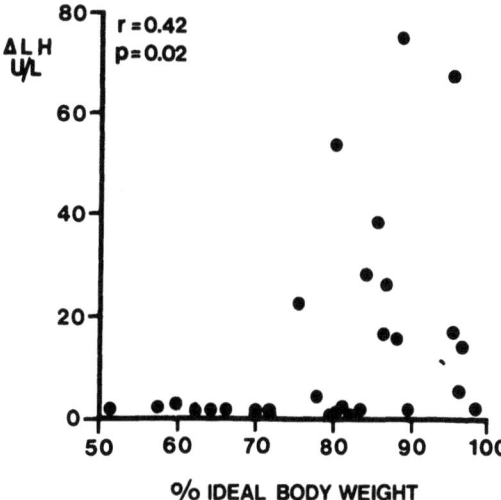

FIGURE 4. Correlation of maximal LH increment following LRH with percentage of ideal body weight. (From Isaacs et al., 1980.)

et al., 1974). The data of Isaacs et al. (1980) suggest that below 75% of ideal body weight (IBW), no LH response to LRH occurs (see Fig. 4). Two groups, however, did not find a consistently blunted LH response to LRH stimulation at low body weights (Aona et al., 1975; Mecklenburg et al., 1974).

The FSH response to LRH in AN tends to be more variable with reports ranging from blunted to supranormal (Aona et al., 1975; Beumont et al., 1976; Isaacs et al., 1980; Jeuniewic et al., 1978; Palmer et al., 1975; Sherman et al., 1975; Vigersky et al., 1976). Although Isaacs et al. (1980) found a positive correlation between FSH responsiveness to LRH and body weight, Jeuniewic et al. (1978) and Palmer et al. (1975) did not.

### 2.3. Estrogen Feedback

Estrogen either directly (pituitary gonadotropes) or indirectly (hypothalamic LHRH) affects gonadotropin secretion (Kastin et al., 1972; Yen et al., 1974; Jaffee and Keye, 1974). In states of high FSH/LH output (i.e., menopause), a negative feedback action is demonstrable. On the other hand, during the normal cycle, and in particular at midcycle, a positive effect is documented by the rising LH values.

Aona et al. (1975) and Wakeling et al. (1977) studied estrogen feedback in AN. Aona and co-workers gave 20 mg Premarin i.v. to nine normal women. At 8 hr plasma LH was suppressed (negative feedback). Later, at 48 hr, it rose to above pretreatment levels (positive feedback). In females with AN who were less than 75% IBW, the plasma LH was also suppressed at 8 hr in all nine patients studied but elevation at 48 hr occurred in only three (see Fig. 5). In summary, during the acute state of AN, negative feedback appeared to be intact but positive feedback was defective in the majority of patients studied.

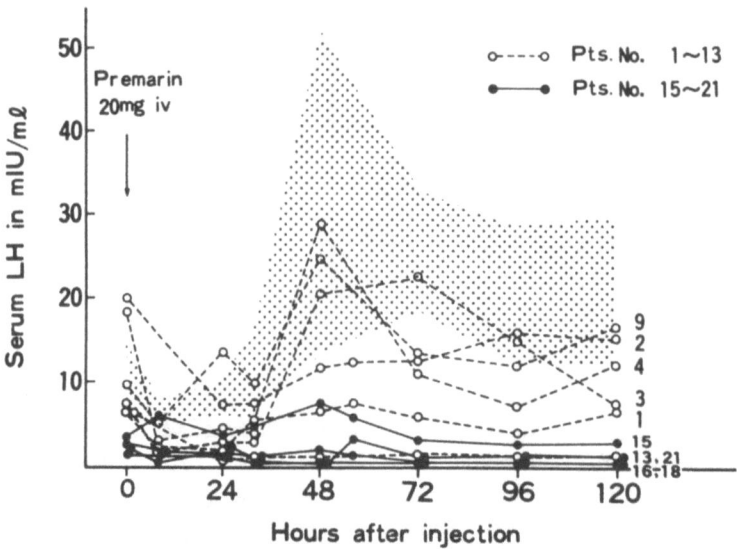

FIGURE 5. Serum levels after i.v. injection of conjugated estrogen (Premarin) in patients with anorexia nervosa. Shaded area represents mean ± S.D. response of LH in normal cyclic women in the mid-follicular phase ($D^{7-9}$). (From Aona et al., 1975.)

Wakeling et al. (1977) got similar results giving 100 μg ethinyl estradiol b.i.d. for 3 days. In five menstruating women studied in the follicular phase, there was an initial lowering of LH on day 1 (negative feedback). This was followed by a significant rise in plasma LH to above baseline values on day 3 or 4 (positive feedback). When 11 patients in the acute stage of AN (64–87% IBW) were compared, plasma LH suppressed normally on day 1 (negative feedback) but failed to rise on day 3 or 4 (lack of positive feedback).

The use of clomiphene is another way in which estrogen feedback can be assessed. Clomiphene acts as an antiestrogen, blocking E receptors in the hypothalamus. This causes an initial rise in LH (i.e., negative feedback is blocked) which in turn stimulates the ovary to produce more estrogen. The ovarian estrogen, in turn, acts by a positive feedback mechanism at the hypothalamus causing a second LH peak 5 to 7 days after clomiphene is stopped (see Fig. 6). Menses occurs 14 days later.

In acute AN, clomiphene either fails to elicit a rise in plasma LH or produces only the initial LH rise without the second peak which normally occurs 5 to 7 days after the clomiphene is discontinued (Aona et al., 1975; Jeuniewic et al., 1978; Marshall and Fraser, 1971; Mortimer et al., 1973; Rudolf et al., 1979; Travaglini et al., 1976; Wakeling et al., 1976).

## 2.4. LRH Priming

The findings of low basal estrogens, low basal gonadotropins with a poor response to LRH, and failure to elicit a positive feedback response to either estrogen or clomiphene, support the hypothesis that the functional defect in AN is at the level of the hypothalamus which probably secretes insufficient LRH. Unfortunately, this cannot be measured directly. The work of Yoshimoto *et al.* (1975) and Nillius *et al.* (1975) provides indirect support for this hypothesis. Yashimoto *et al.* (1975) studied six acute AN patients who had low basal LH and FSH levels and little response to LRH. After treating them with 400 µg LRH i.v. over 2 hr for 3 to 5 days, both LH and FSH responses to LRH became either normal or exaggerated. It was concluded that the pituitary requires adequate previous exposure to LRH before it can respond normally to an acute LRH stimulus, i.e., it needs to be primed. From this it can be extrapolated that patients in the acute stage of AN respond poorly to LRH because

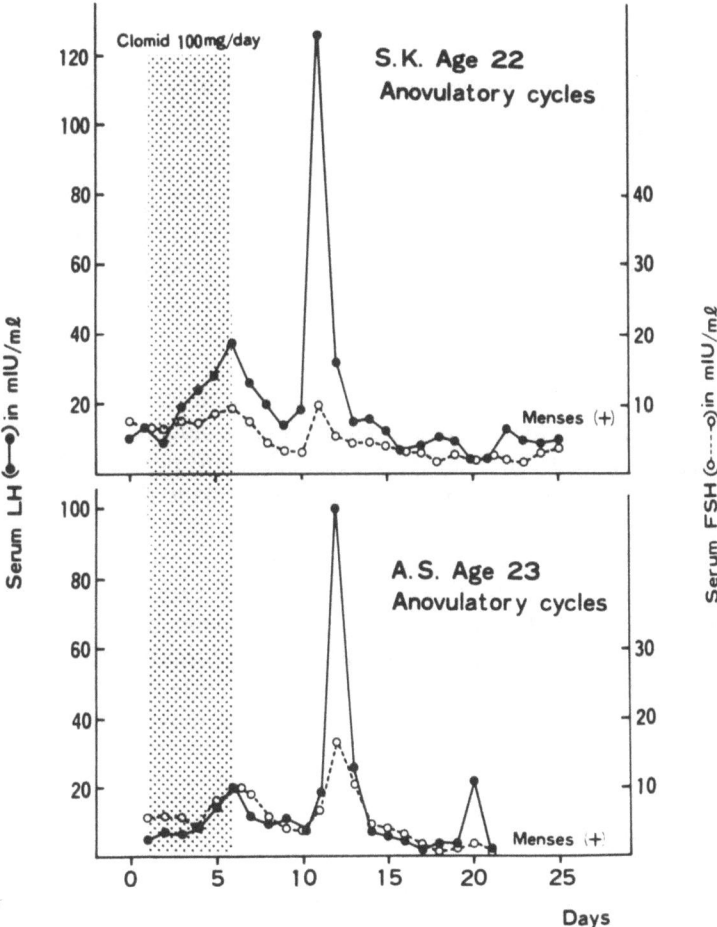

FIGURE 6. Serum LH and FSH levels during and after clomiphene treatment in two patients with anovulatory cycles. Shaded area represents duration of clomiphene treatment. (From Aona *et al.*, 1975.)

they have not been adequately exposed to it in the preceding days, weeks, or months.

The patients of Nillius *et al.* (1975), on the other hand, although they too had low basal LH and estrogen levels, had a normal LH response and a supranormal FSH response to LRH ($N = 4$, weights 36–47 kg). After 4 weeks of 500 μg LRH (i.m.) t.i.d., urinary and plasma estrogen levels returned to normal and all four showed evidence of ovulation as indicated by serum progesterone levels.

These two studies clearly show that although the pituitary in acute AN may need to be primed with LRH, it is capable of responding normally, as is the ovary.

The low estrogen levels found in acute AN may contribute to the blunting of the gonadotropin response to LRH, since estrogens have been shown to augment the pituitary response to LRH (Jaffe and Keye, 1974; Yen *et al.*, 1974).

## 3. OTHER HYPOTHALAMIC AND ENDOCRINE DYSFUNCTION IN ANOREXIA NERVOSA

The hypothalamic–pituitary–ovarian axis is not the only hypothalamic or endocrine system affected in AN. Vigersky *et al.* (1977) and Mecklenburg *et al.* (1974) reported abnormal temperature regulation and partial diabetes insipidus. As well, there is often dysfunction in the thyroid, adrenal, GH, and prolactin systems (Jarrell *et al.*, 1979). Serum $T_3$ levels are consistently found to be low, the levels correlating with the degree of weight loss. The basal plasma TSH is not elevated, as would be expected, and the TSH response to TRH is either normal or delayed. These patients have been described as "hypo-metabolic but with an intact hypothalamic–pituitary–thyroidal axis which has been reset" (Jarrell *et al.*, 1979). The hypo-metabolic state protects the organism from excessive tissue breakdown in the face of an inadequate calorie intake. GH levels are normal or high, rising normally with slow-wave sleep and responding normally to arginine or glucagon. The response to insulin may be blunted. Somatomedin levels, however, are low. Occasionally GH responds abnormally to TRH, LRH, or dopamine agonists. Plasma cortisol is frequently elevated but the circadian rhythm is maintained and urinary cortisol levels are normal. The elevated plasma cortisol values are attributed to a prolonged half-life of cortisol. Serum prolactin levels are normal but may fail to rise normally during sleep.

## 4. ALTERATIONS IN NEUROTRANSMITTERS

Norepinephrine has been found in many experimental situations to stimulate a gonadotropin surge. α-Blockers inhibit both ovulation in rats and the gonadotropin rise usually seen following oophorectomy in monkeys (Ojeda and McCann, 1978). In humans, catecholamine metabolism was studied in 15 women with AN who were all less than 85% IBW. During the acute illness, plasma norepinephrine and the urinary excretion of 3-methoxy-4-hydroxyphenylglycol (MHPG, a urinary metabolite of norepineophrine) were significantly below values found in normal controls. These levels returned to normal with weight gain. Although plasma norepinephrine and its urinary metabolites are chiefly indicators of peripheral metabolism, it would not be unreasonable to postulate that similar changes might occur in the hypothalamus, thereby suppressing LRH levels.

Dopamine, under different experimental conditions, both stimulates and suppresses gonadotropin release. Recent clinical data support a suppressive role in humans (Leblanc et al., 1976 Lachelin et al., 1977; Judd et al., 1978; Quigley et al., 1979). Opiate-like substances also inhibit LH release in normal men (Stubbs et al., 1978). This effect is blocked by naloxone (Bruni et al., 1977). Quigley et al. (1980) assessed the response of LH to both dopamine antagonists and opiate antagonists in patients with hypothalamic amenorrhea. Four of eight patients had a significant increase in plasma LH following both inhibitors, an effect not seen in normals. They postulated that an increased opiate and/or dopamine activity in these patients suppressed LH secretion. One might extrapolate from this that a similar derangement of dopamine/opiates could occur in AN patients. This would be at variance, however, with the observation of Gross et al. (1979) that AN patients had decreased homovanillic acid excretion. Since 50% of urinary homovanillic acid is believed to come from dopamine metabolism in the brain, this finding would seem to indicate a decreased rather than an increased brain dopaminergic activity in AN.

Estrogen metabolites also influence gonadotropin secretion. Fishman et al. (1975) reported decreased estriol levels and increased 2-hydroxyestrone levels in acute AN patients. 2-Hydroxyestrone is a catechol estrogen which may act at the hypothalamus or pituitary as an antiestrogen. Because of its catechol nature it may also compete with brain catecholamines, altering their metabolism. The exact role that catechol estrogens play in influencing gonadotropin secretion is unknown.

In summary, although neurotransmitters such as norepinephrine, dopamine, endorphins, or catechol estrogens may be altered in AN, their exact role, be it causative of the clinical syndrome or simply a consequence of the weight loss, has yet to be defined.

## 5. HYPOTHALAMIC AND ENDOCRINE ALTERATIONS IN SIMPLE WEIGHT LOSS (SWL)

Amenorrhea and infertility occur in times of famine when the infertility serves to protect women from the added nutritional demands of a fetus. Dieting for aesthetic reasons, but without the abnormal attitude of food seen in AN, is also frequently accompanied by secondary amenorrhea. Although less well documented, the hypothalamic and endocrine manifestations of these two situations seem to be the same as in AN. In a large study of severe malnutrition in Mexico (Zubiran and Gomez-Mont, 1953),the estrogen effects on vaginal cytology, urinary estrogens, and urinary gonadotropins were all very low. In dieters, plasma estrogens have also been found to be low (McArthur et al., 1976). Basal plasma LH has been reported as both low (McArthur et al., 1976) and normal (Vigersky et al., 1977)—as has basal FSH (Vigersky et al., 1977; McArthur et al., 1976). The response of LH and FSH to 10 μg LRH was normal but delayed in Vigersky and co-workers' study (1977); the same observation was made in those suffering from AN. In the same study, abnormal thermoregulation and partial diabetes insipidus were observed in SWL but to a lesser degree than in AN. This could be explained by the fact that the AN patients weighed less.

$T_3$ levels are low in starvation (Chopra and Smith, 1975): GH levels and plasma free cortisol levels are elevated (Pimstone, 1976). These changes have all been described in acute AN.

## 6. OTHER FACTORS

It is unlikely that weight loss alone is responsible for the decreased LRH secretion in AN. In eight studies with a total of 400 patients, 255 had an onset of amenorrhea prior to our concurrent with the onset of weight loss (Beck and Brochner-Mortensen, 1954; Crisp and Stonehill, 1971; Halmi, 1974; Hurd et al., 1977; Kanis et al., 1974; Kay and Leigh, 1954; Starkey and Lee, 1969; Warren and Vande Wiele, 1973).

In normal women, gonadotropin secretion can be exquisitely sensitive to emotional or physical trauma. Menses is frequently interrupted by travel or emotional upsets—especially in the teen years. Studies in menopausal depressed women revealed a low LH pattern (Altman et al., 1975). Both plasma LH and FSH may be low in the severely ill with or without weight loss (Warren et al., 1977) or following surgery (Charters et al., 1969).

Even vigorous exercise has been shown to alter menstrual function. Malina et al. (1973) found a significant delay in age at menarche in track and field athletes when compared with nonathletic controls. Feicht et al. (1978) noted a high incidence of secondary amenorrhea among long-distance runners with a direct correlation between the number of miles ran per week and the incidence of secondary amenorrhea which ranged from 6% in those running the least to 43% in those who ran the most ($p < 0.01$). In comparing runners with joggers or controls, Dale et al. (1979) also found a significantly increased incidence of oligomenorrhea or secondary amenorrhea. Only 50% of the runners ovulated. Those with anovulatory cycles had noncyclic plasma LH and FSH values which were in the low to low-normal range. Mean estradiol levels were also below those of controls. Of importance, the percent body fat was much less in runners than in joggers or controls.

Whether the amenorrhea so frequently observed in athletes and the accompanying low levels of gonadotropins and estrogens are related to exercise per se, body fat, or a combination of the two, is not clear. Frisch and McArthur (1974) reported data to illustrate their hypothesis that a certain percentage of body weight as fat had to be attained before menarche could occur or menses resume in secondary amenorrhea due to weight loss. In a recent study of ballerinas, Frisch et al. (1980) found that those with menses were above the critical weight for height (as defined in her previous study) while those without menses were below it. In another study of ballerinas, Warren (1980) obtained different results. The amenorrhea that occurred during the training period appeared to be more closely related to exercise than to body fat. Menses tended to occur during periods of inactivity but without a change in body weight. There was a recurrence of amenorrhea in 11 of 13 girls when exercise was resumed—again without a weight change. All with delayed menarche had attained a body weight within the range where menses should occur according to the criteria of Frisch and McArthur (1974). At the time of amenorrhea, whether primary or secondary, prepubertal LH, FSH, and estrogen levels were reported.

Since part of the clinical syndrome of AN is a compulsion to strenuous exercise, this may play an important role in the suppression of gandotropins observed in these patients.

## 7. AFTER WEIGHT GAIN

The endocrine changes that occur in AN are reversible. However, a return to normal ovulatory menses may not occur for a prolonged period after weight

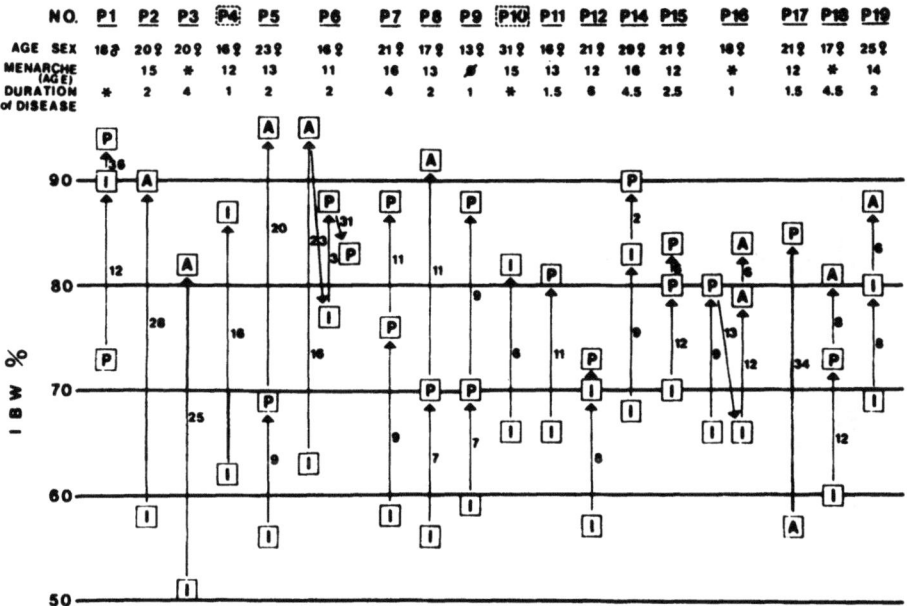

FIGURE 7. Twenty-four-hour sleep–wake pattern of LH in 16 patients with anorexia nervosa, in one patient with schizophrenia (P.4), and one patient with a gastric ulcer (P10). The 24-hr studies are arranged according to the percent of ideal body weight (IBW) at the time of the study. The numbers beside the arrows connecting the studies indicate the time in weeks which elapsed between studies. I infantile, P pubertal, A adult LH pattern. The duration of the disease is given in years. Asterisks indicate missing data; ø indicates primary amenorrhea. (From Pirke et al., 1979.)

has reached the normal range. Thus, studies of estrogen levels at the time when near IBW has been achieved report both normal (Beumont *et al.*, 1976) and subnormal levels (Sherman *et al.*, 1975; Wakeling *et al.*, 1977). Similarly, plasma LH levels have been reported as both normal (Beumont *et al.*, 1976; Marshall and Fraser, 1971; Palmer *et al.*, 1975; Rudolf *et al.*, 1979) and low (Isaacs *et al.*, 1980; Sherman *et al.*, 1975). Basal plasma FSH levels have consistently been reported as normal (Beumont *et al.*, 1976; Isaacs *et al.*, 1980; Palmer *et al.*, 1975; Sherman *et al.*, 1975). It is not possible from the data as presented in these studies to assess whether the duration that IBW is maintained is an important factor in determining whether or not estrogen or gonadotropin secretion returns to normal.

The LH and FSH responses to LRH improve with weight gain—the FSH response often being supranormal during the intermediate stages (Beumont *et al.*, 1976; Sherman *et al.*, 1975). Response to clomiphene (Clomid) is usually (Beumont *et al.*, 1976; Marshall and Fraser, 1971; Travaglini *et al.*, 1976; Wakeling *et al.*, 1976) but not always (Rudolf *et al.*, 1979) normal.

Pirke *et al.* (1979) showed that, in general in AN, the secretory pattern of LH becomes more mature with weight gain (see Fig. 7). However, it is important to note that three cases maintained an infantile pattern in spite of being

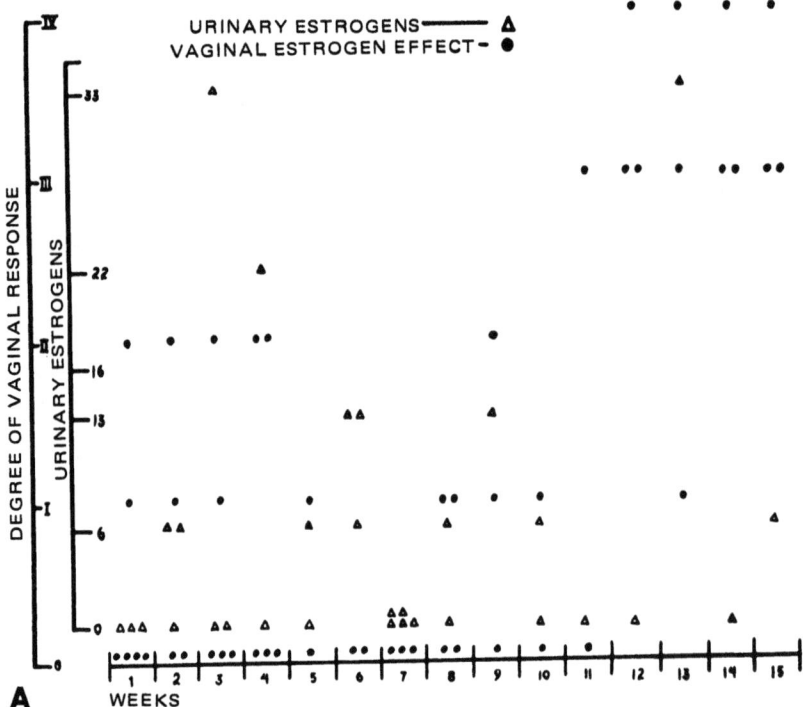

FIGURE 8. (A) Estrogen activity during recovery in several women of menstrual age. Time in weeks is represented along the abscissa. The ordinates correspond to the degree of vaginal response (grades I to IV) or to urinary estrogen excretion expressed in mouse units. (B) Pituitary gonado-

more than 80% IBW (cases 1, 10, 14). The duration that weight was maintained did not appear to be a factor. Case 19 passed from an infantile pattern at 80% IBW to an adult pattern at 90% IBW in a time of 6 weeks. Case 1 reverted from a pubertal to an infantile pattern in 12 weeks in spite of being 90% IBW. Thirty-six weeks later she still had a pubertal pattern.

In a similar study of eight patients who had gained weight after AN and who approached IBW, Katz *et al.* (1978) observed that an adult secretory pattern occurred only in those who were symptomatically improved with respect to their ideation about food, i.e., two of eight patients. However, both remained amenorrheic while three patients with mid- or late-pubertal patterns had irregular menses. There was no correlation between the percentage of body weight as fat and either maturity of the gonadotropin secretory pattern or onset of menses. Estrogen levels were not reported.

After recovery from malnutrition or dieting, gonadal endocrine function also returns to normal. In the Mexican study, Zubiran and Gomez-Mont (1953) demonstrated a progressive rise in both estrogen activity and urinary gonadotropins during refeeding (Fig. 8). Studies of dieters also show a rise in estrogen levels (Lev-Ran, 1974; McArthur *et al.*, 1976) and plasma LH (Lev-Ran, 1974). Lev-Ran (1974), however, observed that although these patients may respond

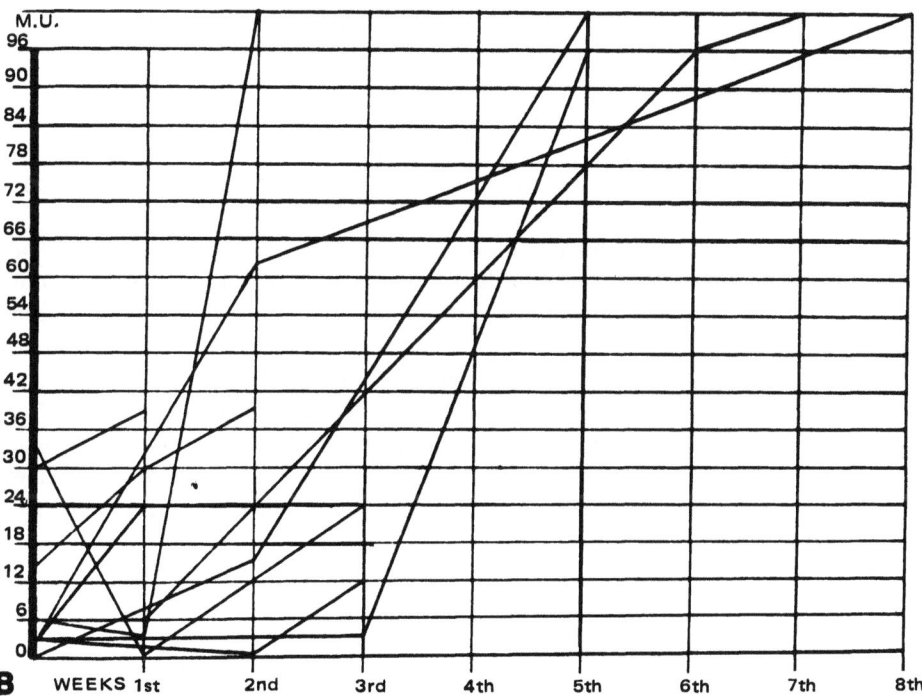

tropin excretion during recovery from malnutrition in women in menstrual age. Each line graph represents a different case. Time in weeks after admission to the hospital is represented along the abscissa. Gonadotropin excretion in mouse units per 24 hr is expressed along the ordinates. (From Zubiran and Gomez-Mont, 1953.)

normally to Clomid, spontaneous menses can be delayed for as long as 20–26 months after weight gain. McArthur and co-workers reported a lag of 2–11 months before menses returned in nine dieters after refeeding. Although Zubiran and Gomez-Mont (1953) did not report when menses returned in the 25 malnourished women they studied, the example they gave as typical had a return of menses with 14 weeks of refeeding at a weight of approximately 45 kg (99 lb).

Thus, from the published data, it is difficult to determine how long it takes for menses to resume after inadequate caloric intake in those who do not have the abnormal food attitudes of the anorectic.

## 8. SUMMARY

AN is a disease of self-imposed starvation which occurs mainly in adolescent girls. (Ten percent of cases occur in boys with similar endocrine manifestations, i.e., testicular atrophy and impotence.) Most of the clinical and laboratory features are also found in simple starvation, i.e., low metabolic rate, abnormal thermoregulation, partial diabetes insipidus, low $T_3$, high GH, low somatomedins, high plasma cortisols, and gonadal atrophy. The special features found in those suffering from AN which do not occur in SWL are the obsession with thinness and the distorted image these patients have of themselves. Even when frankly cachectic, they are convinced that they are still too fat. They are proud of the control they have over themselves in achieving this extraordinary feat and are terrified of losing control and becoming obese. In spite of the label given to the disease, they are hungry and obsessed with food rather than being anorectic.

The fact that 50–90% have been reported as having amenorrhea before or concurrent with the onset of weight loss indicates that weight loss alone is not responsible for all the manifestations. It seems not unreasonable to postulate that at the time of puberty, when there is a changing endocrine and metabolic milieu, the central neurotransmitters would be vulnerable to minor alterations which with time would mature to a normal level. This would account for the variable severity of disease which ranges from mild to fatal. It would also account for the fact that after prolonged follow-up (i.e., 15–25 years), the majority become normal with respect to body weight and fertility (Beck and Brochner-Mortensen, 1954).

## 9. REFERENCES

Altman, N., Sachar, E. J., Gruen, P. H., Halpern, F. S., and Eto, S., 1975, Reduced plasma LH concentration in postmenopausal depressed women, *Psychosom. Med.* **37**:274.

Aona, T., Kinugasa, T., Yamamoto, T., Miyake, A., and Kurachi, K., 1975, Assessment of gonadotrophin secretion in women with anorexia nervosa, *Acta Endocrinol.* (Copenh)**80**:630.

Beumont, P. J. V., George, G. C. W., Pimstone, B. L., and Vanik, A. I., 1976, Body weight and the pituitary response to hypothalamic releasing hormones in patients with anorexia nervosa, *J. Clin. Endocrinol. Metab.* **43**:487.

Beck, J. C., and Brochner-Mortensen, K., 1954, Observations on the prognosis in anorexia nervosa, *Acta Med. Scand.* **149**:409.

Bell, E. T., Harkness, R. A., Loraine, J. A., and Russell, G. F. M., 1966, Hormone assay studies in patients with anorexia nervosa *Acta Endocrinol.* (Copenh) **51**:140.

Boyar, R. M., and Katz, J., 1977, Twenty-four hour gonadtropin secretory patterns in anorexia nervosa, in: *Anorexia Nervosa* (R. A. Vigersky, ed.), p. 186, Raven Press, New York.

Boyar, R. M., Katz, J., Finkelstein, J. W., Kapen, S., Weiner, H., Weitzman, E. D., and Hellman, L., 1974, Immaturity of the 24-hour luteinizing hormone secretory pattern, *N. Engl. J. Med.* **291**:861.

Bruni, J. F., Van Vugt, D., and Marshall, S., 1977, Effects of naloxone, morphine and methionine enkephalin on serum prolactin, luteinizing hormone, follicle stimulating hormone, thyroid stimulating hormone and growth hormone, *Life Sci.* **21**:461.

Charters, A. C., Odell, W. D., and Thompson, J. C., 1969, Anterior pituitary function during surgical stress and convalescence: Radioimmunoassay measurement of blood TSH, LH, FSH, and growth hormone, *J. Clin. Endocrinol. Metab.* **29**:63.

Chopra, I. J., and Smith, S. R., 1975, Circulating thyroid hormones and thyrotropin in adult patients with protein-calorie malnutrition, *J. Clin. Endocrinol. Metab.* **40**:221.

Crisp, A. H., and Stonehill, E., 1971, Relation between aspects of nutritional disturbance and menstrual activity in primary anorexia nervosa, *Br. Med. J.* **3**:149.

Dale, E., Gerlach, D. H., and Wilhite, A. L., 1979, Menstrual dysfunction in distance runners, *Obstet. Gynecol.* **54**:47.

Feicht, C. B., Johnson, T. S., Martin, B. J., Sparkes, K. E., and Wagner, W. W., Jr., 1978, Secondary amenorrhea in atheletes, *Lancet* **II**:1145.

Fishman, J., Boyar, R. M., and Hellman, L., 1975, Influence of body weight on estradiol metabolism in young women, *J. Clin. Endocrinol. Metab.* **41**:989.

Frisch, R. E., and McArthur, J. W., 1974, Menstrual cycles: Fatness as a determinant of minimum weight for height necessary for thier maintenance or onset, *Science* **185**:949.

Frisch, R. E., Wyshak, G., and Vincent, L., 1980, Delayed menarche and amenorrhea in ballet dancers, *N. Engl. J. Med.* **303**:17.

Gross, H. A., Lake, C. R., Ebert, M. H., Ziegler, M. G., and Kopin, I. J., 1979, Catecholamine metabolism in primary anorexia nervosa, *J. Clin. Endocrinol. Metab.* **49**:805.

Grumbach, M. M., Roth, J. C., Kaplan, S. L., and Kelch, R. P., 1974, Hypothalamic–pituitary regulation of puberty in man: Evidence and concepts derived from clinical research, in: *Control of the Onset of Puberty* (M. M. Grumbach, G. D. Grave, and F. E. Mayer, eds.), pp. 115–166, Wiley, New York.

Halmi, K., 1974, Anorexia nervosa: Demographic and clinical features in 94 cases, *Psychosom. Med.* **36**:18.

Hurd, H. P., II, Palumbo, J., and Gharib, H., 1977, Hypothalamic–endocrine dysfunction in anorexia nervosa, *Mayo Clin. Proc.* **52**:711.

Issacs, A. J., Leslie, R. D., Gomez, J., and Bayliss, R., 1980, The effect of weight gain on gonadotrophins and prolactin in anorexia nervosa, *Acta Endocrinol.* (Copehn) **94**:145.

Jaffe, R. B., and Keye, W. R., Jr., 1974, Estradiol augmentation of pituitary responsiveness to gonadotropin-releasing hormone in women, *J. Clin. Endocrinol. Metab.* **39**:850.

Jarrell, J., Meltzer, S., and Tolis, G., 1979, Anorexia nervosa: A review of the endocrine abnormalities in the hypothalamic–pituitary axis, in: *Clinical Neuroendocrinology: A Pathological Approach* (G. Tolis, F. Labrie, J. B. Martin, and F. Naftolin, eds.), pp. 355–365, Raven Press, New York.

Jeuniewic, N., Brown, G. M., Garfinkel, P. E., and Moldofsky, H., 1978, Hypothalamic function as related to body weight and body fat in anorexia nervosa, *Psychosm. Med.* **40**:187.

Judd, S. J., Rakoff, J. S., and Yen, S. S. C., 1978, Inhibition of gonadotropin and prolactin release by dopamine: Effect of endogenous estradiol levels, *J. Clin. Endocrinol. Metab.* **47**:494.

Kanis, J. A., Brown, P., Fitzpatrick, K., Hibbart, D. J., Horn, D. B., Nairn, I. M., Shirling, D.,

Strong, J. A., and Walton, H. J., 1974, Anorexia nervosa: A clinical, psychiatric, and laboratory study. I. Clinical and laboratory investigation, *Q. J. Med.* **43**:321.

Kastin, A. J., Gual, C., and Schally, A. V., 1972, Clinical experience with hypothalamic releasing hormones, II. Luteinizing hormone-releasing hormone and other hypophysiotropic releasing hormones, *Recent Prog. Horm. Res.* **28**:201.

Katz, J. L., Boyar, R., Roffwarg, H., Hellman, L., and Weiner, H., 1978, Weight and circadian luteinizing hormone secretory pattern in anorexia nervosa, *Psychosom. Med.* **40**:549.

Kay, D. W. K., and Leigh, D., 1954, The natural history, treatment and prognosis of anorexia nervosa based on a study of 38 patients, *J. Ment. Sci.* **100**:411.

Lachelin, G. C. L., Leblanc, H., and Yen, S. S. C., 1977, The inhibitory effect of dopamine agonists on LH release in women, *J. Clin. Endocrinol. Metab.* **44**:728.

Leblanc, H., Lachelin, G. C. L., Abu-Fadil, S., and Yen, S. S. C., 1976, Effects of dopamine infusion on pituitary hormone secretion in humans, *J. Clin. Endocrinol. Metab.* **43**:668.

Lev-Ran, A., 1974, Secondary amenorrhea resulting from uncontrolled weight-reducing diets, *Fertil. Steril.* **25**:459.

McArthur, J. W., O'Loughlin, K. M., Beitins, I. Z., and Johnson, L., 1976, Endocrine studies during the refeeding of young women with nutritional amenorrhea and infertility, *Mayo Clin. Proc.* **51**:607.

Malina, R. M., Harper, A. B., Avent, H. H., and Campbell, D. E., 1973, Age at menarche in atheletes and nonatheletes, *Med. Sci. Sports* **5**:11.

Marshall, J. C., and Fraser, T. R., 1971, Amenorrhea in anorexia nervosa: Assessment and treatment with clomiphene citrate, *Br. Med. J.* **4**:590.

Mecklenberg, R. S., Loriaux, L., Thompson, R. H., Andersen, A. E., and Lipsett, M. B., 1974, Hypothalamic dysfunction in patients with anorexia nervosa, *Medicine (Baltimore)* **53**:147.

Mortimer, C. H., Besser, G. M., McNeilly, A. S., Marshall, J. C., Harsoulis, P., Tunbridge, W. M. G., Gomez-Pan, A., and Hall, R., 1973, Luteinizing hormone and follicle stimulating hormone-releasing hormone test in patients with hypothalamic–pituitary–gonadal dysfunction, *Br. Med. J.* **4**:73.

Nillius, S. J., Fries, H., and Wide, L., 1975, Successful induction of follicular maturation and ovulation by prolonged treatment with LH-releasing hormone in women with anorexia nervosa, *Am. J. Obstet. Gynecol.* **122**:921.

Ojeda, S. R., and McCann, S. M., 1978, Control of LH and FSH release by LHRH: Influence of putative neurotransmitters, in: *Clinics in Obstetrics and Gynecology: Neuroendocrinology of Reproduction*, Vol. 5, No. 2 (J. E. Tyson, ed.), pp. 283–303, Saunders, Philadelphia.

Palmer, R. L., Crisp, A. H., MacKinnon, P. C. B., Franklin, M., Bonnar, J., and Wheeler, M., 1975, Pituitary sensitivity to 50 µg LH/FSH-RH in subjects with anorexia nervosa and acute and recovery stages, *Br. Med. J.* **1**:179.

Pimstone, B., 1976, Endocrine function in protein-calorie malnutrition, *Clin. Endocrinol. (Oxford)* **5**:79.

Pirke, K. M., Fichter, M. M., Lund, R., and Doerr, P., 1979, Twenty-four hour sleep–wake pattern of plasma LH in patients with anorexia nervosa, *Acta Endocrinol.* (Copenh) **92**:193.

Quigley, M. E., Judd, S. J., Gilliland, G. B., and Yen, S. S. C., 1979, Effects of a dopamine antagonist on the release of gonadotropin and prolactin in normal women and women with hyperprolactinemic anovulation, *J. Clin. Endocrinol. Metab.* **48**:718.

Quigley, M. E., Sheehan, K. L., Casper, R. F., and Yen, S. S. C., 1980, Evidence for increased dopaminergic and opioid activity in patients with hypothalamic hypogonadotropic amenorrhea, *J. Clin. Endocrinol. Metab.* **50**:949.

Rudolf, K., Göretzlehner, G., and Kunkel, S., 1979, The gonadotropin releasing hormone stimulation and the clomiphene tests in female patients with anorexia nervosa, *Endokrinologie* **73**:287.

Sherman, B. M., Halmi, K. A., and Zumudio, R., 1975, LH and FSH response to gonadotropin-releasing hormone in anorexia nervosa: Effect of nutritional rehabilitation, *J. Clin. Endocrinol. Metab.* **41**:135.

Starkey, T. A., and Lee, R. A., 1969, Menstruation and fertility in anorexia nervosa, *Am. J. Obstet. Gynecol.* **105**:374.

Stubbs, W. A., Delitala, G., Jones, A., Jeffcoate, W. J., Edwards, C. R. W., and Ratter, S. J., 1978, Hormonal and metabolic responses to an enkephalin analogue in normal man, *Lancet* **II**:1225.

Travaglini, P., Beck-Peccoz, P., Ferrari, C., Ambrosi, B., Paracchi, A., Severgnini, A., Spada, A., and Faglia, G., 1976, Some aspects of hypothalamic–pituitary function in patients with anorexia nervosa, *Acta Endocrinol.* (Copenh) **81**:252.

Vigersky, R. A., Loriaux, D. L., Andersen, A. E., Mecklenburg, R. S., and Vaitukaitis, J. L., 1976, Delayed pituitary hormone response to LRF and TRF in patients with anorexia nervosa and with secondary amenorrhea associated with simple weight loss, *J. Clin. Endocrinol. Metab.* **43**:893.

Vigersky, R. A., Andersen, A. E., Thompson, R. H., and Loriaux, D. L., 1977, Hypothalamic dysfunction in secondary amenorrhea associated with simple weight lose, *N. Engl. J. Med.* **297**:1141.

Wakeling, A., Marshall, J. C., Beardwood, C. J., De Souza, V. F. A., and Russell, G. F. M., 1976, The effects of clomiphene citrate on the hypothalamic–pituitary–gonadal axis in anorexia nervosa, *Psychol. Med.* **6**:371.

Wakeling, A., De Souza, V., and Beardwood, C. J., 1977, Effects of administered estrogen on luteinizing hormone release in subjects with anorexia nervosa in acute and recovery stages, in: *Anorexia Nervosa* (R. A. Vigersky, ed.), pp. 199–209, Raven Press, New York.

Warren, M. P., 1980, The effects of exercise on pubertal progression and reproductive function in girls, *J. Clin. Endocrinol. Metab.* **51**:1150.

Warren, M. P., and Vande Wiele, R. L., 1973, Clinical and metabolic features of anorexia nervosa, *Am. J. Obstet. Gynecol.* **117**:435.

Warren, M. P., Siris, E. S., and Petrovich, C., 1977, The influence of severe illness on gonadotropin secretion in the postmenopausal female, *J. Clin. Endocrinol. Metab.* **45**:99.

Yen, S. S. C., Vandenberg, G., and Siler, T. M., 1974, Modulation of pituitary responsiveness to LRF by estrogen, *J. Clin. Endocrinol. Metab.* **39**:170.

Yoshimoto, Y., Moridera, K., and Imura, H., 1975, Restoration of normal pituitary gonadotropin reserve by administration of luteinizing-hormone-releasing hormone in patients wtih hypogonadotropic hypogonadism, *N. Engl. J. Med.* **292**:242.

Zubiran, S., and Gomez-Mont, F., 1953, Endocrine disturbances in chronic human malnutrition, in: *Vitamins and Hormones: Advances in Research and Applications* (R. S. Harris, G. F. Marrian, and K. V. Thimann, eds.), Vol. II, pp. 97–129, Academic Press, New York.

CHAPTER 6

# Abnormalities in Anorexia Nervosa of Dexamethasone Suppression Test and Urinary MHPG Suggest a Norepinephrine Hypothesis

### HARRY E. GWIRTSMAN, ROBERT H. GERNER, and HARVEY A. STERNBACH

## 1. INTRODUCTION

### 1.1. Etiology of Primary Anorexia Nervosa

The illness of primary anorexia nervosa (PAN) has been considered from several etiological standpoints. Some authors have suggested that the syndrome is secondary to psychological conflicts within the patient or their family (Bruch, 1977; Minuchin *et al.*, 1978), others that it is a psychosomatic disorder (Weiner, 1977) with multiple etiologies, and yet others have considered it as a primary hypothalamic disturbance (Vigersky and Anderson, 1977; Minuchin *et al.*, 1978). There have been numerous studies and reviews (Walsh *et al.*, 1978) showing endocrinological abnormalities related to hypothalamic dysfunction during the acute phase of AN, although there is controversy about whether the endocrinological changes precede or follow the behaviors of PAN. A large group of anorexic patients have not previously been studied with the two biochemical tests that are currently in widespread use in clinical and research psychiatry: the dexamethasone suppression test (DST) and urine 3-methoxy-4-hydroxypenylglycol (MHPG).

---

HARRY E. GWIRTSMAN • National Institutes of Health, Bethesda, Maryland 20205. ROBERT H. GERNER • University of California, Irvine, California, and Long Beach Veterans Administration Hospital, Long Beach, California 90822. HARVEY A. STERNBACH • Department of Psychiatry, University of California at Los Angeles, Neuropsychiatric Institute, Los Angeles, California 90024, and Department of Psychiatry, Brentwood Veterans Administration Hospital, Los Angeles, California 90073.

## 1.2. Dexamethasone Suppression Test

The overnight DST is the most widely studied neuroendocrine test in the psychiatric disorders. The test involves the administration of an oral dose of dexamethasone, a potent synthetic glucocorticoid, at the "critical period" for circadian HPA (hypothalamic–pituitary–adrenal) organization, i.e., at 2300 hr or midnight, and measurement of plasma cortisol at some time during the next day. The normal circadian rhythm of cortisol has been studied by a number of investigators (Sachar, 1975; Carroll et al., 1976b). Throughout the day, cortisol secretion occurs in bursts, which taper off from 1600 hr → midnight and reach a trough at that point. During the morning hours plasma cortisol reaches a peak at approximately 0900 hr. The DST can demonstrate normal feedback inhibition, i.e., the presence of an increased amount of glucocorticoid will shut off the hypothalamic production of corticotropin-releasing factor (CRF), and this will cause a diminution of ACTH and, hence of cortisol production by the adrenal gland (see Fig. 1). An abnormal failure to suppress is generally accepted to be present if a plasma cortisol of greater than or equal to 5 μg/dl (done by competitive protein binding) or 4 μg/dl (by radioimmunoassay) occurs within a 24-hr period after the dose of dexamethasone has been given (Carroll et al., 1976b). The long plasma half-life of dexamethasone (3–6 hr) probably explains why a single dose of the substance can have prolonged action on the HPA axis (Carroll et al., 1976a).

The DST has been studied extensively by several groups of investigators; they have shown that between 39 and 53% of patients with endogenous types of depression fail to suppress their cortisol after dexamethasone (Carroll et al., 1981; Schlesser et al., 1980; Gold et al., 1981). Failure of depressed patients to suppress cortisol is consistent with a CNS deficit of norepinephrine (NE)

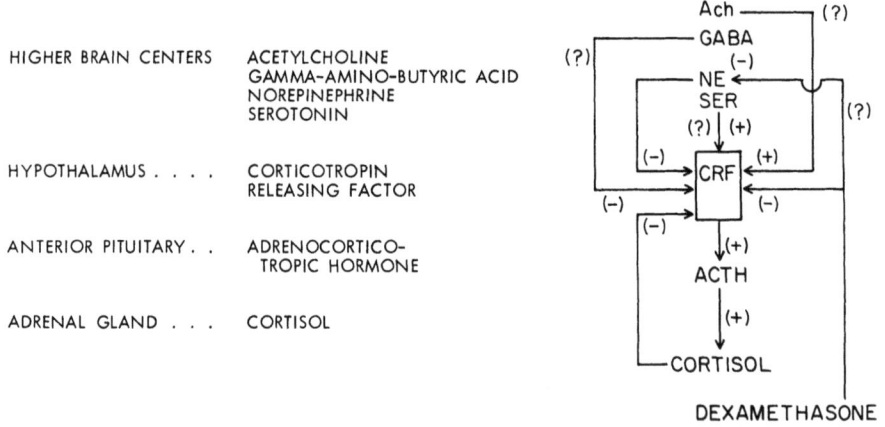

FIGURE 1. Relationships within the hypothalamic–pituitary–adrenal axis. Excitatory relationships, (+); inhibitory relationships, (−); relationships that are as yet insufficiently studied, (?).

(at least in the hypothalamus), since CRF is inhibited by NE and therefore low or inadequate levels of NE would result in increased production of CRF with consequent higher levels of cortisol (Schlesser *et al.*, 1979).However, other neurotransmitters are also involved in CRF regulation and thus an abnormal DST cannot by itself be used to argue for a decrease in central NE. As proposed by Jones *et al.* (1976), γ-aminobutyric acid (GABA) also inhibits CRF secretion, while serotonin and acetylcholine stimulate CRF. The relationship of these transmitters to the DST and depression has been addressed by Sachar *et al.* (1980a).

In AN, it has been demonstrated that the circadian rhythm of cortisol production is probably preserved (Boyar *et al.*, 1977). However, the actual rate of cortisol secretion and the ability of the adrenal gland to respond to an ACTH stimulus is enhanced (Van de Weile, 1977; Walsh *et al.*, 1978). Two preliminary studies have been published to date suggesting that anorexic patients may show nonsuppression of their adrenal response when challenged with dexamethasone. Bethge *et al.* (1970) studied 66 patients with PAN, of whom 14 had a DST using 3 mg of dexamethasone with a cortisol level at 0900 hr. They found nonsuppression of cortisol in 4 of 8 patients 51–55% underweight and in 1 of 6 patients 20–31% underweight. They reported an increase in plasma cortisol significantly correlated with weight and concluded that their findings are most consistent with a disturbance of central regulation of ACTH secretion.

Walsh *et al.* (1978) carried out the DST using 1 mg of dexamethasone and measured cortisol at 0800 hr in 3 of 19 patients with PAN and found nonsuppression of cortisol in all 3. They also reported the cortisol production rate was increased in PAN and that these cortisol changes are *not* found in uncomplicated malnutrition and are intrinsic to PAN. They hypothesized that this could be due to a "functional" hypothyroidism although this was not evaluated. Although these authors did not use the DST, Boyar *et al.* (1977)found a prolonged half-life of cortisol from 50 to 78 min in 10 PAN patients. His group did not find a change in cortisol production. They did find low $T_3$ in this group and speculated that the change in cortisol metabolism may be due to a malnourished individual's decrease in $T_3$. The latter investigators were able to normalize cortisol production rate by administering $T_3$ in five of these "hypothyroid" patients with PAN.

### 1.3. Urinary MHPG

More recently, urinary MHPG, a metabolite of NE, has been measured both during the acute phase and after refeeding in two studies with a total of 35 patients. Patients with PAN appear to have lower MHPG during the acute phase of illness with a return to higher levels when body weight attains 85–90% of ideal. However, neither all patients nor the group as a whole achieve normal levels after regaining weight (Halmi *et al.*, 1978; Gross *et al.*, 1979). Bipolar depressed patients also appear to have low urine MHPG (Edwards *et al.*, 1980). Although a significant amount of urinary MHPG does not come from brain NE (Blombery et al., 1980), this test remains the most clinically applicable measure of central NE turnover.

### 1.4. Relationship of Primary Anorexia Nervosa and Affective Disorder

It has been observed that many patients with PAN are clinically depressed. This depression has usually been thought to be secondary to the primary illness. Since many patients with PAN also develop depression in later life (Cantwell et al., 1977), it is possible that these preliminary DST and MHPG results are related more to a state of depression than to PAN itself, as both abnormal DST and low MHPG are found in depression. If PAN patients who are not depressed had altered MHPG and DST, this would suggest that abnormalities are related to the PAN and not a state of depression. This is the first study to investigate the relationship of the DST and urinary MHPG to the acute phase of illness and secondary characteristic of mood.

## 2. METHODS

Thirty-seven patients with PAN who were admitted to the Neuropsychiatric Institute were subjects for this investigation which was part of their standard evaluation. The patients ranged in age from 13 to 41 with a mean age of 24.4 ($\pm$ 1 S.E.M.). Only one was male. All patients fulfilled the Feighner criteria for PAN, and had an onset before age 30 with development of symptoms in the absence of medical illness or psychosis (Feighner et al., 1972). Ideal body weight (IBW) was calculated by using standard tables (Diem, 1962). Prior to initiation of a behavioral treatment program, patients were assessed on multiple parameters while stabilized on a locked inpatient unit for a week in order to establish baseline behavioral data. During the time, their weight and eating behavior were monitored. None of the patients at this time continued to lose weight but were either stable or had slight increases in their weight. Thyroid function tests (free $T_4$, $T_4$ index, $T_3$, TSH) as well as SMA-6, 12, CBC, and urinalysis were obtained. At the end of the week, when patients were not in a catabolic state, 20 patients collected a 24-hr urine for MHPG in polyethylene containers to which 500 mg of sodium metabisulfite had been added. Specimens were refrigerated during collection and then aliquots were frozen and sent for assay to the National Psychopharmacology Laboratory, Knoxville, Tennessee, where assays in duplicate were performed using previously described techniques (Dekirmenjian and Maas, 1970). Patients were kept on the ward in order to ensure reliable collection. Two patients had low volume and are not included in the data.

All patients had a creatinine/weight ratio of 14 mg/kg or greater and five patients had two collections and the average of MHPG values was used. The creatinine/weight ratio has previously been accepted as an indication of complete collection (Edwards et al., 1980). We included three patients without creatinine data because close staff scrutiny and their urine volumes (1–2.5 liters) convinced us that the collections were complete. Following this, a DST modified from the method of Carroll et al. (1976b) was obtained (1 mg of dexamethasone was given orally at 2300 hr and plasma cortisol measured by radioimmunoassay was obtained between 1600 and 1700 hr the following day). Patients were drug-free for a week or more, except as mentioned. Thorough family histories were taken from the patient and at least one of their close relatives. Patients were either diagnosed as having a minor or major depression using the DSM-III, or were subclassified as hyperactive-type if they did not have depressed symptomatology and exhibited a compulsive desire to jog, swim, do calisthenics, run stairs, etc. Patients who were neither depressed nor hyperactive were called "other." No patients were both depressed and hyperactive. Three drug-free schizophrenics who had starved themselves to 70–80% of IBW also received the DST to determine if weight loss alone might cause a positive DST. Data were analyzed by $t$ test and Pearson's $r$.

## 3. RESULTS

### 3.1. Dexamethasone Suppression Test

Thirty-three patients completed the DST. Two of these suppressed to < 1 μg but were on drugs which probably invalidated the results (preludin and thyroid in one patient and short-term steroid therapy in another) and these patients were not included in the biochemical data analysis. The 26 female primary anorexics who were then < 80% of IBW demonstrated a uniform and marked abnormality in the DST (cortisol level ≥ 5 μg). The average cortisol of these patients after DST was 15.4 ± 1.45 μg and the average percent of

TABLE 1
DST and Urinary MHPG in Anorexia Nervosa and Schizophrenia: Relationship to Body Weight

|  | PAN[a] < 80% IBW | PAN[a] ≥ 80% IBW | Schizophrenia < 80% IBW |
|---|---|---|---|
| DST $N = 35$ |  |  |  |
| Positive ($\geq 5$ μg/dl) | 26 | 2 | 0 |
| Negative ($< 5$ μg/dl) | 2 | 2 | 3 |
| MHPG $N = 18$ |  |  |  |
| Low ($< 1196$ μg/24 hr) | 13 | 2 | — |
| Normal ($\geq 1196$ μg/24 hr) | 1 | 2 | — |

[a] Females only.

IBW in these patients was 66 ± 1.4. Two patients had cortisol levels below threshold, i.e., 1.3 μg (71% IBW) and 4.6 μg (79% IBW) (see Table 1). When the patients were divided according to whether they were depressed ($N = 12$), hyperactive ($N = 5$), or other ($N = 9$), there were no significant differences of cortisol after DST between groups. When the DST cortisol was correlated with the percent of IBW, no significant relationship was found. However, when 13 patients who were less than 65% of IBW were compared to 12 with greater than 76% of IBW (this includes the male and two patients retested after weight gain; see below), cortisol levels were significantly higher in patients with lower weight (17.1 ± 2.2 vs. 5.2 ± 1.5; $p < 0.001$).

The three schizophrenics all had normal suppression after dexamethasone to < 2 μg.

### Follow-up DST

DST was repeated after some degree of weight gain in six patients, three of whom also had DSTs when < 65% of IBW. On retesting, four patients had cortisol levels of less than 1 (e.g., normal). Their weights were 70%, 70%, 81%, and 93%. One patient at 80% had a cortisol of 5.3 μg and another at 90% had a cortisol of 7.4 μg. The two patients with 70% of IBW had gained back 10% of their ideal weight at the time of the repeat DST. However, they were not exhibiting any symptoms of anorexia at the time such as obsessions with food, vomiting, running, bulimia, etc., even though their weight was low. The patient with a cortisol of 7.4 μg (90% of IBW) was still having obsessions about food and was not free of concern over her body weight and configuration.

### 3.2. Thyroid Function

Thyroid function tests were obtained in 33 patients. A mild decrease in thyroid function was found in 5 patients as measured by either $T_4$ RIA or free $T_4$ index, and $T_3$ was just below normal in 2 of the 18 patients where it was determined. However, in every instance of low thyroid function test, one of the other indices of thyroid function was normal ($T_4$, $T_3$, or TSH). TSH levels in three of the mildly hypothyroid patients were normal (<2) and TSH was not measured in the other "low-thyroid" patients. All patients with low thyroid function were also of low weight (58% to 71%). Four of the five had high cortisol (17–19 μg) and the other had received steroids. MHPG obtained on three of the drug-free "low-thyroid" patients was not different from that of the group as a whole.

### 3.3. Urinary MHPG

Complete 24 hr urine MHPG levels were obtained on 18 of the 37 patients. For the 14 female patients whose weight was less than 80% of ideal ($\bar{x} = 65\%$ ± 2.0), the MHPG was 877 ± 101 μg. This is substantially lower than published

### TABLE 2
### Relationship between DST and Urinary MHPG in Anorexia Nervosa

| DST | MHPG | |
|---|---|---|
| | Low (< 1196 μg/24 hr) | Normal (≥ 1196 μg/24 hr) |
| Positive (≥ 5 μg/dl) | 14 | 1 |
| Negative (< 5 μg/dl) | 0 | 1 |

female controls assayed by the same laboratory (1196 μg) (Halmi *et al.*, 1978). All of these patients were nonsuppressors of cortisol in the DST ($\bar{x} = 16.0 \pm 1.6$) (see Table 2). Within this group, the mean MHPG for four hyperactive types was 881 μg and for six depressed, 880 μg. The urinary MHPG for the male (77% IBW) was low (883 μg). There was no significant correlation of urine MHPG levels and postdexamethasone cortisol levels nor between 24-hr urinary MHPG and percent of IBW for the total group. Urine MHPG was obtained in four patients who had gained to > 80% of IBW; their levels were: 1938, 1425, 1026, and 687 μg. One depressed type who gained from 50% to 70% did not show an increase in MHPG.

### 3.4. Family History

Family history was obtained for 31 patients and revealed psychiatric illness in first-degree relatives in 19 of the patients. Five of these had two or more members with psychiatric illness. Nine patients had a first-degree relative who was an alcoholic (eight fathers and one mother) and six patients had a first-degree relative with primary affective disorder. Four patients had a first-degree relative hospitalized for unknown psychiatric illness. No patients had first- or second-degree relatives with schizophrenia.

## 4. DISCUSSION

This investigation extends previous studies suggesting abnormalities in patients with PAN of the DST and of NE metabolism as measured by urinary MHPG. We have demonstrated the occurrence of these in the same patients. These abnormalities were not related to subtyping of anorexics as "hyperactive," "depressed," or "other." Thus, we conclude that these abnormalities are intrinsic to PAN and not to a concomitant mood state. We also found these abnormalities in patients regardless of our measures of thyroid function, which was marginally low in only 6 of 37 patients. All but 1 of the 24 patients who were less than 75% of IBW had abnormal DSTs, while the DST was abnormal in 5 of 13 patients who were greater than 75% of IBW and was marginal (4.6 µg) in one. Similar to previous reports, urinary MHPG was low for the group of female patients with weights < 80% of ideal compared to norms assayed identically (Halmi et al., 1978). Although we do not have data of a group of truly recovered PANs, two of four who gained to > 80% normalized on DST and two others normalized on urine MHPG.

Our investigation is consistent with other studies, suggesting an association of affective disorders (Bethge et al., 1970) to AN by the high incidence of their first-degree relatives of either affective disorders or alcoholism, both of which are increased in relatives of patients with primary affective disorders, especially depressive spectrum disease (Winokur, 1978).

Regulation of the HPA system, which involves CRF, ACTH, and cortisol, seems to have many fascinating relationships with the catecholamines and indoleamines. NE is known to exert a tonic inhibition upon the HPA axis, probably acting via CRF (Carroll, 1978; Schlesser et al., 1979; Sachar et al., 1980a,b; Jones et al., 1976) and there have been attempts to explain the inability of the system to respond to negative feedback by dexamethasone as a disinhibition caused by a deficiency of NE in the limbic–hypothalamic pathways (Collu, 1977). (See Fig. 1.) There is evidence that the circadian rhythm for NE secretion is a mirror image of cortisol rhythms, as would be expected if NE were tonically inhibiting cortisol secretion (Sachar, 1975).

Work by Sachar confirms a major NE role in regulating cortisol secretion. The infusion of $D$-amphetamine into monkeys and depressed patients caused a prompt normalization (within 90 min) of cortisol hypersecretion (Sachar et al., 1980b). It is unlikely that DA was involved because this effect was not abolished in monkeys by pretreatment with the DA blocker pimozide. We think it probable that the abnormal DSTs were mediated through functionally low NE levels in the hypothalamus since NE in the hypothalamus inhibits CRF, which itself stimulates ACTH release resulting in increased cortisol production as reported by Walsh et al.(1978) and Bethge et al. (1970). Such a mechanism is also consistent with the catecholamine hypothesis of depression, since some depressions are associated with abnormal DST and low MHPG. However, our results were found in both depressed and nondepressed PANs, strongly suggesting that we were not merely finding effects due to a depressive affective state. Additionally, one cannot exclude a contribution to the DST results from a possible alteration of other neuromodulators of CRF (i.e., increased serotonin or ACh, or decreased GABA).

This cannot be considered a conclusive study for several reasons. Our patients did not usually collect multiple urine samples for MHPG analysis. Although specific efforts were made to ensure that samples were complete and day-to-day variability in urine MHPG may not be great, it would be important to do multiple urine samples for MHPG in order to confirm our findings. Another reason for abnormal DST could be due to increased half-life of cortisol in these patients rather than to any abnormality in CNS metabolism. We think that this is unlikely since the half-life of cortisol in the body of anorexic patients has been shown to be 78 min (Boyar et al., 1977) and that over the 17–18 hr after the administration of dexamethasone there would be ample time for cortisol levels to suppress if synthesis of cortisol had truly been shut off, as pointed out by Walsh et al. (1978). One might suppose that the finding of increased and deregulated cortisol production is due to "malnutrition" but this has been previously shown not to be the case (Walsh et al., 1978). In point of fact, studies of patients with protein-calorie malnutrition have shown that cortisol production is either normal (Alleyne and Young, 1967) or decreased (Smith et al., 1975). Further, our patients were not starving, but were either at stable low weights or had gained a few pounds from time of admission to the hospital. In addition, we found normal DST in the three drug-free schizophrenics who had starved themselves, suggesting that low weight per se does not necessarily cause DST nonsuppression. Finally, the DST and MHPG abnormalities were found in patients both at moderate as well as low weights. We did not find that the abnormalities in DST and MHPG were correlated with patient weight although some patients who achieved near-normal weights had normalization of these parameters. Both biochemical parameters vary considerably, and thus significant associations could be missed by this methodology. Further studies should carry out repeated DST and MHPG in order to more clearly determine their relationship to weight, behavioral change, family history, and neuroendocrine parameters of PAN.

Although we believe PAN to be a distinct entity and not invariably associated with depressed mood, we think it likely that the high incidence of affective disorders in family members (Winokur et al., 1980), a high liability for depression in recovered patients, abnormal DST, and low MHPG during the anorexic illness, are all consistent with the vulnerability to dysfunction of NE regulation in these patients. This hypothesis is complemented by the finding that decrements in central NE in monkeys have been shown to interfere with satiety (Redmond et al., 1977). Further, Gerner et al. (1983) have found low tyrosine, but not another diet-dependent amino acid, tryptophan, in anorexic CSF. Since NE synthesis is dependent to some extent on tyrosine levels (Gibson and Wurtman, 1978; Carlsson and Lindqvist, 1978), low tyrosine may be implicated in a hypothesis of low central NE in PAN. We do not know why a decrease in NE would be associated with depression at one time and PAN at another, and indeed, we have not demonstrated an etiologic relationship between low NE and PAN. It is interesting, though, that the peak incidence of PAN and the depressive disorders occur at different points in the life cycle, and one follow-up study of PAN has shown that many of these patients go on to develop full-blown depressive disorders (Cantwell et al., 1977). One is led to speculate that these two disorders, linked by genetic, biochemical, and behavioral parameters, may represent the expressed variants of a common diathesis in a population which is highly vulnerable to both a depressive and an eating disorder.

In view of the poor prognosis for full recovery of many patients with PAN

(Hsu, 1980; Cantwell *et al.*, 1977; Pertshuk, 1977; Crisp *et al.*, 1977), it would be important to evaluate whether return to normal endocrine and MHPG values are possible predictors of a stable remission. It is important to consider that recovery from PAN may not be absolutely linked to weight. Some patients who gain weight still have many behavioral symptoms of AN (and abnormal MHPG and DST), while others who may not fully reestablish their ideal weights have recovered from the multitude of behavioral abnormalities associated with this disorder (and may have normal MHPG and DST). Future studies should assess both behavioral manifestations of the disorder, using sophisticated rating scales, as well as weight, not limiting their independent variable to weight alone. As we have suggested here, biological parameters of the "illness" of AN may not be correlated with weight per se, and a focus on the latter may hide important findings in these patients.

## 5. SUMMARY

Thirty-seven patients with PAN were assessed on variables of weight, DST, urine MHPG, thyroid function, mood, and family history. Patients at less than 75% of IBW had low urine MHPG and abnormally high cortisol after DST regardless of their mood state. Thyroid dysfunction was found in only 6 of 35 patients. Forty percent of the patients had a first-degree relative with primary affective disorder or alcoholism. It is hypothesized that low NE is intrinsic to PAN and is responsible for the urine MHPG and DST abnormalities.

## 6. REFERENCES

Alleyne, G. A., and Young, V. H., 1967, Adrenocortical function in children with severe protein–calorie malnutrition, *Clin. Sci.* **33**:189.

Bethge, H., Nagel, A. M., Solbach, H. G., Wiegelmann, W., and Zimmerman, H., 1970, Disturbance of cortisol regulation of adrenocorticol function in anorexia nervosa, parallels to endogenous depression and Cushing's syndrome, *Mat. Med. Nordmark* **22**:204.

Blombery, P. A., Kopin, I. J., Gordon, E. K., Markey, S. P., and Ebert, M. H., 1980, Conversion of MHPG to vanillylmandelic acid, *Arch. Gen. Psychiatry* **37**:1095.

Boyar, R. M., Hellman, L. D., Roffwang, H., Katz, J., Zumoff, B., O'Connor, J., Bradlow, H. L., and Fukushima, D. K., 1977, Cortisol secretion and metabolism in anorexia nervosa, *N. Engl. J. Med.* **296**:190.

Brunch, H., 1977, Psychological antecedents of anorexia nervosa, in: *Anorexia Nervosa* (R. A. Vigersky, ed.), pp. 1–11, Raven Press, New York.

Cantwell, D. P., Sturzenberger, S., Burroughs, J., Salkin, B., and Green J. K., 1977, Anorexia nervosa: An affective disorder, *Arch. Gen. Psychiatry* **34**:1087.

Carlsson, A., and Lindqvist, M., 1978, Dependence of 5-HT and catecholamine synthesis on concentrations of precursor amino acids in rat brain, *Naunyn-Schmiedebergs Arch. Pharmacol.* **303**:157.

Carroll, B. J., 1978, Neuroendocrine function in psychiatric disorders, in: *Psychopharmacology: A Generation of Progress* (M. A. Lipton, A. DiMascio, and K. F. Killam, eds.), pp. 487–497, Raven Press, New York.

Carroll, B. J., Curtis, G. C., and Mendels, J., 1976a Neuroendocrine regulation in depression. I. Limbic system adrenocorticol dysfunction, *Arch. Gen. Psychiatry* **33**:1039.

Carroll, B. J., Curtis, C. G., and Mendels, J., 1976b, Neuroendocrine regulation in depression, II. Discrimination of depression from nondepressed patients, *Arch. Gen. Psychiatry* **33**:1051.

Carroll, B. J., Feinberg, M., Greden, J. F., Tarika, J., Albala, A., Haskett, R. F., James, N. M., Kronfol, Z., Lohr, N., Steiner, M., de Vigne, J. P., and Young, E., 1981, A specific laboratroy test for the diagnosis of melacholia: Standardiaation, validation, and clinical utility, *Arch. Gen. Psychiatry* **38**:15.

Collu, R., 1977, Role of central cholinergic and aminergic neurotransmitters in the control of anterior pituitary hormone secretion, in: *Clinical Neuroendocrinology* (L. Martini and G. M. Besseri, eds.), pp. 43–65, Academic Press, New York.

Crisp, A. H., Kalucy, R. S., Lacey, J. H., and Harding, B., 1977, The long-term prognosis in anorexia nervosa: Some factors predictive of outcome, in: *Anorexia Nervosa* (R. A. Vigersky, ed.), pp. 55–65, Raven Press, New York.

Dekirmenjian, H., and Maas, J. W., 1970, An improved procedure of 3-methoxy-4-hydroxyphenyl-ethylene glycol determination by gas–liquid chromatography, *Anal. Biochem.* **35**:113.

Diem, K. (ed.), 1962, *Scientific Tables: Documenta Geigy*, 6th ed., p. 623, Geigy, Ardsley, N.Y.

Edwards, D. J., Spiker, D. G., Neil, J. F., Kupfer, D. J., and Rizk, M., 1980, MHPG excretion in depression, *Psychiatry Res.* **2**:295.

Feighner, J. P., Robins, E., Guze, S. B., Woodruff, R. A., Winokur, G., and Munoz, R., 1972, Diagnostic criteria for use in psychiatric research, *Arch. Gen. Psychiatry* **26**:57.

Gerner, R. H., Cohen, D. J., Fairbanks, J. G., Anderson, L., Young, G. M., Scheinin, M., Linnoila, M., Schaywitz, B. A., and Hare, T. A., 1983, CSF in anorexia nervosa compared to normal and depressed women, *Am. J. Psychiatry*, in press.

Gibson, C. J., and Wurtman, R. J., 1978, Physiological control of brain norepinephrine synthesis by brain tyrosine concentration, *Life Sci.* **22**:1399.

Gold, M. S., Pottash, A. L. C., Extein, I., and Sweeney D. R., 1981, Diagnosis of depression in the 1980's, *J. Am. Med. Assoc.* **245**:1562.

Gross, H. A., Lake, C. R., Ebert, M. H., Ziegler, M. G., and Kopin, I. J., 1979, Catecholamine metabolism in primary anorexia nervosa, *J. Clin. Endocrinol. Metab.* **49**:805.

Halmi, K. A., Dekirmenjian, H., Davis, J. M., Casper, R., and Goldberg, S., 1978, Catecholamine metabolism in anorexia nervosa, *Arch. Gen. Psychiatry* **35**:458.

Hsu, L. K. G., 1980, Outcome of anorexia nervosa, *Arch. Gen. Psychiatry* **37**:1041.

Jones, M. T., Hillhouse, E., and Burden, J., 1976, Secretion of corticotropin releasing hormone in vitro, in: *Frontiers in Neuroendocrinology*, Vol. 4 (L. Martini and W. F. Ganong, eds.),pp. 194–226, Raven Press, New York.

Minuchin, S., Rosman, B. L., and Baker, L., 1978, *Psychosomatic Families: Anorexia Nervosa in Context*, Harvard University Press, Cambridge, Mass.

Pertshuk, M. J., 1977, Behavior therapy: Extended follow-up, in: Anorexia Nervosa (R. A. Vigersky, ed.), pp. 305–313, Raven Press, New York.

Redmond, D. E., Jr., Huang, Y. H., Snyder, D. R., Maas, J. W., and Baulu, J., 1977, Norepinephrine and satiety in monkeys, in: *Anorexia Nervosa* (R. A. Vigersky, ed.), pp. 81–96, Raven Press, New York.

Sachar, E. J., 1975, Twenty-four-hour cortisol secretory patterns in depressed and manic patients, *Prog. Brain Res.* **42**:81.

Sachar, E. J., Asnis, G., Halbreich, U., Nathan, R. S., and Halpern, F. S., 1980a, Recent studies in the neuroendocrinology, in: *Psychiatric Clinics of North America*, Vol. 3 (E. J. Sachar, ed.), pp. 313–326, Saunders, Philadelphia.

Sachar, E. J., Asnis, G., Nathan, R. S., Harlbreich, U., Tabrizi, M. A., and Halpern, F. S., 1980b, Dextroamphetamine and cortisol in depression, morning plasma cortisol levels suppressed, *Arch. Gen. Psychiatry* **37**:755.

Schlesser, M. A., Winokur, G., and Sherman, B. M., 1979, Genetic subtypes of unipolar primary depressive illness distinguished by hypothalamic–pituitary–adrenal axis activity, *Lancet* **I**:739.

Schlesser, M. A., Winokur, G., and Sherman, B. M., 1980, Hypothalamic–pituitary–adrenal axis activity in depressive illness: Its relationship to classification, *Arch. Gen. Psychiatry* **37**:737.

Smith, S. R., Beldsoe, T., and Chhetri, M. K., 1975, Cortisol metabolism and pituitary–adrenal axis in adults with protein–calorie malnutrition, *J. Clin. Endocrinol. Metab.* **40**:43.

Van de Weile, R. L., 1977, Anorexia nervosa and the hypothalamus, *Hosp. Pract.* **12**:45.

Vigersky, R. A., and Anderson, A. E., 1977, Conclusion, in: *Anorexia Nervosa* (R. A. Vigersky, ed.), pp. 383–384, Raven Press, New York.

Walsh, B. T., Katz, J. L., Levin, J., Kream, J., Fukushima, D. K., Hellman, L. D., Weiner, H., and Zumoff, B., 1978, Adrenal activity in anorexia nervosa, *Psychosom. Med.* **40**:499.

Weiner, H., 1977, The psychobiology of human disease: An overview, in: *Psychiatric Medicine* (G. Usdin, ed.), pp. 3–72, Brunner/Mazel, New York.

Winokur, G., 1978, Mania and depression: Family studies and genetics in relation to treatment, in: *Psychopharmacology: A Generation of Progress* (M. A. Lipton, A. DiMascio, and K. F. Killam, eds.), pp. 1213–1221, Raven Press, New York.

Winokur, G., March, V., and Mendels, J., 1980, Primary affective disorders in relatives of patients with anorexia nervosa, *Am. J. Psychiatry* **137**:695.

CHAPTER 7

# Role of Brain Monoamines and Peptides in the Regulation of Male Sexual Behavior

## G. SERRA and G. L. GESSA

### 1. INTRODUCTION

Sexual behavior in male rats is stimulated by treatments which either decrease brain serotonin (5HT) concentration or increase that of brain dopamine (DA). Conversely, this behavior is suppressed by treatments which either increase 5HT transmission or impair DA function. Recent findings indicate that besides monoamines, the brain neuropeptides also play an important role in the control of male sexual behavior, e.g., endorphins have been shown to inhibit and ACTH-like peptides and LH-RH to stimulate this behavior.

Although the majority of these studies have investigated male rats, some of the results obtained in this species also apply to rabbits, cats, and humans.

The copulatory pattern of the adult male rat with a receptive female consists of repetitive mountings with or without intromission of the penis into the vagina. After each mounting or intromission, the male dismounts from its partner and, after a certain number of intromissions, ejaculation occurs. This is followed by a postejaculatory interval of a few minutes, after which the copulatory sequence is resumed. Sexually active rats ejaculate five to seven times before reaching sexual exhaustion.

Moreover, drugs which arouse sexual activity increase the percentage of animals exhibiting male-to-male mounting behavior, which occurs spontaneously in a certain number of rats in the absence of females in estrus.

Fortunately, male rats often show disturbances in sexual behavior. Therefore, this species offers a useful model to study not only the basic mechanisms controlling such behavior but also possible reasons for sexual disturbances in humans. In fact, male rats may be impotent, or may ejaculate either prema-

G. SERRA and G. L. GESSA • Institute of Pharmacology, University of Cagliari, 09100 Cagliari, Italy.

turely or sluggishly, and some animals may require fewer intromissions than others to ejaculate.

Drugs may influence various measures of sexual pattern: for example, they may increase or decrease the percentage of animals mounting, ejaculating, or exhibiting male-to-male mounting behavior, may shorten or prolong the ejaculation latency, or may increase or decrease the number of intromissions necessary to achieve ejaculation.

## 2. MONOAMINES

The finding that *p*-chlorophenylalanine (PCPA), an inhibitor of 5HT synthesis, stimulates copulatory behavior in rats and the clinical observation that L-dopa, the direct precursor of DA, may produce an aphrodisiac effect in humans, stimulated a great number of studies on the possible role of brain 5HT and catecholamines in the control of male sexual behavior.

### 2.1. 5HT

The hypothesis that brain 5HT plays an inhibitory role in the control of male sexual behavior is supported by the fact that different treatments which decrease brain 5HT levels or block brain 5HT receptors stimulate copulatory behavior. Conversely, treatments which increase brain 5HT levels or stimulate brain 5HT receptors inhibit such behavior. This hypothesis is supported by the following experimental evidence:

#### 2.1.1. Treatments Which Decrease Brain 5HT Levels

PCPA induces male-to-male mounting behavior (Sheard, 1969; Tagliamonte et al., 1969; Shillito, 1970; Malmnäs and Meyerson, 1971).

PCPA increases the percentage of animals reaching ejaculation when administered to sexually inexperienced male rats (Tagliamonte et al., 1971), rats with a low baseline of sexual activity (Mitler et al., 1972), and castrated rats treated with suboptimal doses of testosterone (Malmnäs and Meyerson, 1971; Malmnäs, 1973). On the contrary, PCPA causes no further increase in the number of ejaculations in rats with a high baseline of sexual activity (Whalen and Luttge, 1970), but shortens the ejaculation latency in such animals (Salis and Dewsbury, 1971; Ahlenius et al., 1971).

Stimulation of sexual behavior may be produced by the administration of 5,6-dihydroxytryptamine (5,6DHT) or a tryptophan-free diet, which have the common ability to deplete brain 5HT content but act through different mechanisms.

After injection into the lateral ventricles or into the raphe nuclei 5,6DHT produces male-to-male mounting behavior (Da Prada et al., 1972a) and increases the percentage of sexually inexperienced rats reaching ejaculation (Gessa and Tagliamonte, 1975), respectively. The latter effect is observed 2 weeks after treatment, when brain 5HT is maximally depleted.

The administration of a tryptophan-free diet produces a long-lasting decrease in brain tryptophan levels and brain 5HT synthesis (Biggio et al., 1974). A large percentage of rats or rabbits fed on this diet display repeated episodes of male-to-male mounting behavior (Fratta et al., 1977).

Such behavior is observed also in rats treated with different 5HT antagonists, such as methysergide, mesorgydine, and WA-335-BS in combination with testosterone (Benkert and Eversman, 1971; Soulairac and Soulairac, 1971).

### 2.1.2. Treatments Which Increase Brain 5HT Levels

The administration of 5HTP, the direct precursor of 5HT, abolishes both spontaneous and PCPA-activated copulatory behavior (Sheard, 1969; Tagliamonte et al., 1969; Shillito, 1970; Gessa, 1970; Perez-Cruet et al., 1971; Ferguson et al., 1970; Tagliamonte et al., 1972). The 5HTP inhibitory effect is potentiated by pretreatment with benserazide (Ro 4-4602), an inhibitor of peripheral decarboxylase which causes a much greater 5HT accumulation in the CNS than 5HTP alone.

In contrast to the effect of 5HTP, the administration of L-tryptophan fails to inhibit copulatory behavior in male rats (Tagliamonte et al., 1972), suggesting that 5HT levels accumulating in brain after L-tryptophan loading are insufficient to inhibit copulatory behavior.

The inhibitory effect of 5HTP on male sexual behavior is potentiated by 5HT uptake inhibitors, such as zimelidine or alaproclate, which increase the amount of 5HT available at central 5HT receptors. In fact, these treatments produce a prolongation of the ejaculatory latency and of the postejaculatory interval in rats treated with a subthreshold dose of 5HTP (Ahlenius et al., 1980a).

Quipazine, a direct stimulant of 5HT receptors, inhibits male-to-male mounting behavior induced by PCPA plus testosterone and that by PCPA plus pargyline. The inhibitory effect of quipazine is counteracted by methysergide, a drug believed to block central 5HT receptors (Grabowska, 1975).

Inhibition of copulatory behavior in male rats has been observed after the administration of LSD (Malmnäs, 1973), which is considered to be a direct stimulant of central 5HT receptors.

Finally, the monoamine oxidase inhibitors (MAOI) pargyline, isoniazide, and tranylcypromine suppress the spontaneous copulatory behavior of male rats with receptive females (Tagliamonte et al., 1971; Gessa and Tagliamonte, 1973).

The inhibitory effect of MAOI on sexual behavior seems to be mediated by the accumulation of brain 5HT, since this effect is not only prevented but also reversed by PCPA (Tagliamonte et al., 1971). In fact, the finding that the combination of MAOI and PCPA causes a selective accumulation of brain catecholamines and also a marked sexual stimulation led to the suggestion that brain catecholamines might play an excitatory role in the regulation of male sexual behavior.

## 2.2. DA

A great deal of experimental evidence indicates that brain DA stimulates male sexual behavior.

Male-to-male mounting behavior induced by PCPA in rats is potentiated not only by MAOI but also by L-dopa (Benkert et al., 1973), the direct precursor of DA. Moreover, L-dopa, in combination with the peripheral decarboxylase inhibitor Ro 4-4602, produces male-to-male mounting behavior in rats (Da Prada et al., 1972b). The administration of this combination to sexually sluggish male rats markedly improves their copulatory behavior with receptive females (Tagliamonte et al., 1974).

Similar improvement is obtained with apomorphine, a selective stimulant of DA receptors (Tagliamonte et al., 1974).

Both the administration of apomorphine and of L-dopa (with Ro 4-4602) to sexually experienced male rats decrease the number of penile intromissions necessary to reach ejaculation and accelerate the achievement of ejaculation (Paglietti et al., 1978).

Lisuride, a semisynthetic ergot derivative which is believed to stimulate brain DA receptors, induces male-to-male and female-to-female mounting behavior (Da Prada et al., 1977) and decreases the number of intromissions preceding ejaculation and the ejaculatory latency (Ahlenius et al., 1980b).

The effect of L-dopa and apomorphine on sexual behavior is prevented by neuroleptics, such as pimozide or haloperidol, which selectively block DA receptors in the CNS (Tagliamonte et al., 1974; Paglietti et al., 1978).

Moreover, both neuroleptics, which block DA receptors, and α-methyltyrosine, which inhibits DA synthesis, suppress copulatory behavior in male rats paired with females in estrus (Tagliamonte et al., 1974; Malmnäs, 1973).

Finally, in isolated or grouped male rats, drugs which directly or indirectly stimulate DA receptors induce repeated episodes of penile erection, eventually followed by ejaculation. This effect is observed after the administration of apomorphine (Benassi-Benelli et al., 1979), L-dopa (Baraldi and Benassi-Benelli, 1975), amantadine (Baraldi and Bertolini, 1974), amphetamine (Baraldi and Benassi-Benelli, 1975, 1977), and N-n-propyl-norapomorphine (Benassi-Benelli and Ferrari, 1979; Benassi-Benelli et al., 1979). The latter results suggest that DA plays an important role in the control of erection and ejaculation. These findings are in agreement with the fact that the stimulation of DA receptors accelerates the achievement of ejaculation and have led to the suggestion that dopaminergic hyperactivity might be involved in premature ejaculation in man (Paglietti et al., 1978).

## 3. NEUROPEPTIDES

### 3.1. ACTH

The injection of ACTH and MSH peptides into the CSF of different mammals causes sexual excitation (Bertolini *et al.*, 1975).

In isolated animals, the sexual stimulant effect of ACTH is characterized by recurrent episodes of penile erection accompanied by copulatory movements, each episode often culminating in ejaculation. In rabbits, which are the most sensitive to the sexual excitatory effect of ACTH, sexual stimulation may be so intense that, during the first 2 or 3 hr following treatment, the animals may ejaculate up to a dozen times. It is important to point out that ACTH peptides, unlike drugs which act on brain monoamines such as PCPA or apomorphine, do not enhance social interaction. This is particularly evident in the male rabbit which does not seek to copulate with either male or female partners during episodes of sexual stimulation. In addition, the administration of $ACTH_{1-24}$ into the lateral ventricle of sexually sluggish male rats does not increase the percentage of males copulating with receptive females. However, in sexually experienced animals, the hormone markedly shortens the ejaculation latency and also decreases the number of mounts and intromissions prior to ejaculation.

In contrast to the above findings, it has been reported that the intraventricular injection of $ACTH_{4-10}$ decreases the responsiveness of the male to the female rat, suggesting that the peptidic sequence needed for sexual arousal does not coincide with that of $ACTH_{4-10}$ (Bohus *et al.*, 1975).

Testosterone plays a permissive role in the sexual response to ACTH. In fact, castration eliminates this effect in rats and rabbits. Moreover, the sexual response is prevented in intact rabbits by cyproterone acetate, a substance which antagonizes the testosterone effect (Bertolini *et al.*, 1975).

A lesion of the preoptic area eliminates the capacity of ACTH to induce erection and ejaculation in rabbits, suggesting that this area is essential for the sexual stimulant effect of ACTH (Bertolini *et al.*, 1969).

### 3.2. Endorphins

Chronic users of morphine and other narcotic analgesics often complain of impotence and recent evidence suggests that endorphins, a generic name for endogenous peptides with morphine-like activity, play an inhibitory role in the control of male sexual behavior.

Meyerson and Terenius (1977) reported that the percentage of male rats mounting receptive females decreases when the males receive an intraventricular injection of 1 µg of β-endorphin. This effect of β-endorphin is abolished by pretreatment with the opiate antagonist naltrexone. Similarly, an intraventricular injection of 6 µg of [D-Ala$^2$]-Met-enkephalinamide (DALA), an enkephalin derivative resistant to enzymatic destruction, suppresses copulatory behavior in rats, while 3 μg increases mounting and intromission latencies. Pretreatment with the opiate antagonist naloxone completely prevents the inhibitory effect of DALA (Pellegrini-Quarantotti et al., 1978). Finally, the intraperitoneal administration of morphine produces a complete loss of copulatory behavior in male rats (McIntosh et al., 1980).

The administration of naloxone to vigorous sexually active rats causes no further enhancement of the percentage of rats reaching ejaculation but shortens the ejaculation latency and reduces the number of intromissions occurring before ejaculation (Pellegrini-Quarantotti et al., 1979; McIntosh et al., 1980).

Moreover, naloxone induces copulatory behavior in sexually inactive male rats (Gessa et al., 1979). This finding is of great theoretical and practical interest. In fact, it may indicate that sexual inadequacy in rats may be due, among other causes, to the stimulation of brain opiate receptors by endorphins. Moreover, opioid antagonists may be potentially useful therapeutic agents for sexual impotence in man.

Naloxone modifies copulatory behaviour in a similar manner as L-dopa or apomorphine. This similarity might be explained by an enhanced dopaminergic transmission by naloxone. In fact, it has been reported that enkephalins decrease DA release by an action on opiate receptors at the DA nerve terminals (Loh et al., 1976; Biggio et al., 1978). Since the administration of naloxone or naltrexone increases serum LH levels, it has also been suggested that these compounds stimulate sexual behavior by the central release of luteinizing hormone-releasing hormone (LH-RH) (Myers and Baum, 1980).

## 3.3. LH-RH

LH-RH administration has been reported to facilitate male sexual behavior. In sexually experienced male rats, a single subcutaneous injection of LH-RH significantly reduces the time to the first intromission and ejaculation, but fails to affect the number of mounts or intromissions prior to ejaculation (Moss, 1978; Myers and Baum, 1980). In castrated male rats treated with suboptimal doses of testosterone, the administration of LH-RH decreases (Moss, 1978) or causes no change in (Myers and Baum, 1980) the ejaculatory latency.

Intracerebroventricular injection of LH-RH facilitates mounting behavior in intact rats deprived of genital sensory information by local anesthesia of the penis, without inducing alterations in sex hormone secretion (Dorsa and Smith, 1980).

Indirect evidence that LH-RH is involved in sexual arousal is the observation that exposure of the male rat to a female in estrous stimulates the secretion of LH-RH into the hypophyseal portal system (Kamel et al., 1977; Kamel and Frankel, 1978).

It is important to note that LH-RH alone does not facilitate sexual behavior in rats unless testosterone is present. It has been proposed that LH-RH increases sexual activity by interacting with some neuronal substrate involved in sexual arousal (Dorsa and Smith, 1980).

### 3.4. Prolactin

It is well known that chronic pathological hyperprolactinemia is commonly associated with a decrease or loss of sexual potency in humans (Thorner et al., 1974).

DA exerts a tonic inhibitory role on pituitary prolactin (PRL) release and, vice versa, inhibition of dopaminergic activity results in the stimulation of PRL release as well as in the inhibition of sexual behavior.

Thus, the question has been raised as to whether hyperprolactinemia is the cause of impotence or both these phenomena are the consequence of a diminshed dopaminergic activity.

Several studies have been carried out in laboratory animals to clarify the direct influence of PRL on male sexual behavior.

It has been reported that short-term hyperprolactinemia induced by grafting two pituitary glands under the kidney capsule reduces mount and intromission latencies (Drago et al., 1981), suggesting that PRL may have a stimulant effect on sexual behavior.

Recent findings from our laboratory (Fratta et al., 1981) have shown that subcutaneous or intracerebroventricular PRL treatment (twice daily for 15 days) fails to decrease copulatory behavior in male rats with receptive females. Moreover, in contrast to haloperidol, which not only increases PRL levels but also blocks central DA receptors, domperidone, a peripheral DA receptor blocker which is more potent than haloperidol in increasing blood PRL levels, has no effect on sexual behavior. These results support the hypothesis that inhibition of central dopaminergic activity reduces sexual behavior in male rats, and demonstrate that hyperprolactinemia per se has no effect on sexual behavior.

Thus, it is tempting to speculate that when sexual inadequacy is associated with hyperprolactinemia, the two events are secondary to an impairment of central dopaminergic activity.

## 4. DRUG TREATMENT OF SEXUAL DYSFUNCTION IN MAN

Despite current knowledge on the neurochemistry of sexual behavior in animals, only little and often contrasting clinical data are available on drug treatment of human sexual disorders.

Pharmacological treatments have been attempted for human sexual dysfunctions, such as impotency and premature ejaculation.

### 4.1. Impotency

#### 4.1.1. Drugs Which Influence Brain 5HT Transmission

The finding of the aphrodisiac effect of PCPA in animals led to the clinical study of this compound as a potential treatment of sexual deficiency in man.

Some sexually impotent patients in an uncontrolled study showed improvement after PCPA treatment (3 g/day for 4 weeks) (Benkert, 1973a). Sicuteri *et al.* (1975) studied the effect of PCPA alone or associated with testosterone in an open controlled study carried out in 16 men suffering from severe migraine and complaining of sexual deficiency. The administration of PCPA alone (15 mg/kg/day, orally) resulted in mild sexual stimulation, whereas the authors claim that the coadministration of PCPA with testosterone induced a "clear aphrodisiac effect."

However, in a double-blind study in 10 patients with sexual impotency, Benkert (1975) observed no therapeutic effect of PCPA (1 g/day for 4 weeks) as compared to placebo. In another double-blind study, the effect of PCPA (0.5 g/day for 4 weeks) in combination with testosterone was tested against placebo in 10 patients with sexual impotency. The PCPA plus testosterone combination was found to be no more effective than placebo plus testosterone (Benkert, 1980).

It may be that the lack of effectiveness of PCPA observed in these studies is due to inadequate dosage. In fact, PCPA, in doses over 1 g/day, causes severe side effects which discourage the use of higher doses of this drug.

Negative results have been obtained with the use of 5HT blockers in the treatment of sexual impotency.

In an uncontrolled study, 10 impotent patients were treated with the 5HT blocker methysergide for 4 weeks. Two patients showed increased sex drive; this effect, however, was no better than that of placebo (Benkert, 1973b).

The effect of 5HT blockers has been studied by Bara *et al.* (1978) in 20 impotent patients. In an uncontrolled study, 5 patients were given methysergide in combination with mesterolone for 30–35 days, 4 received methysergide with bromocriptine for 30–50 days, and 11 were treated with metergoline for 30–50 days. None of these treatments showed any therapeutic effect.

### 4.1.2. Drugs Which Influence Brain DA Transmission

Since the introduction of L-dopa in the treatment of Parkinson's disease, increased sexual interest and/or erection capacity has been repeatedly observed in patients under treatment with this drug (Hyyppä *et al.*, 1970; Jenkins and Groh, 1970; Calne and Sandler, 1970; Bowers *et al.*, 1971; Brown *et al.*, 1978). This effect does not seem to be related to improvement of locomotor function (Brown *et al.*, 1978).

No therapeutic effect was seen in the administration of L-dopa alone or in combination with mesterolone to 10 impotent patients, although some of these patients showed a slight increase in erection capacity (Benkert *et al.*, 1972).

Bromocriptine, a direct stimulant of central DA receptors, has been reported to possess therapeutic effect in patients suffering from impotency associated with endocrinological dysfunction. In impotent patients with hyperprolactinemia, the administration of bromocriptine lowered plasma PRL levels and restored sexual potency (Thorner *et al.*, 1974, 1977; Rocco *et al.*, 1983). Increase of libido and sexual activity was observed in acromegalic patients under bromocriptine treatment (Wass *et al.*, 1977).

In contrast, no therapeutic effect was seen when bromocriptine was given at a dose of 5 mg/day alone or in combination with mesterolone to 15 impotent patients (Cooper, 1977). Negative results have also been reported by Ambrosi *et al.* (1977), who examined the effect of bromocriptine (7.5 mg/day) versus placebo in a double-blind study on 30 patients.

Finally, it has recently been reported that the administration of lisuride, a semisynthetic ergot derivative which stimulates DA receptors, to patients suffering from headache and loss of libido induces a significant improvement in sexual performance (Sicuteri and Del Bene, 1980).

### 4.1.3. Opioid Antagonists

While it is well known that opioid addicts frequently suffer from impotency, there have been no controlled studies on the use of opiate antagonists in the treatment of this disorder.

However, cyclazocine, an opioid antagonist used in the treatment of opiate addicts, causes an increase of libido in a certain number of patients (Kaplan, 1974).

Moreover, in a single male subject, naloxone greatly reduced the time needed to achieve ejaculation by masturbation (Goldstein and Hansteen, 1977), and naltrexone has been reported to cause penile erections in several men in the absence of any erotic context (Mendelson *et al.*, 1978). These observations suggest that these compounds may have a stimulatory effect on human sexual activity as observed in laboratory animals.

### 4.1.4. LH-RH

Conflicting results have been reported on the effect of LH-RH on sexual inadequacy in humans. Thus, despite some promising early reports, more recent investigations have been less optimistic.

LH-RH injection, given for several weeks, has been reported to increase sexual potency (Mortimer *et al.*, 1974). Benkert *et al.* (1975) reported increased sexual activity in previously impotent men, lasting 4 to 6 weeks after interruption of LH-RH treatment. Moss (1978) observed increased sexual activity in men after LH-RH administration.

In contrast, the administration of LH-RH has been found to produce no effect in healthy adult men (Ehrensing and Kastin, 1976) and slight and clinically insignificant effects in men suffering from secondary impotency (Davies *et al.*, 1976). In a recent double-blind study on patients suffering from decreased ability to obtain and maintain penile erections, no clinically significant effects were seen after treatment with large doses of LH-RH (Mauk *et al.*, 1980).

### 4.1.5. Androgens

Androgen therapy is indicated in impotent patients with low testosterone levels. In contrast, the effectiveness of the administration of exogenous androgens to sexually impotent men with normal testosterone levels has not been shown (Schmidt, 1980).

### 4.1.6. Antidepressants and Antianxiety Drugs

Sexual dysfunction, such as lowered libido, impotency, and premature ejaculation, are frequently associated with depression. Usually, in such cases, the administration of antidepressant drugs improves the sexual disturbances along with the other symptoms of depression (Renshaw, 1974, 1975). Similarly to antidepressants, the administration of antianxiety drugs can be useful in the treatment of sexual disorders caused by anxiety.

## 5. PREMATURE EJACULATION

The hypothesis that dopaminergic hyperactivity may be involved in the pathogenesis of premature ejaculation could explain the effectiveness of DA receptor blockers in the treatment of this disorder and prompted recent clinical trials with metoclopramide in the treatment of premature ejaculation.

In 1961, several authors reported the first observations of inhibition of ejaculation as a side effect of the neuroleptic thioridazine. These observations prompted Singh (1963) to use thioridazine in small doses in the treatment of premature ejaculation, a treatment which gave very encouraging results.

Recently, Falaschi et al. (1981) carried out a double-blind cross-over trial of metoclopramide, a DA receptor blocker, and placebo in patients with premature ejaculation. These patients took metoclopramide at a dose of 10 mg, twice daily, or an identical placebo preparation. After 90 days of treatment, the patients who received metoclopramide showed a significant improvement. These stimulating results obtained with chronic treatment led these authors to study the acute effect of the drug. The administration of metoclopramide, 2 hr before intercourse, produced an improvement in premature ejaculation.

Positive results in the treatment of premature ejaculation have also been reported with the use of the antidepressant chlorimipramine. In an open study, Eaton (1973) treated 13 men with chlorimipramine.

These results have recently been repeated by Porto (1981) in a double-blind study carried out in 20 patients suffering from premature ejaculation, who received chlorimipramine for 35 days.

The positive effect of this compound in the treatment of "eiaculatio praecox" has been attributed to blockade of peripheral $\alpha$-adreno-receptors in the sympathetic nervous system. However, it is important to point out that premature ejaculation may be associated with depressive illness and that successful treatment of depression with antidepressant drugs usually restores the patients to their premorbid level of sexual function.

## 6. CONCLUSION

Considerable experimental evidence indicates that sexual behavior in male animals is inhibited by brain 5HT and stimulated by brain DA. Other than monoamines, neuropeptides such as ACTH, LH-RH, and endorphins seem to play an important role in the control of such behavior: ACTH-like peptides and LH-RH stimulate while endorphins inhibit male sexual behavior.

This chapter reviews the main experimental observations which have led to the above hypotheses and the results of the first clinical attempts at pharmacological treatment of impotency and premature ejaculation in man.

## 7. REFERENCES

Ahlenius, S., Eriksson, H., Larsson, K., Modigh, K., and Södersten, P., 1971, Mating behaviour in the male rat treated with p-chlorophenylalanine methyl ester alone and in combination with pargyline, *Psychopharmacologia* **20**:383.

Ahlenius, S., Larsson, K., and Svensson, L., 1980a, Further evidence for an inhibitory role of central 5HT in male rat sexual behavior, *Psychopharmacology* **68**:217.

Ahlenius, S., Larsson, K., and Svensson, L., 1980b, Stimulating effects of lisuride on masculine sexual behaviour of rats, *Eur. J. Pharmacol.* **64**:47.

Ambrosi, B., Bara, R., and Faglia, G., 1977, Bromocriptine in impotence, *Lancet* **II**:987.

Bara, R., Ambrosi, B., Travaglini, P., Elli, R., Rondena, M., Moriondo, P., Gaggini, M., and Faglia, G., 1978, Effetti del trattamento con agenti antiserotoninergici in pazienti con impotenza sessuale, Atti XVII Congresso Nazionale S.I.E., Abstract C, P. 180.

Baraldi, M., and Benassi-Benelli, A., 1975, Induzione di erezioni ripetute nel ratto adulto mediante apomorfina, *Riv. Farmacol. Ter.* **6**:147.

Baraldi, M., and Benassi-Benelli A., 1977, Sexual excitement induced in the adult male rat by low doses of d-amphetamine or apomorphine: Suppression of penile erections by severe stereotyped behaviour, *Riv. Farmacol. Ter.* **8**:49.

Baraldi, M., and Bertolini, A., 1974, Penile erections induced by amantadine in male rats, *Life Sci.* **14**:1231.

Benassi-Benelli, A., and Ferrari, F., 1979, Comparazione tra apomorfina e N-n-propyl-norapomorfina per la induzione della sindrome di stiramento e sbadiglio, di erezione peniena e di stereotipie nel ratto, *Riv. Farmacol. Ter.* **10**:121.

Benassi-Benelli, A., Ferrari, F., and Pellegrini-Quarantotti, B., 1979, Penile erection induced by apomorphine and N-n-propyl-norapomorphine in rats, *Arch. Int. Pharmacodyn. Ther.* **242**:241.

Benkert, O., 1973a, Wirkung von serotonin: Antagonisten bei sexueller impotenz, *Pharmakopsychiat. Neuropsychopharm.* **6**:218.

Benkert, O., 1973b, Pharmacological experiments to stimulate human sexual behavior, in: *Psychopharmacology, Sexual Disorders and Drug Abuse* (T. A. Ban, J. R. Boissier, G. L. Gessa, H. Heimann, L. Hollister, H. E. Lehmann, I. Munkvad, H. Steinberg, F. Sulser, A. Sundwall, and O., Vinar, eds.), pp. 489–495, North-Holland, Amsterdam.

Benkert, O., 1975, Clinical studies on the effects of neurohormones on sexual behavior, in: *Sexual Behaviour: Pharmacology and Biochemistry* (M. Sandler and G. L. Gessa, eds.), pp. 297–305, Raven Press, New York.

Benkert, O., 1980, Pharmacology of sexual impotence in the male, in: *Modern Problems of Pharmacopsychiatry* (T. A. Ban, F. A. Freyhan, P., Pichat, and W. Poldinger, eds.), Vol. 15, pp. 158–173, Karger, Basel.

Benkert, O., and Eversman, T., 1971, Importance of the antiserotonin effect for mounting behavior in rats, *Experientia* **28**:532.

Benkert, O., Crombach, G. and Kockott, G., 1972, Effect of L-DOPA on sexually impotent patients, *Psychopharmacologia* **23**:91.

Benkert, O., Renz, A., and Matussek, N., 1973, Dopamine, noradrenaline and 5-hydroxytryptamine in relation to motor activity fighting and mounting behaviour. II. L-DOPA and DL-threodihydroxyphenylserine in combination with Ro-4-4602 and parachlorophenylalanine, *Neuropharmacology* **12**:187.

Benkert, O., Jordan, R., Dahlen, H. G., Schneider, H. P. G., and Gammel, G., 1975, Sexual impotence: A double-blind study of LH-RH nasal spray versus placebo, *Neuropsychobiology* **1**:203.

Bertolini, A., Vergoni, W., and Bernardi, M., 1969, Perdita della capacità dell'ACTH di indurre eccitazione sessuale per lesione delle zone cerebrali accumulanti testosterone, *Boll. Soc. Ital. Biol. Sper.* **45**:1139.

Bertolini, A., Gessa, G. L., and Ferrari, W., 1975, Penile erection and ejaculation: A central effect of ACTH-like peptides in mammals, in: *Sexual Behavior: Pharmacology and Biochemistry* (M. Sandler and G. L. Gessa, eds.), pp. 247–257, Raven Press, New York.

Biggio, G., Fadda, F., Fanni, P., Tagliamonte, A., and Gessa, G. L., 1974, Rapid depletion of serum tryptophan, brain tryptophan, serotonin and 5-hydroxyindoleacetic acid by a tryptophan-free diet, *Life Sci.* **14**:1321.

Biggio, G., Casu, M., Corda, M. G., Di Bello, C., and Gessa, G. L., 1978, Stimulation of dopamine synthesis in caudate nucleus by intrastriatal enkephalins and antagonism by naloxone, *Science* **200**:552.

Bohus, B., Hendricks, H. H. L., Von Kolfschoten, A. A., and Krediet, T. G., 1975, Effect of $ACTH_{4-10}$ on copulatory and sexually motivated approach behaviour in the male rat, in: *Sexual Behaviour: Pharmacology and Biochemistry* (M. Sandler and G. L. Gessa, eds.), pp. 269–275, Raven Press, New York.

Bowers, M. B., Van Woert, M. D., and Davis, L., 1971, Sexual behaviour during L-DOPA treatment for Parkinsonism, *Am. J. Psychiatry* **127**:1691.

Brown, E., Brown, G. M., Kofman, O., and Quarrington, B., 1978, Sexual function and affect in Parkinsonian men treated with L-DOPA, *Am. J. Psychiatry* **135**:1552.

Calne, D. B., and Sandler, M., 1970 L-DOPA and Parkinsonism, *Nature (London)* **226**:21.

Cooper, A. J., 1977, Bromocriptine in impotence, *Lancet* **II**:567.

Da Prada, M., Carruba, M., O'Brien, R. A., Saner, A., and Pletscher, A., 1972a, The effect of 5,6-dihydroxytryptamine on sexual behavior of male rats, *Eur. J. Pharmacol.* **19**:288.

Da Prada, M., Carruba, M., O'Brien, R. A., Saner, A., and Pletscher, A., 1972b, L-DOPA and sexual activity of male rats, *Psychopharmacologia* **26**(Suppl.):135.a Prada, M., Bonetti, E. P., and Keller, H. H., 1977, Induction of mounting behavior in female and male rats by lisuride, *Neurosci. Lett.* **6**:349.

Davies, T. F., Mountjoy, C. Q., Gomez-Pan, A., Watson, M. J., Hanker, J. P., Besser, G. M., and Holl, R., 1976, A double-blind cross over trial of gonadotropin releasing hormone (LH-RH) in sexually impotent men, *Clin. Endocrinol.* **5**:601.

Dorsa, D. M., and Smith, E. R., 1980, Facilitation of mounting behaviour in male rats by intracranial injections of luteinizing hormone-releasing hormone, *Regul. Peptides* **1**:147.

Drago, F., Pellegrini-Quarantotti, B., Scapagnini, U., and Gessa, G. L., 1981, Short-term endogenous hyperprolactinemia and sexual behaviour of male rats, *Physiol. Behav.* **26**:277.

Eaton, H. J., 1973, Chlorimipramine (Anafranil) in the treatment of premature ejaculation, *J. Int. Med. Res.* **1**:432.

Ehrensing, R. H., and Kastin, A. J., 1976, Clinical investigations for emotional effects of neuropeptide hormones, *Pharmacol. Biochem. Behav.* **5**:89.

Falaschi, P., Rocco, A., De Giorgio, G., Frajese, G., Fratta, W., and Gessa, G. L., 1981, Brain dopamine and premature ejaculation: Results of treatment with dopamine antagonists, in:

*Apomorphine and Other Dopaminomimetics*, Vol. I, *Basic Pharmacology* (G. L. Gessa and G. U. Corsini, eds.), pp. 117–121, Raven Press, New York.

Ferguson, J., Henriksen, S., Cohen, H., Mitchell, G., Barchas, J., and Dement, W., 1970, "Hypersexuality" and behavioral changes in cats caused by administration of *p*-chlorophenylalanine, *Science* **168**:499.

Fratta, W., Biggio, G., and Gessa, G. L., 1977, Homosexual mounting behaviour induced in male rats and rabbits by a tryptophan-free diet, *Life Sci.* **21**:379.

Fratta, W., Napoli-Farris, L., Falaschi, P., Bruno, R., Rocco, A., D'Urso, M. R., and Gessa, G. L., 1981, Short and long term effects of hyperprolactinemia on sexual behaviour in male rats, in: 2nd Capo Boi Conference on Neuroscience, Abstract 25.

Gessa, G. L., 1970, Serotonin now: Clinical implications of inhibiting its synthesis with parachlorophenylalanine, *Ann. Intern. Med.* **73**:607.

Gessa, G. L., and Tagliamonte, A., 1973, Role of brain monoamines in controlling sexual behaviour in male animals, in: *Psychopharmacology, Sexual Disorders and Drug Abuse* (T. A. Ban, J. R. Boissier, G. L. Gessa, H. Heimann, L. Hollister, H. E. Lehmann, I. Munkvad, H. Steinberg, F. Sulser, A Sundwall, and O. Vinar, eds.), pp. 451–462, North-Holland, Amsterdam.

Gessa, G. L., and Tagliamonte, A., 1975, Role of brain serotonin and dopamine in male sexual behaviour, in: *Sexual Behaviour: Pharmacology and Biochemistry* (M. Sandler and G. L. Gessa, eds.), pp. 117–128, Raven Press, New York.

Gessa, G. L., Paglietti, E., and Pellegrini-Quarantotti, B., 1979, Induction of copulatory behaviour in sexually inactive rats by naloxone, *Science* **204**:203.

Goldstein, A., and Hansteen, R. W., 1977, Evidence gainst involvement of endorphins in sexual arousal and orgasm in man, *Arch. Gen. Psychiatry* **34**:1179.

Grabowska, M., 1975, Influence of quipazine on sexual behaviour in male rats, in: *Sexual Behaviour: Pharmacology and Biochemistry* (M. Sandler and G. L. Gessa, eds.), pp. 59–62, Raven Press, New York.

Hyyppä, M., Rinne, U. K., and Sonninen, V., 1970, The activating effect of L-DOPA treatment on sexual functions and its experimental background, *Acta Neurol. Scand.* **43**(Suppl. 46):223.

Jenkins, R. B., and Groh, R. H., 1970, Mental symptoms in parkinsonian patients treated with L-DOPA, *Lancet* **II**:177.

Kamel, F., and Frankel, A. I., 1978, Hormone release during mating in the male rat: Time course, relation to sexual behavior and interaction with handling procedures, *Endocrinology* **103**:2172.

Kamel, F., Wright, W. W., Mock, E. J., and Frankel, A. I., 1977, The influence of mating and related stimuli on plasma levels of luteinizing hormone, follicle-stimulating hormone, prolactin and testosterone in the male rat, *Endocrinology* **101**:421.

Kaplan, H., 1974, *Nuove terapie sessuali*, Bompiani Editore.

Loh, H., Brase, D. A., Sampath-Khanna, S., Mar, J. B., May, E. L., and Li, C. H., 1976, β-Endorphin *in vitro* inhibition of striatal dopamine release, *Nature (London)* **264**:567.

McIntosh, T. K., Vallano, M. L., and Barfield, R. J., 1980, Effects of morphine, β-endorphin and naloxone on catecholamine levels and sexual behaviour in the male rat, *Pharmacol. Biochem. Behav.* **13**:435.

Malmnäs, C. O., 1973, Monoaminergic influence of testosterone activated copulatory behaviour in the castrated male rat, *Acta Physiol. Scand. Suppl.* **395**:1.

Malmnäs, C. O., and Meyerson, B. J., 1971, *p*-Chlorophenylalanine and copulatory behaviour in the male rat, *Nature (London)* **232**:398.

Mauk, M. D., Olson, G. A., Kastin, A. J., and Olson, M. D., 1980, Behavioural effects of LH-RH, *Neurosci. Behav. Rev.* **4**:1.

Mendelson, J. H., Ellingboe, J., Keuhnle, J. C., and Mello, N. K., 1978, Effects of naltrexone on mood and neuroendocrine function in normal adult males, *Psychoneuroendocrinology* **3**:231.

Meyerson, B. J., and Terenius, L., 1977, β-Endorphin and male sexual behaviour, *Eur. J. Pharmacol.* **42**:191.

Mitler, M. M., Morden, B., Levine, S., and Dement, W., 1972, The effects of parachlorophenylalanine on the mating behaviour of male rats, *Physiol. Behav.* **8**:1147.

Mortimer, C. H., McNielly, A. S., Fisher, R. A., Murray, M. A. E., and Besser, G. M., 1974, Gonadotrophin-releasing hormone therapy in hypogonadal males with hypothalamic or pituitary dysfunctions *Br. Med. J.* **4**:617.

Moss, R. L., 1978, Effects of hypothalamic peptide on sex behaviour in animals and man, in: *Psychopharmacology: A Generation of Progress* (M. A. Lipton, A. DiMascio, and K. F. Killam, eds.), pp. 431–440, Raven Press, New York.

Myers, B. M., and Baum, M. J., 1980, Facilitation of copulatory performance in male rats by naloxone: Effects of hypophysectomy, 17-α-estrodiol and luteinizing hormone releasing hormone, *Pharmacol. Biochem. Behav.* **12**:365.

Paglietti, E., Pellegrini-Quarantotti, B. P., Mereu, G., and Gessa, G. L., 1978, Apomorphine and L-DOPA lower ejaculation threshold in the male rat, *Physiol. Behav.* **20**:559.

Pellegrini-Quarantotti, B., Corda, M., Paglietti, E., Biggio, G., and Gessa, G. L., 1978, Inhibition of copulatory behaviour in male rats by D-Ala$^2$-Met-enkephalinamide, *Life Sci.* **23**:673.

Pellegrini-Quarantotti, B., Paglietti, E., Bonanni, A., Petta, M., and Gessa, G. L., 1979, Naloxone shortens ejaculation latency in male rats, *Experientia* **35**:524.

Perez-Cruet, J., Tagliamonte, A., Tagliamonte, P., and Gessa, G. L., 1971, Differential effect of *p*-chlorophenylalanine (PCPA) on sexual behaviour and on sleep patterns of male rabbits, *Riv. Farmacol. Ter.* **2**:27.

Porto, R., 1981, Chlorimipramine versus "eiaculatio praecox," *Med. Hyg.* **39**:1249.

Renshaw, D. C., 1974, Sexual disfunctions, in: *Somatic Manifestations of Depressive Disorders* (K. Kiev, ed.), pp. 86–106, Excerpta Medica, Amsterdam.

Renshaw, D. C., 1975, Doxepin treatment of sexual dysfunctions associated with depression, in: *Sinequam (Doxepin. HCl): A Monograph of Recent Clinical Studies*, pp. 23–31, Excerpta Medica, Amsterdam.

Rocco, A., Falaschi, P., Pompei, P., D'Uzzo, R., and Frajese, G., 1983, Reproductive parameters in prolactinemic man, *Arch. Androl.* **10**:179–183. in press.

Salis, P. J., and Dewsbury, D. A., 1971, *p*-Chlorophenylalanine facilitates copulatory behaviour in male rats, *Nature (London)* **232**:400.

Schmidt, C. W., Jr., 1980, Biochemical methods in the treatment of sexual disorders, in: Psychiatric Clinics of North America (J. K. Meyer, ed.), Vol. 3, No. 1, pp. 189–199, Saunders, Philadelphia.

Sheard, M. H., 1969, The effect of *p*-chlorophenylalanine on behaviour in rats: Relation to brain serotonin and 5-hydroxyindoleacetic acid, *Brain Res.* **15**:524.

Shillito, E. E., 1970, The effect of *p*-chlorophenylalanine on social interaction of male rats, *Br. J. Pharmacol.* **38**:305.

Sicuteri, F., and Del Bene, E., 1980, Sexuality and headache, Int. Congr. Headache, Abstract 32.

Sicuteri, F., Del Bene, E., and Anselmi, B., 1975, Aphrodisiac effect of testosterone in parachlorophenylalanine-treated sexually deficient men, in: *Sexual Behaviour: Pharmacology and Biochemistry* (M. Sandler and G. L. Gessa, eds.), pp. 335–339, Raven Press, New York.

Singh, H., 1963, Therapeutic use of thioridazine in premature ejaculation, *Am. J. Psychiatry* **119**:891.

Soulairac, A., and Soulairac, M. L., 1971, Action de la sérotonine sur le comportement sexuel du rat male, *Soc. Biol.* **165**:253.

Tagliamonte, A., Tagliamonte, P., Gessa, G. L., and Brodie, B. B., 1969, Compulsive sexual activity induced by *p*-chlorophenylalanine in normal and pinealectomized male rats, *Science* **166**:1433.

Tagliamonte, A., Tagliamonte, P., and Gessa, G. L., 1971, Reversal of pargyline-induced inhibition of sexual behaviour in male rats by *p*-chlorophenylalanine, *Nature (London)* **230**:24.

Tagliamonte, A., Fratta, W., Mercuro, G., Gibbio, G., Camba, R. C., and Gessa, G. L., 1972, 5-Hydroxytryptophan, but not tryptophan, inhibits copulatory behaviour in male rats, *Riv. Farmacol. Ter.* **3**:405.

Tagliamonte, A., Fratta, W., Del Fiacco, M., and Gessa, G. L., 1974, Possible stimulatory role of brain dopamine in the copulatory behaviour of male rats, *Pharmacol. Biochem. Behav.* **2**:257.

Thorner, M. O., McNeilly, A. S., Hagan, C., and Besser, G. M., 1974, Long-term treatment of galactorrhea and hypogonadism with bromocriptine, *Br. Med. J.* **II**:419.

Thorner, M. O., Edwards, C. R. W., Hanker, J. P., Abraham, G., and Besser, G. M., 1977, Prolactin and gonadotropin interaction in the male, in: *The Testis in Normal and Infertile Men* (P. Troen and S. Nankin, eds.), pp. 351–366, Raven Press, New York.

Wass, J. A. H., Thorner, M. O., Morris, D. V., Rees, L. H., Stuart-Mason, A., Jones, A. E., and Besser, G. M., 1977, Long-term treatment of acromegaly with bromocriptine, *Br. Med. J.* **I**:875.

Whalen, R. E., and Luttge, W. G., 1970, *p*-Chlorophenylalanine-methyl ester: An aphrodisiac?, *Science* **169**:1000.

CHAPTER 8

# Plasma Catecholamines as an Index of a Neuroendocrine Response

IRWIN J. KOPIN

## 1. INTRODUCTION

Epinephrine, the pressor substance released into the blood from the adrenal medulla, was the first hormone to be chemically characterized. In 1897, Abel and Crawford isolated and identified epinephrine (which they called "adrenaline") as $N$-methyl-3,4-dihydroxyphenylethanolamine (Fig. 1). Because the effects of administered epinephrine were so similar to those of sympathetic nerve stimulation (Langley, 1901), one of Langley's students proposed that an epinephrine-like substance is released from sympathetic nerve endings (Elliott, 1905). He later showed that epinephrine released from the adrenals is evoked by splanchnic nerve stimulation (Elliott, 1912), but his hypothesis regarding sympathetic nerves remained unproven until Loewi (1921) and Cannon and Uridil (1921) were able to demonstrate that a sympathomimetic substance was discharged when sympathetic nerves were stimulated. The nature of the chemical substance that was released was established over 25 years later when Von Euler (1948) found that norepinephrine (Fig. 1) is present in sympathetic nerve endings. Shortly thereafter, Peart (1949) showed that stimulation of the splenic nerves evoked release of norepinephrine from the spleen, whereas epinephrine was the catecholamine released from the adrenal medulla (Cannon and Rosenbleuth, 1937). Thus, these catecholamines might be considered the first examples of neurohumoral substances and the adrenal medulla the first discovered neuroendocrine gland.

---

IRWIN J. KOPIN • Laboratory of Clinical Science, National Institute of Mental Health, Bethesda, Maryland 20205.

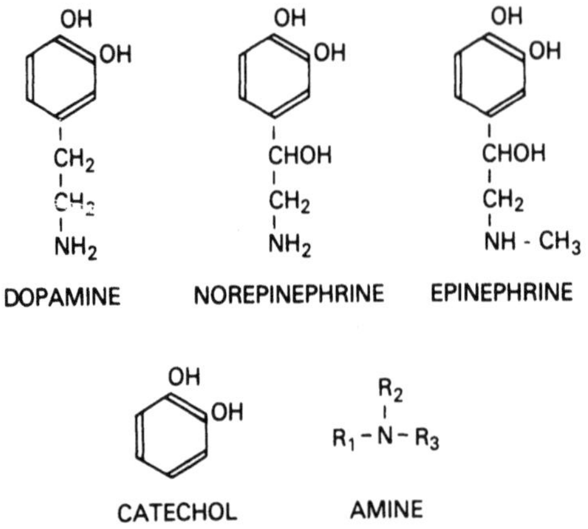

FIGURE 1. Chemical structure of catecholamines. The name for these compounds is derived from the terms used for the distinguishing functional components of their chemical properties—catechol and the family of amines ($R_1$, $R_2$, $R_3$ are functional groups; $R_2$ and/or $R_3$, hydrogen atoms).

## 2. ASSAYS OF PLASMA CATECHOLAMINES

Characteristically colored oxidation products of substances, probably catecholamines, in adrenal venous blood were first discribed over 125 years ago by Vulpian (1856), but such crude methods cannot be used for assay of catecholamine levels in plasma, even after their secretion is evoked by stimuli. The low levels of catecholamines normally present in plasma could barely be detected by bioassay. They could be determined by fluorimetric methods only by use of relatively large volumes of plasma and exacting chemical procedures. The recent development of convenient, sensitive, and specific radioenzymatic and other sophisticated assays has resulted in a striking proliferation of studies on plasma catecholamines.

### 2.1. Bioassay

Oliver and Schafer (1895) noted the striking cardiovascular effects of adrenal medullary extracts and such responses of blood pressure and heart rate were used by Abel and Crawford (1897) as a guide to their purification and identification of epinephrine. Elliott (1912) used the blood pressure responses of cats, after preventing vasomotor reflexes by destruction of the spinal cord down to the fourth spinal segment, to assay the sympathomimetic substances released from the adrenal gland. The pithed rat, however, is far more sensitive to pressor agents and has been used for bioassay of both epinephrine and norepinephrine. Several isolated tissues, e.g., rat uterus, rabbit ear blood vessels, intestinal smooth muscle of rats or chickens, the cat nictitating membrane, etc., are extremely sensitive to catecholamines and have been used to assay their levels in plasma (see Gaddum, 1959). These methods are tedious and lack precision, but if properly carried out they are sensitive and specific.

### 2.2. Fluorimetric Assay

The first of the chemical methods that were sufficiently sensitive to assay the minute quantities of catecholamines in normal human plasma were the fluorimetric methods described by Lund (1950) and Weil-Malherbe and Bone (1953). In both, the catecholamines are adsorbed from serum or plasma onto alumina. They are then eluted from the alumina with acid and converted to derivatives that can be assayed fluorimetrically. The method of Lund (1950) is based on oxidation of the catecholamines to their corresponding chrome derivatives. This is followed by strong alkaline treatment in the presence of an autoxidant, which results in rearrangement of the molecular structures to highly fluorescent derivatives. Weil-Malherbe and Bone (1953) used ethylenediamine to form a highly fluorescent condensation product with the catecholamines. Although the ethylenediamine method is more sensitive, it is relatively nonspecific and yields levels in plasma that are severalfold greater than those obtained by the trihydroxyindole method of Lund. Details of modifications of the extraction and derivatization procedure enhanced the specificity and sensitivity of this method so that it could be used to assay plasma catecholamine levels (see Von Euler, 1959). More recent modifications (Renzini et al., 1970) have reduced the volumes of plasma needed for the assay, but with the introduction of less tedious and more precise methods, most investigators have abandoned the fluorimetric methods.

### 2.3. Radioenzymatic Methods

These methods employ specific enzymes to catalyze addition of a radioactive methyl group from labeled S-adenosylmethionine to either the nitrogen using phenylethanolamine-N-methyltransferase (PNMT), or the 3-hydroxy position, using catechol-O-methyltransferase (COMT), of norepinephrine (Fig. 2). When COMT is used, epinephrine and dopamine are also O-methylated and can be assayed simultaneously.

#### 2.3.1. PNMT Assay

Saelens et al. (1967) introduced the use of PNMT isolated from adrenal medullae of animals to enzymatically N-methylate norepinephrine using [$^{14}$C]-S-adenosylmethionine as methyl donor. The product is isolated by alumina adsorption, eluted, and the isotope content assayed by liquid scintillation spectrometry. Subsequent modifications and use of [$^3$H]- instead of [$^{14}$C]-S-adenosylmethionine have increased the sensitivity of this assay to provide a rapid convenient method for assay of plasma norepinephrine (Lake et al., 1976). Because PNMT is not entirely specific and partially N-methylates epinephrine,

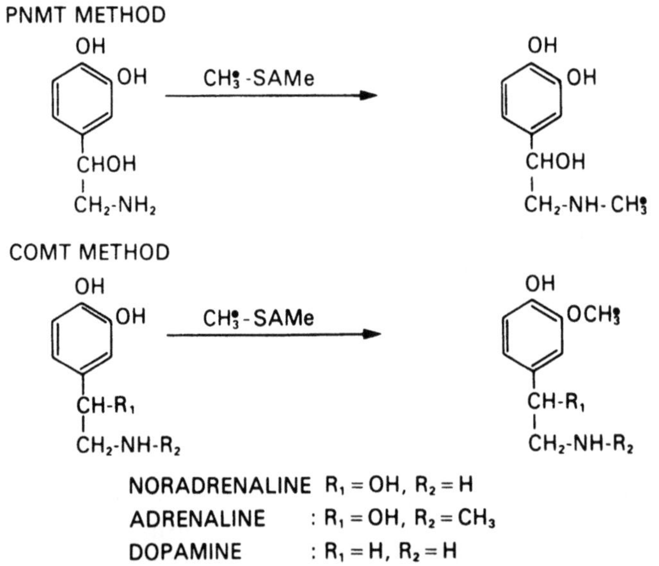

FIGURE 2. Radioenzymatic assays for catecholamines. PNMT, phenylethanolamine-$N$-methyltransferase; COMT, catechol-$O$-methyltransferase.

the values obtained are slightly higher than those obtained by more specific methods.

### 2.3.2. COMT Assay

Engleman and Portnoy (1970) introduced a double-label method in which [$^3$H]catecholamines were added to plasma as internal standards and these were $O$-methylated, along with the endogenous catecholamines, using COMT and [$^{14}$C]-$S$-adenosylmethionine. The labeled products were extracted and separated using thin-layer chromatography, converted to vanillin, and assayed for $^3$H, which provided an index of recovery, and $^{14}$C, which indicated the amount of catecholamine originally present. Subsequent modifications (Passon and Peuler, 1973; Cryer, 1976; Weise and Kopin, 1976; Da Prada and Zürcher, 1976; Peuler and Johnson, 1977) have resulted in a method that eliminates the requirement for an internal standard and allows use of [$^3$H]-$S$-adenosylmethionine of high specific activity. The single-isotope method requires less time and only 1/200th of the sample size as the double-isotope method.

### 2.4. High-Performance Liquid Chromatography with Electrochemical Detection (HPLC-ED)

Catechols are readily oxidized and along with many other compounds can be electrochemically oxidized. By use of alumina adsorption to separate catechols from other oxidized substances and HPLC to rapidly resolve the eluted compounds, it has been possible, after the development of electrochemical detectors of sufficiently high sensitivity, to devise a relatively simple, sensitive, and specific assay for catecholamines (Kissinger et al., 1973) which has been successfully applied to plasma. The effluent from the HPLC column flows through the electrochemical detector which is set at a predetermined potential to oxidize catechols and the generated current (picoamps) recorded as a measure of the quality of catechol oxidized. This method yields results which are comparable to those obtained by radioenzymatic methods (Goldstein et al., 1981) although the relatively large volumes (1 ml) of plasma which are required limit its use for studies in small experimental animals.

### 2.5. Gas Chromatography with Mass Spectrometry (GC-MS)

In this method, the catecholamines are separated by adsorption on alumina or an ion-exchange resin and, after elution and removal (by evaporation) of the aqueous solvent, converted to a volatile substance which can be dissolved in a volatile organic solvent. The solution is injected into a hot chamber where the derivatives and solvent are converted into gases and washed through a porous column in a flow of relatively inert gases. The outflow from this column is led into an ionization chamber where the compounds are bombarded with electrons and converted to ions (chemical ionization) or fragments of the molecules. These ions are accelerated into a magnetic field and resolved on the basis of their mass. The retention times on the column and the fragmentation patterns of known mass ensure absolute specificity. Use of deuterium-labeled or unnatural catecholamines (e.g., *N*-propylnorepinephrine) as internal standards affords exact determination of recoveries. Thus, GC-MS, while tedious and requiring expensive equipment, is exact and provides a standard against which other methods can be evaluated. It is sufficiently sensitive for application to plasma (Ehrhardt and Schwartz, 1978) but is not widely used because the other methods are almost equally sensitive and specific, but less tedious and less expensive.

## 3. VARIATIONS IN PLASMA CATECHOLAMINE LEVELS

Plasma catecholamine levels are maintained by their secretion from the adrenal medulla (mostly epinephrine) or from overflow of norepinephrine released from peripheral sympathetic nerve endings. Because the catecholamines are rapidly removed from plasma, the levels are low in a basal state, but with anxiety, excitement, or stress, the levels may increase markedly and are subject to wide fluctuations. The levels may vary with the method and site of blood collection, time of the day, etc. It is necessary, therefore, to first establish values for a basal state, usually taken in reclining awake subjects at least 20–30 min after insertion into the antecubital vein of an indwelling needle for atraumatic blood sampling.

TABLE 1
Normal Plasma Catecholamine Levels[a]

| Method | Norepinephrine | Epinephrine |
|---|---|---|
| Fluorimeter | 0.89 ± 0.12 | 0.26 ± 0.09 |
| PNMT | 1.94 ± 0.62 | — |
| COMT | 1.32 ± 0.25 | 0.24 ± 0.05 |
| HPLC-ED | 1.84 ± 0.12 (7) | 0.34 ± 0.02 (7) |
| GC-MS | 1.18 ± 0.23 (9) | 0.32 ± 0.06 (9) |

[a] Values are expressed in pmoles/ml and are the mean levels ± S.E.M. from at least five different laboratories or for the number of subjects indicated in parentheses.

### 3.1. Normal Basal Levels

As indicated above, the various methods yield similar, extremely low values for basal catecholamines in plasma (Table 1). The fluorimetric method of Renzini et al. (1970) yields slightly lower levels for plasma norepinephrine, possibly because blank values are overestimated. As indicated earlier, the PNMT method may include some labeled N-methylated epinephrine with the labeled epinephrine measured, so that values obtained from plasma norepinephrine are slightly higher than those found using other methods. The COMT method yields values which are not significantly different from those obtained by GC-MS and when the same plasma samples were assayed by the COMT and HPLC-ED methods, the values were almost identical. Thus, either the COMT or the HPLC-ED method can be used with equal reliability when sufficient volumes of plasma are available.

### 3.2. Determinants of Plasma Catecholamine Levels

Plasma catecholamine levels are determined by their rates of entry and removal from the circulation. Plasma clearance rates for norepinephrine, measured during constant infusion of tritiated norepinephrine, range from 2.4 to 3.5 liters/min (Esler et al., 1979; Ghione et al., 1978) with a half-life of about 2 min for the initial decline in levels of the labeled compound after cessation of the infusion. The levels of norepinephrine found in plasma are elevated when sympathetic nerve activity increases; such increases occur during usual activities (Fig. 3). Even relatively minor stimuli, such as postural changes (Christensen and Brandsborg, 1973; Cryer et al., 1974; Lake et al., 1976), drinking coffee (Robertson et al., 1978), or smoking (Cryer et al., 1976), evoke significant increases in plasma norepinephrine. Common stresses, such as public speaking, elicit adrenal medullary discharge which is reflected by increased plasma epinephrine levels (Dimsdale and Moss, 1980). Plasma norepinephrine increases with age (Ziegler et al., 1976), but this is a minor factor compared to the physiologic variations even within a single individual.

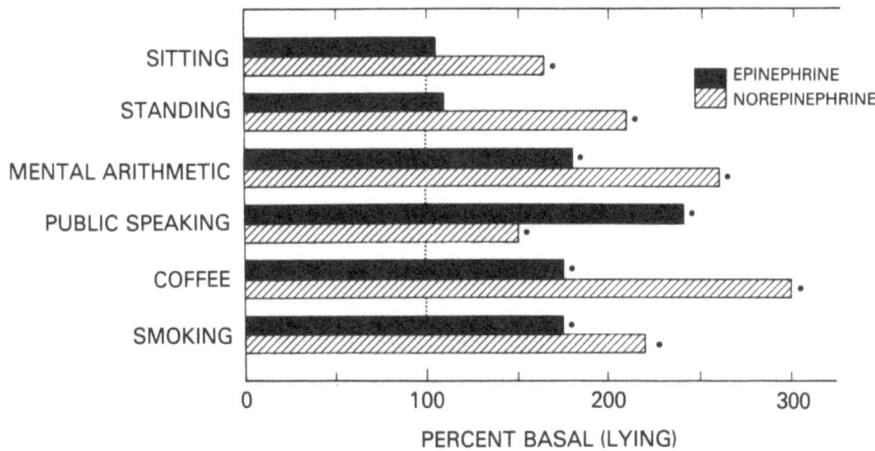

FIGURE 3. Effects of usual activities on plasma catecholamine levels.

### 3.3. Physiological Significance of Plasma Catecholamines

Plasma levels of the catecholamines are usually indices of the rate of their entry into the circulation. At sympathetic neuroeffector junctions, norepinephrine is mostly taken up into the tissues and enzymatically inactivated or returned to the vesicular storage sites in the nerve terminals. Only a relatively small portion of the norepinephrine is released into the circulation. This portion varies, depending upon the width of the neuroeffector junction; reuptake into the neuron is most important for inactivation of the released transmitter at narrow effector junctions, such as those in the vas deferens, than at wide junctions, such as those in blood vessels (Bevan, 1977). The levels of norepinephrine in plasma reflect the sum total of overflow at sympathetic neuroeffector junctions and are much below the level required to elicit a physiological response (Silverberg et al., 1978). Only when there are relatively massive increases in norepinephrine release are attained levels in excess of the 11 pmoles/liter required to evoke increases in blood pressure. In most circumstances, plasma levels of norepinephrine reflect sympathetic discharge. In pithed rats, there is a direct relationship between the rate of sympathetic nerve discharge and the increase in plasma norepinephrine (Yamaguchi and Kopin, 1979). The pressor response is related to the logarithm of the rise in plasma norepinephrine. When sympathetic nerve, but not adrenal medullary, discharge is prevented by administration of bretylium, increases in blood pressure are abolished. Adrenal medullectomy abolishes the stimulation-evoked rise in plasma epinephrine, but has no effect on the pressor response.

The situation is different for epinephrine. Plasma levels of this catecholamine are lower than those of norepinephrine, but relatively small increases in the level of this catecholamine evoke responses. Thus, Clutter et al. (1980) found that the plasma epinephrine thresholds were 0.28–0.58 mmole/liter for

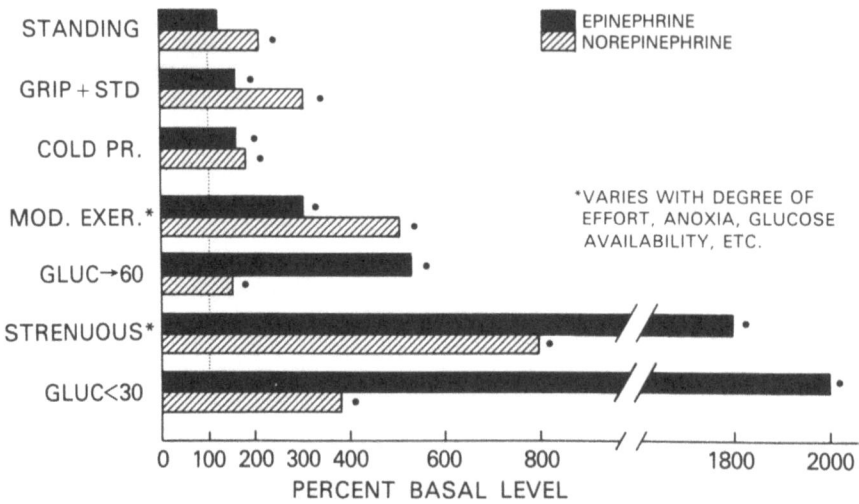

FIGURE 4. Evoked increases in plasma catecholamines.

increasing heart rate, 0.41–0.69 for increment in glycerol or systolic pressure, and 0.84–1.12 for increments in diastolic pressure, glucose, and lactic acid. The epinephrine discharged into the blood from the adrenal medulla is more efficient as a neurohumor than is norepinephrine, which acts specifically within the neuroeffector junction at which it was released.

Increases in catecholamine levels may be evoked by a variety of physiologic stimuli (Fig. 4). As indicated above, postural change increases mainly norepinephrine. Mild exertion, such as maintaining for 5 min a grip on a hand dynameter at 30% of maximal effort, evokes further increases in norepinephrine and some rise in epinephrine (Lake et al., 1976). Immersion of a hand in ice cold water for 3 min also evokes rises in both catecholamines. Decreases in blood glucose, particularly to hypoglycemic levels, evoke striking increases in adrenal medullary discharge of epinephrine and corresponding elevation of plasma levels to that catecholamine. Adrenal medullary responses are also seen with strenuous exercise. Thus, the sympathetic nerve responses appear to be primarily concerned with alterations in individual vascular bed or particular organ activity, whereas epinephrine release from the adrenal medulla is a hormonal response to more severe stresses and appears to be aimed at metabolic and emergency responses.

### 3.4. Effects of Drugs on Plasma Catecholamines

Drugs may alter levels of catecholamines by direct actions on the processes involved in the formation, storage, release, or termination of action of the catecholamines or produce an indirect effect by altering the rate of sympathetic impulse inflow. Decreases in rat plasma norepinephrine levels which occur 7 hr after administration of α-methyltyrosine, an inhibitor of tyrosine hydroxylase which is the rate-limiting step in catecholamine biosynthesis, are attended by compensatory release of adrenal epinephrine (which is normally stored and released very slowly) to cause a threefold increase in levels of the adrenal medullary hormone (Avakian and Horvath, 1980). In such animals, the ability to elevate catecholamines in response to stress is compromised. Drugs, such as reserpine, which prevent storage of catecholamines, decrease plasma levels of norepinephrine (Reid and Kopin, 1975) as does bretylium, which prevents release of the transmitter at sympathetic nerve terminals (McCarty and Kopin, 1979). Ganglionic blocking agents or drugs such as clonidine, which diminish outflow of sympathetic inpulses from the CNS, also diminish plasma norepinephrine levels.

Inhibitors of the enzymes involved in the metabolism of catecholamines do little to potentiate the effects of sympathetic nerve stimulation or to elevate plasma catecholamine levels whereas drugs such as cocaine or desipramine, which prevent uptake by sympathetic nerve endings of norepinephrine, potentiate stimulation-induced responses and elevate plasma catecholamine levels.

Drugs which block α-adrenergic receptors increase plasma norepinephrine by a combination of several actions. Presynaptic α-adrenoceptors normally modulate release of the transmitter and when this auto-inhibition is prevented by a drug, release of norepinephrine is enhanced (see review by Langer, 1977). Nerve impulses are required, however, since the effects on plasma norepinephrine are prevented by ganglionic blocking agents (Reid and Kopin, 1975). Thus, the block of α-receptors reflexly induces an increase in sympathetic nerve activity as well as enhancing release of the catecholamine. Furthermore, some α-adrenergic blocking agents inhibit uptake of catecholamines into the nerve endings.

Drugs which act in the CNS to decrease sympathetic nerve activity diminish plasma catecholamine levels. Clonidine is an $\alpha_2$-adrenergic agonist which, in low doses, appears to act in the brain to decrease sympathetic outflow and thereby lowers plasma norepinephrine levels.

Decreases in blood pressure evoke reflexly-induced sympathetic nerve responses with attendant rises in plasma norepinephrine. Thus, drugs which have direct actions on vascular smooth muscle may alter catecholamine levels indirectly through compensatory reflex recruitment of the sympathoadrenal medullary system to support blood pressure at satisfactory levels.

## 4. PLASMA CATECHOLAMINES IN DISEASE STATES

Abnormally high basal plasma levels of catecholamines are the result of either tumors of the sympatho-adrenal medullary system or excessive sympatho-adrenal medullary activity evoked by reflex stimulation of the overflow of nerve impulses from the sympathetic nervous system, whereas abnormally low basal catecholamines are found with diseases involving degeneration of sympathetic nerves or in conditions in which there is a reduction of sympathetic outflow.

## 4.1. Pheochromocytoma

The highest plasma levels of catecholamines which have been found occur in association with tumors of the adrenal medulla or other sympathetic–chromaffin tissue. These tumors discharge either epinephrine or norepinephrine continuously or episodically and produce symptoms of a hyperadrenergic state. The levels of catecholamines found in plasma vary widely, but are almost invariably elevated. The very high levels of catecholamines in venous blood from the area of the tumor can be used for localization of the tumor.

## 4.2. Plasma Catecholamines in Hypertension

Basal levels of norepinephrine and epinephrine are normal in most patients with essential hypertension. Some young persons with labile hypertension may appear to have high basal levels of norepinephrine, but it is unclear as to whether this represents a response to the procedure for obtaining blood, is pathogenetically related to the development of sustained increases in blood pressure, or is a biochemical marker for hyperreactive subjects who will later develop hypertension (see review by Kopin *et al.*, 1981). It has become apparent, however, that hospital personnel, medical students, etc. are not appropriate controls and appear to have lower basal plasma norepinephrine levels than normotensive civil servants or normotensive outpatients.

Of the various forms of experimental hypertension in animals, the genetic variant, spontaneously hypertensive (SHR) rats appear to provide the most satisfactory model for human hypertension. These animals are hyperreactive to stress. A variety of different stresses have been shown to evoke greater plasma catecholamine responses and pressor responses than in normotensive rats of the same strain. The strain of normotensive rats from which the hypertensive rats were derived, the Wistar–Kyoto (WKY) strain, also appear to have hyperractive sympathoadrenal responses relative to other normotensive strains. The WKY rats, however, are more susceptible to stress-induced ulcers than are the SHR rats. This had led to the hypothesis that both SHR and WKY rats have generally determined abnormal autonomic responses to stress, but that another, perhaps peripheral heritable factor is responsible for the disease state which becomes manifest; ulcer in WKY, hypertension in SHR rats (see Kopin *et al.*, 1980). This concept, of course, can be extended to other disease states, including psychosomatic diseases in humans (see below).

### 4.3. Orthostatic Hypotension

Changes in posture are normally attended by compensatory cardiovascular reflexes which maintain blood pressure within physiological limits. Failure of the baroreflex acts (baroreceptors, CNS, sympathetic nerves) to activate appropriately release of norepinephrine or a deficient receptor effector system when standing, results in a fall in blood pressure, to a point where the blood flow to the brain is insufficient and lightheadedness or fainting occurs. Reflex arc failure is attended by an absence of an increase in plasma norepinephrine, called hypoadrenergic orthostatic hypotension (Cryer *et al.*, 1978), in contrast to receptor-effector failure, which evokes reflexly induced sympathetic responses and elevation in plasma norepinephrine levels. When reflex arc failure is due to a disorder in the CNS, the patients have other signs of a central deficit characterized by incoordination or parkinsonian-like tremors (Shy–Drager syndrome). Other patients appear to have degeneration of peripheral sympathetic nerves, e.g., in diabetic neuropathy. Ziegler *et al.* (1977) distinguished between the peripheral (primary or idiopathic as well as diabetic) orthostatic hypotension and the Shy–Drager syndrome (or multiple system atrophy) on the basis of the basal plasma norepinephrine levels. In diseases which result in destruction of sympathetic nerves, the plasma norepinephrine levels are low, whereas in autonomic failure due to CNS disorders, the plasma levels of norepinephrine are normal during reclining. In both forms of reflex failure there is almost no change in plasma norepinephrine levels with standing.

### 4.4. Cardiovascular Disorders

Because the sympathoadrenal system is important in the regulation of the cardiovascular system, as well as in the response to stress, it is not surprising that cardiovascular diseases may be attended by changes in plasma catecholamine levels. In heart failure (Fig. 5), the degree of elevation of plasma norepinephrine is related to the severity of the failure (Thomas and Marks, 1978). The mechanisms responsible for this are not clear, but it is likely that reflexes initiated from dilated atria or veins may be responsible, although decreased clearance rates of the catecholamines from the plasma have not been ruled out.

Myocardial infarction, as other catastrophic illnesses, evokes stress–pain-induced increases in plasma norepinephrine and epinephrine. The severity of the infarct occurrence of pulmonary edema, and increases in plasma catecholamines are interrelated.

### 4.5. Metabolic Disorders

Since catecholamines are also involved in regulation of metabolism, it might be expected that their plasma levels might vary in different metabolic

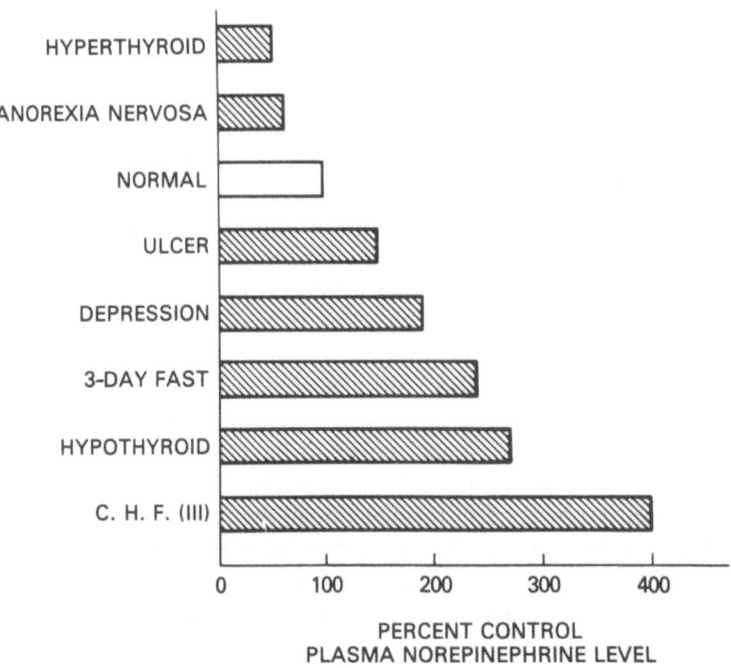

FIGURE 5. Plasma norepinephrine levels in disease states which are not related to primary disorder of the sympathetic nervous system and/or are not associated with hypertension. The blood pressure is usually low in anorexia nervosa. In hyperthyroidism the systolic pressure may be elevated, but the diastolic pressure is normal or depressed.

states. While oral glucose is attended by increases in plasma norepinephrine (Young *et al.*, 1980), the stress of a 3-day fast also increases plasma norepinephrine levels (Fig. 5). In anorexia nervosa, however, plasma norepinephrine (and blood pressures) are low (Fig. 5) and there is a blunting of the catecholamine responses to postural change and exertion (Gross *et al.*, 1979). These abnormalities are reversed when the patients regain weight.

In hypothyroidism (Fig. 5), plasma norepinephrine levels are elevated and in hyperthyroid patients the levels are depressed (Christensen, 1973). These changes may be related to compensatory responses and/or the alterations in β-receptors which attend abnormalities in thyroid state (Williams *et al.*, 1977).

### 4.6. Psychosomatic Disorders

Anxiety is frequently attended by signs of increased sympatho-adrenal medullary activity, and plasma catecholamines are increased in depressions and in patients with ulcer.

## 5. SUMMARY

There have been developed several highly sensitive and specific methods for assay for plasma catecholamine levels. These levels are useful indices of sympatho-adrenal medullary activity and rapidly reflect the physiological responses to postural changes, temperature alterations, plasma glucose levels, or a variety of stresses.

Abnormal basal plasma catecholamine levels are found in a variety of diseases in which there are compensatory sympathoadrenal responses as well as in disorders which involve the sympatho-adrenal medullary system or its control centers in the CNS.

## 6. REFERENCES

Abel, J. J., and Crawford, A. C., 1897, On the blood pressure raising constituent of the suprarenal capsule, *Bull. Johns Hopkins Hos.* **8**:151.

Avakian, E. V., and Horvath, S. M., 1980, Plasma catecholamine responses to tyrosine hydroxylase inhibition and cold exposure, *Life Sci.* **26**:1691.

Bevan, J. A., 1977, Some functional consequences of variation in adrenergic synaptic cleft width and in nerve density and distribution, *Fed. Proc.* **36**:2439.

Cannon, W. B., and Rosenbleuth, A., 1937, *Autonomic Neuroeffector Systems*, Macmillan Co., New York.

Cannon, W. B., and Uridil, J. E., 1921, Studies on the conditions of activity in endocrine glands, VIII. Some effects on the denervated heart of stimulating the nerves of the liver, *Am. J. Physiol.* **58**:353.

Christensen, N. J., 1973, Plasma noradrenaline and adrenaline in patients with thyrotoxicosis and myxoedema, *Clin. Sci. Mol. Med.* **45**:163.

Christensen, N. J., and Brandsborg, O., 1973, The relationship between plasma catecholamine concentration and pulse rate during exercise and standing, *Eur. J. Clin. Invest.* **3**:299.

Clutter, W. E., Bier, D. M., Shah, S. D., and Cryer, P. E., 1980, Epinephrine plasma metabolic clearance rates and physiologic thresholds for metabolic and hemodynamic actions in man, *J. Clin. Invest.* **66**:94.

Cryer, P. E., 1976, Isotope-derivative measurements of plasma norepinephrine and epinephrine in man, *Diabetes* **25**:1071.

Cryer, P. E., Santiago, J. V., and Shah, S. D., 1974, Measurement of norepinephrine and epinephrine in small volumes of human plasma by a single isotope derivative method: Response to the upright posture, *J. Clin. Endocrinol. Metab.* **39**:1025.

Cryer, P. E., Haymond, M. W., Santiago, J. V., and Shan, S. D., 1976, Norepinephrine and epinephrine release and adrenergic mediation of smoking associated hemodynamic and metabolic events, *N. Engl. J. Med.* **295**:573.

Cryer, P. E., Silverberg, A. B., Santiago, J. V., and Shah, S. D., 1978, Plasma catecholamines in diabetes: The syndromes of hypoadrenergic and hyperadrenergic postural hypotension, *Am. J. Med.* **64**:407.

Da Prada, M., and Zürcher, G., 1976, Simultaneous radio-enzymatic determination of plasma and tissue adrenaline, noradrenaline and dopamine within the femtomole range, *Life Sci.* **19**:1161.

Dimsdale, J. E., and Moss, J., 1980, Plasma catecholamines in stress and exercise, *J. Am. Med. Assoc.* **243**:340.

Ehrhardt, J. D., and Schwartz, J., 1978, A gas chromatography–mass spectrometry assay of human plasma catecholamines, *Clin. Chim. Acta* **88**:71.

Elliott, T. R., 1905, The action of adrenaline, *J. Physiol. (London)* **32**:401.
Elliott, T. R., 1912, The control of the suprarenal glands by the splanchnic nerve, *J. Physiol. (London)* **44**:374.
Engleman, K., and Portnoy, B., 1970, A sensitive double-isotope derivative assay for norepinephrine and epinephrine, *Circ. Res.* **26**:53.
Esler, M., Jackman, G., Bobik, A., Kelleher, D., Jennings, G., Leonard, P., Skews, H., and Korner, P., 1979, Determination of norepinephrine apparent release rate and clearance in humans, *Life Sci.* **25**:1461.
Gaddum, J. H., 1959, Bioassay procedures, *Pharmacol. Rev.* **11**:241.
Ghione, S., Palombo, C., Pellegrini, M., Fommei, E., Pilo, A., and Donato, L., 1978, The kinetics of plasma noradrenaline in normal and hypertensive subjects, *Clin. Sci. Mol. Med.* **55**:89s.
Goldstein, D. S., Feuerstein, G., Izzo, J. L., Kopin, I. J., and Keiser, H. R., 1981, Validity and reliability of high pressure liquid chromatography with electrochemical detection for measuring plasma levels of norepinephrine and epinephrine in man, *Life Sci.* **28**:467.
Gross, H. A., Lake, C. R. Ebert, M. H., Ziegler, M. G., and Kopin, I. J., 1979, Catecholamine metabolism in primary anorexia nervosa, *J. Clin. Endocrinol. Metab.* **49**:805.
Kissinger, P. T., Refshauge, C., Dreiling, R., and Adams, R. N., 1973, An electrochemical detector for liquid chromatography with picogram sensitivity, *Anal. Lett.* **6**(5):465.
Kopin, I. J., McCarty, R., and Yamaguchi, I., 1980, Plasma catecholamines in human experimental hypertension, *Clin. Exp. Hypertens.* **2**:379.
Kopin, I. J., Goldstein, D. S., and Feuerstein, G. Z., 1981, A position paper: The sympathetic nervous system and hypertension. in: *Frontiers in Hypertension Research* (J. J. Laragh, F. R. Buhler, and D. W. Seldin, eds.), pp. 283–289, Springer-Verlag, New York.
Lake, C. R., Ziegler, M. G., and Kopin, I. J., 1976, Use of plasma norepinephrine for evaluation of sympathetic neuronal function in man, *Life Sci.* **18**:1315.
Langer, S. Z., 1977, Presynaptic receptors and this role in the regulation of transmitter release. *Br. J. Pharmacol.* **60**:481.
Langley, J. N., 1901, The difference of behavior of central and peripheral pilomotor nerve cells, *J. Physiol. (London)* **27**:224.
Loewi, O., 1921, Ober humorale ubertragberkeit der herznervenwirkung, *Pfluegers Arch. Ges. Physiol.* **189**:239.
Lund, A., 1950, Simultaneous fluorimetric determinations of adrenaline and noradrenaline in blood, *Acta Pharmacol. Toxicol.* **6**:137.
McCarty, R., and Kopin, I. J., 1979, Stress-induced alterations in plasma catecholamines and behavior of rats: Effects of chlorisondamine and bretylium, *Behav. Neural Biol.* **27**:249.
Oliver, G., and Schafer, E. A., 1895, The physiological effects of extracts on the suprarenal capsules, *J. Physiol. (London)* **18**:230.
Passon, P. G., and Peuler, J. D., 1973, A simplified radiometric assay for plasma norepinephrine and epinephrine, *Anal. Biochem.* **51**:618.
Peart, W. S., 1949, The nature of splenic sympathin, *J. Physiol. (London)* **108**:491.
Peuler, J. D., and Johnson, G. A., 1977, Simultaneous single isotope radioenzymatic assay of plasma norepinephrine, epinephrine and dopamine, *Life Sci.* **21**:625.
Reid, J. L., and Kopin, I. J., 1975, The effects of ganglionic blockade, reserpine and vinblastine on plasma catecholamines and dopamine-β-hydroxylase in the rat, *J. Pharmacol. Exp. Ther.* **193**:748.
Renzini, V., Brunori, C. A., and Valori, C., 1970, A sensitive and specific fluorimetric method for the determination of noradrenaline and adrenaline in human plasma, *Clin. Chim. Acta* **30**:587.
Robertson, D., Jürgen, C., Frölich, J. C., Carr, R. K., Watson, J. T., Hollifield, J. W., Shand, D. G., and Oates, J. A., 1978, Effects of caffeine on plasma renin activity catecholamines and blood pressure, *N. Engl. J. Med.* **298**:181.
Saelens, J. K., Schoen, M. S., and Kovacsics, G. B., 1967, An enzyme assay for norepinephrine in brain tissue, *Biochem. Pharmacol.* **16**:1403.
Silverberg, A. B., Shah, S. D., Haymond, M. W., and Cryer, P. E., 1978, Norepinephrine: Hormone and neurotransmitter in man, *Am. J. Physiol.* **234**(3):E252.

Thomas, J. A., and Marks, B. H., 1978, Plasma norepinephrine in congestive heart failure, *Am. J. Cardiol.* **41**(2):233.

Von Euler, U. S., 1948, Identification of the sympathomimetic Ergone in adrenergic nerves of cattle (sympathin N) with laevo-noradrenaline, *Acta Physiol. Scand.* **16**:63.

Von Euler, U. S., 1959, The development and applications of the trihydroxyindole method for catecholamines, *Pharmacol. Rev.* **11**:262.

Vulpian, M., 1856, Note sur quelque reactions propes a la substance des capsules surrenales, *C.R. Acad. Sci.* **13**:663.

Weil-Malherbe, H., and Bone, A. D., 1953, The adrenergic amines of human blood, *Lancet* **I**:974.

Weise, V. K., and Kopin, I. J., 1976, Assay of catecholamines in human plasma: Studies of a single isotope radioenzymatic procedure, *Life Sci.* **19**:1673.

Williams, L. T., Lefkowitz, R. J., Watanabe, A. M., Hathaway, D. R., and Besch, H. R., 1977, Thyroid hormone regulation of beta-adrenergic receptor number, *J. Biol. Chem.* **252**:2787.

Yamaguchi, I., and Kopin, I. J., 1979, Plasma catecholamine and blood pressure responses to sympathetic stimulation in pithed rats, *Am. J. Physiol.* **237**(3):H305.

Young, J. B., Rowe, J. W., Pallotta, J. A., Sparrow, D., and Landsberg, L., 1980, Enhanced plasma norepinephrine response to upright posture and oral glucose administration in elderly human subjects, *Metabolism* **29**:532.

Ziegler, M. G., Lake, C. R., and Kopin, I. J., 1976, Plasma noradrenaline increases with age, *Nature (London)* **261**:333.

Ziegler, M. G., Lake, C. R., and Kopin, I. J., 1977, The sympathetic nervous system defect in primary orthostatic hypotension, *N. Engl. J. Med.* **296**:293.

CHAPTER 9

# Nonspecific Pituitary Responses to Hypothalamic Hormones in Basic and Clinical Research

## DANIELA COCCHI, VITTORIO LOCATELLI, and EUGENIO E. MÜLLER

## 1. INTRODUCTION

In endocrinology the concept of neurohormonal specificity, e.g., that a specific hypothalamic regulatory hormone (RH) selectively affects the secretion of an anterior pituitary (AP) hormone, has been a tenet for many years. However, in the early 1970s the demonstration by Tashjian *et al.* (1971) that thyrotropin-releasing hormone (TRH) was capable of affecting prolactin (PRL) release in cultured rat pituitary tumor cells provided strong evidence against that notion. Another hypophysiotropic hormone, e.g., somatostatin, demonstrated an even more evident example of the lack of neurohormonal specificity; in addition to blocking growth hormone (GH) release, it inhibits the secretion of thyrotropin (TSH) but not PRL stimulated by TRH and of an array of gastrointestinal and kidney hormones, e.g., gastrin, secretin, insulin, glucagon, renin, by acting directly on the cells secreting these peptides (for review see Guillemin and Gerich, 1976). Another RH which can be added to this list is luteinizing hormone-releasing hormone (LH-RH), since indisputable evidence has been presented for its ability to induce concurrent release of LH and follicle-stimulating hormone (FSH) from pituitary gonadotropes, and proof for the existence of a distinct FSH-releasing hormone is lacking (Schally *et al.*, 1976). The aforementioned and other endocrine effects of RHs by no means can be referred to as nonspecific, for they occur in either animals or humans under physiologic conditions. What is intended here as nonspecific AP responses is rather the ability of RHs to affect under certain circumstances the secretion of some AP hormones which are normally refractory to stimulation.

---

DANIELA COCCHI, VITTORIO LOCATELLI, and EUGENIO E. MÜLLER • Department of Pharmacology, School of Medicine, University of Milan, 20129 Milan, Italy.

In this contribution many pathologic conditions of basic and clinical research in which altered AP responsiveness to RHs occurs are reviewed and discussed. It was our aim to focus especially on the potential of these neuroendocrine responses to advance knowledge into the pathophysiology of the underlying human diseases.

## 2. NONSPECIFIC GH-RELEASING EFFECT OF NEUROHORMONES

### 2.1. Animal Studies

#### 2.1.1. TRH

Search of the elusive GH-releasing factor or hormone (GRF or GH-RH) (Müller et al., 1979) only recently has culminated in the identification, characterization, and synthetic replicate of a specific peptide (Guillemin et al., 1982; Rivier et al., 1982) but has prompted in the past studies on the potential GRF activity of identified RHs (Table 1). TRH was reported capable of inducing GH release from in vitro incubated sheep (Takahara et al., 1974b), calf (Machlin et al., 1974), and rat (see below) APs. In vivo, an increased release of GH after TRH was first described in the cow infused with a 100-μg dose of the peptide (Convey et al., 1973).

In the rat positive responses to TRH dealt with direct infusions into the hypophyseal portal vasculature (Takahara et al., 1974a), or perfused hemipituitary glands in vitro (Carlson et al., 1974). It soon became apparent that the TRH-induced GH rise in the rat may be affected by various factors, a major one referring to anesthesia. Wakabayashi et al. (1979) failed to show any effect of TRH (0.2 μg/100 g body wt) on the physiologic pulsatile release of GH in unanesthetized freely moving male rats, but in their hands the same TRH dose increased plasma GH in urethane-anesthetized rats. Similar results were reported by other authors in male rats anesthetized with urethane (Kato et al., 1975) or ether (Chihara et al., 1976a).

A sex-related difference may also be envisaged. In our own studies the minimal effective doses of TRH for increasing GH release in urethane-anesthetized rats were 0.6 and 1.2 μg/rat for males and females, respectively. Panerai et al. (1977b) and Ojeda et al. (1977) reported a greater GH rise after TRH (0.85 μg/100 g body wt) in male than female rats under tribromoethanol anesthesia.

However, susceptibility to make a GH secretory response to TRH in the rat is even more related to the state of anatomical and/or functional connections between the CNS and the AP gland. In an extensive series of studies, we have shown that TRH administered to intact, urethane-anesthetized female rats increased baseline GH levels only at the highest dose used (1.2 μg/rat), but was completely ineffective at doses of 0.15, 0.30, and 0.6 μg/rat. In contrast, TRH increased GH levels also when injected at the lower doses into hypophysec-

## TABLE 1
### Hypothalamic Releasing and Inhibiting Hormones

| Hypothalamic hormone (or factor) | Abbreviation | Structure[a,b] |
|---|---|---|
| Corticotropin (ACTH)-releasing factor | CRF | H-Ser-Gln-Glu-Pro-Pro-Ile-Ser-Leu-Asp-Leu-Thr-Phe-His-Leu-Leu-Arg-Gln-Val-Leu-Glu-Met-Thr-Lys-Ala-Asp-Glu-Leu-Ala-Gln-Gln-Ala-His-Ser-Asn-Arg-Lys-Leu-Leu-Asp-Ile-Ala-NH$_2$ |
| Thyrotropin (TSH)-releasing hormone | TRH | pGlu-His-Pro-NH$_2$ |
| Luteinizing hormone (LH)-releasing hormone | LH-RH | pGlu-His-Trp-Ser-Tyr-Gly-Leu-Arg-Pro-Gly-NH$_2$ |
| Follicle-stimulating hormone (FSH)-releasing hormone | FSH-RH | pGlu-His-Trp-Ser-Tyr-Gly-Leu-Arg-Pro-Gly-NH$_2$ |
| Growth hormone (GH)-releasing factor | GRF | H-Tyr-Ala-Asp-Ala-Ile-Phe-Thr-Asn-Ser-Tyr-Arg-Lys-Val-Leu-Gly-Gln-Leu-Ser-Ala-Arg-Lys-Leu-Leu-Gln-Asp-Ile-Met-Ser-Arg-Gln-Gln-Gly-Glu-Ser-Asn-Gln-Glu-Arg-Gly-Ala-Arg-Ala-Arg-Leu-NH$_2$* |
| Growth hormone (GH) release-inhibiting hormone | GH-RIH[c] | H-Ala-Gly-Cis-Lys-Asn-Phe-Phe-Trp-Lys-Thr-Phe-Thr-Ser-Cys-OH |
| Prolactin-inhibiting factor | PIF | — |
| Prolactin-releasing factor | PRF | — |
| Melanocyte-stimulating hormone (MSH) release-inhibiting hormone | MR-IH | H-Pro-Leu-Gly-NH$_2$ |
| Melanocyte-stimulating hormone (MSH)-releasing hormone | MRH | H-Cys-Tyr-Ile-Gln-Asn-OH |

[a] The three-letter symbols for amino acids are those recommended in *Biochem. J.* **102**:23 (1967), except for pGlu, pyroglutamic acid.
[b] Asterisk indicates the 44 amino acid peptide with growth hormone releasing activity isolated from a human tumor of the pancreas (Guillemin *et al.*, 1982).
[c] Or somatostatin.

tomized rats bearing a pituitary transplant under the kidney capsule (Udeschini *et al.*, 1976). As in rats bearing an ectopic AP, TRH proved to be a more effective GH releaser than in sham-operated controls in rats with hypothalamic ablation or bearing bilateral electrolytic lesions in the median eminence (ME) (Chihara *et al.*, 1976b; Müller *et al.*, 1977a).

The susceptibility of the somatotropes to the releasing action of TRH increases with time from CNS–pituitary disconnection. Figure 1 shows that in hypophysectomized rats with an ectopic AP, the TRH-induced GH rise was absent at the dose of 1.2 µg 1 day after transplantation; it occurred at the dose of 0.6 µg 3 and 8 days after transplantation and was clearly elicitable also at the low dose of 0.15 µg from 15 up to 60 days from transplantation (Panerai *et al.*, 1977b). Supporting these data, light microscopic and ultrastructural studies showed that with time, somatotropes of the ectopic gland acquired an en-

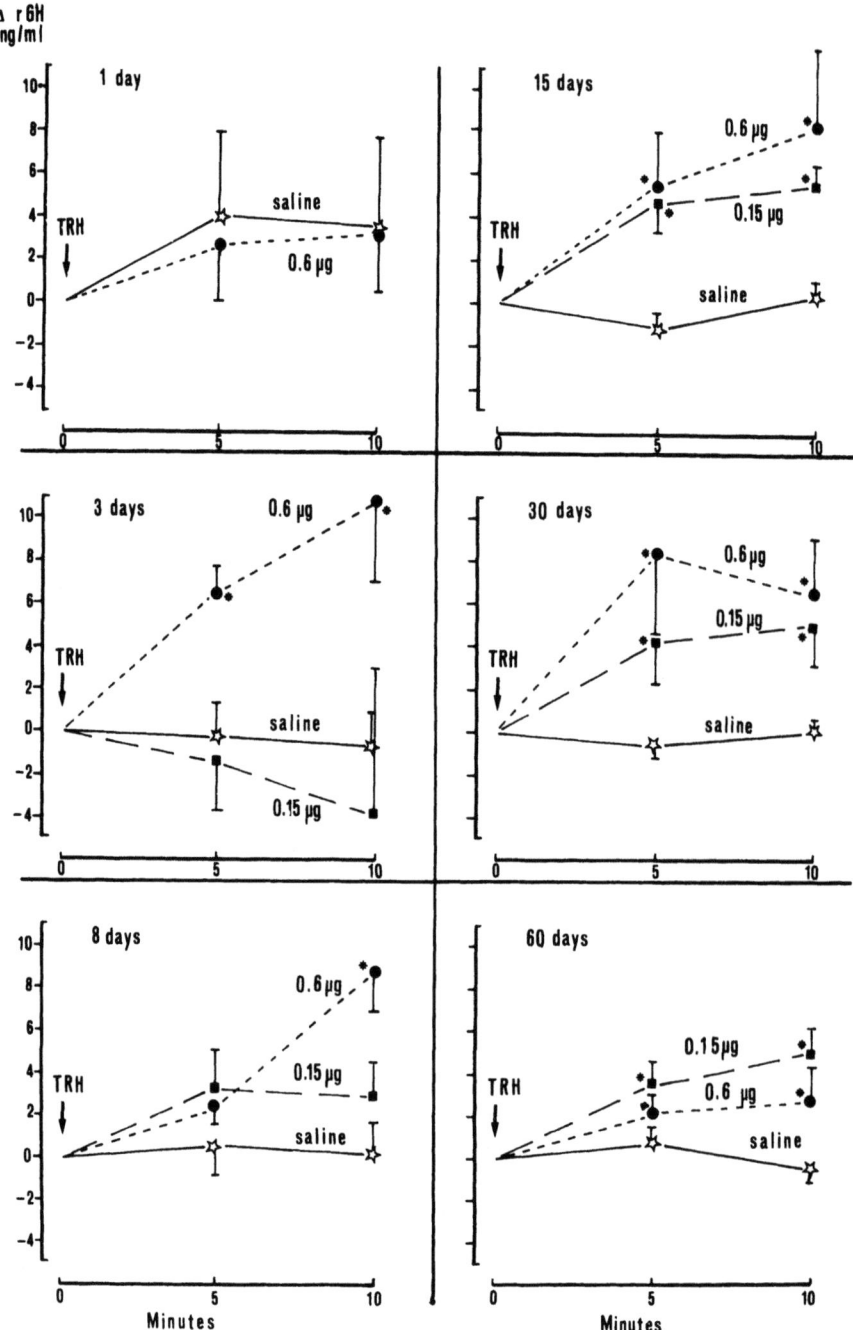

FIGURE 1. GH-releasing effect of TRH in hypophysectomized female rats bearing one anterior pituitary gland transplanted under the kidney capsule: effect of the time interval after transplantation. Data are expressed as means ± S.E.M. (vertical lines) of the change in values from baseline. In each group 6 animals were used, except in the experiments 15 days after transplantation, in which 9 animals per group were used. Asterisks indicate a statistically significant difference ($p < 0.01$) from saline-injected controls. Number of days after transplantation, and the dose of TRH (μg) are indicated. From Panerai et al. (1977b) J. Endocrinol., courtesy of the English Society for Endocrinology.

hanced sensitivity to TRH and responded with exocytosis at doses ineffective on *in situ* APs (Rossi *et al.*, 1977). Extensive biochemical studies conducted on normotopic and 30-day-old rat pituitary grafts have demonstrated that TRH infused *in vivo* or added *in vitro* to the incubation fluid produced a large increase in GH biosynthesis in and release from the grafts, being instead ineffective on the normotopic glands (Giannattasio *et al.*, 1979). More recent ultrastructural findings have confirmed the aforementioned results. In the study of McComb *et al.* (1981) both ectopic lactotropes and somatotropes of hypophysectomized rats showed subcellular changes indicative of stimulation following chronic TRH treatment.

In preliminary reports (Shani and Siegel, 1980; Tonon *et al.*, 1981) TRH has been shown to be a potent releaser of ACTH in rats and α-MSH in frogs, both of which species are known to have a well-developed intermediate lobe. Since *in vitro* the cells of the rat AP do not show ACTH release in response to TRH (Gershengorn *et al.*, 1980), the pars intermedia is the most likely source of ACTH release by TRH.

### 2.1.2. Other Neurohormones

Induction of GH release from APs disconnected from the CNS is not unique to TRH. Administration of LH-RH at doses (0.6 and 1.2 μg/rat) ineffective to raise GH levels in anesthetized female rats increased significantly GH levels in rats bearing an ectopic gland. In the latter, melanocyte-stimulating hormone release-inhibiting hormone (MIF) (1.2 μg) was active as GH releaser, though its action was less consistent than that of LH-RH. In contrast, a peptide of pituitary origin such as α-MSH proved to be ineffective, indicating that the above is not a general phenomenon (Panerai *et al.*, 1976). Hormonal and/or nonhormonal factors such as lack or reduction of the inhibitory effect of estrogens (Szabo and Frohman, 1975) or thyroid hormones (Chihara *et al.*, 1976b) or reduced clearance rate of the RHs following hypophysectomy (Kastin *et al.*, 1974) cannot be dismissed as responsible for the GH rise induced by TRH, LH-RH, and MIF. However, the ability of TRH to evoke GH release in many *in vitro* pituitary systems (Machlin *et al.*, 1974; Takahara *et al.*, 1974b) and also from a pituitary grafted under the kidney capsule of intact rats (Panerai *et al.*, 1977c) mitigates against this view (see Section 4.5.1 for a further discussion).

In conclusion, TRH and other RHs are poor GH releasers in several animal species, except in the cow, when they are injected into intact animals. They are instead more potent GH releasers when acting on APs *in vitro* or anatomically and/or functionally disconnected from CNS influences. In this vein, facilitation of the TRH-induced GH rise in intact animals by anesthesia likely reflects a pharmacologically induced disruption of CNS–AP links due to an impairment of CNS transmitter function. Therefore, it may be suggested that the secretory response to TRH present in rat "ectopic" AP does not represent the appearance of a new feature but rather that amplification of a feature which is genetically programmed in normotopic somatotropes. However, as long as these pituitary cells are under control by specific RHs this potential is largely repressed. In this connection, of interest is the observation that in intact rats administration of an antiserum to somatostatin allowed a GH response to TRH to be evidenced at a dose (0.6 μg) ineffective in control rats (Panerai et al., 1977a). Alternatively, abnormal somatotrope responsiveness to TRH may denote reduction of endogenous TRH production and, hence, diminished downregulation of TRH receptors (see Section 4.4.1). The quantitative regulation of the TRH receptor level was first demonstrated in a pituitary tumor line cultured in the presence of the neurohormone (Hinkle and Tashjian, 1975). Obviously, more studies are needed to clarify the mechanism(s) underlying inappropriate responsiveness of somatotropes to RHs in experimental animals.

## 2.2. Anomalous Responses in Fetuses and Newborns

Due to incomplete development of the portal hypophyseal system (Glydon, 1957; Campbell, 1966) and the ontogenetic immaturity of the brain neurotransmitter system (Loizou, 1972), regulatory mechanisms for the control of some AP hormones may not become fully operative until infancy. For GH, data in the fetus and newborn are consistent with the existence of such immaturity (see Kaplan et al., 1972), functionally reflected by the lack of pulsatile release (Cocchi et al., 1976). Hence, animal and human fetuses and newborns represent suitable models for evaluating GH responses to TRH.

Table 2 shows that in 12-day-old female and male pups, TRH at all doses used (0.15, 0.3, 0.6 and 1.5 μg/100 g body wt i.p.) induced a significant, though not dose-related, increase in plasma GH. The neurohormone instead released

TABLE 2
Effect of Various Doses of TRH on Plasma GH and PRL Levels in 13-Day-Old Pups Pretreated with Vehicle or 6-OHDA When 5 Days Old[a]

| Pretreatment[b] | Treatment (μg/100 g body wt) | Plasma GH (ng/ml) | % of control | Plasma PRL (ng/ml) | % of control |
|---|---|---|---|---|---|
| Vehicle | Saline (10)[c] | 15.2 ± 1.7[d] | 100 | 5.4 ± 0.6 | 100 |
| 6-OHDA | Saline (10) | 11.9 ± 1.6 | 100 | 14.1 ± 2.5[f] | 100 |
| Vehicle | TRH, 0.15 (6) | 31.9 ± 4.0[e] | 210 | 6.2 ± 1.4 | 114 |
| 6-OHDA | TRH, 0.15 (6) | 29.5 ± 6.0[e] | 248 | 13.8 ± 2.5[f] | 97 |
| Vehicle | TRH, 0.3 (8) | 35.0 ± 7.0[e] | 230 | 5.8 ± 0.8 | 107 |
| 6-OHDA | TRH, 0.3 (7) | 34.5 ± 8.0[e] | 290 | 19.8 ± 6.4[f] | 140 |
| Vehicle | TRH, 0.6 (8) | 30.0 ± 3.5[e] | 198 | 5.4 ± 1.5 | 100 |
| 6-OHDA | TRH, 0.6 (8) | 35.5 ± 3.6[e] | 282 | 12.2 ± 1.4[f] | 86 |
| Vehicle | TRH, 1.5 (7) | 26.0 ± 2.5[e] | 170 | 18.9 ± 4.0[e] | 350 |
| 6-OHDA | TRH, 1.5 (7) | 36.3 ± 4.4[e] | 305 | 8.4 ± 5.0 | 59 |

[a] From Gil-Ad et al. (1976), Neuroendocrinology, courtesy of Karger, Basel.
[b] 10 μl 0.1% ascorbic acid was used as vehicle; 6-OHDA was given in a dose of 60 μg in 10 μl vehicle.
[c] Number of rats in parentheses.
[d] Mean ± S.E.M.
[e] $p < 0.01$ vs. saline-treated (same pretreatment).
[f] $p < 0.01$ vs. vehicle-pretreated (same treatment).

PRL only at the highest dose. The TRH-induced GH rise was even more evident if pups were previously subjected to central sympathectomy by intraventricularly injected 6-hydroxydopamine (6-OHDA). In these animals, TRH failed to release PRL at any dose level. All these data showing that neonatal pituitaries are highly susceptible to the GH-releasing action of TRH, and that their susceptibility is increased further by 6-OHDA-induced disruption of catecholaminergic neurotransmission, support the proposition that functional CNS–AP disconnection is the cause of the deranged TRH responsiveness (see above). Data reminiscent of those obtained in the rat have been reported recently in human fetuses (Roti et al., 1982). Administration of 400 μg TRH to 59 mothers during labor increased significantly GH levels in the blood collected by the funiculum 60–90 min later, indicating that TRH stimulated GH release in the newborn. In contrast to these findings, however, no rise in GH was evoked by a lower TRH dose (8 μg/kg) directly injected into 2-day-old newborns (Hiba et al., 1977). The GH-releasing effect of TRH was not observed in either ovine fetuses between 88 and 133 days of gestation (term 147 days) or in 5- to 17-day-old lambs (Thomsett et al., 1980) suggesting that also species-related differences are involved in this effect.

With regard to other RHs, clonal strains of pituitary cells derived from Rathke's pouch epithelium of 11- to 13-day-old fetal rats responded to low doses of LH-RH by releasing PRL; at high doses the opposite effect was present (Herbert and Rennels, 1977). Cultures of dispersed pituitary cells from adult animals did not respond to LH-RH (Blackwell et al., 1973; Tang and Spies, 1976).

## 3. OTHER ANOMALOUS AP RESPONSES TO RHs

To the wide spectrum of inhibitory actions of somatostatin on the secretory activity of a variety of endocrine and exocrine glands inhibition of PRL release also can be listed. Because this activity is not invariably manifested and a dissociation between *in vitro* and *in vivo* data is frequently present (see below), connotation of this response as anomalous seems warranted.

A number of reports have been published on the inhibition of PRL release by somatostatin from cultured rat pituitary cells either normal (Grant et al., 1974; Vale et al., 1974) or tumoral (Schonbrunn and Tashjian, 1978; Maruyama and Ishikawa, 1977). The action of somatostatin on the release of PRL was considerably less than its effect on the secretion of GH, and was increased with AP cells derived from castrated or from estrogen–progesterone-primed rats (Vale et al., 1974). In *in vivo* studies in the rat, somatostatin was devoid of PRL inhibitory activity in conditions of either normo- or hyperprolactinemia due to neuroleptics or to grafting an additional pituitary under the kidney capsule. Somatostatin instead lowered PRL levels dose-dependently when intact rats treated with neuroleptics or grafted had been previously primed with estradiol (Cooper and Shin, 981). Thus, from *in vitro* and *in vivo* studies it would appear that estrogen priming in the rat unmasks or induces receptors which may be acted upon by somatostatin. How these findings may be reconciled with the inhibitory action of somatostatin on lactotropes from gonadectomized rats is presently unknown (see Section 4.5.3d).

At odds with these data is the finding that in the cow, administration of somatostatin increased basal and premilking serum concentrations of PRL (Gorewit, 1980), though the mechanism subserving this action is presently obscure.

## 4. HUMAN STUDIES

### 4.1. Normal Subjects

Before dealing with anomalous AP responses to RHs in human pathologic states a brief review of some of those responses in healthy subjects is proper, for also in the latter nonspecific responses have at times been reported.

According to some reports, TRH would be competent to release GH in normal individuals (Karlberg et al., 1971; Saito et al., 1971; Eastman and Lazarus, 1972; Torjesen et al., 1973). However, (1) in some of these studies GH responded to administration of rather huge amounts of TRH, which resembles the GH responsiveness of intact rats to high TRH doses (see above), (2) GH responses were evident in women but not in men, and (3) finally, proper control experiments were not performed. Hence, the possibility of a nonspecific stress effect could not be excluded. In our own experience TRH (400–500 µg i.v.) administered to normal subjects failed to elicit a rise in GH levels (Brambilla et al., 1978; Salerno et al., 1980), so that we regard that neuroendocrine response as qualitatively abnormal.

A small increase in serum FSH levels has been reported following injection or infusion of TRH in normal males but not pre- and postmenopausal females. Administration of estrogens suppressed this response (Mortimer et al., 1973, 1974). A transient stimulation of LH secretion by TRH has been instead described in normal women at midcycle (Franchimont, 1972). No effect of somatostatin has been detected on basal PRL levels in five subjects when the peptide was infused at doses of 750 µg in 1 hr (Barreca et al., 1979). In the study of Bratusch-Marrain and Waldhäusl (1979) somatostatin was effective in inhibiting the increase in serum PRL induced by arginine and phenylalanine loading, a finding not confirmed by Siler et al. (1973) in a study with arginine. Somatostatin also was unable to inhibit PRL release stimulated by hypoglycemia or TRH (Hall et al., 1973; Copinschi et al., 1976; Carr et al., 1975).

### 4.2. Pathologic States

Abnormal AP responsiveness to RHs has been described in various human pathologic states comprising psychiatric, metabolic, and endocrine disorders. While the true mechanism(s) underlying the anomalous response patterns has not been clarified yet, some of the data obtained offer considerable promise as neurobiological markers, especially in psychiatric diseases, for differentiating patient groups on the basis of biological differences.

### 4.3. Psychiatric Diseases

#### 4.3.1. Primary Affective Disorders

Almost paralleling animal studies on abnormal somatotrope's responsiveness to some RHs in ectopically located APs (see above), anomalous response to RHs were reported in patients presenting with primary affective disorders (PAD). In the study of Maeda et al. (1975b), administration of 500 μg of TRH induced an increase in plasma GH in 8 of 13 patients with PAD, irrespective of whether they were unipolars or bipolars. GH responses were found 90–120 min after TRH in 3 of 8 positively responding depressive patients. Since then many studies have confirmed these findings, and a nonspecific GH release after TRH has been found in approximately half of the patients with PAD in the studies of Chazot et al. (1974), Takahashi et al. (1975), and Brambilla et al. (1978).

However, abnormal GH responsiveness to TRH is not the only qualitative alteration of AP responsiveness to RHs in PAD. In fact, by extending to five hormones the evaluation of the AP responsiveness to TRH and in addition to LH-RH, we succeeded in disclosing a much broader alteration than previously suspected (Brambilla et al., 1978). In 7 of the 16 patients considered, in fact, the TRH-induced GH response was concomitant with a TRH-induced rise in FSH and/or LH, and an isolated FSH responsiveness to TRH was present in one patient. In addition, one patient responded to LH-RH by releasing GH and three patients by releasing PRL. In only three patients was conventional AP responsiveness to RHs found. Saline administration did not evoke nonspecific hormone release in the five patients in whom it was administered (Table 3).

Disorders of brain monoamine transmitters are thought to play an important role in the pathogenesis of PAD, and alterations of pre- and postsynaptic monoamine receptors have been evidenced in depressive patients (Schildkraut, 1974; Sulser, 1978). Thus, the most parsimonious hypothesis to explain deranged pituitary responsiveness to neurohormones is that it may reflect a neuroendocrine impairment, e.g., a functional CNS–pituitary disconnection. This would remove the neurohormonal influences for the secretion of pituitary hormones, e.g., for GH, GRF, and somatostatin, or, alternatively, be responsible for chronic deprivation of other RHs, e.g., TRH, with resulting hyperresponsiveness to the exogenous peptide. Alternatively, it has been proposed that RHs can modify nonspecifically AP hormone secretion in man through a CNS-mediated mechanism. For instance, TRH and LH-RH may act as neurotransmitters to affect monoamine transmission (Keller et al., 1974, Breese et al., 1975) and proof has been given that pharmacologic doses of TRH inhibit GH secretion in pentobarbital; morphine; or chlorpromazine-treated rats (Collu et

TABLE 3
Altered Anterior Pituitary Responses in Depressed Patients after Regulatory Hormone Administration[a]

| Case/Sex | Regulatory hormone testing | Pituitary responsiveness[b] | | | | |
|---|---|---|---|---|---|---|
| | | GH | PRL | FSH | LH | TSH |
| 1/M | TRH | + + | + | + + | + + | Normal |
| | LH-RH | − | − | Delayed | + | |
| 2/M | TRH | + + | + | − | − | Low |
| | LH-RH | − | − | + | + | |
| 3/M | Saline only | | | | | |
| 4/F post[c] | TRH | − | + | − | − | Normal |
| | LH-RH | − | − | Delayed | + | |
| 5/F post | TRH | + + | + | + + | − | Normal |
| | LH-RH | − | + + | + | + | |
| 6/F post | TRH | + + | + | − | − | Low |
| | LH-RH | − | − | + | + | |
| 7/F post | TRH | − | + | − | − | High |
| | LH-RH | − | − | + | + | |
| 8/F post | TRH | − | + | + + | − | High |
| | LH-RH | | | | | |
| 9/F post | Saline only | | | | | |
| 10/F post | TRH | + + | + | − | + + | Low |
| | LH-RH | − | − | + | + | |
| 11/F | TRH | + + | + | + + | − | Normal |
| | LH-RH | − | + + | + | + | |
| 12/F | TRH | + + | + | − | + + | Normal |
| | LH-RH | + + | + + | + | + | |
| 13/F | TRH | − | + | − | − | Normal |
| | LH-RH | − | − | − − | + | |
| 14/F | TRH | + + | + | − | + + | Normal |
| | LH-RH | | | | | |
| 15/F | TRH | + + | + | − | − | Low |
| | LH-RH | | | | | |
| 16/F | TRH | + + | + | + + | + + | Normal |
| | LH-RH | | | | | |

[a] Modified from Brambilla *et al.* (1978), *Arch. Gen. Psychiatry*, courtesy of The American Medical Association.
[b] +, responsiveness; + +, abnormal responsiveness; −, unresponsiveness; − −, abnormal unresponsiveness.
[c] Post, postmenopausal.

*al.*, 1975; Chihara *et al.*, 1976a), DA-induced GH release in baboons (Steiner *et al.*, 1977), and arginine, insulin (Maeda *et al.*, 1976b), L-dopa (Maeda *et al.*, 1975b), clonidine (Zanoboni *et al.*, 1979), and sleep-induced (Chihara *et al.*, 1977) GH release in man. Consistent with an effect of RHs on AP hormone secretion mediated via the hypothalamus or upper brain structures, through changes in brain monoamine metabolism, are the delayed pituitary responses to TRH and LH-RH stimulation found in some PAD patients (see above).

However, against the pathophysiological significance of a CNS-mediated mechanism for RHs in humans is the observation that the TRH doses shown capable of affecting the stimulated GH release, in either animals or men (see above), are considerably higher than those effective in inducing nonspecific GH release. In a careful study in unanesthetized calves, Hedlung et al. (1977) demonstrated that TRH was ineffective in stimulating GH secretion at doses which elicited striking rises in plasma GH by intrajugular route. Nevertheless, though with these reservations, the existence in PAD of an abnormal hypothalamic or suprahypothalamic neurotransmitter responsiveness to RHs cannot be excluded.

Whatever the true mechanism(s) of RHs, many endocrine findings in depressive illness substantiate the view of an impairment of neuroendocrine function. They include deficient GH response to CNS-acting stimuli, such as insulin hypoglycemia (Sachar et al., 1972) or clonidine (Checkley, 1980), absent or delayed sleep-entrained GH secretory peaks (Carroll and Mendels, 1976), disinhibiton of the hypothalamic–pituitary–adrenal axis resulting in cortisol hypersecretion, defective suppression of cortisol levels by dexamethasone, and blunting of the normal circadian rhythm of cortisol secretion. Finally, reduced TSH secretion after TRH, absent nocturnal PRL rises, and reduced LH secretion in postmenopausal women also have been reported (see Rubin and Poland, 1982).

Though the sensitivity of the overall abnormal pituitary responsiveness to RHs in PAD is only of about 30–40%, i.e., there are about 60–70% of true depressives who are false-negatives, its specificity seems to be very high (Brambilla et al., 1982), suggesting that anomalous AP responses may be helpful in the differential diagnosis between PAD and secondary affective disorders (SAD). In a furtherance of our previous studies, it was shown that in none of five patients with SAD did RHs elicit aberrant pituitary responses (Brambilla et al., 1980), a finding in keeping with those of Loosen et al., (1979) in depressed alcoholic men.

In patients with PAD no correlation could be evidenced between neuroendocrine pathology and type of depression, unipolar or bipolar, duration of the illness and number of depressive episodes, presence of anxiety and agitation, type of previous therapy, age of the patient or age of onset of the disease. The most prominent feature of abnormal AP responsiveness seems to be its heterogeneity; the abnormalities were not the same in all subjects, even though they tended to be constant in the same patient. This pattern is reminiscent of the heterogeneity of depressed patients with regard to course and polarity of the disease, biochemical alterations, responses to therapy, and prognosis (Murphy et al., 1978; Goodwin et al., 1978). Thus, the heterogeneity of the hormonal responses may be related to the large spectrum of individual entities which are diagnosed as depression. Persistence of the neuroendocrine pathology in PAD patients either in the depressive or normothymic phase (Brambilla et al., 1980) lends support to the proposition that it may be a continuous underlying process throughout the course of the illness, divorced from the various clinical phases. It should therefore be viewed as a trait of the disease. In this vein it is noteworthy that in two of six phenotypically normal relatives of an affected proband at high risk for the disease, the same type of neuroendocrine impairment was present (Brambilla et al., 1980). While this suggests that the latter is heritable, it remains to be shown that "impaired" relatives develop a higher incidence of depression than "nonimpaired" relatives. At variance with some of the above data, Kirkegaard et al. (1981) reported reduction of the anomalous TRH-induced GH rise in PAD patients after recovery from depression, thus suggesting that it may be a state phenomenon.

With regard to the biochemical mechanism(s) underlying abnormal AP responsiveness, scant information is available. A trend toward a positive correlation was found in PAD patients between urinary levels of 3-methoxy-4-hydroxyphenylglycol, a norepinephrine (NE) metabolite, partly reflecting brain NE turnover (Maas et al., 1979), and the GH response to TRH. This would imply that central NE activity plays a role in the deranged GH response. In this respect, indices of reduced 5HT function (Shaw et al., 1967; Coppen et al., 1973) may be related in the depressive illness to the anomalous AP responsiveness to RHs (see Section 4.4.1).

### 4.3.2. Anorexia Nervosa

Although in their study Lundberg et al. (1972) disregarded the increase in GH levels present after TRH in two of seven patients with anorexia nervosa (AN) such finding has been subsequently confirmed by many authors (Maeda et al., 1976a; Macaron et al., 1978; Gold et al., 1980). AN is a mental disease which shares with PAD some psychogenic determinants (Beumont, 1970; Casper and Davis, 1977). In AN patients as in PAD (see above) a host of endocrine findings indicates the existence of hypothalamic dysfunction (Mecklenburg et al., 1974). Similarly to PAD, plasma levels of cortisol are elevated in many patients with AN and adrenocortical activity is not as readily suppressed by dexamethasone as in normal subjects (Warren and Vande Wiele, 1973). It is therefore not surprising that patients with AN share with PAD patients a broad alteration of AP responsiveness to RHs. In fact, of the 10 patients investigated by Brambilla et al. (1981), four released GH after TRH and three after LH-RH; the anomalous GH responses coexisted in three subjects with TRH-induced rises in FSH and/or LH, in one subject with an LH-RH-induced PRL rise, and there was isolated PRL responsiveness to LH-RH in two subjects. In addition, one patient responded to LH-RH by releasing TSH. Collectively, conventional AP responsiveness to RHs was found in only 1 of the 10 patients, who was in a phase of initial nutritional rehabilitation.

Analysis of anomalous AP responsiveness to RHs as related to resting hormone concentrations might give insight into the pathophysiology of the disease. Macaron et al. (1978) reported significant increases in circulating GH titers following TRH only in AN patients with elevated basal GH levels, an interesting finding since it might underlie absence of an inhibitory somatostatinergic tone. However, no such correlation was present in the studies of

Gold et al. (1980) or our own studies, either after TRH or LH-RH stimulation. Reportedly, low baseline FSH and/or LH levels are present in many patients with AN (Beumont et al., 1973; Lundberg et al., 1972; Marshall and Fraser, 1971; Brambilla et al., 1981), while PRL levels higher than controls have been occasionally found (Travaglini et al., 1976). It has been postulated that an increased activity of DA in the CNS might play a role in the pathophysiology of the disorder (Barrry and Klawans, 1976; Sherman and Halmi, 1977). Turning to the endocrine changes, the increased DA tone might account for the defective gonadotropin secretion (Fuxe et al., 1975a), the presence of high resting GH levels (Müller et al., 1977b), and, perhaps, the blunted GH response to L-dopa or apomorphine (Sherman and Halmi, 1977). On the other hand, enhanced dopaminergic tone in the CNS can hardly reconcile with basal PRL levels within the normal range (Maeda et al., 1976a) or elevated (Travaglini et al., 1976).

Caution should be exercised in relating endocrine findings present in AN to a well-defined pathophysiology (see above) since they may simply reflect the degree of caloric deprivation in the preceding few days. Thus, underweight patients who have good caloric intake and are gaining weight have lower levels of GH than do patients with poor caloric intake at the time of study (Casper et al., 1977; Garfinkel et al., 1975). A variety of abnormalities have been reported in AN in the GH response to provocative stimuli, such as insulin hypoglycemia (Brauman and Gregoire, 1975; Casper et al., 1977). Similar disturbances are observed in other forms of malnutrition and most of the abnormalities of GH in AN disappear with restoration of normal diet and weight (Samuel and Deshpande, 1972; Casper et al., 1977; Frankel and Jenkins, 1975). Also, abnormal AP responsiveness to RHs in AN might merely reflect the status of malnutrition. In this view, a low TRH dose (100 µg) was capable of increasing the already high GH levels in six of eight children with protein calorie malnutrition (Becker et al., 1975). With restoration of weight, those patients in whom basal GH levels had returned to normal showed either no GH increment or more delayed responses. Similarly, Brown et al. (1978) suggested that the high GH levels and the abnormal GH response to TRH of anorectic patients are related to the low caloric intake rather than to the amount of weight loss. Consistent with this view, Tolis et al. (1982) found that in patients with AN the TRH-induced GH rise was abolished by normal weight recovery. At odds with these findings, however, we have observed that in one anorectic patient who had restored either normal weight or regular menses, abnormal GH responsiveness to TRH and LH-RH persisted (Brambilla et al., unpublished results). Similarly, a paradoxical PRL response has been observed in AN patients during the course of LH-RH infusion, and, interestingly, the rise was more prominent in those patients who had gained weight during the week before the infusion (Beumont et al., 1980). It must be recalled, however, that the abnormal PRL rise after LH-RH might be related to the anovulatory state of these subjects. In fact, a marked increase in serum PRL levels has been reported after LH-RH (100 µg i.v.) in 13 normoprolactinemic anovulatory women and ascribed to hyperresponsiveness of pituitary lactotropes (Giampietro et al., 1979).

In conclusion, further studies are mandatory to ascertain whether abnormal AP responsiveness to RHs is in AN patients merely related to the low caloric intake or instead reflects an underlying CNS disturbance persisting throughout the course of the illness. With such an approach further analogies between AN and PAD may become apparent or be refuted.

### 4.3.3. Schizophrenia

Recently, an abnormal GH response to TRH and LH-RH has been reported also in adolescent schizophrenic boys (Gil-Ad *et al.*, 1981). The anomalous GH response after LH-RH was present in 8 of 10 patients and after TRH in 4 of 6 patients; no effect on GH was found in normal aged-matched controls. As in PAD, in schizophrenia the presence of alterations in brain neurotransmitters— i.e., an excess of DA and endorphin activity (Snyder, 1977; Gruen, 1978) or a deficiency of NE and prostaglandin activity (Hartmann, 1976; Horrobin, 1977) which could impair the regulatory mechanisms of GH secretion—has also been suggested. Treatment of schizophrenic boys with antipsychotic drugs, e.g., haloperidol, thioridazine, and chlorpromazine, which strongly interact with catecholamine systems likely by reducing the excessive dopaminergic function, was able to abolish the abnormal GH response to LH-RH in 5 of the 6 patients tested. This fact and why the treatment failed to abolish the TRH-induced GH response in 3 of 4 subjects have no obvious explanation (Gil-Ad *et al.*, 1981).

## 4.4. Diseases with Metabolic Alterations

### 4.4.1. Renal Failure and Liver Cirrhosis

In addition to psychiatric disorders, abnormal AP responsiveness to RHs occurs frequently in patients with alterations of metabolism resulting from chronic diseases. In 1973 Gonzales-Barcena and his colleagues reported first an unexpected GH rise after TRH in 7 of 8 patients with renal failure. The paradoxical GH increase also was found in uremic patients on chronic hemodialysis treatment (Gonzales-Barcena *et al.*, 1973; Czernichow *et al.*, 1976). The neuroendocrine disturbance would be related to the severity of the disease; patients with mild renal insufficiency behaved like controls while patients at a more advanced stage or at the end stage of the disease exhibited proportionally higher GH responses (Weissel *et al.*, 1979).

A TRH-induced GH release also has been described by our group (Panerai *et al.*, 1977d) and later confirmed by others (Zanoboni and Zanoboni-Muciaccia, 1977; Van Thiel *et al.*, 1978) in patients with liver cirrhosis. The anomalous response to TRH occurs in about 60% of cirrhotic patients and, at odds with renal failure, can be correlated neither with the different etiology nor to any of the biochemical indices of hepatic failure, reflecting the severity of the disease (albumin, bilirubin, cholinesterase, etc.) (Salerno *et al.*, 1979).

In either severe renal disease or in hepatic cirrhosis, the GH response to TRH is coupled to high basal GH levels and a paradoxical GH rise after glucose administration (Samaan and Freeman, 1970; Becker et al., 1969), a stimulus inhibitory to GH secretion in normal subjects. Such elevations could be, at least in part, due to altered GH metabolic disposal, as suggested by the studies of Cameron et al. (1972) in patients with hepatic and renal failure. Also, variations in the production of GH-dependent somatomedins could play a role in the paradoxical and/or anomalous GH responsiveness to stimuli and the presence of high basal GH levels. However, since patients with severe liver disease and AN have reduced production of somatomedins whereas uremic patients have serum levels higher than normal (Hintz et al., 1978), this possibility seems to be unlikely. Finally, studies on the disappearance rate of plasma GH after i.v. somatostatin infusion suggest that in cirrhotic and uremic subjects actual GH hypersecretion is present (Pimstone et al., 1975).

In addition to the paradoxical GH elevation after glucose, many other neuroendocrine disturbances are present in liver cirrhosis and uremia, a likely reflection of an impaired CNS neurotransmitter function. Hyperprolactinemia occurs frequently in patients with renal disease coupled to a decrease or absent PRL suppression by dopaminergic agents (Ramirez et al., 1977) and blunted PRL and TSH responses to TRH (Czernichow et al., 1976). Cirrhotic patients also have often high resting PRL levels associated with gynecomastia (Van Thiel et al., 1975; Panerai et al., 1977d). However, at variance with uremic patients, PRL and TSH responses to TRH may even be exaggerated (Panerai et al., 1977d; Van Thiel et al., 1982).

Patients with altered AP responsiveness to RHs and deranged CNS neurotransmitter function provide a naturally occurring experimental model for investigating possible causal interrelationships between the two events. An increased brain 5HT turnover due to a facilitated transport of tryptophan from plasma into the brain has been postulated to occur in patients with chronic hepatic failure (Munro et al., 1975). Indeed, very high concentrations of free tryptophan, 5HT, and its main catabolic product, 5-hydroxyindoleacetic acid, are found in brain and cerebrospinal fluid of cirrhotic patients (Lal et al., 1975; Jellinger and Riederer, 1977). Based on these findings, an interrelationship could be envisaged between altered 5HT turnover and anomalous GH responses (to TRH and glucose). Figure 2 shows that pretreatment with metergoline (MCE), a potent blocker of 5HT receptors (Fuxe et al., 1975b; Sastry and Phillips, 1977), did not modify basal GH levels in a group of cirrhotic patients but potentiated the TRH-induced GH rise in 4 of 9 patients. In addition, 2 of 8 TRH-nonresponder patients developed the anomalous GH response after MCE. In addition, MCE potentiated in 5 of 9 patients the paradoxical GH response to glucose, even though there was no significant difference between pre- and posttreatment GH values (Salerno et al., 1980). All these findings lend credence to the view that in cirrhotic patients, central 5HT neurotransmission is causally related to the anomalous responses to GH secretion. However, since

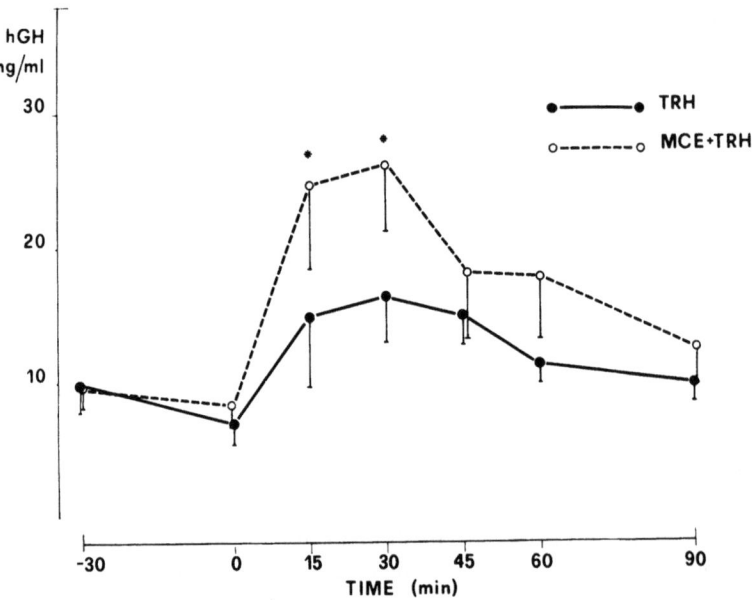

FIGURE 2. GH response to TRH before and after pretreatment with metergoline (MCE) in nine patients with severe liver disease. MCE was administered at a dose of 2 mg every 4 hr for 48 hr. The mean values and S.E.M. are given.* $p < 0.05$ vs. pretreatment values. From Salerno et al. (1980), J. Clin. Endocrinol. Metab., courtesy of Williams & Wilkins.

MCE blockade of 5HT receptors did not counteract but instead magnified the deranged GH responsiveness, it may be foreseen that in cirrhotic patients the postulated enhancement in brain 5HT turnover is actually coupled to a reduction of 5HT neuronal function. Other findings obtained with TRH are consistent with the above proposition. In fact, administration of another 5HT receptor blocker (methysergide) to six normal volunteers caused an abnormal GH response to TRH in two of them (Collu, 1979).

Pharmacologically induced alterations in brain 5HT function are not the only way to modify AP responsiveness to TRH in liver cirrhosis. Infusion of DA to five TRH-responder patients increased basal GH levels per se in three subjects and allowed an earlier GH rise after TRH to be seen. An earlier GH rise also was present in one of the remaining two patients in whom DA per se did not affect basal GH levels (Müller et al., 1979). An interaction between TRH and DA on GH secretion, with TRH enhancing the GH-releasing effect of DA when given "after" but not "before" DA, had been reported in healthy subjects by Burrows et al. (1977). Though these findings are difficult to interpret, it may be proposed that in healthy subjects pretreatment with DA removed a hypothalamic influence (possibly somatostatin) normally hindering the potential GH-releasing ability of TRH (Giannattasio et al., 1979). The same mechanism when applied to cirrhotic patients, who respond to TRH by releasing GH also in the resting state, may account for the earlier GH rise after the tripeptide. Presence of a somatostatinergic "tone" which restrains the potential GH-releasing action of TRH would be supported, albeit inferentially, by the finding that a somatostatin infusion strikingly suppresses the GH rise in TRH-responder cirrhotics (Salerno et al., 1982). To our knowledge, no data exist on the ability of somatostatin to abolish the TRH-induced GH rise in either the uremic subjects or patients with psychiatric disorders. In contrast to subjects with psychiatric disorders, the GH release after TRH seems to be the unique deranged response to hypophysiotropic stimuli. In fact, administration of LH-RH to 12 cirrhotic patients, of whom 7 were TRH responders, failed to stimulate a rise in plasma GH or PRL and in none of the 7 patients investigated was TRH capable of inducing an increase of LH or FSH levels (Salerno et al., 1982). The reason for the selective alteration in AP responsiveness of cirrhotic patients is presently obscure; it may suggest selective disturbance of CNS neurotransmitter–neurohormone function. A similar type of disarrangement in brain function may underlie the nonspecific GH release following administration of TRH present in patients with homozygous β-thalassemia. Out of 12 subjects investigated, 8 responded with a paradoxical rise of GH levels to administration of TRH, none to administration of LH-RH (Masala et al., 1982).

### 4.4.2. Diabetes

Abnormalities in GH secretion and metabolism exist in patients with insulin-dependent diabetes which are reminiscent of those present in uremic and cirrhotic patients. For example, diurnal and mean 24-hr GH levels are higher in diabetics than in controls (Molnar et al., 1972), an increased GH response to exercise has been reported in poorly controlled insulin-dependent diabetes (Hansen, 1970), and hyperglycemia does not inhibit the response of GH to provocative maneuvers (arginine infusion or L-dopa) (Ajlouni et al., 1975; Cremer et al., 1973).

It should not be surprising, therefore, that 6 of 13 insulin-dependent diabetics exhibited a significant GH rise after TRH (Dasmahapatra et al., 1981). However, analysis of these data reveals absence of placebo studies, particularly important in the diabetic state where GH secretion is a highly labile function (Johansen and Hansen, 1971), and the presence of only inconsistent GH rises (2–4 ng/ml) in at least 3 of the "responder" patients. Results in accordance with those of Dasmahapatra et al. (1981), but of more clear-cut evidence, were recently obtained by Ceda et al. (1982). They showed that administration of TRH induced an unequivocal GH rise in 7 of 15 males and in 13 of 16 females with insulin-dependent diabetes. Interestingly, in the study of Dasmahapatra et al. (1981), all TRH responders were females. Moreover, there was in either male or female subjects a positive correlation between basal values and peak values of GH and a negative correlation between GH peaks and age. The positive correlation between basal values and GH peaks following TRH found by Ceda et al. (1982) poses the interesting problem of the existence in responder

### 4.5. Endocrine Diseases

#### 4.5.1. Hypothyroidism

One of the endocrine disorders not due to adenomatous pituitary hyperfunction (see below) in which paradoxic AP responses to RHs have been evidenced is primary hypothyroidism. Hamada *et al.* (1976) first reported that TRH was capable of releasing GH in 6 of 13 adult hypothyroid subjects. Later on, it was shown that TRH, which did not affect GH release in 7 euthyroid children, induced a significant GH increment in 7 of 11 hypothyroid children. There was no consistent GH increase present after administration of LH-RH in any of the subjects (Collu, 1979). At odds with these data are the findings of Tolis *et al.* (1979); in an extensive study in 29 adult hypothyroid subjects, they reported that the concentration of GH was variable, with the amplitude and frequency of the secretory pattern similar to those reported by others for normal individuals. After administration of TRH, GH values did not differ from the values observed as spontaneous surges, leading the authors to conclude that TRH is not a reliable GH secretagogue in human subjects with hypothyroidism. Irrespective of whether the TRH-induced GH release is an important biologic marker of hypothyroidism, other neuroendocrine alterations are evident in this disease. Of 12 children with primary hypothyroidism, in whom baseline PRL levels were high as was the mean 24-hr LH concentration, administration of TRH evoked a nonspecific LH release in five cases and FSH release in four cases. In three children tested again after 6 months of thyroxine therapy, responses to TRH reverted to normal though two continued to show significant LH responses to the peptide (Daneman *et al.*, 1982).

Though the actual ability of TRH as GH releaser in human hypothyroidism has not been definitely established, some of the reported findings pose the important question of whether the aberrant AP responses to TRH present in animals and humans may, totally or in part, be due to a defective negative feedback inhibition by thyroid hormones on pituitary somatotropes, a mechanism which operates to control TSH secretion from thyrotropes (Morley, 1981).

Along this line, it is noteworthy recalling that the scarce GH-releasing effect of TRH in urethane-anesthetized rats was greatly enhanced by surgical or pharmacological thyroidectomy, while, conversely, induction of hyperthyroidism by a $T_4$ or $T_3$ regimen led to an opposite effect (Chihara *et al.*, 1976a; Kato *et al.*, 1975). Presence of a secondary hypothyroidism can be envisaged in hypophysectomized rats bearing an ectopic pituitary or in rats bearing electrolytic lesions in the ME, two experimental models in which a nonspecific GH release to TRH is present (see Section 2.1).

In man, serum $T_4$ production rate may be low in uremic subjects and increases after successful renal transplantation or intensive dialysis (Joasoo et al., 1974; Lim et al., 1977). Both in cirrhotic patients and in patients with AN or malnutrition, significant decreases in plasma levels of $T_4$ with shift of the peripheral metabolism of the latter away from the production of $T_3$ and toward the production of reverse $T_3$ have been reported (Gastineau, 1979; Burman et al., 1977). The possibility should also be considered that hypothyroidism induced by lithium therapy (Shopsin et al., 1969; Emerson et al., 1973) may be responsible for the paradoxical TRH-induced GH rise reported by Yamaguchi et al. (1980) in depressive patients in the manic state under but not before lithium therapy. However, that absence of the negative feedback by thyroid hormones on pituitary cells may play a key role in the occurrence of the anomalous AP responses to TRH which would be denied by the finding that the ectopic AP gland shows a higher susceptibility to TRH also when placed in a physiologic endocrine environment, e.g., following transplantation into intact euthyroid rats (Panerai et al., 1977c). Another point with respect to TRH and GH secretion in hypothyroidism deals with the down-regulating effect of RHs on their own pituitary receptors, e.g., their ability to control in an inhibitory fashion the number of binding sites on target pituitary cells (see Section 2.1). Data have been presented that in situations of primary hypothyroidism, when the secretion of TSH is increased, there is a lowering of endogenous TRH production (Sinha and Meites, 1966; Wilberg and Seibel, 1973), likely due to the existence of a short-loop feedback regulation.

A down-regulating mechanism of TRH on TSH secretion has been demonstrated in both humans and rats. Chronic administration of oral TRH significantly blunts the TSH response to acute administration of the tripeptide both in euthyroid and in primary hypothyroid men (Staub et al., 1978). Therefore, another mechanism which may account for the TRH-induced GH rise in hypothyroidism would be a defective TRH production. This possibility has been investigated by our group in patients with chronic liver cirrhosis. A group of these patients was treated with oral TRH (20 mg twice daily) for 6 days and the effect of the treatment was evaluated on the GH response to acutely administered TRH (400 μg i.v.). Of nine cirrhotic patients who were TRH responders, chronic TRH treatment reduced the GH peak to acute TRH administration in six while the response was unaffected in two and potentiated in one (Cocchi et al., 1982). Thus, it would appear that at least in some cirrhotic patients, defective endogenous TRH production and/or function may play a role in the development of the abnormal GH responsiveness (see also Section 4.4.1).

### 4.5.2. Primary Hypogonadism

Recently, an anomalous GH response after a low dose of LH-RH (50 μg) has been reported in 56.3% of a group of 16 young patients with Klinefelter's syndrome (Dickerman *et al.*, 1981). Stimulation with TRH, however, led to a GH response in only 1 of the 14 patients investigated. Though interpretation of these findings is not obvious, the authors hypothesize that failure of testis function may lead through the accumulation of the corresponding RH (e.g., LH-RH) in the hypothalamus or absence of the negative feedback mechanism of gonadal hormones (with ensuing short-loop feedback inhibition of gonadotropins on LH-RH production), to an unmasking of LH-RH receptors located on other pituitary cells (e.g., somatotropes). The close analogy existing between this alleged mechanism and that possibly underlying the abnormal GH release after TRH, seen in hypothyrodism (see above), does not escape attention.

### 4.5.3. Pituitary Secreting Tumors

*4.5.3a. Acromegaly.* First demonstration of an aberrant response to a hypothalamic hormone in acromegaly was given by Irie and Tsushima (1972) and Schalch *et al.* (1972) by showing that in a large proportion of patients, TRH administration induced an increase in plasma GH. It was soon demonstrated that LH-RH also was effective, though the latter appeared in general a less powerful GH stimulator than TRH (Faglia *et al.*, 1973a; Rubin *et al.*, 1973). Repeating the TRH or LH-RH testing on different occasions in the same patients revealed a markedly constant response even after intervals of several months between testing. The aberrant GH responsiveness was not related to the activity of the disease (it persisted after pituitary irradiation in spite of lowered GH plasma titers) and was independent from basal GH levels, tumor size, or the dose of RHs administered. Even higher doses of RHs were ineffective in unresponsive patients (Cantalamessa *et al.*, 1976). Discordant findings have been reported on the suppressibility of the GH release by TRH operated by thyroid hormones. While Pawlikowski and Owczarczyk (1977) demonstrated that treatment with $T_3$ (60 μg daily for 1 day) inhibited in 4 of 5 patients the TRH-induced GH rise, Faglia *et al.* (1973b) and Carlson *et al.* (1977) did not observe such effect when $T_3$ was administered to patients with active acromegaly even at higher doses (100–120 μg daily for 1 week). Regarding the mechanism(s) underlying the paradoxical GH release, either an action of the RHs exerted on the hypothalamus (via release of GRF?) (see also Section 4.3) or a direct action on pituitary somatotropes has been proposed (Faglia *et al.*, 1973a). The second possibility, which relies on a state of immaturity of the adenomatous cells of some acromegalics with ensuing susceptibility to nonspecific RHs, appears to be the most likely. Compelling evidence for a pituitary site of action has been provided by Matsukura *et al.* (1977), who showed in GH-producing pituitary adenomas *in vitro* activation of adenylate cyclase (AC) by TRH, LH-RH, and some biogenic amines. The conclusion of these authors was that the presence of multiple hormone receptors in those tissues proved that the paradoxical GH releases are due to an alteration of the cellular membrane receptors of the tumors. Results of *in vivo* studies also are consistent with this proposition. Centrally acting GH-suppressing agents such as dexamethasone, a receptor blocker such as phentolamine, or hyperglycemia, in spite of their ability to lower basal GH levels, failed to counteract the TRH-induced GH release in acromegalic patients (Nakagawa and Obara, 1971). Instead, selective adenomectomy in TRH-responder acromegalics was followed by restoration of normal regulatory responses, suggesting location of the anomalous receptor in the excised tissue (Hoyte and Martin, 1975; Faglia et al., 1977). Finally, a homogeneity has been evidenced in acromegaly in the GH response to combined application of TRH and direct DA agonists, which lower GH levels in about 60% of acromegalics by a direct pituitary mechanism (see Müller, 1979), e.g., most of the subjects either responded to both stimuli or to none (Liuzzi et al., 1974; Faglia et al., 1975).

Also, comparison of the anomalous TRH-induced GH rise and other neuroendocrine responses points to a direct pituitary site of action for the peptide in acromegaly. In hyperprolactinemic acromegalic patients, a high correlation was found between GH and PRL responsiveness to TRH (Moriondo et al., 1980). The most likely explanation for these findings is that TRH in these instances was acting on populations of mixed GH- and PRL-secreting adenomatous cells, though the possibility that a single stem cell may produce either hormone cannot be ruled out (Zimmerman et al., 1974).

At variance with the ability of somatostatin to counteract the TRH-induced GH rise in patients with liver cirrhosis (Salerno et al., 1982) stands its ineffectiveness to do the same in acromegalics. In all the five patients studied by Giustina et al. (1974), the inhibition of GH release induced by somatostatin was rapidly removed by TRH or LH-RH injection, which was followed by a striking increase in plasma GH levels. Despite its ineffectiveness in blunting the abnormal AP responsiveness to RHs, somatostatin holds promise of enriching our understanding of the underlying pathophysiology. In fact, in the study of Pieters et al. (1982), infusion of increasing doses of the peptide disclosed the presence of two groups of acromegalics: a group in which the elevated GH levels decreased to normal values, and a group in which GH levels decreased but still remained elevated. Serum GH levels showed a significant rebound phase above baseline values after stopping somatostatin only in group 1, suggesting that in the latter endogenous somatostatin deficiency way account for the higher sensitivity. Importantly, administration of LH-RH induced a marked GH increase only in the patients who normalized their GH levels following somatostatin, which suggests a close connection between the action of these RHs. In contrast, the GH response to TRH did not differ among the two groups, pointing to different modes of action of TRH and LH-RH at the somatotropes (Pieters et al., 1982). Somewhat at variance with these data are those of Hanew et al. (1980), who demonstrated that acromegalics who exhibit a large GH increase to at least one of the stimuli (TRH, LH-RH, arginine) show the highest responsiveness to somatostatin.

In conclusion, the mechanism(s) subserving the aberrant responses to TRH (and LH-RH?) in acromegaly is still unknown, though a (primary?) alteration in the cellular membrane of pituitary somatotropes seems to be the most likely explanation. Studies of TRH–dopaminergic drug interaction already mentioned for cirrhotic patients (see above) also point to this direction. In fact, infusion of DA, which in acromegaly acts directly to inhibit GH release from somatotropes (Verde et al., 1976; Camanni et al., 1977), completely inhibited the TRH-induced GH rise (Fig. 3). Oddly, the TRH-induced GH release is only scarcely

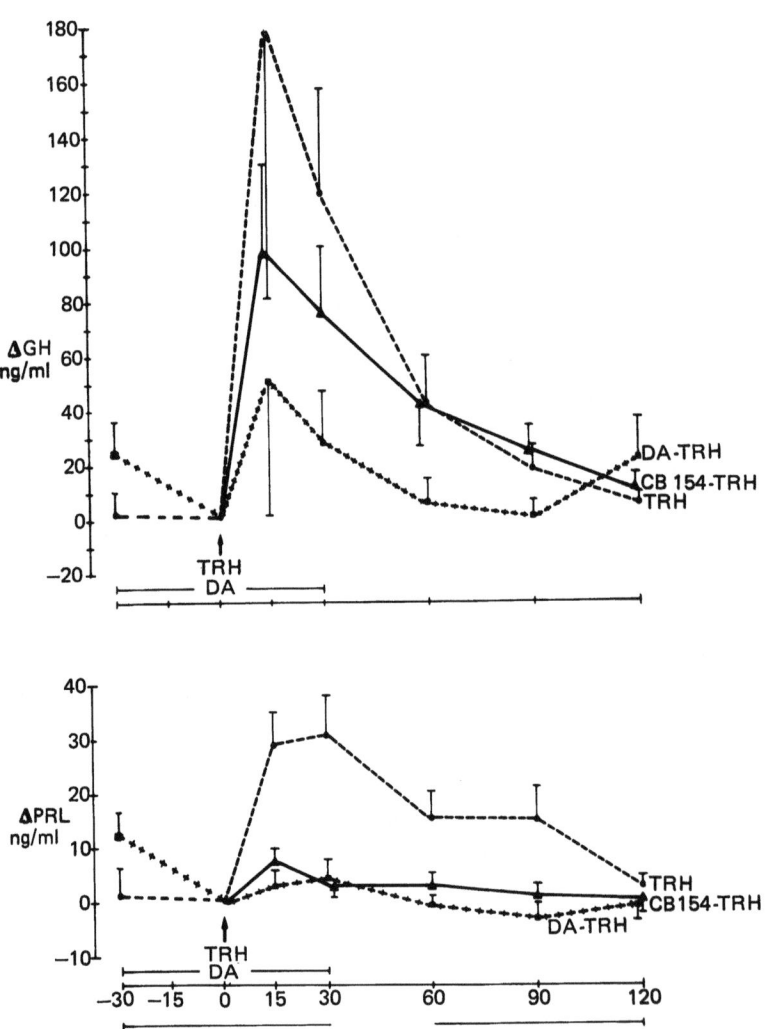

FIGURE 3. Effect of TRH on plasma GH (upper graph) and PRL (lower graph) in 12 acromegalic subjects before and after treatment with dopaminergic drugs. Bromocriptine (CB-154) was administered to the patients at a dose of 10 mg daily (2.5 mg every 6 hr) p.o. (last administration 2 hr before). Dopamine hydrochloride was infused at a dose of 30 mg/60 min. In the DA experiment, GH and PRL values at 0 min were 21 ± 6.7 and 35.0 ± 19.0 ng/ml, respectively. From Müller et al. (1979), Clin. Endocrinol., courtesy of Blackwell Scientific Publications.

affected by concurrent administration of bromocriptine [Fig. 3 and Ishibashi et al. (1977)] whereas the GH-releasing effect of LH-RH is suppressed by this drug (Ishibashi et al., 1978). These data reinforce the idea that different mechanisms for TRH and LH-RH are operative at the level of the somatotropes. The nonspecific effect of LH-RH in acromegaly is not confined to GH release; LH-RH, in fact, also can release PRL and, though few acromegalics were studied in this context, a partial concordance between GH and PRL responses to LH-RH was reported. In fact, most of the PRL responders to LH-RH were also GH responders (Catania et al., 1976; Ishibashi et al., 1978).

*4.5.3b. PRL-Secreting Pituitary Tumors.* As alluded to before (see Section 3) somatostatin cannot modify plasma PRL titers either when they are within a normal range or are pharmacologically increased. In contrast, in 2 of 5 patients bearing a PRL-secreting adenoma (prolactinoma) infusion of somatostatin (750 µg) reduced promptly, though transiently, PRL levels. Similar results were obtained in acromegalic hyperprolactinemic subjects [3 of 6 patients in the study of Barreca et al. (1979); 2 of 5 in the study of Yen et al. (1974)]. However, no placebo studies were performed in any of the patients with prolactinoma, so that the possibility is not ruled out that spontaneous fluctuations in the high basal PRL level could account for some of the observed effects (Genazzani et al., 1980).

*4.5.3c. ACTH-Secreting Tumors.* Somatostatin's inability to affect the release of some AP hormones under resting conditions but its competence to act on a tumorous cell (see above) extends to include ACTH. In fact, the peptide, which does not influence ACTH secretion in normal men (Hall et al., 1979), is effective in this context in patients with Nelson's syndrome (Tyrrell et al., 1975), when a pituitary adenoma untied from normal feedback control by circulating glucocorticoids secretes higher amounts of the hormone.

Somatostatin's ability to act on tumorous cells (see also above) calls for the presence of a common mechanism underlying the action of the peptide. Reportedly, the action of somatostatin rests on the inhibition of cAMP production by the pituitary gland (Labrie et al., 1975). Inhibition of AC by DA is hardly seen in normal pituitaries but occurs in prolactinomas (De Camilli et al., 1979), a condition in which the AC linked to lactotropes would be increased (G. Giannattasio, personal communication). For analogy, elevated enzyme activity also in GH- and ACTH-secreting tumors would explain their susceptibility to the inhibitory action of somatostatin.

Inhibition of ACTH secretion by somatostatin is not the only aberrant response of ACTH-producing tumors. A rise in plasma ACTH and cortisol after TRH was first reported by Krieger and Luria (1977) in patients with Nelson's syndrome and Cushing's disease, though other authors (Connell et al., 1975; Matsukura et al., 1977) did not confirm these data in untreated and treated patients with Cushing's disease. Consistent with Krieger and Luria's findings, Pieters et al. (1979) described a paradoxical rise in plasma cortisol in 3 of 6 patients with Cushing's disease after either TRH or LH-RH administration, and, more recently, a paradoxical ACTH and cortisol release to TRH and LH-

RH in 9 of 22 consecutive patients with this disease (Pieters *et al.*, 1982c). A nonspecific release of ACTH induced by the neurohormones would be the mechanism. More recently, Oki *et al.* (1980) have reported an increase in ACTH and β-endorphin plasma levels following TRH in one patient with Nelson's syndrome; somatostatin had the opposite effect.

The anomalous ACTH response to TRH of patients with ACTH hypersecretion subsided after treatment with the antiserotoninergic agent cyproheptadine (Krieger and Luria, 1977), a finding reminiscent of the effect of another antiserotoninergic drug MCE on the TRH-induced GH rise in cirrhotic patients [though in that instance an opposite effect was seen (see Section 4.4.1)]. Since cyproheptadine can inhibit ACTH secretion from trypsin-dispersed rat pituitary tumor cells (Lamberts *et al.*, 1980), a direct action of the drug on the tumor cells can be envisaged.

Regarding the source of the ACTH secreted in response to TRH the pars intermedia-like human pituitary cells are likely candidates. Supporting this proposition is the finding that like in many patients with Cushing's disease paradoxical rises in plasma ACTH and cortisol after TRH are also present in pregnant women (Pieters *et al.*, 1982b), in whom typical intermediate lobe peptides are secreted (Clark *et al.*, 1978) (see Section 2.1.1).

*4.5.3d. FSH-Secreting Pituitary Tumors.* Recently, the ability of TRH to increase LH and FSH levels in men with pituitary adenomas associated with FSH hypersecretion has been reported (Snyder *et al.*, 1980). Serum LH concentrations after TRH increased to 136% of resting levels in comparison to a low increase (about 50%) in normal men or men with pituitary adenomas without FSH hypersecretion. Serum FSH concentrations did not increase either in any of the normal men considered or in those with pituitary adenomas without FSH hypersecretion, but did increase, though slightly (38.6%), in 5 of the 10 men with FSH hypersecretion. The possibility that gonadotropin hypersecretion per se could result in the nonspecific gonadotropin responses to TRH was excluded, since there were no LH and FSH responses to TRH in men with primary hypogonadism. Supporting these *in vivo* data is the finding that TRH added to four "functionless" tumors containing and secreting gonadotrophins in cell cultures elicited release of either LH or FSH in three of them (Mashiter *et al.*, 1982).

In conclusion, the mechanism(s) whereby RHs elicit paradoxical AP responses in patients bearing pituitary adenomas is obscure (see also above). In at least some instances, possible paracrine effects between adenomatous cells and the normal cell population of the pituitary may play a role. Thus, a release of PRL after addition of LH-RH has been shown by rat pituitary lactotropes only when cultured in presence of a highly purified population of gonadotropes (Denef, 1981). A similar mechanism might explain the abnormal inhibition of PRL secretion by somatostatin present in acromegaly (see above), since the adenomatous pituitary can be viewed as a somatotrope-enriched cell population. Such mechanism, however, cannot account satisfactorily for most of the altered responses present in other pituitary adenomas. It can only be said that,

## 5. CONCLUSIONS

Though not exhaustive, this review has attempted to list and discuss most of the pathologic states in which abnormal AP responses to RHs take place. Of these responses, the most frequently occurring and investigated is the GH increase after TRH, which is present in psychiatric, metabolic, and endocrine disturbances. Studies in laboratory animals support the view that defective CNS control of GH release may underlie this neuroendocrine disorder and that the action of TRH is exerted directly on the pituitary somatotropes, though a CNS-mediated mechanism for TRH and other RHs cannot be ruled out. In diseases with metabolic alterations, e.g., liver cirrhosis, the abnormal GH increase after TRH is the unique alteration is AP responsiveness present, suggesting selective impairment in the CNS function. In psychiatric disorders, more extensive derangement of AP responsiveness to RHs would rather reflect complex alterations in CNS neurotransmitter function. The trait character of the responses in PAD holds promise for a better diagnostic understanding of depressive patients.

While in the aforementioned conditions abnormal AP responsiveness to RHs would depend upon the appearance or unmasking of new receptor sites on a "normal" target pituitary cell, possibly due to impaired secretion of the specific endogenous neurohormone, a different mechanism would operate in the case of adenomatous pituitary cells. Here, the occurrence of nonspecific receptors would be linked to the development of the tumor itself, though the biochemical events underlying the nonspecific release are poorly understood.

Whatever the mechanism(s) of action may be, it is evident that evaluation of abnormal AP responses to regulatory neurohormones may represent a tool for getting insight into the pathophysiology of many human diseases.

## 6. REFERENCES

Ajlouni, K., Martinson, D. R., and Hagen, T. C., 1975, Effect of glucose on the growth hormone response to L-dopa in normal and diabetic subjects, *Diabetes* **24**:633.

Barreca, T., Cicchetti, V., Perria, C., Masturzo, P., and Rolandi, E., 1979, Effets de la somatostatine sur la sécrétion de prolactine: Etude chez des sujets normaux et chez des patients ayant un adénome hyophysaire, *Nouv. Press. Med.* **8**:331.

Barry, V. C., and Klawans, H. L., 1976, On the role of dopamine in the pathophysiology of anorexia nervosa, *J. Neural Transm.* **38**:107.

Becker, M. D., Cook, G. C., and Wright, A. D., 1969, Paradoxical elevation of growth hormone in active chronic hepatitis, *Lancet* **II**:1035.

Becker, D., Kronheim, S., and Pimstone, B., 1975, Serum growth hormone responses to thyrotrophin releasing hormone in children with protein-calorie malnutrition, *Horm. Metab. Res.* **7**:358.

Beumont, P. J. V., 1970, Anorexia nervosa: A review, *S. Afr. Med. J.* **44**:911.

Beumont, P. J. V., Friesen, H. G., Gelder, M. G., and Kolakowska, T., 1973, Plasma prolactin and luteinizing hormone levels in anorexia nervosa: Response to clomiphene citrate, *Psychol. Med.* **4**:219.

Beumont, P. J. V., Abraham, S. F., and Turtle, J., 1980, Paradoxical prolactin response to gonadotropin releasing hormone during weight gain in patients with anorexia nervosa, *J. Clin. Endocrinol. Metab.* **51**:1283.

Blackwell, R., Vale, W., Amoss, M., Burgus, R., Monahan, M., Rivier, J., Ling, N., and Guillemin, R., 1973, Lack of effect of native or synthetic LRF on secretion of prolactin in vitro, *Am. J. Physiol.* **224**:176.

Brambilla, F., Smeraldi, E., Sacchetti, E., Negri, F., Cocchi, D., and Müller, E. E., 1978, Deranged anterior pituitary responsiveness to hypothalamic hormones in depressed patients, *Arch. Gen. Psychiatry* **35**:1231.

Brambilla, F., Smeraldi, E., Bellodi, L., Sacchetti, E., and Müller, E. E., 1980, Neuroendocrine correlates and monoaminergic hypothesis in primary affective disorders (PAD), in: *Progress in Psychoneuroendocrinology* (F. Brambilla, G. Racagni, and D. de Wied, eds.), pp. 235–245, Elsevier/North-Holland, Amsterdam.

Brambilla, F., Cocchi, D., Nobile, P., and Müller, E. E., 1981, Anterior pituitary responsiveness to hypothalamic hormones in anorexia nervosa, *Neuropsychobiology* **7**:225.

Brambilla, F., Smeraldi, E., Sacchetti, E., Bellodi, L., Genazzani, A. R., Facchinetti, F., and Müller, E. E., 1982, Neuroendocrine abnormalities in depressive illness, in: *Typical and Atypical Antidepressants: Clinical Practice* (E. Costa and G. Racagni, eds.), pp. 329–340, Raven Press, New York.

Bratusch-Marrain, P., and Waldhäusl, W., 1979, The influence of amino acids and somatostatin on prolactin and growth hormone release in man, *Acta Endocrinol. (Kbh.)* **90**:403.

Brauman, H., and Gregoire, F., 1975, The growth hormone response to insulin induced hypoglycemia in anorexia nervosa and control underweight or normal subjects, *Eur. J. Clin. Invest.* **5**:289.

Breese, G. R., Cott, J. M., Cooper, B. R., Prange, A. J., Jr., Lipton, M. A., and Plotnikoff, N. P., 1975, Effects of thyrotropin releasing hormone (TRH) on the actions of pentobarbital and other centrally acting drugs, *J. Pharmacol. Exp. Ther.* **193**:11.

Brown, G. M., Seggie, J. A., Chambers, J. W., and Ettigi, P. G., 1978, Psychoneuroendocrinology and growth hormone: A review, *Psychoneuroendocrinology* **3**:131.

Burman, K. D., Lukes, Y., Wright, F. D., and Wartofsky, L., 1977, Reduction in hepatic triiodothyronine binding capacity induced by fasting, *Endocrinology* **101**:1331.

Burrows, G. N., May, P. B., Spaulding, S. W., and Donabedian, R. K., 1977, TRH and dopamine interactions affecting pituitary hormone secretion, *J. Clin. Endocrinol. Metab.* **45**:65.

Camanni, F., Massara, F., Belforte, L., Rosatello, A., and Molinatti, G. M., 1977, Effect of dopamine on plasma growth hormone and prolactin levels in normal and acromegalic subjects, *J. Clin. Endocrinol. Metab.* **44**:465.

Cameron, D. P., Burger, H. G., Catt, K. J., Gordon, E., and Watts, J. M., 1972, Metabolic clearance of human growth hormone in patients with hepatic and renal failure and in the isolated perfused pig liver, *Metabolism* **21**:895.

Campbell, H. J., 1966, The development of the primary portal plexus in the median eminence of the rabbit, *J. Anat. Physiol. London* **100**:381.

Cantalamessa, L., Reschini, E., Catania, A., and Giustina, G., 1976, Pituitary hormone responses to hypothalamic releasing hormones in acromegaly, *Acta Endocrinol. (Kbh)* **83**:673.

Carlson, H. E., Mariz, I. K., and Daughaday, W. H., 1974, Thyrotropin-releasing hormone stimulation and somatostatin inhibition of growth hormone secretion from perfused rat adenohypophyses, *Endocrinology* **94**:1709.

Carlson, H. E., Sowers, J. R., and Rand, R. W., 1977, Lack of effect of thyroid hormones on the growth hormone response to thyrotropin-releasing hormone in acromegaly, *Metabolism* **26**:801.

Carr, D., Gomez-Pan, A., Weightman, D. R., Roy, V. C. M., Hall, R., Besser, G. M., Thorner, M. O., McNeilly, A. S., Schally, A. V., Kastin, A. J., and Coy, D. H., 1975, Growth-hormone release inhibiting hormone: Actions on thyrotropin and prolactin secretion after thyrotropin-releasing hormone, *Br. Med. J.* **3**:67.

Carroll, B. J., and Mendels, J., 1976, Neuroendocrine regulation in affective disorders, in: *Hormones, Behaviour and Psychopathology* (E. J. Sachar, ed.), pp. 193–224, Raven Press, New York.

Casper, R. C., and Davis, J. M., 1977, On the course of anorexia nervosa, *Am. J. Psychiatry* **134**:974.

Casper, R. C., Davis, J. M., and Pandey, G. N., 1977, The effect of nutritional status and weight changes on hypothalamic function tests in anorexia nervosa, in: *Anorexia Nervosa* (R. A. Vigersky, ed.), pp. 137–147, Raven Press, New York.

Catania, A., Cantalamessa, L., and Reschini, E., 1976, Plasma prolactin response to luteinizing hormone releasing hormone in acromegalic patients, *J. Clin. Endocrinol. Metab.* **43**:689.

Ceda, G. P., Speroni, G., Dall'Aglio, E., Valenti, G., and Butturrini, U., 1982, Nonspecific growth hormone responses to thyrotropin-releasing hormone in insulin-dependent diabetes: Sex- and age-related pituitary responsiveness, *J. Clin. Endocrinol. Metab.* **55**:180.

Chazot, G., Chalumeau, A., Aimard, G., Mornex, R., Garde, A., Schott, B., and Girard, P. F., 1974, Facteur-de liberation de l'hormone thyreotrope et etats depressifs: Des acroagonismes au TRH, *Lyon Med.* **231**:831.

Checkley, S. A., 1980, Neuroendocrine tests of monoamine function in man: A review of basic theory and its application to the study of depressive illness, *Psychol. Med.* **10**:35.

Chihara, K., Kato, Y., Ohgo, S., Iwasaki, Y., Abe, H., Maeda, K., and Imura, H., 1976a, Stimulating and inhibiting effects of thyrotropin-releasing hormone on growth-hormone release in rats, *Endocrinology* **98**:1047.

Chihara, K., Kato, Y., Ohgo, S., Iwasaki, Y., Maeda, K., Miyamoto, Y., and Imura, H., 1976b, Effects of hyperthyroidism and hypothyroidism on rat growth-hormone release induced by thyrotropin-releasing hormone, *Endocrinology* **98**:1396.

Chihara, K., Kato, Y., Maeda, K., Abe, H., Furumoto, M., and Imura, H., 1977, Effect of thyrotropin-releasing hormone on sleep and sleep-related growth hormone release in normal subjects, *J. Clin. Endocrinol. Metab.* **44**:1094.

Clark, D., Thody, A. J., Shuster, S., Bowers, H., 1978, Immunoreactive α-MSH in human plasma in pregnancy, *Nature* **273**:163.

Cocchi, D., Gil-Ad, I., Panerai, A. E., Locatelli, V., and Müller, E. E., 1976, Circadian variations in plasma growth hormone and prolactin in the infant rat: Comparison with the adult pattern, *Life Sci.* **19**:825.

Cocchi, D., Locatelli, S., and Salerno, F., 1982, Effect of repeated oral administration of TRH on TRH-induced anterior pituitary hormone release in patients with severe liver disease, *J. Endocrinol. Invest.* **5**(Suppl. 1):222 (abstract).

Collu, R., 1979, Abnormal pituitary hormone response to thyrotropin-releasing hormone: An index of central nervous system dysfunction, in: *Central Neuroendocrinology: A Pathophysiological Approach* (G. Tolis, F. Labrie, J. B. Martin, and F. Naftolin, eds.), pp. 129–137, Raven Press, New York.

Collu, R., Clermont, M. J., and Letarte, J., 1975, Inhibition of pentobarbital-induced release of growth hormone by thyrotropin releasing hormone, *Endocrinol. Res. Commun.* **2**:123.

Connell, G. M., Garcia, J. F., and Linfoot, J. A., 1975, Response to TRF in Cushing's disease and Nelson's syndrome: Absence of any effect on ACTH secretion, Abstracts of the 57th Annual Meeting of the Endocrine Society, New York, p. 216.

Convey, E. M., Tucker, H. A., Smith, V. G., and Zolman, J., 1973, Bovine prolactin, growth-hormone, thyroxine and corticoid response to thyrotropin-releasing hormone, *Endocrinology* **92**:471.

Cooper, G. R., and Shin, S. H., 1981, Somatostatin inhibits prolactin secretion in the estradiol primed male rat, *Can. J. Physiol. Pharmacol.* **59**:1082.

Copinschi, G., Leclercq-Meyer, V., Virasoro, E., L'Hermite, M., Vanhaelst, L., Golstein, J., Leclercq, R., Fery, F., and Robyn, C., 1976, Pituitary and extrapituitary effects of somatostatin in normal man, *Horm. Metab. Res.* **8**:226.

Coppen, A., Eccleston, E., and Peet, M., 1973, Total and free tryptophan concentration in the plasma of depressive patients, *Lancet* **II**:60.

Cremer, G. M., Molnar, G. D., Taylor, W. F., Rosevear, J. W., and Ackermann, E., 1973, Growth hormone release in unstable diabetes: Test with saline, arginine, glucagon and epinephrine, *Acta Diabetol. Lat.* **10**:1216.

Czernichow, P., Dauzet, M. C., Broyer, M., and Rappaport, R., 1976, Abnormal TSH, PRL and GH response to TSH releasing factor in chronic renal failure, *J. Clin. Endocrinol. Metab.* **43**:630.

Daneman, D., Gutai, J. P., Foley, T. P., Jr., Johnson, L., and Winters, S. J., 1982, Hypothalamic pituitary gonadal (HPG) axis in children with severe hypothyroidism (HT), 64th Annual Meeting of the Endocrine Society, San Francisco, p. 86 (Abstract).

Dasmahapatra, A., Urdaniva, E., and Cohen, M. P., 1981, Growth hormone response to thyrotropin-releasing hormone in diabetes, *J. Clin. Endocrinol. Metab.* **52**:859.

De Camilli, P., Macconi, D., and Spada, A., 1979, Dopamine inhibits adenylate cyclase in human prolactin secreting adenomas, *Nature (London)* **278**:252.

Denef, C., 1981, LH-RH stimulates prolactin release from rat pituitary lactotrophs cultured with a highly purified population of gonadotrophs, *Ann. Endocrinol.* **42**:65.

Dickerman, Z., Rachmel, A., Gil-Ad, I., Prager-Lewin, R., Galatzer, A., and Laron, Z., 1981, Rise in plasma growth hormone in response to exogenous LRH in Klinefelter's syndrome, *Clin. Endocrinol.* **15**:403.

Eastman, C. J., and Lazarus, L., 1972, The effect of orally administered synthetic thyrotropin releasing factor on adenohypophyseal function, *Horm. Metab. Res.* **4**:58.

Emerson, G. H., Dyson, W. L., and Utiger, R. D., 1973, Serum thyrotropin and thyroxine concentrations in patients receiving lithium carbonate, *J. Clin. Endocrinol. Metab.* **36**:338.

Faglia, G., Beck-Peccoz, P., Travaglini, P., Paracchi, A., Spada, A., and Lewin, A., 1973a, Elevations in plasma growth hormone concentration after luteinizing hormone-releasing hormone (LRH) in patients with active acromegaly, *J. Clin. Endocrinol. Metab.* **37**:338.

Faglia, G., Beck-Peccoz, P., Ferrari, C., Travaglini, P., Ambrosi, B., and Spada, A., 1973b, Plasma growth-hormone response to thyrotropin-releasing hormone in patients with active acromegaly, *J. Clin. Endocrinol. Metab.* **36**:1259.

Faglia, G., Paracchi, A., Beck-Peccoz, P., and Ferrari, C., 1975, An explanatory hypothesis for plasma GH response to non-specific releasing hormone and for "paradoxical" GH inhibition after dopaminergic drugs in acromegaly, *Acta Endocrinol. (Kbh.)* **199**(Suppl.):323 (abstract).

Faglia, G., Paracchi, A., Beck-Peccoz, P., and Ferrari, C., 1977, Assessment of the results of transphenoidal hypophysectomy in acromegaly by means of TRH and L-dopa tests, in: *Treatment of Pituitary Adenomas* (R. Fahlbusch and K. von Werder, eds.), pp. 91–94, Thieme, Stuttgart.

Franchimont, P., 1972, Thyrotrophin releasing hormone, in: *Frontiers in Hormone Research, 1972* (R. Hall, I. Werner, and H. Holgate, eds.), Vol. 1, pp. 139–140, Karger, Basel.

Frankel, R. J., and Jenkins, J. S., 1975, Hypothalamic pituitary function in anorexia nervosa, *Acta Endocrinol. (Kbh.)* **78**:209.

Fuxe, K., Agnati, L. F., Corrodi, H., Everitt, B. J., Hökfelt, T., Löfstrom, A., and Ungerstedt, U., 1975a, Action of dopamine receptor agonists in forebrain and hypothalamus: Rotational behaviour, ovulation, and dopamine turnover, in: *Advances in Neurology* (J. Calne, T. Chase, and L. Barbeau, eds.), Vol. 9, pp. 223–242, Raven Press, New York.

Fuxe, K., Agnati, L, and Everitt, B., 1975b, Effects of metergoline on central monoamine neurons: Evidence for a selective blockade of central 5-HT receptors, *Neurosci. Lett.* **1**:283.

Garfinkel, P. E., Brown, G. M., Stancer, H. C., and Moldofsky, H., 1975, Hypothalamic–pituitary function in anorexia nervosa, *Arch. Gen. Psychiatry* **32**:739.

Gastineau, C. F., 1979, Alcohol and the endocrine system in: *Metabolic Effects of Alcohol* (P. Avogaro, C. R. Sirtori, and E. Tremoli, eds.), pp. 103–110, Elsevier/North-Holland, Amsterdam.

Genazzani, A. R., Camanni, F., Massara, F., Picciolini, E., Cocchi, D., Belforte, L., and Müller, E. E., 1980, A new pharmacological approach to the diagnosis of hyperprolactinemic states: The nomifensine test, *Acta Endocrinol. (Kbh.)* **93**:139.

Giampietro, O., Moggi, G., Chisci, R., Coluccia, A., Dalle Luche, A. D., Simonini, N., and Brunori, I., 1979, Unusual prolactin response to luteinizing hormone-releasing hormone in some anovulatory women, *J. Clin. Endocrinol. Metab.* **49**:141.

Gershengorn, M., Arevalo, C., Geras, E., Rebecchi, M., 1980, Thyrotropin-releasing hormone stimulation of adrenocorticotropin production by mouse pituitary tumor cells in culture, *J. Clin. Invest.* **65**:1294.

Giannattasio, G., Zanini, A., Panerai, A. E., Meldolesi, J., and Müller, E. E., 1979, Studies on rat pituitary homografts. II. Effects of thyrotropin-releasing hormone on *in vitro* biosynthesis and release of growth hormone and prolactin, *Endocrinology* **104**:237.

Gil-Ad, I., Cocchi, D., Panerai, A. E., Locatelli, V., Mantegazza, P., and Müller, E. E., 1976, Altered growth-hormone and prolactin responsiveness to TRH in the infant rat, *Neuroendocrinology* **21**:366.

Gil-Ad, I., Dickerman, Z., Weizman, R., Weizman, A., Tyano, S., and Laron, Z., 1981, Abnormal growth-hormone response to LRH and TRH in adolescent schizophrenic boys, *Am. J. Psychiatry* **138**:357.

Giustina, G., Reschini, E., Peracchi, M., Cantalamessa, L., Cavagnini, F., Pinto, M., and Bulgheroni, P., 1974, Failure of somatostatin to suppress thyrotropin releasing factor and luteinizing hormone releasing factor-induced growth-hormone release in acromegaly, *J. Clin. Endocrinol. Metab.* **38**:906.

Glydon, R. S. T. J., 1957, The development of the blood supply to the pituitary in the albino rat, with specific reference to the portal vessels, *J. Anat. London* **91**:237.

Gold, M. S., Pottash, A. L. C., Sweeney, D. R., Martin, D. M., and Davies, R. K., 1980, Further evidence of hypothalamic–pituitary dysfunction in anorexia nervosa, *Am. J. Psychiatry* **137**:101.

Gonzales-Barcena, D., Kastin, A. J., Schalch, D. S., Torres-Zamora, M., Perez-Pasten, E., Kato, A., and Schally, A. V., 1973, Responses to thyrotropin-releasing hormone in patients with renal failure and after infusion in normal men, *J. Clin. Endocrinol. Metab.* **36**:117.

Goodwin, F. K., Cowdry, R. W., and Wester, M. H., 1978, Predictors of drug response in the affective disorders: Towards an integrated approach, in: *Psychopharmacology: A Generation of Progress* (M. A. Lipton, A. DiMascio, and K. F. Killam, eds.), pp. 1277–1288, Raven Press, New York.

Gorewit, R. C., 1980, Influence of somatostatin on serum prolactin concentrations of cow during rest and milking, *Experientia* **36**:359.

Grant, N. H., Sarantakis, D., and Yardley, P., 1974, Action of growth hormone release inhibitory hormone on prolactin release in rat pituitary cell cultures, *J. Endocrinol.* **61**:163.

Gruen, P. H., 1978, Endocrine changes in psychiatric diseases, *Med. Clin. North Am.* **62**:285.

Guillemin, R., and Gerich, J. E., 1976, Somatostatin: Physiological and clinical significance, *Annu. Rev. Med.* **27**:379.

Guillemin, R., Brazeau, P., Böhlen, P., Esch, F., Ling, N., and Wehrenberg, W. B., 1982, Growth hormone-releasing factor from a human pancreatic tumor that caused acromegaly, *Science* **218**:585.

Hall, R., Schally, A. V., Evered, D., Kastin, A. J., Mortimer, C. H., Turnbridge, W. M. G., Besser, G. M., Coy, D. H., Goldie, D. J., McNeilly, A. S., Phenekos, C., and Weightman, D., 1973, Action of growth-hormone-release inhibitory hormone in healthy men and in acromegaly, *Lancet* **II**:581.

Hamada, N., Uoi, K., Nishizawa, Y., Okamoto, T., Hasagawa, K., Mirii, H., and Wada, H., 1976, Increase of serum GH concentration following TRH injection in patients with primary hypothyroidism, *Endocrinol. Jpn.* **23**:5.

Hanew, K., Kokubun, M., Sasaki, A., Mouri, T., and Yoshinaga, K., 1980, The spectrum of pituitary growth hormone responses to pharmacological stimuli in acromegaly, *J. Clin. Endocrinol. Metab.* **51**:292.

Hanse, A. P., 1970, Abnormal serum growth hormone responses to exercise in juvenile diabetics, *J. Clin. Invest.* **49**:1467.

Hartmann, E., 1976, Schizophrenia, a theory, *Psychopharmacol. Bull.* **49**:1.

Hedlund, L., Doelger, S. G., Tollerton, A. J., Lischko, M. M., and Johnson, H. D., 1977, Plasma growth hormone concentrations after cerebroventricular and jugular injection of thyrotropin-releasing hormone, *Proc. Soc. Exp. Biol. Med.* **156**:422.

Herbert, D. C., and Rennels, E. G., 1977, Effect of synthetic luteinizing hormone releasing hormone on prolactin secretion from clonal pituitary cells, *Biochem. Biophys. Res. Commun.* **79**:133.

Hiba, J., Del Pozo, E., Genazzani, A., Pusterla, E., Lancranjan, I., Sidiropoulos, D., and Gunti, J., 1977, Hormonal mechanism of milk secretion in the newborn, *J. Clin. Endocrinol. Metab.* **44**:973.

Hinkle, P. M., and Tashjian, A., Jr., 1975, Thyrotropin releasing hormone regulates the number of its own receptors in the $GH_3$ strain of pituitary cells in culture, *Biochemistry* **14**:3845.

Hintz, R. L., Suskind, R., Amatayakul, K., Thanangkul, O., and Olson, R., 1978, Plasma somatomedin and growth hormone values in children with protein-calorie malnutrition, *J. Pediatr.* **92**:153.

Horrobin, D. F., 1977, The role of prostaglandins and prolactin in depression, mania and schizophrenia, *Postgrad. Med. J.* **53**(Suppl. 4):160.

Hoyte, K. M., and Martin, J. B., 1975, Recovery from paradoxical growth hormone responses in acromegaly after transphenoidal selective adenomectomy, *J. Clin. Endocrinol. Metab.* **41**:656.

Irie, M., and Tsushima, T., 1972, Increase of serum growth hormone concentration following thyrotropin releasing hormone injection in patients with acromegaly or gigantism, *J. Clin. Endocrinol. Metab.* **35**:97.

Ishibashi, M., Yamaji, T., and Kosaka, K., 1977, Effect of bromocriptine on TRH induced growth-hormone and prolactin release in acromegalic patients, *J. Clin. Endocrinol. Metab.* **45**:275.

Ishibashi, M., Yamaji, T., and Kosaka, K., 1978, Induction of growth hormone and prolactin secretion by luteinizing hormone-releasing hormone and its blockade by bromocriptine in acromegalic patients, *J. Clin. Endocrinol. Metab.* **47**:418.

Jellinger, K., and Riederer, P., 1977, Brain monoamines in metabolic (endotoxic) coma: A preliminary biochemical study in human post-mortem material, *J. Neural Transm.* **41**:275.

Joasoo, A., Murray, I. P. C., Parkin, J., Robenson, M. R., and Jeremy, D., 1974, Abnormalities of in vitro thyroid function tests in renal disease, *Q. J. Med.* **43**:245.

Johansen, K. and Hansen, A. P., 1971, Diurnal serum growth hormone levels in poorly and well-controlled juvenile diabetics, *Diabetes* **20**:239.

Kaplan, S. L., Grumbach, M. M., and Shepard, T. H., 1972, The ontogenesis of human fetal hormones. I. Growth hormone and insulin, *J. Clin. Invest.* **51**:3080.

Karlberg, B., Almqvist, S., and Werner, S., 1971, Effects of synthetic pyroglutamyl-histidyl-proline-amide on serum levels of thyrotropin, cortisol, growth hormone, insulin and PBI in normal subjects and patients with pituitary and thyroid disorders, *Acta Endocrinol. (Kbh.)* **67**:288.

Kastin, A. J., Nissen, C., Redding, T. W., Nair, R. M. G., and Schally, A. V., 1974, Delayed disappearance of $^{14}C$-labeled Pro-Leu-Gly-$NH_2$ from the blood of hypophysectomized rats, *Neuroendocrinology* **16**:36.

Kato, Y., Chihara, K., Maeda, K., Ohgo, S., Okanishi, Y., and Imura, H., 1975, Plasma growth hormone responses to thyrotropin-releasing hormone in the urethane-anesthetized rat, *Endocrinology* **96**:1114.

Keller, H. H., Bartolini, G., and Pletscher, A., 1974, Enhancement of cerebral noradrenaline turnover by thyrotropin releasing hormone, *Nature (London)* **248**:528.

Kirkegaard, C., Eskildsen, P. C., and Bjørum, N., 1981, Parallel changes of the responses of thyrotropin, growth hormone and prolactin to thyrotropin-releasing hormone in endogenous depression, *Psychoneuroendocrinology* **6**:253.

Krieger, D. T., and Luria, M., 1977, Plasma ACTH and cortisol responses to TRF, vasopressin or hypoglycemia in Cushing's disease and Nelson's syndrome, *J. Clin. Endocrinol. Metab.* **44**:361.

Labrie, F., Borgeat, P., Ferland, L., Dupont, A., Lamaire, S., Pelletier, G., Barden, N., Drouin, J., DeLéan, A., Belanger, A., and Jolitoeur, P., 1975, Mechanism of action and modulation of activity of hypothalamic hypophysiotropic hormones, in: *Hypothalamic Hormones* (M. Motta, P. G. Crosignani, and L. Martini, eds.), pp. 109–123, Academic Press, New York.

Lal, S., Young, S. N., and Sourkes, T. L., 1975, 5-Hydroxytryptamine and hepatic coma, *Lancet* II:979.

Lamberts, S. W. J., Klijn, J. G. M., de Quijada, M., Timmermans, H. A. T., Uitterlinden, P., and Birkenhäger, J. C., 1980, Bromocriptine and the medical treatment of Cushing's disease, in: *Neuroactive Drugs in Endocrinology* (E. E. Müller, ed.), pp. 371–382, Elsevier/North-Holland, Amsterdam.

Lim, V. S., Fang, V. S., Kara, A. I., and Refetoff, S., 1977, Thyroid dysfunction in chronic renal failure: A study of the pituitary–thyroid axis and peripheral turnover kinetics of thyroxine and triiodothyronine, *J. Clin. Invest.* **60**:522.

Liuzzi, A., Chiodini, P. G., Botalla, L., Silvestrini, F., and Müller, E. E., 1974, Growth hormone (GH) releasing activity of TRH and GH-lowering effect of dopaminergic drugs in acromegaly: Homogeneity in the two responses, *J. Clin. Endocrinol. Metab.* **39**:871.

Loizou, L. A., 1972, The postnatal ontogeny of monoamine-containing neurones in the central nervous system of albino rat, *Brain Res.* **40**:395.

Loosen, P. T., Prange, A. J., Jr., and Wilson, I. C., 1979, TRH (Protirelin) in depressed alcoholic men: Behavioral changes and endocrine responses, *Arch. Gen. Psychiatry* **36**:540.

Lundberg, P. O., Walinder, J., Werner, J., and Wide, L., 1972, Effects of thyrotropin-releasing hormone on plasma levels of TSH, FSH, LH and GH in anorexia nervosa, *Eur. J. Clin. Invest.* **2**:150.

Maas, J. W., Hattox, S. E., Greeve, N. M., and Landis, D. H., 1979, 3-Methoxy-4-hydroxyphenethyleneglycol production by human brain in vivo, *Science* **205**:1025.

Macaron, C., Wilber, J. F., Green, O., and Freinkel, N., 1978, Studies of growth hormone (GH), thyrotropin (TSH) and prolactin (PRL) secretion in anorexia nervosa, *Psychoneuroendocrinology* **3**:181.

McComb, D. J., Ryan, N., Ryder, D., Horvath, E., Kovacs, K., Domokos, I., and Làszlò, F. A., 1981, Response to thyrotropin-releasing hormone (TRH) of rat lactotrophs and somatotrophs deprived of hypothalamic control, *Endokrinologie* **3**:303.

Machlin, L. J., Jacobs, L. S., Cirulis, N., Kimes, R., and Miller, R., 1974, An assay for growth hormone and prolactin-releasing activities using a bovine pituitary cell culture system, *Endocrinology* **95**:1350.

Maeda, K., Kato, Y., Ohgo, S., Chihara, K., Yoshimoto, Y., Yamaguchi, N., Kuromaru, S., and Imura, H., 1975a, Growth hormone and prolactin release after injection of thyrotropin releasing hormone in patients with depression, *J. Clin. Endocrinol. Metab.* **40**:501.

Maeda, K., Kato, Y., Chihara, K., Ohgo, S., Iwasaki, Y., and Imura, H., 1975b, Suppression by thyrotropin-releasing hormone (TRH) of human growth hormone release induced by L-dopa, *J. Clin. Endocrinol. Metab.* **41**:408.

Maeda, K., Kato, Y., Yamaguchi, N., Chihara, K., Ohgo, S., Okanishi, Y., Yoshimoto, Y., Moridera, K., Kuromaru, S., and Imura, H., 1976a, Growth hormone release following thyrotropin releasing hormone injection into patients with anorexia nervosa, *Acta Endocrinol. (Kbh.)* **81**:1.

Maeda, K., Kato, Y., Chihara, K., Ohgo, S., Iwasaki, Y., Abe, H., and Imura, H., 1976b, Suppression by thyrotropin-releasing hormone (TRH) of growth hormone release induced by arginine and insulin-induced hypoglycemia in man, *J. Clin. Endocrinol. Metab.* **43**:453.

Marshall, J. C., and Fraser, T. R., 1971, Amenorrhea in anorexia nervosa: Assessment and treatment with clomiphene citrate, *Br. Med. J.* **IV**:590.

Maruyama, T., and Ishikawa, H., 1977, Somatostatin: Its inhibiting effect on the release of hormones and IgG from clonal cell strains. Its Ca-influx dependence, *Biochem. Biophys. Res. Commun.* **3**:1083.

Masala, A., Meloni, T., Galisai, D., Alagna, S., Rovasio, P. P., Rassu, S., and Milia, A. F., 1982, The effects of thyrotropin-releasing hormone and luteinizing hormone-releasing hormone on growth hormone release in patients with homozygous β-thalassaemia, *J. Clin. Endocrinol. Metab.* **54**:1271.

Mashiter, R., White, M. O., Adams, E. F., Loizou, M., Winslow, C., and Surmont, D. W. A., 1982, Peptide regulation of hormone secretion by human pituitary tumours in cell culture, in: *Pituitary Hyperfunction, Physiopathology and Clinical Aspects* (D. Cocchi and F. Massara, eds.), September 9–11, p. 37 (abstract), Milan, Italy.

Matsukura, S. Kakita, Y., Hirata, Y., Yoshimi, M., Fukase, Y., Iwasaki, Y., Kato, Y., and Imura, H., 1977, Adenylate cyclase of GH and ACTH producing tumors of human: Activation by non-specific hormones and other bioactive substances, *J. Clin. Endocrinol. Metab.* **44**:392.

Mecklenburg, R. S., Loriaux, D. L., Thompson, R. H., Andersen, A. E., and Lipsett, M. B., 1974, Hypothalamic dysfunction in patients with anorexia nervosa, *Medicine (Baltimore)* **53**:147.

Molnar, G. D., Taylor, W. F., Langworthy, A., and Fatourechi, A., 1972, Diurnal growth hormone and glucose abnormalities in unstable diabetics: Studies of ambulatory-fed subjects during continuous blood glucose analysis, *J. Clin. Endocrinol. Metab.* **34**:837.

Moriondo, P., Travaglini, P., Ronderna, M., Beck-Peccoz, P., Conti-Puglisi, F., Ambrosi, B., and Faglia, G., 1980, Prolactin secretion in acromegaly, in: *Pituitary Microadenomas* (G. Faglia, M. A. Giovanelli, and R. M. MacLeod, eds.), pp. 247–255, Academic Press, New York.

Morley, J. E., 1981, Neuroendocrine control of thyrotropin secretion, *Endocrine Rev.* **2**:396.

Mortimer, C. H., Besser, G. M., McNeilly, A. S., Tumbridge, W. M. G., Gomez-Pan, A., and Hall, R., 1973, Interaction between secretion of the gonadotropins, prolactin, growth hormone, thyrotropin and corticosteroids in man: The effect of LH/FSH-RH, TRH and hypoglycemia alone and in combination, *Clin. Endocrinol.* **2**:317.

Mortimer, C. H., Besser, G. M., Goldie, D. J., Hook, J., and McNeilly, A. S., 1974, The TSH, FSH and PRL responses to continuous infusions of TRH and the effects of oestrogens administration in normal males, *Clin. Endocrinol.* **3**:97.

Müller, E. E., 1979, The control of somatotropin secretion, in: *Hormonal Proteins and Peptides* (C. H. Li, ed.), Vol. VII, pp. 124–204, Academic Press, New York.

Müller, E. E., Panerai, A. E., Cocchi, D., Gil-Ad, I., Rossi, G. L., and Olgiati, V. R., 1977a, Growth hormone releasing activity of thyrotropin-releasing hormone in rats with hypothalamic lesions, *Endocrinology* **100**:1663.

Müller, E. E., Liuzzi, A., Cocchi, D., Panerai, A. E., Oppizzi, G., Locatelli, V., Silvestrini, F., Mantegazza, P., and Chiodini, P. G., 1977b, Role of dopaminergic receptors in the regulation of growth hormone secretion, in: *Non-Striatal Dopaminergic Neurons* (E. Costa and G. L. Gessa, eds.), pp. 127–138, Raven Press, New York.

Müller, E. E., Salerno, F., Cocchi, D., Locatelli, V., and Panerai, A. E., 1979, Interaction between the thyrotropin-releasing hormone-induced growth hormone rise and dopaminergic drugs: Studies in pathologic conditions of the animal and man, *Clin. Endocrinol.* **11**:645.

Munro, H. N., Fernstrom, J. D., and Wurtman, R. J., 1975, Insulin, plasma amino acid imbalance, and hepatic coma, *Lancet* **I**:722.

Murphy, D. L., Schilin, D. J., and Murray, R. N., 1978, Psychoactive drug responder subgroups: Possible contributions to psychiatric classification, in: *Psychopharmacology: A Generation of Progress* (M. A. Lipton, A. D. Mascio, and K. F. Killam, eds.), pp. 807–820, Raven Press, New York.

Nakagawa, K., and Obara, T., 1977, Failure of growth-hormone suppressing agents to affect TRH-releasing hormone and LH releasing hormone-induced growth-hormone release in acromegaly, *J. Clin. Endocrinol. Metab.* **44**:189.

Ojeda, S. R., Castro Vasquez, A., and McCann, S. M., 1977, TRH-induced growth hormone (GH) release in rats of both sexes: Changes in pituitary response after gonadectomy and during estrous cycle, *Proc. Soc. Exp. Biol. Med.* **54**:254.

Oki, S., Nakai, Y., Nakao, K., and Imura, H., 1980, Plasma-endorphin responses to somatostatin,

thyrotropin-releasing hormone or vasopressin in Nelson's syndrome, *J. Clin. Endocrinol. Metab.* **50:**194.
Panerai, A. E., Cocchi, D., Gil-Ad, I., Locatelli, V., Rossi, G. L., and Müller, E. E., 1976, Stimulation of growth hormone release by luteinizing hormone-releasing hormone and melanocyte stimulating hormone-release inhibiting hormone in the hypophysectomized rat bearing an ectopic pituitary, *Clin. Endocrinol.* **5:**717.
Panerai, A. E., Giannattasio, G., Zanini, A., Meldolesi, J., Rossi, G. L., and Müller, E. E., 1977a, TRH: A potential growth-hormone releasing factor, Abstracts, 59th Annual Meeting of the Endocrine Society, Chicago, p. 128.
Panerai, A. E., Gil-Ad, I., Cocchi, D., Locatelli, V., Rossi, G. L., and Müller, E. E., 1977b, Thyrotropin releasing hormone-induced growth hormone and prolactin release: Physiological studies in intact rats and hypophysectomized rats bearing an ectopic pituitary gland, *J. Endocrinol.* **72:**301.
Panerai, A. E., Rossi, G. L., Cocchi, D., Gil-Ad, I., Locatelli, V., and Müller, E. E., 1977c, Release of growth hormone by TRH in intact rats or in intact or hypophysectomized rats bearing a heterotopic pituitary, *Proc. Soc. Exp. Biol. Med.* **154:**573.
Panerai, A. E., Salerno, F., Manneschi, M., Cocchi, D., and Müller, E. E., 1977d, Growth hormone and prolactin responses to thyrotropin releasing hormone in patients with severe liver disease, *J. Clin. Endocrinol. Metab.* **45:**134.
Pawlikowski, M., and Owczarczyk, I., 1977, Effect of triiodothyronine on thyroliberin-induced somatostatin release in patients with acromegaly, *Endokrynol. Pol.* **6:**497.
Pieters, G. F. F. M., Smals, A. G. H., Benraad, T. J., and Kloppenborg, P. W. C., 1979, Plasma cortisol response to thyrotropin releasing hormone and luteinizing hormone-releasing hormone in Cushing's disease, *J. Clin. Endocrinol. Metab.* **48:**874.
Pieters, G. F. F. M., Romeijn, J. E., Smals, A. G. H., and Kloppenborg, P. W. C., 1982a, Somatostatin sensitivity and growth hormone reponses to releasing hormones and bromocriptine in acromegaly, *J. Clin. Endocrinol. Metab.* **54:**942.
Pieters, G. F. F. M., Smals, A. G. H., Goverde, H. J. M., Kloppenborg, P. W. C., 1982b, Paradoxical responsiveness of adrenocorticotropin and cortisol to thyrotropin releasing hormone (TRH) in pregnant women: Evidence for intermediate lobe activity, *J. Clin. Endocrinol. Metab.* **55:**387.
Pieters, G. F. F. M., Smals, A. G. H., Goverde, H. G. M., Pesman, G. J., Meyer, E., Kloppenborg, P. W. G., 1982c, Adrenocorticotropin and cortisol responsiveness to thyrotropin-releasing hormone and luteinizing hormone-releasing hormone discloses two subsets of patients with Cushing's disease, *J. Clin. Endocrinol. Metab.* **55:**1188.
Pimstone, B. L., LeRoith, D., Epstein, S., and Kronheim, S., 1975, Disappearance rates of plasma growth hormone after intravenous somatostatin in renal and liver disease, *J. Clin. Endocrinol. Metab.* **41:**392.
Ramirez, G., O'Neill, W. M., Jr., Bloomer, H. A., and Jubiz, W., 1977, Abnormalities in the regulation of prolactin in patients with chronic renal failure, *J. Clin. Endocrinol. Metab.* **45:**658.
Rivier, J., Spiess, J., Thorner, M., Vale, W., 1982, Characterization of a growth-hormone-releasing factor from a human pancreatic islet tumour, *Nature* **300:**276.
Rossi, G. L., Probst, D., Panerai, A. E., Gil-Ad, I., Cocchi, D., and Müller, E. E., 1977, Light and electron microscopic studies of thyrotropin releasing hormone-induced growth hormone and prolactin release in hypophysectomized rats bearing an ectopic pituitary gland, *J. Endocrinol.* **72:**313.
Roti, E., Gnudi, A., Robuschi, G., Rossella, P., Benassi, L., and Brawerman, L. E., 1982, Response of growth hormone to thyrotropin-releasing hormone during fetal life, *J. Clin. Endocrinol. Metab.* **54:**1255.
Rubin, A. L., Levin, S. R., Bernstein, R. L., Tyrrell, J. B., Noacco, C., and Forsham, P. H., 1973, Stimulation of growth hormone by luteinizing hormone-releasing hormone in active acromegaly, *J. Clin. Endocrinol. Metab.* **37:**160.
Rubin, R. T., and Poland, R. E., 1982, The chronoendocrinology of endogenous depression, in:

*Neuroendocrine Perspectives* (E. E. Müller and R. M. MacLeod, eds.), pp. 305–331, Elsevier/North-Holland, Amsterdam.

Sachar, E. J., Mushrush, G., Perlow, M., Weitzman, E. D., and Sassin, J., 1972, Growth hormone responses to L-dopa in depressed patients, *Science* **178**:1304.

Saito, S., Abe, K., Yoshida, H., Kaneko, T., Nakamura, E., Shimizu, T. N., and Yamaihara, N., 1971, Effects of synthetic thyrotropin-releasing hormone on plasma thyrotropin, growth hormone and insulin levels in man, *Endocrinol. Jpn.* **18**:101.

Salerno, F., Cocchi, D., Zanardi, P., Casanueva, F., and Müller, E. E., 1979, Growth hormone and prolactin secretion in cirrhotic patients, in: *Metabolic Effects of Alcohol* (P. Avogaro, C. R. Sirtori, and E. Tremoli, eds.), pp. 77–88, Elsevier/North-Holland, Amsterdam.

Salerno, F., Cocchi, D., Frigerio, C., Colombo, A. M., and Müller, E. E., 1980, Anomalous growth hormone responses to thyrotropin releasing hormone and glucose in cirrhotic patients: The effect of metergoline, *J. Clin. Endocrinol. Metab.* **51**:641.

Salerno, F., Cocchi, D., Monza, G., Lampertico, M., and Müller, E. E., 1982, Growth hormone response to thyrotropin-releasing hormone in liver cirrhosis: Unique alteration in anterior pituitary responsiveness to hypothalamic hormones, *Horm. Metab. Res.* **14**:482.

Samaan, N. A., and Freeman, R. M., 1970, Growth hormone levels in severe renal failure, *Metabolism* **19**:102.

Samuel, A. M., and Deshpande, U. R., 1972, Growth hormone levels in protein calorie malnutrition, *J. Clin. Endocrinol. Metab.* **35**:863.

Sastry, B. S. R., and Phillips, J. W., 1977, Metergoline as a selective 5-hydroxytryptamine (5-HT) antagonist in the cerebral cortex, *Can. J. Physiol. Pharmacol.* **55**:130.

Schalch, D. S., Gonzalez-Barcena, D., Kastin, A. J., Schally, A. V., and Lee, L. A., 1972, Abnormalities in the release of TSH in response to thyrotropin-releasing hormone (TRH) in patients with disorders of the pituitary, hypothalamus and basal ganglia, *J. Clin. Endocrinol. Metab.* **35**:609.

Schally, A. V., Arimura, A., Redding, T. W., Debeljuk, U., Carter, W., Dupont, A., and Vilchez-Martinez, J. A., 1976, Re-examination of porcine and bovine hypothalamic fractions for additional luteinizing hormone and follicle stimulating hormone-releasing activity, *Endocrinology* **98**:380.

Schildkraut, J. J., 1974, Catecholamines and affective disorders: New concepts in brain research, Health Learning Systems Inc, Bloomfield.

Schonbrunn, A., and Tashjian, A. H., Jr., 1978, Characterization of functional receptors for somatostatin in rat pituitary cells in culture, *J. Biol. Chem.* **253**:6473.

Shani, J., and Siegel, R. A., 1980, ACTH release by TRH in conscious and anesthetized rats, 62nd. Annu. Meet. Endocr. Soc., Washington, P. 234 (abstract).

Shaw, D. M., Camps, F. E., and Eccleston, E. C., 1967, 5-Hydroxytryptamine in the hind brain of depressive suicides, *Br. J. Psychiatry* **113**:1407.

Sherman, B. M., and Halmi, K. A., 1977, Effect of nutritional rehabilitation on hypothalamus-pituitary function in anorexia nervosa, in: *Anorexia Nervosa* (R. A. Vigersky, ed.), pp. 211–233, Raven Press, New York.

Shopsin, B., Blum, M., and Gershon, S., 1969, Lithium-induced thyroid disturbance: Case report and review, *Compr. Psychiatry* **10**:215.

Siler, T. M., Van Den Berg, G., and Yen, S. S. C., 1973, Inhibition of growth hormone release in humans by somatostatin, *J. Clin. Endocrinol. Metab.* **37**:632.

Sinha, D., and Meites, J., 1966, Effects of thyroidectomy and thyroxine on hypothalamic concentration of thyrotropin releasing factor and pituitary content of thyrotropin in rats, *Neuroendocrinology* **1**:4.

Snyder, P. J., Muzyka, R., Johnson, J., and Utiger, R. D., 1980, Thyrotropin-releasing hormone provokes abnormal follicle-stimulating hormone (FSH) and luteinizing hormone responses in men who have pituitary adenomas and FSH hypersecretion, *J. Clin. Endocrinol. Metab.* **51**:744.

Snyder, S. H., 1977, Opiate receptors in the brain, *N. Engl. J. Med.* **296**:266.

Staub, J. J., Girard, J., Mueller-Brand, J., Noelpp, B., Werner-Zodrow, I., Baur, U., Heitz, P.

H., and Gemsenjaeger, E., 1978, Blunting of TSH response after repeated oral administration of TRH in normal and hypothyroid subjects, *J. Clin. Endocrinol. Metab.* **46**:260.

Steiner, R. A., Illner, P., Marques, P., Williams, D., Shen, L., Edwards, L., and Gale, C. C., 1977, Inhibition of dopamine-induced release of growth hormone by thyrotropin-releasing hormone, *Am. J. Physiol.* **233**:430.

Sulser, F., 1978, Functional aspects of the norepinephrine receptor coupled adenylate cyclase system in the limbic forebrain and its modification by drugs which precipitate or alleviate depression: Molecular approaches to an understanding of affective disorders, *Pharmakopsychiatr. Neuro-Psychopharmakol.* **11**:43.

Szabo, M., and Frohman, L. A., 1975, Effects of porcine stalk median eminence and prostaglandin $E_2$ on rat growth hormone secretion in vivo and their inhibition by somatostatin, *Endocrinology* **96**:955.

Takahara, J., Arimura, A., and Schally, A. V., 1974a, Stimulation of prolactin and growth hormone release by TRH infused into a hypophysial portal vessel, *Proc. Soc. Exp. Biol. Med.* **146**:831.

Takahara, J., Arimura, A., and Schally, A. V., 1974b, Effect of catecholamines on the TRH-stimulated release of prolactin and growth hormone from sheep pituitaries in vitro, *Endocrinology* **95**:1490.

Takahashi, S., Kondo, H., and Yoshimura, M., 1975, Enhanced growth hormone responses to TRH injection in bipolar depressed patients, *Folia Psychiatr. Neurol. Jpn.* **29**:215.

Tang, L. K. L., and Spies, H. G., 1976, Effects of hypothalamic-releasing hormones on LH, FSH and prolactin in pituitary monolayer cultures, *Proc. Soc. Exp. Biol. Med.* **151**:189.

Tashjian, A. H., Jr., Barowsky, N. J., and Jensen, D. K., 1971, Thyrotropin releasing hormone: Direct evidence for stimulation of prolactin production by pituitary cells in culture, *Biochem. Biophys. Res. Commun.* **43**:516.

Thomsett, M. J., Marti-Henneberg, C., Gluckman, P. D., Kaplan, S. L., Rudolph, A. M., and Grumbach, M. M., 1980, Hormone ontogeny in the ovine fetus. VIII. The effect of thyrotropin-releasing factor on prolactin and growth hormone release in the fetus and neonate, *Endocrinology* **106**:1074.

Tolis, G., Banovac, K., Kleissl, P., Martin, J. B., and McKenzie, J. M., 1979, Episodic and TRH induced growth hormone release in primary hypothyroidism of man and rat, *Endocrinol. Res. Commun.* **6**:213.

Tolis, G., Richardson, S., and Cripsin, J., 1982, Anorexia nervosa–Hormonal profiles, in: *Brain Peptides and Hormones* (R. Collu, J. Ducharme, A. Barbeau, and G. Tolis, eds.), pp. 343–356, Raven Press, New York.

Torjesen, P. A., Haug, E., and Sand, T., 1973, Effect of thyrotropin-releasing hormone on serum levels of pituitary hormones in men and women, *Acta Endocrinol. (Kbh.)* **73**:455.

Travaglini, P., Beck-Peccoz, P., Ferrari, C., Ambrosi, B., Paracchi, A., Severgnini, A., Spada, A., and Faglia, G., 1976, Some aspects of hypothalamic–pituitary function in patients with anorexia nervosa, *Acta Endocrinol. (Kbh.)* **81**:252.

Tyrrell, J. B., Lorenzi, M., Gerich, J. E., and Forshman, P. H., 1975, Inhibition by somatostatin of ACTH secretion in Nelson's syndrome, *J. Clin. Endocrinol. Metab.* **40**:1125.

Udeschini, G., Cocchi, D., Panerai, A. E., Gil-Ad, I., Rossi, G., Chiodini, P. G., Liuzzi, A., and Müller, E. E., 1976, Stimulation of growth hormone release by thyrotropin releasing hormone in the hypophysectomized rat bearing an ectopic pituitary, *Endocrinology* **98**:807.

Vale, W., Rivier, C., Brazeau, P., and Guillemin, R., 1974, Effects of somatostatin on the secretion of thyrotropin and prolactin, *Endocrinology* **95**:968.

Van Thiel, D. H., Gavaler, J. S., and Lester, R., 1975, Plasma estrone, prolactin, neurophysin and sex-steroid-binding globulin in chronic alcoholic men, *Metabolism* **24**:1015.

Van Thiel, D. H., Gavaler, J. S., Wright, C., Smith, W. J., Jr., and Abuid, J., 1978, Thyrotropin-releasing hormone (TRH)-induced growth hormone (hGH) responses in cirrhotic men, *Gastroenterology* **75**:66.

Tonon, M. C., Leroux, P., Jegon, S., Lihrmann, I., Leboulanger, F., Delarue C., and Vaudry, H., 1981, Multineuronal control of the intermediate lobe of the pituitary gland in amphibia, 63rd Annu. Meet. Endocr. Soc., p. 244 (abstract).

Van Thiel, D. H., Gavaler, J. S., and Sanghvi, A., 1982, Lack of dissociation of prolactin responses to thyrotropin releasing hormone and metoclopramide in chronic alcoholic men, *Endocrinol. Invest.* **5:**281.

Verde, G., Oppizzi, G., Colussi, G., Cremascoli, G., Botalla, L., Müller, E. E., Silvestrini, F., Chiodini, P. G., and Liuzzi, A., 1976, Effect of dopamine infusion on plasma levels of growth hormone in normal subjects and in acromegalic patients, *Clin. Endocrinol.* **5:**419.

Wakabayashi, I., Kanda, M., Miki, N., Demura, R., and Shizume, K., 1979, Plasma growth hormone and thyrotropin responses to thyrotropin releasing hormone in freely behaving and urethane anesthetized rats, *Life Sci.* **24:**2119.

Warren, M. P., and Vande Wiele, R. L., 1973, Clinical and metabolic features of anorexia nervosa, *Am. J. Obstet. Gynecol.* **117:**435.

Weissel, M., Stummvoll, H. K., Kolbe, M., and Höfer, R., 1979, Basal and TRH stimulated thyroid and pituitary hormones in various degrees of renal insufficiency, *Acta Endocrinol. (Kbh.)* **90:**23.

Wilberg, J. F., and Seibel, M. J., 1973, Thyrotropin-releasing hormone interactions with an anterior pituitary membrane receptor, *Endocrinology* **92:**888.

Yamaguchi, N., Tanimoto, K., and Kuromaru, S., 1980, Growth hormone (GH) release following thyrotropin-releasing hormone (TRH) injection in manic patients receiving lithium carbonate, *Psychoneuroendocrinology* **5:**253.

Yen, S. S. C., Siler, T., and DeVane, G., 1974, Effect of somatostatin in patients with acromegaly: Suppression of growth hormone, prolactin, insulin and glucose levels, *N. Engl. J. Med.* **290:**935.

Zanoboni, A., and Zanoboni-Muciaccia, W., 1977, Elevated basal growth hormone levels and growth hormone response to TRH in alcoholic patients with cirrhosis, *J. Clin. Endocrinol. Metab.* **45:**576.

Zanoboni, A., Zanoboni-Muciaccia, W., Zanussi, C., and Baraldi, R., 1979, Suppression of clonidine-induced release of growth hormone by thyrotropin releasing hormone in humans, *J. Endocrinol. Invest.* **2:**347.

Zimmerman, E. A., Defendini, R., and Frantz, A. G., 1974, Prolactin and growth hormone in patients with pituitary adenomas: A correlative study of hormone in tumor and plasma by immunoperoxidase technique and radioimmunoassay, *J. Clin. Endocrinol. Metab.* **38:**577.

CHAPTER 10

# Neuroendocrine Studies in Huntington's Disease

RAYMON DURSO, STEFANO A. RUGGIERI,
ALESSANDRO DENARO, and
CAROL A. TAMMINGA

## 1. INTRODUCTION

During the past 15 years, the neuroendocrine system has been studied in Huntington's disease (HD) for the purpose of examining underlying hormonal and neurotransmitter abnormalities within the hypothalamic–pituitary axis. Although a clinically overt endocrinopathy has not been described, symptoms of progressive weight loss, disturbances in sweating, and reports of increased fertility in HD women have indeed been observed (Reed and Neel, 1959; Bruyn, 1968; Marx, 1973), and suggest hypothalamic dysfunction in affected individuals. Moreover, postmortem studies of the hypothalamus in HD patients have demonstrated a number of pathological findings including atrophy associated with neuronal cell loss (Vogt and Vogt, 1952; Bruyn, 1973) and alterations in monoamine metabolites (Bernheimer and Hornykiewicz, 1973).

The investigation of neuroendocrine function in HD has emerged in two directions. The first relates to eliciting evidence of endocrine abnormalities in HD patients which might explain clinical symptomatology. The second uses the hypothalamic–pituitary axis as an *in vivo* system in which to define existing neurochemical defects in HD. Knowing that specific monoamine neurotransmitters modulate anterior pituitary hormone secretion through pathways within the hypothalamus, this research strategy is designed to examine the integrity of such interactions. This method, as it is generally applied, tests hormonal response to specific drugs which act as pharmacologic probes to stimulate or

---

RAYMON DURSO • Department of Neurology, Boston University School of Medicine, Boston, Massachusetts 02130.  STEFANO A. RUGGIERI and ALESSANDRO DENARO • Department of Neurology, University of Rome, Rome, Italy.  CAROL A. TAMMINGA • Department of Psychiatry, University of Maryland, Baltimore, Maryland 21228.

inhibit activity of a single neurotransmitter system. Such investigations, including our own data reported here, have attempted to unveil a neurotransmitter imbalance representative of pathology throughout the choreic brain.

This chapter will review the neuroendocrine research in HD to date and add our own data to the existing literature. Our investigations replicate earlier work as well as expand the spectrum of pharmacologic probes previously used in evaluating the hypothalamic–pituitary axis. These novel probes include a GABAergic agonist, muscimol, and a muscarinic cholinergic drug, arecoline. Additionally, we have gathered 24-hr endocrine measures in HD patients to examine spontaneous daily secretion of growth hormone (GH), prolactin (PRL), and luteinizing hormone (LH). Abnormalities in neuroendocrine activity present in HD patients will be discussed including possible explanations for such observed defects.

## 2. NEUROTRANSMITTER MODULATION OF HORMONE RELEASE: BACKGROUND FOR HD STUDIES

HD has been characterized clinically as a hyperdopaminergic disorder. Support for this contention comes from the observation that choreiform movements frequently improve following administrations of drugs known to inhibit dopaminergic activity. In an effort to further evaluate this hypothesis, HD neuroendocrine research has investigated the regulation of GH and PRL—two hormones whose secretion is predominantly influenced by dopaminergic neural activity.

In man, dopaminergic manipulation clearly influences GH levels. This hormone increases in response to a number of direct dopamine receptor agonists including bromocriptine, piribedil, apomorphine, lisuride, and lergotrile (Lal *et al.*, 1972; Brown *et al.*, 1973; Camanni *et al.*, 1975; Thorner *et al.*, 1976; Liuzzi *et al.*, 1978; Thorner *et al.*, 1978). In contrast, the role of dopaminergic stimulation in altering GH levels of other mammalian species appears limited (Lovinger *et al.*, 1975; Steiner *et al.*, 1976; Ruch *et al.*, 1977) as regulation is predominantly controlled by noradrenergic mechanisms (Müller *et al.*, 1968; Toivola and Gale, 1972; Toivola *et al.*, 1972; Lal *et al.*, 1975; Chambers and Brown, 1976).

The pathways which subserve dopamine-induced GH release are not well defined. Present evidence would suggest that the relevant dopamine neurons lie within the blood–brain barrier. Support for this comes from (1) the failure of carbidopa, a peripheral decarboxylase inhibitor, to modify L-dopa-induced GH elevation (Lovinger *et al.*, 1976), (2) the absence of dopamine receptors in hypothalamic areas lying outside the blood–brain barrier (e.g., medial basal hypothalamus, including the median eminence) (Brown *et al.*, 1976), and (3) the demonstration that dopamine does not act directly on pituitary somatotrophs to release GH; *in vitro* experiments with cultured pituitary cells fail to demonstrate any GH alteration with dopamine agonists (Birge *et al.*, 1970).

Those areas lying within the blood–brain barrier which have been postulated to mediate GH release through dopaminergic mechanisms include both hypothalamic and extrahypothalamic locations. Within the hypothalamus, there is evidence that the arcuate and ventral medial nuclei influence GH secretion (Martin, 1972) in a manner which may involve hypothalamic dopaminergic pathways (Nemeroff et al., 1977). More attention, however, has been focused on extrahypothalamic areas. Specifically, electrical stimulation of the limbic system in both the rat (Martin, 1972) and human (Luedecke et al., 1976) results in the release of GH. Dopamine is likely to play a role influencing the response since (1) the limbic system receives a major neural input from dopamine mesolimbic tracts arising from the $A_{10}$ area of the midbrain and (2) stimulation of the $A_{10}$ area releases GH from the pituitary (Martin, 1972).

With PRL secretion, dopamine clearly plays a pivotal role. In numerous pharmacologic studies involving multiple mammalian species, dopaminergic stimulation consistently results in PRL inhibition. Additionally, there is good evidence to indicate that a site for dopaminergic action lies outside the blood–brain barrier. The tuberoinfundibular dopamine neurons in the median eminence are thought to secrete dopamine into the pituitary portal venous system. From there, it is transported to receptors located on lactotrophs in the adenohypophysis. Evidence supporting the adenohypophysis as the site for dopamine-mediated PRL secretions includes (1) the finding that apomorphine, a dopamine agonist, suppresses PRL in animals with large medial basal hypothalamic lesions (Cheung and Weiner, 1976), (2) suppression of plasma PRL by apomorphine occurs in animals with donor transplantations of the pituitary gland separated from brain and placed under the kidney capsule (Horowski and Gräf, 1976), (3) dopamine infusion into a portal vessel suppresses PRL release (Takahara et al., 1974), and (4) *in vitro* studies demonstrate that dopamine agonists can inhibit spontaneous PRL secretion from cultured anterior pituitary cells (Clemens et al., 1975; Smalstig et al., 1974; MacLeod and Lehmeyer, 1974; Caron et al., 1977; Cheung and Weiner, 1976). Further evidence in support of a pituitary site of action for dopamine-mediated PRL inhibition is the demonstration that this monoamine, when infused into normal subjects, both reduces basal levels of PRL (Leblanc et al., 1976) as well as inhibits the PRL release seen after TRH administration (Besses et al., 1975).

Very little information is available on the role of GABA in human GH and PRL secretion. Data on existing GABA pathways in the hypothalamus are similarly scarce (Tappaz et al., 1976; Tappaz and Brownstein, 1977). Muscimol, a GABA agonist, has been demonstrated to elevate human GH and PRL levels (Tamminga et al., 1978). At variance with this report is the finding that the GABA analog baclofen significantly reduces arginine- or hypoglycemia-induced GH release (Cavagnini et al., 1977) although this latter drug may act via mechanisms other than GABA stimulation (Andén and Wachtel, 1977). It is not known whether GABA affects GH secretion through (1) inhibition or stimulation of dopaminergic pathways (Koulu et al., 1980; Lahti and Losey, 1974; Walters et al., 1973), (2) inhibition of acetylcholine (Nistri and Constanti, 1975), or (3) release of norepinephrine (Philippu et al., 1973). A physiologic role of GABA-mediated PRL regulation is also poorly defined. While GABA does inhibit PRL release from in vitro incubated pituitary, much greater than physiologic concentrations are needed (Schally et al., 1977). Furthermore, the concentration of GABA in the median eminence is the lowest found in the brain (Tappaz et al., 1977). A recent report, however, suggests that there may be dual GABA control of PRL secretion in the rat (Locatelli et al., 1979); one stimulatory, exerted through the central nervous system, the other inhibitory occurring peripherally at the level of the adenohypophysis. The demonstration that intraventricularly administered GABA in rats elevates PRL (Vijayan and McCann, 1978) is supportive of the central stimulatory role for GABA. Although further clarification of GABA-mediated PRL regulation is most certainly needed, it is inviting to explain, at least in part, this central action of GABA on PRL through a GABA–dopamine interaction. The demonstration of GABAergic terminals in the arcuate and ventromedial nuclei (Tappaz and Brownstein, 1977) makes such an interaction possible, since it is from these areas of the hypothalamus that the dopaminergic tuberoinfundibular tract arises.

The role of cholinergic transmission in GH and PRL regulation has been only marginally investigated. Atropine was found to lack an affect on hypoglycemia-induced GH release in humans (Blackard and Waddell, 1969), although, in rats, inhibition of normal episodic GH bursts occurred after administration of this same agent (Martin et al., 1978). Eserine, a cholinesterase inhibitor, abolished the GH rise in dogs seen after enkephalin administration (Casaneuva et al., 1980). With regard to PRL, there are some data to indicate that cholinergic inhibition of PRL release occurs and is mediated by dopamine (Grandison and Meites, 1976).

## 3. NEUROENDOCRINE STUDIES IN HD

### 3.1. GH

#### 3.1.1. Literature Review

Over the last decade, GH secretion has been evaluated in HD both under baseline conditions and after stimulation with various agents known to release this hormone. Podolsky and Leopold (1974) were the first to demonstrate that HD patients failed to normally suppress GH after oral glucose administration and that they exhibited abnormally elevated GH levels after L-dopa administration. Baseline GH levels were slightly higher in patients with HD although this difference was significant only at a $p < 0.10$ level. Later studies showed that secretion of GH after insulin infusion was significantly higher in HD patients than in normal controls (Keogh et al., 1976; Phillipson and Bird, 1976). Baseline levels of GH in HD subjects were significantly elevated in Phillipson and Bird's study but not in Keogh and co-workers' report. Abnormally exaggerated GH release in HD patients was also found in response to arginine

(Leopold and Podolsky, 1975). Finally, administration of the dopamine agonists bromocriptine (Caraceni et al., 1977) and apomorphine (Müller et al., 1979a) have both resulted in significantly greater GH secretion by HD Patients as compared to controls. In these latter cited studies, baseline values for HD subjects were normal.

In contrast to the aforementioned reports, two investigations have not been able to show similar elevation of GH stimulation. In fact, both have demonstrated impaired GH release in HD patients; one in response to bromocriptine (Chalmers et al., 1978), the other after apomorphine administration (Levy et al., 1979). Some points can be raised to account, at least in part, for these apparent discrepancies. Chalmer and co-workers' study contained two HD patient groups; the first was discontinued from neuroleptic medication only 72 hr prior to study; the second consisted of nonmedicated patients, but was markedly older than the control group (HD patients' mean age 59, controls' mean age 44). In the first situation, a persistence of neuroleptic effect, i.e., dopamine antagonist activity, may have contributed to the blunted GH response seen in HD subjects. Similarly, the relatively older age of the HD subjects in the second group may also have contributed to the reduced GH response. That GH secretion diminishes with age is well documented (Finkelstein et al., 1972). Levy and co-workers' study can likewise be criticized as being poorly controlled for both sex and age. Here the HD group was both older and contained a greater number of males as compared to controls. In addition to the relationship between GH secretion and age, the observation that female human subjects secrete higher levels of GH than their male counterparts is also well described (Frantz and Rabkin, 1965; Merimee et al., 1966).

Despite previously mentioned reports of normal baseline GH levels in HD, recent investigations have suggested that under closer scrutiny there is indication that daily spontaneous secretion of GH is also abnormally elevated in HD. The earlier rationale of using single or few values to estimate GH baseline secretion is clearly inadequate in evaluating a hormone which has several secretory episodes during the course of a day. By sampling at frequent intervals over the course of a night, Murri et al. (1980) found that spontaneous GH release was significantly increased in HD patients as compared to controls during the early morning hours.

### 3.1.2. Spontaneous GH Secretion in HD

Our own studies (Durso et al., 1983a) have included an analysis of 24-hr plasma GH secretion in HD. Nine female patients with HD were compared to nine normal female age-matched controls (Table 1). Blood was sampled every 30 min during a 24-hr period from an indwelling heparinized venous catheter and sleep was monitored by electroencephalography. Patients refrained from strenuous exercise although they were allowed to ambulate during the day, usually within the confines of their room. Meals were served at 8:00 a.m., 12:30 p.m., and 5:00 p.m. Ovulating women were studied within 10 days of their last

TABLE 1
Circadian Rhythm Study

|  | HD patients (N = 9) | Normal controls (N = 9) |
| --- | --- | --- |
| Age (years)[a] | 43 ± 12 (20–61) | 44.8 ± 14 (20–60) |
| Sex | 9 females | 9 females |
| Neuroleptic history[b] | 5 never treated | 9 never treated |
|  | 2 haloperidol (2 years) |  |
|  | 1 haloperidol (6 months) |  |
|  | 1 thioridazine (2 weeks) |  |
| Other medications[b] | 6 none | 9 none |
|  | 1 L-dopa (2 weeks) |  |
|  | 1 imipramine (2 weeks) |  |
|  | 1 hydroxyzine and diazepam (2 weeks) |  |
| Duration of symptoms (years)[a] | 4 ± 1.5 (2–7) |  |
| Total sleep during study (min)[c] | 332 ± 29 | 394 ± 13 |
| Slow-wave sleep (min)[c] (stages 3 and 4) | 79.4 ± 14.5 | 92.7 ± 18.5 |

[a] Mean ± S.D. (range).
[b] Number of patients, drugs, withdrawal time.
[c] Mean ± S.E.M.

day of menses. The blood samples were immediately put on ice after being drawn, then centrifuged and separated within 1 hr. The resultant plasma was frozen and stored at −70°C until assay. Plasma GH levels were analyzed in duplicate using a double-antibody radioimmunoassay technique with materials and methods as described by the National Pituitary Agency.

Our results demonstrated increased levels of GH secretion over 24 hr in HD patients compared with normal controls. The mean (± S.E.M.) 24-hr GH level in HD subjects (4.44 ± 0.39 ng/ml) as compared to that of controls (2.66 ± 0.34 ng/ml) was significantly higher ($p < 0.01$, unpaired $t$ test). Furthermore, repeated measures analysis of variance between the two groups demonstrated abnormally high GH secretion in the HD subjects over the 24-hr period ($F = 10.8, p < 0.01$). Closer examination of the 24-hr curve (Fig. 1) reveals that HD patients consistently demonstrate higher peaks of GH relative to controls throughout the day while retaining a normal pattern of secretion. Such results are consistent with an exaggerated response in HD to endogenous cyclic induction of GH release. Stress, known to stimulate GH release (Brown and Reichlin, 1972), was evaluated as a possible explanation for the difference in GH secretion observed between the HD subjects and controls. As a mechanism to investigate this factor, cortisol samples were analyzed at 7:30 a.m., 8:00 a.m., 8:30 a.m., and then again at 7:30 p.m., 8:00 p.m., and 8:30 p.m. The mean (± S.E.M.) morning and nighttime cortisol levels in HD subjects were 207 ± 23 and 99 ± 16 ng/ml, respectively. In controls, the same measures were 151 ± 22 and 61 ± 7 ng/ml. The difference between HD and normal

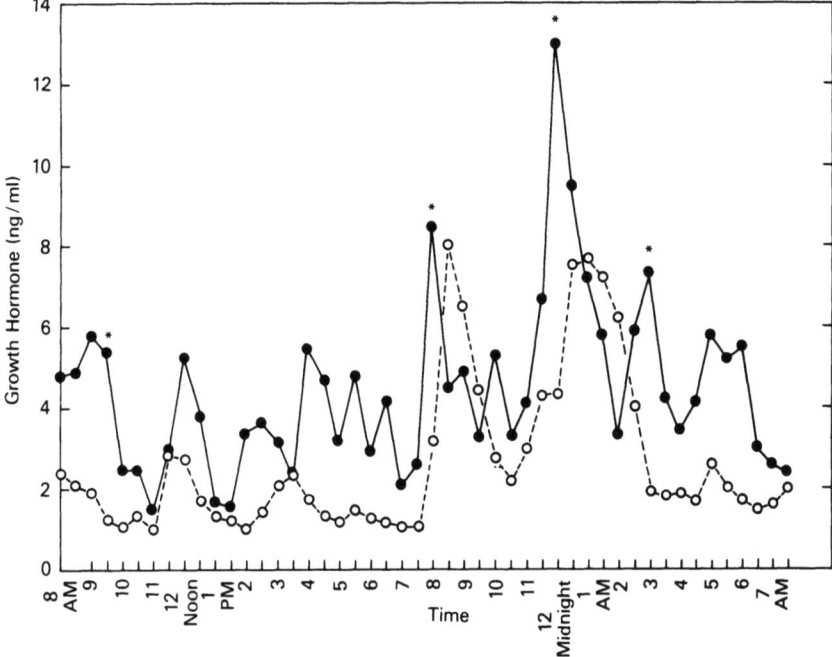

FIGURE 1. Twenty-four-hour secretion patterns for growth hormone are depicted for nine female HD patients (closed circles) and nine female age-matched controls (open circles). The quantitative difference seen is significant ($F = 10.8$, $p < 0.01$—analysis of variance).

cortisol secretion for either peiod of time was not significant by repeated measures analysis of variance. Because exercise releases GH (Hansen, 1971), it was considered that excessive secretion found in HD could conceivably be attributed to the presence of choreiform movements. Closer inspection, however, revealed that those subjects with the rigid variant of HD and who therefore lacked chorea, (three patients) demonstrated similar elevated GH levels (mean 24-hr plasma level $4.03 \pm 0.28$ ng/ml). Furthermore, the elevations in all HD patients were maintained during sleep, the time when movements were virtually absent. Hence, it is not likely that exercise is responsible for the observed results. Finally, other factors which might alter GH secretion such as sex, time of menstrual cycle, or amount of slow-wave sleep were either carefully controlled for or did not significantly differ between the two groups.

### 3.1.3. Pharmacologic Studies

To further evaluate GH secretion in HD, we also studied patients after administration of (1) a dopamine agonist (apomorphine), (2) a dopamine antagonist (haloperidol), (3) a GABAergic agonist (muscimol), and (4) a cholinergic muscarinic agonist (arecoline) (Durso et al., 1983b). In these patients

TABLE 2
Pharmacologic Studies

|  | HD patients (N = 8) | Normal controls (N = 8) |
|---|---|---|
| Age (years)[a] | 39 ± 16 (18–58) | 42.5 ± 18 (20–63) |
| Sex | 3 males, 5 females | 3 males, 5 females |
| Neuroleptic history | 3 never treated<br>3 haloperidol (3 years)<br>1 haloperidol (6 months)<br>1 thioridazine (2 weeks) | 8 never treated |
| Other medications[b] | 6 none<br>1 L-dopa (2 weeks)<br>1 hydroxyzine and diazepam (2 weeks) | 8 none |
| Duration of symptoms (years)[a] | 4.75 ± 2.12 (2–7) |  |

[a] Mean ± S.D. (range).
[b] Number of patients, drugs, withdrawal time.

and controls (Table 2), all experiments were conducted in the morning after an all-night fast. Not all participants were used for each pharmacologic study, although within any single drug investigation HD subjects and controls were closely age- and sex-matched. After receiving a single administration of one of the aforementioned drugs, blood was sampled from an indwelling catheter at a predetermined schedule. Subjects remained in bed throughout the course of the blood drawing and meals were withheld. The drug doses and routes of administration were as follows: (1) apomorphine 0.75 mg s.c., (2) haloperidol 1 mg i.m., (3) muscimol 5 mg p.o., (4) arecoline 5 mg s.c., and (5) saline 1 ml i.m. Studies were done at least 48 hr apart.

Our results demonstrate that HD patients secrete abnormally high GH amounts in response to a number of different pharmacologic stimuli (Table 3). Exaggerated GH concentrations were seen after administration of the dopamine agonist apomorphine (Fig. 2). GH levels were almost six times as high in HD subjects than controls shortly after injection of this drug. Similarly, after oral administration of the GABAergic drug muscimol, HD individuals again secreted significantly higher amounts of GH as compared to normal subjects (Fig. 3). Finally, haloperidol injection as well was associated with abnormally elevated GH amounts in HD patients (Fig. 4). No GH difference was found between the two groups after arecoline or saline administration. In these latter studies, HD levels of GH, although higher, were not statistically significant (Table 3).

One might have expected that saline administration would show significant elevations of GH in the HD group since higher levels of this hormone were seen in patients during the 24-hr study. Here, the nature of procedures for the two different studies is probably responsible for the observed differences. Participants in the saline experiments remained supine and fasted throughout the duration of the study. In contrast, the 24-hr procedure attempted to best approximate patients' daily activities and, hence, allowed for ambulation and

TABLE 3
Mean (± S.E.M) GH Levels (ng/ml)[a]

| | Time (min) | | | | | | | | | Significance HD vs. controls |
|---|---|---|---|---|---|---|---|---|---|---|
| | 0 | 30 | 60 | 90 | 120 | 150 | 180 | 240 | 300 | |
| HD | | | | | | | | | | |
| Saline (N = 6) | 4.3 ± 1.0 | 4.0 ± 1.6 | 3.7 ± 1.1 | 3.1 ± 0.6 | 3.2 ± 0.47 | 3.5 ± 0.66 | 3.8 ± 0.58 | 6.7 ± 3.6 | 4.9 ± 0.60 | ns |
| Apomorphine (N = 8) | 2.2 ± 0.45 | 34.9 ± 10 | 31 ± 9.4 | 12.6 ± 3.5 | 5.6 ± 1.4 | 3.4 ± 0.65 | 2.9 ± 0.51 | 2.8 ± 1.5 | | $p < 0.025$ |
| Haloperidol (N = 6) | 4.7 ± 0.47 | 5.6 ± 0.73 | 6.6 ± 1.0 | 9.9 ± 3.8 | 7.8 ± 3.1 | 9.3 ± 5.5 | 5.1 ± 0.61 | 4.1 ± 0.52 | 3.7 ± 0.51 | $p < 0.025$ |
| Muscimol (N = 7) | 5.1 ± 0.57 | 21.6 ± 15 | 18.6 ± 4.4 | 22.8 ± 9.9 | 18.3 ± 8.8 | 12.3 ± 5.1 | 8.0 ± 2.5 | 6.5 ± 1.4 | 7.5 ± 1.7 | $p < 0.05$ |
| Arecoline (N = 8) | 4.1 ± 0.40 | 8.3 ± 2.9 | 7.8 ± 2.6 | 4.0 ± 0.78 | 3.7 ± 0.28 | 5.2 ± 1.1 | 4.6 ± 0.7 | | | ns |
| Controls | | | | | | | | | | |
| Saline (N = 6) | 3.5 ± 0.85 | 3.6 ± 1.3 | 3.0 ± 0.64 | 3.4 ± 0.74 | 4.0 ± 1.6 | 4.4 ± 1.4 | 3.6 ± 0.68 | 3.3 ± 0.48 | 4.6 ± 0.62 | |
| Apomorphine (N = 8) | 2.7 ± 0.49 | 6.3 ± 1.7 | 7.0 ± 2.2 | 4.2 ± 1.2 | 2.3 ± 0.39 | 2.1 ± 0.14 | 3.1 ± 1.3 | 4.1 ± 1.2 | | |
| Haloperidol (N = 6) | 3.5 ± 0.72 | 3.4 ± 0.32 | 2.8 ± 0.15 | 3.0 ± 0.26 | 3.1 ± 0.43 | 4.1 ± 1.3 | 3.4 ± 0.61 | 3.1 ± 0.16 | 3.2 ± 0.16 | |
| Muscimol (N = 7) | 4.6 ± 0.89 | 3.9 ± 0.62 | 6.1 ± 2.9 | 8.7 ± 2.7 | 5.3 ± 1.2 | 5.4 ± 2.1 | 6.2 ± 2.4 | 5.3 ± 1.8 | 4.4 ± 1.1 | |
| Arecoline (N = 8) | 2.3 ± 1.1 | 4.8 ± 1.2 | 5.5 ± 2.2 | 3.4 ± 1.2 | 2.3 ± 0.53 | 2.0 ± 0.31 | 1.9 ± 0.25 | | | |

[a] The number of persons participating in each study is shown in parentheses. Statistical evaluation was done by analysis of variance using repeated measures and a general linear model.

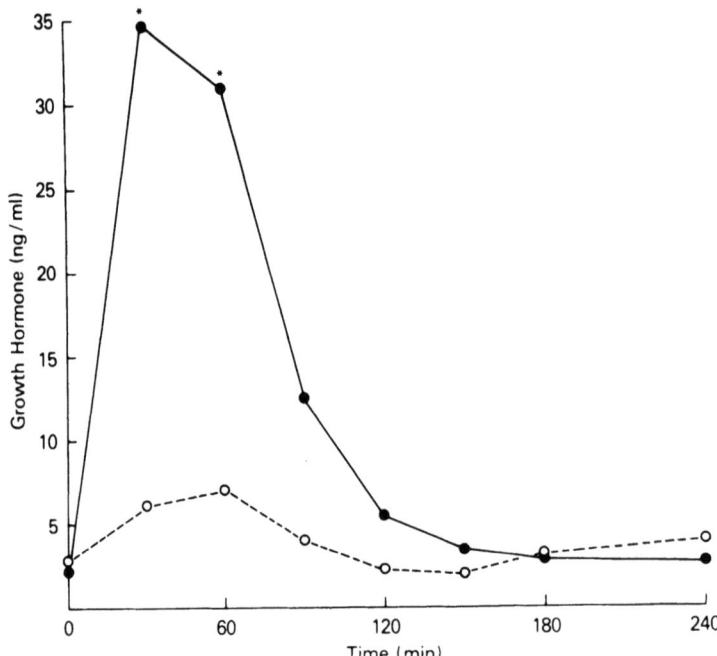

**FIGURE 2.** Growth hormone levels after 0.75 mg apomorphine s.c. are shown for eight HD patients (closed circles) and eight age- and sex-matched controls (open circles). The difference in levels of growth hormone secreted by the two groups is significant ($F = 6.47$, $p < 0.025$—analysis of variance).

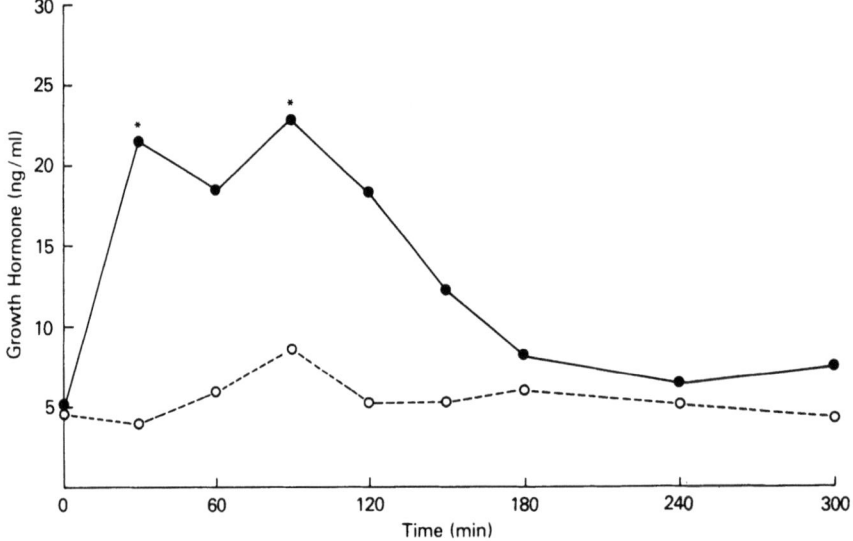

**FIGURE 3.** Growth hormone levels after 5 mg muscimol p.o. are shown for seven HD patients (closed circles) and seven age- and sex-matched controls (open circles). The difference in levels of growth hormone secreted by the two groups is significant ($F = 6.37$, $p < 0.05$—analysis of variance).

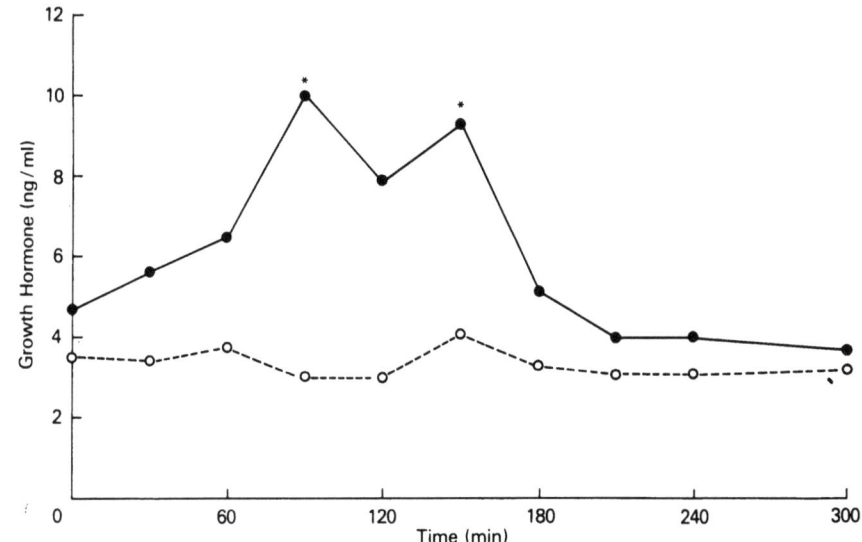

FIGURE 4. Growth hormone levels after 1 mg haloperidol i.m. are shown for six HD patients (closed circles) and six age- and sex-matched controls (open circles). The difference in levels of growth hormone secreted by the two groups is significant ($F = 8.95$, $p < 0.025$—analysis of variance).

meals—stimuli known to release GH. Thus, it is likely that the GH abnormalities seen in the 24-hr studies were the result of additional GH-releasing factors not present in the saline experiments.

## 3.2. PRL

### 3.2.1. Literature Review

Studies of PRL secretion in HD are conflicting. Baseline levels have been reported as increased (Caraceni et al., 1977; Caine et al., 1978), decreased (Hayden et al., 1977), or within normal limits (Chalmers et al., 1978; Levy et al., 1979). More complete evaluations of spontaneous PRL release measuring secretion over a 24-hr period (Polleri et al., 1980) or during the course of a night (Murri et al., 1980) both have reported PRL secretion in HD to be similar to controls. Pharmacologic studies have also failed to demonstrate reproducible abnormalities of PRL release in HD. Thus, while some authors report a lack of normal PRL suppression to dopamine agonists (Caraceni et al., 1977; Caine et al., 1978), others demonstrate normal responses to these same drugs (Chalmers et al., 1978; Levy et al., 1979; Müller et al., 1979b; Leopold and Podolsky, 1979). Similarly, PRL release in HD patients after administration of the dopamine antagonist chlorpromazine has been reported as both impaired (Hayden et al., 1977; Caine et al., 1978) and within normal limits (Levy et al., 1979). Finally, HD subjects have shown normal PRL responses after TRH administration (Levy et al., 1979; Müller et al., 1979b).

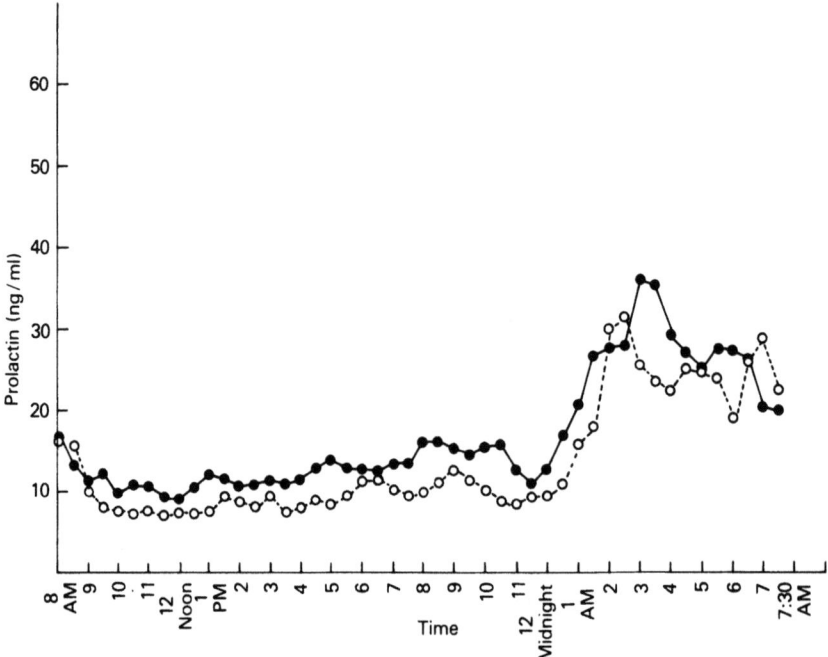

FIGURE 5. Twenty-four-hour secretion patterns for prolactin are depicted for nine female HD patients (closed circles) and nine female age-matched controls (open circles). There is no significant difference noted between the two (analysis of variance).

### 3.2.2. Spontaneous PRL Secretion in HD

Using the clinical and laboratory methodology previously described (Section 3.1.2), we report plasma PRL levels in HD and normal individuals sampled every 30 min over 24 hr (Durso et al., 1983a). These results confirm earlier reports that spontaneous secretion of this hormone in HD patients is similar to that in age- and sex-matched normal volunteers. The mean ($\pm$ S.E.M.) 24-hr PRL level in HD patients did not significantly differ from that of controls (16.9 $\pm$ 2.0 ng/ml vs. 13.9 $\pm$ 1.8 ng/ml). Similarly, analysis of variance using repeated measures failed to find a significant difference. Furthermore, it can be seen that the pattern of secretion in both groups (Fig. 5) were alike with highest secretory levels occurring in the early morning hours (3:00 a.m.–8:00 a.m.). Within the HD group, mean ($\pm$ S.E.M.) PRL levels in patients with the rigid variant (three patients) were no different from those seen in chorea subjects (16.1 $\pm$ 1.7 ng/ml vs. 17.3 $\pm$ 2.7 ng/ml).

### 3.2.3. Pharmacologic Studies

We further investigated PRL secretion in HD using the same clinical and laboratory procedures as outlined in Section 3.1.3 (Durso et al., 1983b). It was

TABLE 4
Mean (± S.E.M.) PRL Levels (ng/ml)[a]

| | Time (min) | | | | | | | | | Significance HD vs. controls |
|---|---|---|---|---|---|---|---|---|---|---|
| | 0 | 30 | 60 | 90 | 120 | 150 | 180 | 240 | 300 | |
| HD | | | | | | | | | | |
| Saline ($N = 8$) | 10.6 ± 2.4 | 7.6 ± 1.3 | 8.0 ± 1.6 | 6.8 ± 1.3 | 6.5 ± 1.6 | 7.1 ± 1.1 | 7.5 ± 1.1 | 7.9 ± 2.1 | 9.0 ± 2.3 | ns |
| Bromocriptine ($N = 8$) | 14.3 ± 3.6 | 12.7 ± 2.8 | 10.4 ± 2.4 | 8.6 ± 2.0 | 7.3 ± 1.6 | 6.4 ± 1.5 | 5.3 ± 1.2 | 4.8 ± 1.1 | 4.7 ± 0.83 | ns |
| Apomorphine ($N = 8$) | 12.2 ± 2.2 | 8.5 ± 1.1 | 6.8 ± 1.0 | 7.7 ± 1.4 | 7.9 ± 1.0 | 7.8 ± 0.92 | 8.1 ± 1.0 | 12.2 ± 2.8 | 11.2 ± 3.3 | ns |
| Haloperidol ($N = 8$) | 10.1 ± 2.2 | 20.7 ± 4.0 | 31.2 ± 5.7 | 41.9 ± 10.6 | 47.0 ± 12.3 | 45.9 ± 13.6 | 36.9 ± 10.2 | 32.3 ± 8.4 | 30.7 ± 9.7 | $p < 0.05$ |
| Muscimol ($N = 8$) | 17.2 ± 7.2 | 15.3 ± 6.3 | 16.4 ± 7.2 | 18.6 ± 9.1 | 18.7 ± 10.1 | 27.1 ± 17.4 | 25.7 ± 14.3 | 20.7 ± 10.3 | 14.3 ± 5.0 | ns |
| Controls | | | | | | | | | | |
| Saline ($N = 8$) | 8.2 ± 1.0 | 6.7 ± 1.0 | 7.0 ± 1.0 | 5.7 ± 0.60 | 5.8 ± 0.63 | 5.4 ± 0.56 | 5.5 ± 0.45 | 6.4 ± 0.58 | 7.1 ± 0.80 | |
| Bromocriptine ($N = 8$) | 10.7 ± 1.3 | 10.4 ± 1.7 | 8.6 ± 0.59 | 6.9 ± 2.0 | 5.8 ± 0.52 | 5.1 ± 0.46 | 5.5 ± 0.39 | 4.9 ± 0.49 | 4.3 ± 0.33 | |
| Apomorphine ($N = 8$) | 10.8 ± 1.8 | 8.8 ± 1.2 | 8.0 ± 1.2 | 7.5 ± 1.6 | 7.5 ± 1.6 | 6.9 ± 1.2 | 7.3 ± 1.2 | 9.4 ± 2.4 | 11.7 ± 4.4 | |
| Haloperidol ($N = 8$) | 8.3 ± 0.80 | 9.5 ± 0.78 | 10.6 ± 1.4 | 14.3 ± 2.4 | 15.0 ± 2.0 | 17.8 ± 3.8 | 18.1 ± 2.7 | 19.8 ± 3.5 | 20.6 ± 3.1 | |
| Muscimol ($N = 8$) | 16.0 ± 5.2 | 17.5 ± 4.2 | 21.3 ± 4.4 | 21.7 ± 5.4 | 28.5 ± 6.4 | 24.6 ± 7.0 | 27.9 ± 9.6 | 27.8 ± 9.0 | 24.4 ± 7.8 | |

[a] The number of persons participating in each study is shown in parentheses. Statistical evaluation was done by analysis of variance using repeated measures and a general linear model.

found that PRL secretion was similar for patients and controls in response to a number of different pharmacologic stimuli including saline, bromocriptine, apomorphine, and muscimol (Table 4). Significant alteration, however, of PRL after haloperidol administration was observed in the HD group. Greater than a sixfold increase in PRL concentrations occurred in HD subjects as compared to age- and sex-matched controls following i.m. haloperidol (Fig. 6). Three previous investigations report data in seeming contradiction with our finding. However, Hayden and co-workers' (1977) study, which reported an impaired PRL response in HD patients to chlorpromazine, analyzed the absolute levels of PRL rise as opposed to utilizing relative increases in the face of prominent differences in baseline levels of PRL between patients and controls. The two additional studies which have reported either normal (Levy et al., 1979) or diminished (Caine et al., 1978) PRL responses in HD patients differ from ours in the duration of prestudy neuroleptic-free period. Our patients, with the exception of a single subject, had never been treated with neuroleptics or were withdrawn at least 6 months before testing; the two aforementioned reports utilized a 2-week washout period. After 2 weeks, while most drug tissue stores may be metabolized, it is still likely that neuroleptic-induced changes in sensitivity of dopamine receptors persist (Smith and Davis, 1976; Clow et al., 1980). Such an increase in dopmaine receptor sensitivity could blunt the effect of chlorpromazine, a drug designed to block the action of dopamine at its receptor. Hence, the effect of premature testing of dopaminergic control of PRL following neuroleptic withdrawal would be to achieve an artificially low

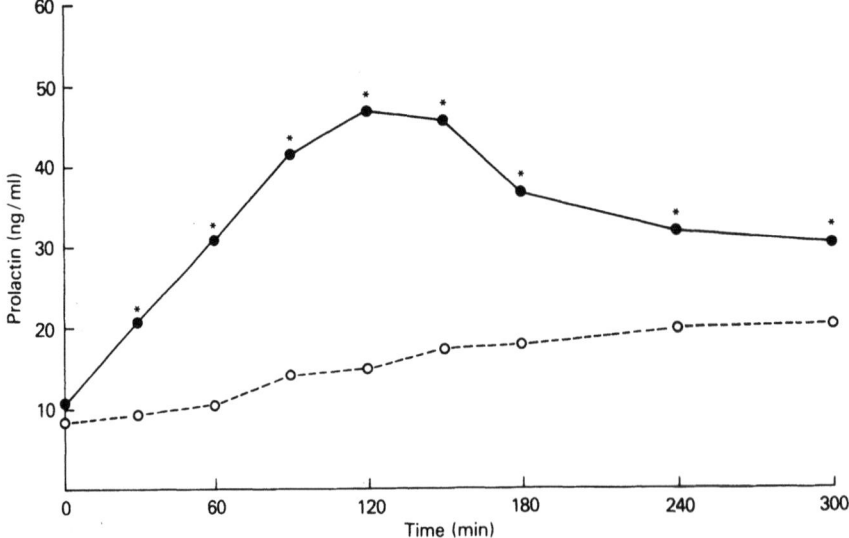

FIGURE 6. Prolactin levels after 1 mg haloperidol i.m. are shown for eight HD patients (closed circles) and eight age- and sex-matched controls (open circles). The difference in levels of prolactin secreted by the two groups is significant ($F = 4.77$, $p < 0.05$—analysis of variance).

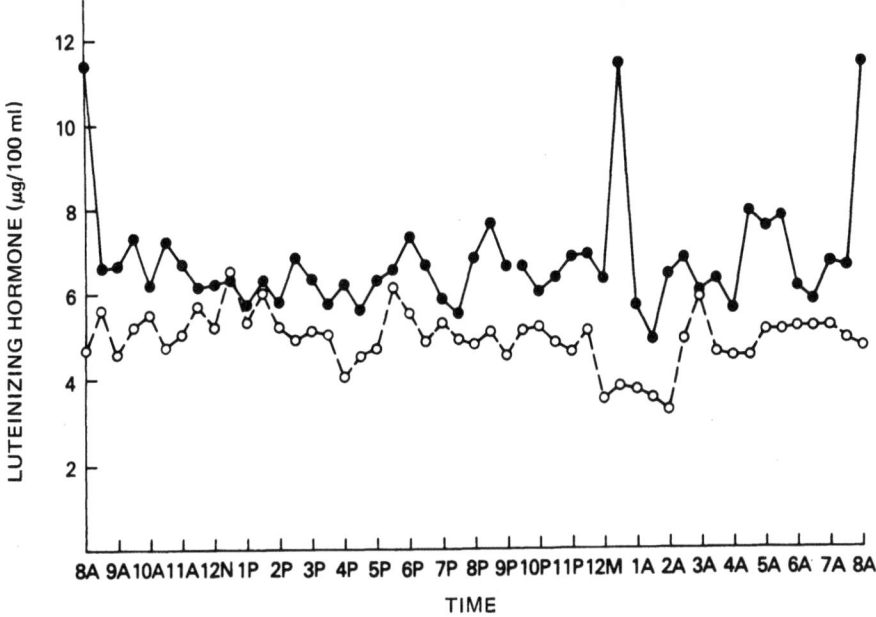

FIGURE 7. Twenty-four-hour secretion patterns for luteinizing hormone are shown for four premenopausal female HD patients (closed circles) and age-matched female controls (open circles). There is no significant difference noted between the two (analysis of variance).

PRL level in response to a dopamine antagonist, whereas without chronic drug effect, as in our studies, an exaggerated PRL response might be obtained.

### 3.3. LH

Epidemiologic studies have suggested increased fertility of choreic women in comparison to their nonaffected siblings (Reed and Neel, 1959; Marx, 1973). That hypothalamic dysfunction might be the cause of this observation is supported by Bird et al. (1976) who demonstrated abnormalities of gonadotropin-releasing factor (GrRF) in postmortem hypothalamic tissue from HD women. A fourfold increase in the median eminence concentration of this peptide was found in the female choreic patients as compared with controls. No such difference was observed in affected males.

Since GrRF serves as a releasing factor for LH, we evaluated the latter in four HD females (mean age 31.3) and four female normal controls (mean age 33.0). All subjects were studied within 10 days after the last day of menses and none were taking any medication. Sampling was done over a 24-hr period in the manner previously described (Section 3.1.2). The 24-hr curves are illustrated in Fig. 7. Using analysis of variance, the apparently elevated LH secretion by HD patients is not statistically significant. It should be noted, however, that the high variability of LH values in the HD group as well as the small sample size do not allow for reliable statistical evaluation. Any definite conclusions await further study.

## 4. PROPOSED DEFECTS OF NEUROENDOCRINE CONTROL IN HD

Exaggerated GH secretion in HD is well substantiated in both the literature as well as the data reported here. This abnormality, however, appears not the result of any specific neurotransmitter defect; multiple pharmacologic studies involving noradrenergic (insulin and arginine infusion), dopaminergic (L-dopa, dopamine agonists and antagonists), and GABAergic (muscimol) stimuli all result in excessive GH secretion in HD patients. Furthermore, as evident in our 24-hr endocrine study, circadian GH-stimulating mechanisms also result in similar increased secretion of this hormone in HD subjects.

We could not demonstrate in our HD patients a consistent dopaminergic defect in neuroendocrine regulation. While increase GH levels noted after administration of either apomorphine or haloperidol could be consistent with the existence of supersensitive dopamine receptors, the levels seen during our PRL studies involving bromocriptine, apomorphine, and haloperidol failed to substantiate any such supersensitivity. In fact, the abnormally elevated HD PRL levels noted after haloperidol are consistent with a reduced number of dopamine receptors since a given dose of haloperidol more effectively blocked the action of dopamine in HD patients. Similarly, pharmacologic testing using muscimol or arecoline also did not produce consistent results indicating altered GABA or cholinergic neuroendocrine interaction.

Hence, the diverse spectrum of stimuli to which GH is excessively secreted in HD in combination with an inability to demonstrate a consistent neurotransmitter defect through our endocrine studies may indicate that a general factor influencing neurotransmitter–hormone regulation in HD is abnormal. From the data available, it is reasonable to speculate that somatostatin may be such an influence. This 14-amino-acid peptide produces inhibition of GH secretion and is distributed through the CNS. Concentrations in the hypothalamus are highest (Brownstein *et al.*, 1975), but other extrahypothalamic areas contain significant amounts, including the septum–preoptic area (Brownstein *et al.*, 1975), the amygdala, and other limbic areas referable to basal-medial cortical structures (Epelbaum *et al.*, 1977). In man, somatostatin inhibits GH release associated with L-dopa and arginine (Siler *et al.*, 1973). Plasma GH elevations in response to exercise or sleep are similarly inhibited by this substance (Prange Hansen *et al.*, 1973; Parker *et al.*, 1974). Evidence that somatostatin influences basal GH secretion derives from the demonstration that antisomatostatin serum significantly elevates basal GH secretion in rats (Chihara *et al.*, 1978). In addition, immunization of baboons with antisomatostatin serum results in a similar increase in basal GH secretion (Steiner *et al.*, 1978). Thus, it is conceivable that reduced levels of somatostatin in HD patients could result in the elevated GH levels observed in our circadian rhythm study as well as those seen in the pharmacologic investigations.

The demonstration of markedly increased PRL levels after haloperidol in our HD patients is more difficult to explain on the basis of diminished somatostatin levels. Although somatostatin can influence PRL release in special instances, the vast majority of reports show that this substance affects neither basal levels of PRL secretion (Siler *et al.*, 1973), TRH-induced PRL release (Hall *et al.*, 1973; Siler *et al.*, 1974), nor hypoglycemia-induced PRL secretion (Hall *et al.*, 1973) in human subjects. In addition, PRL is unaffected in rats given antisomatostatin serum (Chihara *et al.*, 1978). The possibility remains, however, that a subtle effect which somatostatin may have in inhibiting PRL release is masked by the dominant inhibitory role of dopamine originating from tracts in the hypothalamus. In situations where the latter influence is removed, the effect of somatostatin may become more evident. That this mechanism might occur is strengthened by the demonstration that somatostatin significantly inhibits PRL release in pituitary cell cultures (Vale *et al.*, 1974; Drouin *et al.*, 1976). Similarly, the observation that somatostatin can reduce basal secretion of PRL in acromegalic patients (Yen *et al.*, 1974) may be related to a similar disruption of hypothalamic–pituitary dopamine control. In our study, dopaminergic inhibition of PRL was blocked by haloperidol administration. Such a blockade may have unmasked other factors regulating PRL secretion which differed between patient and control groups. PRL inhibition by somatostatin may have been one of these factors.

Studies evaluating somatostatin in HD patients have been both scarce and contradictory. While one investigation reports somatostatin as significantly reduced in the spinal fluid of HD subjects (Cramer *et al.*, 1981), another has cited elevated concentrations in postmortem HD basal ganglia tissue (Cooper *et al.*, 1981). Consistent with these reports as well as with the results presented in this chapter would be a possible defect in HD patients of somatostatin release. However, further investigations regarding the role this peptide plays in observed HD neuroendocrine abnormalities are needed. Specifically, postmortem measures of somatostatin content in the hypothalamus would seem most helpful, since levels in this area are likely to be a more sensitive indicator of neuroendocrine pathology than sites more remote from the hypothalamic–pituitary axis. Additionally, this area would be subject to less experimental error than the more severely diseased basal ganglia as the presence of severe atrophy in the latter could result in artifactually elevated somatostatin concentrations.

## 5. SUMMARY

Multiple studies over the past 15 years have demonstrated the existence of neuroendocrine defects in Huntington's chorea. Abnormalities in GH secretion are described by the authors in this text as well as supported in previous medical literature. Evidence of dysfunction in release of other hormones in this disease awaits further research. Such investigation may allow us to uncover a selective defect in neuroendocrine regulation indicative of hypothalamic catecholamine imbalance or lack of a specific releasing or inhibitory peptide. The existence of a consistent pattern of catecholamine imbalance could be extrapolated to other parts of the choreic brain. Similarly, evidence for absence of a particular peptide would have implications that extend outside the endocrine system since such peptides are now believed to function as neurotransmitters or neuromodulators of synaptic transmission throughout the CNS.

## 6. REFERENCES

Andén, N. E., and Wachtel, H., 1977, Biochemical effects of baclofen (beta-parachlorophenyl-GABA) on the dopamine and the noradrenaline in the rat brain, *Acta Pharmacol. Toxicol.* :opKbh.) **40**:310.

Bernheimer, H., and Hornykiewicz, O., 1973, Brain amines in Huntington's chorea, in: *Advances in Neurology*, Vol. 1, *Huntington's Chorea 1872–1972* (A. Barbeau, T. N. Chase, and G. W. Paulson, eds.), pp. 525–531, Raven Press, New York.

Besses, G. S., Burrow, G. N., Spaulding, S. W., and Donabedion, R. K., 1975, Dopamine infusion acutely inhibits the TSH and prolactin response to TRH, *J. Clin. Endocrinol. Metab.* **41**:985.

Bird, E. D., Chiappa, S. A., and Fink, G., 1976, Brain immunoreactive gonadotropin-releasing hormone in Huntington's chorea and non-choreic subjects, *Nature (London)* **260**:536.

Birge, C. A., Jacobs, L. S., Hammer, C. T., and Daughaday, W. H., 1970, Catecholamine inhibition of prolactin secretion by isolated rat adenohypophyses, *Endocrinology* **86**:120.

Blackard, W. G., and Waddell, C. C., 1969, Cholinergic blockade and growth hormone responsiveness to insulin hypoglycemia, *Proc. Soc. Exp. Biol. Med.* **131**:192.

Brown, G. M., and Reichlin, S., 1972, Psychologic and neural regulation of growth hormone secretion, *Psychosom. Med.* **34**:45.

Brown, G. M., Seeman, P., and Lee, T., 1976, Dopamine/neuroleptic receptors in basal hypothalamus and pituitary, *Endocrinology* **99**:1407.

Brown, W. A., Von Woert, M. H., and Ambani, L. M., 1973, Effect of apomorphine on growth hormone release in humans, *J. Clin. Endocrinol. Metab.* **37**:463.

Brownstein, M., Arimura, A., Sato, H., Schally, A. V., and Kizer, J. S., 1975, The regional distribution of somatostatin in the rat brain, *Endocrinology* **96**:1456.

Bruyn, G. W., 1968, Huntington's chorea, historical, clinical, and laboratory synopsis, in: *Handbook of Clinical Neurology*, Vol. 6, *Diseases of the Basal Ganglia* (P. J. Vinken and G. W. Bruyn, eds.), pp. 298–378, North-Holland, Amsterdam.

Bruyn, G. W., 1973, Neuropathological changes in Huntington's chorea, in: *Advances in Neurology*, Vol. 1, *Huntington's Chorea 1872–1972* (A. Barbeau, T. N. Chase, and G. W. Paulson, eds.), pp. 399–403, Raven Press, New York.

Caine, E., Kartzinel, R., Ebert, M., and Carter, A. C., 1978, Neuroendocrine function in Huntington's disease: Dopaminergic regulation of prolactin release, *Life Sci.* **22**:911.

Camanni, F., Massara, F., Belforte, L., and Molinatti, G. M., 1975, Changes in plasma growth hormone levels in normal and acromegalic subjects following administration of 2-bromo-α-ergocryptine, *J. Clin. Endocrinol. Metab.* **40**:363.

Caraceni, T., Panerai, A. E., Parat, E. A., Coechi, D., and Muller, E. E., 1977, Altered growth hormone and prolactin responses to dopaminergic stimulation in Huntington's chorea, *J. Clin. Endocrinol. Metab.* **44**:870.

Caron, M., Raymond, V., Lefkowitz, R., and Labrie, F., 1977, Identification of dopaminergic receptors in anterior pituitary: Correlation with the dopaminergic control of prolactin release, *Fed. Proc.* **36**:278.

Casaneuva, F., Betti, R., Frigerio, C., Cocchi, D., Mantegazza, P., and Müller, E. E., 1980, Growth hormone-releasing effect of an enkephalin analog in the dog: Evidence for cholinergic mediation, *Endocrinology* **106**:1239.

Cavagnini, F., Invitti, C., Di Landro, A., Tenconi, L., Maraschini, C., and Girotti, G., 1977, Effects of gamma-aminobutyric acid (GABA) derivative, baclofen, on growth hormone and prolactin secretion in man, *J. Clin. Endocrinol. Metab.* **45**:579.

Chalmers, R. J., Johnson, R. H., Keogh, H. J., and Nanda, R. N., 1978, Growth hormone and prolactin response to bromocriptine in patients with Huntington's chorea, *J. Neurol. Neurosurg. Psychiatry* **41**:135.

Chambers, J. W., and Brown, G. M., 1976, Neurotransmitter regulation of growth hormone and ACTH in the rhesus monkey: Effects on biogenic amines, *Endocrinology* **98**:420.

Cheung, C. Y., and Weiner, R. I., 1976, Supersensitivity of anterior pituitary dopamine receptors involved in the inhibition of prolactin secretion following destruction of the medial basal hypothalamus, *Endocrinology* **99**:914.

Chihara, K., Arimura, A., Chihara, M., and Schally, A. V., 1978, Studies on the mechanism of growth hormone and thyrotropin responses to somatostatin antiserum in anesthetized rats, *Endocrinology* **103**:1916.

Clemens, J. A., Smalstig, E. G., and Shaar, C. J., 1975, Inhibition of prolactin secretion by lergotrile mesylate: Mechanism of action, *Acta Endocrinol.* (Kbh) **79**:230.

Clow, A., Theodorou, A., Jenner, P., and Marsden, C. D., 1980, Changes in cerebral dopamine function induced by a year's administration of trifluoperazine or thioridazine and their subsequent withdrawal, in: *Advances in Biochemical Psychopharmacology*, Vol. 24, *Long-term Effects of Neuroleptics* (F. Cattabeni, P. F. Spano, G. Racagni, and E. Costa, eds.), pp. 335–340, Raven Press, New York.

Cooper, P. E., Aronin, N., Bird, E. D., Leeman, S. E., and Martin, J. B., 1981, Increased somatostatin in basal ganglia of Huntington's disease, *Neurology* **31**:64 (abstract).

Cramer, H., Kohler, J., Oepen, G., Schomburg, G., and Schroter, E., 1981, Huntington's chorea—Measurements of somatostatin, substance P and cyclic nucleotides in the cerebrospinal fluid, *J. Neurol.* **225**:183.

Drouin, J., De Léan, A., Rainville, D., LaChance, R., and Labrie, F., 1976, Characteristics of the interaction between thyrotropin-releasing hormone and somatostatin for thyrotropin and prolactin release, *Endocrinology* **98**:514.

Durso, R., Tamminga, C. A., Ruggieri, S., Denaro, A., Gillespie, M., and Chase, T. N., 1983a, Circadian release of growth hormone, prolactin, and luteinizing hormone in Huntington's disease, *J. Neuro. Neurosurg. Psychiatry* in press.

Durso, R., Tamminga, C. A., Ruggieri, S., Denaro, A., Gillespie, M., and Chase, T. N., 1983b, Neuroendocrine defects in Huntington's disease, *Neurology* in press.

Epelbaum, J., Brazeau, P., Tsang, D., Brawer, J., and Martin J. B., 1977, Subcellular distribution of radioimmunoassayable somatostatin in rat brain, *Brain Res.* **126**:309.

Finkelstein, J. W., Boyar, R. M., Roffwarg, H. P., Kream, J., and Hellman, L., 1972, Age-related change in the twenty-four hour spontaneous secretion of growth hormone. *J. Clin. Endocrinol. Metab.* **35**:665.

Frantz, A. G., and Rabkin, M. T., 1965, Effects of estrogen and sex difference on secretion of human growth hormone, *J. Clin. Endocrinol. Metab.* **25**:1470.

Grandison, L., and Meites, J., 1976, Evidence for adrenergic mediation of cholinergic inhibition of prolactin release, *Endocrinology* **99**:775.

Hall, R., Besser, G. M., Schally, A. V., Coy, D. H., Evered, D., Goldie, D. J., Kastin, A. J., McNeilly, A. S., Mortimer, C. H., Phenekos, C., Tunbridge, W. M. G., and Weightman, D., 1973, Action of growth-hormone-release inhibitory hormone in healthy men and in acromegaly, *Lancet* **II**:581.

Hansen, A. P., 1971, The effect of adrenergic receptor blockade on the exercise-induced serum growth hormone rise in normals and juvenile diabetics, *J. Clin. Endocrinol. Metab.* **33**:807.

Hayden, M. R., Paul, M., Vinik, A. I., and Beighton, P., 1977, Impaired prolactin release in Huntington's chorea: Evidence for dopaminergic excess, *Lancet* **II**:423.

Horowski, R., and Gräf, H. J., 1976, Influence of dopaminergic agonists and antagonists on serum prolactin concentrations in the rat, *Neuroendocrinology* **22**:273.

Keogh, H. J., Johnson, R. H., Nanda, R. N., and Sulaiman, W. R., 1976, Altered growth hormone release in Huntington's chorea, *J. Neurol. Neurosurg. Psychiatry* **39**:244.

Koulu, M., Lammintausta, R., and Dahlström, S., 1980, Effects of some gamma-aminobutyric acid (GABA)-ergic drugs on the dopaminergic control of human growth hormone secretion, *J. Clin. Endocrinol. Metab.* **51**:124.

Lahti, R. A., and Losey, E. G., 1974, Antagonism of the effects of chlorpromazine and morphine on dopamine metabolism by GABA, *Res. Commun. Chem. Pathol. Pharmacol.* **7**:31.

Lal, S., De La Vega, C. E., Sourkes, T. L., and Friesen, H. G., 1972, Effect of apomorphine on human-growth-hormone secretion, *Lancet* **II**:661.

Lal, S., Tolis, G., Martin, S. B., Brown, G. M., and Guyda, H., 1975, Effect of clonidine on growth hormone, prolactin, luteinizing hormone, follicle-stimulating hormone, and thyroid-stimulating hormone in the serum of normal men, *J. Clin. Endocrinol. Metab.* **41**:827.

Leblanc, H., Lachelin, G. C. L., Abu-Fadil, S., and Yen, S. S. C., 1976, Effects of dopamine infusion on pituitary hormone secretion in humans, *J. Clin. Endocrinol. Metab.* **43**:668.

Leopold, N. A., and Podolsky, S., 1975, Exaggerated growth hormone response to arginine infusion in Huntington's disease, *J. Clin. Endocrinol. Metab.* **41**:160.

Leopold, N. A., and Podolsky, S., 1979, Levodopa and glucose influence on prolactin secretion in Huntington's disease, in: *Advances in Neurology*, Vol. 23, *Huntington's Disease* (T. N. Chase, N. S. Wexler, and A. Barbeau, eds.), pp. 299–304, Raven Press, New York.

Levy, C. L., Carlson, H. F., Sowers, J. R., Goodlett, R. E., Tourtellotte, W. W., and Hershman, J. M., 1979, Growth hormone and prolactin secretion in Huntington's disease, *Life Sci.* **24**:743.

Liuzzi, A., Chiodini, P. G., Oppizzi, G., Botalla, L., Verde, G., DeStefano, L., Colussi, G., Gräf, K. J., and Horowski, R., 1978, Lisuride hydrogen maleate: Evidence for a long lasting dopaminergic activity in humans, *J. Clin. Endocrinol. Metab.* **46**:196.

Locatelli, V., Cocchi, D., Frigerio, C., Betti, R., Krogsgaard-Larsen, P., Racagni, G., and Müller, E. E., 1979, Dual $\gamma$-aminobutyric acid control prolactin secretion in the rat, *Endocrinology* **105**:778.

Lovinger, R. D., Rose, J., Boryczka, A. T., Shackelford, R., Kaplan, S. L., Ganong, W. F., and Grumbach, M. M., 1975, Effect of altered brain amine content and selective electrical stimulation in the control of growth hormone in the dog, Int. Symp. Growth Hormone Related Peptides, Milan, p. 59.

Lovinger, R., Holland, L. Jr., Kaplan, S., Grumbach, M., Boryczka, A. T., Shackleford, R., Salmon, J., Reid, I. A., and Ganong, W. F., 1976, Pharmacological evidence for stimulation of growth hormone secretion by a central noradrenergic system in dogs, *Neuroscience* **1**:443.

Luedecke, G., Mueller, D., and Patino, J., 1976, 5th Int. Congr. Endocrinol. (Abstract Book), Hamburg, p. 272.

MacLeod, R. M., and Lehmeyer, J. E., 1974, Studies on the mechanism of the dopamine-mediated inhibition of prolactin secretion, *Endocrinology* **25**:1077.

Martin, J. B., 1972, Plasma growth hormone (GH) response to hypothalamic or extrahypothalamic electrical stimulation, *Endocrinology* **91**:107.

Martin, J. B., Durand, D., Gurd, W., Faille, G., Audet, J., and Brazeau, P., 1978, Neuropharmacological regulation of episodic growth hormone and prolactin secretion in the rat, *Endocrinology* **102**:106.

Marx, R. N., 1973, Huntington's chorea in Minnesota, in: *Advances in Neurology*, Vol. 1, *Huntington's Chorea 1872–1972* (A. Barbeau, T. N. Chase, and G. W. Paulson, eds.), p. 237, Raven Press, New York.

Merimee, T. J., Bergess, J. A., and Rabinowitz, D., 1966, Sex-determined variation in serum insulin and growth hormone response to amino acid stimulation, *J. Clin. Endocrinol. Metab.* **26**:791.

Müller, E. E., Dal Pra, P., and Pecile, A., 1968, Influence of brain neurohumors injected into the lateral ventricle of the rat on growth hormone release, *Endocrinology* **83**:893.

Müller, E. E., Parati, E. A., Panerai, A. E., Cocchi, D., and Caraceni, T., 1979a, Growth hormone hyperresponsiveness to dopaminergic stimulation in Huntington's chorea, *Neuroendocrinology* **28**:313.

Müller, E. E., Parati, E. A., Cocchi, D., Zanardi, P., and Caraceni, T., 1979b, Dopaminergic drugs on growth hormone and prolactin secretion in Huntington's disease, in: *Advances in Neurology*, Vol. 23, *Huntington's Disease* (T. N. Chase, N. S. Wexler, and A. Barbeau, eds.), pp. 319–334, Raven Press, New York.

Murri, L., Iudice, A., Muratorio, A., Polleri, A., Barreca, T., and Murialdo, G., 1980, Spontaneous nocturnal plasma prolactin and growth hormone secretion in patients with Parkinson's disease and Huntington's chorea, *Eur. Neurol.* **19**:198.

Nemeroff, C. B., Konkol, R. J., Bissette, G., Youngblood, W., Martin, J. B., Brazeau, P., Rone, M. S., Prange, A. J., Breese, G. R., and Skizer, J., 1977, Analysis of the disruption in hypothalamic–pituitary regulation in rats treated neonatally with monosodium L-glutamate (MSG): Evidence for the involvement of tuberoinfundibular cholinergic and dopaminergic systems in neuroendocrine regulation, *Endocrinology* **101**:613.

Nistri, A., and Constanti, A., 1975, Some observations on the mechanism of action of baclofen (beta-chlorophenyl-gamma-amino-butyric acid), *Experientia* **31**:64.

Parker, D. C., Rossman, L. G., Siler, T. M., Rivier, J., Yen, S. S. C., and Guillemin, R., 1974, Inhibition of the sleep-related peak in physiologic human growth hormone release by somatostatin, *J. Clin. Endocrinol. Metab.* **38**:496.

Philippu, A., Przuntek, H., and Rosenberg, W., 1973, Superfusion of the hypothalamus with gamma-aminobutyric acid: Effect on release of noradrenaline and blood pressure, *Naunyn-Schmiedebergs Arch. Pharmacol.* **276**:103.

Phillipson, O. T., and Bird, E. D., 1976, Plasma growth hormone concentrations in Huntington's chorea, *Clin. Sci. Mol. Med.* **50**:551.

Podolsky, S., and Leopold, N. A., 1974, Growth hormone abnormalities in Huntington's chorea: Effect of L-dopa administration, *J. Clin. Endocrinol. Metab.* **39**:36.

Polleri, A., Savoldi, F., Muratorio, A., Masturzo, P., Murialdo, G., Martignori, E., Nappi, G., Iudice, A., and Murri, L., 1980, Circadian rhythmicity of prolactin secretion in Huntington's chorea, *Life Sci.* **26**:1609.

Prang Hansen, A., Ørskov, H., Seyer-Hansen, K., and Lundbaek, K., 1973, Some actions of growth hormone release inhibiting factor, *Br. Med. J.* **3**:523.

Reed, T. E., and Neel, J. V., 1959, Huntington's chorea in Michigan. 2. Selection and mutation, *Am. J. Hum. Genet.* **11**:107.

Ruch, W., Mixter, R. C., Russell, R. M., Garcia, G. F., and Gale, C. C., 1977, Aminergic and thermoregulatory mechanisms in hypothalamic regulation of growth hormone in cats, *Am. J. Physiol.* **223**:E61.

Schally, A. V., Redding, T. W., Arimura, A., Dupont, A., and Linthicum, G. L., 1977, Isolation of gamma-aminobutyric acid from pig hypothalami and demonstration of its prolactin release-inhibiting (PIF) activity *in vivo* and *in vitro*, *Endocrinology* **100**:681.

Siler, T. M., Vanden Berg, G., and Yen, S. S. C., 1973, Inhibition of growth hormone release in humans by somatostatin, *J. Clin. Endocrinol. Metab.* **37**:632.

Siler, T. M., Yen, S. S. C., Vale, W., and Guillemin, R., 1974, Inhibition by somatostatin on the release of TSH induced in man by thyrotropin-releasing factor, *J. Clin. Endocrinol. Metab.* **38**:742.

Smalstig, E. B., Sawyer, B. D., and Clemens, J. A., 1974, Inhibition of rat prolactin release by apomorphine *in vivo* and *in vitro*, *Endocrinology* **95**:123.

Smith, R. C., and Davis, J. M., 1976, Behavioral evidence for supersensitivity after chronic administration of haloperidol, clozapine, and thioridazine, *Life Sci.* **19**:725.

Steiner, R. A., Rolfs, A., Shen, L., Illner, P., and Gale, C. C., 1976, Neuroendocrine control of prolactin and growth hormone secretion in the baboon, *Clin. Res.* **24**:101 (abstract).

Steiner, R. A., Stewart, J. K., Barber, J., Koerker, D., Goodner, C. J., Brown, A., Illner, P., and Gale, C. C., 1978, Somatostatin: A physiological role in the regulation of growth hormone secretion in the adolescent male baboon, *Endocrinology* **102**:1587.

Takahara, J., Arimura, A., and Schally, A. V., 1974, Suppression of prolactin release by a purified porcine PIF preparation and catecholamines infused into a rat hypophyseal portal vessel, *Endocrinology* **96**:462.

Tamminga, C. A. Neophytides, A., Chase, T. N., and Frohman, L. A., 1978, Stimulation of prolactin and growth hormone secretion by muscimol, a γ-aminobutyric acid agonist, *J. Clin. Endocrinol. Metab.* **47**:1348.

Tappaz, M. L., and Brownstein, M. J., 1977, Origin of glutamate-decarboxylase (GAD)-containing cells in discrete hypothalamic nuclei, *Brain Res.* **132**:95.

Tappaz, M. L., Brownstein, M. J., and Palkovits, M., 1976, Distribution of glutamate decarboxylase in discrete brain nuclei, *Brain Res.* **108**:371.

Tappaz, M. L., Brownstein, M. J., and Kopin, I. J., 1977, Glutamate decarboxylase (GAD) and gamma-aminobutyric acid (GABA) in discrete nuclei of the hypothalamus and substantia nigra, *Brain Res.* **125**:109.

Thorner, M. O., Wass, J. A. H., Jones, A., Bloom, S. R., and Macleod, R. M., 1976, 5th Int. Congr. Endocrinol. (Abstract Book), Hamburg, p. 24.

Thorner, M. O., Ryan, S. M., Wass, J. A. H., Jones, A., Bouloux, P., Williams, S., and Besser, G. M., 1978, Effect of the dopamine agonist, lergotrile mesylate, on circulating anterior pituitary hormones in man, *J. Clin. Endocrinol. Metab.* **47**:372.

Toivola, P. T. K., and Gale, C. C., 1972, Stimulation of growth hormone release by microinjection of norepinephrine into hypothalamus of baboons, *Endocrinology* **90**:895.

Toivola, P. T. K., Gale, C. C., Goodner, C. J., and Werrbach, J. H., 1972, Central-adrenergic regulation of growth hormone and insulin, *Hormones* **3**:192.

Vale, W., Rivier, C., Brazeau, P., and Guillemin, R., 1974, Effects of somatostatin on the secretion of thyrotropin and prolactin, *Endocrinology* **95**:968.

Vijayan, E., and McCann, S. M., 1978, The effects of intraventricular injection of gamma-aminobutyric acid (GABA) on prolactin and gonadotropin release in conscious female rats, *Brain Res.* **155**:35.

Vogt, C., and Vogt, O., 1952, Precipitating and modifying agents in chorea, *J. Nerv. Ment. Dis.* **116**:601.

Walters, J. R., Roth, R. H., and Aghajanian, G. K., 1973, Dopaminergic neurons: Similar biochemical and histochemical effects of gamma-hydroxybutyrate and acute lesions of the nigro-neostriatal pathway, *J. Pharmacol. Exp. Ther.* **186**:630.

Yen, S. S. C., Siler, T. M., and DeVane, G. W., 1974, Effect of somatostatin in patients with acromegaly: Suppression of growth hormone, prolactin, insulin and glucose levels, *N. Engl. J. Med.* **290**:935.

CHAPTER 11

# $ACTH_{4-10}$
## Effects on Pavlovian Conditioning

D. A. POWELL and SHIRLEY L. BUCHANAN

## 1. INTRODUCTION

The behavioral effects of a variety of hypothalamic neuropeptides have long been known (e.g., Applezweig and Baudry, 1955; Levine and Jones, 1965). Early experiments by de Wied and his colleagues (e.g., de Wied, 1965, 1969) revealed that adenohypophysectomy produced active avoidance deficits; others have shown that hypophysectomy also impairs passive avoidance behavior (e.g., Weiss et al., 1970). Later experiments revealed that the performance of these animals can be restored by treatment with ACTH peptide fragments with little or no adrenocortical effect, as reviewed recently by de Wied and Gispen (1977) and de Wied (1980).

In contrast to adenohypophysectomy, it was found that removal of the neurohypophysis including the intermediate lobe did not interfere with avoidance learning, but extinction of a previously learned avoidance response was found to be markedly facilitated (de Wied and Gispen, 1977). Similarly, it was reported that treatment of intact rats with ACTH or ACTH fragments, as well as vasopressin, did not alter the rate of acquisition of shock-motivated avoidance behavior, but increased resistance to extinction (Kelsey, 1975; de Wied, 1980). Apparently extinction of avoidance behavior is more sensitive to treatment with ACTH than with vasopressin (Murphy and Miller, 1955; de Wied et al., 1968). Vasopressin and ACTH also facilitate passive avoidance behavior in intact animals (Levine and Jones, 1965; Ader and de Wied, 1972). In addition, deficits in acquisition of conditioned avoidance behaviors due to (1) hypophysectomy, (2) inherited vasopressin deficiency, or (3) experimental amnesia are reversed by vasopressin administration, as recently reviewed (de Wied, 1980).

These peptide influences on learning and memory have been shown to be due to central effects, since analogs which have little or no peripheral hormonal

---

D. A. POWELL and SHIRLEY L. BUCHANAN • Neuroscience Laboratory, Wm. Jennings Bryan Dorn Veterans' Hospital, and University of South Carolina, Columbia, South Carolina 29201.

effects (i.e., $ACTH_{4-10}$ and des-Gly-Arg vasopressin) produce results identical to those produced by the naturally occurring peptides. These and similar findings (e.g., Bohus, 1975; Sandman et al., 1971) have thus firmly established the fact that vasopressin, ACTH (as well as α-MSH) are related to learning and memory in subhuman animals. The most frequently reported finding is that these peptides do not influence acquisition of new response repertoires, but instead retard the response decrement associated with extinction when reinforcement contingencies are removed.

The CNS sites for the behavioral action of ACTH and vasopressin have been studied by injection or implantation of these substances into the brain. Microinjections of doses that are systemically ineffective were placed in various limbic midbrain structures by van Wimersma Greidanus and de Wied (1971) and van Wimersma Greidanus et al. (1974). Application of $ACTH_{4-10}$ to posterior thalamic areas, including the parafascicular nucleus, resulted in increased resistance to extinction of an active avoidance response; vasopressin had a similar but smaller effect. Other areas, including the ventral, medial, and posterior hypothalamus, substantia nigra, and reticular formation, were ineffective sites. Moreover, lesions of the parafascicular nucleus prevented the behavioral effects of $ACTH_{4-10}$ on conditioning and extinction (Bohus and de Wied, 1967). Recent experiments by Dogterom and Buijs (1980), using both radioimmunoassy and immunohistochemical techniques, have revealed a network of vasopressin- as well as oxytocin-containing fibers in lateral septum, habenula, and mesencephalic areas. These findings, taken together, thus suggest that limbic structures may mediate the behavioral effects of vasopressin and ACTH through receptors located on cells in limbic system structures.

There is an extensive literature relating limbic system structures to learning and memory formation. Both electrophysiological studies (e.g., Segal and Olds, 1973; Berger et al., 1980; Thompson et al., 1980; Gabriel et al., 1980) as well as lesion studies (Cohen and MacDonald, 1976; Powell et al., 1976, 1978; Moore, 1979; Solomon, 1980) have revealed that a variety of forebrain structures are involved in various stages of the learning process. Prominent among these structures are those with neurons containing behaviorally active peptides, as recently demonstrated by Dogterom and Buijs (1980). These studies have employed both operant and classical (Pavlovian) conditioning paradigms. However, with a single exception (Bohus, 1975), all of the experiments which have studied the effects of neuropeptides on learning and memory formation have employed operant conditioning paradigms.

In the present experiment we sought to determine the effects of several different doses of $ACTH_{4-10}$ on Pavlovian conditioning of both somatomotor and autonomic responses. We have postulated that the development of classically conditioned autonomic responses represents an early stage of engram formation associated with "attentional processes," whereas later occurring somatomotor responses represent an attempt by the organism to contend with biologically significant environmental contingencies (Powell et al., 1978). It is conceivable that the various neuropeptides which influence learning and memory may have differential effects on various stages of the learning process. The purpose of a recent series of investigations begun in our laboratory has thus been to investigate this problem using the heart rate (HR) and eyeblink (EB) Pavlovian conditioned responses (CRs) in the rabbit, as indices of early and late stages of associative learning, respectively. In the present report we describe the effects of $ACTH_{4-10}$ on extinction of HR and EB CRs.

## 2. METHODS

### 2.1. Animals

Sixty experimentally naive New Zealand albino rabbits were used as experimental subjects. The animals were approximately 150 days old at the beginning of the experiment. After receipt from a local supplier, the animals were housed in a climate-controlled room with a 7 a.m./7 p.m. light/dark cycle and maintained on *ad lib*. food and water for the duration of the experiment. All animals were run during the daylight portion of the light/dark cycle. Twelve animals each received subcutaneous injections of 0, 125, 250, 500, or 1000 µ/kg $ACTH_{4-10}$ (Organon), dissolved in a volume of physiological saline equal to 1 $cm^3$/kg body wt, prior to extinction training (see below). No injections were employed during acquisition.

### 2.2. Apparatus

During the experiment, animals were restrained in standard Plexiglas restrainers (Gormezano, 1966) and placed in Industrial Acoustics Co. experimental chambers. EB, HR, and electromyographic (EMG) activity were recorded on a Grass Model 7D polygraph. EB and EMG responses were measured by a Grass Model 7P-3 preamplifier and integrator set in the integrator mode. The output of the polygraph was input to a Digital Equipment Co. PDP-11/10 computer which was used to collect and summarize all data in real time. Tone conditional stimuli (CSs) were produced by TTL square-wave audio oscillators. The CSs were presented through 10.16-cm-diameter speakers located 30 cm above the animal's head. The PDP-11/10 computer in conjunction with TTL programming equipment automatically programmed delivery of the shock, CS–US intervals, etc. ECG electrodes were stainless steel safety pins inserted beneath the skin on the right front leg and left haunch. Stainless steel eyeclips inserted beneath the lower and upper eyelids were used to record the EB response (see Powell and Joseph, 1974). No. 4 insect pins inserted into the neck muscles served as EMG electrodes. The electric shock unconditioned stimulus (US) was produced by a TTL AC constant-current shock source. Shock intensity was monitored constantly for each animal during the experiment by in-line milliammeters.

### 2.3. Procedure

A simple classical conditioning procedure was used with a constant 45-sec intertrial interval. The CS was a 1216-Hz tone with an intensity of 75 db

(SPL). The CS duration was 0.5 sec and its cessation coincided with the onset of a 3-mA, 0.25-sec-duration AC electric shock train presented across the upper and lower eyelids through chronically implanted Michel stainless steel wound clips. The first 3 days of training consisted of 100 conditioning trials per session. Four were unreinforced test trials for HR assessment (see below). During these sessions a twofold learning criterion was employed consisting of (1) the occurrence of five consecutive EB CRs no later than the end of the second session, and (2) responding of greater than 80% on the third day of acquisition. Only animals that met both aspects of this criterion were placed on extinction. Animals that met this criterion received three extinction sessions. Animals were injected 1 hr prior to the beginning of extinction with the appropriate concentration of $ACTH_{4-10}$. The first 15 trials of extinction consisted of "reacquisition" trials during which the tone was followed by shock. Following these 15 trials, subjects received 100 trials of tone alone (i.e., extinction trials). An EB extinction criterion was employed which consisted of the number of trials during which the animals failed to show an EB response during CS presentation for five consecutive extinction trials.

## 2.4. Response Measurement

EB and EMG CRs were recorded during each tone. These CRs were defined as a 200-$\mu$V change from baseline during the 0.5-sec CS. The computer sampled baseline values for each response channel on the polygraph immediately prior to tone onset. If a response was in progress at this point, the trial was discarded. If not, five samples were taken (at 100-msec intervals) during the tone interval. The largest value obtained (above 200-$\mu$V) was recorded as EB or EMG amplitude. HR CRs were recorded only during four evenly spaced test trials during which the CS was not followed by shock. During these trials the computer sampled tachograph voltage from each channel 16 times/sec beginning 4 sec prior to tone onset and continuing for 8 sec after tone onset. These 16 samples were converted to instantaneous HR and averaged within each 1-sec interval. Mean HR from the four pre-CS 1-sec intervals was averaged to yield one baseline value for each test trial. HR change from baseline was obtained by subtracting the mean HR associated with each 1-sec block after tone onset from the pre-CS measure.

## 3. RESULTS

### 3.1. Acquisition

#### 3.1.1. Eyeblink

In order to achieve an $N$ of 12 per experimental group, 81 animals were subjected to acquisition training; however, 21 of these animals failed to achieve the previously established EB acquisition criterion. Thus, approximately 75%

of the rabbits initially studied during acquisition were subjected to extinction training. The EB acquisition criterion employed was thus a fairly rigorous one, which ensured that, prior to extinction training, both the level of performance and duration of training were comparable for all animals. Analyses of variance (ANOVA) of the EB acquisition data thus revealed no significant differences between the groups which later received different $ACTH_{4-10}$ doses during extinction.

### 3.1.2. Heart Rate

ANOVA of baseline (viz., pretone) HR revealed no significant effects. Mean baseline HR was 219 beats/min. The HR conditioned response consisted of bradycardia which began with tone (CS) onset and reached a maximum of 15 to 16 beats/min during the second or third 1-sec interval following tone onset. ANOVA procedures were also utilized to analyze the HR changes which occurred during acquisition. This analysis was a mixed design analysis which employed groups as a major independent dimension with five levels, and sessions (three levels) and blocks of post-CS heart beats in 1-sec intervals (nine levels, including one pre-CS measurement and eight post-CS measurements) as repeated measures. This analysis revealed that during acquisition there were significant post-CS blocks ($F = 10.45$, $df = 8, 440$, $p < 0.001$), and sessions ($F = 2.2$, $df = 32, 856$, $p < 0.0003$) effects; however, no group differences were significant, nor did groups interact with any of the other major dimensions.

### 3.1.3. EMG Responses

Few EMG CRs occurred. The mean numbers of CRs for the saline, 125, 250, 500, and 1000 μg/kg groups were 1.75, 0.40, 0.20, 5.10, and 0.50, respectively. Analysis of these data revealed no significant effects.

## 3.2. Extinction

### 3.2.1. Eyeblink

$ACTH_{4-10}$ had a dose-related effect on the eyelid conditioned response during extinction. These data are shown in Fig. 1. This figure shows the mean trials ($\pm$ 1 S.E.M.) to attain the extinction criterion for each of the five groups of animals. These data thus represent the mean number of trials required for the animals to receive five successive trials during extinction on which a CR (as previously defined) did not occur. The three highest dosages of $ACTH_{4-10}$ resulted in a considerably greater number of trials to reach this criterion than did the two lower doses, i.e., the 125 μg/kg group and the vehicle group. Many animals in the former three groups as well as one animal in the saline group and three animals in the 125 μg/kg group failed to reach this criterion throughout any of the three daily extinction sessions. These data were thus extremely

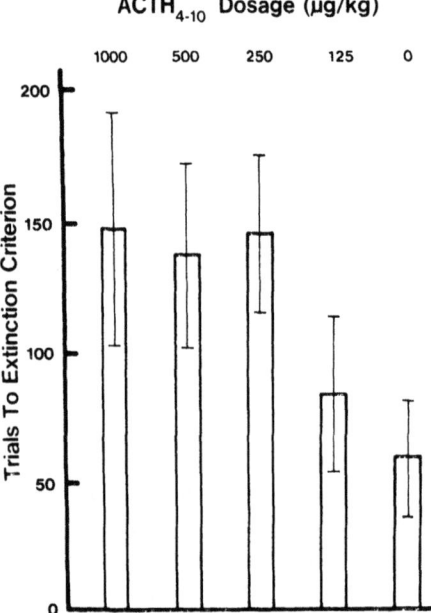

FIGURE 1. Mean trials to achieve an EB response criterion of groups of rabbits administered $ACTH_{4-10}$ during extinction as shown. The criterion consisted of five consecutive trials during extinction in which a CR did not occur. $N = 12$ rabbits per group previously exposed to 300 trials of Pavlovian conditioning training.

skewed and consequently could not be appropriately subjected to parametric statistical analysis. The data were thus analyzed by $\chi^2$ analysis in which animals which met or did not meet the criterion served as the major dependent variable. The number of animals which did not meet the extinction criterion throughout the three extinction session were 7, 5, 6, 3, and 1 for the 1000, 500, 250, 125, and µg/kg groups, respectively. An overall analysis of these data resulted in $\chi^2$ of 8.95 ($df = 3$, $0.05 < p < 0.10$), which did not quite attain normally accepted levels of confidence; however, individual $\chi^2$, comparing each of the groups with the saline control group, revealed that the 1000 µg/kg group was significantly different from the saline group ($\chi^2 = 6.67$, $df = 1$, $p < 0.01$). In addition, differences between 1000 µg/kg and 125 µg/kg group and the 500 µg/kg and 0 µg/kg group approached but did not reach the α level, $p < 0.05$ ($\chi^2 = 3.84$ and 2.54 respectively, $df = 1$, $0.05 < p < 0.10$). No other differences were significant. Analysis of EB percent CRs, EB response frequency, and EB response amplitude revealed no significant differences between the groups. With the exception of CR amplitude, these data were, like the trials to criterion data, extremely skewed due to the large number of responses exhibited by the animals in the higher dosage groups.

### 3.2.2. Heart Rate

ANOVA of baseline HR again revealed no significant effects. Mean baseline HR during extinction was 217.2 beats/min, not significantly different from that obtained during acquisition. Figure 2 illustrates the HR change which occurred in response to the CS averaged over the four test trials for each of the five groups and each of the eight post-CS 1-sec intervals during the first session of extinction. Figure 3 shows similar data pooled over the three experimental sessions. As can be seen in these figures, relatively little HR change occurred in the saline vehicle group until the last two to three 1-sec intervals, during which an HR acceleration of 3–5 beats/min occurred. On the other hand, relatively large HR decelerations occurred in the $ACTH_{4-10}$ groups during the tone CS. The magnitude of these HR decelerations appeared to be dose related, especially during the first extinction session (Fig. 2). The greatest change was exhibited by the animals in the largest dosage groups; the largest magnitude change averaged 8–9 beats/min, occurring in the 1000 μg/kg group, whereas the 125 μg/kg group showed a change of only slightly greater than 3 beats/min. All animals exhibited a return to baseline with minimum HR accelerations occurring, except those shown by the saline group. However, these HR decelerations became somewhat attenuated during later extinction sessions, and

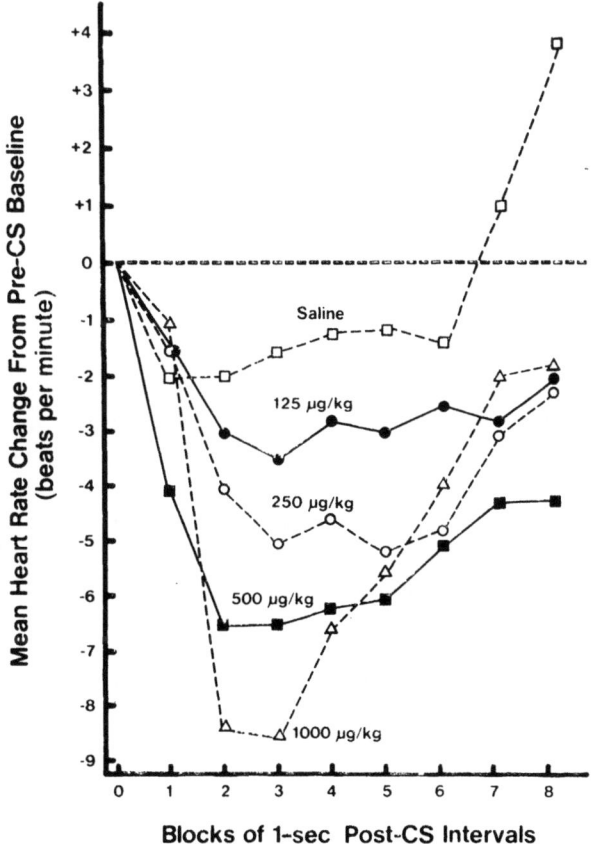

FIGURE 2. Mean heart rate change from pre-CS baseline of rabbit groups administered $ACTH_{4-10}$ as indicated during the first of three extinction sessions. $N = 12$ rabbits per group previously exposed to 300 trials of Pavlovian conditioning training.

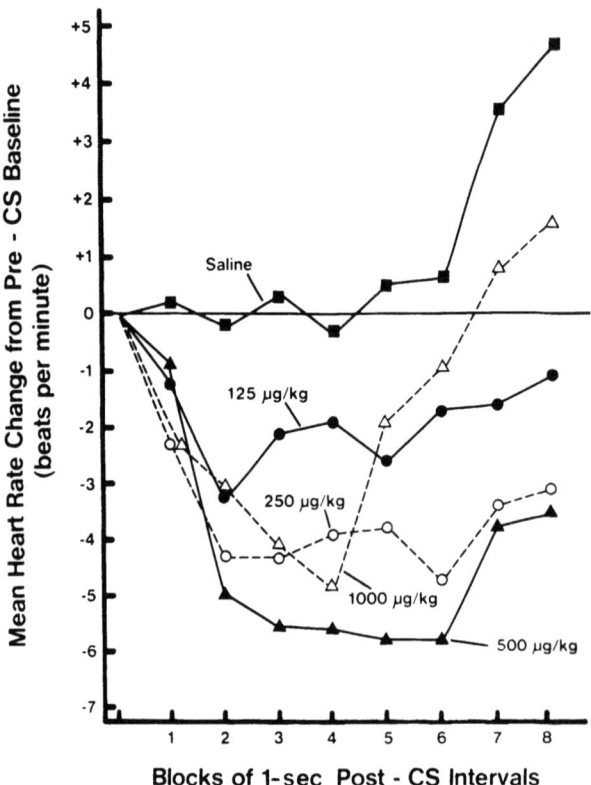

FIGURE 3. Mean heart rate change from pre-CS baseline of rabbit groups administered $ACTH_{4-10}$ as indicated, averaged over three sessions of extinction training. $N = 12$ rabbits per group previously exposed to 300 trials of Pavlovian conditioning training.

the systematic dose–response relationship appeared to break down, although $ACTH_{4-10}$ continued to potentiate the bradycardiac response throughout all three sessions. Figure 3, which shows HR change from baseline averaged over the three extinction sessions, illustrates these findings. A comparison of Figs. 2 and 3 thus suggests that although the effects of $ACTH_{4-10}$ on the HR response decreased as a function of sessions, it continued to attenuate extinction of the response throughout training.

The HR data during extinction were analyzed using a mixed design ANOVA identical to that employed to analyze the acquisition data. This analysis revealed that the group difference in mean HR change from baseline (viz. pooled over the eight post-CS 1-sec intervals) did not quite reach normally accepted levels of statistical significance ($F = 2.11$, $df = 4, 55$, $p < 0.09$). However, the group × blocks of post-CS intervals effect was highly significant ($F = 4, 20$, $df = 32, 400$, $p < 0.0003$). The blocks main effect ($F = 10.69$, $df = 8, 440$, $p < 0.0001$) was also significant. However, the sessions effect and interactions with sessions were not significant. Separate ANOVAs were also calculated for each post-CS interval using groups and sessions as independent and repeated dimensions, respectively. This analysis revealed a significant groups effect for post-CS intervals 1 ($p < 0.01$), 5 ($p < 0.01$), 6 ($p < 0.03$), 7 ($p < 0.003$), and 8 ($p < 0.0005$). These analyses thus suggest that the group differences shown in Figs. 2 and 3, indicating a dose-related retardation in extinction of the Pavlovian conditioned HR response, are reliable.

### 3.2.3. EMG Responses

As was obtained during acquisition, few EMG responses occurred during extinction. Mean number of EMG responses during extinction were 0.80, 0.30, 3.0, 3.5, and 0.90 for the saline, 125, 250, 500, and 1000 µg/kg groups, respectively. Analysis of these data revealed no significant effects.

## 4. DISCUSSION

The present experiment demonstrated that the administration of $ACTH_{4-10}$ during extinction training of Pavlovian conditioned somatomotor and autonomic responses affects classically conditioned behaviors in a manner similar to its effects on operant conditioning. Thus, higher dosages of $ACTH_{4-10}$ significantly increased the numbers of trials to achieve a previously established EB extinction criterion compared to lower dosages. Similarly, the typically obtained cardiac decelerations associated with Pavlovian EB conditioning extinguished completely in a saline control group, whereas dose-related increases in the bradycardiac CR occurred in animals which received $ACTH_{4-10}$ during extinction. Thus, administration of $ACTH_{4-10}$ appears to extend to the extinction period the normally occurring parasympathetic bias of the cardiovascular system in response to Pavlovian CSs, even though the CS is no longer serving as a signal.

It is notable that this effect occurred in spite of the fact that fairly high rates of EB responding occurred during extinction in all groups. Asymptotic EB conditioning has been associated with a biphasic classically conditioned cardiac response (Powell et al., 1974). This response typically consists of an initial cardiac deceleration followed by an acceleration, which occurs only after the eyelid conditioned response has reached asymptotic levels (Powell et al., 1974; Powell and Kazis, 1976). It has been determined that the later-occurring accelerative response is related to the execution of the somatomotor (i.e., EB) CR and is almost always accompanied by increases in various other somatomotor behavior (Powell and Joseph, 1974). The initial decelerative component of the response, however, is associated with relative stimulus ambiguity or uncertainty, and occurs in spite of the fact that somatomotor responses, as well as the conditioned EB responses, are concomitantly occurring. The effects of $ACTH_{4-10}$ are thus to enhance the initial cardiac deceleration during extinction and to prevent the somatomotor biasing of the cardiac response which occurs under normal circumstances (Powell et al., 1974). In fact, this later-occurring tachycardiac response was evident in the saline animals but absent in the animals receiving $ACTH_{4-10}$. Assuming that $ACTH_{4-10}$ affects primarily mnemonic processes, these findings thus suggest that the learned cardiac response accompanying classical conditioning consists of cardiac decelerations and that the later-occurring HR acceleration is a somatomotor bias produced by the occurrence of concomitantly conditioned somatomotor responses.

It is also of note that the ACTH fragment $ACTH_{4-10}$, which has no peripheral hormonal effects, was utilized in the present experiments. Since this fragment acts only on the CNS (de Wied, 1980), it may be concluded that the effects of ACTH in the present experiment on both the eyelid and cardiac CRs were on CNS mechanisms. Further experiments in which these agents are implanted or injected into different limbic system and forebrain structures should provide information regarding possible CNS substrates for the neuropeptide mediation of various stages of the learning process.

As noted above, we have suggested that the learning process consists of several discrete and probably experimentally discriminable stages which almost certainly involve different CNS structures and mechanisms as physiological substrates (Powell *et al.*, 1978). At its simplest level, at least three such stages are required for learning to occur: (1) The first involves "attention" to relevant signals; (2) the second involves the formation of an association between these signals and their consequences (viz., reinforcers); and (3) the third involves the elaboration of a somatomotor response to deal effectively with such contingencies. Prior lesion studies in the rabbit have shown that neostriatal mechanisms, as well as other extrapyramidal motor structures (e.g., cerebellum), are possibly involved in the third of these three stages (McCormick *et al.*, 1981; Powell *et al.*, 1978). Experiments in which the septal–hippocampal circuit has been studied suggest that limbic system mechanisms are most probably involved in the early "attention" stage (Solomon and Moore, 1975; Vinogradova, 1975; Powell *et al.*, 1976; Thompson *et al.*, 1976; Solomon, 1977, 1980). The results of the present experiment suggest that $ACTH_{4-10}$ affects the early attentional stage of this process, assuming that the bradycardia associated with CS presentation is indicative of such processes, but that it also affects the third stage involving the development of somatomotor execution, since EB responding during extinction was also influenced by $ACTH_{4-10}$ administration.

## 5. SUMMARY

Graded doses of peripherally acting ACTH analog $ACTH_{4-10}$ were administered to New Zealand albino rabbits during extinction of Pavlovian conditioned EB and HR responses. During 3 days of classical conditioning, 1-sec tones served as the CS and a brief paraorbital electric shock train was the US. The animals were trained to an EB CR criterion, followed by 3 days of extinction training during which the tone CS was not paired with the shock US. During this time $ACTH_{4-10}$ was administered. $ACTH_{4-10}$ produced increases in responding during extinction in both the EB and HR response systems. However, this effect appeared to be much stronger for HR than for EB, since significant differences were obtained only for the number of animals which failed to reach criterion on the EB task. Significant differences were not obtained for total number of CRs, CR latency, percent CRs, trials to criterion, etc. However, dose-related increases in the magnitude of the bradycardia associated with classical conditioning (viz., the HR CR) occurred during extinction. Although this effect was quite clear-cut during the first day of extinction, the effects related to dose tended to disappear with subsequent extinction training. These findings are compatible with earlier reports of the effects of $ACTH_{4-10}$ on operant conditioning and suggest that neuropeptides influence associative processes as well as instrumental learning.

ACKNOWLEDGMENTS. This research was supported by VA Institutional Research funds awarded to the Wm. Jennings Bryan Dorn VA Hospital, Columbia South Carolina.

We thank Organon International for their generous donation of the $ACTH_{4-10}$.

## 6. REFERENCES

Ader, R., and de Wied, D., 1972, Effects of vasopressin on active and passive avoidance learning, *Psychon. Sci.* **29**:46.

Applezweig, M. H., and Baudry, F. D., 1955, The pituitary adrenocortical system in avoidance learning, *Psychol. Rep.* **1**:417.

Berger, T. W., Clark, G. A., and Thompson, R. F., 1980, Learning dependent neuronal responses recorded from limbic system brain structures during classical conditioning, *Physiol. Psychol.* **8**:155.

Bohus, B., 1975, Pituitary peptides and autonomic responses, in: *Progress in Brain Research*, Vol. 42, *Hormones, Homeostasis and the Brain* (W. H. Gispen, T. B. van Wimersma Greidanus, B. Bohus, and D. de Wied, eds.), pp. 275–283. Elsevier, Amsterdam.

Bohus, B., and de Wied, D., 1967, Avoidance and escape behavior following medial thalamic lesions in rats, *J. Comp. Physiol. Psychol.* **64**:26.

Cohen, D. H., and MacDonald, R. L., 1976, Involvement of the avian hypothalamus in defensively conditioned heart rate change, *J. Comp. Neurol.* **167**:465.

de Wied, D., 1965, The influence of posterior and intermediate lobe of the pituitary and pituitary peptides on the maintenance of a conditioned avoidance response in rats, *Int. J. Neuropharmacol.* **4**:157.

de Wied, D., 1969, Effects of peptide hormones on behavior, in: *Frontiers in Neuroendocrinology* (W. F. Ganong and L. Martini, eds.), pp. 97–140, Oxford University Press, London.

de Wied, D., 1980, Behavioral actions of neurohypophysial peptides, in: *Neuroactive Peptides* (A. Burgen, H. W. Kosterlitz, and L. L. Iversen, eds.), pp. 183–194, The Royal Society, London.

de Wied, D., and Gispen, W. H., 1977, Behavioral effects of peptides, in: *Peptides in Neurobiology* (H. Gainer, ed.), pp. 379–448, Plenum Press, New York.

de Wied, D., Bohus, B., and Greven, H. M., 1968, Influence of pituitary and adrenocortical hormones on conditioned avoidance behavior in rats, in: *Endocrinology and Human Behavior* (R. P. Michael, ed.), pp. 188–199, Oxford University Press, London.

Dogterom, J., and Buijs, R., 1980, Vasopressin and oxytocin distribution in rat brain: Radioimmunoassay and immunocytochemical studies, in: *Neuropeptides and Neural Transmission* (C. Marsan and W. Tarczyk, eds.), pp. 307–314, Raven Press, New York.

Gabriel, M., Foster, K., Orona, E., and Lambert, R. W., 1980, Early and late acquisition of discriminative neuronal activity during differential conditioning in rabbit: Specificity within the laminae of cingulate, *J. Comp. Physiol. Psychol.* **94**:1069.

Gormezano, I., 1966, Classical conditioning, in: *Experimental Methods and Instrumentation in Psychology* (J. B. Sidowski, ed.), pp. 181–196, McGraw-Hill, New York.

Kelsey, J. E., 1975, Role of pituitary–adrenocortical system in mediating avoidance behavior of rats with septal lesions, *J. Comp. Physiol. Psychol.* **88**:271.

Levine, S., and Jones, L. E., 1965, Adrenocorticotrophic hormone (ACTH) and passive avoidance learning, *J. Comp. Physiol. Psychol.* **59**:357.

McCormick, D. A., Lavand, D. G., Clark, G. A., Kettner, R. E., Rising, C. E., and Thompson, R. F., 1981, The engram found? Role of the cerebellum in classical conditioning of nictitating membrane and eyelid responses, *Bull. Psychon. Soc.* **18**:103.

Moore, J. W., 1979, Brain processes and conditioning, in: *Mechanisms of Learning and Motivation: A Memorial Volume to Jerzy Konorski* (A. Dickinson and R. A. Boakes, eds.), pp. 111–142, Erlbaum, Hillsdale, N.J.

Murphy, J. V., and Miller, R. E., 1955, The effect of adrenocorticotropic hormone (ACTH) on avoidance conditioning in the rat. *J. Comp. Physiol. Psychol.* **48**:47.

Powell, D. A., and Joseph, J. A., 1974, Autonomic–somatic interaction and hippocampal theta activity, *J. Comp. Physiol. Psychol.* **87**:978.

Powell, D. A., and Kazis, E., 1976, Blood pressure and heart rate changes accompanying classical eyeblink conditioning in the rabbit (*Oryctolagus cuniculus*), *Psychophysiology* **13**:441.

Powell, D. A., Lipkin, M., and Milligan, W. L., 1974, Concomitant changes in classically conditioned heart rate and corneoretinal potential discrimination in the rabbit (*Oryctolagus cuniculus*), *Learn. Motiv.* **5**:532.

Powell, D. A., Milligan, W. L., and Buchanan, S., 1976, Orienting and classical conditioning in the rabbit (*Oryctolagus cuniculus*): Effects of septal area lesions, *Physiol. Behav.* **17**:855.

Powell, D. A., Mankowski, D., and Buchanan, S., 1978, Concomitant heart rate and corneoretinal potential conditioning in the rabbit (*Oryctolagus cuniculus*): Effects of caudate lesions, *Physiol. Behav.* **20**:143.

Sandman, C. A., Kastin, A. J., and Schally, A. V., 1971, Behavioral inhibition as modified by melanocyte-stimulating hormone (MSH) and light–dark conditions, *Physiol. Behav.* **6**:45.

Segal, M., and Olds, J., 1973, Activity of units in the hippocampal circuit of the rat during differential classical conditioning, *J. Comp. Physiol. Psychol.* **82**:195.

Solomon, P. R., 1977, Role of the hippocampus in blocking and conditioned inhibition of the rabbit's nictitating membrane response, *J. Comp. Physiol. Psychol.* **91**:407.

Solomon, P. R., 1980, A time and place for everything? Temporal processing views of hippocampal function with special reference to attention, *Physiol. Psychol.* **8**:254.

Solomon, P. R., and Moore, J. W., 1975, Latent inhibition and stimulus generalization of the classically conditioned nictitating membrane response in rabbits (*Oryctolagus cuniculus*) following dorsal hippocampal ablation, *J. Comp. Physiol. Psychol.* **84**:145.

Thompson, R. F., Berger, T. W., Cegavske, C. F., Patterson, M. M., Roemer, R. A., Teyler, T. J., and Young, R. A., 1976, The search for the engram, *Am. Psychol.* **31**:209.

Thompson, R. F., Berger, T. W., Berry, S. D., Hoehler, F. K., Kettmer, R. E., and Weisz, D. J., 1980, Hippocampal substrates of classical conditioning, *Physiol. Psychol.* **8**:262.

van Wimersma Greidanus, T. B., and de Wied, D., 1971, Effects of systemic and intracerebral administration of two opposite acting ACTH-related peptides on extinction of conditioned avoidance behavior, *Neuroendocrinology* **7**:291.

van Wimersma Greidanus, T. B., Bohus, B., and de Wied, D., 1974, Differential localization of the influence of lysine vasopressin and of $ACTH_{4-10}$ on avoidance behavior: A study in rats bearing lesions in the parafascicular nuclei, *Neuroendocrinology* **14**:280.

Vinogradova, O. S., 1975, The hippocampus and the orienting reflex, in: *Neuronal Mechanisms of the Orienting Reflex* (E. N. Sokolov and O. S. Vinogradova, eds.), pp. 128–169, Wiley, New York.

Weiss, J. M., McEwen, B. S., Silva, M., and Kalkut, M., 1970, Pituitary–adrenal alterations and fear responding, *Am. J. Physiol.* **218**:864.

CHAPTER 12

# $ACTH_{4-10}$ and Memory in Psychiatric Patients

GIACOMO D'ELIA and
SVEND-OTTO FREDERIKSEN

## 1. INTRODUCTION

Historically the pituitary–adrenal axis has been implicated in the stress response (Selye, 1956), which was thought to be mediated by a target organ away from the brain. Early observations by Thompson and McConnell (1955) and Ungar (1973) indicated, however, that amino acids may directly influence the CNS. The early work of Mirsky et al. (1953) and Miller and Ogawa (1962) indicated a relationship between ACTH and psychological processes. de Wied and Bohus (1966) proposed that ACTH, MSH, and their fragments improved memory.

Much recent evidence strongly indicates that $ATCH_{4-10}$ influences aspects of behavior by direct action on the CNS (Rigter and van Riezen, 1978). The most convincing effects are on attention-arousal features, but effects on memory are unimpressive. For example, it has been argued that the effect of $ACTH_{4-10}$ on retrieval of memory in animal studies may be secondary to higher attention-arousal.

Most human studies have been carried out on healthy people. Psychiatric patients treated with electroconvulsive therapy (ECT) for depressive disorders seem to be uniquely suited for the study of attention, consolidation, and retrieval of memory. Such patients have attentional disturbances secondary to the depressive state, together with transient disturbance of consolidation and/or decreased availability of memory items to retrieval secondary to ECT (Williams, 1977).

Since there already exist a number of excellent reviews on the behavioral effects of the pituitary peptides in animals and man (Marx, 1975; Rigter and van Riezen, 1978; Miller et al., 1977), we will confine ourselves to published

---

GIACOMO D'ELIA and SVEND-OTTO FREDERIKSEN • Department of Psychiatry, University of Linköping, Linköping, Sweden.

data relevant to research on psychiatric patients. We will propose some lines for further clinical research with ACTH analogs in psychiatric patients.

## 2. ATTENTION-AROUSAL

There is some controversy about whether the effects of ACTH peptide on animal behavior consist of increase in motility or enhancement of arousal, motivation (Bohus *et al.*, 1977), vigilance (Wolthus and de Wied, 1976), or selective attention (Sandman *et al.*, 1972, 1975, 1976; Rigter and van Riezen, 1978). The distinctions implied by these numerous terms seem largely artificial, and may simply indicate that early stages of information processing are influenced by $ACTH_{4-10}$.

Miller *et al.* (1976) found that $ACTH_{4-10}$ improved performance on the digit symbol substitution test in man. Monitoring of evoked potentials disclosed that the peptide altered the late component of these potentials. Gaillard and Sanders (1975) demonstrated improved performance on a continuous reaction-time task, a test which demands a sustained level of attention. These human data are compatible with findings in animal studies. It should be kept in mind, however, that these observations were made on healthy people.

In a double-blind intraindividual cross-over comparison (d'Elia and Frederiksen, 1980a), patients suffering from depressive disorders of such severity that ECT was indicated were administered the Bourdon Test 150 min after s.c. injection of 30 mg of $ACTH_{4-10}$ or placebo (Fig. 1, Study I). Half of the patients were started at random with $ACTH_{4-10}$, and the other half with placebo. The observations were done in connection with the second and third treatments in a unilateral ECT series.

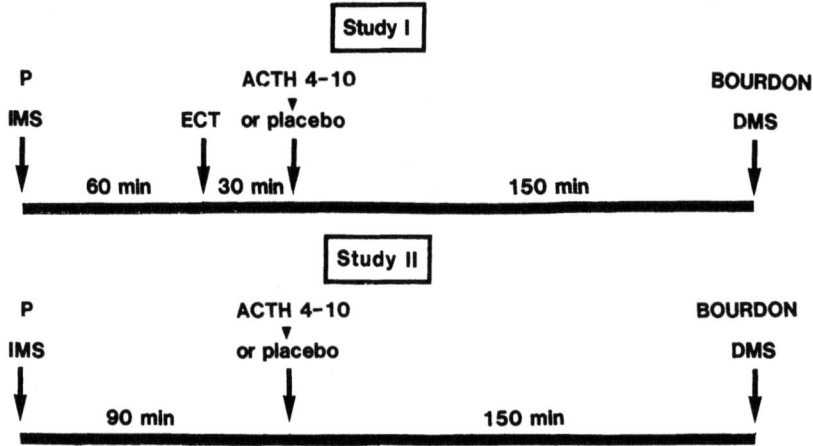

FIGURE 1. Time intervals in Studies I and II. P, presentation of the memory test; IMS, immediate memory score; DMS, delayed memory score; ECT, electroconvulsive treatment.

## TABLE 1
### The Bourdon Test Score

| Variable | $ACTH_{4-10}$ $\bar{X}$ | Placebo $\bar{X}$ | F |
|---|---|---|---|
| Study I (N = 20) | | | |
| Time (sec) | 309.3 | 318.9 | 0.91 |
| Deleted | 68.8 | 66.6 | 0.16 |
| Not deleted | 19.2 | 21.4 | 0.51 |
| Study II (N = 20) | | | |
| Time (sec) | 324.9 | 294.9 | 1.01 |
| Deleted | 77.4 | 77.7 | 0.04 |
| Not deleted | 11.7 | 10.3 | 0.04 |
| Study III (N = 18) | | | |
| Time (sec) | 469.4 | 523.5 | 1.49 |
| Deleted | 126.6 | 127.1 | 0.07 |
| Not deleted | 29.4 | 29.5 | 0.01 |
| Study IV (N = 20) | | | |
| Time (sec) | 268.5 | 250.8 | 3.95 |
| Deleted | 79.9 | 75.5 | 0.07 |
| Not deleted | 8.1 | 9.0 | 0.07 |

The Bourdon Test is a cancellation test with meaningless text. The patient is asked to delete the consonant *d* as fast and exactly as possible. Three scores are obtained, the time employed to perform the task, the number of deleted *d*'s, and the number of undeleted *d*'s. The test is assumed to give a measure of sustained attention, or concentration. Two parallel forms of the test, A and B, were administered according to the ABBA principle. Unilateral ECT was administered 30 min before giving $ACTH_{4-10}$ or placebo (Fig. 1, Study I). No difference was found between $ACTH_{4-10}$ and placebo (Table 1, Study I), nor

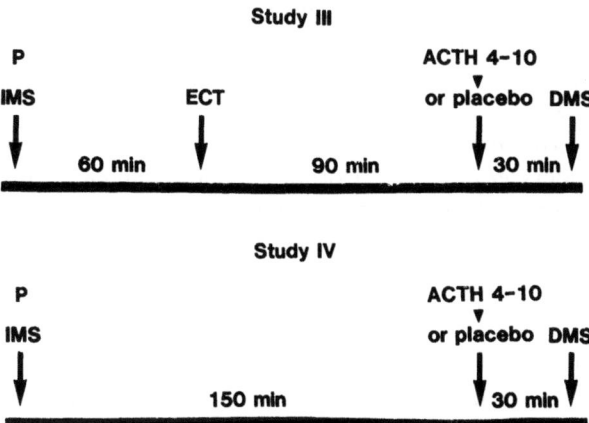

FIGURE 2. Time intervals in Studies III and IV. P, presentation of the memory test; IMS, immediate memory score; DMS, delayed memory score; ECT, electroconvulsive treatment.

was any difference found in a group of psychiatric patients suffering from neurotic depression and receiving light psychopharmacological treatment (Table 1, Study II). The design of Study II was identical to that of Study I on ECT-treated patients.

To investigate the significance of the time interval, $ACTH_{4-10}$ or placebo was administered 30 min before the test and 90 min after ECT in a new series of patients (Fig. 2, Study III) (d'Elia and Frederiksen, 1980b). No difference was found (Table 1, Study III), and no difference was seen in a group of healthy volunteers (Table 1, Study IV, and Fig. 2, Study IV). Intergroup differences are not compared because the groups did not match with regard to a number of essential variables such as age, treatment, etc.

## 3. CONSOLIDATION

In spite of improvement in avoidance response in animals, suggesting that relevant information was stored in memory, a progressive decline in avoidance to placebo level suggested that consolidation of memory may not have been greatly affected (Bohus et al., 1977). This is in contrast to the results of posttrial administration of ACTH and $ACTH_{4-10}$, which indicate that these substances may influence consolidation of memory. Flood et al. (1976) found that $ACTH_{4-10}$ injected s.c. in mice after acquisition of an active or passive avoidance response improved performance in a retrieval test 1 week later. Similar results have been reported by others (Gold and van Buskirk, 1976; Gold and McGaugh, 1978). Thus, an enhancing effect of $ACTH_{4-10}$ on the consolidation of memory cannot be ruled out.

In human volunteers and patients, $ACTH_{4-10}$ improved performance in the Benton Visual Retention Test (Miller et al., 1974; Sandman et al., 1975, 1976). However, in other studies using the same test, no improvement could be shown (Rapaport et al., 1976; Veith et al., 1978; Draper et al., 1978). Other visual, nonverbal memory tests (mainly subtests of the Wechsler Memory Scale ) have also given contradictory results, some authors claiming beneficial effects (Ferris et al., 1976; Miller et al., 1976) and others reporting no effect at all (Draper et al., 1978). Verbal memory has in general been claimed to be unaffected by $ACTH_{4-10}$ (Sanders et al., 1975; Dornbursh and Volavka, 1976; Ferris et al., 1976; Branconnier et al., 1979; Draper et al., 1978), but beneficial effects have also been reported (Veith et al., 1978). On the whole, human studies have failed to give consistent support to a consolidation facilitating effect of $ACTH_{4-10}$ in single doses of 10–20 mg administered s.c. A possible explanation is that different populations have been studied. Well-rested, healthy volunteers with intact attentional and amnestic functions probably leave little scope for improvement. Also, the dose of $ACTH_{4-10}$ may have been too low. Further, the tests may have been insensitive to memory changes. In fact, the Wechler Memory Scale and the Benton Visual Retention Test mainly measure learning, and probably do not adequately reflect consolidation.

We investigated our depressive patients with a memory test battery suited for the study of consolidation and comprising the 30-word-pair test, the 30-figure test, the 30-geometrical-figure test, and the 30-face test. A higher dose of $ACTH_{4-10}$ was used. The design of the study is shown in Fig. 1 (Study I).

The 30-word-pair test (Cronholm and Molander, 1957) is constructed according to the traditional Ranschburg model. Thirty word-pairs are read, ten at a time, at the rate of one pair about every 3 sec. The subject is then asked to complete the pairs after being presented with the first word of each pair but in a new order. The procedure is repeated after 4 hr with the word-pairs in yet another order.

In the 30-figure test (Cronholm and Molander, 1957), the subject is shown a picture of 30 common objects, which are pointed out and named at the rate of roughly 1 item/sec. Immediately after presentation, the subject is asked to point out the objects which he recognizes on a new picture in which the 30 original objects are mixed with 30 others. After 4 hr the procedure is repeated, with the first 30 objects distributed differently and mixed with the other 30 objects. The 30-geometrical-figure test (d'Elia and Frederiksen, 1982) involves relatively complex geometrical figures. The subject is shown a picture of 30 figures, one item every 2 sec. Immediately afterwards, using a new picture where the original 30 figures are mixed with 30 others, the subject is asked to point out those he recognizes. After 4 hr the procedure is repeated, as in the 30-figure test. The 30-face test (d'Elia and Frederiksen, 1982) comprises black-and-white frontal-view photographs (4 × 5 cm) of 30 human faces (medical students), 15 men and 15 women, showing the head and neck with little distinctive clothing. Some of the men have moustaches or glasses, and the same

TABLE 2
Study I. Memory Score[a]

| Memory test | $ACTH_{4-10}$ $\bar{X}$ | Placebo $\bar{X}$ | F | p |
|---|---|---|---|---|
| IMS | | | | |
| 30-word-pair test | 16.6 | 16.3 | 0.1 | ns |
| 30-figure test | 24.3 | 23.1 | 1.9 | ns |
| 30-geometrical-figure test | 16.8 | 15.9 | 0.6 | ns |
| 30-face test | 16.5 | 18.4 | 4.2 | ns |
| DMS | | | | |
| 30-word-pair test | 10.4 | 7.4 | 7.0 | 0.05 |
| 30-figure test | 19.5 | 19.3 | 0.1 | ns |
| 30-geometrical-figure test | 13.5 | 13.7 | 0.0 | ns |
| 30-face test | 12.8 | 14.0 | 0.9 | ns |
| FS | | | | |
| 30-word-pair test | 6.3 | 8.9 | 7.3 | 0.05 |
| 30-figure test | 4.8 | 3.8 | 1.2 | ns |
| 30-geometrical-figure test | 3.3 | 2.3 | 1.5 | ns |
| 30-face test | 3.7 | 4.6 | 0.3 | ns |

[a] IMS = immediate memory score; DMS = delayed memory score; FS = (IMS − DMS) forgetting score.

## TABLE 3
### Study II. Memory Score

| Memory test | ACTH$_{4-10}$ $\bar{X}$ | Placebo $\bar{X}$ | F | p |
|---|---|---|---|---|
| IMS | | | | |
| 30-word-pair test | 20.2 | 22.1 | 4.99 | 0.05 |
| 30-figure test | 25.9 | 25.2 | 0.63 | ns |
| 30-geometrical-figure test | 20.9 | 20.0 | 0.13 | ns |
| 30-face test | 19.8 | 19.8 | 0.01 | ns |
| DMS | | | | |
| 30-word-pair test | 14.1 | 14.7 | 6.36 | ns[a] |
| 30-figure test | 21.8 | 22.7 | 0.96 | ns |
| 30-geometrical-figure test | 18.7 | 17.0 | 2.17 | ns |
| 30-face test | 18.0 | 18.2 | 0.01 | ns |
| FS | | | | |
| 30-word-pair test | 6.1 | 7.4 | 2.64 | ns[b] |
| 30-figure test | 4.1 | 2.5 | 0.06 | ns |
| 30-geometrical-figure test | 2.2 | 3.1 | 3.13 | ns |
| 30-face test | 1.8 | 1.6 | 0.01 | ns |

[a] ANACOVA IMS/DMS, $F = 1.36$, ns.
[b] ANACOVA IMS/FS, $F = 0.35$, ns.

number of moustaches and glasses are included in the second set of 30 faces. The test is carried out in the same way as the 30-geometrical-figure test.

In each test, three scores are obtained: immediate memory score (IMS), immediately after the presentation of the items; delayed memory score (DMS) (Figs. 1 and 2); and forgetting score (FS), which is the difference IMS − DMS. FS is assumed to represent an indicator of consolidation. Parallel forms of the tests were used. This posttrial study was aimed to show whether ACTH$_{4-10}$ counteracts a retrograde dysmnesic effect of unilateral ECT during the phase of consolidation.

A beneficial effect was found in one subtest in the ACTH$_{4-10}$ group (Table 2) but on the whole the peptide in a dose of 30 mg s.c. did not improve test performance in ECT-treated depressive patients or in patients with mild depressive disorders (Table 3).

## 4. RETRIEVAL

Preretrieval administration of ACTH$_{4-10}$ in animals has been shown to reduce amnesia induced by carbon dioxide or electroshock, irrespective of the interval between learning and testing or of the conditioning situation (Rigter and van Riezen, 1978). These effects have been attributed to a peptide-induced change in motivation, arousal, or attention. However, it has also been argued that ACTH$_{4-10}$ may improve retrieval by counteracting factors that interfere with retrieval (Quinton, 1972). In healthy volunteers, ACTH$_{4-10}$ given before

TABLE 4
Study III. Memory Score

| Memory test | ACTH$_{4-10}$ $\bar{X}$ | Placebo $\bar{X}$ | F | p |
|---|---|---|---|---|
| IMS | | | | |
| 30-word-pair test | 17.9 | 18.0 | 0.012 | ns |
| 30-figure test | 23.7 | 23.6 | 0.03 | ns |
| 30-geometrical-figure test | 17.9 | 16.7 | 1.66 | ns |
| 30-face test | 18.3 | 17.7 | 0.29 | ns |
| DMS | | | | |
| 30-word-pair test | 11.6 | 10.8 | 0.48 | ns |
| 30-figure test | 19.8 | 20.7 | 1.02 | ns |
| 30-geometrical-figure test | 15.0 | 14.7 | 0.18 | ns |
| 30-face test | 14.2 | 14.2 | 0.00 | ns |
| FS | | | | |
| 30-word-pair test | 6.3 | 7.2 | 0.46 | ns |
| 30-figure test | 3.8 | 2.9 | 0.57 | ns |
| 30-geometrical-figure test | 2.9 | 2.0 | 0.45 | ns |
| 30-face test | 4.1 | 3.6 | 0.18 | ns |

TABLE 5
Study IV. Memory Score

| Memory test | ACTH$_{4-10}$ $\bar{X}$ | Placebo $\bar{X}$ | F | p |
|---|---|---|---|---|
| IMS | | | | |
| 30-word-pair test | 23.2 | 23.7 | 0.53 | ns |
| 30-figure test | 25.7 | 25.2 | 0.50 | ns |
| 30-geometrical-figure test | 19.2 | 22.0 | 7.90[a] | 0.01 |
| 30-face test | 19.9 | 21.9 | 10.02[b] | 0.01 |
| DMS | | | | |
| 30-word-pair test | 16.0 | 16.5 | 0.17 | ns |
| 30-figure test | 22.2 | 21.9 | 0.29 | ns |
| 30-geometrical-figure test | 16.7 | 18.5 | 2.70[a] | ns |
| 30-face test | 16.6 | 19.3 | 98.50[b] | 0.001 |
| FS | | | | |
| 30-word-pair test | 7.3 | 7.2 | 0.02 | ns |
| 30-figure test | 3.5 | 3.3 | 0.01 | ns |
| 30-geometrical-figure test | 2.5 | 3.5 | 1.20 | ns |
| 30-face test | 3.3 | 2.6 | 0.37 | ns |

[a] ANACOVA IMS/DMS, $F = 1.18$, ns.
[b] ANACOVA IMS/DMS, $F = 0.72$, ns.

retrieval and 7 days after learning did in fact enhance memory (Sanders et al., 1975). However, $ACTH_{4-10}$ apparently did not improve test performance in brain-damaged chronic alcoholics with regard to visuo-motor-spatial function and visual memory (Draper et al., 1978).

In elderly subjects with mild cognitive impairment, memory retrieval in two tests on nonverbal memory was impaired on the day following injection of 30 mg $ACTH_{4-10}$. In patients with severe cognitive impairment, on the other hand, test performance improved after a single dose of $ACTH_{4-10}$ (Ferris et al., 1976).

In depressive patients treated with bilateral ECT, no significant differences between $ACTH_{4-10}$ and placebo were observed in terms of memory retrieval (Small et al., 1977). We gave $ACTH_{4-10}$ or placebo 30 min before retrieval to psychiatric patients treated with unilateral ECT for depressive disorders (d'Elia and Frederiksen, 1980b). The time schedule for testing and treatment is shown in Fig. 2 (Study III). Like Studies I and II, this was an intraindividual double-blind cross-over comparison. No differences in test performance were found, indicating that $ACTH_{4-10}$ did not facilitate retrieval (Table 4). No effect was found in healthy volunteers (Fig. 2 and Table 5, Study IV) (d'Elia and Frederiksen, 1980b). In conclusion, whereas animal experiments indicate that $ACTH_{4-10}$ may facilitate memory retrieval and reduce amnesia, studies on healthy volunteers and psychiatric patients do not support this hypothesis.

## 5. CONCLUDING REMARKS

The results of the human studies are inconsistent, and provide no convincing evidence that $ACTH_{4-10}$ improves memory function. An effect on attention seems probable but is not sufficiently supported by clinical observations. In depressive patients treated with ECT or light psychopharmacological treatment, we found no clinically significant effects on sustained concentration, consolidation of engrams, or enhancement of memory retrieval. Different sources of error may be discussed, however. Little is known about the pharmacodynamic and pharmacokinetic properties of $ACTH_{4-10}$: A single 30-mg dose may be too small; s.c. injection may be inappropriate; and the dose–response curve may be nonmonotonic for the behavioral indices considered (Kimble, 1977).

Witter et al. (1975) found that the half-life of $ACTH_{4-10}$ is less than $1\frac{1}{2}$ min. The effect of the drug on the CNS is probably rapidly evanescent. The time schedules of clinical studies, including ours, are therefore probably nonsuitable. Determination of plasma concentration levels could help when designing clinical trials.

Attention is a complex variable comprising neurophysiological as well as emotional–motivational aspects. Reports apparently suggest that only the former are influenced by $ACTH_{4-10}$. Miller et al. (1974) found that $ACTH_{4-10}$ reduces EEG activity in the 3- to 7-Hz band, and increases it in bands with frequencies of 7 and over. $ACTH_{4-10}$ appears also to diminish the habituation of the α-blocking response. However, these results are not consistent with those of Sannita *et al.* (1976) and Small (1977).

Miller *et al.* (1976) report also an N350 peak of visual evoked response that did not appear under placebo conditions, and discuss the possibility that the finding could reflect the activation of highly specific central alerting mechanisms. Though inconsistent and open to speculative interpretation, data from neurophysiological studies may indicate that the action of $ACTH_{4-10}$ takes place during the earliest stages of information processing.

## 6. SUGGESTED LINES FOR FUTURE CLINICAL RESEARCH

With the availability of more potent peptides, testing under more favorable conditions becomes feasible. The analog $ACTH_{4-9}$, for instance, can be given my mouth and over a relatively long period. It has been suggested that $ACTH_{4-9}$ may have CNS-stimulating, possibly mood-elevating effects (Dornbush *et al.*, 1981), which opens interesting perspectives. Since normal, well-rested young people are very good at attentional tasks, drugs should be tested on subjects with different baselines in order to avoid ceiling effects. Sleep-deprived or fatigued people and subjects with pathological weakened attention should be studied (depressed, schizophrenic, brain-damaged, etc.).

Motivational and neurophysiological aspects of attention should be kept apart. Orientation responses, habituation, reaction time, EEG α-blocking response, and evoked potentials may be useful in testing subchronic effects of the drugs; these variables are probably less tied to motivation than are other aspects of attention that are dependent on positive or negative reinforcement and on learned behavior.

Sensory information remains for only 1–2 sec in the sensory–neural registers (i.e., immediate memory). During this stage images are formed and materials are identified. In an experiment in which an array of 12 letters was presented by tachistoscope, Sperling (1960) found that immediately after presentation subjects were able to remember about 75% of the letters arranged in three subsets. At intervals ranging up to 2 sec he gave an acoustic signal that asked the subject to report particular subsets of the array. Sperling has argued that the experiment provides evidence for the existence of a visual memory system (iconic memory) in which items decay rapidly and are lost within about 1–2 sec (Sperling, 1963). If neuropeptides have effect at this level of "learning," it ought to postpone the decay or enhance recollection as a result of a stronger initial registration.

Much of the input to immediate memory is lost within a very short time, and only a part finds its way from immediate memory to short-term memory where it remains for another 15–30 sec. At this stage the information is highly sensitive to interference. If we present a list of words the length of which exceeds the subject's memory span (about 7), some of them will be transferred from short-term to long-term memory. When each recalled word is plotted in its original serial position of presentation, it emerges that items at the ends of the list are recalled more frequently than words in the middle of the list; the final and early items are characteristically recalled slightly more successfully than the central items ("recency" and "primacy" effects). It is assumed that recency effect represents recall from short-term memory, i.e., the items still exist in consciousness and need not be transferred to more permanent storage in order to be recalled. On the other hand, central and early items are retrieved from long-term memory, which is a more permanent form of storage (Underwood, 1976). With a relatively simple method it should be possible to test effects of neuropeptides on these types of recall.

## 7. REFERENCES

Bohus, B., van Wimersma Greidanus, T. J. B., Urball, I., and de Wied, D., 1977, Hypothalamo-neurohypophyseal hormone effects on memory and related functions in the rat, in: *Neurobiology of Sleep and Memory* (R. R. Drucker-Colin and J. L. McGaugh, eds.), pp. 333–346, Academic Press, New York.

Branconnier, R. J., Cole, J. O., and Gardos, G., 1979, $ACTH_{4-10}$ in the amelioration of neuropsychological symptomatology associated with senile organic brain syndrome, *Psychopharmacology* **61**:161.

Cronholm, B., and Molander, L., 1957, Memory disturbances after electroconvulsive therapy. I. Conditions 6 hours after electroshock treatment, *Acta Psychiatr. Neurol. Scand.* **32**:280.

d'Elia, G., and Frederiksen, S. O., 1980a, $ACTH_{4-10}$ and memory in ECT-treated and untreated patients. I. Effect on consolidation, *Acta Psychiatr. Scand.* **62**:418.

d'Elia, G., and Frederiksen, S. O., 1980b, $ACTH_{4-10}$ and memory in ECT-treated patients and untreated controls. II. Effect on retrieval, *Acta Psychiatr. Scand.* **62**:429.

d'Elia, G., and Frederiksen, S. O., 1982, Reliability and validity of two memory tests, unpublished data.

de Wied, D., and Bohus, B., 1966, Long-term and short-term effects on retention of a conditioned avoidance response in rats by treatment with long-acting pitressin and alpfa-MSH, *Nature (London)* **212**:1484.

Dornbush, R. L., and Volavka, J., 1976, $ACTH_{4-10}$: A study of toxicological and behavioral effects in an aging sample, *Neuropsychobiology* **2**:350.

Dornbush, R. L., Shapiro, B., and Freedman, A. M., 1981, Effects of an ACTH short chain neuropeptide in man, *Am. J. Psychiatry* **138**:962.

Draper, R. J., Feldman, B., and Haughton, H., 1978, Trial of a peptide fraction of ACTH ($ACTH_{4-10}$) in the alcoholic brain damage syndrome, President at the 2nd World Congress of Biological Psychiatry, Barcelona.

Ferris, S. H., Sathananthan, G., Gershon, S., Clark, C., and Moshinsky, J., 1976, Cognitive effects of $ACTH_{4-10}$ in the elderly, *Pharmacol. Biochem. Behav.* **5**(Suppl. 1):73.

Flood, J. F., Jarvik, M. E., Bennett, E. L., and Orme, A. E., 1976, Effects of ACTH peptide fragments on memory formation, *Pharmacol. Biochem. Behav.* **5**(Suppl. 1):41.

Gaillard, A. W. K., and Sanders, A. R., 1975, Some effects of $ACTH_{4-10}$ on performance during a serial reaction task, *Psychopharmacologia* **42**:201.

Gold, P. E., and McGaugh, J. L., 1978, cited by Rigter, H., and van Riezen, H., 1978, Hormones and memory, in: *Psychopharmacology: A Generation of Progress* (M. A. Lipton, A. DiMascio, and K. F. Killam, eds.), pp. 677–689, Raven Press, New York.

Gold, P. E., and van Buskirk, R., 1976, Enhancement and impairment of memor processes with post-trial injections of adrenocorticotrophic hormone, *Behav. Biol.* **16**:387.

Kimble, G. A., 1977, Is learning involved in neuropeptide effects on behavior?, in: *Neuropeptide Influence on the Brain and Behavior* (L. H. Miller, C. A. Sandman, and A. J. Kastin, eds.), pp. 189–200, Raven Press, New York.

Marx, J. L., 1975, Learning and behavior. I. Effects of pituitary hormones, *Science* **190**:367.

Miller, L. H., Kastin, A. J., Sandman, C. A., Fink, M., and van Tien, W. T., 1974, Polypeptide influences on attention, memory and anxiety in man, *Pharmacol. Biochem. Behav.* **2**:663.

Miller, L. H., Harris, L. C., van Riezen, H., and Kastin, A. J., 1976, Neuroheptapeptide influence on attention and memory in man, *Pharmacol. Biochem. Behav.* **5**(Suppl. 1):17.

Miller, L. H., Sandman, C. A., and Kastin, A. J. (eds.), 1977, Neuropeptide influences on the brain and behavior, in: *Advances in Biochemical Psychopharmacology*, Vol. 17, Raven Press, New York.

Miller, R. E., and Ogawa, N., 1962, The effect of adrenocorticotrophic hormone (ACTH) on avoidance conditioning in the adrenalectomized rat, *J. Comp. Physiol. Psychol.* **55**:211.

Mirsky, I. A., Miller, R., and Stein, M., 1953, Relation of adrenocortical activity and adaptive behavior. *Psychosom. Med.* **15**:574.

Quinton, E., 1972, cited by Rigter, H., and van Riezen, H., 1978, Hormones and memory, in: *Psychopharmacology: A Generation of Progress* (M. A. Lipton, A. DiMascio, and K. F. Killam, eds.), pp. 677–689, Raven Press, New York.

Rapaport, J. L., Quinn, P. O., Capeland, A. P., and Burg, C., 1976, $ACTH_{4-10}$: Cognitive and behavioral effects in hyperactive, learning-disabled children, *Neuropsychobiology* **2**:291.

Rigter, H., and van Riezen, H., 1978: Hormones and memory, in: *Psychopharmacology: A Generation of Progress* (M. A. Lipton, A. DiMascio, and K. F. Killam, eds.), pp. 677–689, Raven Press, New York.

Sanders, A. F., Truijens, C. L., and Bunt, A. A., 1975, $ACTH_{4-10}$ and learning, TNO Institute for Perception.

Sandman, C. A., Miller, L. H., Kastin, A. J., and Schally, A. V., 1972, Neuroendocrine inference on attention and memory, *J. Comp. Physiol. Psychol.* **80**:54.

Sandman, C. A., George, J. M., Nolan, J. D., van Riezen, H., and Kastin, A. J., 1975, Enhancement of attention in man with $MSH/ACTH_{4-10}$, *Physiol. Behav.* **15**:427.

Sandman, C. A., George, J., Walker, B. B., Nolan, J. D., and Kastin, A. J., 1976, $MSH/ACTH_{4-10}$ enhances attention in the mentally retarded, *Pharmacol. Biochem. Behav.* **5**(Suppl. 1):23.

Sannita, W. G., Irwin, P., and Fink, M., 1976, Effects of $ACTH_{4-10}$ on quantitative EEG and memory function in normal male volunteers, Presented at Conf. on Neuropeptides, Philadelphia.

Selye, H., 1956, *The Stress of Life*, McGraw-Hill, New York.

Small, J., 1977, cited by Shagass, C., 1977, EEG and evoked potential approaches to the study of neuropeptides, in: *Neuropeptide Influences on the Brain and Behavior: Advances in Biochemical Psychopharmacology* (L. H. Miller, C. A. Sandman, and A. J. Kastin, eds.), pp. 29–60, Raven Press, New York.

Small, J. G., Small, I. F., Milstein, V., and Dian, D. A., 1977, Effects of $ACTH_{4-10}$ on ECT induced memory dysfunctions, *Acta Psychiatr. Scand.* **55**:241.

Sperling, G., 1960, The information available in brief visual presentations, Psychologic Monograph No. 74, American Psychological Association, New York.

Sperling, G., 1963, A model for visual tasks, in: *Human Factors*, Vol. 5, pp. 19–31.

Thompson, R., and McConnell, J. V., 1955, Classical conditioning in the planarian, *Dugesia dorotocephlia*, *J. Comp. Physiol. Psychol.* **48**:65.

Underwood, G., 1976, *Attention and Memory*, Pergamon Press, Elmsford, N.Y.

Ungar, G., 1973, The problem of molecular coding of neural information: A critical review, *Naturwissenschaften* **60**:307.

Veith, J. L., Sandman, C. A., George, J. M., and Stevens, V. C., 1978, Effects of $MSH/ACTH_{4-10}$ on memory, attention and endogenous hormone levels in women, *Physiol. Behav.* **20**:43.

Williams, M., 1977, Memory disorders associated with electroconvulsive therapy, in: *Amnesia: Clinical, Psychological, and Medicolegal Aspects* (C. W. M. Whytty and O. L. Zangwill, eds.), pp. 183–198, Butterworths, London.

Witter, A., Greven, H. M., and de Wied, D., 1975, Correlation between structure, behavioral activity and rate of biotransformation of some $ACTH_{4-9}$ analogs, *J. Pharmacol. Exp. Ther.* **193:**853.

Wolthus, O. L., and de Wied, D., 1976, The effect of ACTH analogues on motor behavior and visual evoked response in rats, *Pharmacol. Biochem. Behav.* **4:**273.

CHAPTER 13

# Vasopressin in Neuropsychiatric Disorders

## J. J. LEGROS and IOANA LANCRANJAN

### 1. INTRODUCTION

Increasing evidence has been accumulated in the past decade indicating that the neurohormones and/or neuropeptides synthesized by neurons located in the hypothalamus and in extrahypothalamic areas of the brain influence the brain functions, the "peptidergic" neurons being an integral part of the working of the mammalian CNS (Guillemin, 1977).

A large number of neuropeptides and neurohormones have been reported to be present in the CNS. Their regional distribution in the brain has been studied in various mammalian species including man by means of radioimmunoassay and immunohistochemical methods. The first peptidergic neurons to be discovered in the brain were the magnocellular neurons of the hypothalamic supraoptic (SO) and paraventricular (PV) nuclei (Bargmann and Scharrer, 1951; Scharrer and Scharrer, 1954). These neurons synthesize vasopressin or oxytocin plus their respective "carrier proteins," the neurophysins. Vasopressin is synthesized also by the parvocellular neurons of the suprachiasmatic nucleus.

The aim of this chapter is to highlight selected interesting aspects of the synthesis and release of vasopressin, its behavioral effect in experimental animals and in man, and its clinical applications. Finally, the possible mechanisms of action of vasopressin in the brain are discussed and new mechanisms are postulated in order to stimulate research in this field.

### 2. VASOPRESSIN: SYNTHESIS AND RELEASE

The nonapeptides vasopressin and oxytocin and their associated proteins, the neurophysins, derive from a common precursor polypeptide for each non-

---

J. J. LEGROS • Neuroendocrinology Section, CHU University of Liège, Liège, Belgium.    IOANA LANCRANJAN • Department of Clinical Research, Sandoz Ltd., Basel, Switzerland.

apeptide and its neurophysin, a given neuron making only one nonapeptide–neurophysin pair. Gainer et al. (1977) have shown that vasopressin and its respective neurophysin are manufactured from a common precursor, a glycopeptide with a molecular weight of 20,000. The hypothesis—one hormone, one neurophysin, one cell—gained support when Burford et al. (1971) and Sunde and Sokol (1975) demonstrated biochemically the absence of the neurophysin, which is the carrier of vasopressin in the brain of Brattleboro rats homozygous for diabetes insipidus. The synthesis of the precursor occurs on ribosomes in the perikarya of neurons, then the neuropeptides and neurophysins are incorporated in neurosecretory granules in the Golgi zone and transported through the axoplasm by fast axoplasmic flow (Sachs et al., 1971; McKelvy et al., 1980) to terminals projecting to:

1. The posterior pituitary, where vasopressin is stored and released into the general circulation.
2. The organum vasculosum of the lamina terminalis and the third ventricle.
3. The external zone of the median eminence in contact with the portal vessels which reach the anterior pituitary (only fibers originating in the PV nucleus).

The release of the nonapeptides and their neurophysins by exocytosis is a $Ca^{2+}$ entry-dependent process (McKelvy et al., 1980).

Berl et al. (1973) have shown that actin-like molecules can be found in nerve terminal-contining subcellular fractions from whole brain, with the implication that neurohormone release may occur by a contractile process. McKelvy and Epelbaum (1977) proposed that $Ca^{2+}$ promotes this contractile process by inhibiting "neurosecretory granule" membrane myosin phosphorylation and stimulating troponin–tropomyosin interaction, but further studies are necessary to substantiate the "contractile" mechanism for neurohormone release. Several groups demonstrated the concomitant peripheral release of the nonapeptide and its neurophysin (Robinson, 1978) triggered by physiologic stimuli such as suckling for oxytocin and increased osmolarity or decrease blood pressure for vasopressin. Besides the release in the peripheral blood stream, indirect evidence has been given that vasopressin is released also in the pituitary portal blood (Oliver et al., 1977; Zimmerman et al., 1973) where it might work as a corticotropin-releasing factor (Stillman et al., 1977). Vasopressin is released also in the CSF and most probably in various regions of the CNS (Zimmerman, 1981).

Indeed, it has been reported that in man and in the rat, vasopressin and oxytocin fibers from magnocellular neurons are also projected via the stria terminalis to the central nucleus of the amygdala, while descending fibers are distributed to the nucleus tractus solitarius and the dorsal nucleus of the vagus in the medulla oblongata to the substantia gelatinosa of the central gray, and to the lateral parts of the spinal cord. In addition, vasopressin fibers originating in the parvocellular neurons of the suprachiasmatic nucleus are directed to the lateral septum, lateral habenulae, posterior hypothalamus, interpeduncular nucleus, mediodorsal thalamus, and periventricular gray of the brain stem. Also, fine fibers are present in the medial amygdala, ventral hippocampus, arcuate nucleus, organum vasculosum lamina terminalis, subfornical organ, and the pineal gland (Buijs et al., 1978; Dierickx and Vandersande, 1977; Nilaver et al., 1980; Weindl and Sofroniew, 1980).

The wide distribution of vasopressinergic fibers in the brain and their connection with the limbic system strongly suggest that vasopressin could have important neuromodulatory and behavioral effects in mammals. Indeed, a great deal of evidence has accumulated supporting the role of vasopressin in learning and memory processes in rats (de Wied, 1965; de Wied et al., 1977; van Wimersma Greidanus and Versteeg, 1982).

## 3. BEHAVIORAL EFFECTS OF VASOPRESSIN

de Wied was the first to report that the extinction of a conditioned avoidance response in rats is markedly accelerated after the removal of the posterior lobe of the pituitary. Subsequent studies have shown that a posterior pituitary extract or vasopressin restores the disturbed extinction process (de Wied, 1965; de Wied et al., 1977).

In intact rats, the s.c. injection of a posterior pituitary extract preparation or lysine-vasopressin (LVP) had long-term effects on the extinction of the conditioned avoidance response. Studies with vasopressin analogs have shown that the behavioral effects of vasopressin are independent of its endocrine action. Indeed, desglycinamide-lysine-8-vasopressin (DGAVP), which has almost no pressor or antidiuretic activity, is still centrally active, its central potency being 50% of that of arginine-vasopressin.

Administered i.v., LVP increased also avoidance latencies and retention in passive avoidance behavior studies. Furthermore, the fact that the intracerebroventricular (i.c.v.) administration of vasopressin inhibited the extinction of the pole-jumping avoidance response at doses 200-fold less than were needed for its systemic administration suggests that vasopressin may be physiologically involved in memory processes (de Wied et al., 1977).

The intraventricular administration of vasopressin antiserum after a learning trial leads to a deficit in passive avoidance retention whereas the peripheral administration of large amounts of vasopressin antiserum has no effects on behavior, which also supports the physiological implication of vasopressin in memory processes.

Further evidence for the role of vasopressin in memory processes was given by studies showing that the hormone protects against puromycin-induced memory loss, CO and electroshock-induced amnesia (van Wimersma Greidanus and Versteeg, 1982).

Therefore, a large body of convincing evidence has accumulated showing that vasopressin is involved in memory processes in experimental animals.

On the contrary, the evidence that vasopressin has a physiological role in modulating learning and memory processes in humans is still missing. Indicative of such a role are data showing a decrease of neurophysin levels beyond the age of 50 in man and of the vasopressin response to the water deprivation test in old age (Legros, 1975), and a relationship between circulating neurophysins and some psychometric memory tests (item 7 of Rey's PRM) (Legros and Gilot, 1979). Moreover, Gilot et al. (1979) and de Wied (personal communication) reported that patients with hereditary diabetes insipidus have a deficit of attention and memory which may be improved after treatment with vasopressin. Laszlo et al. (1981) reported also that attention and short- and long-term memories are poorer in patients with diabetes insipidus in comparison with healthy controls. The 7-day DGAVP 32.5 µg/day nasal spray treatment improved significantly the attention and short- and long-term memory in both patients and controls.

On the other hand, vasopressin levels in the spinal fluid of 10 elderly normal subjects, 9 patients with multi-infarct dementia and 5 patients with Alzheimer's type of dementia, were all in the same range (Lancranjan et al., unpublished data). The lack of decrease with age of vasopressin levels in the spinal fluid was reported by Jenkins et al. (1980) and Luerssen and Robertson (1980). Furthermore, the vasopressin content of postmortem human brains of patients with Alzheimer's type of dementia failed to show statistically significant differences in comparison with controls matched for age, in spite of an average 33% decrease of vasopressin levels in the hypothalamus, locus coeruleus, substantia nigra—pars compacta and reticulata—and globus pallidus (Rossor et al., 1980).

Despite the fact that the physiological role of vasopressin in the control of memory processes in man has not yet been fully established, important evidence has accumulated showing that vasopressin has pharmacological central effects in humans. These effects were described in (1) electrophysiological studies carried out in normal volunteers using the spontaneous EEG and the contingent negative variation (CVN) methods and (2) clinical therapeutical studies carried out in elderly subjects and in patients with either memory disturbances or psychiatric diseases.

## 4. CENTRAL EFFECTS OF VASOPRESSIN IN HUMANS

### 4.1. Electrophysiological Studies in Normal Volunteers

#### 4.1.1. Influence of LVP on the Spontaneous EEG in Man

Timsit-Berthier et al. (1978a) have reported the effects of intramuscular (i.m.) and intranasal administration of LVP on the spontaneous EEG of young healthy volunteers. A dose of 10 IU i.m. decreased the alpha activity and increased the delta activity 1–2 hr after LVP injection. This effect lasted for 4–6 hr and was accompanied by clinical sedation. On the other hand 5.4 IU

LVP nasal spray did not influence the EEG patterns, but three out of four subjects reported a "feeling of elation and well-being." In a cross-over, randomized placebo-controlled study in eight normal volunteers, Matejcek (personal communication) confirmed the effects of LVP on the spontaneous EEG. Indeed, after placebo and 1 IU LVP i.m., no effect on EEG has been noticed, whereas after 2 and 4 IU LVP i.m., a dose–response effect has been recorded. This effect—a reduction of the alpha activity and an increase of the delta and theta activity—was significant after 4 hr and lasted for up to 6 hr after the administration of LVP. On the other hand, 7.5, 15, and 30 IU LVP administered intranasally showed only a late effect, namely an increase in the alpha activity which was statistically significant only 6 and 8 hr after the administration of 15 and 30 IU LVP.

### 4.1.2. Influence of LVP on the CNV

The CNV is considered as a sensitive and reliable indicator of the central effects of drugs with stimulant and depressant actions. It is a slow cortical potential recorded at the vertex, extracted from the spontaneous EEG by averaging several EEG fragments in the same phase in relation to the stimuli. The CNV is obtained by warning the subjects, by way of a first, conditioned stimulus (S1), of the arrival of a second, unconditioned stimulus (S2) to which he must give a response. Between S1 and S2 there is a negative slow potential, the CNV, which is developed in the frontal lobes of the brain and stopped by the occurrence of S2.

In the studies carried out by Timsit-Berthier *et al.* (1978b) the EEG was recorded for 30 min with the subject's eyes closed and 100 pairs of stimuli were delivered; S1 was a brief tone, and S2 a long tone that the subject had to stop by pressing a button. The interval between the pair of stimuli varied at random from 10 to 20 sec. Each CNV was obtained by averaging 12 artifact-free vertex EEG fragments in phase-relation to 12 successive pairs of stimuli. So, for each subject a set of six to eight successive CNV was collected and analyzed. It was thus possible to appreciate the evolution of CNV in time. Fifteen minutes after placebo administration a progressive decrease of CNV amplitude was noticed, whereas the administration of LVP induced a resistance to this habituation phenomenon and the amplitude of CNV remained constant. Further evidence for the central effects of LVP has been given by Devos *et al.* (personal communication) also using the CNV recording. In a cross-over, double-blind study with placebo and a single 15-IU dose of LVP administered intranasally in 10 young healthy male volunteers, a significant increase of the wave P 300, 2 hr and of the CNV 6 hr after the administration of LVP has been registered. Recent studies carried out by Timsit-Berthier (personal communication) confirmed these consistent long-term effects of LVP and, moreover, showed that the effect on CNV pattern of a single intranasal dose of 15 IU LVP lasted 1 week.

## 4.2. Influence of LVP on Memory in Healthy Man

The decrease of neurophysins levels beyond the age of 50 in man prompted Legros et al. (1978) to carry out a double-blind study in 23 male patients (aged 50–65 years) with no metabolic, cardiovascular, or psychiatric diseases. In a randomized order 12 men received approximately 16 IU LVP intranasally per day (divided in three doses) and 11 men received placebo for 3 days. Besides clinical (weight, pulse rate, blood pressure) and biochemical investigation (urinary volume, serum and urinary osmolarity, blood urea nitrogen, protein and sodium plasma levels) a psychological investigation testing mood, attention (KT Attention Test, Wais Digit Symbol Test) and memory (PRM, 15 words, 30 figures, complex figure of Rey) was carried out before and after 3 days' treatment with LVP/placebo. No statistically significant changes were noted in the clinical, biological, and mood evaluation. However, the psychometric tests revealed statistically significant differences (Median-Test-21 statistics) between the two groups of subjects. The subjects receiving LVP performed better in tests of attention, concentration, and motor rapidity (KT Attention Test and Wais Digit Symbol Test) and in memory tests using visual graphic material for the measurement of visual retention (30 figures of Rey), recognition (subtests 2, 5, 3, and 4 of the PRM of Rey) and immediate and delayed free recall (subtests 6 and 7 of the PRM and complex figure of Rey). Using audioverbal material, it has been found that LVP improved attention and immediate memory (Wais Digit Span) as well as learning and recognition (15 words of Rey).

In another double-blind study made in a group of 20 men aged 80 ± 4 years tested with either LVP (16 IU for 2 weeks, 13 cases) or placebo (8 cases) we failed to find such a significant improvement of psychometric tests (Legros et al., 1980). However, a clinical impression of well-being was clearly noticed by the medical staff (in blind condition) in 5 out of the 12 patients receiving LVP as early as at the third day of the treatment whereas such an improvement was never noted in the placebo group.

In six young volunteers, Weingartner et al. (1981a) found in a double-blind, placebo cross-over study an improvement of serial learning, free recall, and recall of semantically related words after 2- to 3-week treatment with DDAVP 30–60 µg daily. Similarly, Laszlo et al. (1981) in a no-control study reported that in healthy volunteers a 7-week treatment with DGAVP 32.5 µg/day was followed by an increase of attention and short- and long-term memory.

## 4.3. Influence of LVP in Amnesic Patients

Oliveros et al. (1978) were the first to report the therapeutic effects of LVP administered intranasally in four patients with posttraumatic retrograde amnesia. Negative results in two patients (treated with LVP 16 IU for 2 weeks) and positive results in another patient (43 years old) with Korsakoff psychosis treated with LVP 22 IU as nasal spray were subsequently reported (Le Boeuf et al., 1978). Recently, the therapeutic effect of LVP in posttraumatic amnesia was confirmed in another open study in six patients (Timsit-Berthier et al., 1980). An improvement of memory and mood was noticed in five of the six patients treated with LVP 15 IU intranasally for 15 days. The effects occurred either during the treatment or 1-2 months after the end of the treatment. The effect lasted for the 6-month observation period. The only case that did not show an improvement was a woman suffering from anteroretrograde amnesia which occurred 8 years previously after a brain trauma. Moreover, positive results were also obtained in five patients in a double-blind cross-over placebo-controlled study carried out in seven patients with posttraumatic amnesia. All patients were treated with LVP 14 IU daily for 15 days and five out of six patients improved after LVP.

Jenkins et al. (1981) reported negative results in five patients with posttraumatic amnesia treated with either DDAVP or DGAVP whereas Drago et al. (1981) reported a significant improvement of memory, mainly of retrieval in six patients (three depressed, two patients with Korsakoff syndrome, and one with posttraumatic anterograde amnesia) in a placebo-controlled study with LVP (25 IU daily for 2 weeks). The administration of DDAVP has been shown to improve the ability of patients with Lesch-Nyhan syndrome to learn a passive avoidance response (Anderson et al., 1979) and the psychological status of children under DDAVP therapy for central diabetes insipidus. Based on the analysis of a 61-item questionnaire, Waggoner et al. (1978) concluded that DDVAP treatment "affords the child an enhanced energy level and out-look on life and permits the child freer use of imaginative and creative abilities."

## 5. INFLUENCE OF LVP TREATMENT IN PSYCHIATRIC PATIENTS

### 5.1. Schizophrenia

As early as 1952, Forizs published remarkable clinical results in schizophrenic patients treated with pitressin 10 IU every 24 or 48 hr for 6 to 12 months. In two-thirds of the 62 patients treated, most of them with chronic residual schizophrenia, there was a notable effect and in one-third of the patients the improvement was spectacular. The following beneficial effects have been described: increased sleeping time, reduced psychomotor activity and better adequacy of the emotions, increased awareness of the environment, disappearance of ward problems, and the patients showing more interest in their families and more insight. More recently, Vranckx et al. (1979) in an open clinical trial have reported moderate to marked improvement in 8 out of 16 schizophrenic patients treated for 28 days with 7.5 to 45 IU LPV nasal spray daily. Fourteen of the patients had chronic schizophrenia with deterioration and the other two had chronic schizophrenia with delusions and hallucinations.

A statistically significant change in the following items was noticed: somatic concern anxiety, emotional withdrawal, depressive mood, uncooperativeness, motor retardation, and blunted affect. There was a tendency for the item "excitement" to deteriorate and this was related to discontinuation of the drug in one case, because of insomnia. The effects observed took place very rapidly, i.e., from days 3 and 7 of treatment, reaching a plateau on day 14. The final global assessment reveals highly positive results in 3 cases, positively in 5, inadequate in 6, and negative in 2. Compared to previous treatment, LVP was noted superior in 8 cases. Twelve of the 16 patients had an unsatisfactory response to their previous treatment: in 5 out of these 12 cases LVP produced moderate to marked therapeutic response. In 4 patients the improvement observed as sufficient to allow discharge from hospital.

Similar results were reported in another open study (24 patients) by Alterwain (personal communication) and by Korsgaard et al. (1980) and Nuyts (personal data) in double-blind placebo-controlled studies.

Korsgaard et al. (1980) reported that the thinking disorder factor of the BPRS was also significantly decreased after 3 weeks of LVP treatment (22.5–67.5 IU/day) in a double-blind cross-over study in 16 chronic anergic schizophrenic patients. A decrease of the BPRS anergic factor was also found after 2 and 3 weeks of LVP treatment. On the other hand, no effect on the depression factor of BPRS, on the BPRS total, and on tardive dyskinesia was recorded by Korsgaard et al. (1980).

In conclusion, the studies carried out with LVP in schizophrenic patients showed that treatment with vasopressin can improve mainly those symptoms of schizophrenia which are not influenced or are aggravated by neuroleptics, namely emotional withdrawal, anergia, and blunted affect.

## 5.2. Depression

Therapeutic effects of vasopressin were reported in depression by Gold et al. (1979). Two out of four patients with primary affective disorder previously refractory to at least two "classical" antidepressive treatments improved after DDAVP treatment (Gold et al., 1978, 1979; Weingartner et al., 1981a).

Three of these four patients also demonstrated a significant cognitive enhancement beginning 2 days and continuing for 2 weeks during DDAVP treatment and measured by recall tasks. Moreover, the retrograde amnesia after ECT was substantially reversed in two depressed patients treated with DDAVP 40–60 µg daily, as indicated by the threefold increase in the recall of items (Weingartner et al., 1981a). The same group reported that the concentrations of vasopressin in the CSF were significantly lower in drug bipolar patient in the depressed compared with the manic phase. Moreover, CSF vasopressin levels were lower in bipolar depressed patients compared to normal controls ($p < 0.1$). A substantial lower release of vasopressin was also reported after hypertonic saline infusion in three depressed patients compared with mania. Vranckx (personal data) was unable to confirm the therapeutic efficacy of vasopressin (LVP) in severely depressed patients refractory to the classical treatment. Therapeutic effects were, however, recorded in a few patients with mode rate endogenous depression treated with 7.5–22.5 IU LVP daily (Vranckx and Benghezal, unpublished data).

### 5.3. Opiate Addiction

Preliminary results showed that vasopressin (DGAVP) facilitates the methadone detoxification of heroin addicts (van Beek-Verbeek et al., 1979) and significantly reduces the heroin consumption in six of nine toxicomanic patients investigated in an open study with LVP (Kryspin-Exner, personal communication).

### 5.4. Senile Dementia

#### *5.4.1. Senile Dementia of Alzheimer's Type*

Mainly negative results have been reported so far in patients with Alzheimer's type of dementia treated with LVP (7.7 or 22.5 IU daily for 2–4 weeks) or DDAVP (30 µg intranasally daily) (Chase et al., 1981; Tinklenberg et al., 1981, 1982).

Therapeutic transient effects were, however, noticed in some Alzheimer's patients (Agnoli, personal communication; Nishimura, personal communication; Adolfsson et al., personal communication) treated with either LVP or pitressin. Delwaide et al. (1980) reported better results of the Rey Auditory Verbal Learning Test in senile patients after single application of LVP (15 IU as nasal spray) than after piracetam (2000 mg i.v.) given randomly in a crossover design. Moreover, Weingartner et al. (1981a) reported that patients with mild senile dementia treated in a cross-over placebo/DDAVP study showed cognitive enhancement by facilitating access to semantic memory structures after DDAVP administration.

Finally, data collected in a multiple cross-over study in nine patients with a mild to moderate mental impairment due to Alzheimer's disease showed a consistent positive effect of LVP treatment (15 IU daily for 7 days). The best improvements were observed in the verbal associative memory test and the visual recognition memory test and consistent trends indicative for positive cognitive effects were shown by the verbal memory retrieval and the simple motor speed tests (Ferris and Reisber, 1981).

#### *5.4.2. Multi-infarct Dementia*

A mild but significant improvement of memory (retrieval and recall failure) and social behavior was reported in 5 of 11 LVP-treated patients in a placebo, double-blind controlled study (Bucht et al., 1983). The diagnosis of multi-infarct dementia was based on the history of either hypertension, transitory ischemic attack, or stroke, a stepwise mental deterioration, and an ischemic score of 7 score of 7 on the Hachinski score; focal neurological symptoms and CT-scan and EEG abnormalities.

The treatment with LVP 15 IU twice daily for 3 weeks was followed by an improvement of the social behavior in 8 of 11 LVP-treated patients. The patients became more active, mentally more alert, and socially better integrated. The relatives observed an improvement in the daily life activities of the patients. Two of nine patients treated with placebo were also clinically improved at the end of the 3-week treatment period. A significant improvement of observed items on the BPRS (scale) of disorientation and long-term memory was also found. Retrieval and recall failure on the Fuld object memory test were also significantly improved in the LVP-treated patients.

### 5.4.3. Dementia in Parkinsonian Patients

A "significant improvement in cerebral function" was observed also in 5 of 20 parkinsonian patients with incipient dementia in an open study carried out by Rinne (personal communication). The patients felt more lively and memory function improved. In the Philadelphia Geriatric Center Mental Status Questionnaire (PGCMQ) a better score was obtained after 4-week treatment with LVP in 1 patient with on–off phenomenon and dementia and in 4 cases with Parkinson's disease and incipient dementia.

After withdrawal of LVP treatment the "cerebral activation" decreased in all five patients and very dramatically so in two of them.

All these preliminary results reported in patients with various types of dementia need confirmation in large double-blind controlled studies.

## 6. CENTRAL EFFECT OF VASOPRESSIN IN MAN: MECHANISMS OF ACTION

Experimental and clinical data suggest that vasopressin exerts behavioral and cognitive effects in man such as:

1. Stimulation of attention, concentration, and memory in young and elderly normal volunteers as well as in patients with diabetes insipidus.
2. Improvement of retrograde and anterograde amnesia in some patients with posttraumatic amnesia as well as in some patients with Korsakoff psychosis.
3. Slight improvement of memory (or memory tests) and social behavior in some demented patients.
4. Antianergic effects and improvement of the social behavior in schizophrenic patients.
5. Antidepressant effects in endogenous depressed patients and, finally, facilitation of detoxification of opiate-addicted patients.

The major criticism of the clinical data published so far is that most of the findings are not consistent, possibly due to the heterogeneity of the patients included in the studies and to the lack of objective criteria for selecting the presumed responders.

The EEG and CVN data support, on the other hand, the fact that vasopressin has central effects in man. Vasopressin's mechanisms of action within the CNS in man are still unknown. In the rat, vasopressin has a major role in memory consolidation and retrieval, and central noradrenergic pathways have been implicated in mediating its actions on learning and memory (Kovács et al., 1979; Tanaka et al., 1977; van Wimersma Greidanus and Versteeg, 1982).

The improvement of attention and memory in patients with diabetes insipidus treated with vasopressin might be due to the replacement of their hormonal insufficiency, although it is not yet proven that patients with diabetes insipidus have a central vasopressinergic deficit. It has been reported that depressed patients have a functional vasopressin insufficiency to osmotic stimulation (Weingartner et al., 1981a); thus, neither a correlation between the degree of vasopressin insufficiency and depression nor a relation between this deficit and the clinical improvement after vasopressin therapy has been established so far.

## 6.1. Vasopressin: Modulation of Neurotransmitter Turnover in Specific Brain Nuclei

Extrapolating the animal data one might postulate that the structures concerned with noradrenergic neurotransmission could have an important role in the effect of vasopressin on memory and behavior in man. Indeed, Kovács et al. (1979) and Tanaka et al. (1977) showed that in the rat peripheral administration of LVP and i.c.v. microinjection of arginine-vasopressin facilitate the increase of the turnover rate of norepinephrine (NE) in some limbic midbrain and lower brain stem nuclei, such as the dorsal septal nucleus, anterior hypothalamic nucleus, medial forebrain bundle, parafascicular nucleus, dorsal raphe nucleus, locus coeruleus, nucleus ruber, nucleus tractus solitarii, and the Al region. It has previously been shown that the following structures are the most probably involved in the consolidation of memory and the retrieval of stored information: the rostral septal area, the region of the parafascicular nucleus of the thalamus, the dorsal hippocampus, and the locus coeruleus (Kovács et al., 1979). In three of these regions, the dorsal septal nucleus, the parafascicular nucleus, and the locus coeruleus, NE turnover is enhanced by i.c.v. administration. Data recorded in Brattleboro homozygous rats showed that these rats have an altered brain catecholamine turnover, in a direction opposite to that of the changes induced by the i.c.v. administration of vasopressin. Furthermore, the i.c.v. administration of an antiserum antivasopressin induces a decrease of the NE turnover in the dorsal septal nucleus, the parafascicular nucleus, and the rostral part of the nucleus tractus solitarii. A further support that the effect of vasopressin on memory consolidation is mediated by the NE system was given by the work of Kovács et al. (1979) who showed that the destruction of the ascending dorsal NE bundle by microinjection of 6-hydroxydopamine abolished the effect on memory consolidation in rats. Moreover, it has been found that bilateral injection of small amounts of vasopressin (25 pg) in regions containing terminals of the coeruleo-telencephalic noradrenergic system, viz. the dorsal septal nucleus, the dentate gyrus of the hippocampus, and the dorsal raphe nucleus, caused a facilitation of memory consolidation, whereas microinjections in the central amygdaloid nucleus and the locus coeruleus, which contain the cell bodies of this sytem, failed to do so. In contrast with this well-documented evidence in animals supporting the modulatory effect of vasopressin on the NE structures in the limbic system, no data on this subject are available in man. The fact that the NE metabolite MHPG in the CSF of multi-infarct demented patients did not change after 3-week treatment with LVP (Adolfsson et al., 1983) cannot simply rule out the possible increase of NE turnover in specific structures of the limbic system of man following vasopressin treatment.

### 6.2. Vasopressin and ACTH Secretion

The fact that vasopressin releases ACTH after i.m. or i.v. injection (Krieger and Luria, 1977) and the ACTH has central effects in animals and man raises the question whether or not the central effects of vasopressin are mediated by the release of ACTH. In a cross-over study carried out in two groups of six young male volunteers (Langranjan et al., unpublished data), placebo and three doses of LVP were administered either i.m. or intranasally and ACTH plasma levels were measured in the subsequent 2 hr. A dose–response ACTH-releasing effect was noticed after the i.m. administration of 1, 5, and 10 IU, whereas no effect on ACTH release was noticed after the intranasal administration of LVP (7.5, 15, and 30 IU). Moreover, a daily ACTH profile was made before and after 7-day treatment with LVP (15 IU intranasally daily) and showed that LVP had no effect on the ACTH pattern. Based on these data we ruled out the possibility that the central effect of LVP is mediated by the ACTH release. These conclusions are in agreement with de Wied and co-worker's data (1977) showing that ACTH and vasopressin have different central effects, ACTH being involved in attention, motivation, and retrieval and not in memory storage.

### 6.3. Vasopressin and the Circadian Secretion of Pituitary Hormones

Vasopressin is a main constituent of the suprachiasmatic nucleus, and increased evidence supports the idea that the suprachiasmatic nucleus is the site of the control of biological rhythms (Moore, 1978). Indeed, studies in adult and neonatal rats revealed the primary significance of the suprachiasmatic nucleus and the retinohypothalamic projection in the generation and maintenance

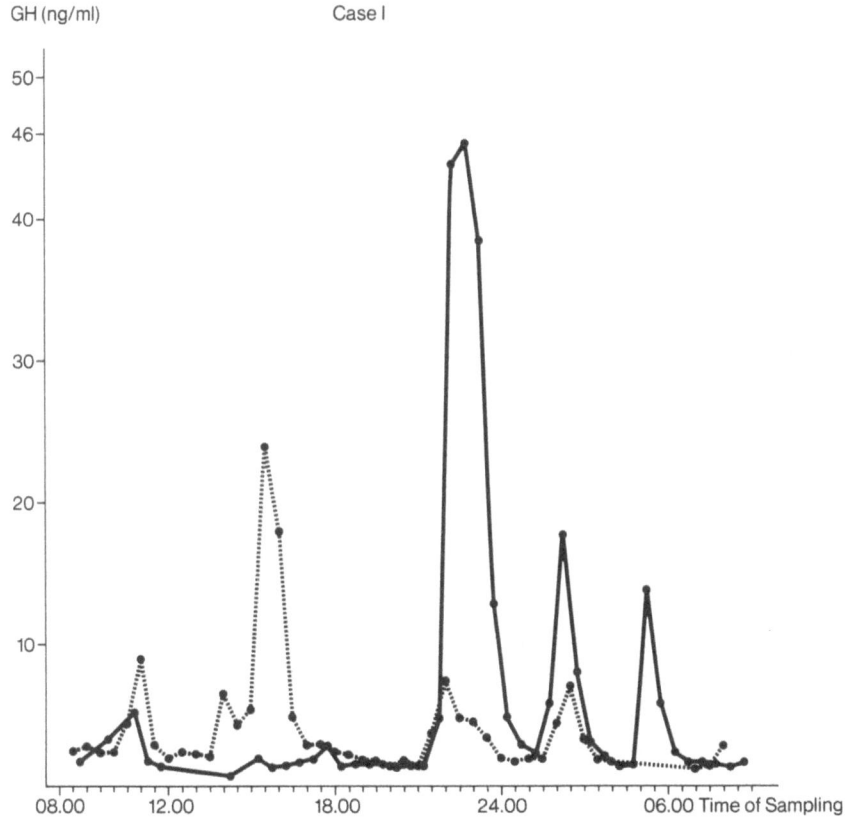

FIGURE 1. The growth hormone 24-hr profile in a schizophrenic patient (case I) before (●---●) and after (●—●) LVP treatment.

of circadian rhythms. It has recently been reported that total ablation of the suprachiasmatic nucleus in 2-day-old rats permanently abolishes circadian rhythms. On the other hand, it has been reported that deterioration in cognitive processes in schizophrenic patients is associated with a flattening of the diurnal rhythm of 17 ketosteroids or with a reversal of it. Moreover, there is a disappearance of the nocturnal GH peak in schizophrenic patients when their slow-wave sleep is absent "as it seems to happen rather frequently in schizophrenia" (Brambilla and Penati, 1978). Consequently we became interested to investigate the effect of LVP on the nyctohemeral rhythm of pituitary hormones in some patients treated with vasopressin. The nyctohemeral rhythm of GH and PRL secretion in two schizophrenic patients (Lancranjan et al., unpublished results) showed a normalization after 3-week treatment with LVP doses built up to 22.5 IU (Figs. 1–4).

The normalization of the GH profile might reflect the normalization of the sleep pattern (stage III–IV of sleep) in these schizophrenic patients treated

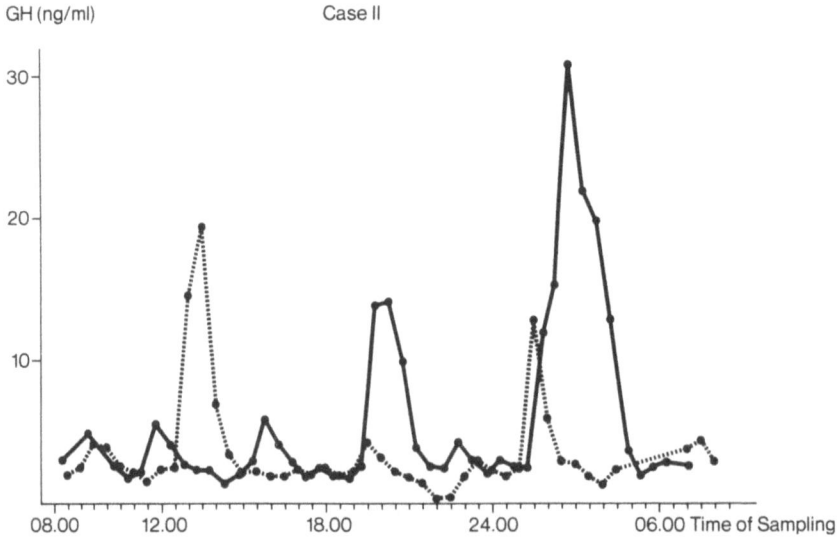

FIGURE 2. The growth hormone 24-hr profile in a schizophrenic patient (case II) before (●---●) and after (●—●) LVP treatment.

with LVP. It is a known fact that the nocturnal GH peak is related with the slow-wave sleep mainly during the first 2 hr. Animal data have also shown that the hippocampal theta activity in Brattleboro rats is significantly lower and could be temporarily normalized by the intraventricular injection of vasopressin (Urban and de Wied, 1978).

In the same patients the ACTH profile was normal before LVP treatment and the pattern did not change after treatment (Figs. 5, 6). On the other hand

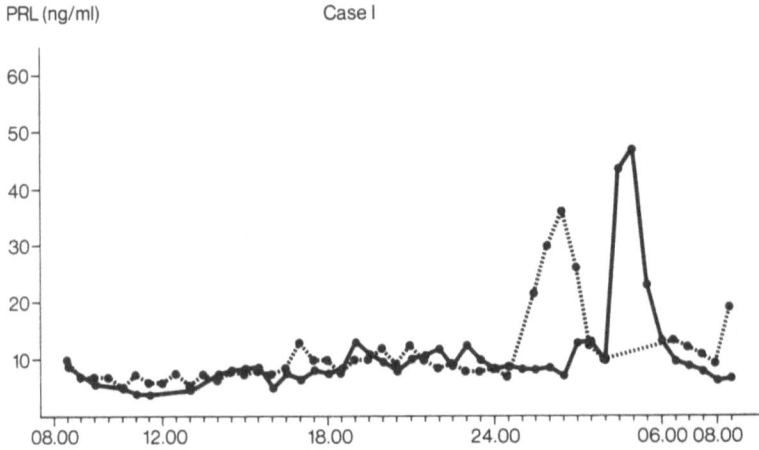

FIGURE 3. PRL 24-hr profile in a schizophrenic patient (case I) before (●---●) and after (●—●) LVP treatment.

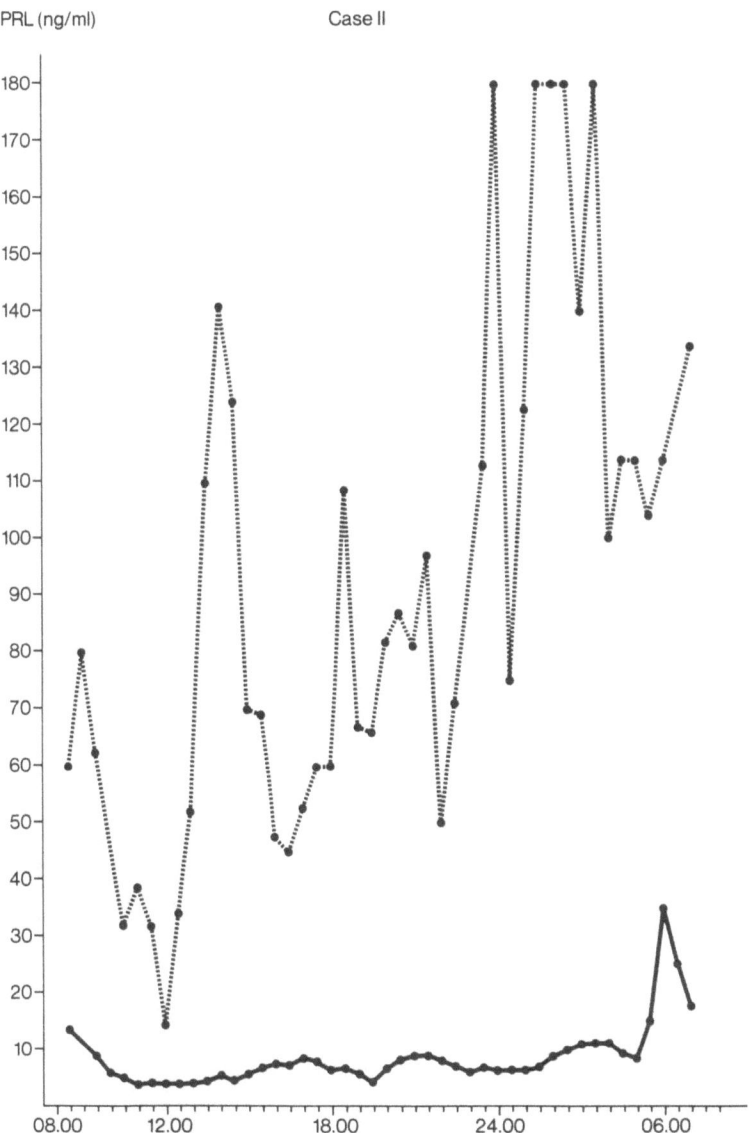

FIGURE 4. PRL 24-hr profile in a schizophrenic patient (case II) before (●---●) and after (●—●) LVP treatment.

the profile of neurophysins and plasma osmolality (Figs. 7, 8) showed a chaotic profile before LVP treatment and a normal aspect thereafter. Highly interesting is that these two schizophrenic patients improved also clinically during the LVP therapy.

The 24-hr profile of GH secretion before and after 3-week treatment with LVP was also investigated in a double-blind study carried out in patients with

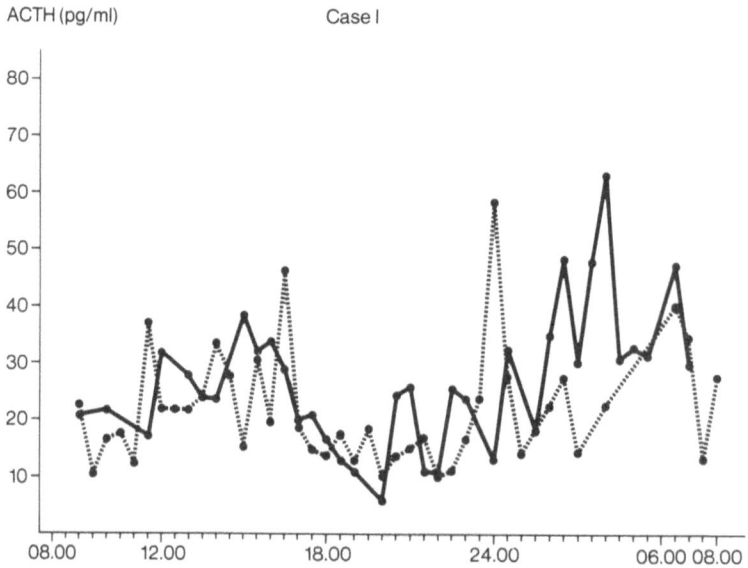

FIGURE 5. ACTH 24-hr profile in a schizophrenic patient (case I) before (●---●) and after (●—●) LVP treatment.

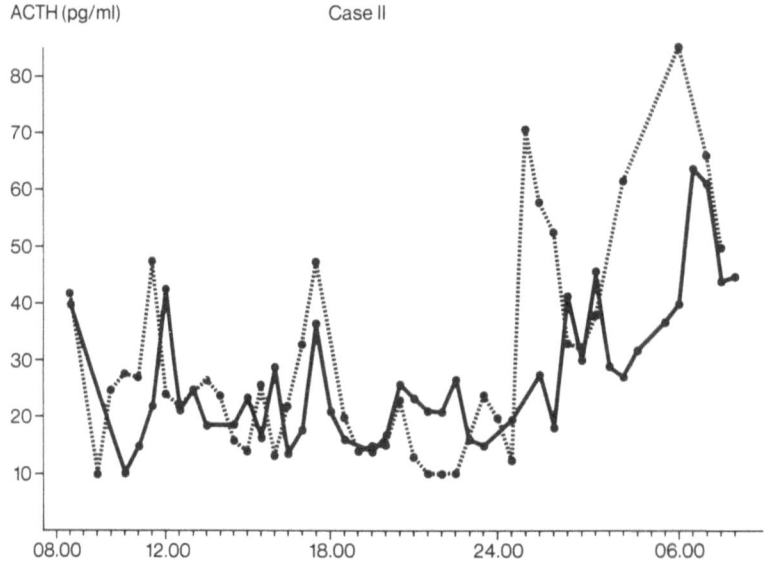

FIGURE 6. ACTH 24-hr profile in a schizophrenic patient (case II) before (●---●) and after (●—●) LVP treatment.

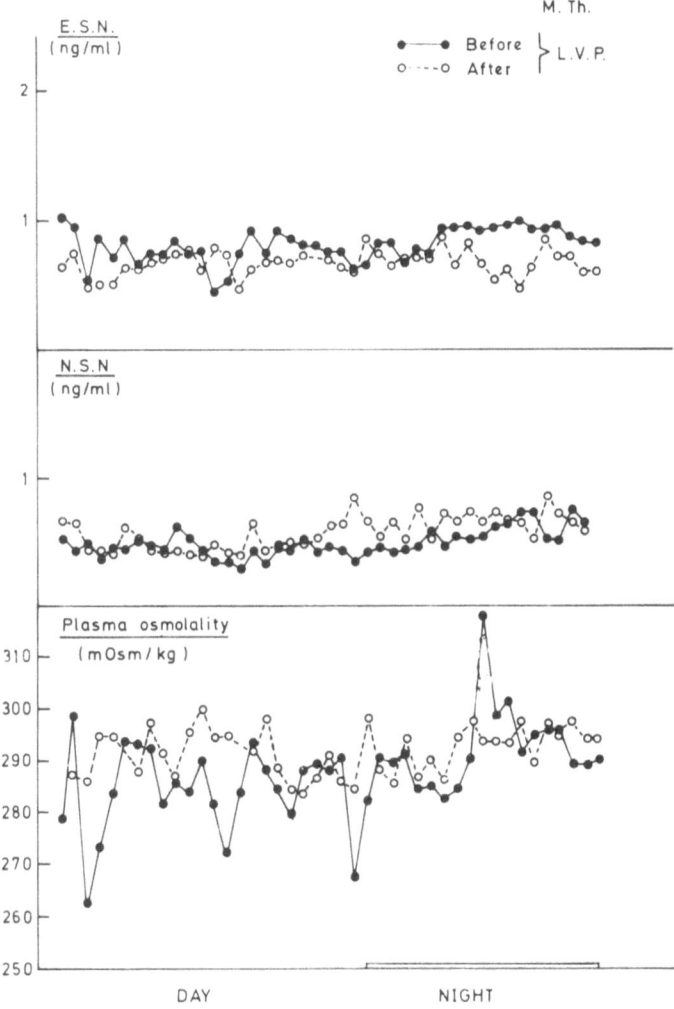

FIGURE 7. Estrogen-stimulated neurophysin (ESN), nicotine-stimulated neurophysin (NSN), and plasma osmolality in a schizophrenic patient (M.Th., case I) before (●—●) and after (○---○) LVP treatment.

multi-infarct dementia (Lancranjan and Adolfsson, unpublished data). Two out of six patients lacked and one patient had only a very small sleep-related GH peak before LVP/placebo treatment. The two patients treated with LVP had a normalization of the GH profile whereas after placebo treatment no changes occurred.

In spite of the interest of these observations, one cannot draw firm conclusions due to the paucity of these data. It seems, however, that vasopressin's central effects can be reflected also by a normalization of the circadian secretion of pituitary hormones.

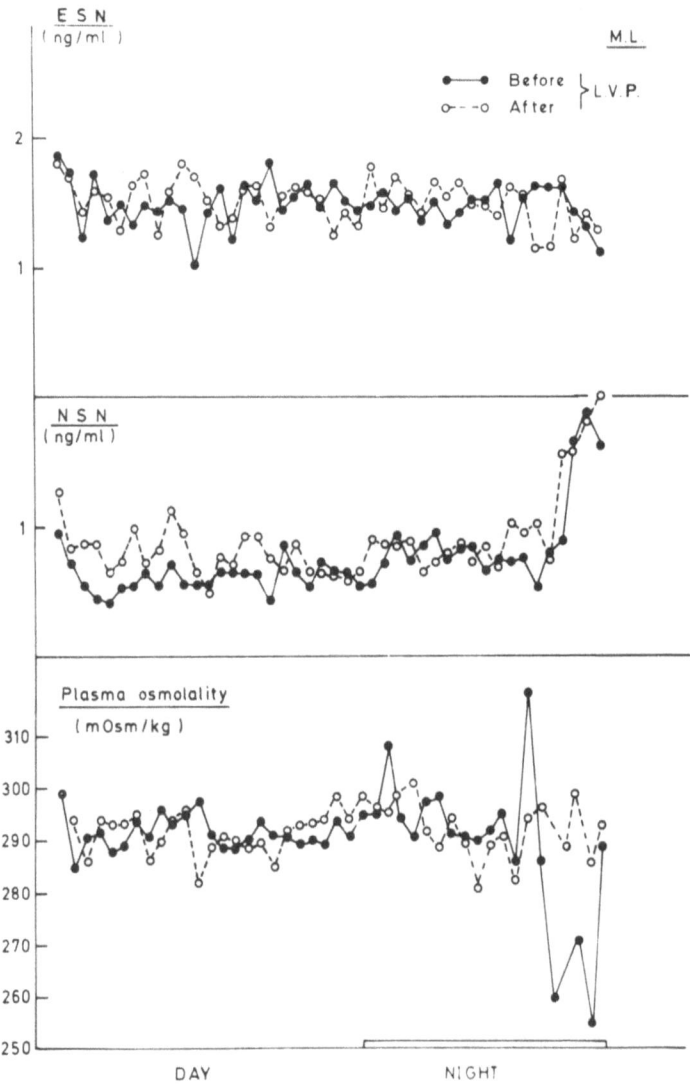

FIGURE 8. Estrogen-stimulated neurophysin (ESN), nicotine-stimulated neurophysin (NSN), and plasma osmolality in a schizophrenic patient (M.L., case II) before (●—●) and after (○---○) LVP treatment.

### 6.4. Vasopressin: The Natural Opioid Peptides

Recent data strongly suggest that pituitary opioid peptides and neurohypophyseal hormones may participate in a regulatory loop in which each affects the other's release. It has been shown that plasma ACTH and β-LPH rose in parallel in response to vasopressin administrated i.v. or i.m. in normal subjects and in patients with Cushing's disease and Nelson's syndrome (Krieger and Luria, 1977).

Moreover, Simantov and Snyder (1977) reported the presence of high concentrations of opiate receptors in membrane preparations of the posterior lobe of the pituitary, which might suggest a physiologic role of β-LPH and β-endorphin in the modulation of the posterior pituitary's secretion. Furthermore, Mata et al.(1977) reported that dehydrated rats, with increased oxytocin and vasopressin release, have significantly decreased levels of pituitary opioid peptides, probably owing to depletion of their endogenous stores. Finally, Rossier et al. (1979) found a close anatomical relationship between enkephalin fibers from the PV and SO nuclei and the neurointermediate pituitary lobe.

They found high amounts of enkephalin-like material in extracts of the neuropituitary enkephalin cell bodies in the PV and SO nuclei and enkephalin fibers in the pars nervosa of the pituitary. Section of the pituitary stalk produced the disappearance of enkephalin fibers in the posterior pituitary whereas lesions of the PV nucleus produced a decrease by only 40% of the enkephalin content in the pituitary. Furthermore, the fact that dehydration decreased also enkephalin concentrations supported the idea that enkephalin fibers modulate the release of vasopressin at the pituitary levels. Finally, the fact that there are opiate receptors in the pars nervosa suggests the enkephalin fibers control the secretion of vasopressin in rats (Rossier et al., 1979).

In order to assess whether or not there are any relationships between endogenous opioid peptides and vasopressin's treatment in humans, a pilot study was carried out in six hebephrenic schizophrenic patients. ACTH and β-LPH were measured in plasma and CSF before and after 3-week intranasal LVP treatment (Lancranjan and Vranckx, unpublished data). No important changes in ACTH and β-LPH plasma levels were found whereas a consistent increase of β-LPH CSF levels was noted (Table 1). These data might suggest that at least in schizophrenic patients, vasopressin might act by stimulating the central release of β-LPH (probably from neurons originating in the arcuate

TABLE 1
ACTH and β-LPH, CSF and Plasma Levels in Six Hebephrenic Schizophrenic Patients before (A) and after (B) LVP Treatment

| | | CSF | | | | Plasma | | | |
|---|---|---|---|---|---|---|---|---|---|
| | | ACTH (ng/l) | | β-LPH (ng/l) | | ACTH (ng/l) | | β-LPH (ng/l) | |
| Case | Age (years) | A | B | A | B | A | B | A | B |
| 1 | 29 | 108 | 86 | 9 | 69 | 35 | 65 | 136 | 32 |
| 2 | 22 | 71 | — | 54 | — | 22 | 14 | 90 | 77 |
| 3 | 27 | 63 | 43 | 55 | 184 | 16 | 29 | 173 | 73 |
| 4 | 28 | 115 | — | 21 | — | 39 | 14 | 2091 | 316 |
| 5 | 26 | 123 | 116 | 9 | 18 | 8 | 54 | 131 | 136 |
| 6 | 20 | 88 | 52 | 31 | 100 | 36 | 41 | 329 | 137 |

nucleus). Similar studies ought to be done in a larger number of patients with schizophrenia as well as in depressed and opiate-addicted patients.

### 6.5. Vasopressin: cAMP Formation

The effects of vasopressin on cAMP formation and prostaglandin synthesis in the kidney and the bladder are well documented. Orloff and Handler (1962) showed that cAMP mediates the antidiuretic action of vasopressin, and Bentley (1969) showed that vasopressin stimulates the adenylcyclase whereas valinomycin blocks both the adenylcyclase and the action of vasopressin on the bladder. Furthermore, Grantham and Burg (1966) reported that AVP stimulates also the adenylcyclase in the renal medulla. A number of investigators showed that cAMP activates a protein kinase which catalyzes the phosphorylation of membrane proteins which in turn regulate membrane permeability (Dousa *et al.*, 1977). Finally, Barker (1976) found that the iontophoretic application of vasopressin into snail neurons produced changes in the permeability of the neuronal membranes. Thus, it is likely that vasopressin, via synthesis of cAMP, can also change the permeability of neurons in the mammalian brain, but this aspect has not yet been investigated in relation to the central effects of vasopressin.

### 6.6. Vasopressin: Stimulation of Critical Enzymes in the Brain

Knowing that hormones exert metabolic effects by stimulating key enzymes, one may wonder whether vasopressin has selective effects on one of the critical enzymes involved in the synthesis or metabolism of neurotransmitters and/or on enzymes involved in RNA and protein synthesis. If one agrees with Rose and Haywood (1977) that the biochemical events involved in learning and memory are increased RNA polymerase, increased RNA and protein synthesis, and increased axonal flow to synapses, the effects of vasopressin on all these processes deserve further investigation.

### 6.7. Vasopressin: Modulation of the Phosphorylation of Myelin

Another new biochemical finding is that in aged rats the *in vivo* phosphorylation of myelin in the cortex and hippocampus is increased. The administration in both young and old rats of 0.05 IU LVP/rat is followed by a reduction of myelin phosphorylation *in vivo* in the cortex and hypothalamus (Hiestand, to be published). Additional experimental work is, however, required for correlating the phosphorylation of myelin with memory processes in mammals.

### 6.8. Vasopressin: Increased Permeability of the Blood–Brain Barrier

It has recently been reported that the i.c.v. administration of vasopressin specifically influences the brain permeability to water (and neurotransmitter precursors?) in primates (Raichle and Grubb, 1978). Whether or not this effect is correlated with the effect of vasopressin on memory processes is not known.

## 7. CONCLUSIONS

The wide distribution of vasopressinergic fibers in the brain and the well-documented behavioral effects of vasopressin in experimental animals suggest that vasopressin might be an important neuromodulator with multiple effects on brain functions.

Evidence has accumulated that vasopressin has cognitive and behavioral effects in man, but further controlled studies are required to ascertain its physiological role and the pharmacological effects of vasopressin in psychoneurology.

## 8. REFERENCES

Adolfsson, R., Bucht, G., Lancranjan, I., Said, S., Seif, S., Svennerholm, L., and Winblad, B., 1983, Endocrine and metabolic effects of vasopressin in patients with multiinfarct dementia, in: *Neuropeptide and Hormone Modulation of Brain Function and Homeostasis* (J. M. Orday, J. R. Sladek, and B. Reisberg, eds.), Raven Press, New York, in press.

Anderson, L. T., David, R., Bonnet, K., and Dancis, J., 1979, Passive avoidance learning in Leasch–Nyhan disease: Effect of 1-desamino-8-arginine-vasopressin, *Life Sci.* **24**:905.

Bargmann, W., and Scharrer, E., 1951, The site of origin of the hormones of the posterior pituitary, *Am. Sci.* **39**:255.

Barker, J. L., 1976, Peptides: Roles in neuronal excitability, *Physiol. Rev.* **56**:435.

Bentley, P. J., 1969, The effect of valinomycin on the toad bladder: Antagonism to vasopressin and aldosterone, *J. Endocrinol.* **45**:287.

Berl, S., Puszkin, S., Nicklas, W. J., 1973, Actomyosin-like protein may function in the release of transmitter material at synaptic endings. *Science* **179**:441.

Brambilla, F., and Penati, G., 1978, Schizophrenia: Endocrinological review, in: *Perspectives in Endocrine Psychobiology* (F. Brambilla, P. K. Bridges, E. Endroczi, and G. Heuser, eds.), pp. 309–421, Wiley, New York.

Bucht, G., Adolfsson, R., Lancranjan, I., and Winblad, B., 1983, Vasopressin in multiinfarct dementia and other neuropsychiatric disorders, in: *Neuropeptide and Hormone Modulation of Brain Function and Homeostasis* (J. M. Ordy, J. R. Sladek, and B. Reisberg, eds.), Raven Press, New York, in press.

Buijs, R. M., Swaab, D. F., Dogterom, J., and van Leeuwen, F. W., 1978, Intra- and extrahypothalamic vasopressin and oxytocin pathways in the rat, *Cell Tissue Res.* **186**:423.

Burford, G. D., Jones, C. W., and Pickering, B. T., 1971, Tenative identification of a vasopressin-neurophysin and an oxytocin neurophysin in the rat, *Biochem. J.* **124**:809.

Chase, T. N., Durso, R., Fedio, P., and Tamminga, C. A., 1981, Vasopressin treatment of cognitive deficits in Alzheimer's disease, in: *Alzheimer's Disease: A Report of Progress in Research* (S. Corkin, K. L. Davis, J. H. Growdon, E. Usdin, and R. J. Wurtman, eds.), Raven Press, New York.

Delwaide, P. J., Devoitille, J. M., and Ylieff, M., 1980, Acute effect of drugs upon memory of patients with senile dementia, *Acta Psychiatr. Belg.* **80**:748.

de Wied, D., 1965, The influence of the posterior and intermediate lobe of the pituitary and pituitary peptides on the maintenance of a conditioned avoidance response in rats, *Int. J. Neuropharmacol.* **4**:157.

de Wied, D., Bohus, B., Gispen, W. H., van Ree, J. M., Urban, I., and van Wimersma Greidanus, T. B., 1977, Neurohypophyseal hormones and behavior, in: *Neurohypophysis* (A. M. Moses and L. Share, eds.) pp. 201–210, Karger, Basel.
Dierickx, K., and Vandersande, F., 1977, Immunocytochemical localization of the vasopressinergic and the oxytocinergic nerves in the human hypothalamus, *Cell Tissue Res.* **184**:15.
Dousa, T. P., Barnes, L. D., and Kim, J. K., 1977, The role of the cyclic AMP-dependent protein phosphorylations and microtubules in the cellular action of vasopressin in mammalian kidney, in: *Neurohypophysis* (A. M. Moses and L. Share, eds.), pp. 220–235, Karger, Basel.
Drago, F., Rapisarda, V., Calandra, C., Filetti, S., and Scapagnini, U., 1981, A clinical evaluation of the effects of vasopressin on memory disorders, *Acta Ther.* **7**:345.
Ferris, S. H., and Reisber, B., 1981, Treating dementia with neuropeptides and piracetam, in: *Alzheimer's Disease: A Report of Progress in Research* (S. Corkin, K. L. Davis, J. H. Growdon, E. Usdin, and R. J. Wurtman, eds.), Raven Press, New York.
Forizs, L., 1952, The use of pitressin in the treatment of schizophrenia with deterioration, *N.C. Med. J.* **13**:76.
Gainer, H., Sarne, Y., and Brownstein, M. J., 1977, Neurophysin biosynthesis: Conversion of a putative precursor during axonal transport, *Science* **195**:1354.
Gilot, P., Crabbe, J., and Legros, J. J., 1979, Bilan mnesique chez 5 sujets presentant und diabète insipide, central idiopathique familial, *Acta. Psychiatr. Belg.* **80**:755.
Gold, P. W., Reuss, V. I., and Goodwin, F. K., 1978, Vasopressin in affective illness, *Lancet* **I**:1233.
Gold, P. W., Weingartner, H., Ballenger, J. C., Goodwin, F. K., and Post, R. M., 1979, Effects of 1-desamino-8-D-arginine vasopressin on behaviour and cognition in primary affective disorder, *Lancet* **II**:992.
Grantham, J. J., and Burg, M. B., 1966, Effect of vasopressin and cyclic AMP on permeability of isolated collecting tubules, *Am. J. Physiol.* **211**:255.
Guillemin, R., 1977, The expanding significance of hypothalamic peptides, or, is endocrinology a branch of neuroendocrinology?, *Recent Prog. Horm. Res.* **33**:1.
Jenkins, J. S., Mather, H. M., and Ang, V., 1980, Vasopressin in human cerebrospinal fluid, *J. Clin. Endocrinol. Metab.* **50**:364.
Jenkins, J. S., Mather, H. M., Coughlan, A. K., and Jenkins, D. G., 1981, Desmopressin and desglycinamide vasopressin in post-traumatic amnesia, *Lancet* **I**:39.
Korsgaard, S., Casey, D. E., Pacersen, N. E. D., Jorgensen, A., and Gerlach, G., 1980, Vasopressin in anergic schizophrenia, Commun. XIth Int. Congr. Int. Soc. Psychoneuroendocrinol., Florence.
Kovács, G. L., Bohus, B., Versteeg, D. H., de Kloet, E. R., and de Wied, D., 1979, Effect of oxytocin and vasopressin on memory consolidation: Sites of action and catecholaminergic correlates after local microinjection into limbic-midbrain structures, *Brain Res.* **175**:303.
Krieger, D. T., and Luria, M., 1977, Plasma ACTH and cortisol responses to TRF, vasopressin of hypoglycemia in Cushing's disease and Nelson's syndrome, *J. Clin. Endocrinol. Metab.* **44**:361.
Laszlo, F. A., Laczi, F., Valkusz, Z., Jardanhazy, T., Wagner, A., Szilard, J., and de Wied, D., 1981, Effects of desglycinamide-arginine-vasopressin (DGAVP) on the memory in diabetes insipidus patients, *Acta Endocrinol. Suppl.* **243**:97.
Le Boeuf, A., Lodge, J., and Eames, P. G., 1978, Vasopressin and memory in Korsakoff syndrome, *Lancet* **II**:1370.
Legros, J. J., 1975, The radioimmunoassay of human neurophysins: Contribution to the understanding of the physiopathology of neurohypophysial function, *Ann. N.Y. Acad. Sci.* **248**:281.
Legros, J. J., and Gilot, P., 1979, Vasopressin and memory in the human, in: *Brain Peptides: A New Endocrinology* (A. M. Gotto, Jr., E. J. Peck, Jr., and A. E. Boyd, III, eds.), pp. 347–364, Elsevier/North-Holland, Amsterdam.
Legros, J. J., Gilot, P., Seron, X., Claessens, J. J., Adam, A., Moeglen, J. M., Audibert, A., and Berchier, P., 1978, Influence of vasopressin on learning and memory, *Lancet* **I**:41.
Legros, J. J., Schmitz, S., Bruwier, M., Mantanus, H., and Timsit-Berthier, M., 1980, Neuro-

hypophyseal peptides and cognitive function: A clinical approach, in: *Progress in Psychoneuroendocrinology* (F. Brambilla, G. Racagni, and D. de Wied, eds.), pp. 325–337, Elsevier/North-Holland, Amsterdam.

Luerssen, T. G., and Robertson, G. L., 1980, Cerebrospinal fluid vasopressin and vasotocin in health, in: *Neurobiology of Cerebrospinal Fluid* (J. H. Wood, ed.), pp. 613–623, Plenum Press, New York.

McKelvy, J. F., and Epelbaum, J., 1977, Biosynthesis, packaging, transport and release of brain peptides, in: *The Hypothalamus* (S. Reichlin, R. Baldessarini, and J. Martin, eds.), pp. 195–211, Raven Press, New York.

McKelvy, J. F., Glasel, J. A., and Foreman, M., 1980, Biochemical aspects of hypothalamic function, in: *Handbook of the Hypothalamus: Physiology of the Hypothalamus* (P. J. Morgane, and J. Panksepp, eds.), pp. 1–61, Dekker, New York.

Mata, M. M., Gainer, H., and Klee, W. A., 1977, Effect of dehydration on the endogenous opiate content of the rat neuro-intermediate lobe, *Life Sci.* **21**:1159.

Moore, R. Y., 1978, Central neural control of circadian rhythms, in: *Frontiers in Neuroendocrinology*, Vol. 5 (W. F. Ganong and L. Martini, eds.), pp. 185–206, Raven Press, New York.

Nilaver, G., Zimmerman, E. A., Wilkins, J., Michaels, J., Hoffman, D., and Silverman, A.-J., 1980, Magnocellular hypothalamic projections to the lower brain stem and spinal cord of the rat: Immunocytochemical evidence for predominance of the oxytocin–neurophysin system compared to the vasopressin–neurophysin system, *Neuroendocrinology* **30**:150.

Oliver, C., Mical, R. S., and Porter, J. C., 1977, Hypothalamic–pituitary vasculature: evidence for retrograde blood flow in the pituitary stalk, *Endocrinology* **100**:598.

Oliveros, J. C., Jandali, M. K., Timsit-Berthier, M., Remy, R., Benghezal, A., Audibert, A., and Moeglen, J. M., 1978, Vasopressin in amnesia, *Lancet* **I**:42.

Orloff, J., and Handler, J. S., 1962, The similarity of effects of vasopressin, adenosine 3′-5′ phosphate (cyclic AMP) and theophylline on the toad bladder, *J. Clin Invest.* **41**:702.

Raichle, M. E., and Grubb, R. L., Jr., 1978, Regulation of brain water permeability by centrally-released vasopressin, *Brain Res.* **143**:191.

Robinson, A. G., 1978, Neurophysins, an aid to understanding the structure and function of the neurohypophysis, in: *Frontiers in Neuroendocrinology*, Vol. 5 (W. F. Ganong and L. Martini, eds.), pp. 35–59, Raven Press, New York.

Rose, S. P. R., and Haywood, J., 1977, Experience, learning and brain metabolism, in: *Biochemical Correlates of Brain Structure and Function* (A. N. Davison, ed.), pp. 249–292, Academic Press, New York.

Rossier, J., Battenberg, E., Pittman, Q., Bayon, A., Koda, L., Miller, R., Guillemin, R., and Bloom, F., 1979, Hypothalamic enkephalin neurones may regulate the neurohypophysis, *Nature (London)* **277**:653.

Rossor, M. N., Iversen, L. L., Mountjoy, C. Q., Roth, M., Hawthorn, J., Ang, V. Y., and Jenkins, J. S., 1980, Arginine vasopressin and choline acetyltransferase in brains of patients with Alzheimer type senile dementia, *Lancet* **II**:1367.

Sachs, H., Goodman, R., Osinchak, J., and McKelvy, J., 1971, Supraoptic neurosecretory neurons of the guinea pig in organ culture: Biosynthesis of vasopressin and neurophysin, *Proc. Natl. Acad. Sci. USA* **68**:2782.

Scharrer, E., and Scharrer, B., 1954, Hormones produced by neurosecretory cells, *Recent Prog. Horm. Res.* **10**:183.

Simantov, R., and Snyder, S. H., 1977, Opiate receptor binding in the pituitary gland, *Brain Res.* **124**:178.

Stillman, M. A., Recht, L. D., Rosario, S. L., Seif, S. M., Robinson, A. G., and Zimmerman, E. A., 1977, The effects of adrenalectomy and glucocorticoid replacement of vasopressin and vasopressin-neurophysin in the zona externa of the median eminence of the rat, *Endocrinology* **101**:42.

Sunde, D. A., and Sokol, H. W., 1975, Quantification of rat neurophysins by polyacrylamide gel electrophoresis (PAGE): Application to the rat with hereditary hypothalamic diabetes insipidus, *Ann. N.Y. Acad. Sci.* **248**:345.

Tanaka, M., de Kloet, E. R., de Wied, D., and Versteeg, D. H. G., 1977, Arginine$^8$-vasopressin affects cathecholamine metabolism in specific brain nuclei, *Life Sci.* **20**:1799.

Timsit-Berthier, M., Audibert, A., and Moeglen, J. M., 1978a, Influence de la lysine-vasopressine sur l'EEG chez l'homme: Resultats préliminaires, *Neuropsychobiology* **4**:129.

Timsit-Berthier, M., Audibert, A., and Moeglen, J. M., 1978b, Modification de l'EEG et de la VCN sous l'effect d'une hormone post-hypophysaire: la lysine vasopressine (LVP), IIème Congrès Mondial de psychiatrie Biologique, Barcelona, Abstract 438.

Timsit-Berthier, M., Mantanus, H., Jaques, M. C., and Legros, J. J., 1980, Use of lysine-vasopressine in the treatment of post-traumatic amnesia, *Acta Psychiatr. Belg.* **80**:353.

Tinklenberg, J. R., Pfefferbaum, A., and Berger, P. A., 1981, l-Desamino-D-arginine vasopressin (DDAVP) in cognitively impaired patients *Psychopharmacol. Bull.* **17**:206.

Tinklenberg, J. R., Pigache, R., Pfefferbaum, A., and Berger, P. A., 1982, Vasopressin peptides and dementia, in: *Alzheimer's Disease: A Report of Progress in Research* (S. Corkin, K. L. Davis, J. H. Growdon, E. Usdin, and R. J. Wurtmann, eds.), pp. 463–468, Raven Press, New York.

Urban, I., and de Wied, D., 1978, Neuropeptides: Effects on paradoxical sleep and theta rhythm in rats, *Pharmacol. Biochem. Behav.* **8**:51.

van Beek-Verbeek, G., Fraenkel, M., Geerlings, P. J., van Ree, J. M., and de Wied, D., 1979, Vasopressin analogue in methadone detoxification of heroin addicts, *Lancet* **II**:738.

van Wimersma Greidanus, T., and Versteeg, D. H. G., 1982, Neurohypophysial hormones: Their role in endocrine function and behavioral homeostasis, in: *Behavioral Neuroendocrinology* (C. B. Nemeroff and A. J. Dunn, eds.), Spectrum, New York.

Vranckx, C., Minne, P., Benghezal, A., Moeglen, J. M., and Audibert, A., 1979, Vasopressin and schizophrenia, in: *Biological Psychiatry Today* (J. Obiols, C. Ballus, E. Gonzales Monclus, and J. Pujol, eds.), pp. 753–758, Elsevier/North-Holland, Amsterdam.

Waggoner, R. W., Jr., Slonim, A. E., and Armstrong, S. H., 1978, Improved psychological status of children under DDAVP therapy for central diabetes insipidus, *Am. J. Psychiatry* **135**:361.

Weindl, A., and Sofroniew, M. V., 1980, Immunohistochemical localization of hypothalamic peptide hormones in neural target areas, Brain and Pituitary Peptides, Ferring Symposium, Munich, 1979, pp. 97–109, Karger, Basel.

Weingartner, H., Gold, P., Ballenger, J. C., Smallberg, S. A., Summers, R., Rubinow, D. R., Post, R. M., and Goodwin, F. K., 1981a, Effects ov vasopressin on human memory functions, *Science* **211**:601.

Weingartner, H., Kaye, W., Gold, P. H., Smollberg, S., Peterson, R., Gillin, J. C., and Ebert, M., 1981b, Vasopressin treatment of cognitive dysfunction in progressive dementia, *Life Sci.* **29**:2721.

Zimmerman, E. A., 1981, The organization of oxytocin and vasopressin pathways, in: *Neurosecretion and Brain Peptides* (J. B. Martin, S. Reichlin, and K. L. Bick, eds.), pp. 63–75, Raven Press, New York.

Zimmerman, E. A., Carmel, P. W., Husain, M. K., Ferin, M., Tannenbaum, M., Frantz, A. G., and Robinson, A. G., 1973, Vasopressin and neurophysin: High concentrations in monkey hypophyseal portal blood, *Science* **182**:925.

CHAPTER 14

# Endogenous Opiate Systems May Modulate Learning and Memory

GAYLE A. OLSON, RICHARD D. OLSON,
ABBA J. KASTIN, and DAVID H. COY

## 1. INTRODUCTION

It is becoming increasingly realized that the endogenous opiate peptides have numerous effects on a wide range of behaviors and physiological conditions. Some of these effects may reflect the multiple physiological roles these peptides might play other than the ones related to pain perception; these additional roles may, in fact, be of more importance to the organism than the antinociceptive function with which the peptides were originally associated. Learning and memory processes might represent important functions modulated by the peptides.

In this chapter we shall review the existing literature on the results of the opiate peptides and opiate antagonists that were injected before or after the learning process in an attempt to influence the acquisition and memory of the experience. The use of opiate antagonists to modulate learning might allow inferences about inhibition of the effects being produced by the endogenous opiate peptides. We shall look at the effects of the opiates and their antagonists on the acquisition process itself, using both appetitive and aversive conditioning, as well as their effects on retention and extinction of that which was learned. In addition, we shall review opiate modulation of brain areas involved in learning and shall attempt to assess the hypothesis that the opiate peptides are natural reward mediators and thus influence reinforcement. Finally, we shall speculate briefly on the possible role of the opiate peptides in the development of and treatment of senile dementia, a memory deficit of the aged.

---

GAYLE A. OLSON and RICHARD D. OLSON • Department of Psychology, University of New Orleans, New Orleans, Louisiana 70148.    ABBA J. KASTIN • Veterans Administration Medical Center and Tulane University School of Medicine, New Orleans, Lousiana 70146.    DAVID H. COY • Tulane University School of Medicine, New Orleans, Louisiana 70146.

## 2. APPETITIVE ACQUISITION

The first study to investigate the effects of the opiate peptides on learning was published in 1976 by Kastin and co-workers. They reported that i.p. administration of Met-enkephalin and its D-Ala$^2$ analog facilitated acquisition in a complex maze for food reward in rats. An analog with essentially no opiate activity, [D-Phe$^4$]-Met-enkephalin, had the same facilitating effect as the D-Ala$^2$ analog, suggesting to the authors that the facilitation they noted for the opiate peptides was probably dissociated from their narcotic properties. A pentafluorinated analog of Met-enkephalin was found to improve reversal learning in monkeys after s.c. administration, although it had no significant effect on the acquisition of the original discrimination (Olson et al., 1979). Met-enkephalin, when injected into rats in their first week of life, facilitated maze performance when tested 3 months later, even though no subsequent injections were given at testing (Kastin et al., 1980a). This suggested that the enkephalin might have had its effect on the organization of the developing brain, thereby influencing its ability to process new information.

On the other hand, Met-enkephalin and several of its analogs, as well as the potent pentafluorinated analog of β-endorphin, were reported not to affect the acquisition of discrimination or reversal learning in monkeys after s.c. injections (Olson et al., 1981), perhaps due to some methodological variations. Furthermore, [D-Ala$^2$]-β-endorphin increased running times and errors when given at a dose of 80 μg/kg but not at 800 μg/kg, suggesting an inverted U-shaped function (Kastin et al., 1980b). Thus, it is possible that Met-enkephalin and its analogs sped up performance as a nonopiate effect, whereas β-endorphin showed the typical opiate response of slowing behavioral activity. Naloxone given by itself, however, also produced slower running times and did not antagonize the effects of the endorphin (Kastin et al., 1980b), raising the question of whether these changes are indeed narcotic in nature.

## 3. AVERSIVE ACQUISITION

Attempts to modulate the acquisition process with opiate peptides have largely used the active avoidance paradigm, often with electric shock as the noxious stimulus. One should be somewhat cautious in interpreting such studies since some of the peptides might have been producing some minimal analgesic effects, thereby reducing the aversiveness of the stimuli, and thus not influencing the learning process itself, but the motivation to perform. In addition, the stress of the aversive situation might cause release of endogenous opiates, which in turn might influence learning. Indeed, Izquierdo et al. (1980b) reported release of β-endorphin during training sessions with four different aversive conditioning tasks.

Most reports have indicated that the opiate peptides either impaired acquisition of the avoidance response or had no effect. Leu-enkephalin, [D-Ala-

D-Leu]-enkephalin (Rigter et al., 1980a,b), and Met-enkephalin (Rigter et al., 1980a) produced interference, but [D-Ala$^2$]-Met-enkephalinamide did not affect learning (Rigter et al., 1980a). β-Endorphin had no effect on learning (Gorelick et al., 1978; Izquierdo, 1980a,b), except at a dose of 20 µg/kg at which it slowed acquisition (Izquierdo, 1980b). β-Endorphin, although it had no effect on trials to criterion, did increase latencies (Gorelick et al., 1978), suggesting that it might be exerting a slightly sedative effect.

Since the opiate peptides generally impaired acquisition, it might have been expected that naloxone would facilitate learning, but such results were not reported. Naloxone alone was found to interfer with acquisition (Izquierdo, 1980a,b) or have no effect at all (Rigter et al., 1980a). Naloxone, however, also antagonized the interference of the enkephalins (Rigter et al., 1980a) and β-endorphin (Izquierdo, 1980b), thus presenting a confusing picture as to its role in the modulation of learning. Its effect on the retention of the learned response, however, is clearer, as will be discussed in a later section of this chapter.

The acquisition of a passive avoidance task was facilitated by Leu-enkephalin and [D-Ala-D-Leu]-enkephalin (Rigter et al., 1980b), in contrast to their effect on the active avoidance paradigm. It was postulated that these influences were due to a general suppression of behavior by the peptides. The passive avoidance task requires that the organism not make an active response but rather remain relatively inactive, so that a lowering of activity level would facilitate that response. On the other hand, the active paradigm requires that the organism make an active response, such as jumping to the other side of a shuttle box, so that the sedative properties of the peptides would interfere with that response. Further support for the notion of opiate suppression of behavior is provided by a report that naloxone facilitated escape learning in one strain of mice for all escape tasks whereas in another strain facilitated it in one task and inhibited it in the others (Castellano, 1981). This suggests that the antagonist might be blocking the suppression produced by endogenous opiates released during training of the escape paradigm, with strain differences due to previously documented differences in responsiveness to opiates (e.g., Oliverio and Castellano, 1974). However, Leu-enkephalin and [D-Ala-D-Leu]-enkephalin had no effect on a swim escape response when injected i.p. (Rigter et al., 1980b), so that any suppression possibly exerted by these peptides was not sufficient to influence that particular response. Thus, it appears that variables such as task and subject are important considerations when studying the effects of the opiates on learning.

Habituation of a rearing response to a tone in rats was not influenced by i.p. administration of β-endorphin (Izquierdo, 1980a,b) except at a dose of 20 µg/kg, which produced interference (Izquierdo, 1980b). Since this was a similar finding to that in the active avoidance paradigm, it was postulated that there might be an inverted U-shaped dose–response curve for β-endorphin with respect to its influence on learning (Izquierdo, 1980b). Although the interference reported here is inconsistent with the notion that the opiates might work through suppression of behavior, other studies do support it. A number of enkephalin and endorphin analogs were found to facilitate learning of the habituation response to a loud buzzer by goldfish by increasing the latency for responding, probably through the immobilizing influence of the peptides (Olson *et al.*, 1978). In addition, naloxone impaired the habituation of a rearing response to a 70-db tone in rats (Izquierdo, 1980a,b), presumably by antagonizing the opiate peptides that are endogenously released during training. Naloxone also blocked the interference produced by the 20 µg/kg dose of β-endorphin (Izquierdo, 1980b).

The classical conditioning paradigm has also been used to study the effects of the opiate peptides on learning. Intracerebral administration of the opiate agonist levorphanol significantly impaired the acquisition of a conditioned heart rate response to a tone–shock pairing. Naloxone facilitated the learning and blocked the interference of the agonist (Gallagher *et al.*, 1981). However, although i.v. injection of naloxone was found to have no effect on acquisition of the eyeblink response using shock as the unconditioned stimulus, the extinction of the response was prolonged and conditioning of the heart rate response was attenuated by naloxone (Hernandez and Powell, 1980). Variations in subjects, routes of administration, and dose levels might account for the differences. These findings, however, suggest that the endogenous opiates might be involved in some phase of the learning of this aversively motivated task.

## 4. RETENTION

Although more research has been done measuring retention than acquisition, little work has centered on retention of an appetitive task. Olson *et al.* (1981) reported that [D-Phe$^4$]-Met-enkephalin, an analog with little or no opiate activity, interfered with retention in a delayed response task for reward of food in monkeys. Since the delays involved 0, 15, or 30 sec, it was concluded that the enkephalin analog impaired short-term memory. However, in a short-term memory task involving retention of word lists in man, naloxone had no effect on either immediate or delayed recall (Volavka *et al.*, 1979). Naloxone had variable effects on performance on the Graham–Kendall Memory for Designs Test in humans, with three of eight subjects showing impaired retention, one slightly improving, and four having no effect (Morley *et al.*, 1980). Thus, more work needs to be done to unravel the confusion from these quite different studies.

A number of researchers have used the aversive conditioning paradigms to investigate retention as a function of opiate modulation. The results are relatively consistent with respect to memory in active avoidance paradigms. In general, the opiate agonists interfere with retention, suggesting that they are natural amnesics, and the antagonists facilitate retention. β-Endorphin (Izquierdo, 1980a,b) impairs memory, although the dose is important, since 2 µg/kg is effective only when administered before learning (Izquierdo, 1980b). Morphine (Jensen *et al.*, 1978; Izquierdo, 1979), β-endorphin (Izquierdo *et al.*, 1980a,b; Izquierdo and Dias, 1981), Leu-enkephalin (Izquierdo *et al.*, 1980a;

Izquierdo and Dias, 1981), and Met- and [des-Tyr]-Met-enkephalin (Izquierdo and Dias, 1981) produced retrograde amnesia. The interference of the enkephalins was blocked by the injection of naloxone (Izquierdo and Dias, 1981). Naloxone facilitated retention when given either before or after training (Izquierdo, 1979, Izquierdo, 1980a,b; Jensen et al., 1978; Messing et al., 1979), presumably by antagonizing the endogenous opiates released during training (Izquierdo, 1980b).

Reports from studies using passive avoidance paradigms are similar to those for active avoidance but are less consistent. The opiate agonist levorphanol (Gallagher and Kapp, 1978), and β-endorphin, when given after training but not if given before training (Martinez and Rigter, 1980), impaired memory. Interestingly, α- and γ-endorphin have opposite results, although the two studies of them reported different findings; Kovács et al. (1981) found that γ-endorphin interfered with retention and α-endorphin facilitated it, but Martinez and Rigter (1980) reported that γ-endorphin facilitated memory and α-endorphin had no effect at a wide range of dose levels. The two used comparable doses and systemic administration, eliminating those variables as factors in the discrepancy. The former study, however, is consistent with results reported concerning resistance to extinction that will be discussed later. Met-enkephalin was found to have no effect on retention at any dose tested when given either before or after training (Martinez and Rigter, 1980).

Met- and Leu-enkephalin were shown to have some antiamnesic effects, since they were able to reduce $CO_2$-induced amnesia in rats at low doses (Rigter et al., 1977; Rigter, 1978). The diminished amnesia was found after systemic injection of Met-enkephalin either before training or before retrieval, but Leu-enkephalin reduced amnesia only if given before retrieval. It was suggested that the preacquisition effect might have been a result of facilitation of consolidation or protection from the negative effects of the $CO_2$. The preretrieval effect might have been due to strengthening of the memory trace or reversal of the effects of $CO_2$. Subsequent studies reviewed above, although not directly related to $CO_2$-induced amnesia, have suggested that the opiate peptides probably do not facilitate consolidation or strengthen the memory trace. Thus, their effect on $CO_2$-induced amnesia may be an interaction with $CO_2$.

Naloxone has relatively consistent effects on the retention of passive avoidance and escape. It facilitates memory (Castellano, 1975; Gallagher and Kapp, 1978; Messing et al., 1979), although its effects are partially dependent on the age of the animal studied and other task variables. Naloxone facilitated retention in young rats but inhibited it in old rats (Jensen et al., 1980a,b), possibly due to different numbers or activity of opiate receptors in the two populations. Naloxone had the same differential effect on age for memory of an escape task (Jensen et al., 1980a,b). The level of shock interacted with age, too, since in old rats naloxone inhibited retention when the shock was at a low level but facilitated it when the shock was high; for young animals, naloxone regularly facilitated memory (Vasquez et al., 1979).

One study reported results opposite from the bulk of the literature, in that morphine and Met-enkephalin were found to facilitate retention of the passive avoidance response (Stein and Belluzzi, 1979). The dose levels used here were high relative to those used in other studies, so the discrepancy may be due to the nonmonotonic function of dose or due to compensation of endogenous mechanisms for the large doses. Stein and Belluzzi (1979) suggested that the endogenous opiates might be natural reward mediators, accounting for the effect of reinforcement on behavior.

Retention of a habituation response is affected in the same way by the opiates and their antagonists. Morphine (Izquierdo, 1979) and β-endorphin (Izquierdo, 1980a,b; Izquierdo et al., 1980a,b) inhibited memory or produced retrograde amnesia. The same inverted U-shaped function for dose was reported for β-endorphin on this task as in previously mentioned tasks (Kastin et al., 1980b; Izquierdo, 1980b). Naloxone facilitated retention of the habituation response to a tone (Izquierdo, 1979, 1980a; Izquierdo and Graudenz, 1980), so that the investigators concluded that this nonpainful stimulus produced the same effects as the electric shock of the avoidance response.

## 5. EXTINCTION

Another method for studying retention of a learned task is to measure resistance to extinction. The stronger the memory is, the longer it should take to extinguish the response. Task variables are particularly important here, since different responses extinguish differentially with the same peptides. In the only appetitive paradigm studied, extinction of running down an alley for food by rats was delayed after administration of both α- and γ-endorphin (Le Moal et al., 1979). However, γ-endorphin facilitated extinction of an active avoidance response (de Wied et al., 1978a,c; Kiraly et al., 1980; Le Moal et al., 1979), whereas it delayed extinction of a passive avoidance task (de Wied et al., 1978a). [Des-Tyr]-γ-endorphin (DTγE) produced the same effects as the parent peptide (de Wied et al., 1978a,c), suggesting that it is probably not a narcotic effect, since DTγE has little or no opiate activity.

In the active avoidance paradigm, α-endorphin produced opposite results of γ-endorphin, prolonging extinction (de Wied et al., 1978a,b; Le Moal et al., 1979). β-Endorphin also delayed extinction of the active avoidance response (de Wied et al., 1978a,b), except in one instance in which it had no effect (Kiraly et al., 1980). Met-enkephalin, likewise, delayed extinction (de Wied et al., 1978b). Naltrexone facilitated extinction of the pole-jumping avoidance response, as might be expected, and both α- and β-endorphin restored the behavior that had been rapidly extinguished under the influence of naltrexone (de Wied et al., 1978b). Extinction of an aversive Pavlovian conditioned response, however, was delayed by administration of naloxone before the extinction trials (Hernandez and Powell, 1980), thus perhaps emphasizing the importance of procedural variables.

## 6. NEUROPHYSIOLOGICAL MEASUREMENTS

It has long been thought that the hippocampus is at least partially responsible for the transfer of memory from short-term storage to long-term storage. This function was confirmed in the tragic surgical case reported by Scoville and Milner (1957), in which the patient, after bilateral removal of the mesial temporal lobes and hippocampus, was unable to establish any new memories. Since the transfer from short-term to long-term memory is the process often considered learning and since the hippocampus is known to possess opiate receptors (Di Giulio et al., 1979; Fry et al., 1979; Lindberg et al., 1979; Zakarian and Smyth, 1979), it seems reasonable to discuss the activity of the hippocampus and its response to opiate administration in this review of memory and learning.

In general, the opiate peptides tend to excite the hippocampus. β-Endorphin (Taylor et al., 1979; Botticelli and Wurtman, 1979), Met-enkephalin (Gähwiler, 1980; Taylor et al., 1979), Leu-enkephalin (Gähwiler, 1980), and several enkephalin analogs (Botticelli and Wurtman, 1979; Dingledine, 1981; Gähwiler, 1980; Martinez et al., 1979) were reported to increase the firing of the pyramidal cells and the amplitude of their spikes. Naloxone blocked this excitation (Botticelli and Wurtman, 1979; Martinez et al., 1979; Taylor et al., 1979), although when given alone it had no effect on a majority of the cells (Botticelli and Wurtman, 1979; Fry et al., 1979). Naloxone did, however, have some excitatory and some inhibitory effects on a few of the pyramidal cells (Fry et al., 1979). The relationship between the increased activity and the effects on retention due to opiate administration is not yet established, however.

The amygdala also contains high concentrations of opiate receptors (Kuhar et al., 1973; Law et al., 1979; Lindberg et al., 1979) and seems to be involved in the process of acquisition. Administration of the opiate agonist levorphanol into the amygdala before learning inhibited the acquisition of a classically conditioned response (Gallagher et al., 1981) and when given after training interfered with retention (Gallagher and Kapp, 1978; Jensen et al., 1980a). Naloxone does the opposite, facilitating learning when injected into the amygdala before the start of the session (Gallagher et al., 1981) and improving retention when given after training (Gallagher and Kapp, 1978; Jensen et al., 1980a). The paraventricular area of the brain appears to have a totally different reaction to the application of opiate agonists, producing enhanced rather than impaired performance (Jensen et al., 1980a).

The brain apparently releases β-endorphin during the training process, since the amount of the peptide found in the brain tissue except in the hypothalamus decreases as a result of the experience (Izquierdo et al., 1980b). The tasks involved aversive conditioning, pseudoconditioning, and presentation of tone or shock alone, suggesting that it is stimulation rather than learning that produced release of the endorphin. The fact that naloxone generally facilitates acquisition might be due, then, to the antagonism of the presumed endogenous amnesic that is released during training.

## 7. REWARD MEDIATION

Stein and Belluzzi (1979) proposed a role for the endogenous opiate peptides in the mediation of behavioral reinforcement. They reported that (1) central injections of enkephalin served as reinforcement in self-stimulation studies, (2) that enkephalin might be the basis for reinforcement for electrical stimulation since naloxone blocked the effect, and (3) that retention is facilitated by injections of the opiates after training. These findings have produced controversy and widespread experimentation to test them. Particularly at question is the third conclusion of Stein and Belluzzi (1979) of memory facilitation, since most of the work reviewed above has produced opposite results. The other findings have also been heuristic, but have more support for them.

There have been several reports supporting the finding that the opiate peptides can provide reinforcement when self-administered to brain areas or even peripherally. The potent enkephalin analog FK-33-824 was found to maintain i.v. self-administration established with a reward of morphine in monkeys, suggesting that the enkephalin had the same reinforcing properties as did morphine (Mello and Mendelson, 1978). Drug-naive monkeys also exhibited self-administration of FK-33-824 i.v. with the mean amount increasing gradually until addiction occurred (Roemer *et al.*, 1977), indicating that the enkephalin was a quite powerful reinforcer. Rats were found to self-administer another potent enkephalin analog, [D-Ala$^2$]-Met-enkephalinamide (Hudson *et al.*, 1980), and when given a choice of self-administering that enkephalin analog or cerebrospinal fluid (which should have no reinforcing properties since it is present all the time), the rats chose the enkephalin (Olds and Williams, 1980). Thus, apparently animals will learn new behaviors to receive opiate peptides as their only reinforcement, attesting to the reward properties of these peptides.

The hypothesis that electrical stimulation of the brain is reinforcing and establishes and maintains self-stimulation for it because it releases endogenous opiates has not received unequivocal support. Stein and Belluzzi (1979) found that naloxone decreased the amount of such self-stimulation, suggesting that the antagonist was blocking the reinforcing properties of the endogenous opiates. Support for that result came from Stapleton *et al.* (1979b), who also reported that naloxone lowered the amount of self-stimulation, and from Broekkamp *et al.* (1979), who found that Met- and Leu-enkephalin produced increases in self-stimulation. However, contradictory findings have also been reported. Naloxone failed to alter self-stimulation in some cases (Esposito *et al.*, 1980; Stilwell *et al.*, 1980; van der Kooy *et al.*, 1977), as did naltrexone (Weibel and Wolf, 1979). In addition, morphine (Weibel and Wolf, 1979) and DTγE (Dorsa *et al.*, 1979) decreased self-stimulation rather than enhancing it, thus greatly confusing the issue. More research needs to be done here with close attention to location of stimulation and dose levels of the peptides and antagonists in order to attempt to understand the phenomenon.

The hypothesized reward properties of the opiate peptides have been tested in other ways also. Stapleton *et al.* (1979a) demonstrated that rats prefer to go to areas of an alley in which they had previously been given injections of [D-Ala$^2$]-Met-enkephalin. The enkephalin did not, however, produce any conditioned taste aversion, as morphine did in the same situation, suggesting to the investigators that the peptide had only reward properties. Rats also changed their preference from the dark side of a shuttle box to the light side (Katz and Gormezano, 1979), attesting to the reward value of the opiates. On the negative side, however, Chipkin et al. (1980) reported that neither enkephalin analogs nor naloxone altered responding for sweetened milk in an ethanol discrimination test, thus concluding that the opiates are not general reward mediators. Indeed, although the opiate peptides can be reinforcing in some situations, they are not rewarding in all situations and do not mediate all reward.

## 8. SENILE DEMENTIA

Since memory deficits are a prominent symptom of senile dementia, it is possible that the endogenous opiate peptides might be involved in its development. A particular problem in senile dementia is that memory for recent events is severely impaired. although memory of past events remains relatively intact. It is possible, then, that senile dementia involves decreased ability to transfer memory from short-term storage to long-term storage or to establish long-term memories. Since we have reviewed numerous studies indicating that the opiate peptides can modulate that process, it is possible that the peptides play a role in the etiology of senile dementia.

Consistent with that notion are results indicating that there are age-related changes in opiate receptors and in reactivity to the opiates. There appear to be changes in the number and affinity of receptors with increasing age, and these are apparently regionally specific and sex-related. In aged female rats, there was a decrease in the densities of receptors in the thalamus, midbrain, and anterior cortex relative to young female rats, and a decrease in binding in the anterior cortex (Messing et al., 1980). In male rats, there was a decrease with age in the concentrations of receptors in the frontal poles, the anterior cortex, and the striatum, and there was higher affinity for binding in the frontal poles but not in other areas in the old rats, possibly as a compensatory mechanism for the loss of receptors (Jensen et al., 1980a,b). The significance of the sex differences is unclear, although they are consistent with an interaction between the opiates and gonadal steroids.

Age-related differences in the effect of naloxone on retention have also been reported, presumably reflecting changes that occur with the endogenous opiate peptides. Injections of naloxone were found to facilitate retention of an inhibitory avoidance response and of a swim-escape response in young rats, with the opposite effect occurring in old rats. These differences were not due to differential sensitivity to shock or reactivity with age, since those behaviors were unchanged (Jensen et al., 1980a,b). The direction of the naloxone effect was also dependent on the level of footshock, since the above finding was reported with a footshock of 500 µA, but with a 750-µA footshock there was a naloxone-induced facilitation in both ages (Vasquez *et al.*, 1979). Thus, with a strong shock the age differences were eliminated.

It is possible, then, that these decreases in the functioning of the endogenous opiate system might be related in some way to the behavioral changes that accompany aging, especially with respect to the memory deficits that develop in some aged individuals. The severe impairment of recent memory that characterizes senile dementia may be a result of an inability of the endogenous opiates to operate in their usual way in the modulation of memory.

## 9. SUMMARY

It appears then that the endogous opiate systems may modulate learning and memory, although in quite complex ways with such factors as task, specific opiate, and subject variables being important. Enkephalins were shown to facilitate appetitive acquisition, but β-endorphin interfered with it; naloxone also produced interference on that task but did not antagonize the effect of the endorphin, suggesting that the reaction was nonnarcotic. The opiates impaired acquisition of active avoidance tasks but facilitated passive avoidance responses, supporting the notion that the opiate peptides generally suppressed behavior in the presence of aversive stimuli. Naloxone facilitated escape responses, further supporting that suggestion, although it interfered with acquisition of the active avoidance paradigm. The effects of the opiates on habituation were confusing, with reports of facilitation, impairment, and no effect after administration of the peptide and of facilitation after naloxone. Acquisition of a classical conditioning paradigm was impaired by opiates, and naloxone facilitated it.

Retention of an appetitive task has been affected by injections of opiate peptides and naloxone, but in no predictable way as can be seen from the conflicting results. Retention of an aversive paradigm, however, produced more consistent results, with most investigators reporting that the opiate agonists interfered with retention and that the antagonists facilitated it. This is pretty much the case with retention of both active and passive avoidance, escape, and habituation tasks, with some minor exceptions. Most opiate agonists also delayed extinction, and their antagonists facilitated it.

Opiate peptides tended to excite the hippocampus, an area of the brain thought to have a role in the formation of memory, and antagonists inhibited it. The relationship between the activity of hippocampal cells and retention is still not known, however. Administration of agonists into another area of the brain involved in learning, the amygdala, impaired learning and retention, but naloxone injected there facilitated them. When opiate agonists were applied to the paraventricular area, the opposite resulted, however.

It has been suggested that part of the role of the opiate peptides in learning might involve reward mediation. Animals have learned to self-administer opiates in preference to control substances and to choose stimuli associated with previous administration of the opiates. There is contradictory evidence for the hypothesis that electrical self-stimulation is rewarding because it releases endogenous opiate peptides.

A possible link between memory deficits associated with senile dementia and the endogenous opiates has been postulated. The number and affinity of receptors as well as the effect of naloxone on retention change with age, supporting a role for the opiates in senile dementia, but this area has not been well tested. Therefore, the possible effects of the opiate peptides on learning and memory are very complex.

## 10. REFERENCES

Botticelli, L. J., and Wurtman, R. J., 1979, Beta-endorphin administration increases hippocampal acetylcholine levels, *Life Sci.* **24**:1799.

Broekkamp, C. L., Phillips, A. G., and Cools, A. R., 1979, Facilitation of self-stimulation behavior following intracerebral microinjections of opioids into the ventral tegmental area, *Pharmacol. Biochem. Behav.* **11**:289.

Castellano, C., 1975, Effects of morphine and heroin on discrimination learning and consolidation in mice, *Psychopharmacologia* **42**:235.

Castellano, C., 1981, Strain-dependent effects of naloxone on discrimination learning in mice *Psychopharmacology* **73**:152.

Chipkin, R. E., Stewart, J. M., and Channabasavaiah, K., 1980, The effects of peptides on the stimulus properties of ethanol, *Pharmacol. Biochem. Behav.* **12**:93.

de Wied, D., Bohus, B., van Ree, J. M., Kovács, G. L., and Greven, H. M., 1978a, Neuroleptic-like activity of [des-Tyr$^1$]-$\gamma$-endorphin in rats, *Lancet* **I**:1046.

de Wied, D., Bohus, B., van Ree, J. M., and Urban, I., 1978b, Behavioral and electrophysiological effects of peptides related to lipotropin ($\beta$-LPH), *J. Pharmacol. Exp. Ther.* **204**:570.

de Wied, D., Kovács, G. L., Bohus, B., van Ree, J. M., and Greven, H. M., 1978c, Neuroleptic activity of the neuropeptide $\beta$-LPH$_{62-77}$ ([des-Tyr$^1$]-$\gamma$-endorphin; DT$\gamma$E]), *Eur. J. Pharmacol.* **49**:427.

Di Giulio, A. M., Majane, E. M., and Yang, H. Y., 1979, On the distribution of (Met 5)- and (Leu 5)-enkephalins in the brain of the rat, guinea-pig and calf, *Br. J. Pharmacol.* **66**:297.

Dingledine, R., 1981, Possible mechanisms of enkephalin action on hippocampal CAl pyramidal neurons, *J. Neurosci.* **1**:1022.

Dorsa, D. M., van Ree, J. M., and de Wied, D., 1979, Effects of [des-Tyr$^1$]$\gamma$-endorphin and $\alpha$-endorphin on substantia nigra self-stimulation, *Pharmacol. Biochem. Behav.* **10**:899.

Esposito, R. U., Perry, W., and Kornetsky, C., 1980, Effects of d-amphetamine and naloxone on brain stimulation reward, *Psychopharmacology* **69**:187.

Fry, J. P., Zieglgänsberger, W., and Herz, A., 1979, Specific versus nonspecific actions of opioids on hippocampal neurones in the rat brain, *Brain Res.* **163**:295.

Gähwiler, B. H., 1980, Excitatory action of opioid peptides and opiates on cultured hippocampal pyramidal cells, *Brain Res.* **194**:193.

Gallagher, M., and Kapp, B. S., 1978, Manipulation of opiate activity in the amygdala alters memory processes, *Life Sci.* **23**:1973.

Gallagher, M., Kapp, B. S., McNall, C. L., and Pascoe, J. P., 1981, Opiate effects in the amygdala central nucleus on heart rate conditioning in rabbits, *Pharmacol. Biochem. Behav.* **14**:497.

Gorelick, D. A., Catlin, D. H., George, R., and Li, C. H., 1978, Beta-endorphin is behaviorally active in rats after chronic intravenous administration, *Pharmacol Biochem. Behav.* **9**:385.

Hernandez, L. L., and Powell, D. A., 1980, Effects of naloxone on Pavlovian conditioning of eyeblink and heart rate responses in rabbits, *Life Sci.* **27**:863.

Hudson, R., King, A., Singer, G., Tucker, A., Coy, D. H., and Kastin, A. J., 1980, Schedule-induced self-injection of enkephalin and heroin by the rat, in: *Problems in Pain* (C. Peck and M. Wallace, eds.), pp. 73–77, Pergamon Press, Elmsford, N.Y.

Izquierdo, I., 1979, Effect of naloxone and morphine on various forms of memory in the rat: Possible role of endogenous opiate mechanisms in memory consolidation, *Psychopharmacology* **66**:199.

Izquierdo, I., 1980a, Effect of beta-endorphin and naloxone on acquisition, memory and retrieval of shuttle avoidance and habituation learning in rats, *Psychopharmacology* **69**:115.

Izquierdo, I., 1980b, Effects of a low and a high dose of β-endorphin on acquisition and retention in the rat, *Behav. Neural Biol.* **30**:460.

Izquierdo, I., and Dias, R. D., 1981, Retrograde amnesia caused by Met-, Leu- and des-Tyr-Met-enkaphalin in the rat and its reversal by naloxone, *Neurosci. Lett.* **22**:189.

Izquierdo, I., and Graudenz, M., 1980, Memory facilitation by naloxone is due to release of dopaminergic and beta-adrenergic systems from tonic inhibition, *Psychopharmacology* **67**:265.

Izquierdo, I., Paiva, A. C., and Elisabetsky, E., 1980a, Post-training intraperitoneal administration of Leu-enkephalin and β-endorphin causes retrograde amnesia for two different tasks in rats, *Behav. Neural Biol.* **28**:246.

Izquierdo, I., Souza, D. O., Carrasco, M. A., Dias, R. D., Perry, M. L., Eisinger, S., Elisabetsky, E., and Vendite, D. A., 1980b, Beta-endorphin causes retrograde amnesia and is released from the rat brain by various forms of training and stimulation, *Psychopharmacology* **70**:173.

Jensen, R. A., Martinez J. L., Jr., Messing, R. B., Spiehler, V. R., Vasquez, B. J., Soumireu-Mourat, B., Liang, K. C., and McGaugh, J. L., 1978, Morphine and naloxone alter memory in rats, *Soc. Neurosci. Abstr.* **4**:260.

Jensen, R. A., Messing, R. B., Martinez, J. L., Jr., Vasquez, B. J., and McGaugh, J. L., 1980a, Opiate modulation of learning and memory in the rat, in: *Aging in the 1980's: Psychological Issues* (L. W. Poon, ed.), pp. 191–200, American Psychological Association, Washington, D.C.

Jensen, R. A., Messing, R. B., Spiehler, V. R., Martinez, J. L., Jr., Vasquez, B. J., and McGaugh, J. L., 1980b, Memory, opiate receptors, and aging, *Peptides* **1**(Suppl. 1):197.

Kastin, A. J., Scollan, E. L., King, M. G., Schally, A., and Coy, D., 1976, Enkephalin and a potent analog facilitate maze performance after intraperitoneal administration in rats, *Pharmacol. Biochem. Behav.* **5**:691.

Kastin, A. J., Kostrzewa, R. M., Schally, A. V., and Coy, D. H., 1980a, Neonatal administration of Met-enkephalin facilitates maze performance of adult rats, *Pharmacol. Biochem. Behav.* **13**:883.

Kastin, A. J., Mauk, M. D., Schally, A. V., and Coy, D. H., 1980b, Unusual dose-related effect of an endorphin analog in a complex maze, *Physiol. Behav.* **25**:959.

Katz, R. J., and Gormezano, G., 1979, A rapid and inexpensive technique for assessing the reinforcing effects of opiate drugs, *Pharmacol. Biochem. Behav.* **11**:231.

Kiraly, I., Borsy, J., Tapfer, M., and Graf, L., 1980, Study on the neuroleptic activity of endorphins, in: *Opiate Receptors and the Neurochemical Correlates of Pain* (S. Furst, ed.), pp. 93–99, Pergamon Press, Elmsford, N.Y.

Kovács, G. L., Bohus, B., and de Wied, D., 1981, Retention of passive avoidance behavior in rats following α- and γ-endorphin administration: Effects of post-learning treatments, *Neurosci. Lett.* **22**:79.

Kuhar, M. J., Pert, C. B., and Snyder, S. H., 1973, Regional distribution of opiate receptor binding in monkey and human brain, *Nature (London)* **245**:447.

Law, P.-Y., Loh, H. H., and Li, C. H., 1979, Properties and localization of beta-endorphin receptor in rat brain, *Proc. Natl. Acad. Sci. USA* **76**:5455.

Le Moal, M., Koob, G. J., and Bloom, F. E., 1979, Endorphins and extinction: Differential actions on appetitive and adversive tasks, *Life Sci.* **24**:1631.

Lindberg, I., Smythe, S. J., and Dahl, J. L., 1979, Regional distribution of enkephalin in bovine brain, *Brain Res.* **168**:200.

Martinez, J. L., Jr., and Rigter, H., 1980, Endorphins alter acquisition and consolidation of an inhibitory avoidance response in rats, *Neurosci. Lett.* **19**:197.

Martinez, J. L., Jr., Jensen, R. A., Craeger, R., Veliquette, J., Messing, R. B., McGaugh, J. L., and Lynch, G., 1979, Selective effects of enkephalin on electrical activity of the *in vitro* hippocampal slice, *Behav. Neural Biol.* **26**:128.

Mello, N. K., and Mendelson, J. H., 1978, Self-administration of an enkephalin analog by rhesus monkey, *Pharmacol. Biochem. Behav.* **9**:579.

Messing, R. B., Jensen, R. A., Martinez, J. L., Jr., Spiehler, V. R., Vasquez, B. J., Soumireu-Mourat, B., Liang, K. C., and McGaugh, J. L., 1979, Naloxone enhancement of memory, *Behav. Neural Biol.* **27**:266.

Messing, R. B., Vasquez, B. J., Spiehler, V. R., Martinez, J. L., Jr., Jensen, R. A., Rigter, H., and McGaugh, J. L., 1980, $^3$H-Dihydromorphine binding in brain regions of young and aged rats, *Life Sci.* **26**:921.

Morley, J. E., Baranetsky, N. G., Wingert, T. D., Carlson, H. E., Hershman, J. M., Melmed, S., Levin, S. R., Jamison, K. R., Weitzman, R., Chang, R. J., and Varner, A. A., 1980, Endocrine effects of naloxone-induced opiate receptor blockade, *J. Clin. Endocrinol. Metab.* **50**:251.

Olds, M. E., and Williams, K. N., 1980, Self-administration of D-Ala$^2$-Met-enkephalinamide at hypothalamic self-stimulation sites, *Brain Res.* **194**:155.

Oliverio, A., and Castellano, C., 1974, Genotype-dependent sensitivity and tolerance to morphine and heroin: Dissociation between opiate-induced running and analgesia in the mouse, *Psychopharmacologia* **39**:13.

Olson, G. A., Olson, R. D., Kastin, A. J., Green, M. T., Roig-Smith, R., Hill, C. W., and Coy, D. H., 1979, Effects of an enkephalin analog on complex learning in the rhesus monkey, *Pharmacol. Biochem. Behav.* **11**:341.

Olson, G. A., Roig-Smith, R., Mauk, M. D., LaHoste, G. J., Coy, D. H., Hill, C. W., and Olson, R. D., 1981, Differential effects of neuropeptides on short-term memory in primates, *Peptides* **2**(Suppl. 1):131.

Olson, R. D., Kastin, A. J., Michell, G. F., Olson, G. A., Coy, D. H., and Montalbano, D. M., 1978, Effects of endorphin and enkephalin analogs on fear habituation in goldfish, *Pharmacol. Biochem. Behav.* **9**:111.

Rigter, H., 1978, Attenuation of amnesia in rats by systemically administered enkephalins, *Science* **200**:83.

Rigter, H., Greven, H., and van Riezen, H., 1977, Failure of naloxone to prevent reduction of amnesia by enkephalins, *Neuropharmacology* **16**:545.

Rigter, H., Hannan, T. J., Messing, R. B., Martinez, J. L., Jr., Vasquez, B. J., Jensen, R. A., Veliquette, J., and McGaugh, J. L., 1980a, Enkephalins interfere with acquisition of an active avoidance response, *Life Sci.* **26**:337.

Rigter, H., Jensen, R. A., Martinez, J. L., Jr., Messing, R. B., Vasquez, B. J., Liang, K. C., and McGaugh, J. L., 1980b, Enkephalin and fear-motivated behavior, *Proc. Natl. Acad. Sci. USA* **77**:3729.

Roemer, D., Buescher, H. H., Hill, R. C., Pless, J., Bauer, W., Cardinaux, F., Closse, A., Hauser, D., and Huquenin, R., 1977, A synthetic enkephalin analogue with prolonged parenteral and oral analgesic activity, *Nature (London)* **268**:547.

Scoville, W. B., and Milner, B., 1957, Loss of recent memory after bilateral hippocampal lesions, *J. Neurol. Neurosurg. Psychiatry* **20**:11.

Stapleton, J. M., Lind, M. D., Merriman, V. J., Bozarth, M. A., and Reid, L. D., 1979a, Affective consequences and subsequent effects on morphine self-administration of D-Ala$^2$-methionine enkephalin, *Physiol. Psychol.* **7**:146.

Stapleton, J. M., Merriman, V. J., Coogle, C. L., Gelbard, S. D., and Reid, L. D., 1979b, Naloxone reduces pressing for intracranial stimulation of sites in the periaqueductal gray area, accumbens nucleus, substantia nigra, and lateral hypothalamus, *Physiol. Psychol.* **7**:427.

Stein, L., and Belluzzi, J. D., 1979, Brain endorphins: Possible role in reward and memory formation, *Fed. Proc.* **38**:2468.

Stilwell, D. J., Levitt, R. A., Horn, C. A., Irvin, M. D., Gross, K., Parsons, K. S., Scott, R. H., and Bradley, E. L., 1980, Naloxone and shuttlebox self-stimulation in the rat, *Pharmacol. Biochem. Behav.* **13**:739.

Taylor, D., Hoffer, B., Zieglgänsberger, W., Siggins, G., Ling, N., Seiger, A., and Ohlson, L., 1979, Opioid peptides excite pyramidal neurons and evoke epileptiform activity in hippocampal transplants *in oculo, Brain Res.* **176**:135.

van der Kooy, D., LePiane, F. G., and Phillips, A. G., 1977, Apparent independence of opiate reinforcement and electrical self-stimulation systems in rat brain, *Life Sci.* **20**:981.

Vasquez, B. J., Jensen, R. A., Messing, R. B., Martinez, J. L., Jr. Rigter, H., and McGaugh, J. L., 1979, Naloxone impairs memory in aged rats, *Pharmacologist* **21**:269.

Volavka, J., Dornbush, R., Mallya, A., and Cho, D., 1979, Naloxone fails to affect short-term memory in man, *Psychiatry Res.* **1**:89.

Weibel, S. L., and Wolf, H. H., 1979, Opiate modification of intracranial self-stimulation in the rat, *Pharmacol. Biochem. Behav.* **10**:71.

Zakarian, S., and Smyth, D., 1979, Distribution of active and inactive forms of endorphins in rat pituitary and brain, *Proc. Natl. Acad. Sci. USA* **76**:5972.

CHAPTER 15

# Possible Role of Opioids in Mental Disorders
## Present State of Knowledge

### H. M. EMRICH

## 1. INTRODUCTION

The discovery of endorphinergic systems (for a review see Herz, 1978; Snyder, 1978) initiated a new field of research in biological psychiatry, the importance of which is, as yet, incompletely assessible. There are, however, some indications of the importance of opioid peptides in stress phenomena, in affective disorders, and, possibly, in schizophrenic psychoses. The aim of this chapter is to evaluate the present state of knowledge in this field.

## 2. THE ROLE OF ENDORPHINS IN STRESS

The prima vista astonishing fact that brains normally contain high amounts of substances which have opiate-like activity promoted a diversity of investigations aimed at the elucidation of the conditions under which these opioids may be of functional significance (reviewed by Emrich, 1981). In both animals and humans, under normal conditions, an injection of the specific opiate antagonist naloxone induces only few changes in either subjective performance and objective behavior data or such physiological variables as temperature, pain thresholds, respiratory rate, and so on (e.g., a reduction in grooming and activity and a slight hyperalgesia). From these observations it must be concluded that under normal conditions, endorphins are probably functionally active to only a minimal degree from nerve terminals or the pituitary. The situation changes completely, however, upon exposure to stress factors, such as inescapable pain, cold water immersion, or other intensely stressful stimuli (reviewed by Millan, 1981).

---

H. M. EMRICH • Max-Planck-Institut für Psychiatrie, Munich, FRG.

From these findings it may be concluded that, under particular conditions, endorphins are released and exert actions similar to those of opiates, i.e., endorphins appear to play a modulatory role in the regulation of specific vegetative functions and possibly also in cognition. They might be interpreted as endogenous anxiolytic or tranquilizing and analgetic regulators and modulators which serve as a reserve system for emergency situations. The mechanisms by which this release is elicited are not as yet known. There are some experimental results, however, which provide some insights into this process. One is the demonstration that electrical stimulation of the periaqueductal gray and other centers of the brain stem in man induces a release of endorphins and a simultaneous elevation in nociceptive threshold (Millan and Emrich, 1981). Such results have been obtained in humans by Hosobuchi (1981). The analgesic effect of acupuncture is, moreover, apparently explicable at a pharmacological level as an action of endorphins, since Mayer *et al.* (1977; see also Mayer and Watkins, 1981) have shown that the antinociception evoked by acupuncture is reversible by the opiate antagonist naloxone. Endorphins may, additionally, represent mediators of the anxiolytic and tranquilizing actions of benzodiazepines (reviewed by Millan and Duka, 1981) and neuroleptic drugs (Höllt, 1981; Höllt and Bergmann, 1982; Hong *et al.*, 1979).

## 3. THE POSSIBLE SIGNIFICANCE OF OPIOID PEPTIDES IN AFFECTIVE DISORDERS

Based on the observation that opiates and endorphins exert euphorogenic effects in normal volunteers (Kline *et al.*, 1977; Martin, 1977), the hypothesis of an involvement of endorphins in different types of affective disorders may be derived. Endogenous depression, according to such a concept, would represent a state of hypoactive endorphinergic systems, whereas mania would be considered as a state of endorphin hyperactivity (Byck, 1976; Emrich, 1982a).

An evaluation of the hypothesis of an opioid dysfunction in depression might be achieved by the employment of the following different types of approach:

1. Measurement of levels of different types of endorphins in the CSF and plasma of patients suffering from affective disorders.
2. An evaluation of the clinical effects of the opiate antagonist naloxone in mania and depression.
3. An evaluation of the possible therapeutic effects of different types of opioids in depressive syndromes.

### 3.1. Levels of Endorphins in CSF and Plasma of Patients Afflicted with Affective Disorders

By use of a radioreceptor assay (RRA), Terenius *et al.* (1976) succeeded in determining levels of endorphins in the CSF of patients afflicted with different types of psychiatric disturbances. They observed an elevation in radioreceptor activity in chronic schizophrenics and, although less pronounced, in patients displaying manic depression, as compared to a control group of healthy volunteers. An intraindividual analysis of the data from patients with affective disorders revealed a trend toward higher values in manic as compared to depressive phases. Pickar et al. (1981), however, employing a similar RRA, detected no differences between the CSF endorphin levels of a group of 41 depressed patients, 13 manic patients, and 41 normal probands. These authors (Pickar et al., 1980) did, on the other hand, observe a significant decrease in plasma opioid activity in a patient with a history of severe manic depressive illness upon a switch from mania to depression.

Levels of β-endorphin immunoreactivity in plasma of patients suffering from affective disorders have been evaluated by Höllt et al. (1978a) and Emrich et al. (1979a) by the use of a radioimmunoassay (RIA) (Höllt et al., 1978b). No differences between levels of immunoreactive β-endorphin in manic, depressed, and neurotic patients were observed in two consecutive studies. Similar data have been reported recently by Pickar et al. (1981) with respect to β-endorphin immunoreactivity in CSF.

### 3.2. Effect of Naloxone in Affective Disorders

It may be anticipated from the hypothesis of a possible hyperactivity of endorphinergic systems in mania and of a hypoactivity in depressive syndromes that naloxone may possess antimanic properties and should, moreover, exacerbate depressive symptomatology. A depressiogenic action of naloxone in normal probands would, additionally, have to be postulated. Grevert and Goldstein (1978) did not, however, find any significant change in the mood of 30 healthy volunteers treated, in a double-blind procedure, with i.v. doses of up to 10 mg naloxone, Emrich et al. (1977, 1979b, 1980a), while treating a total number of 48 schizophrenic patients with a dose of naloxone of up to 24.8 mg in a double-blind design, did not observe a tendency toward a depressed mood as an effect of this treatment. Furthermore, Watson et al. (1978), while treating 9 schizophrenic patients, in a double-blind fashion, with 10 mg naloxone reported decreased anxiety and an improvement in mood after this therapy, a result which is at variance with the hypothesis, since anxiety is an important symptom in depression. A depressiogenic effect of naloxone in normal probands and in schizophrenic patients is, thus, very unlikely, However, chronic naltrexone treatment may induce a depression-like symptomatology in normals (Hollister et al., 1981).

The therapeutic action of naloxone in mania is, on the other hand, apparently minimal: Davis et al. (1979) reported a positive effect in 1 of 4 manic patients upon administration of 10.0–30.0 mg naloxone, and Janowsky et al. (1977) found, in a group of 12 patients, 4 naloxone responders after i.v. injection of 20 mg (double-blind study). There are several reports, however, with data which are inconsistent with the hypothesis of an antimanic property of naloxone: Emrich (1978; Emrich et al., 1979c) treating 2 manic patients with 4.0 mg naloxone (one open, one double-blind trial) found no improvement, Davis et al. (1980) in 10 manic patients (double-blind, 20 mg) observed no significant change, and a recently performed collaborative WHO study (25 manic patients, 0.3 mg/kg naloxone; Pickar and Bunney, 1981) showed no beneficial effects of this therapy. An antimanic property of naloxone appears, therefore, highly questionable. In depressive patients the influence of naloxone is, similarly, minimal (Emrich, 1978; Emrich et al., 1979c; Terenius et al., 1977).

### 3.3. Action of Opioid Agonists in Affective Disorders

The therapeutic use of opiate agonists in depression was initiated in the "opium cure," recommended by Emil Kraepelin as early as 1901, to be taken in slowly increasing and later decreasing doses, especially in patients with agitated depression. The clinical reports, although no standardized evaluations of the clinical effects have been performed, point to a genuine antidepressive effect of tinctura opii, of morphine and other opiates (Kraepelin, 1901, 1927; Weygandt, 1935). Subsequently, Fink et al. (1970), in an open study, treated 10 severely depressed patients with the partial agonist cyclazocine (1.0–3.0 mg) and observed, within 3 weeks, a remarkable reduction of "depressed mood" and "apathy." Kline et al. (1977) were the first to apply β-endorphin to depressed patients and reported beneficial effects (open study, 2 patients, 1.5–6.0 mg). Angst et al. (1979) observed a switch to hypomania/mania after β-endorphin infusions (10.0 mg i.v.) in 3 of 6 depressed patients. Gorelick et al. (1981) and Gerner et al. (1980) using a double-blind design in 10 patients (1.0–11.5 mg β-endorphin i.v.) found unequivocal antidepressive effects of this medication. A therapeutic trial with the synthetic enkephalin analog FK 33-824 in 10 patients with endogenous depression achieved a remarkable improvement in 3 patients and a tranquilizing effect in 4 others (Nedopil and Rüther, 1979).

Extein et al. (1981), in contrast, reported only a slight antidepressive effect of 5.0 mg morphine in 10 inpatients with major depressive disorders (open study), whereas, in a double-blind investigation of 6 depressed inpatients, employing 5.0 mg methadone, no difference to placebo was observed.

A possible therapeutic action of opioids is furthermore suggested by the observation that electroconvulsive therapy (ECT), which is probably the most effective rapidly acting somatic therapy for endogenous depression, apparently activates endorphinergic systems. Holaday et al. (1981) showed that ECT in rats induces a spectrum of vegetative, naloxone-reversible changes, which apparently reflect an ECT-induced activation of particular endorphinergic systems. Consonant with this finding are the observations of Emrich et al. (1979a) who demonstrated that levels of β-endorphin immunoreactivity in the plasma of patients exhibiting endogenous depression are increased upon ECT. It may, thus, be speculated that the therapeutic efficacy of ECT in depression is mediated via a mobilization of endorphins and that, possibly, a future therapy of depressed patients may lie in the development of a pharmacological means of mimicking these neurobiological effects of ECT.

The body of evidence, summarized above, provides no substantive evidence for the hypothesis of a central endorphinergic deficiency in depression and hyperactivity in mania. It appears very probable, however, that the activation of special central opiate receptors (e.g., by "opiate cure," β-endorphin, partial agonists, and possibly release of endorphins via ECT) has beneficial effects in endogenous depression. It may, thus, be hypothesized that depression represents a neurochemical defect independent of endorphinergic systems, but that this defect can partially be compensated for by an activation of opiate receptors. Data favoring this concept have recently been obtained by Emrich *et al.* (1981a) in a clinical trial using the partial opiate-receptor agonist with potent analgesic properties, buprenorphine (Temgesic; Reckitt and Colman, Co.), 0.4 mg/day in 10 patients with endogenous depression, showing about 40% mean reduction of depression scores within 3 days of treatment (double-blind ABA design).

## 4. A POSSIBLE INVOLVEMENT OF ENDORPHINS IN SCHIZOPHRENIA

The hypothesis that endorphins may play a causative role in the pathogenesis of schizophrenia was based on the following observations: Certain partial opiate agonists [cyclazocine and nalorphine (Jasinski *et al.*, 1967)] induce immediately naloxone-reversible hallucinations in healthy volunteers. Terenius *et al.* (1976), on the other hand, found elevated endorphin levels in the CSF of chronic schizophrenics and observed a normalization of these accompanying the clinical improvement produced by neuroleptic therapy. From these observations it was concluded that productive symptoms (e.g., hallucinations, delusions) may be induced in schizophrenic patients by hyperactivity of normally or abnormally functioning opioids (Wahlström and Terenius, 1981).

### 4.1. Action of Naloxone in Schizophrenia

Gunne *et al.* (1977) initiated the search for an antipsychotic effect of the opioid antagonist naloxone. They observed, in a single-blind study, an immediate reversal of auditory hallucinations in 4 schizophrenic patients upon i.v. injection of 0.4. mg of this drug. In two cases with hallucinatory behavior, however, no change occurred. Volavka *et al.* (1977) failed to reproduce these effects in 7 schizophrenic patients under similar conditions, by use of a double-blind design. Furthermore, in a study by Davis *et al.* (1977), employing 0.4–10.0 mg naloxone in 14 schizophrenic patients, no unequivocal naloxone-induced improvement of psychotic symptoms was demonstrable. Only the item "unusual thought content" showed a statistically significant improvement. Negative findings with low doses of naloxone (0.4–1.6 mg) have also been communicated by other authors (Gunne et al., 1977; Janowsky et al., 1977; Kurland et al., 1977).

In studies with naloxone performed by Emrich et al. (1977, 1979c,d), 40 schizophrenic patients have been examined, 20 with a dose of naloxone of 4.0 mg and the remaining 20 with a dose of 24.8 mg. In both cases a double-blind placebo-controlled cross-over design was undertaken. Only patients reporting frequent hallucinations and/or actual delusional experience were included in the studies. Psychopathological changes were registered by use of the IMPS and a self-constructed "VBS Scale" which can especially be adapted to hallucinations and actual delusional experience. Both investigations (employing 4.0 or 24.8 mg) established that i.v. injection of naloxone produced a reduction in psychotic symptomatology (especially hallucinations). As compared to placebo, this effect reached statistical significance within 2–7 hr postinjection. Berger et al. (1979; see also Watson et al., 1978), employing 10 mg naloxone, found a statistically significant reduction of hallucinations and of psychotic anxiety in 10 schizophrenics. The maximal effect, in accordance with the observations of Emrich et al. (1977), occurred about 2 hr after injection. Similar results have been acquired in a collaborative study sponsored by WHO, reviewed by Pickar and Bunney (1981).

From the studies performed with naloxone during the last 5 years, it may be concluded that low doses, sufficient to precipitate withdrawal symptoms in heroin addicts, are completely ineffective in schizophrenic patients, whereas higher dosages (4.0–25.0 mg) exert a relatively small, but statistically significant, antipsychotic effect. These results are not easily interpreted: On the one hand, these findings may constitute support for the hypothesis of Terenius et al. (1976) that endorphins may play a causative role in schizophrenia; on the other hand, the data may be interpreted in terms of interactions between endorphinergic and dopaminergic neuronal systems (Deyo et al., 1979; Extein et al., 1979; Ferland et al., 1977), possibly involving an activation of counterregulatory processes due to the very short, rapidly decaying pulse of naloxone blockade of opiate receptors.

### 4.2. Measurement of Endorphins in CSF and Plasma of Schizophrenic Patients

Terenius et al. (1976) and Lindström et al. (1978) were the first to measure elevated amounts of endorphins in the CSF of schizophrenic patients by use of an RRA. These authors analyzed two CSF fractions, separated by gel chromatography, and observed, in particular in fraction I, an increase of endorphin levels in patients suffering from schizophrenia. Recently this group (Rimón et al., 1980) in 6 of 9 acute schizophrenic patients, in 4 of 6 schizophrenics reentering the hospital, and in 2 of 9 chronic cases found CSF endorphin levels above the range exhibited by healthy volunteers. Pickar et al. (1981), however, likewise using an opioid RRA, found no difference in CSF levels of opioid activity in comparing the data of 41 normals, 27 schizophrenic patients, and

14 patients with schizoaffective psychosis. In agreement with this observation are the findings of Emrich et al. (1979a) and Höllt et al. (1982) that the β-endorphin immunoreactivity content of CSF obtained from 15 schizophrenic patients was not elevated in comparison with a group of 16 patients with mixed neurological diagnoses and another group of 8 medical patients who were punctured owing to the suspicion of meningitis, although the CSF was found to be normal ("normals").

### 4.3. Endorphins and Hemodialysis

Measurements of the β-endorphin immunoreactivity content of plasma have been undertaken with special reference to the finding of Palmour et al. (1979) that in the dialysate of schizophrenic patients there exists a high amount of Leu-β-endorphin. Since the antibody, employed by Höllt et al. (1978b), displays a 100% cross-reactivity with Leu-β-endorphin, a metabolic abnormality in schizophrenic patients involving elevated levels of Leu-β-endorphin in plasma should have proved detectable in this radioimmunoassay. Levels of β-endorphin-like immunoreactivity in the plasma of 14 patients with schizophrenia (not receiving neuroleptics) showed, however, only a small tendency toward an elevation in comparison with the data of 13 patients exhibiting different types of neuroses (Emrich et al., 1979a). Similar results have been communicated by Ross et al. (1979). The contention that abnormally high levels of endorphins can be eliminated by hemodialysis (Palmour et al., 1979) is, furthermore, at variance with the data of Höllt et al. (1979) who have demonstrated that in vitro hemodialysis/hemoperfusion is ineffective in clearing β-endorphin and $Leu^5$-β-endorphin from plasma, a finding which may be explicable by the fact that β-endorphin might be bound to carrier proteins in plasma. Furthermore, Lewis et al. (1979) were not able to detect Leu-β-endorphin in hemodialysates of schizophrenic patients.

### 4.4. Activation of Endorphins by Neuroleptic Drugs

A more important hint for an involvement of endorphinergic systems in the pathobiochemistry of schizophrenia may be derived from observations of Costa's group (Hong et al., 1979) and of Höllt and Bergmann (1982) that chronic treatment with haloperidol and other neuroleptic drugs elevates levels of endorphins in the striatum, pituitary, and plasma. Based on the activating effect of neuroleptic drugs upon endorphin biosynthesis (Höllt and Bergmann, 1982), the hypothesis might be constructed that the curative actions of neuroleptic drugs are mediated via an increase in endorphin levels (Hong et al., 1979). In an attempt to estimate this hypothesis, a clinical trial was performed in order to determine whether the antipsychotic effects of neuroleptic drugs can be antagonized by administration of the opioid antagonist naloxone. The study was performed in a double-blind fashion with placebo control employing a dose of naloxone of $2 \times 20$ mg per day for 2 days in 8 schizophrenic patients revealing productive psychotic symptoms. The results showed no blockade by naloxone of the antipsychotic effect of neuroleptic drugs and apparently do not support the idea that a neuroleptic-effected "activation of endorphins" represents the causative factor underlying the beneficial effects of neuroleptic therapy. However, if the short pulse of i.v. naloxone treatment has merely to be interpreted as a trigger, initiating counterregulatory processes (cf. discussion in Emrich, 1982b), this finding does not contradict the view of an endorphin-mediated antipsychotic action of neuroleptics. Thus, a clinical study employing chronic naltrexone–neuroleptic interaction, apparently, would be of high interest.

### 4.5. Action of Des-Tyr-γ-Endorphin in Schizophrenia

An endorphin hypothesis substantially differing from that enunciated by Terenius et al. (1976) is the concept of the existence of an "endogenous neuroleptic" (de Wied, 1979), assuming that sufficient quantities of such a substance are normally present in the brain, whereas schizophrenic patients possess an inborn error in the metabolism of this substance. This hypothesis emerged from the findings that γ-endorphin, and in particular its des-Tyr derivative, which is devoid of opiate-like activity, displays a profile of action similar to that of neuroleptic drugs in several animal tests (de Wied et al., 1978). It was hypothesized that des-Tyr-γ-endorphin (DTγE) and possibly also des-enkephalin-γ-endorphin may act as "endogenous neuroleptics," being functionally defective in schizophrenia. The use of these substances would, consequently, represent the most rational therapy for this illness. Indeed, Verhoeven et al. (1978), in an open study, treated 6 schizophrenic patients undergoing continuous neuroleptic therapy, with a daily dose of 1 mg DTγE (i.m.) and observed an improvement in all of them, although this effect was only short-lived in three cases. Subsequently, a double-blind placebo-controlled cross-over study was undertaken by the same authors (Verhoeven et al., 1979) in 6 more patients exhibiting chronic or frequently relapsing schizophrenia and similarly under neuroleptic treatment, and additionally in 2 patients not receiving neuroleptics. Once more, a remarkable therapeutic efficacy of DTγE (1 mg/day, i.m.) involving a change of about 60–80% within a period of 3–4 days was observed in three-point rating scales comprising hallucinations, delusions, train of thought disorders, emotional flatness, orientation and motor activity; injection of placebo was practically ineffective. In view of the importance of the theoretical and therapeutic implications of these findings, several groups have attempted to reproduce such effects. Emrich et al. (1980b,c), by use of a double-blind placebo-controlled cross-over investigation, treated 13 patients undergoing continuous neuroleptic therapy and suffering from either chronic or acute, frequently relapsing schizophrenia displaying persistent productive symptoms (hallucinations, acute delusions). After 1 day of single-blind injection of placebo, two successive double-blind treatment periods of 4 days each followed, viz. 4 days with i.m. injections of 2 mg DTγE preceding 4 days of placebo injections or vice versa. Psychopathological evaluation was performed twice daily by use of the IMPS and an eight-point scale appropriate for the estimation of special target symptoms (VBS). The mean data obtained from the whole sample of 13 patients showed that placebo and DTγE produced a reduction in symptomatology of an approximately equal magnitude. The results provide no support for the hypothesis of an antipsychotic efficacy of DTγE in the treatment of chronic schizophrenic patients. In the subgroup of acute cases, however, a therapeutic action of DTγE appears possible. Similar findings, showing no effect in chronic cases, have also been communicated by Bourgeois *et al.* (1980), Casey *et al.* (1981), Manchanda and Hirsch (1981), and Tamminga *et al.* (1981) (for reviews see Emrich *et al.*, 1981b; van Ree *et al.*, 1981). In the group of acute schizophrenic patients, further evaluation of a possible antipsychotic action of DTγE is necessary.

## 5. ROLE OF OPIOIDS IN OTHER DISORDERS

It has been hypothesized by Goldstein (1976) that opioids may be of possible significance in heroin addicts. According to this concept a constitutional predisposition to heroinism may lie in a genetic deficiency, the self-administration of exogenous opioid material constituting an attempt to counterbalance this condition. As discussed by Herz (1981), there are numerous methodological problems to the evaluation of this idea. On the other hand, there are several findings (reviewed by Herz, 1981; Höllt *et al.*, 1978c) showing that chronic administration of morphine results in a large (50%) decrease in the β-endorphin content of the intermediate/posterior pituitary lobe. Höllt *et al.* (1980) were able to demonstrate that the reduction of β-endorphin formation after long-term treatment with morphine in the intermediate pituitary of the rat is due to a reduction in the activity of mRNA coding for the β-endorphin ACTH precursor—whereas the opposite is true for chronic haloperidol treatment which produces, in contrast, elevated levels in this lobe (Höllt and Bergmann, 1982). From these findings one may hypothesize that some component of the addictive processes characterizing heroinism may relate to an opiate-induced inhibition of endorphinergic systems, a concept (Herz, 1981) consistent with the hypothesis of Dole and Nyswander (1967) that heroin addiction is a consequence of a disturbance of brain metabolism.

Another field of pathophysiology in which endogenous opioids may be of central importance is in different types of shock state (reviewed by Faden and Holaday, 1981), e.g., endotoxic, hypovolemic, and "spinal shock." It has been shown by these authors that intracerebroventricular administration of naloxone specifically antagonizes the vegetative changes induced by spinal transection (Holaday *et al.*, 1981). Furthermore, Holaday and Faden (1978) have demonstrated the ability of naloxone to reverse the early hypotension manifested in endotoxin shock and hypovolemic shock (Faden and Holaday, 1979), findings pointing to the suggestion that a hyperactivity of endorphinergic systems may have fatal medical consequences and that, therefore, naloxone therapy aimed at the antagonism of endogenous opioid systems may be developed as a novel central feature in emergency medicine in the immediate future, as has been revealed in the life-saving action of naloxone in a child suffering from "irreversible shock" (Holaday, personal communication).

There are, additionally, some hints that ethanol-induced coma may be due, to some extent, to an opioid hyperactivity (Naber *et al.*, 1981) and that naloxone may also be beneficial in these states of intoxication (Mackenzie, 1979; Schenk *et al.*, 1978).

## 6. SUMMARY

Opioid peptides are apparently of considerable functional importance in the regulation of particular vegetative reactions upon exposure to stress and might be interpreted as endogenous anxiolytics or tranquilizing and analgetic regulators and modulators, which serve as a reserve system for the response to emergency situations. In affective disorders, a possible deficiency in the operation of endorphinergic mechanisms has been hypothesized, although biochemical data and the effect of the specific opiate antagonist naloxone have provided no really compelling support for such a hypothesis. On the other hand, the activation of opiate receptors apparently exerts antidepressant effects. Since electroconvulsion induces an activation of endorphinergic systems, it is plausible to postulate that at least a component of the beneficial effects of electroconvulsion in endogenous depression may be constituted by an endorphinergic type of mechanism. In schizophrenic patients, several studies have shown an antipsychotic effect of high-dosage naloxone treatment, although initial findings of endorphin hyperactivity in schizophrenia have not been replicated. The mechanistic basis of the naloxone effect in schizophrenia is herein discussed. Hemodialysis in schizophrenia has been shown not to be effective via an elimination of endorphins from the circulation. On the contrary, hemodialysis induces an elevation in levels of plasma $\beta$-endorphin, possibly due to an unspecific stress effect. Chronic application of neuroleptic drugs induces an increased biosynthesis of $\beta$-endorphin and an increased turnover and release of Met- and Leu-enkephalin. The concept of the existence of an "endogenous neuroleptic," namely the des-Tyr derivative of $\gamma$-endorphin, has critically been evaluated in different clinical studies. According to these results, a therapeutic effect in patients with acute schizophrenia appears possible, although it has not really been, as yet, convincingly demonstrated. The possible functional significance of endorphins in heroinism and psychosomatic disorders and, also, in different types of "shock state" is discussed.

## 7. REFERENCES

Angst, J., Autenrieth, V., Brem, F., Koukkou, M., Meyer, H., Stassen, H. H., and Storck, U., 1979, Preliminary results of treatment with $\beta$-endorphin in depression, in: *Endorphins in Mental*

*Health Research* (E. Usdin, W. E. Bunney, Jr., and N. S. Kline, eds.), pp. 518–528, MacMillan Press, London.

Berger, P. A., Watson, S. J., Akil, H., and Barchas, J. D., 1979, Naloxone administration in chronic hallucinating schizophrenic patients, in: *Endorphins in Mental Health Research* (E. Usdin, W. E. Bunney, Jr., and N. S. Kline, eds.), pp. 423–434, MacMillan Press, London.

Bourgeois, M., Laforge, E., Muyard, J., Blayac, J., and Lemoine, J., 1980, Endorphines et schizophrénies. II. Les essais de traitement des schizophrénies par la Des-tyr-γ-endorphine (DTγE) (dont un essai personnel), *Ann. J. Med. Psychiatr.* **138**:1112.

Byck, R., 1976, Peptide transmitters: A unifying hypothesis for euphoria, respiration, sleep, and the action of lithium, *Lancet* **II**:72.

Casey, D. E., Korsgaard, S., Gerlach, J., Jørgensen, A., and Simmelsgaard, H., 1981, Effect of des-tyrosine-γ-endorphin in tardive dyskinesia, *Arch. Gen. Psychiatry* **38**:158.

Davis, G. C., Bunney, W. E., Jr., De Fraites, E. G., Kleinman, J. E., van Kammen, D. P., Post, R. M., and Wyatt, R. J., 1977, Intravenous naloxone administration in schizophrenia and affective illness, *Science* **197**:74.

Davis, G. C., Bunney, W. E., Jr., Buchsbaum, M. S., De Fraites, E. G., Duncan, W., Gillin, J. C., van Kammen, D. P., Kleinman, J., Murphy, D. L., Post, R. M., Reus, V., and Wyatt, R. J., 1979, Use of narcotic antagonists to study the role of endorphins in normal and psychiatric patients, in: *Endorphins in Mental Health Research* (E. Usdin, W. E., Bunney, Jr., and N. S. Kline, eds.), pp. 393–406, MacMillan Press, London.

Davis, G. C., Extein, I., Reus, V. I., Hamilton, W., Post, R. M., Goodwin, F. K., and Bunney, W. E., Jr., 1980, Failure of naloxone to reduce manic symptoms, *Am. J. Psychiatry* **137**:1583.

de Wied, D., 1979, Schizophrenia as an inborn error in the degradation of β-endorphin—A hypothesis, *Trends Neurosci.* **2**:79.

de Wied, D., Kovács, G. L., Bohus, B., van Ree, J. M., and Greven, H. M., 1978, Neuroleptic activity of the neuropeptide β-LPH$_{62-77}$ ([des-Tyr$^1$]-γ-endorphin; DTγE), *Eur. J. Pharmacol.* **49**:427.

Deyo, S. N., Swift, R. M., and Miller, R. J., 1979, Morphine and endorphins modulate dopamine turnover in rat median eminence, *Proc. Natl. Acad. Sci. USA* **76**:3006.

Dole, V. P., and Nyswander, M. E., 1967, Heroin addiction—A metabolic disease, *Arch. Intern. Med.* **120**:19.

Emrich, H. M., 1978, Ueber eine moegliche Rolle von Endorphinen bei psychischen Krankheiten, *Arzneim. Forsch.* **28**:1270.

Emrich, H. M. (ed.), 1981, *The Role of Endorphins in Neuropsychiatry*, Karger, Basel.

Emrich, H. M., 1982a, A possible role of opioid substances in depression, in: *Typical and Atypical Antidepressants*, Vol. 1, *Molecular Mechanisms* (E. Costa and G. Racagni, eds.), pp. 77–84 Raven Press, New York.

Emrich, H. M., 1982b, A possible role of endorphinergic systems in schizophrenia, in: *Psychobiology of Schizophrenia* (M. Namba and H. Kaiya, eds.), pp. 291–297. Pergamon Press, Elmsford, N.Y.

Emrich, H. M., Cording, C., Pirée, S., Kölling, A., von Zerssen, D., and Herz, A., 1977, Indication of an antipsychotic action of the opiate antagonist naloxone, *Pharmakopsychiatr. Neuro-Psychopharmakol.* **10**:265.

Emrich, H. M., Höllt, V., Kissling, W., Fischler, M., Laspe, H., Heinemann, H., von Zerssen, D., and Herz, A., 1979a, β-Endorphin-like immunoreactivity in cerebrospinal fluid and plasma of patients with schizophrenia and other neuropsychiatric disorders, *Pharmakopsychiatr. Neuro-Psychopharmakol.* **12**:269.

Emrich, H. M., Höllt, V., Laspe, H., Fischler, M., Heinemann, H., Kissling, W., von Zerssen, D., and Herz, A., 1979b, Studies on a possible pathological significance of endorphins in psychiatric disorders, in: *Neuro-psychopharmacology* (B. Saletu, P. Berner, and L. Hollister, eds.), pp. 527–534, Pergamon Press, Elmsford, N.Y.

Emrich, H. M., Cording, C., Pirée, S., Kölling, A., Möller, H.-J., von Zerssen, D., and Herz, A., 1979c, Actions of naloxone in different types of psychoses, in: *Endorphins in Mental Health Research* (E. Usdin, W. E. Bunney, Jr., and N. S. Kline, eds.), pp. 452–460, MacMillan Press, London.

Emrich, H. M., Möller, H.-J., Laspe, H., Meisel-Kosik, I., Dwinger, H., Oechsner, R., Kissling, W., and von Zerssen, D., 1979d, On a possible role of endorphins in psychiatric disorders: Actions of naloxone in psychiatric patients in: *Biological Psychiatry Today* (J. Obiols, C. Ballús, E. Gonzáles Monclús, and J. Pujol, eds.), pp. 798–805, Elsevier/North-Holland, Amsterdam.

Emrich, H. M., Höllt, V., Bergmann, M., Kissling, W., Schmid, W., von Zerssen, D., and Herz, A., 1980a, Plasma levels of β-endorphin under chronic neuroleptic treatment in schizophrenic patients: Failure of naloxone to counteract curative effects of neuroleptic drugs, in: *Neural Peptides and Neuronal Communication* (E. Costa and M. Trabucchi, eds.), pp. 489–502, Raven Press, New York.

Emrich, H. M., Zaudig, M., Kissling, W., Dirlich, G., von Zerssen, D., and Herz, A., 1980b, Des-tyrosyl-γ-endorphin in schizophrenia: A double-blind trial in 13 patients, *Pharmakopsychiatr. Neuro-Psychopharmakol.* **13**:290.

Emrich, H. M., Zaudig, M., von Zerssen, D., Herz, A., and Kissling, W., 1980c, Des-tyr-γ-endorphin in schizophrenia, *Lancet* **II**:1364.

Emrich, H. M., Vogt, P., and Herz, A., 1981a, A possible role of opioids in depression: Significant improvement after buprenorphine, in: *Biological Psychiatry 1981* (C. Perris, G. Struwe, and B. Jansson, eds.), pp. 380–385, Elsevier/North-Holland, Amsterdam.

Emrich, H. M., Zaudig, M., von Zerssen, D., Kissling, W., Dirlich, G., and Herz, A., 1981b, Action of (des-Tyr$^1$)-γ-endorphin in schizophrenia, in: *The Role of Endorphins in Neuropsychiatry* (H. M. Emrich, ed.), pp. 279–286, Karger, Basel.

Extein, I., Lo, C., Goodwin, F., and Schoenfeld, R. I., 1979, Dopamine-mediated behavior produced by the enkephalin analogue FK 33-824, *Psychiatry Res.* **1**:333.

Extein, I., Pickar, D., Gold, M. S., Gold, P. W., Pottash, A. L. C., Sweeney, D. R., Ross, R. J., Rebard, R., Martin, D., and Goodwin, F. K., 1981, Methadone and morphine in depression, *Psychopharmacol. Bull.* **17**:29.

Faden, A. I., and Holaday, J. W., 1979, Opiate antagonists: A role in the treatment of hypovolemic shock, *Science* **205**:317.

Faden, A. I., and Holaday, J. W., 1981, Endorphins in traumatic spinal injury: Pathophysiological studies and clinical implications, in: *The Role of Endorphins in Neuropsychiatry* (H. M. Emrich, ed.), pp. 158–174, Karger, Basel.

Ferland, L., Fuxe, K., Eneroth, P., Gustafsson, J.-A., and Skett, P., 1977, Effects of methionine-enkephalin on prolactin release and catecholamine levels and turnover in the median eminence, *Eur. J. Pharmacol.* **43**:89.

Fink, M., Simeon, J., Itil, T. M., and Freedman, A. M., 1970, Clinical antidepressant activity of cyclazocine—a narcotic antagonist, *Clin. Pharmacol. Ther.* **11**:41.

Gerner, R. H., Catlin, D. H., Gorelick, D. A., Hui, K. K., and Li, C. H., 1980, β-Endorphin: Intravenous infusion causes behavioral change in psychiatric inpatients, *Arch. Gen. Psychiatry* **37**:642.

Goldstein, A., 1976, Opioid peptides (endorphins) in pituitary and brain, *Science* **193**:1081.

Gorelick, D. A., Catlin, D. H., and Gerner, R. H., 1981, β-Endorphin studies in psychiatric patients, in: *The Role of Endorphins in Neuropsychiatry* (H. M. Emrich, ed.), pp. 236–245, Karger, Basel.

Grevert, P., and Goldstein, A., 1978, Endorphins: Naloxone fails to alter experimental pain or mood in humans, *Science* **199**:1093.

Gunne, L. M., Lindström, L., and Terenius, L., 1977, Naloxone-induced reversal of schizophrenic hallucinations, *J. Neural Transm.* **40**:13.

Herz, A. (ed.), 1978, *Developments in Opiate Research*, Dekker, New York.

Herz, A., 1981, On the role of endorphins in addiction, in: *The Role of Endorphins in Neuropsychiatry* (H. M. Emrich, ed.), pp. 175–180, Karger, Basel.

Holaday, J. W., and Faden, A. I., 1978, Naloxone reversal of endotoxin hypotension suggests role of endorphins in shock, *Nature (London)* **275**:450.

Holaday, J. W., Tortella, F. C., and Belenky, G. L., 1981, Electroconvulsive shock (ECS) results in a functional activation of endorphin systems, in: *The Role of Endorphins in Neuropsychiatry* (H. M. Emrich, ed.), pp. 142–157, Karger, Basel.

Hollister, L. E., Johnson, K., Boukhabza, D., and Gillespie, H. K., 1981, Aversive effects of naltrexone in subjects not dependent on opiates *Drug Alcohol Depend.* **7**:1.

Höllt, V., 1981, Effects of neuroleptic drugs on endogenous peptides in the rat, in: *The Role of Endorphins in Neuropsychiatry* (H. M. Emrich, ed.), pp. 1–18, Karger, Basel.

Höllt, V., and Bergmann, M., 1982, Effect of acute and chronic haloperidol treatment on the concentrations of immunoreactive β-endorphin in plasma, pituitary and brain of rats, *Neuropharmacology* **21**:147.

Höllt, V., Emrich, H. M., Müller, O. A., and Fahlbusch, R., 1978a, β-Endorphin-like immunoreactivity (β-ELI) in human plasma and cerebrospinal fluid (CSF), in: *Characteristics and Function of Opioids* (J. M. van Ree and L. Terenius, eds.), pp. 279–280, Elsevier/North-Holland, Amsterdam.

Höllt, V., Przewłocki, R., and Herz, A., 1978b, Radioimmunoassay of β-endorphin: Basal and stimulated levels in extracted rat plasma, *Naunyn-Schmiedebergs Arch. Pharmacol.* **303**:171.

Höllt, V., Przewłocki, R., and Herz, A., 1978c, β-Endorphin-like immunoreactivity in plasma, pituitaries and hypothalamus of rats following treatment with opiates, *Life Sci.* **23**:1057.

Höllt, V., Hillebrand, G., Schmidt, B., and Gurland, H.-J., 1979, Endorphins in schizophrenia: Hemodialysis/hemoperfusion are ineffective in clearing β-Leu$^5$-endorphin and β-endorphin from human plasma, *Pharmakopsychiatr. Neuro-Psychopharmakol.* **12**:399.

Höllt, V., Haarmann, I., Przewłocki, R., and Jerlicz, M., 1980, Long-term treatment of rats with morphine decreases *in vitro* biosynthesis in and release of β-endorphin from intermediate/posterior lobes of pituitary, in: *Neural Peptides and Neuronal Communication* (E. Costa and M. Trabucchi, eds.), pp. 399–405, Raven Press, New York.

Höllt, V., Emrich, H. M., Bergmann, M., Nedopil, N., Dieterle, D., Gurland, H.-J., Nusselt, L., von Zerssen, D., and Herz, A., 1982, β-Endorphin-like immunoreactivity in CSF and plasma of neuropsychiatric patients, in: *Endorphins and Opiate Antagonists in Psychiatric Research: Clinical Implications* (N. S. Shah and A. G. Donald, eds.), pp. 231–243, Plenum Press, New York.

Hong, J. S., Yang, H. Y. T., Gillin, J. C., Di Giulio, A. M., Fratta, W., and Costa, E., 1979, Chronic treatment with haloperidol accelerates the biosynthesis of enkephalins in rat striatum, *Brain Res.* **160**:192.

Hosobuchi, Y., 1981, Periaqueductal gray stimulation in humans produces analgesia accompanied by elevation of β-endorphin and ACTH in ventricular CSF, in: *The Role of Endorphins in Neuropsychiatry* (H. M. Emrich, ed.), pp. 109–122, Karger, Basel.

Janowsky, D. S., Segal, D. S., Bloom, F., Abrams, A., and Guillemin, R., 1977, Lack of effect of naloxone on schizophrenic symptoms, *Am. J. Psychiatry* **134**:926.

Jasinski, D. R., Martin, W. R., and Haertzen, C. A., 1967, The human pharmacology and abuse potential of N-allylnoroxymorphone (naloxone), *J. Pharmacol. Exp. Ther.* **157**:420.

Kline, N. S., Li, C. H., Lehmann, H. E., Lajtha, A., Laski, E., and Cooper, T., 1977, β-Endorphin-induced changes in schizophrenic and depressed patients, *Arch. Gen. Psychiatry* **34**:1111.

Kraepelin, E., 1901, *Einfuehrung in die Psychiatrische Klinik*, pp. 11, Joh. Ambrosius Barth-Verlag, Leipzig.

Kraepelin, E., 1927, *Psychiatrie I*, pp. 817–819, Joh. Ambrosius Barth-Verlag, Leipzig.

Kurland, A. A., McCabe, O. L., Hanlon, T. E., and Sullivan, D., 1977, The treatment of perceptual disturbances in schizophrenia with naloxone hydrochloride, *Am. J. Psychiatry* **134**:1408.

Lewis, R. V., Gerber, L. D., Stein, S., Stephen, R. L., Grosser, B. I., Velick, S. F., and Udenfriend, S., 1979, On βH-Leu$^5$-endorphin and schizophrenia, *Arch. Gen. Psychiatry* **36**:237.

Lindström, L. H., Widerlöv, E., Gunne, L. M., Wahlström, A., and Terenius, L., 1978, Endorphins in human cerebrospinal fluid: Clinical correlations to some psychotic states, *Acta Psychiatr. Scand.* **57**:153.

Mackenzie, A. I., 1979, Naloxone in alcohol intoxication, *Lancet* **I**:733.

Manchanda, R., and Hirsch, S. R., 1981, (Des-Tyr$^1$)-γ-endorphin in the treatment of schizophrenia, *Psychol. Med.* **11**:401.

Martin, W. R. (ed.), 1977, *Drug addiction*. I. *Morphine, Sedative-Hypnotic and Alcohol Dependence*, Handbuch der experimentellen Pharmakologie, Vol. 45, Springer, Berlin.

Mayer, D. J., and Watkins, L. R., 1981, Role of endorphins in endogenous pain control systems, in: *The Role of Endorphins in Neuropsychiatry* (H. M. Emrich, ed.), pp. 68–96 Karger, Basel.

Mayer, D. J., Price, D. D., and Rafii, A., 1977, Antagonism of acupuncture analgesia in man by the narcotic antagonist naloxone, *Brain Res*. **121**:368.

Millan, M. J., 1981, Stress and endogenous opioid peptides: A review, in: *The Role of Endorphins in Neuropsychiatry* (H. M. Emrich, ed.), pp. 49–67, Karger, Basel.

Millan, M. J., and Duka, Th., 1981, Anxiolytic properties of opiates and endogenous opioid peptides and their relationship to the actions of benzodiazepines, in: *The Role of Endorphins in Neuropsychiatry* (H. M. Emrich, ed.), pp. 123–141, Karger, Basel.

Millan, M. J., and Emrich, H. M., 1981, Endorphinergic systems and the response to stress, *Psychother. Psychosom*. **36**:43.

Naber, D., Soble, M. G., and Pickar, D., 1981, Ethanol increases opioid activity in plasma of normal volunteers, *Pharmakopsychiatr. Neuro-Psychopharmakol*. **14**:160.

Nedopil, N., and Rüther, E., 1979, Effects of the synthetic analogue of methionine enkephalin FK 33-824 on psychotic symptoms, *Pharmakopsychiatr. Neuro-Psychopharmakol*. **12**:277.

Palmour, R. M., Ervin, F. R., Wagemaker, H., and Cade, R., 1979, Characterization of a peptide from the serum of psychotic patients, in: *Endorphins in Mental Health Research* (E. Usdin, W. E. Bunney, Jr., and N. S. Kline, eds.), pp. 581–593, MacMillan Press, London.

Pickar, D., Vartanian, F., Bunney, W. E., Jr., Maier, H. P., Gastpar, M. T., Prakash, R., Sethi, B. B., Lideman, R., Belyaev, B. S., Tsutsulkovskaja, M. V. A., Jungkunz, G., Nedopil, N., Verhoeven, W., and van Praag, H., 1982, Short-term naloxone administration in schizophrenic and manic patients, *Arch. Gen. Psychiatry* **39**:313–319.

Pickar, D., Cutler, N. R., Naber, D., Post, R. M., Pert, C. B., and Bunney, W. E., Jr., 1980, Plasma opioid activity in manic-depressive illness, *Lancet* **I**:937.

Pickar, D., Naber, D., Post, R. M., van Kammen, D. P., Ballenger, J., Kalin, N., and Bunney, W. E., Jr., 1981, Measurement of endorphins in CSF: Relationship to psychiatric diagnosis, in: *The Role of Endorphins in Neuropsychiatry* (H. M. Emrich, ed.), pp. 246–262, Karger, Basel.

Rimón, R., Terenius, L., and Kampman, R., 1980, Cerebrospinal fluid endorphins in schizophrenia, *Acta Psychiat. Scand*. **61**:395.

Ross, M., Berger, P. A., and Goldstein, A., 1979, Plasma β-endorphin immunoreactivity in schizophrenia, *Science* **205**:1163.

Schenk, G. K., Enders, P., Engelmeier, M. P., Ewert, T., Herdemerten, S., Köhler, K. H., Lodemann, E., Matz, D., and Pach, J., 1978, Application of the morphine antagonist naloxone in psychic disorders, *Arzneim. Forsch*. **28**:1274.

Snyder, S. H., 1978, The opiate receptor and morphine-like peptides in the brain, *Am. J. Psychiatry* **135**:645.

Tamminga, C. A., Tighe, P. J., Chase, T. N., De Fraites, E. G., and Schaffer, M. H., 1981, Destyrosine-γ-endorphin administration in chronic schizophrenics, *Arch. Gen. Psychiatry* **38**:167.

Terenius, L., Wahlström, A., Lindström, L., and Widerlöv, E., 1976, Increased CSF levels of endorphins in chronic psychosis, *Neurosci. Lett*. **3**:157.

Terenius, L., Wahlström, A., and Agren, H., 1977, Naloxone (Narcan®) treatment in depression: Clinical observations and effects on CSF endorphins and monoamine metabolites, *Psychopharmacology* **54**:31.

van Ree, J. M., Verhoeven, W. M. A., van Praag, H. M., and de Wied, D., 1981, Neuroleptic-like and antipsychotic effects of γ-type endorphins, in: *The Role of Endorphins in Neuropsychiatry* (H. M. Emrich, ed.), pp. 266–278, Karger, Basel.

Verhoeven, W. M. A., van Praag, H. M., Botter, P. A., Sunier, A., van Ree, J. M., and de Wied, D., 1978, (Des-Tyr$^1$)-γ-endorphin in schizophrenia, *Lancet* **I**:1046.

Verhoeven, W. M. A., van Praag, H. M., van Ree, J. M., and de Wied, D., 1979, Improvement of schizophrenic patients treated with (des-Tyr$^1$)-γ-endorphin (DTγE), *Arch. Gen. Psychiatry* **36**:294.

Volavka, J., Mallya, A., Baig, S., and Perez-Cruet, J., 1977, Naloxone in chronic schizophrenia, *Science* **196:**1227.
Wahlström, A., and Terenius, L., 1981, Endorphin hypothesis of schizophrenia, in: *The Role of Endorphins in Neuropsychiatry* (H. M. Emrich, ed.), pp. 181–191, Karger, Basel.
Watson, S. J., Berger, P. A., Akil, H., Mills, M. J., and Barchas, J. D., 1978, Effects of naloxone on schizophrenia: Reduction in hallucinations in a subpopulation of subjects, *Science* **201:**73.
Weygandt, W., 1935, *Lehrbuch der Nerven-und Geisteskrankheiten*, p. 507, Marhold-Verlagsbuchhandlung, Halle.

CHAPTER 16

# Endogenous Opioid Peptides in Schizophrenia and Affective Disorders

## FRĂNCESCA BRAMBILLA, ANDREA GENAZZANI, and FABIO FACCHINETTI

### 1. INTRODUCTION

The involvement of endogenous opioid peptides in the nurture of mental diseases is suggested by the following observations:

1. β-Endorphin (β-EP) and enkephalins may act as neurotransmitters in the central nervous system (CNS) in a mediatory process that involves the inhibition of postsynaptic adenylate cyclase and the activation of guanylate cyclase (Gispen *et al.*, 1977; Minneman and Iversen, 1976), or in a modulatory function that alters the response of neurons to the stimulatory effects of classical nonpeptide neurotransmitters. This last effect could be obtained by endorphin-induced modifications of the neuronal membranes, resulting in altered activity of specific enzymes involved in neuronal function. In this context β-EP could interfere with the activity of classical neurotransmitters, and therefore changes in the levels of these natural opioids could be responsible for the appearance and development of mental diseases. Enkephalins have been shown to inhibit the release of acetylcholine (ACh) in the hippocampus, of noradrenalin (NE), dopamine (DA) and adenylate cyclase in the entire brain and especially in the striatum of rats (Kosterlitz and Hughes, 1975).

2. A specific behavioral effect has been demonstrated in animals injected intracerebrally with β-EP, represented by a catatonialike condition with motor retardation and muscular rigidity, which is considered to be an anologue of human catatonia or of the extrapyramidal rigidity elicited in schizophrenics by neuroleptics (Bloom *et al.*, 1976; Jacquet and Marks, 1976). To reinforce the

---

FRĂNCESCA BRAMBILLA • Ospedale Psichiatrico Paolo Pini, Milano Affori, Italy. ANDREA GENAZZANI and FABIO FACCHINETTI • Cattedra di Patologia Ostetrica-Ginecologica, Università di Cagliari, Cagliari, Italy.

significance of this last observation, it has been shown the β-EP-induced catalepsy and neuroleptic-induced rigidity are mediated by similar metabolic alterations, mainly a blockade of DA receptors in the nigrostriatal neurons (Berney and Hornykiewicz, 1977; Van Loon and Kim, 1978). However, the effects of β-EP differ from those of neuroleptics in that the first induces muscle rigidity and loss of the righting reflex followed by a period of excitation. Belluzzi and Stein (1977) and Stein and Belluzzi (1979) suggest that in rats drive-inducing reward functions may be mediated by catecholamines, but drive-reducing reward function, facilitation of learned responses and memory consolidation may be mediated by the natural opioid peptides. Impairment of reward function, learning and memory are frequent symptoms of mental disorders. β-EP administration to normal human volunteers produces mild cognitive impairments, feelings of perplexity and drowsiness (Kline and Lehman, 1978).

3. Morphine and morphinelike compounds, have been anecdotally reported to be effective in improving the symptomatology of mental disorders (Berger, 1978), even though no double-blind studies have ever been done to check this out. On the other hand opiate agonists, such as cyclozocine and nalorphine, have been observed to induce hallucinations and derealization experiences in healthy volunteers (Jasinski et al., 1967). Even though contradictory, the above-mentioned clinical experiences point out a possible influence of opiates in the development and process of mental diseases, or at least of some of their symptoms.

4. Patients suffering from schizophrenia or primary affective disorders (PAD) have been reported to have reduced sensitivity to painful stimuli. It has been suggested that this phenomenon may be related to high levels of β-EP in the CNS, since a therapeutic approach with naloxone (8–20 mg) in a few schizophrenics has decreased the pain threshold (Buchsbaum et al., 1981).

A substantial body of data has been accumulating in recent years from studies based on the hypothesis that β-EP levels are increased or decreased in the major psychoses, schizophrenia, or PAD, or even that a disturbance in the metabolism of this natural opioid is involved in their nurture. The results obtained up to now are not yet convincing for one or another hypothesis, probably due to methodological and technical difficulties.

The following approaches have been used to investigate these theories: administration of opioid antagonists to reduce putative high levels of β-EP; administration of β-EP to increase low levels of the opioid; assay of β-EP-like materials in cerebrospinal fluid (CSF), blood, post-mortem brain, and hemodialysate of patients. The first two approaches were used both to demonstrate putative pathogenetic influences of the opioid peptides in mental diseases and to suggest new therapeutic approaches.

## 1.1. Schizophrenia

### 1.1.1. Opioid Antagonist Administration

The data in this area are frankly contradictory. Gunne et al., (1977) were the first to suggest that naloxone, a typical opioid antagonist, was effective in the treatment of the disease. Administered acutely in an open study to a small group of patients in a very small dose (0.4 mg), naloxone reduced or abolished auditory hallucinations between 2 and 6 hr after the injection. Since then, both positive and negative results have been reported, in open and in double-blind placebo-controlled studies using different doses of the substance. Emrich *et al.* (1977, 1979), Davis *et al.* (1977), Akil *et al.* (1978), Watson *et al.* (1978), Herz *et al.* (1978), Orr *et al.* (1978), Berger *et al.* (1979, 1981), Lehman *et al.* (1979), Barchas *et al.* (1980), Pickar *et al.* (1981a, 1982), Jorgensen *et al.* (1982), and Kleinman *et al.* (1982), using doses of naloxone from 0.4 to 10 mg or 300 mg of naltrexone, described a general improvement in psychotic behavior, mainly represented by reduced or abolished auditory and visual hallucinations, and even improvement in the unusual thought content, for time spans ranging from 2 to 48 hr, and with naloxone administered from 1 to 15 times. Lipinski *et al.* (1979), even though reporting in general no positive results, observed an improvement in the continuous performance task, a measure of vigilance and attention which are supposed to be highly abnormal in schizophrenia (Davis *et al.*, 1980). Schenk *et al.*(1978) observed a striking improvement in catatonic patients given naloxone (0.4–30 mg). Conflicting results in this syndrome have been reported by Emrich *et al.* (1979), who gave 4–10 mg of naloxone.

On the other hand, other studies have failed to demonstrate any significant symptomatological improvement after either acute treatment with naloxone (0.4–20 mg) or chronic therapy with naltrexone (50–250 mg) (Volavka *et al.*, 1977; Janowsky *et al.*, 1977, 1979; Kurland *et al.*, 1977; Simpson *et al.*, 1977; Mielke and Gallant, 1977; Davis *et al.*, 1977; Gitlin and Rosenblatt, 1978; Judd *et al.*, 1978; Gunne *et al.*, 1979; Perez-Cruet *et al.*, 1979; Verhoeven *et al.*, 1981a).

The conflicts between the data could be due to methodological differences, including the variable doses of the antagonists used, the duration of administration (as a bolus or as a slow infusion), whether treatment was acute or prolonged, the selection of the patients (acute or chronic, with or without florid symptomatology, in particular visual and auditory hallucinations, delusion, and psychomotor excitement) and the simultaneous administration of neuroleptics. In this regard, it has been suggested that neuroleptics and naloxone, when administered together, have addictive therapeutic effects. In fact, the patients of Gunne *et al.* (1977) and some of the patients of Watson *et al.* (1979) and Davis *et al.* (1980) who received naloxone and neuroleptics simultaneously were those to show the best results. However, no effects were observed in other subjects given the combined therapy (Janowksy *et al.*, 1977; Mielke and Gallant, 1977; Volavka *et al.*, 1977).

### 1.1.2. β-EP Administration

Kline and Lehman (1979) report up to 2 weeks improvement in the schizophrenic symptomatology after one to four injections of β-EP (1.5–9 mg). The effects were divided into two phases: immediate with paresthesia and autonomic symptoms, and delayed, subdivided again into stage I (5 min–6 hr) defined as antidysphoric with disinhibiting, anxiolitic, and antidepressant activity; stage II (2–3 hr), defined as inhibitory with sedation, drowsiness, and impairment of cognitive processes; stage III (1–10 days), defined as therapeutic with improvement of various schizophrenic symptoms.

According to Barchas et al. (1980), both single bolus and slow infusion of β-EP (20 mg) induced small but significant general improvements in the schizophrenic symptoms plus EEG modifications similar to those elicited by morphine. This was confirmed by Nedopil and Ruther (1979a), using a stabilized enkephalin analogue, and by Krebbs and Roubicek (1979) and by Jorgenson et al. (1979), using a Met-enkephalin analogue. On the contrary, Gerner et al. (1980), Catlin et al. (1981), Petho et al. (1981), Berger et al. (1981), and Pickar et al. (1981b) observed no improvement after β-EP [10 mg in slow infusion]), some of the patients indeed showing psychological deterioration.

According to Jaquet (1980), β-EP fails to cross the blood–brain barrier (BBB) in significant amounts. In fact, it has been demonstrated that in subjects with chronic pain, intravenous injections of the substance elicited no analgesia while intraventricular injections did. The CNS effects reported in psychiatric patients might be secondary, or might be due to fragments of the peptide which can possibly cross the BBB. In this context the results of Verhoeven et al. (1978, 1981, 1982) and Van Ree et al. (1981) are interesting. They observed that the administration of either des-Tyr-gamma-EP (DTγEP) (1 mg a day i.m. for 8 days-3 weeks) or Des-Tyr-gamma-enkephalin (3 mg for 10 days), the first being a peptide probably derived by cleavage of β-EP, devoid of opiate activity, ameliorated the schizophrenic symptomatology. It was pointed out that in rats γ-EP has an effect on behavior similar, even though not identical, to that of haloperidol (De Wied 1980). Both the neuroleptic and β-EP inhibit acquisition and facilitate extinction of conditioned behavior and induce a positive grip test in rats. However, haloperidol decreases locomotor activity and causes sedation, while DTγEP does not. Moreover, DTγEP does not displace [$^3$H]-haloperidol or [$^3$H]-spiperone from their specific binding sites in membrane preparations from rat striatum, frontal cortex or nucleus accumbens and does not antagonize the effects of apomorphine and amphetamine. Therefore, DTγEP does not act as a postsynaptic dopamine antagonist, even though it increases dopamine turnover in certain brain areas by an as yet unknown mechanism. Verhoeven et al. (1982) suggest that in schizophrenia there may be an inborn enzymatic deficit in cleavage of the natural opioid peptides with resulting deficit in γ-EP, or with an accelerated conversion to α-EP, the former metabolite likely representing a natural neuroleptic, the lack of which could influence the clinical symptomatology. These results were not confirmed by Tamminga et al. (1981) and Manchanda et al. (1981), while Meltzer et al. (1982) reported improvement only in 3 out of 8 patients treated.

The data of Berger et al. (1980) and Pfefferbaum et al. (1979) go against the suggestion that β-EP does not cross the BBB. They observed significant improvement of the schizophrenic symptomatology in parallel with modification of the EEG activity, similar to that induced by morphine, when β-EP was injected intravenously either as a bolus or as a slow infusion (20 mg). Moreover, it has been demonstrated (Pezalla et al., 1978; Rapoport et al., 1980) that β-EP given intravenously does cross the mammalian BBB and increases the β-EP levels in the CNS. Therapeutic trials have also been made with FK-33-824 (0.5–3 mg/day for 3 days), a Met-enkephalin derivative, obtaining strong positive effects on hallucinations and an increasing sense of well-being (Jorgensen et al., 1979; Nedopil and Ruther, 1979a).

### 1.1.3. β-EP Levels

Levels of β-EP-like material in blood, CSF, postmortem brain, or hemodialysate have been assayed with variable results. Terenius et al. (1976) and Gunne et al. (1979), using a radioreceptor method, reported that endorphinlike activity, fraction I, is increased in the CSF with a strict relationship between the levels of the peptide and the severity of the disease. Neuroleptic therapy induced positive clinical effects and a decrease of the opioid fraction I, with a tendency of fraction II to increase in parallel.

Domschke et al. (1979) assayed β-EP-like materials by RIA in the CSF and found high levels in acute patients of the paranoid type, and low levels in chronic subjects. Emrich et al. (1979) reported contrasting results of RIA β-EP-like material in CSF and blood: normal values were observed in the CSF and high values in blood, especially in the patients on concomitant neuroleptic therapy. Lindström et al. (1978) examined endorphinlike material in CSF by radioreceptor assay and observed a trend toward elevated levels of fraction I, again with a reduction to normal levels during neuroleptic therapy, in parallel with the clinical improvement. On the contrary, Ross et al. (1979), Höllt et al. (1979), Berger et al. (1981), and Van Kammen et al. (1981) observed comparable β-EP-like immunoreactivity in schizophrenics and controls. Naber et al. (1981) observed that male schizophrenics have nearly a twofold lower CSF opioid activity in comparison to both female patients and controls, and Bianco et al. (1981) reported low plasma β-EP levels in newly schizophrenic patients. Hole et al. (1979) found peptidelike factors in the urine of schizophrenics which have a strong opiate receptor stimulating effect and induce dopaminergic stimulation when injected intracerebroventricularly in rats. Kleine et al. (1981) found increased CSF levels of Met-enkephalin and Leu-enkephalin in six schizophrenics.

It has been occasionally observed that schizophrenics on dialysis for renal diseases tend to improve in their mental symptomatology, and dialysis was also reported to be effective in a group of schizophrenics without renal diseases (Wagemaker and Cade 1977). It was proposed that, through dialysis, a possible low molecular weight schizophrenogenic substance was removed from the body of the patients. This substance was suggested to be Leu-β-EP (Palmour et al. [1979] or a low molecular weight opiatelike material (Lewis et al., 1979). This was not confirmed by Höllt et al. (1979), Nedopil et al. (1979b), Schulz et al. (1981), Diaz-Buxo et al. (1980), and Rorsman et al. (1981), using both hemodialysis and hemoperfusion.

β-EP concentration in five different areas of postmortem human brains (hypothalamus, ventrolateral nucleus of the thalamus, hyppocampus, anterior cingulate gyrus, premotor cortex) has been radioimmunologically measured in schizophrenics and controls by Lightman et al. (1979). No differences in the concentration of the opioid between the two groups of subjects were observed. On the contrary, Kleinman et al. (1981) observed increased levels of Met-enkephalin concentration in caudate nucleus in chronic undifferentiated schizophrenics.

### 1.1.4. Effects of Psychotropic Drugs on Met-enkephalin and Leu-enkephalin

Classical neuroleptics, such as phenothiazines, butyrophenones, and pimozide, administered chronically to rats increase the enkephalin content of the caudatus, putamen, globus pallidus, and nucleus accumbens, but do not increase it in septum and hypothalamus. Clozapine is totally devoid of the above-mentioned activity (Hong et al., 1979; Wise and Stein, 1979). It was inferred that in the former structures a blockade of DA receptors regulate Met-enkephalin production or storage, perhaps transsynaptically.

## 1.2. Primary Affective Disorders

A series of observations seems to link mood impairment to variations in β-EP levels. First, exogenous opiates induce in humans a peculiar state of indifference, an emotional detachment from the experience of suffering (Jaffe and Martin, 1975) or a euphoria similar to that observed in mania (Byck 1976; Kline et al., 1977). Moreover, it has been observed that manic patients seem to feel less pain than normal subjects on administration of experimentally painful stimuli (Davis et al., 1978). Thus, it has been hypothesized that β-EP is elevated in mania and that naloxone therapy might be the treatment of choice of this syndrome (Byck, 1976; Belluzzi and Stein, 1977).

### 1.2.1. Narcotic Antagonist Administration

Naloxone (0.4–6 mg) does not seem to significantly affect depressive mood in PAD patients (Davis et al., 1977; Emrich et al., 1979), even though Terenius et al. (1977) observed a worsening of the symptomatology on withdrawal of the antagonist.

Instead, PAD patients in the manic phase treated with naloxone (2–30 mg in 3–9 trials) showed improvement of the florid symptomatology, even though the change was not always statistically significant (Judd et al., 1978; Janowsky et al., 1978, 1979; Davis et al., 1979). In two manic patients Emrich et al. (1979) found no improvement, and even a worsening, of the florid symptomatology. No significant modification of the manic mood was reported by Davis et al.

(1980). Pickar *et al.* (1982) observed no significant behavioral effects in 12 manic patients given naloxone (10 mg) therapy.

### 1.2.2. β-EP Administration

Kline and Lehman (1979), Gerner *et al.* (1980), Gold *et al.* (1979), and Pickar *et al.* (1981b) have reported that β-EP (1.8–9 mg) improved depressive mood in a limited number of PAD patients. Angst *et al.* (1979) administered β-EP (10 mg) as a single injection and observed an improvement in depressive mood persisting for 2 hr, and in few cases a switch from depression to hypomania or mania. Instead no improvement of depression was observed by Catlin *et al.* (1980) after injection of 4.3–10 mg of β-EP.

### 1.2.3. β-EP Levels

Terenius *et al.* (1977) observed increased CSF levels of β-EP-like material, fraction I and II, in PAD patients both in the depressive and manic phases of the disease. This was confirmed by Lindström *et al.* (1978). Risch (1982) observed high plasma β-EP levels and an increased response to physostigmine stimulation as compared to that of controls. Instead, Emrich (1982), Catlin *et al.* (1981) and Post *et al.* (1981) observed no differences in β-EP immunoreactivity between depressed and manic PAD patients and controls. Naber *et al.* (1981), too, observed no differences in CSF opioid activity between PAD patients and controls, but four patients showed higher opioid levels during the manic than the depressive phase of the disease. Post *et al.* (1981) observed that β-EP levels in PAD patients even though being normal in the CSF tend to be significantly positively correlated with the severity of anxiety, while in normal volunteers they are negatively correlated.

The contradiction in the data reported above might result from many variants that can interfere with the results obtained. These include heterogeneity of the diseases, age and sex of the patients, age of onset and duration of illness, previous treatments and whether or not patients were on psychotropic therapy during the examination. Moreover, the investigative approach using either therapy with the opioid or with the antagonists to demonstrate the existence of an opioid peptide pathology in mental diseases is questionable. In fact, doses and duration of treatment vary from investigation to investigation since we do not know the optimum doses for specific cases. We know, for instance, that both neuroleptics and antidepressants show a latency time from a few days to a few weeks before being effective: it may be that the agonist or antagonist opioid therapies also need longer time or higher doses than those that had been used to correct putative matabolic impairments in specific brain areas.

The direct approach of measuring opioid levels in blood or CSF should be more conclusive for assessing real metabolic patterns of these peptides and their relationship with mental disorders, even though endorphin blood levels may be more directly related to pituitary than brain secretion. However, the studies performed up to now have used radioreceptor assay or radioimmunoassay which measure all of the endorphinlike material, without differentiating between β-EP and β-LPH, or other opioidlike substances. Therefore, they may miss a specific pathology of one or the other peptide.

We decided to reinvestigate blood levels of each of the two opioids, β-EP and β-LPH, in schizophrenics before and during neuroleptic therapy and in PAD and secondary affective disorders (SAD) patients, also assaying the related ACTH levels and calculating the molar ratio (MR) for the two peptides.

## 2. METHODS AND MATERIALS

### 2.1. Subjects

Twenty-eight chronic schizophrenics, 23 hebephrenics and 5 paranoids, were investigated, including 19 men and 9 women (7 fertile and 2 menopausal), aged 23 to 64 years and with 9- to 36-year histories of schizophrenia. All had delusion and severe thought disorders. Only 17 had visual or auditory hallucinations. The hebephrenics had severe mental deterioration, apathy, stolidity, withdrawal from reality, and incapacity for interpersonal relating. No subjects had been on drug therapy for the 10 days prior to study. Previous treatments had included phenothiazines, haloperidol, and benzodiazepines. All subjects were inpatients in our Institute during the investigation, receiving the same common hospital diet and being subjected to the same hygienic rules.

The PAD subjects were 15, 2 men and 13 women (3 fertile and 10 menopausal), aged 33 to 65 years, with 2- to 32-year histories of the disorder; 13 had unipolar disorder, while 2 had bipolar disorder with 1 to 20 depressive episodes. One of the patients was in a normothimic phase of the disease. Three patients had a retarded type of depression, 2 were agitated, 4 strongly anxious. They had all been off therapy for at least 10 days before our investigation; 12 were inpatients while the remaining 3 were investigated as outpatients. They had been previously treated with tricyclic antidepressants, lithium salts, ECT, and benzodiazepines, and 6 of them, briefly, with phenothiazines, haloperidol, and sulpiride.

The SAD patients were 8, 1 man and 7 women (4 fertile and 3 menopausal), aged 23 to 50 years, with 3-month to 5-year histories of the depressive symptomatology. All of them suffered basically from neuroses. They were all outpatients.

The diagnosis of our patients was based on the Research Diagnostic Criteria. The Wittenborn Rating Scale for the schizophrenics and the Hamilton Rating Scale for the depressed patients were used to monitor current psychological states.

Patients with previous histories of cerebral trauma, encephalopathies, chronic organic diseases, overt endocrinopathies, and obesity were not included in this investigation.

Twenty-one subjects defined as psychologically normal after examination by two psychiatrists, matched for age and sex, were used as controls. Informed consent was obtained from all patients and controls after explanation of the entire research protocol.

## 2.2. Assay Procedures for β-EP, β-LPH, and ACTH

Plasma β-EP, β-LPH, and ACTH levels were assayed in all patients when they had been off therapy for 10 days. The blood samples were collected after 12 hr of fasting. The inpatients were in bed for 12 hr before the collection, while outpatients were given rest in bed for 1 hr in the hospital prior to the collection of blood samples. Heparinized blood samples were drawn at 9 a.m. through a 19-gauge butterfly needle inserted in a forearm vein at 8:30 a.m. and kept patent by a saline infusion. The samples were immediately centrifuged and the plasma stored at $-20°C$.

Plasma ACTH was assayed by a charcoal RIA method with the commercially available kits of Serono.

Human β-LPH from extractive source for standard and labeling purposes was kindly donated by Professor C. H. Li; synthetic β-EP was supplied by BACHEM (Torrance, Calif.). Iodination of both proteins was made using the chloramine-T method: purification of labeled β-LPH or β-EP from $^{125}I$ residual was achieved through a Sephadex G-25 (1 × 20.0 cm) column chromatography.

Anti-β-LPH rabbit serum (N-terminal) and anti-β-EP rabbit serum (C-terminal) were also generously provided by Professor C. H. Li. Characteristics of the two antisera were previously reported (Wiedemann et al., 1977, 1979). Briefly, 0.4% of β-EP was recognized by N-terminal serum, while 16% of β-LPH cross-reacts with C-terminal serum.

β-LPH and β-EP were extracted from plasma (3 ml) with silicic acid (200 mg) through a 1-hr mixing. After two successive water and 0.1 M HCl washing, the two peptides were removed from silicic acid with 2 ml of acetone-HCl (2:1) mixture. The acid–acetone mixture was then dried under nitrogen, the residue redissolved with 0.4 ml of 0.1 M acetic acid, 0.01 M BSA and applied on a column of Sephadex G-25 (1.5 × 45 cm) eluted with the above-mentioned acetic acid. According to the respective elution coefficients, two distinct peaks containing respectively β-LPH and β-EP were collected, freeze-dried, redissolved with 0.4 ml of phosphate buffer 0.04 M, pH 7.4, and submitted to the RIA. Characteristics of the RIA were reported elsewhere (Facchinetti and Genazzani, 1979).

Eight schizophrenics were given haloperidol 0.1 mg/kg body wt per day for 10 days, and the assays were repeated.

Data were analyzed statistically by Student's $t$ test. Moreover, in view of the common origin of β-EP, β-LPH, and ACTH (Meins et al., 1977) and their postulated concurrent secretion (Guillemin et al., 1977), the correlation coefficients and multiple linear regression in their plasma concentrations were checked in the three groups of subjects and compared to those of controls.

## 3. RESULTS

### 3.1. Schizophrenic Patients

Basal β-LPH values were elevated in 13 patients (106–217 pg/ml) and normal in the other 15. The mean values (Fig. 1) for the entire group were significantly higher than in controls ($p < 0.01$), with no differences between patients with and without visual or auditory hallucinations or between paranoid and hebephrenic types of the disease.

Basal β-EP levels were elevated in 19 patients (42–253 pg/ml), low in 2 (2–8 pg/ml), and normal in the others. Again the mean value (Fig. 1) for the entire group was significantly higher than in controls ($p < 0.01$), with no differences between patients with and without visual and auditory hallucinations or between paranoid and hebephrenic subjects.

Basal ACTH levels were elevated in 8 patients (102–250 pg/ml) and normal in the others. No statistically significant difference was observed between the schizophrenics as a group and the controls (Fig. 1).

The β-LPH/β-EP molar ratio (MR) (Fig. 1) was low in 12 patients (0.20–0.50) and elevated in 4 (3.72–29.2), 3 of whom had low basal β-EP levels. When the schizophrenics were considered as a group, there was no significant difference between the MR of schizophrenics and controls, between schizophrenics with and without hallucinations, or between paranoid and hebephrenic subjects.

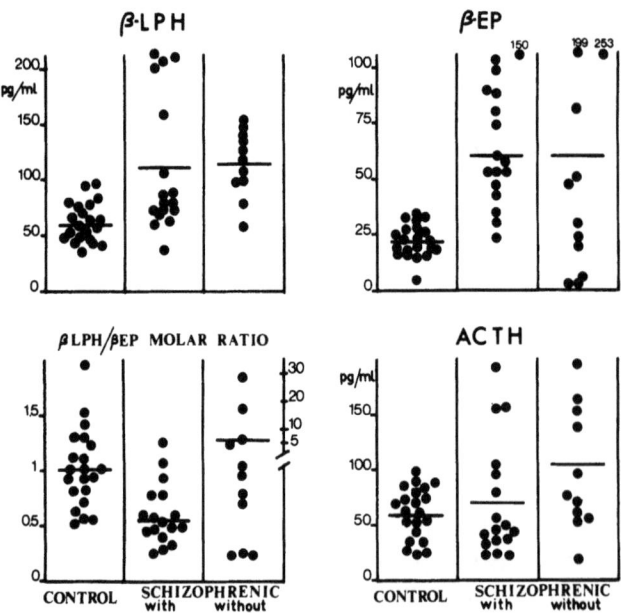

FIGURE 1. β-LPH, β-EP, β-LPH/β-EP MR, and ACTH plasma levels in 21 controls and in 28 chronic schizophrenic patients, 17 with and 11 without hallucinations.

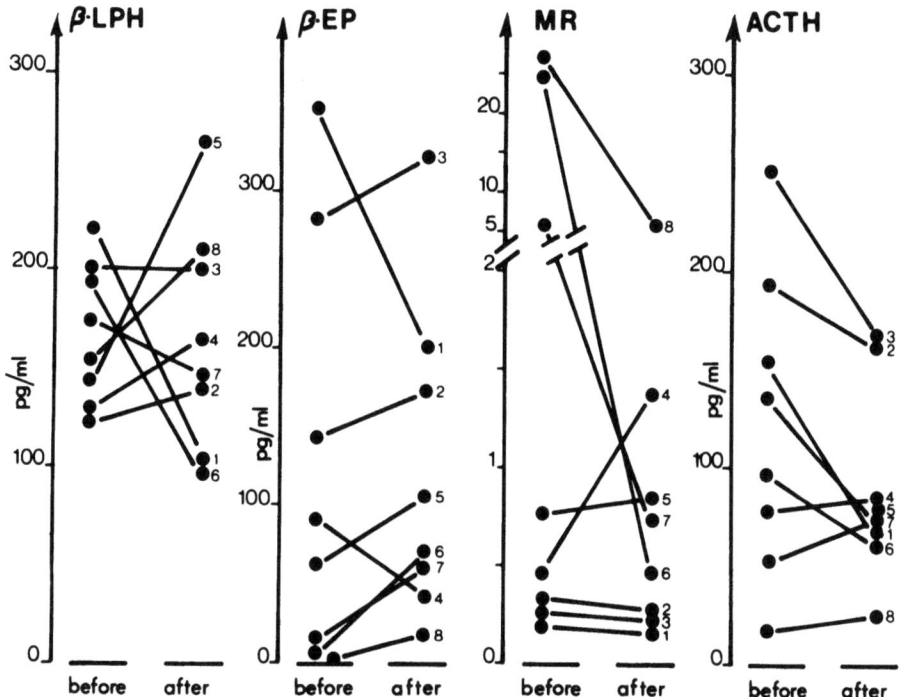

FIGURE 2. β-LPH, β-EP, β-LPH/β-EP MR, and ACTH plasma levels of 8 chronic schizophrenics before and after 10 days of haloperidol treatment (0.1 mg/kg body wt).

Ten patients showed a parallelism in the levels of β-LPH, β-EP, and ACTH (4 with high and 6 with normal values) but in the other 18 there was no paralellism for the levels of the three peptides.

In the entire schizophrenic group ($r = 0.644$; $p < 0.01$) and in the subgroup with hallucinations ($r = 0.704$; $p < 0.01$), β-LPH and β-EP plasma levels were significantly positively correlated. This did not occur in schizophrenics without hallucinations ($r = 0.395$). Moreover, ACTH and β-EP plasma levels were positively correlated in the entire schizophrenic group ($r = 0.523$; $p < 0.01$) and in the subgroup without hallucinations ($r = 0.678$; $p < 0.05$).

A multiple linear regression between β-LPH, β-EP, and ACTH was found in schizophrenics ($r = 0.585$; $p < 0.01$) and in the subgroup without hallucinations.

Haloperidol therapy (Fig. 2) induced no systematic modification in β-LPH levels. An increase in β-EP levels to normal values was observed in three subjects who had low levels before therapy. In the cases with normal to high basal β-EP levels, the effect of the treatment varied. The MR was unchanged in three subjects with very high β-EP levels before and after haloperidol therapy, while it decreased in subjects whose β-EP levels increased from low to normal after therapy. ACTH levels were not modified by the therapy.

## 3.2. PAD Patients

β-LPH values were normal in 6 patients and high in the other 9 (117–258 pg/ml). The mean plasma β-LPH levels (Fig. 3) were significantly higher than in controls ($p < 0.01$).

β-EP levels were normal in 4 cases and elevated in the other 11 (39–137 pg/ml). The mean values (Fig. 3) were significantly higher than in controls ($p < 0.001$). No consistent relationships between β-LPH and β-EP levels was observed ($r = 0.339$).

The MR (Fig. 3) was elevated in 1 subject (4.80), low in 2 (0.36 and 0.46), and normal in the other patients.

ACTH values were normal in 11 cases and elevated in 4 (113–140 pg/ml) (Fig. 3).

A significant multiple linear regression ($r = 0.489$; $p < 0.05$) existed between β-LPH, β-EP, and ACTH plasma levels of PAD patients (ACTH, 44.9 + 0.321β-LPH − 0.318β-EP).

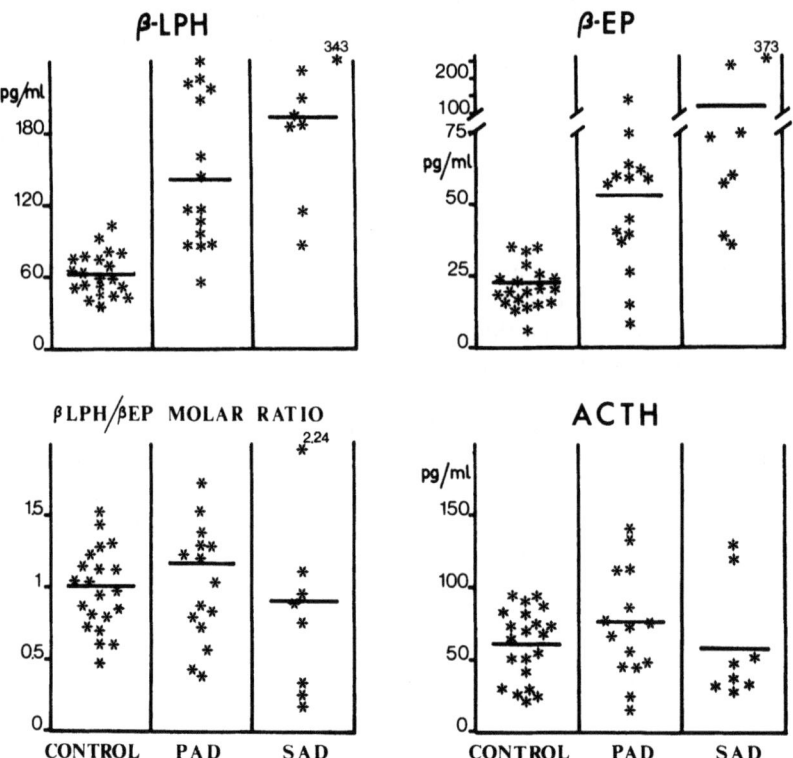

FIGURE 3. β-LPH, β-EP, β-LPH/β-EP MR, and ACTH plasma levels of 21 controls and 15 patients with PAD and 8 patients with SAD.

## 3.3. SAD Patients

β-LPH levels were elevated in 7 cases (116–342 pg/ml) and normal in 1. The mean plasma levels (Fig. 3) of β-LPH were significantly higher than those found in controls ($p < 0.001$).

β-EP levels were high in 6 cases (64–373 pg/ml) and normal in the other 2. The mean value for the entire group (Fig. 3) was significantly higher than in controls ($p < 0.001$). No correlation was observed between β-LPH and β-EP plasma levels ($r = 0.060$).

The MR was high in 1 subject (2.24), low in 2 cases (0.19 and 0.33) and normal or at the lower limit of the normal range in the others. The mean MR (Fig. 3) ($0.90 \pm 0.27$) was significantly lower than in the controls ($p < 0.05$).

ACTH levels were elevated in 2 cases (120–130 pg/ml) and normal in the others (Fig. 3). No significant correlation with β-LPH ($r = 0.420$) or β-EP ($r = 0.525$) was found.

A multiple linear regression between the three peptide plasma levels revealed that ACTH equals $82.6 - 0.234\beta\text{-LPH} + 0.183\beta\text{-EP}$ ($r = 0.687$; $p < 0.01$). This theoretically implies that for equimolar amounts of β-LPH and β-EP, half molar concentrations of ACTH are to be expected.

## 4. CONCLUSION

The results reported above confirm our preliminary data obtained in a small group of schizophrenics, PAD and SAD patients (Brambilla et al., 1981). The schizophrenics, when considered as a group, had significantly higher β-LPH and β-EP levels than the controls, but no strict parallelism was observed between the values of β-LPH, β-EP and ACTH except in few patients. However, the results were heterogeneous, with some of the patients showing frankly elevated levels of the opioids and some normal values. When possible interfering factors were taken into consideration in attempt to explain this variability, we observed no relationship between opioid levels and type of disease, hebephrenic or paranoid, presence or absence of visual and auditory hallucinations, sex and age of the patients, age of onset or duration of the disease or type of previous therapy. At the moment, we have no explanation for the heterogeneity of the results. However, it seems obvious that schizophrenia is not characterized by one common specific pattern of opioid secretion, and that even though the population was strictly selected in term of diagnosis and other possible interfering variants, it includes at least two groups of biochemically different patients, one with normal and one with elevated levels of the peptides. This could be the basis for the conflicting effects of the agonist–antagonist therapy and for the differences in opioid levels reported in the literature.

Haloperidol treatment increased β-EP levels in patients with low basal levels of the opioids and decreased them in two cases with high basal values, with no concomitant changes in β-LPH and ACTH. The effects of the neuroleptic therapy in our patients do not confirm the data of Terenius *et al.* (1976), who observed a decrease in β-EP-like activity in CSF during drug treatment. More extensive studies, in term of number of patients, greater amounts of drug or longer period of treatment might clarify the role of the DA receptor blocking agent on the secretion of the opioids.

PAD patients showed a trend toward higher than normal secretion of the three peptides, especially with regard to β-LPH levels. This finding disagrees with those of Naber *et al.* (1981), Post *et al.* (1981), and Emrich *et al.* (1982), and supports those of Terenius *et al.* (1976), Lindstroöm *et al.* (1978), and Risch *et al.* (1982), who found elevated β-EP-like activity in the CSF and blood. However, here again we observed a wide range of β-LPH, β-EP and ACTH values, with no apparent relationship to age and sex of the patients, duration of the disease, number of depressive episodes, uni- or bipolarity, presence of anxiety and agitation or type of previous therapy. The variability in the results may be part of the heterogeneity of the course, symptomatology, prognosis, and response to therapy which characterizes the disease and may be linked to the large spectrum of individual entities which are all diagnosed as PAD. The multiple linear correlation between the three peptide plasma values suggests that their circulating levels should be present in equimolar amounts, indicating increased secretion of proopiocortin with normal enzymatic cleavage of the precursor molecule.

SAD patients, as a group, showed increased secretion of β-LPH and β-EP and normal-to-low ACTH values. The multiple linear regression between the three peptides showed a significant positive correlation, but the equation (ACTH = 0.1174 + 0.022 β-LPH + 0.171 β-EP ) sustains the observation that, in the physiological range, the molar ratio between each opioid and ACTH is near 2. This suggests an alteration in the enzymatic breakdown of proopiocortin, with hypersecretion of the opioid derivatives independent of ACTH release; or, there may be an alteration in the clearance rate or distribution volumes of the peptides. Our data point out a striking difference in the opioid secretory profiles in PAD and SAD patients.

Even though our findings are preliminary and require more extensive studies in terms of number of patients, effects of stimulation and inhibition tests, circadian rhythms and influence of psychotropic drugs, it seems to us that the patterns of opioid secretion displayed by schizophrenics, PAD, and SAD patients may help to differentiate specific groups of subjects with different metabolic substrata.

At present, it seems difficult to sustain the idea that the alterations in blood opioid levels are, of themselves, responsible for the nurture of schizophrenia and affective disorders. In fact, the impairment seems to be roughly the same in both diseases, with variable increases in β-LPH and β-EP, and the alteration is not present in the 100% of the patients. Moreover, blood levels of the opioids probably reflect mostly the pituitary secretion and not the brain content of the substance. However, we do not see any special reason why β-EP should be hypersecreted by the pituitary in mental disorders, while remaining normal in the brain. Moreover, it is well known that hypophysial hormones can reach the hypothalamus and possibly higher brain areas by reverse flow via the portal vessels or the CSF (Oliver *et al.*, 1977). Therefore, either β-EP hypersecretion is a phenomenon common to the entire neuroendocrine system, including both the pituitary and the brain, or if the pituitary alone be responsible for the high β-EP levels, the opioid may reach specific brain areas and interfere with their functions.

In agreement with the new biological hypothesis of mental disorders, it is likely that opioid alterations are one of the impairments in a complex biochemical dysfunction that includes all the classical neurotransmitters and possibly the related neuromodulators.

With regard to the pathogenesis of schizophrenia, the most widely accepted hypothesis is that the disease is linked to an alteration of DA transmission in specific brain areas. Recently, a theory connecting the DA hypothesis with putative impairments in β-EP levels has been advanced by Volavka *et al.* (1979). Elevated levels of β-EP can cause both inhibition of DA release and decreased phosphorylation of membrane proteins. As a result of the decreased transmitter release, there may be a supersensitivity of the DA postsynaptic receptors, while the decreased membrane phosphorylation can result in an increased cyclic AMP response to DA stimulation. Conversely, a β-EP deficiency would result in a lack of DA inhibition and therefore in an increased secretion of the transmitter. Thus, both decreased and increased β-EP levels could lead to increased dopaminergic tonus in specific brain areas.

The presence of equally high levels of β-EP, β-LPH, and ACTH in PAD might be an expression of the activation of the hypothalamo-pituitary-adrenal axis typical of the disease (Sachar *et al.*, 1980). In this context, it could be another neuroendocrine marker of the illness, possibly related to the same biochemical impairment of the classical neurotransmitter(s), which is the basis of the pathogenesis of PAD.

Whatever the relationship between endogenous opioid levels and nurture of mental disorders may be, it seems worthwhile to extend the investigation of the metabolic patterns of these substances in psychotic patients to identify specific subgroups with peculiar symptomatologic and prognostic aspects of the disease.

## 5. REFERENCES

Akil, H., Richard, D. E., Hughes, J., and Barchas, J. D., 1978, Enkephalin-like material elevated in ventricular cerebrospinal fluid of pain patients after analgetic focal stimulation, *Science* **201**:463.

Angst, J., Autenrieth, V., Brem, F., Koukkou, M., Meyer, H., Stassen, H. H., and Storck, U., 1979, Preliminary results of treatment with β-endorphin in depression, in: *Endorphins in Mental Health Research* (E. Usdin, W. E. Bunney, Jr., and N. S. Kline, eds.), pp. 518–528, Oxford University Press, London.

Barchas, J. D., Berger, P. A., Watson, S. J., Akil, H., and Choh, H. L., 1980, Opioid agonists and antagonists in schizophrenia, in: *Neural Peptides and Neuronal Communication* (E. Costa and M. Trabucchi, eds.), pp. 447–453, Raven Press, New York.

Belluzzi, J. D., and Stein, L., 1977, Enkephalin may mediate euphoria and drive-reduction reward, *Nature (London)* **266**:556.

Berger, P. A., 1978, Investigating the role of endogenous opioid peptides in psychiatric disorders, *Neurosci. Res. Program Bull.* **16**:585.

Berger, P. A., Watson, S. J., Akil, H., and Barchas, J. D., 1979, Naloxone administration in chronic hallucinating schizophrenic patients, in: *Endorphins in Mental Health Research* (E. Usdin, W. E. Bunney, Jr., and N. S. Kline, eds.), pp. 423–434, Oxford University Press, London.

Berger, P. A., Watson, S. J., Akil, H., Elliott, G. R., Rubin, R. T., Pfefferbaum, A., Davis, K. L., Barchas, J. D., and Li, C. H., 1980, Beta-endorphin and schizophrenia, *Arch. Gen. Psychiatry* **37**:635.

Berger, P. A., Watson, S. J., Akil, H., Barchas, J. D., 1981, Clinical studies of the role of endorphins in schizophrenia, *Mod. Probl. Pharmacopsychiat.* **17**:226.

Berney, S., and Hornykiewicz, O., 1977, The effect of beta-endorphin and Met-enkephalin on striatal dopamine metabolism and catalepsy: Comparison with morphine, *Commu. Psychopharmacol.* **1**:597.

Bianco, F., Castro, R., and Sanchez, C., 1981, Endorphins and schizophrenia, *III World Congr. Biological Psychiat.*, Stockholm, Abstracts, F. 1002.

Bloom, F., Segal, D., Ling, N., and Guillemin, R., 1976, Endorphins: Profound behavioral effects in rats suggest new etiological factors in mental illness, *Science* **194**:630.

Brambilla, F., Genazzani, A. R., Facchinetti, F., Parrini, D., Petraglia, F., Sacchetti, E., Scarone, S., Guastalla, A., and D'Antona, N., 1981, Beta-endorphin and beta-lipotropin plasma levels in chronic schizophrenia, primary affective disorders and secondary affective disorders, *Psychoneuroendocrinology* **6**:321.

Buchsbaum, M. S., Davis, G. C., and van Kammen, D. P., 1980, Diagnostic classification and the endorphin hypothesis of schizophrenia: Individual differences and psychopharmacological strategies, in: *Perspectives in Schizophrenia Research* (C. Baxter and T. Melnechuck, eds.), pp. 177–191, Raven Press, New York.

Byck, R., 1967, Peptide transmitters: A unifying hypothesis for euphoria, respiration, sleep, and action of lithium, *Lancet* **II**:72.

Catlin, D. H., Gorelick, D. A., Gerner, R. H., Hui, K. K., and Li, C. H., 1980, Clinical effects of beta-endorphin infusion, in: *Neural Peptides and Neuronal Communication* (E. Costa and M. Trabucchi, eds.), pp. 465–472, Raven Press, New York.

Catlin, D. H., Gerner, R. H., and Gorelick, D. A., 1981, Beta-endorphin:behavioral effects of single and multiple infusions-measurement of CSF levels, III World Congr. Biological Psychiat, Stochkholm, Abstracts S. 56.

Davis, G. C., Bunney, W. E., Jr., De Fraites, E. G., Kleinman, J. E., van Kammen, D. P., Post, R. M., and Wyatt, R. J., 1977, Intravenous naloxone administration in schizophrenia and affective illness, *Science* **197**:74.

Davis, G. C., Buchsbaum, M. S., and Bunney, W. E., Jr., 1978, Naloxone decreases diurnal variation in pain sensitivity and somatosensory evoked potentials, *Life Sci.* **23**:1449.

Davis, G. C., Buchsbaum, M. S., and Bunney, W. E., Jr., 1979, Research in endorphins and schizophrenia, *Schizophr. Bull.* **5**:244.

Davis, G. C., Buchsbaum, M. S., and Bunney, W. E., Jr., 1980, Alterations of evoked potentials link research in attention dysfunction to peptide response symptoms of schizophrenia, in: *Neural Peptides and Neuronal Communication* (E. Costa and M. Trabucchi, eds.), pp. 473–487, Raven Press, New York.

de Wied, D., 1980, Peptides and adaptive behavior, in: *Hormones and the Brain* (D. de Wied and P. A. Van Keep, eds.), pp. 103–113, MTP Press, Lancaster.

Diaz-Buxo, J. A., Caudle, J. A., Chandler, J. T., Farmer, C. D., Holbrook, W. D., 1980, Dyalisis of schizophrenic patients:a double-blind study, *Am. J. Psychiatry* **137**:1220.

Domschke, W., Dickschas, A., and Mitznegg, P., 1979, CSF beta-endorphin in schizophrenia, *Lancet* **II**:425.
Emrich, H. M., 1982, A possible role of opioid substances in depression, in: *Typical and Atypical Antidepressants* (E. Costa and G. Racagni, eds.), pp. 77–86, Raven Press, New York.
Emrich, H. M., Cording, C., Pirée, S., Kölling, A., von Zerssen, D., and Herz, A., 1977, Indication of an antipsychotic action of the opiate antagonist naloxone, *Pharmakopsychiatr. Neuro-Psychopharmakol.* **10**:265.
Emrich, H. M., Cording, C., Pirée, S., Kölling, A., Müller, H.-J. von Zerssen, D., and Herz, A., 1979, Actions of naloxone in different types of psychoses, in: *Endorphins in Mental Health Research* (E. Usdin, W. E. Bunney, Jr., and N. S. Kline, eds.), pp. 452–460, Oxford University Press, London.
Ernst, K., Dorner, I., Horbst, A., Klinkmann, H., Precht, K., and Seidel, M., 1981, Hemodyalisis treatment in schizophrenia, III World Congr. Biological Psychiatry, Stockholm, Abstracts, F. 120.
Facchinetti, F., and Genazzani, A. R., 1979, Simultaneous radioimmunoassay for beta-lipotropin and beta-endorphin in human plasma, in: *Radioimmunoassay of Drugs and Hormones in Cardiovascular Medicine* (A. Albertini, M. Da Prada, and B. A. Peskar, eds.), pp. 347–354, Elsevier/North-Holland Amsterdam.
Gerner, R. H., Catlin, D. H., Gorelick, D. A., Hui, K. K., and Li, C. H., 1980, Beta-endorphin: Intravenous infusion causes behavioral change in psychiatric inpatients, *Arch. Gen. Psychiatry* **37**:642.
Gispen, W. H., van Ree, J. M., and de Wied, D., 1977, Lipotropin and the central nervous system, *Int. Rev. Neurobiol.* **20**:209.
Gitlin, M., and Rosenblatt, M., 1978, Possible withdrawal from endogenous opiates in schizophrenics, *Am. J. Psychiatry* **135**:377.
Gold, M. S., Pottash, A. L. C., Sweeney, D. R., Kleber, H. D., and Redmond, D. E., Jr., 1979, Rapid opiate detoxification: Clinical evidence of antidepressant and antipanic effects of opiates, *Am. J. Psychiatry* **136**:982.
Guillemin, R., Vargo, T., Rossier, J., Minick, S., Ling, N., Rivier, C., Vale, W., and Bloom, F., 1977, Beta-endorphin and adrenocorticotropin are secreted concomitantly by the pituitary gland, *Science* **197**:1367.
Gunne, L.-M., Lindström, L., and Terenius, L., 1977, Naloxone-induced reversal of schizophrenic hallucinations, *J. Neural Transm.* **40**:13.
Gunne, L.-M., Lindström, L., and Widerlöi, E., 1979, Possible role of endorphins in schizophrenia and other psychiatric disorders, in: *Endorphins in Mental Health Research* (E. Usdin, W. E. Bunney, Jr., and N. S. Kline, eds.), pp. 547–552, Oxford University Press, London.
Herz, A., Bläsig, J., Emrich, H. M., Cording, C., Pirée, S., Kölling, A., and von Zerssen, D., 1978, Is there some indications from behavioral effects of endorphins for their involvement in psychiatric disorders?, in: *Advances in Biochemical Psychopharmacology* (E. Costa and M. Trabucchi, eds.), pp. 333–339, Raven Press, New York.
Hole, K., Bergslien, H., Jørgensen, H. A., Berge, O. G., Reichelt, K. L., and Etrygstad, O. E., 1979, A. peptide containing fraction in the urine of schizophrenic patients which stimulates opiate receptors and inhibits dopamine uptake, *Neuroscience* **4**:1883.
Höllt, V., Hillebrand, G., Schmidt, B., and Gurland, H. J., 1979, Endorphins in schizophrenia: Hemodialysis/hemoperfusion are ineffective in clearing beta-Leu$^5$-endorphin and beta-endorphin from human plasma, *Pharmakopsychiatr. Neuro-Psychopharmakol.* **12**:399.
Hong, J. S., Yang, H. Y., Gillin, J. C., Di Giulio, A. M., Fratta, W., and Costa, E., 1979, Chronic treatment with haloperidol accelerates the biosynthesis of enkephalins in rat striatum, *Brain Res.* **160**:192.
Jacquet, Y. F., and Marks, N., 1976, The C-fragment of beta-lipotropin: An endogenous neuroleptic or antipsychotogen, *Science* **194**:632.
Jacquet, Y. F., 1980, Beta-endorphin, blood–brain barrier and schizophrenia, *Lancet (Lett.)* **12**:831.
Jaffe, J. H., 1975, Drug action and drug abuse, in: *Pharmacological Basis of Therapeutics* (L. S. Goodman and A. Gilman, eds.), pp. 284–324, Macmillan Co., New York.

Janowsky, D. S., Segal, D. S., Bloom, F., Abrams, A., and Guillemin, R., 1977, Lack of effects of naloxone on schizophrenic symptoms, *Am. J. Psychiatry* **134**:926.

Janowsky, D. S., Judd, L., Huey, L., Roitman, N., Parker, D., and Segal, D., 1978, Naloxone effects on manic symptoms and growth-hormone levels, *Lancet* **II**:320.

Janowsky, D. S., Judd, L. L., Huey, L., and Segal, D., 1979, Effects of naloxone in normal, manic and schizophrenic patients: Evidence for alleviation of manic symptoms, in: *Endorphins in Mental Health Research* (E. Usdin, W. E. Bunney, Jr., and N. S. Kline, eds.), pp. 435–447, Oxford University Press, London.

Jasinski, D. R., Martin, W. R., and Haertzen, C. A., 1967, The human pharmacology and abuse potential of N-allyl noroxymorphone (naloxone), *J. Pharmacol. Exp. Ther.* **157**:420.

Jorgensen, H. A., and Cappelen, C., Jr., 1982, Naloxone-induced reduction of schizophrenic symptoms: a case report, *Acta Psychiat. Scand.* **65**:370.

Jørgensen, A., Fog, R., and Veilis, B., 1979, Synthetic enkephalin analogue in treatment of schizophrenia, *Lancet* **I**:935.

Judd, L. L., Janowsky, D. S., Segal, D. S., and Huey, L. Y., 1978, Naloxone related attenuation of manic symptoms in certain bipolar depressives, in: *Characteristics and Function of Opioids: Developments in Neurosciences* (J. M. van Ree and L. Terenius, eds.), pp. 173–174, Elsevier/North-Holland, Amsterdam.

Kleine, T. O., Klempel, K. and Funfgeld, E. W., 1981, Methionine enkephalin and leucine enkephalin in CSF of patients suffering from chronic schizophrenia, III World Congr. Biological Psychiatry, Stockholm Abstract F. 238.

Kleinman, J. E., Karoum, F., Rosenblatt, J. E., Hong, J., Gillin, J. C., Bridge, T. P., Zalcman, S., Costa, E., and Wyatt, R. J., 1981, Postmortem studies of catecholamines and endogenous peptides, III World Congr. Biological Psychiatry, Stochkolm, Abstract S. 195.

Kleinman, J. E., Weinberger, D. R., Rogol A., Shilling, D. J., Mendelson, W. B., Davis, G. C., Bunney, W. E., Jr., and Wyatt, R. J., 1982, Naloxone in chronic schizophrenia: neuroendocrine and behavioral effects, *Psychiatry Res.*, **7**:1.

Kline, N. S., and Lehmann, H. E., 1978, Clinical observations with beta-endorphin injections, *Psychopharmacol. Bull.* **14**:12.

Kline, N. S., and Lehmann, H. E., 1979, β-Endorphin therapy in psychiatric patients, in: *Endorphins in Mental Health Research* (E. Usdin, W. E. Bunney, Jr., and N. S. Kline, eds.), pp. 500–517, Oxford University Press, London.

Kline, N. S., Li, C. H., Lehmann, H. E., Lajtha, A., Laski, E., and Cooper, T., 1977, Beta-endorphin-induced changes in schizophrenic and depressed patients, *Arch. Gen. Psychiatry* **34**:1111.

Kosterlitz, H. W., and Hughes, J., 1975, Some thoughts on the significance of enkephalin, the endogenous ligand, *Life Sci.* **17**:91.

Krebs, E., and Rubicek, J., 1979, EEG and clinical profile of a synthetic analogue of methionine-enkephalin—FK 33-824, *Pharmakopsychiatr. Neuro-Psychopharmakol.* **12**:86.

Kurland, A. A., McCabe, O. L., Hanlon, T. E., and Sullivan, D., 1977, The treatment of perceptual disturbances in schizophrenia with naloxone hydrochloride, *Am. J. Psychiatry* **134**:1408.

Lehmann, H. E., Vasavan Nair, W. P., and Kline, N. S., 1979, β-Endorphin and naloxone in psychiatric patients: Clinical and biological effects, *Am. J. Psychiatry* **136**:762.

Lewis, R. V., Gerber, L. D., Stein, S., Stephen, R. L., Grosser, B. I., Velick, S. F., and Undenfriend, S., 1979, on β-Leu$^5$-endorphin and schizophrenia, *Arch. Gen. Psychiatry* **36**:237.

Lightman, S. L., Spokes, E. G., Sagnella, G. A., Gordon, D., and Bird, E. D., 1979, Distribution of beta-endorphin in normal and schizophrenic human brains, *Europ. J. Clin. Invest.* **9**:377.

Lindström, L. H., Widerlöv, E., Gunne, L. M., Wahlström, A., and Terenius, L., 1978, Endorphins in human cerebrospinal fluid: Clinical correlations to some psychotic states, *Acta Psychiatr. Scand.* **57**:153.

Lipinski, J., Meyer, R., Kornetsky, C., and Cohen, B. M., 1979, Naloxone in schizophrenia: Negative result, *Lancet (Lett.)* **I**:1292.

Mains, R. E., Eipper, B. A., and Ling, N., 1977, Common precursor to corticotropins and endorphins, *Proc. Natl. Acad. Sci. USA* **74**:3014.

Manchanda, R., and Hirsch, S. R., 1981, (Des-Tyr$^1$)-gamma-endorphin in the treatment of schizophrenia, *Psychol. Med.* **11**:401.
Meltzer, H. Y., Bush, D. A., Tricon B. J., and Robertson, A., 1982, Effect of (Des-Tyr$^1$)-gamma-endorphin in schizophrenia, Psychiatry Res., **6**:313.
Mielke, D. H., and Gallant, D. M., 1977, An oral opiate antagonist in chronic schizophrenia: A pilot study, *Am. J. Psychiatry* **134**:1430.
Minneman, K. P., and Iversen, L. L., 1976, Enkephalin and opiate narcotics increase cyclic GMP accumulation in slices of rat neostriatum, *Nature (London)* **262**:313.
Naber, D., Pickar, D., Post, R. M., Van Kammen, D. P., Ballenger, J., Rubinow, D., Waters, R. N., and Bunney, W. E., Jr., 1981, CSF opioid activity in psychiatric patients, in: *Biological Psychiatry* (C. Perris, G. Struwe, and B. Jansson, eds.), pp. 372–375, Elsevier/North Holland, Amsterdam.
Nedopil, N., and Ruther, E., 1979a, Effects of the synthetic analogue of methionine enkephalin FK 33-824 on psychotic symptoms, *Pharmakopsychiatr. Neuro-Psychopharmakol.* **12**:277.
Nedopil, N., Dieterle, D., Hillebrand, G., and Gurland H. J., 1979b, Hemoperfusion in chronic schizophrenics, *Klin. Wochenschr.* **57**:1329.
Oliver, C., Mical, R. S., and Porter, J. C., 1977, Hypothalamic–pituitary vasculature: Evidence for retrograde blood flow in the pituitary stalk, *Endocrinology* **101**:598.
Orr, M., and Oppenheimer, C., 1978, Effects of naloxone on auditory hallucinations, Brit. Med. J., **1**:481.
Palmour, R. M., Ervin, R. F., Wagemaker, H., and Cade, R., 1979, Characterization of a peptide from the serum of psychotic patients, in: *Endorphins in Mental Health Research* (E. Usdin, W. E. Bunney, Jr., and N. S. Kline, eds.), pp. 581–593, Oxford University Press, London.
Perez-Cruet, J., Volavka, J., Mallya, A., Baig, S., and Toga, A., 1979, Behavioral effects of naloxone and LSD, in: *Endorphins in Mental Health Research* (E. Usdin, W. E. Bunney, Jr., and N. S. Kline, eds.), pp. 407–415, Oxford University Press, London.
Petho, B., Graf, L., Kaeczag I., Bitter, I., Tolna, J., Baraczka, K., and Li, C. H., 1981, Beta-endorphin and schizophrenia, *Lancet* **1**:212.
Pezalla, P. D., Lis, M., Seidah, N. G., and Chrétien, P. B., 1978, Lipotropin, melanotropin and endorphin: In vivo catabolism and entry into cerebrospinal fluid, *Can. J. Neurol. Sci.* **5**:183.
Pfefferbaum, A., Berger, P. A., Elliott, G. R., Tinkleberg, J. R., Kopell, B. S., Barchas, J. D., and Li, C. H., 1979, Human EEG response to beta-endorphin, *Psychiatry Res.* **1**:83.
Pickar, D., Bunney, W. E., Jr., 1981a, Acute naloxone administration in schizophrenic patients:a world health organization collaborative study, in: *Biological Psychiatry* (C. Perris, G. Struwe, and B. Jansson, eds.), pp. 508–511, Elsevier/North Holland, Amsterdam.
Pickar, D., Cutler, N. R., Naber, D., Post, R. M., Pert, C. B., and Bunney, W. E., Jr., 1980, Plasma-opioid activity in manic-depressive illness, *Lancet* **I**:937.
Pickar, D., Davis, G. C., Schulz, S. C., Extein, I., Wagner, R., Naber, D., Gold, P., Van Kammen, D. P., Goodwin, F. K., Wyatt, R., Li, C. H., and Bunney, W. E., Jr., 1981b, Behavioral and biological effects of acute beta-endorphin injection in schizophrenic and depressed patients, *Am. J. Psychiat.* **138**:160.
Pickar, D., Vartanian, F., Bunney, W. E., Jr., Mayer, H. P., Gastpar, M. T., Prakash, R., Sethi, B. B., Lideman, R., Belayev, B., Tsutsulkovskaja, M. V. A., Jungkunz, G., Nedopil, N., Verhoeven, W., and van Praag, H., 1982, Short-term naloxone administration in schizophrenic and manic patients, *Arch. Gen. Psychiat.* **39**:313.
Post, R., Roubinow, D., Gold, P., Ballenger, F., Goodwin, F., Pickar, D., Naber, D., Uhde, T., Reichlin S. and Bunney W. E. Jr., 1981, Somatostatin and opiate peptides in CSF of affectively ill patients:effects of carbamezapine, III World Congr. Biological Psychiatry, Stockholm, Abstract.
Rapoport, S. I., Klee, W. A., Pettigrew, K. D., and Ohno, K., 1980, Entry of opioid peptides into the central nervous system, *Science* **207**:84.
Risch, S. C., 1982, Beta-endorphin hypersecretion in depression:possible cholinergic mechanisms, Biol. Psychiat., **17**:1071.
Rorsman, B., Franzen, G., Sjöstedt, L., Lindholm, T., Thysell, H., Terenius, L., and Wahlström,

A., 1981, Hemodialysis in schizophrenia: Psychiatric aspects of treatment failure in a pilot study, *Neuropsychobiology* **7**:113.

Ross, M., Berger, P. A., and Goldstein, A., 1979, Plasma beta-endorphin immunoreactivity in schizophrenia, *Science* **205**:1163.

Sachar, E. J., Halbreich, U., Asnis, G., Nathan, S., and Halpern, F. S., 1980, Neuroendocrine disturbance in depression, in: *Progress in Psychoneuroendocrinology* (F. Brambilla, G. Racagni, and D. de Wied, eds.), pp. 263–272, Elsevier/North-Holland, Amsterdam.

Schenk, G. K., Enders, P., Engelmeier, M. P., Ewert, T., Herdemerten, S., Köhler, K. H., Lodmann, E., Matz, D., and Pach, J., 1978, Application of the morphine antagonist naloxone in psychic disorders, *Arzneim. Forsch.* **28**:1274.

Schulz, S. C., Van Kammen, D. P., Balow, J. E., Flye, M. W., and Bunney, W. E., Jr., 1981, Dialysis in schizophrenia: a double-blind evaluation, *Science* **211**:1066.

Simpson, G. M., Branchey, M. H., and Lee, J. H., 1977, A trial of naloxone in chronic schizophrenia, *Curr. Ther. Res. Clin. Exp.* **22**:909.

Stein, L., and Belluzzi, J. D., 1979, Brain endorphins: Possible role in reward and memory formation, *Fed. Proc.* **38**:2468.

Tamminga, C. A., Tighe, P. J., Chase, T. N., De Fraites, E. G., and Schaffer, M. H., 1981, Des-Tyrosine-gamma Endorphin administration in chronic schizophrenics, *Arch. Gen. Psychiat.*, **38**:167.

Terenius, L., Wahlström, A., Lindström, L., and Widerlöw, E., 1976, Increased CSF levels of endorphins in chronic psychosis, *Neurosci. Lett.* **3**:157.

Terenius, L., Wahlström, A., and Agren, H., 1977, Naloxone (Narcan) treatment in depression: Clinical observations and effects on CSF endorphins and monoamine metabolites, *Psychopharmacology* **54**:31.

Van Kammen, D. P., Waters, R. N., Gold, P., Sternberg, D., Robertson, G., Ganten, D., Pickar, D., Naber, D., Ballenger, J. C., Kaye, W. H., Post, R. M., and Bunney W. E., Jr., 1981, Spinal fluid vasopressin, angiotensin I and II, beta-endorphin and opioid activity in schizophrenia: a preliminary evaluation, in: *Biological Psychiatry* (C. Perris, G. Struwe, and B. Jansson, eds.), pp. 339–344, Elsevier/North Holland, Amsterdam.

Van Loon, G. R., and Kim, C., 1978, Dopaminergic mediation of beta-endorphin-induced catalepsy, *Res. Commun. Chem. Pathol. Pharmacol.* **21**:37.

Van Ree, J. M., and de Wied, D., 1981, Endorphins in schizophrenia, *Neuropharmacology* **20**:1271.

Verhoeven, W. M., van Praag, H. M., Botter, P. A., Sunier, A., van Ree, J. M., and de Wied, D., 1978, (Des-Tyr¹)-gamma-endorphin in schizophrenia, *Lancet (Lett.)* **I**:1046.

Verhoeven, W. M. A., van Praag, H. M., and de Jong, J. T. V. M., 1981a, Use of naloxone in schizophrenic psychoses and manic syndromes, *Neuropsychobiology* **7**:159.

Verhoeven, W. M. A., Westenberg, H. G. M., Gerritsen, T. W., Van Praag, H. M., Thijssen, J. H. H., Schwarz, F., Van Ree, J. M., and De Wied, D., 1981b, (Des-Tyrosine¹)-gamma-endorphin in schizophrenia: clinical biochemical and hormonal aspects, *Psychiatry Res.* **5**:293.

Verhoeven, W. M. A., Van Ree, J. M., Van Bentum, A. H., De Wied, D., and Van Praag, H. M., 1982, Antipsychotic properties of Des-Enkephalin-gamma-endorphin in treatment of schizophrenic patients, *Arch. Gen. Psychiat.* **39**:648.

Volavka, J., Mallya, A., Baig, S., and Perez-Cruet, J., 1977, Naloxone in chronic schizophrenia, *Science* **196**:1227.

Volavka, J., Davis, L. G., and Ehrlich, Y. H., 1979, Endorphins, dopamine and schizophrenia, *Schizophr. Bull.* **5**:227.

Wagemaker, H., Jr., and Cade, R., 1977, The use of hemodialysis in chronic schizophrenia, *Am. J. Psychiatry* **134**:684.

Watson, S. J., Berger, P. A., Akil, H., Mills, M. J., and Barchas, J. D., 1978, Effects of naloxone on schizophrenia: Reduction in hallucinations in a subpopulation of subjects, *Science* **201**:73.

Watson, S. J., Akil, H., Berger, P. A., and Barchas, J. D., 1979, Some observations on the opiate peptides in schizophrenia, *Arch. Gen. Psychiatry* **36**:35.

Wiedemann, E., Saito, T., Linfoot, J. A., and Li, C. H., 1977, Radioimmunoassay of human beta-lipotropin in unextracted plasma, *J. Clin. Endocrinol. Metab.* **45**:1108.

Wiedemann, E., Saito, T., Linfoot, J. A., and Li, C. H., 1979, Specific radioimmunoassay of human beta-endorphin in unextracted plasma, *J. Clin. Endocrinol. Metab.* **49**:478.

Wise, C. D., and Stein, L., 1979, Brain endorphin levels increase after long-term chlorpromazine treatment, in: *Endorphins in Mental Health Research* (E. Usdin, W. E. Bunney, Jr., and N. S. Kline, eds.), pp. 115–118, Oxford University Press, London.

CHAPTER 17

# Changes of Brain Monoamines in the Animal Model for Depression

NOBORU HATOTANI, JUNICHI NOMURA, and ISAO KITAYAMA

## 1. INTRODUCTION

Depression is the most fundamental psychosomatic reaction caused by various etiological factors of which some are known to be genetic, some are environmental, and some are of physical nature. The essence of depression, regardless of etiology, is a reduction of vital potency which is directly manifested by symptoms such as depressed mood, retardation of the psychic process, decrease of individual drives and general activity, and impairment of basic biological rhythms. On the basis of these fundamental symptoms, various kinds of individual personality reactions, namely self-depreciation, self-reproach, pessimistic thinking, hypochondriacal preoccupation, phobia, depersonalization and so on, can develop and make up diverse clinical pictures. Despite the variety of its clinical manifestations, depression is the most therapeutically accessible state by such biological treatments as antidepressants or electroconvulsive therapy. Accordingly, it may be indisputable that the basic symptoms of depression are brought about through a final common biological pathway irrespective of etiology.

## 2. PSYCHOENDOCRINE MODEL FOR DEPRESSION

Depression is the principal field upon which biological studies have recently been focused, and neuroendocrine findings, in particular, are accumulating. The significant results of clinical psychoneuroendocrine studies on depression were reviewed by Hatotani *et al.* (1978, 1979). From the results of these studies, it appears that dysfunction of the hypothalamic–pituitary system

NOBORU HATOTANI, JUNICHI NOMURA, and ISAO KITAYAMA • Department of Psychiatry, Mie University School of Medicine, Tsu, Mie, Japan.

and metabolic disturbances of biogenic amines are the most significant findings in depressive illness. The function of the hypothalamic–pituitary system, which is controlled by monoaminergic pathways from the limbic system and the lower brain stem, involves the integration of mood, behavior, consciousness, instinctive drives, autonomic nervous and endocrine functions, and biological rhythms. The basic symptoms of depression closely correlate with the disturbances of these functions. Therefore, a final common biological pathway in the pathogenesis of depression is supposed to be dysfunction of monoaminergic systems which are connected with the hypothalamic–pituitary system.

It is a common experience that the onset of depression in most instances is preceded by a stressful situation of either psychological or physical nature. When the loading of stress exceeds an individual's ability to cope with it, it is probable that the resultant metabolic decompensation of brain monoamines would lead to persistent disturbances of the regulatory mechanisms of the CNS, reflected in the symptomatology of depression. According to such a conceptual model of depression, it is conceivable that stress plays an important part in the pathogenesis of depression, in which a metabolic disturbance of brain monoamines is essential. This stress model of depression is pertinent not only to reactive depression but also to endogenous depression. An individual who is genetically predisposed to depressive illness would be more prone to develop depression precipitated by stress than would an individual with no such predisposition. It is postulated that depression is the most common pattern of decompensation of the function of the CNS.

In order to verify the relevance of such a conceptual model of depression described above, it would be a strategic requirement to produce an animal model for depression by means of stress. It is well known that adult female rats show a regular cycle of spontaneous running activity corresponding to the estrous cycle (Richter, 1965). We took advantage of this species-specific baseline behavior, which is tied to an endocrine function, in order to produce an animal model for depression (Hatotani et al., 1977, 1979).

## 3. ANIMAL MODEL FOR DEPRESSION

### 3.1. Stress-Induced Animal Model for Depression in Female Rats

#### 3.1.1. Method of Producing the Animal Model

*3.1.1a. Selection of Female Rats with a Regular Cycle of Spontaneous Running Activity.* Adult female Wistar rats were raised in revolving cages of 1-m circumference. The cages were set in a room with natural lighting and constant temperature of 22 ± 2°C. The spontaneous running activity was recorded as the number of revolutions of the drum at 10 a.m. every day. The percentage of cornified cells in a vaginal smear, water and food intake, rectal temperature, body weight, and motor responsiveness (running activity in a 10-sec period following a sound stimulus) were recorded at the same time. Only rats that showed a regular cycle of spontaneous running activity corresponding to the estrous cycle were selected for the experiment.

*3.1.1b. Production of the Animal Model for Depression.* The selected rats were moved into drums revolving automatically at 5 rpm, so that they were exposed to stress in terms of forced running. Food and water were freely accessible during the stress. When the fatigue reached its maximum and the rectal temperature came down to 33°C or less, the rats were relieved from the stress and given a rest for 24 hr. This sequence of stress and rest was repeated three times. The duration of each stress was 2–5 days, varying from rat to rat. The rats were then brought back to the revolving cages, and their running activity, vaginal smear, food and water intake, rectal temperature, and motor responsiveness were again recorded daily.

The rats showed no spontaneous running activity for several days after the stress. In about half of the rats, the spontaneous running activity started to reappear in about 2 weeks, and resumed regular cyclicity as shown in Fig. 1. These animals were termed "recovery rats." In the other half of the rats, as shown in Fig. 2, the spontaneous running activity continued to remain very low for more than 2 weeks after the stress and showed no cyclic regularity. These animals were termed "depression-model rats." Depression-model rats were examined from various points of view in comparison with control rats (rats without stress) and recovery rats.

### 3.1.2. Characteristics of the Animal Model

Figure 2 shows an example of the depression-model rat. Before the forced running stress, the rat showed a regular 4-day cycle of spontaneous running

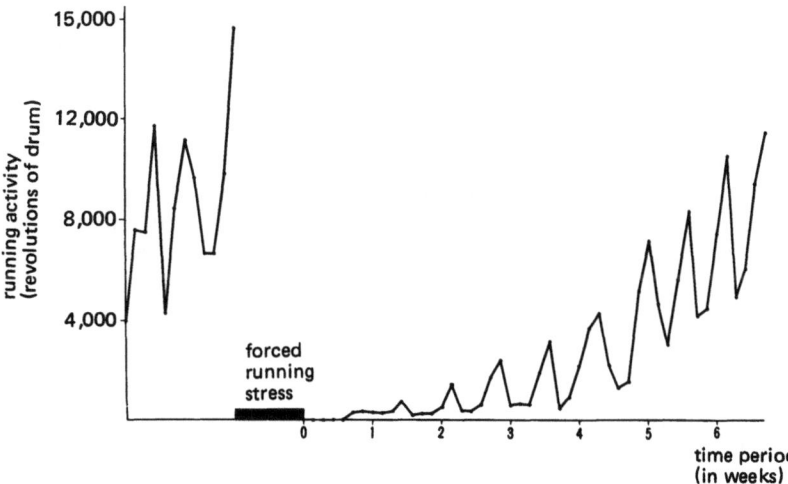

FIGURE 1. Recovery rat. The rat before the stress showed a regular cycle of spontaneous running activity. After the stress, the spontaneous running activity started to reappear in about 2 weeks and resumed regular cyclicity.

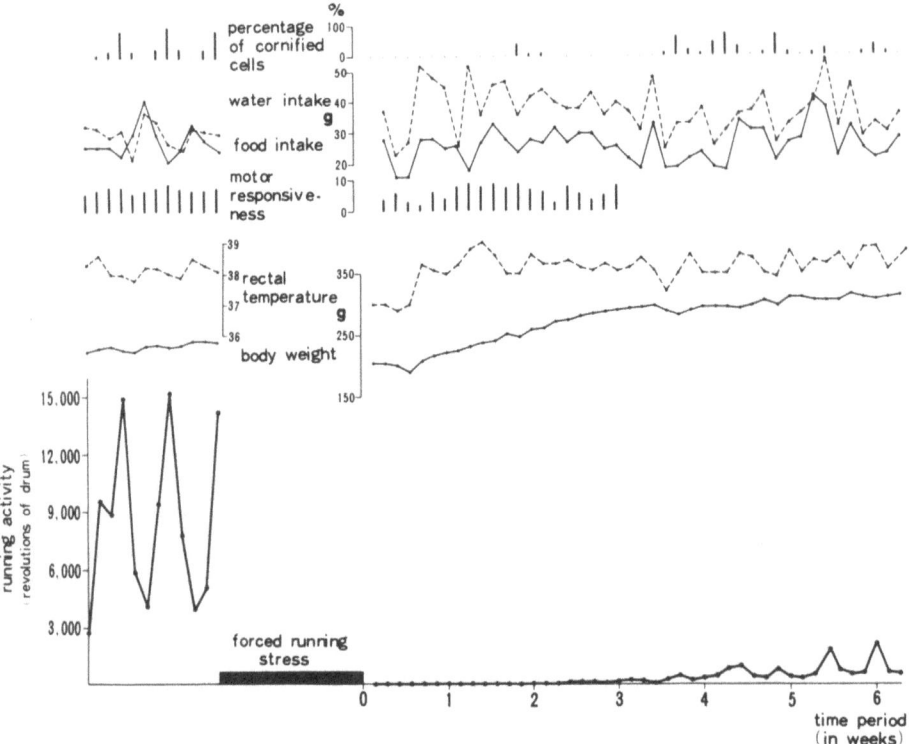

FIGURE 2. Depression-model rat. The rat before the stress showed a 4-day cycle of spontaneous running activity corresponding to the estrous cycle. After the stress, the running activity and its cyclicity almost disappeared for 6 weeks. The vaginal smear showed the pattern of constant diestrus or irregular reappearance of cornified cells. In contrast, food and water intake were almost the same as before the stress. Motor responsiveness and rectal temperature returned to normal levels within several days after the stress. Body weight recovered within 10 days and continued to increase thereafter.

activity corresponding to the estrous cycle. After the stress, the running activity and its cyclicity almost disappeared for 6 weeks. The vaginal smear showed a pattern of constant diestrus or irregular reappearance of cornified cells. The intake of food and water was almost the same as before the stress. Motor responsiveness and rectal temperature returned to normal levels within several days after the stress. Body weight recovered within 10 days and continued to increase thereafter. The rats immediately after the stress were extremely exhausted. As shown in Table 1, immediately after the stress, gastric ulcer was detected in 10 out of 13 rats and the weight of the adrenal glands significantly increased from 66.2 mg to 104.0 mg. However, in depression-model rats as well as recovery rats, gastric ulcer was not detected any more and the weight of the adrenal glands returned to normal levels. Therefore, depression-model animals seemed to have completely recovered from physical exhaustion. From the behavioral point of view, depression-model rats showed ptosis and sleep

TABLE 1
Occurrence of Gastric Ulcer and Changes in Weight of Adrenal Glands in Experimental Animals as Compared with Controls

|  | Control rats | Rats immediately after the stress | Depression-model rats | Recovery rats |
| --- | --- | --- | --- | --- |
| No. of rats with gastric ulcer | 0 (10)[a] | 10 (13) | 0 (14) | 0 (7) |
| Wet weight of adrenal glands (mg) | 66.2 ± 8.8 (10) | 104.0 ± 22.4[b] (9) | 64.1 ± 12.8 (19) | 66.0 ± 6.7 (5) |

[a] Number of rats examined is shown in parentheses.
[b] Statistically significant change ($p < 0.01$).

disturbance, and pronounced aggression though being inactive and seclusive. This stress-induced, prolonged inactive state accompanied by the abolition of hormone-dependent cyclic behavior, a state different from simple exhaustion, was supposed to be a depression analog in rats.

### 3.1.3. Effect of Imipramine on the Animal Model

Reversal of symptoms by clinically effective treatment technique is proposed as one of the criteria that might validate an animal model for depression (McKinney, 1977). Figure 3 shows the effect of imipramine on the running activity and its cyclicity of a depression-model rat. Imipramine hydrochloride

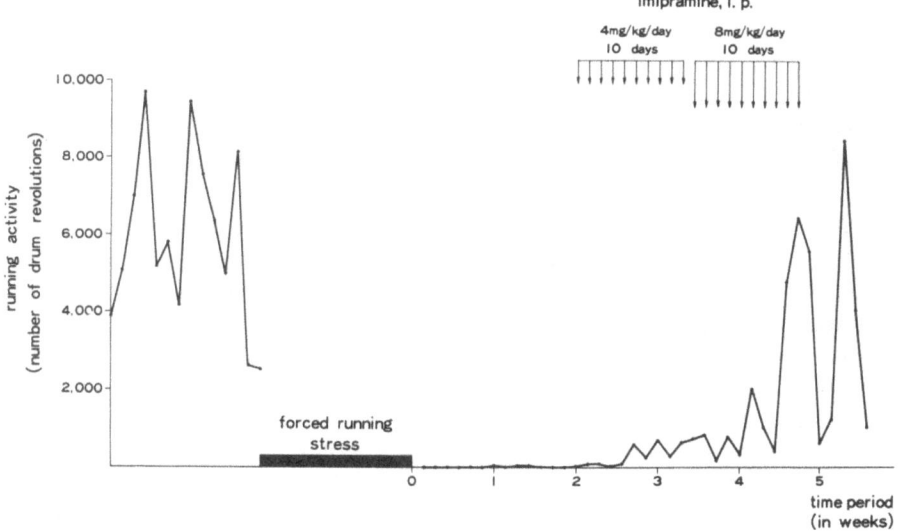

FIGURE 3. Effect of imipramine on the running activity of a depression-model rat. Imipramine was injected i.p. daily from 2 weeks after the stress. The first series of imipramine injections (4 mg/kg for 10 days) was not very effective; however, the rat quickly recovered its normal cyclic running activity during the second series of imipramine injections (8 mg/kg for 10 days).

was injected i.p. daily from 2 weeks after the stress. The first series of imipramine injections (4 mg/kg for 10 days) was not very effective. However, the rat quickly recovered its normal cyclic running activity during the second series of imipramine injections (8 mg/kg for 10 days). When imipramine (8 mg/kg) was injected i.p. for 14 days in seven depression-model rats, all rats showed a similar mode of recovery. It was also noted that the spontaneous running activity began to increase rather suddenly on the fourth or fifth day of imipramine injections. The effectiveness of imipramine was obvious, since the rate of spontaneous recovery without any treatment during the 4 weeks after the stress was approximately 52% (35 of 67 rats). In order to compare the effect of antidepressant drug, the effect of chlorpromazine was also studied. When chlorpromazine hydrochloride (4 mg/kg) was injected i.p. for 14 days in seven depression-model rats, only three rats showed recovery. The results of these studies seemed to further validate this animal model for depression.

## 3.2. Changes of Brain Monoamines in the Animal Model for Depression

### 3.2.1. Studies on Brain Monoamines by the Histochemical Fluorescence Method

Experimental animals, paired with age-matched controls on diestrus day one, were decapitated, and brain tissues were treated according to the histochemical fluorescence method of Falck *et al.* (1962). The 12 regions selected from the main monoaminergic neuronal systems, such as the ascending norepinephrine (NE) system, the nigrostriatal dopamine (DA) system, the tuberoinfundibular DA system, and the ascending serotonin (5HT) system, were examined by means of fluorescence microscopy. In some experiments, the fluorescence intensity of catecholamines was measured semiquantitatively according to a modification of the technique of Lichtensteiger (1970).

The results of semiquantitative study are summarized in Fig. 4. (1) Fluorescence intensity of catecholamines in depression-model rats increased significantly in the cell groups of the ascending NE systems: nucleus reticularis lateralis (A1), ventral part of the nucleus commissuralis (A2), tractus rubrospinalis (A5), locus coeruleus (A6), and nucleus subcoeruleus (A7), which are situated in the medulla oblongata and the pons. (2) In contrast, fluorescence intensity of catecholamines in depression-model rats significantly decreased in the cell bodies of the nucleus arcuatus (A12) and in their nerve terminals in the external layer of the median eminence (EME), which belong to the tuberoinfundibular DA system. (3) The above-described findings were also seen in rats that were examined immediately after the stress, but not in recovery rats. (4) No remarkable change was observed in the fluorescence intensity of the nigrostriatal DA system and of the ascending 5HT system.

Representative fluorescence photomicrographs are shown in Figs. 5 and 6. Figure 5 contains fluorescence photomicrographs of NE-containing cell bodies in the locus coeruleus. The depression-model rat (B) showed much stronger

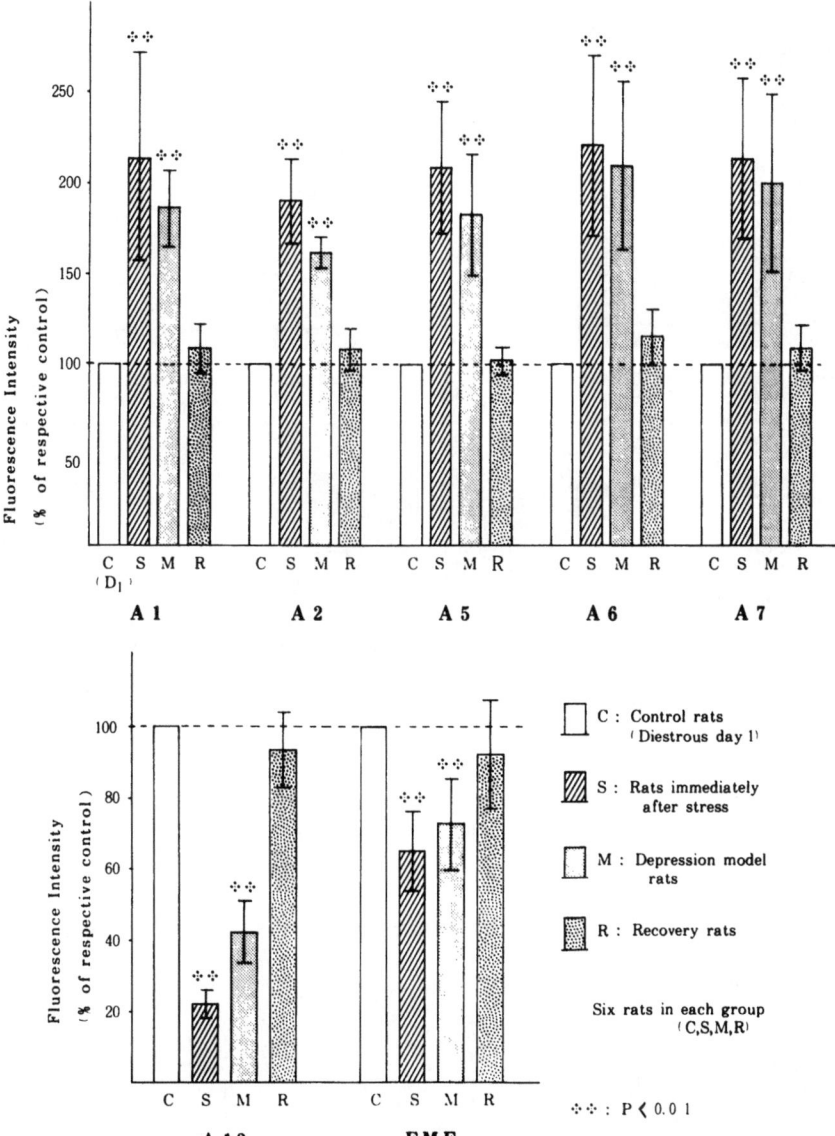

FIGURE 4. Fluorescence intensity of catecholamines in various regions of the monoaminergic neuronal systems in control and experimental rats. The fluorescence intensity was semiquantitatively measured and expressed as percent of respective controls. A1: nucleus reticularis lateralis; A2: ventral part of nucleus commissuralis; A5: tractus rubrospinalis; A6: locus coeruleus; A7: nucleus subcoeruleus; A12: nucleus arcuatus; EME: external layer of median eminence. In rats killed immediately after the stress and in depression-model rats, the fluorescence intensity increased in the cell groups of the ascending NE systems (A1, A2, A5, A6, A7) and decreased in the cell bodies and nerve terminals of the tuberoinfundibular dopamine system (A12, EME).

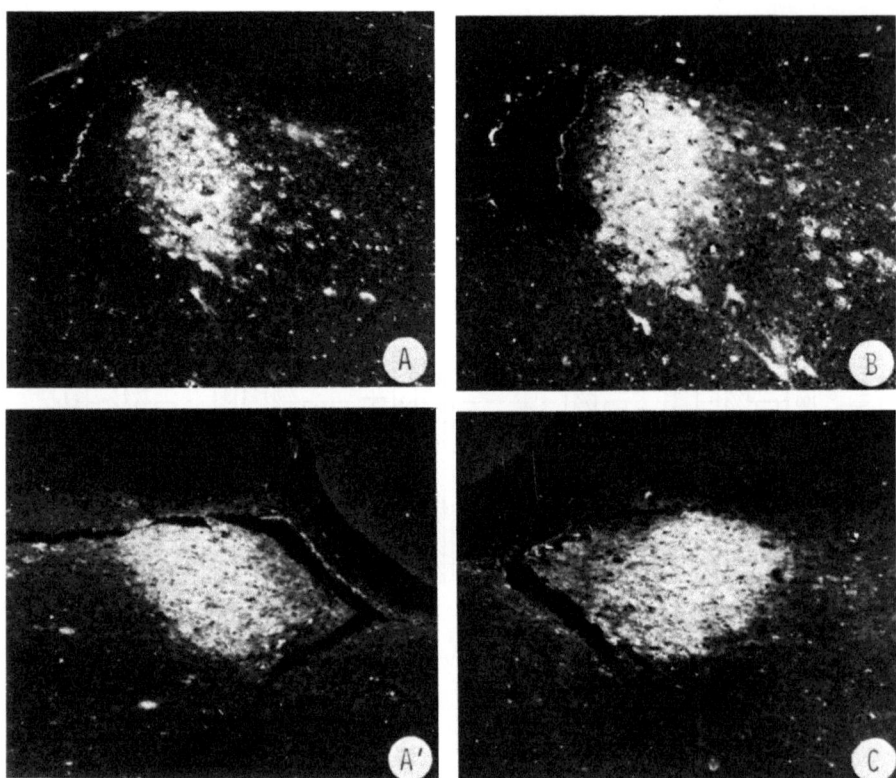

FIGURE 5. Fluorescence photomicrographs of NE-containing cell bodies in the locus coeruleus. The depression-model rat (B) showed much stronger fluorescence than the control (A). The recovery rat (C) displayed almost the same fluorescence intensity as the control (A'). Magnification: × 50.

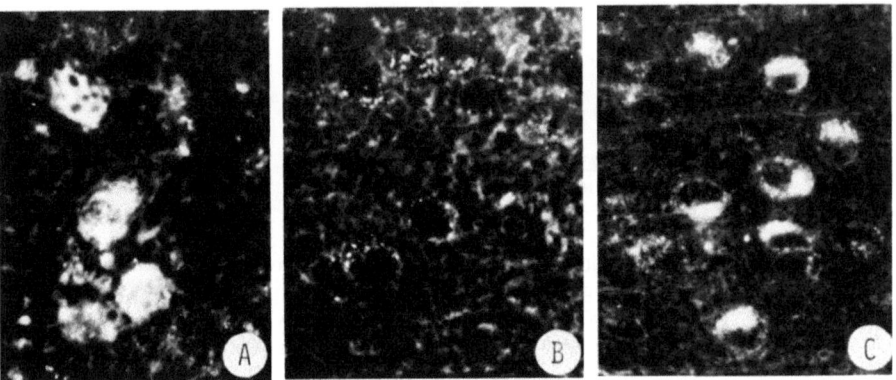

FIGURE 6. Fluorescence photomicrographs of DA-containing cell bodies in the nucleus arcuatus. The fluorescence seen in the control (A) almost vanished in the rat killed immediately after the stress (B). In the depression-model rat (C), the fluorescence recovered only about half the intensity of the control level. × 400.

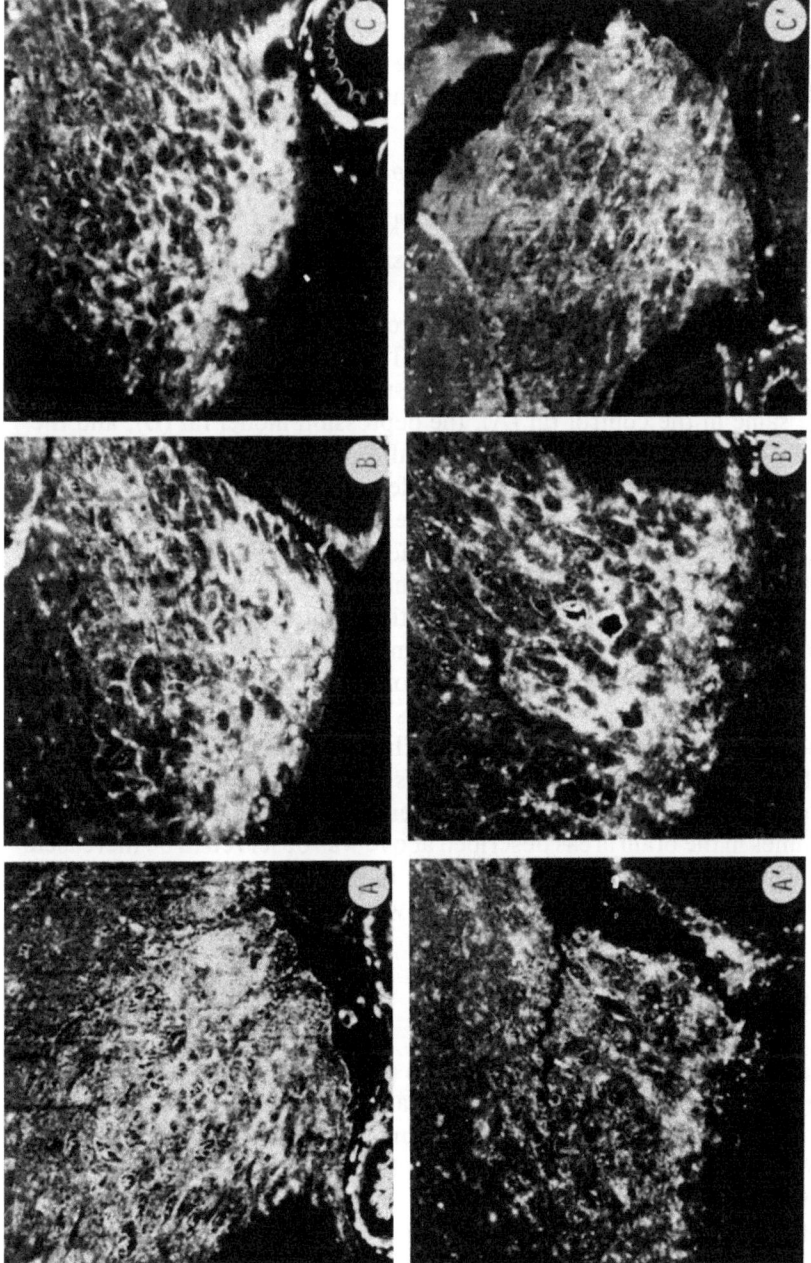

FIGURE 7. Fluorescence photomicrographs of the nucleus supraopticus. The fluorescence intensity of catecholamines in the steady state (without α-methyl-p-tyrosine) was approximately the same in the control rat (A), the depression-model rat (B), and the recovery rat (C). The fluorescence intensity 3 hr after the i.p. injection of α-methyl-p-tyrosine (250 mg/kg) remained high in the depression-model rat (B') as compared with the control (A'). In the recovery rat (C'), the fluorescence intensity after α-methyl-p-tyrosine was almost the same as in the control (A'). × 100.

fluorescence than the control (A), but the recovery rat (C) displayed almost the same fluorescence intensity as the control (A'). Figure 6 shows DA-containing cell bodies in the nucleus arcuatus. Fluorescence seen in the control rat (A) almost disappeared in the rat killed immediately after the stress (B). In the depression-model rat (C), the fluorescence intensity was still low and displayed only about half the intensity of the control level.

### 3.2.2. Studies on the Turnover Rate of Catecholamines

α-Methyl-$p$-tyrosine methylester (250 mg/kg) was injected i.p. 3 hr before the decapitation in order to inhibit the synthesis of catecholamines. The turnover rate of catecholamines was estimated by comparing the fluorescence intensity in control and experimental rats. Figure 7 shows fluorescence photomicrographs of the nucleus supraopticus. The fluorescence intensity of catecholamines in the steady state (without α-methyl-$p$-tyrosine) was approximately the same in the control rat (A), the depression-model rat (B), and the recovery rat (C). The fluorescence intensity after the injection of α-methyl-$p$-tyrosine remained high in the depression-model rat (B') as compared with the control rat (A'). In the recovery rat (C'), the fluorescence intensity after α-methyl-$p$-tyrosine was almost the same as in the control rat (A'). These findings suggest that the turnover rate of catecholamines decreased in the depression-model rat. Similar results were obtained in other regions, in which nerve terminals of the ascending NE system are concentrated, such as the nucleus motorius dorsalis nervi vagi, the internal layer of the median eminence, the retrochiasmatic area, the nucleus dorsomedialis, the nucleus paraventricularis, and the nucleus interstitialis striae terminalis. In order to examine the turnover rate of catecholamines during the acute stress, rats were exposed to forced running for 24 hr. In contrast to the long-term stress, the turnover rate of catecholamines in the brain areas described above was observed to be increased during the acute stress. The period when the turnover rate of catecholamines changes from an increase to a decrease is now under investigation.

## 4. DISCUSSION

Animal models for depression have been produced as a possible research strategy utilizing learned helplessness (Seligman and Maier, 1967), conditioning techniques (Takahashi et al., 1974, 1976), and separation experiments (McKinney and Bunney, 1969; McKinney et al., 1971). At the present stage, it is difficult to compare and discuss the results of these animal models, though changes of brain monoamines have been reported in some of these models. Our stress-induced animal model for depression seems to have the following advantages: (1) the experimental animals have a definite baseline behavior with regular cyclicity corresponding to the estrous cycle; (2) animals were exposed to the long-lasting, uncontrollable stress, which may be an important factor preceding the onset of depression; (3) the stress-induced inactive state of animals accompanied by impairment of the biological rhythm can be differentiated from simple exhaustion; (4) taking into consideration the life span of the rat, the duration of this inactive state is comparable in length to human depression; (5) the changes in animal behavior can be evaluated objectively in terms of running activity and its cyclicity; (6) the recovery rat shows almost the same findings in brain monoamines as the control rat; and (7) the antidepressant drug normalizes the animal behavior.

Taking all these into consideration, it would be reasonable to regard this animal model as a depression analog. The most significant findings concerning brain monoamines in the depression-model rats were: (1) an increase of fluorescence intensity of cell groups in the ascending NE system; (2) a decrease of fluorescence intensity of cell bodies and nerve terminals in the tuberoinfundibular DA system; and (3) a decrease in turnover rate of catecholamines in nerve terminals of the ascending NE system. It is well known that the ventral pathway of the ascending NE system terminates mainly in the hypothalamus (Ungerstedt, 1971) and that the tuberoinfundibular system has a close relation to the regulation of the estrous cycle (Fuxe and Hökfelt, 1969). The function of these brain regions is obviously related to emotion, spontaneity, and the sleep mechanism, as well as to endocrine and autonomic nervous functions. Dysfunction of these areas may be related to the development of depressive symptoms. It is also interesting to note that the turnover rate of catecholamines in nerve terminals of the ascending NE system increased during the acute stress and decreased after the long-lasting stress. One may speculate that the turnover rate of catecholamines, which adaptively increases during the acute stress, changes to a decreased state at a certain critical point in the course of long-lasting stress, and that this alteration of catecholamine metabolism remains unrestored in depression-model animals. The relation between the increase in catecholamine levels in cell bodies and the decrease in catecholamine turnover in nerve terminals of the ascending NE system is also a suggestive finding, and should be studied further.

It is obvious that biochemical studies should also be carried out in order to verify the histochemical findings and to obtain a detailed kinetic analysis. In our preliminary biochemical study, NE, DA, and 5HT were measured in the cerebral cortex, midbrain, hypothalamus, medulla oblongata and pons, striatum, hippocampus, and cerebellum (Nomura *et al.*, 1978). Immediately after the forced running stress, NE levels in the hypothalamus decreased significantly, and remained at a lower level in the depression-model rat. The recovery rat, in contrast, showed no difference from the control. Biochemical investigation of the metabolism of catecholamines in each nucleus is now in progress.

## 5. SUMMARY AND CONCLUSIONS

1. Forced running stress could induce a prolonged inactive state accompanied by the abolition of hormone-dependent cyclic behavior in female rats.

2. This inactive state was different from simple exhaustion and could be reversed by imipramine. The state is supposed to be a depression analog in female rats.

3. Levels and turnover rates of brain monoamines were examined by the histochemical fluorescence method in these animals. In the depression-model rat, an increase of fluorescence intensity in the cell groups of the ascending NE system and a decrease of fluorescence intensity in cell bodies and nerve terminals of the tuberoinfundibular DA system were observed constantly. The turnover rate of catecholamines in nerve terminals of the ascending NE system increased during the acute stress, but decreased in depression-model rats. In the recovery rats, the fluorescence intensity and turnover rate of catecholamines were the same as those in the control rats.

4. A metabolic disorder of brain catecholamines might be a final common biological pathway in the pathogenesis of depression and an appropriate animal model would greatly help in detailed studies of this problem.

## 6. REFERENCES

Falck, B., Hillarp, N.-Å., Thieme, G., and Torp, A., 1962, Fluorescence of catecholamines and related compounds condensed with formaldehyde, *J. Histochem. Cytochem.* **10**:348.

Fuxe, K., and Hökfelt, T., 1969, Catecholamines in the hypothalamus and the pituitary gland, in: *Frontiers in Neuroendocrinology* (W. F. Ganong and L. Martini, eds.), pp. 47–96, Oxford University Press, London.

Hatotani, N., Nomura, J., Yamaguchi, T., and Kitayama, I., 1977, Clinical and experimental studies on the pathogenesis of depression, *Psychoneuroendocrinology* **2**:115.

Hatotani, N., Nomura, J., and Wakoh, T., 1978, Endocrinological studies on periodic psychoses, in: *Perspectives in Endocrine Psychobiology* (F. Brambilla, P. K. Bridges, E. Endröczi, and G. Heuser, eds.), pp. 423–465, Akadémiai Kiadó, Budapest, and Wiley, New York.

Hatotani, N., Nomura, J., Inoue, K., and Kitayama, I., 1979, Psychoendocrine model of depression, *Psychoneuroendocrinology* **4**:155.

Lichtensteiger, W., 1970, Katecholaminhaltige Neurone in der neuroendokrinen Steuerung, Prinzip und Anwendung der Mikrofluorimetrie, *Prog. Histochem. Cytochem.* **1**:185.

McKinney, W. T., Jr., 1977, Biobehavioral models of depression in monkeys, in: *Animal Models in Psychiatry and Neurology* (I. Hanin and E. Usdin, eds.), pp. 117–126, Pergamon Press, Elmsford, N.Y.

McKinney, W. T., Jr., and Bunney, W. E., Jr., 1969, Animal model of depression. I. Review of evidence: Implications for research, *Arch. Gen. Psychiatry* **21**:240.

McKinney, W. T., Jr., Suomi, S. J., and Harlow, H. F., 1971, Depression in primates, *Am. J. Psychiatry* **127**:1313.

Nomura, J., Kitayama, I., and Hatotani, N., 1978, Stress-induced depression model in female rats: Biochemical and histochemical investigation of brain monoamines, *Folia Psychiatr. Neurol. Jpn.* **32**:159.

Richter, C. P., 1965, *Biological Clocks in Medicine and Psychiatry*, Thomas, Springfield.

Seligman, M. E., and Maier, S. F., 1967, Failure to escape traumatic shock, *J. Exp. Psychol.* **74**:1.

Takahashi, R., Nagayama, H., Kido, A., and Morita, T., 1974, An animal model of depression, *Biol. Psychiatry* **9**:191.

Takahashi, R., Tachiki, K. H., Nishiwaki, K., Nakamura, E., Tateishi, T., and Nagayama, H., 1976, Biochemical basis of an animal model of depressive illness: A preliminary report, *Folia Psychiatr. Neurol. Jpn.* **30**:208.

Ungerstedt, U., 1971, Stereotaxic mapping of the monoamine pathway in the rat brain, *Acta Physiol. Scand. Suppl.* **367**:1.

CHAPTER 18

# Psychoendocrinology of Depression

GREGORY M. BROWN, PAUL GROF, and
EVA GROF

## 1. INTRODUCTION

Despite continuing progress in the treatment and prophylaxis of affective disorders, mental depression remains a major world health problem. Statistics on mental depression vary to some extent; however, they all attest to the high frequency of this disorder in the population. The lifetime risk of any one of us suffering from a depression of clinical severity has been estimated at between 10 and 15% (Weissman and Klerman, 1978). According to recent WHO reports, at least one hundred million people are suffering from mental depression of clinical intensity at any time (Sartorius, 1979).

Extensive investigations have been carried out into the biological and psychological dimensions of affective disorder, yet our understanding of the nature of mental depression is still limited. There are some biological abnormalities, however, which have been well documented and which provide promise of further progress. Altered adrenal function in depression has long been known (Sachar, 1967). It has now been established that a variety of endocrine abnormalities exist, involving several different hormones and hormone responses (Sachar, 1975; Carroll, 1976; Ettigi and Brown, 1977; Rubin and Kendler, 1977). With the identification and clinical application of the hypothalamic factors which regulate the pituitary, it has become certain that a variety of neural systems are involved in endocrine regulation.

It has been shown that those neurotransmitter systems which are thought to be implicated in the psychobiology of affective disorders (Schildkraut, 1974; Sourkes, 1977; Maas, 1979) are also involved in the regulation of neuroendocrine function (Frohman and Stachura, 1975; Martin et al., 1977; Brown et al., 1979). If a central neurotransmitter defect is at the heart of the depressive

---

GREGORY M. BROWN • Department of Neurosciences, McMaster University, Hamilton, Ontario, Canada L5H 1X7. PAUL GROF • Department of Psychiatry, McMaster University, and Hamilton Psychiatric Hospital, Hamilton, Ontario, Canada L5H 1X7. EVA GROF • Department of Psychiatry, McMaster University, Hamilton, Ontario, Canada L5H 1X7.

symptoms, then all the phenomena including the various psychopathological characteristics, the visceral symptomatology, and the endocrine abnormalities should coexist as they are all secondary to that central deficit in neurotransmission. Systematic studies of neuroendocrine regulatory mechanisms in affective disorders should help in clarifying the "neurotransmitter dimension" of this important and complex problem (Brown and Seggie, 1980).

It is of particular interest that investigations of neuroendocrine function may provide a means of differentiating various subtypes among the affective disorders, because they may mirror the central activity of biogenic amines in depressed subjects. Different types of neurotransmitter deficit have been postulated in depressive illness. Different underlying abnormalities could lead to syndromes which differ with respect to clinical characteristics, visceral symptomatology, and endocrine abnormalities. Support for this view has already been presented, specifically a lowering of GH response to insulin hypoglycemia in primary unipolar depressions (Gruen et al., 1975), an increase of LH response to LHRH in secondary depressions (Ettigi et al., 1979), and differential findings of cortisol escape in the dexamethasone test in three subgroups of unipolar depressions (Schlesser et al., 1980).

## 2. ENDOCRINE ABNORMALITIES AND DEPRESSION

Abnormalities of neuroendocrine function have been found consistently in plasma and urinary cortisol, in the dexamethasone suppression test, and in the responsiveness of GH and TSH. These findings are covered in the literature in several review articles (Sachar, 1975; Carroll, 1976; Ettigi and Brown, 1977; Rubin and Kendler, 1977) and described in studies focusing on specific hormones as follows:

1. Cortisol in plasma and its derivatives in urine have been found elevated in depressed patients by several groups of investigators (Sachar, 1967; Stancer et al., 1969; Sachar, 1975; Carroll, 1976; Taracha et al., 1978). In mania, lower levels were reported (Stancer et al., 1969). The elevation was originally considered nonspecific and due to anxiety, stress, or hospitalization; however, studies which controlled for these variables have yielded data showing hypersecretion of cortisol (Carroll, 1976). Further evidence for the elevated plasma cortisol levels came from studies measuring the total diurnal pattern of cortisol secretion in depressed patients; alterations in the pattern of secretory bursts have been found (Sachar et al., 1973, 1980). Normalization with clinical improvement has been reported in most of the above changes (Gregoire et al., 1977; Nuller and Ostroumova, 1980). More recently, several workers have shifted to a strategy of examining the dexamethasone suppression of cortisol production.

In most studies depressed patients showed failure to produce a normal response (Shopsin and Gershon, 1971; Stokes et al., 1976; Carroll, 1976; Brown and Shuey, 1980; Schlesser et al., 1980). Carroll (1976) and associates have used the escape from the dexamethasone suppression test as a diagnostic test in an extensive study of depression. One theoretical problem with the dexamethasone suppression test bears discussion. Most investigators examine single blood samples at specific times after administration of dexamethasone, typical times being 8:00 a.m., 4:00 p.m., and 11:00 p.m, and also use an arbitrary concentration (usually 5.0 µg/dl) to define suppression and nonsuppression. Because cortisol is secreted in bursts (Sachar et al., 1980) and blood levels can vary extremely rapidly, a significant amount of error should be introduced. This error could, in theory, be obviated by taking a series of samples bracketed around the times of interest (Goldzieher et al., 1976).

A reduced corticosteroid response to methylamphetamine has been reported in endogenous depression despite a normal GH response, supporting the concept of deficient noradrenergic function at central α-adrenergic receptors (Checkley, 1979).

2. GH secretion and its regulation have also been investigated in depression. There is a complex regulation of GH secretion in man: the effect of a number of stimuli has been reported, from psychological (for instance, sleep or prolonged fasting) to metabolic factors (Brown et al., 1978), and these must be controlled in experimental studies. Several stimuli, such as insulin, hypoglycemia, L-dopa, and 5-hydroxytryptophan, have been useful in testing the GH response. In response to both insulin and 5-hydroxytryptophan, a diminished GH rise has been established in a proportion of depressed patients (Takahashi, et al., 1974; Gruen et al., 1975). Findings of diminished GH response to L-dopa have also been reported in depression, but were not substantiated when properly controlled for sex and age (Sachar et al., 1975). Aberrant GH responses to TRH and LHRH have been reported in patients with primary affective disorders; these responses are not seen in normal subjects (Maeda et al., 1975; Brambilla et al., 1978).

3. TSH has been investigated in depressed patients. Baseline levels in most studies are normal (Ehrensing et al., 1974; Ettigi et al., 1979) but elevated levels in bipolar female patients have been reported (Garfinkel et al., 1979) in response to the infusion of TRH. This stimulation test leads to clearly lower values in a substantial number of depressed patients (Ehrensing et al., 1974; Furlong et al., 1976; Kirkegaard et al., 1978; Linnoila et al., 1979). A flattening of the circadian thyrotropin rhythm has been reported in severe endogenously depressed patients (Weeke and Weeke, 1978). An antidepressant effect of TRH has been reported but has remained equivocal (Ehrensing et al., 1974; Orth et al., 1979).

4. LH has been studied sparingly; however, some interesting findings have been reported. An increase of the LH response to LHRH in secondary depression has been found (Ettigi et al., 1979). In addition, plasma LH levels have been reported lower in depressed postmenopausal women (Altman et al., 1975). At the subordinate level, the relationship between estrogens and progesterone

on the one hand and depression and mood changes on the other hand, remains of considerable interest (Campbell and Whitehead, 1977; Ladisich, 1977; Schneider et al., 1977; Herrmann and Beach, 1978).

5. Prolactin levels in plasma have been investigated in depression reflecting the recent increase of interest in this hormone. In some studies, baseline prolactin levels have been found elevated in both bipolar and unipolar patients and associated with mood changes (Arana et al., 1977; Cuenca et al., 1978), and altered 24-hr rhythm has been reported (Halbreich et al., 1979). Prolactin responses to TRH have been reported as increased (Maeda et al., 1975), decreased (Ehrensing et al., 1974; Linnoila et al., 1979), or normal (Ettigi and Brown, 1979). Recently, considerable attention has also been paid to the changes of prolactin secretion in depressed patients following various antidepressant drugs (Janowsky et al., 1978; Halbreich et al., 1978; Nielsen, 1980) and electroconvulsive therapy (O'Dea et al., 1978). Bipolar depressed female patients have been reported to show a failure of the normal prolactin elevation following carbidopa (Garfinkel et al., 1979).

6. Vasopressin and melatonin are being increasingly studied in affective illness. Reduced nighttime output of melatonin has been reported (Mendlewicz et al., 1979; Wetterberg et al., 1979; Lewy et al., 1979), and a reduction in central vasopressin function postulated (Gold and Goodwin, 1978; Gold et al., 1979).

## 3. RELATION TO CLINICAL SUBTYPES

A number of the above studies indicate that depressed patients fall into subgroups with respect to endocrine abnormalities. Depressed patients of three different genetic subtypes are different with respect to nonsuppressibility on the dexamethasone suppression test (Schlesser et al., 1980). Nonsuppression distinguished patients with primary unipolar depression (45%) from patients with primary bipolar depression (85%), secondary unipolar depression (0%), and controls (6%). Within the group of patients with primary unipolar depression, three familial subtypes were distinguished: familial pure depression disease (76% nonsuppression), sporadic depressive disease (44%), and depression spectrum disease (7%). Nonsuppression following dexamethasone is reported to correlate with a good response to treatment as compared to suppression, providing further evidence for a biological difference in these groups.

Diminished GH responses to hypoglycemia occur in only about 50% of depressed subjects (Gruen et al., 1975; Garver et al., 1975; Ettigi and Brown, 1979). Impaired GH responses in patients with endogenous but not reactive depression have been reported following desmethylimipramine (Carroll et al., 1976), amphetamine (Matussek, 1978), and clonidine (Laakmann and Benkert, 1978). The hyporesponse of TSH to TRH also occurs in only a proportion of patients (Ehrensing et al., 1974; Furlong et al., 1976; Kirkegaard et al., 1978). Bipolar patients are reported to differ from unipolar patients with respect to

TSH response to TRH (Gold et al., 1980), and LH response to LHRH may distinguish primary from secondary depressed patients (Ettigi et al., 1979). Little information, however, is available on the interrelationship of these abnormalities as most studies have examined only one endocrine parameter.

## 4. NEUROTRANSMITTERS AND DEPRESSION

Abnormalities in neurotransmitters, and the catecholamines in particular, have been postulated as the basis of affective disorders during the past two decades.

In original form, the catecholamine hypothesis of depression proposed that some, if not all, depressions are associated with an absolute or relative deficiency of catecholamines, especially norepinephrine, at functionally important receptor sites in the brain; conversely, elation was associated with an excess of such amines (Schildkraut, 1965, 1974). Although further investigations supported a connection between affective disorders and biogenic amines, they have also indicated that the original catecholamine hypothesis probably reflects only one facet of a very complex psychobiological state. It has become clear that other biogenic amines, such as indoleamines and dopamine, as well as neuroendocrine function, are involved in the biologic interplay which provides the substrate of affective disorders. The above suggests the catecholamine hypothesis has so far provided only an incomplete and oversimplified picture (Bunney and Gulley, 1978).

The original impetus for formulation of the catecholamine hypothesis was the finding that various monoamine oxidase inhibitors, notably iproniazid, acted clinically as mood elevators or antidepressants. Shortly thereafter it was found that this class of compounds also produced marked increases in biogenic amine levels. Actually, all three classes of drugs most commonly used to treat various depressive disorders, the tricyclic antidepressants, the monoamine oxidase inhibitors, and the psychostimulants, appeared to interact with catecholamines, in a way which is consistent with the catecholamine hypothesis. Reserpine, lithium, L-dopa, and biogenic amine depletors have also been studied in this context.

A close reexamination of the available experimental data has, however, pointed to some major inconsistencies in the catecholamine hypothesis. Monoamine oxidase inhibitors have not been reported very useful in typical depressions (Sack and Goodwin, 1974), and some psychostimulants (cocaine and amphetamine) that have a very potent effect on catecholamine metabolism are not effective antidepressants (Sack and Goodwin, 1974; Bunney and Gulley, 1978). Reserpine, which produced a major disruption of catecholamine metabolism, leads to depression mainly in predisposed individuals with a previous history of affective problems (Mendels, 1974).

The catecholamine hypothesis proposed that manias and depressions are biochemical opposites; yet, lithium administration has been demonstrated to be therapeutic in both (Sack and Goodwin, 1974). Similarly, experiments with

L-dopa, a precursor of catecholamines, and the biogenic amine depletors AMPT (α-methylparatyrosine) and PCPA (p-chlorophenylalanine) have not been consistent with expectations based on the catecholamine hypothesis (Sack and Goodwin, 1974; Mendels, 1974; Bunney and Gulley, 1978). One may conclude that, although evidence for the involvement of catecholamines in affective disorders is becoming increasingly compelling, their suggested pivotal role in the biochemistry of these disorders is still only speculative.

The serotonin deficiency theory of depression is still in dispute (van Praag, 1977). 5-HIAA in CSF has been reported as reduced as well as normal by different investigators (Aprison et al., 1978). Reduced serotonin levels in brain reported in some studies (e.g., Lloyd et al., 1974) has been disputed by others (Beskow et al., 1976). The serotonin precursors L-tryptophan and L-5-hydroxytryptophan are devoid of antidepressant activity in most studies. Probenecid-induced accumulation of 5-HIAA is reduced in both depression and mania (van Praag, 1977; Aprison et al., 1978), suggesting that there may be a decreased serotonin turnover in both states. This finding has led to the theory that serotonin deficiency underlies bipolar illness and that changes in catecholamines are responsible for the appearance of mania or depression. In contrast to previous theories, Aprison and co-workers have postulated a postsynaptic serotonin hypersensitivity in depression (Aprison et al., 1978).

Of particular interest have been findings linking the treatment response of depressed patients to a particular finding in biogenic amine metabolism. In one study (Maas et al., 1974) a group of depressed patients who responded favorably to imipramine but not to amitriptyline, showed low levels of the norepinephrine metabolite MHPG (3-methoxy-4-hydroxyphenylglycol) and these levels in urine did not change after clinical improvement or D-amphetamine. The patients who responded well to amitriptyline initially had high levels of MHPG, and these decreased either after treatment with tricyclic antidepressants or after a trial of D-amphetamine.

Studies of this type suggest that there are different biochemical subtypes of depression. Further support for this concept comes from findings that effective tricyclic antidepressant drugs differ in their specificity (Brogden et al., 1979; Sulser, 1979). An alternate hypothesis put forward by Sulser, that the tricyclic drugs all affect the postsynaptic noradrenergic receptor, remains to be confirmed (Sulser, 1978, 1979).

Thus, a substantial body of information links depression with alterations in catecholamines and serotonin. However, experimental work in animals makes it clear that most balances in the nervous systems occur via catenary control, that is, via actions of neurons connected like links in a chain (Sourkes, 1977). If this concept is applicable to depression, then different biochemical forms may exist in which different links of the chain are involved. These different biochemical forms may well have clinical, visceral, and neuroendocrine consequences which differ from one another despite the common occurrence of depression as a primary manifestation.

## 5. NEUROENDOCRINE REGULATION AND NEUROTRANSMITTERS

Regulation of a number of hypothalamic and pituitary hormones by neurotransmitters has been documented extensively. In the hypothalamus the neurotransmitters influence the secretion of several regulating factors which, in turn, have impact on the release of the hormones of the anterior pituitary. It is impossible to do justice here to the extensive literature available on this subject; however, several comprehensive articles (Frohman and Stachura, 1975; Brown et al., 1979; Brown and Seggie, 1980) have reviewed these relationships.

Here, only a few typical instances will be quoted. Thus, GH is under a positive control of at least three different neurotransmitter systems: norepinephrine, dopamine, and serotonin. Prolactin is regulated by biogenic amines, particularly by dopamine which exerts an inhibition on the secretion of prolactin. Dopamine appears to be the physiologic prolactin-inhibiting substance (Brown et al., 1976) although originally dopamine was thought to exert its action by increasing the release of prolactin-inhibiting factor (PIF). Melatonin is regulated primarily by norepinephrine (Brown and Niles, 1982).

Table 1 summarizes the relationships of various neurotransmitters and hormones. The data on which it is based are incomplete and in many cases only animal studies have been done and major species differences exist. Nevertheless, it is possible to use this table as a tentative model.

## 6. NEUROENDOCRINE EFFECTS OF LITHIUM

Lithium is a drug which is therapeutic in both mania and depression (Shopsin and Gershon, 1978) but whose mechanism of action is still a matter of controversy (Hendler, 1978). Certain data indicate an absolute or relative decrease of norepinephrine availability at the synapse following lithium treatment (Schildkraut, 1973; Hendler, 1978). It has also been hypothesized that it has a

TABLE 1
Major Effects of Neurotransmitters on Hormones[a,b]

|  | Serotonin | Dopamine | Norepinephrine | Endogenous opiate |
|---|---|---|---|---|
| ACTH | E |  | I | I |
| Prolactin | E | I | I | E |
| GH | E | E | Mixed | E |
| Thyrotropin |  | I | E |  |
| LH |  | I | E | I |

[a] Adapted from Brown et al. (1979).
[b] E: excitatory, facilitates release; I: inhibits release; mixed: α-adrenergic stimulation facilitates release and β-adrenergic stimulation inhibits release.

modulatory effect on serotonin systems (Mandell and Knapp, 1979) or that it may act to limit changes in receptors as it has been shown that it will prevent the development of supersensitivity of dopamine receptors (Bunney et al., 1979).

By far the most of the information on which these theories are based comes from studies in experimental animals. Strategies for investigation of neurobiologic effects of lithium in man require elaboration. We have therefore undertaken studies on the effects of lithium administration on neuroendocrine mechanisms in normal subjects.

Initial studies have been completed in 10 normal male subjects, age 36.9 ± 3.0 (S.E.M.) years, height 176.6 ± 1.9 cm, and body weight 75.7 ± 4.1 kg. All subjects were white, six were married, three single, and one divorced. All had a negative family history of depression and scored zero on the Hamilton and Beck rating scales. All were assessed using the Schedule for Affective Disorders and Schizophrenia. All had normal blood pressure, physical examination, normal routine laboratory examination, and a normal MMPI profile (Minnesota Multiphasic Personality Inventory). None was regularly on any psychotropic medication; at any rate, none received any medication for a minimum of 4 weeks prior to the first testing.

Each normal subject showed a clear-cut adrenal suppression following dexamethasone 2 mg in samples taken the next day at 8:00 a.m., 4:00 p.m., and midnight using the criterion of cortisol <5 μg/dl. Dexamethasone administration preceded the triple bolus test by 57 hr.

Each subject was tested with a modified triple bolus test prior to and following 3 weeks of lithium administration. The modified triple bolus test consisted of the administration of 0.1 U of regular insulin intravenously followed 2 hr later by 100 μg of LHRH and 500 μg of TRH. Responses of cortisol, GH and prolactin were assessed. Following insulin administration, tests were excluded in which hypoglycemia failed to achieve a drop of at least 50%. In such instances subjects were retested with a higher dose of insulin.

With the sample size of 10, no correlation was found of maximum hypoglycemia with the peak levels of GH, prolactin, or cortisol. All subjects showed a clear-cut GH and prolactin response to hypoglycemia and prolactin response to TRH combined with LHRH. The peak GH level showed a significant negative correlation with age.

In six of the subjects, responses have been examined while they were receiving lithium. Both GH (Fig. 1) and PRL (Fig. 2) responses to hypoglycemia were attenuated (GH: $F = 5.57$, $df = 1, 12$, $p < 0.05$; PRL: $F = 5.46$; $df = 1, 10$, $p < 0.05$). PRL (Fig. 3) response to TRH increased ($F = 7.20$, $df = 1, 10$, $p < 0.025$) while cortisol was unaffected ($F = 2.64$, $df = 1, 8$, NS). Despite the attenuated neurohormonal responses, hypoglycemia prior to and during lithium treatment was not significantly different. Thus, lithium treatment particularly after the premedication with dexamethasone has a marked effect on the hormonal response to hypoglycemia.

Recent pilot data suggest that the interaction between the effects of lithium

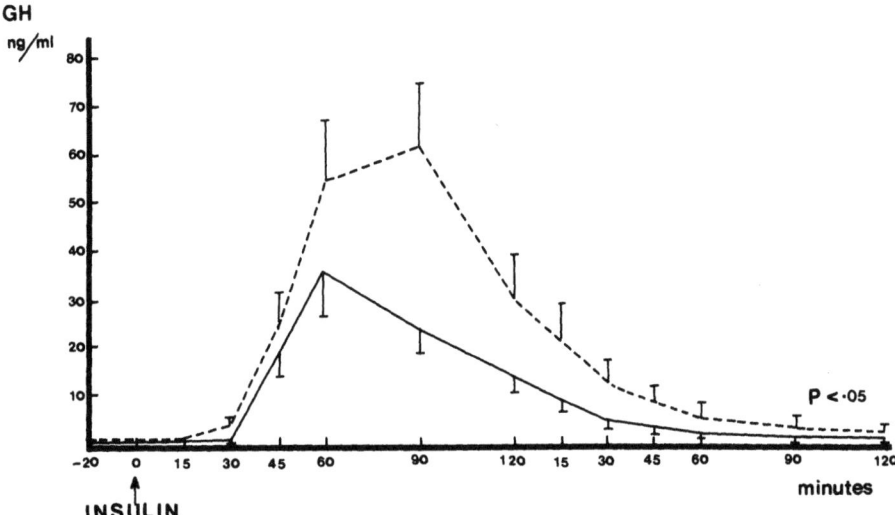

FIGURE 1. Growth hormone response to hypoglycemia in six healthy volunteers before lithium administration (interrupted line) and after 3 weeks on lithium (solid line).

and dexamethasone on neuroendocrine function are quite complex and require further study.

The finding that lithium treatment alters this neuroendocrine reflex provides the opportunity to gain an insight into the neurobiologic effects of lithium.

One possible explanation for this finding is that lithium interferes with one or more of the neurotransmitters which mediate the GH and PRL response to

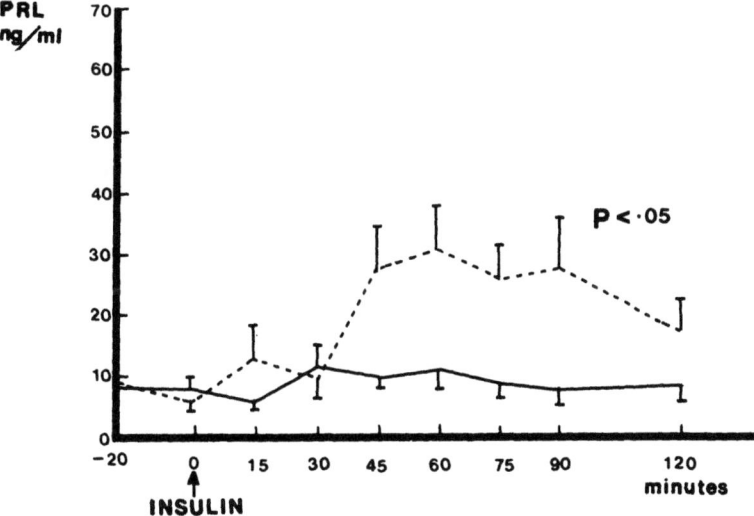

FIGURE 2. Prolactin response to hypoglycemia in six healthy volunteers before lithium administration (interrupted line) and after 3 weeks on lithium (solid line).

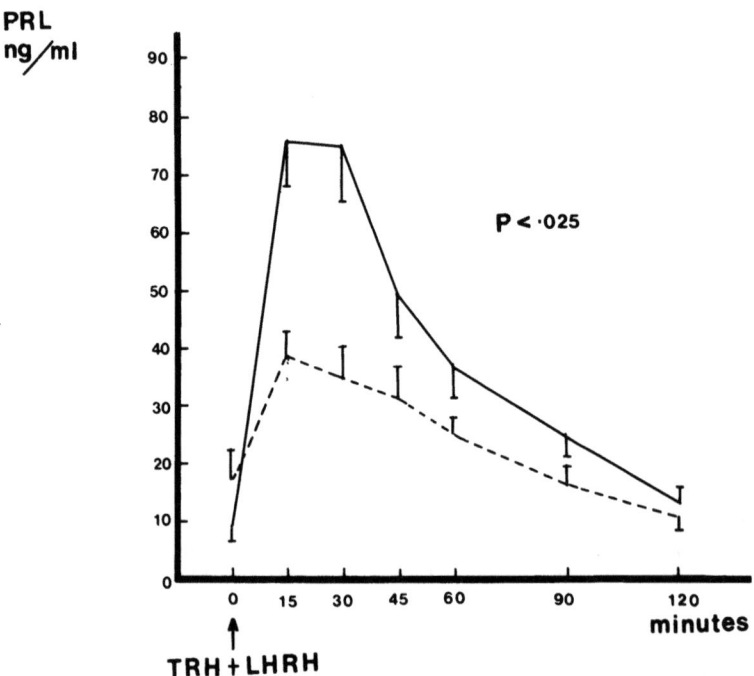

FIGURE 3. Prolactin response to TRH in six healthy volunteers before administration of lithium (interrupted line) and after 3 weeks on lithium (solid line).

hypogycemia. Possible neurotransmitter candidates are serotonin and one of the opioid peptides as these substances are excitatory to both GH and PRL (Brown et al., 1979). Alteration in dopamine is unlikely as a factor because this transmitter has opposite effects on GH and PRL and because lithium treatment does not alter either the PRL response to haloperidol or the GH response to apomorphine (Lal et al., 1978). Another possibility is that lithium alters the sensitivity of the glucoreceptors which mediate the response. Further studies will be necessary to clarify these issues.

## 7. CONCLUSION

Extensive investigation has established that there is a wide variety of endocrine abnormalities in depressive disorder involving the regulation of corticotropin, GH, thyrotropin, LH and perhaps vasopressin and melatonin. There is also evidence that biological subtypes of depressive disorders exist which may be differentiated endocrinologically. To date it has not been possible to establish the neural basis of these endocrine disorders with any certainty but it is probable that a common factor is responsible for both the mood and endocrine changes. The endocrine approach may therefore be useful not only as a clinical test for diagnosis and monitoring of the course of illness but also as a research tool useful in defining the underlying abnormality, and examining the action of antidepressant drugs. For example, marked neuroendocrine changes are produced by lithium administration which provide evidence of potent neurobiologic effects of this agent.

## 8. REFERENCES

Altman, N., Sachar, E. J., Gruen, P. H., Halpern, F. S., and Eto, S., 1975, Reduced plasma LH concentration in post menopausal depressed women, *Psychosom. Med.* **37**:274.

Aprison, M. H., Takahashi, R., and Tachiki, K., 1978, Hypersensitive serotonergic receptors involved in clinical depression: A theory, in: *Neuropharmacology and Behavior* (B. Haber and M. H. Aprison, eds.), pp. 23–53, Plenum Press, New York.

Arana, G., Boyd, A. E., III, Reichlin, S., and Lipsitt, D., 1977, Prolactin levels in mild depression, *Psychosom. Med.* **39**:193.

Beskow, J., Gottfries, C. G., Roose, B. F., and Winblad, B., 1976, Determination of monoamines and monoamine metabolites in the human brain: Post mortem studies in a group of suicides and in a control group, *Acta Psychiatr. Scand.* **53**:7.

Brambilla, F., Smeraldi, E., Sacchetti, E., Negri, F., Cocchi, D., and Muller, E. E., 1978, Deranged anterior pituitary responsiveness to hypothalamic hormones in depressed patients, *Arch. Gen. Psychiatry* **35**:1231.

Brogden, R. N., Keel, R. C., Speight, T. M., and Avery, C. S., 1979, Mianserin: A review of its pharmacological properties and therapeutic efficacy in depressive illness, *Drugs* **16**:273.

Brown, G. M., and Niles, L. P., 1981, Studies on melatonin and other pineal factors, in: *Clinical Neuroendocrinology*, Vol. 2 (G. M. Besser and L. Martini, eds.), pp. 205–264, Academic Press, New York.

Brown, G. M., and Seggie, J. A., 1980, Neuroendocrine mechanisms and their implications for psychiatric research, in: *Psychiatric Clinics of North America*, Vol. 3, No. 2 (E. J. Sachar, ed.), pp. 205–221, Saunders, Philadelphia.

Brown, G. M., Seeman, P., and Lee, T., 1976, Dopamine/neuroleptic receptors in basal hypothalamus pituitary, *Endocrinology* **99**:1407.

Brown, G. M., Seggie, J. A., Chambers, J. W., and Ettigi, P. G., 1978, Psychoendocrinology and growth hormone: A review, *Psychoneuroendocrinology* **3**:131.

Brown, G. M., Friend, W. C., and Chambers, J. W., 1979, Neuropharmacology of hypothalamic-pituitary regulation, in: *Clinical Neuroendocrinology: A Pathophysiological Approach* (G. Tolis, F. Labrie, J. B. Martin, and F. Naftolin, eds.), pp. 47–81, Raven Press, New York.

Brown, W. A., and Shuey, I., 1980, Response to dexamethasone and subtype of depression, *Arch. Gen. Psychiatry* **37**:747.

Bunney, W. E., Jr., and Gulley, B. L., 1978, The current status of research in the catecholamine theories of affective disorder, in: *Biochemistry of Mental Disorders: New Vistas* (E. Usdin and A. Mandell, eds.), pp. 83–100, Dekker, New York.

Bunney, W. E., Jr., Pert, A., Rosenblatt, J., Pert, C. B., and Gallaper, D., 1979, Mode of action of lithium, *Arch. Gen. Psychiatry* **36**:898.

Campbell, S., and Whitehead, M., 1977, Oestrogen therapy and the menopausal syndrome, *Clin. Obstet. Gynecol.* **4**:31.

Carroll, B. J., 1976, Psychoendocrine relationships in affective disorders, in: *Modern Trends in Psychosomatic Medicine*, Vol. 3 (O. W. Hill, ed.), pp. 121–153, Butterworths, London.

Carroll, B. J., Curtis, G. C., and Mendels, J., 1976a, Neuroendocrine regulation in depression. I. Limbic system-adrenocortical dysfunction, *Arch. Gen. Psychiatry* **33**:1039.

Carroll, B. J., Curtis, G. C., and Mendels, J., 1976b, Neuroendocrine regulation in depression. II. Discrimination of depressed from nondepressed patients, *Arch. Gen. Psychiatry* **33**:1051.

Checkley, S. A., 1979, Corticosteroid and growth hormone responses to methylamphetamine in depressive illness, *Psychol. Med.* **9**:107.
Cuenca, E., Galiana, J., and Cañete, J. M., 1978, Prolactin in depression, *Arch. Farmacol. Toxicol.* **4**:47.
Ehrensing, R. H., Kastin, A. J., Schalch, D. S., Friesen, H. G., Vargas, J. R., and Schally, A. V., 1974, Affective state and thyrotropin and prolactin responses after repeated injections of thyrotropin-releasing hormone in depressed patients, *Am. J. Psychiatry* **131**:714.
Ettigi, P. G., and Brown, G. M., 1977, Psychoneuroendocrinology of affective disorder: An overview, *Am. J. Psychiatry* **134**:493.
Ettigi, P. G., and Brown, G. M., 1979, Psychoendocrine correlates in affective disorder, in: *Neuroendocrine Correlates in Neurology and Psychiatry* (E. E. Müller and A. Agnoli, eds.), pp. 225–238, Elsevier/North-Holland, Amsterdam.
Ettigi, P. G., Brown, G. M., and Seggie, J. A., 1979, TSH and LH responses in subtypes of depression, *Psychosom. Med.* **41**:203.
Frohman, L. A., and Stachura, M. E., 1975, Neuropharmacologic control of neuroendocrine function in man, *Metabolism* **24**:211.
Furlong, F. W., Brown, G. M., and Beeching, M. F., 1976, Thyrotropin-releasing hormone: Differential antidepressant and endocrinological effects, *Am. J. Psychiatry* **133**:1187.
Garfinkel, P. E., Brown, G. M., Warsh, J. J., and Stancer, H. C., 1979, Neuroendocrine responses to carbidopa in primary affective disorders, *Psychoneuroendocrinology* **4**:13.
Garver, D. L., Pandey, G. N., Dekirmenjian, H., and Deleon-Jones, F., 1975, Growth hormone and catecholamines in affective disease, *Am. J. Psychiatry* **132**:1149.
Gold, M. S., Pottash, A. L., Ryan, N., Sweeny, D. R., Davies, R. K., and Martin, D. M., 1980, TRH-induced TSH response in unipolar, bipolar and secondary depressions: Possible utility in clinical assessment and differential diagnosis, *Psychoneuroendocrinology* **5**:147.
Gold, P. W., and Goodwin, F. K., 1978, Vasopressin in affective illness, *Lancet* **I**:1233.
Gold, P. W., Weingartner, H., Ballenger, J. C., Goodwin, F. K., and Post, R. M., 1979, Effects of 1-desamino-8-D-arginine vasopressin on behaviour and cognition in primary affective disorder, *Lancet* **II**:992.
Goldzieher, J. W., Dozier, T. S., and Smith, K. D., 1976, Improving the diagnostic reliability of rapidly fluctuating plasma hormone levels by optimized multiple-sampling techniques, *J. Clin. Endocrinol. Metab.* **43**:824.
Gregoire, F., Brauman, H., De Buck, R., and Corvilain, J., 1977, Hormone release in depressed patients before and after recovery, *Psychoneuroendocrinology* **2**:303.
Gruen, P. H., Sachar, E. J., Altman, N., and Sassin, J., 1975, Growth hormone responses to hypoglycemia in post-menopausal depressed women, *Arch. Gen. Psychiatry* **32**:31.
Halbreich, U., Assad, M., and Ben-David, M., 1978, Prolactin secretion during and after noveril infusions to depressive patients, *Psychopharmacologia* **56**:167.
Halbreich, U., Grunhaus, L., and Ben-David, M., 1979, Twenty-four-hour rhythm of prolactin in depressive patients, *Arch. Gen. Psychiatry* **36**:1183.
Hendler, N. H., 1978, Lithium pharmacology and physiology, in: *Handbook of Pharmacology*, Vol. 14, *Affective Disorders: Drug Actions in Animals and Man* (L. L. Iversen, S. D. Iversen, and S. H. Snyder, eds.), pp. 233–273, Plenum Press, New York.
Herrmann, W. M., and Beach, R. C., 1978, The psychotropic properties of estrogens, *Pharmakopsychiatr. Neuro-psychopharmakol.* **11**:164.
Janowsky, D. S., Leichner, P., Parker, D., Judd, L., Huey, L., and Clopton, P., 1978, Methylphenidate and serum prolactin in man, *Psychopharmacologia* **58**:43.
Kirkegaard, C., Bjørum, N., Cohn, D., and Lauridsen, V. B., 1978, Thyrotrophin-releasing hormone (TRH) stimulation test in manic-depressive illness, *Arch. Gen. Psychiatry* **35**:1017.
Laakmann, G., and Benkert, O., 1978, Neuroendokrinologie und Psychopharmaka, *Arnzeim. Forsch.* **28**:1277.
Ladisich, W., 1977, Influence of progesterone on serotonin metabolism: A possible causal factor for mood changes, *Psychoneuroendocrinology* **2**:257.

Lal, S., Nair, N. P. V., and Guyda, H., 1978, Effect of lithium on hypothalamic-pituitary dopaminergic function, *Acta Psychiatr. Scand.* **57**:91.

Lewy, A. J., Wehr, J. A., Gold, P., and Goodwin, F. K., 1979, Plasma melatonin in manic-depressive illness, in: *Catecholamines: Basic and Clinical Frontiers*, Vol. 2 (E. Usdin, I. J. Kopin, and T. Barchas, eds.), pp. 1173-1175, Pergamon Press, Elmsford, N. Y.

Linnoila, M., Lamberg, B. A., Rosberg, G., Karonen, S. L., and Welin, M. G., 1979, Thyroid hormones and TSH, prolactin and LH responses to repeated TRH and LRH injections in depressed patients, *Acta Psychiatr. Scand.* **59**:536.

Lloyd, K. G., Farley, I. J., Deck, J. H., and Hornykiewicz, O., 1974, Serotonin and 5-hydroxyindoleacetic acid in discrete areas of the brainstem of suicide victims and control patients, *Adv. Biochem. Psychopharmacol.* **11**:387.

Maas, J. W., 1979, Biochemistry of the affective disorders, *Hosp. Pract.* **14**:113.

Maas, J. W., Dekirmenjian, H., and Fawcett, J. A., 1974, MHPG excretion by patients with affective disorders, *Int. Pharmacopsychiatry* **9**:14.

Maeda, K., Kato, Y., Ohgo, S., Chihara, K., Yoshimoto, Y., Yamaguchi, N., Muromaru, S., and Imura, H., 1975, Growth hormone and prolactin release after injection of thyrotropin-releasing hormone in patients with depression, *J. Clin. Endocrinol. Metab.* **40**:501.

Mandell, A. J., and Knapp, S., 1979, Asymmetry and mood, emergent properties of serotonin regulation, *Arch. Gen. Psychiatry* **36**:909.

Martin, J. B., Reichlin, S., and Brown, G. M., 1977, *Clinical Neuroendocrinology*, pp. 1-410, Davis, Philadelphia.

Matussek, N., 1978, Effect of amphetamine and clonidine on human growth hormone release in psychiatric patients and controls, in: *Depressive Disorders* (S. Garattini, ed.), pp. 431-46., Schattauer Verlag, Stuttgart.

Mendels, J., 1974, Amine depletion as an experimental approach to the aminergic hypothesis, *Psychopharmacol. Bull.* **10**:54.

Mendelwicz, J., Linkowski, P., Branchey, L., Weinberg, U., Weitzman, E. D., and Branchey, M., 1979, Abnormal 24 hour pattern of melatonin secretion in depression, *Lancet* **II**:1362.

Nielsen, J. L., 1980, Plasma prolactin during treatment with nortriptyline, *Neuropsychobiology* **6**:52.

Nuller, J. L., and Ostroumova, M. N., 1980, Resistance to inhibiting effect of dexamethasone in patients with endogenous depression, *Acta Psychiatr. Scand.* **61**:169.

O'Dea, J. P. K., Gould, D., Hallberg, M., and Wieland, R. G., 1978, Prolactin changes during electroconvulsive therapy, *Am. J. Psychiatr.* **135**:609.

Orth, J. P., Braccini, T., Krebs, B. P., and Darcourt, G., 1979, The psychotropic action of TRH, *Encephale* **5**:375.

Rubin, R. T., and Kendler, K. S., 1977, Psychoneuroendocrinology: Fundamental concepts and correlates in depression, in: *Depression: Clinical, Biological and Psychological Perspectives* (E. Usdin, ed.), pp. 122-138., Brunner/Mazel, New York.

Sachar, E. J., 1967, Corticosteroids in depressive illness. I. A reevaluation of control issues and the literature, *Arch. Gen. Psychiatry* **17**:544.

Sachar, E. J., 1975, Hormonal changes in stress and mental illness, *Hosp. Pract.* **10**:(7):49.

Sachar, E. J., Hellman, L., Roffwarg, H. P., Halpern, F. S., Fukushima, D. K., and Gallagher, T. F., 1973, Disrupted 24-hour patterns of cortisol secretion in psychotic depression, *Arch. Gen. Psychiatry* **28**:19.

Sachar, E. J., Altman, N., Gruen, P. H., Glassman, A., Halpern, F. S., and Sassin, J., 1975, Human growth hormone response to levodopa: Relation to menopause, depression and plasma dopa concentration, *Arch. Gen. Psychiatry* **32**:502.

Sachar, E. J., Asnis, G., Halbreich, U., Nathan, R. S., and Halpern, F., 1980, Recent studies in the neuroendocrinology of major depressive disorders, in: *Psychiatric Clinics of North America*, Vol. 3, No. 2 (E. J. Sachar, ed.), pp. 313-326, Saunders, Philadelphia.

Sack, R. L., and Goodwin, F. K., 1974, Inhibition of dopamine β-hydroxylase in manic patients, *Arch. Gen. Psychiatry* **31**:649.

Sartorius, N., 1979, Depressive disorders, a major public health problem, in: *Mood Disorders* (F. J. Ayd and J. J. Taylor, eds.), pp. 1–8, Ayd Medical Communications, Baltimore.

Schildkraut, J. J., 1965, The catecholamine hypothesis of affective disorders: A review of supporting evidence, *Am. J. Psychiatry* **122**:509.

Schildkraut, J. J., 1973, Pharmacology—The effects of lithium on biogenic amines, in: *Lithium: Its Role in Psychiatric Research and Treatment* (S. Gershon and B. Shopsin, eds.), pp. 51–73, Plenum Press, New York.

Schildkraut, J. J., 1974, Biogenic amines and affective disorders, *Annu. Rev. Med.* **25**:333.

Schlesser, M. A., Winokur, G., and Sherman, B. M., 1980, Hypothalamic–pituitary–adrenal axis activity in depressive illness, *Arch. Gen. Psychiatry* **37**:737.

Schneider, M. A., Brotherton, P. L., and Hailes, J., 1977, The effect of exogenous oestrogens on depression in menopausal women, *Med. J. Aust.* **2**:162.

Shopsin, B., and Gershon, S., 1971, Plasma cortisol response to dexamethasone suppression in depressed and control patients, *Arch. Gen. Psychiatry* **24**:320.

Shopsin, B., and Gershon, S., 1978, Lithium: Clinical considerations, in: *Handbook of Psychopharmacology*, Vol. 14, *Affective Disorders: Drug Actions in Animals and Man* (L. L. Iversen, S. D. Iversen, and S. H. Snyder, eds.), pp. 275–325, Plenum Press, New York.

Sourkes, T. L., 1977, Biochemistry of mental depression, *Can. Psychiatr. Assoc. J.* **22**:487.

Stancer, H. C., Quarrington, B., Cookson, B. A., Brown, G. M., Bonkalo, A., and Lyall, W. A. L., 1969, A longitudinal drug study and central amines, *Arch. Gen. Psychiatry* **20**:290.

Stokes, P. E., Stoll, P. M., Mattson, M. R., and Sollod, R. N., 1976, Diagnosis and psychopathology in psychiatric patients resistant to dexamethasone, in: *Hormones, Behavior and Psychopathology* (E. J. Sachar, ed.), pp. 225–229, Raven Press, New York.

Sulser, F., 1978, Functional aspects of the norepinephrine receptor coupled adenylate cyclase system in the limbic forebrain and its modification by drugs which precipitate or alleviate depression: Molecular approaches to an understanding of affective disorders, *Pharmakopsychiatr. Neuro-Psychopharmakol.* **11**:43.

Sulser, F., 1979, Tricyclic antidepressants: Animal pharmacology (biochemical and metabolic aspects), in: *Handbook of Psychopharmacology*, Vol. 14 *Affective Disorders: Drug Actions in Animals and Man* (L. L. Iversen, S. D. Iversen, and S. H. Snyder, eds.), pp. 157–197, Plenum Press, New York.

Takahashi, S., Kondo, H., Yoshimura, M., and Ochi, Y., 1974, Growth hormone responses to administration of L-5-hydroxytryptophan (L-5-HTP) in manic-depressive psychosis, in: *Psychoneuroendocrinology: Proceedings* (N. Hatotani, ed.), pp. 32–38, Karger, Basel.

Taracha, E., Szukalski, B., and Zaluska, M., 1978, Plasma cortisol level in patients with affective disturbances and its correlation with tryptophan and 5-hydroxyindoleacetic acid [Preliminary communication], *Psychiatr. Pol.* **12**:317.

van Praag, H. M., 1977, Indoleamines in depression, in: *Neuroregulators and Psychiatric Disorders* (E. Usdin, D. A. Hamburg, and J. D. Barchas, eds.), pp. 163–176, Oxford University Press, London.

Weeke, A., and Weeke, J., 1978, Disturbed circadian variation of serum thyrotropin in patients with endogenous depression, *Acta Psychiatr. Scand.* **57**:281.

Weissman, M. M., and Klerman, G. L., 1978, Epidemiology of mental disorders, *Arch. Gen. Psychiatry* **35**:705.

Wetterberg, L., Beck-Friis, J., Aperia, B., and Petterson, U., 1979, Melatonin/cortisol ratio in depression, *Lancet* **II**:1361.

CHAPTER 19
# Neuroendocrine Dysfunction in Subtypes of Depression

DETLEV VON ZERSSEN, MATHIAS BERGER, and PETER DOERR

## 1. OUTLINE OF THE PROBLEM

### 1.1. Endocrine Dysfunction in Depression

As pointed out by G. M. Brown and associates in their contribution to this volume, "neuroendocrine function may provide a means of differentiating various subtypes among the affective disorders, because they may mirror the central activity of biogenic amines in depressed subjects. Different types of neurotransmitter deficit have been postulated in depressive illness. Different underlying abnormalities could lead to syndromes which differ with respect to clinical characteristics, visceral symptomatology, and endocrine abnormalities." They have listed a series of endocrine abnormalities of depressed patients as described in the literature and added information about their own findings on neuroendocrine effects of lithium (see also Prange and Loosen, this volume; Johnson, 1982; Sachar, 1982; Winokur et al., 1982).

As far as subtypes of depression are concerned, neuroendocrine dysfunction was found mainly in primary depression as compared to secondary depression and in endogenous depression as compared to nonendogenous (psychogenic) depression (Carroll, 1982; Carroll et al., 1980a; Czernik, 1982; Matussek et al., 1980). Findings with respect to genetic subtypes have been less consistent. For example, one group (Schlesser et al., 1980; see also Coryell et al., 1982a; Targum et al., 1982) has reported a significantly higher incidence of cortisol nonsuppression or early escape from suppression after a midnight dose of dexamethasone for bipolar patients than for unipolar patients, and within the unipolar group, for the familial subtype of pure depression than for sporadic depression and depression spectrum disease, but these results have not been

---

DETLEV VON ZERSSEN, MATHIAS BERGER, and PETER DOERR • Max-Planck-Institut für Psychiatrie, Munich, FRG.

confirmed by others (Amsterdam *et al.*, 1982; Carroll *et al.*, 1980b; Rudorfer *et al.*, 1982). Similar discrepancies are found with respect to psychopathological subtypes. Thus, indications of increased cortisol production were originally related to specific psychopathological features of depression (e.g., anxiety and psychotic turmoil, on the one hand, and denial of illness, on the other; see Sachar, 1967) but later found to be unrelated to psychopathological characteristics of (the endogenous subtype of) depression. It seems fairly well established that an increased activity of the hypothalamic–pituitary–adrenal system (HPAS) is particularly often found in severely depressed patients exhibiting psychotic (delusional) symptoms (Carroll *et al.*, 1976c; Rudorfer *et al.*, 1982). Within the lower range of severity, however, no consistent correlation with endocrine abnormalities has as yet been ascertained (Carroll, 1982; Johnson, 1982).

"The final general problem is one of interpretation. As psychoendocrinologists, we tend to make use of clinical screening procedures . . . soon after their introduction by clinical endocrinologists and before the mechanisms of these provocative tests are understood. From a little information . . . we are tempted to construct explanatory hypotheses, e.g., about endogenous brain catecholamine defects in depression. . . . Such hypotheses may have heuristic value to workers in the field, but they can be misleading to general readers," as Carroll and Mendels (1976, p. 217) have expressed it with special reference to the psychoendocrine study of the growth hormone system in depression. Their warning should be taken seriously also with respect to research on other hormonal systems and other mental disorders, e.g., neuroendocrine dysfunction in anorexia nervosa.

## 1.2. Endocrine Dysfunction in Depression and Anorexia Nervosa

Although it is well known that malnutrition leads to marked changes in metabolism (see Fichter and Pirke, 1982) including alterations in the production rate as well as clearance rate of almost all hormones, it was suggested that endocrine abnormalities in anorexic patients (Katz *et al.*, 1976) were, in part, reflections of a hypothalamic dysfunction underlying the disorder (Russell, 1977). A more thorough analysis of psychoendocrine relationships with special reference to the nutritional state of anorexic patients (Fichter *et al.*, 1982) and experimental research on rats (Pirke and Spyra, 1982) and human volunteers (Pirke *et al.*, 1982) under starvation have, however, convincingly demonstrated that all the endocrine changes and other somatic abnormalities so far observed in anorexia nervosa (Fichter and Pirke, 1982) can probably be traced back to malnutrition (see also Aro *et al.*, 1978).

Depression is one of the conditions of involuntary weight loss (Marton *et al.*, 1981), and some of the neuroendocrine findings in depression [particularly increased cortisol production, flattening of the diurnal rhythm (see Fig. 1), and nonsuppression or early escape from suppression of cortisol production by dexamethasone] are similar to findings in anorexia nervosa (Doerr *et al.*, 1980;

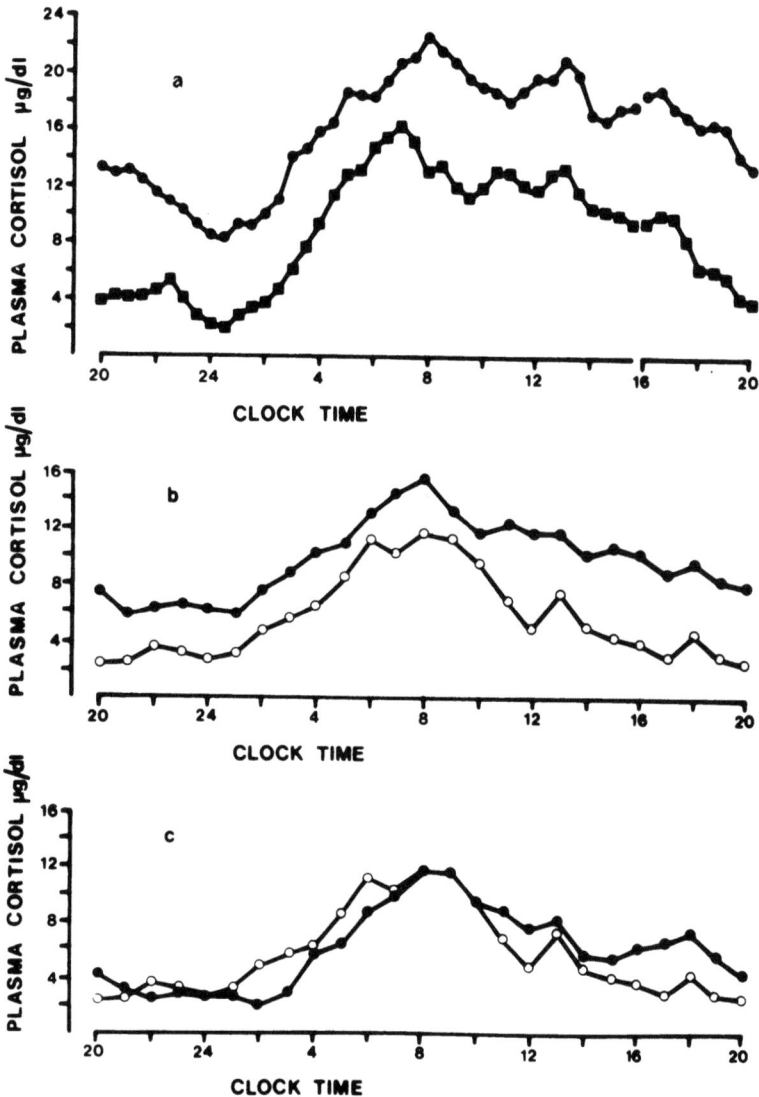

FIGURE 1. Mean plasma cortisol concentration in: (a) Patients with anorexia nervosa on admission to hospital (dots, $N = 18$) and after recovery (squares, $N = 15$). (From Doerr et al., 1979.) (b) "Hypersecreting" patients with endogenous depression (dots, $N = 12$) and normal postmenopausal women (circles, $N = 6$). (c) "Normosecreting" patients with endogenous depression ($N = 12$) and normal postmenopausal women (same as b). (From Sachar, 1982.) [The figures are redrawn from the originals in a simplified manner (without symbols for standard deviations or significance levels, respectively) using equal units in the coordinate system; moreover, in (1) the time interval 16–20 hr is switched from the beginning of the measurement period to its end in order to facilitate visual comparison with curves (b) and (c). Differences in the general level of cortisol values between (a) and (b)/(c) are probably due to methodological differences between the laboratories.]

unchanged in depression in contrast to anorexia nervosa. Therefore, it seems worthwhile to consider how far weight loss may be responsible, at least in part, for neuroendocrine dysfunctions in depressed patients (Fichter et al., 1982). It should be mentioned in this context that changes in the hypothalamic–pituitary–gonadal system in anorexia nervosa are clearly related to the degree to which the anorexic is underweight (Pirke et al., 1980) whereas the activation of the HPAS is induced by fasting and consequently more related to actual weight loss than to the deviation of a patient's real weight from his ideal weight (Doerr et al., 1980). It may be inferred from these findings that in depression, actual weight loss due to a reduced caloric intake may be a factor underlying neuroendocrine dysfunction, particularly with respect to the increased activity of the HPAS.

### 1.3. Endocrine Dysfunction in Depression and Cushing's Syndrome

Even more similarities in neuroendocrine dysfunction than between anorexia nervosa and depression are, however, found between the latter condition and early stages of Cushing's disease (Carroll and Mendels, 1976). These endocrine similarities may sometimes even cause problems in differential diagnosis, particularly if both obesity and depressed mood are present. This is not only because of an increase in cortisol production with a marked activation of the HPAS during the evening and the first half of the night, dexamethasone-nonsuppression, and a blunted cortisol response to insulin-induced hypoglycemia, but also the blunted responses of human growth hormone (hGH) to hypoglycemia and other provocative tests, the blunted response of thyroid-stimulating hormone (TSH) to TSH-releasing factor (TRH) and other neuroendocrine dysfunctions (Krieger, 1982). Insofar as the same hormone analysis and the same provocation or suppression tests have been applied in Cushing's disease and depression, all abnormalities so far repeatedly described in (at least severe forms of) depression are also and even more regularly found in Cushing's disease. It was therefore concluded that similar changes in brain function might underlie Cushing's disease (without autonomous adenoma of the pituitary) and melancholia (Carroll and Mendels, 1976). The high incidence of depressive features in patients with Cushing's disease (Cohen, 1980; Whybrow and Hurwitz, 1976) seems to support this assumption. The basic defect in melancholic patients was supposed to be located within the limbic system (Carroll et al., 1976b). These inferences, however, must be regarded premature in view of the following facts:

1. The cerebral pathology of Cushing's disease (without autonomous adenoma of the pituitary) is not yet well understood. Lesions have been described in the diencephalon and in the cerebral cortex of such patients (see Krieger, 1982), but so far not in the limbic system.
2. Depression does not regularly occur in patients with Cushing's disease and is also, though possibly less often (Cohen, 1980), observed in Cush-

ing's syndrome. Other mental disorders (mainly delusional psychoses and confusional states) are observed in almost the same frequency in both conditions (Whybrow and Hurwitz, 1976). There is no convincing evidence of psychopathological differences between Cushing's disease and Cushing's syndrome (Bleuler, 1964; Jeffcoate et al., 1979) and the psychopathological changes usually disappear after normalization of cortisol levels due to therapy (Jeffcoate et al., 1979; Voigt et al., in press). Moreover, similar mental disorders as in endogenous hyperadrenocorticism are also observed, though relatively less frequently, during pharmacotherapy with ACTH and/or glucocorticoids and tend to respond well to an adequate reduction of exogenous hormone supply (Ling et al., 1981; von Zerssen, 1976).

3. Endocrine changes as described for melancholia are also observed in Cushing's syndrome due to autonomous ACTH-producing tumors and, except for an increased ACTH production, in Cushing's syndrome due to cortisol-producing tumors of the adrenal cortex and, except for the increased production of ACTH (and cortisol), under pharmacotherapy with ACTH or glucocorticoids (Krieger, 1982). Therefore, they are to be regarded as consequences of hyperadrenocorticism (Kendler and Davis, 1977) rather than as reflections of an underlying brain dysfunction as in Cushing's disease (without autonomous adenoma of the pituitary) and in melancholia.

It should be pointed out that a derangement of pituitary function with altered responsiveness to releasing factors develops in Cushing's disease and Cushing's syndrome (Krieger, 1982) as well as during starvation (Fichter and Pirke, 1982). Such dysfunction at the pituitary level has also been reported to occur in depression (Brambilla et al., 1978). Nonetheless, the blunted cortisol response to methylamphetamine and D-amphetamine was interpreted as indicating a reduced α-adrenergic sensitivity in the brains of patients with endogenous depression, supposed to underlie the symptom formation of this disorder (Checkley, 1979; Sachar et al., 1980). However, in vitro studies of cultures and suspensions of rat pituitary cells (Vale et al., 1978; Voigt et al., 1980) favor an alternative interpretation in which this rapidly occurring response is mediated by α-adrenergic receptors in the anterior lobe of the pituitary (physiologically responsible for ACTH release by norepinephrine from the periphery). Cortisol hypersecretion may prevent this reaction in some depressed patients with an increased activity of the HPAS. [There are, however, experimental data in humans indicating that methylamphetamine may directly stimulate cortisol secretion at the adrenal level (Fehm et al., in press).]

### 1.4. The Stress Hypothesis of Endocrine Dysfunction in Depression

The causative factors of cortisol hypersecretion in depression have not yet been clearly elucidated. Besides starvation, the mental stress induced by the disorder has to be taken into account although the following objections have been raised against this stress hypothesis (e.g., Carroll and Mendels, 1976; Sachar, 1975):

1. Patients with a neurotic form of the disorder generally do not show indications of an increased activity of the HPAS unless they experience rather specific situational stresses, e.g., related to particular psychotherapeutic sessions.
2. Among the patients with endogenous depression, no clear relationship has been established between cortisol hypersecretion and clinical features, suspected to reflect the degree of stress experienced by the patient, e.g., anxiety or severity of depression.
3. Cortisol secretion is more enhanced in the evening when depressive mood tends to be alleviated.
4. Increased cortisol secretion persists even during sleep.
5. Diazepam was shown to prevent the adrenocortical stress response in rats; however, the drug does not seem to influence cortisol hypersecretion in depressed patients.
6. The secretion of the other "stress hormones," prolactin and hGH, is usually not enhanced (prolactin) and in some cases can even be decreased (hGH during sleep) in depressed patients.

These objections are, however, not quite convincing:

1. There are some reports in the literature according to which an increased activity of the HPAS may also appear in patients with neurotic depression (e.g., Holsboer et al., 1980; Shulman and Diewold, 1977; Stokes et al., 1976; Träskman et al., 1980).
2. In patients with severe delusional depression, neuroendocrine abnormalities are found with higher frequency and to a greater extent than in patients with less severe forms of endogenous depression (Carroll et al., 1976c). This may be indicative of a higher degree of stress experienced in the context of depressive delusions. The fact that the DST is usually normal in patients with hypomania or euphoric mania but frequently abnormal in severe degrees of mania with dysphoric affect (Carroll, 1979) may point in the same direction. Moreover, among healthy people in predicament, those who are "suffering in silence" (Mason, 1975) seem to be under more stress according to 170H-CS excretion levels than those who express negative emotions openly. Thus, the degree of "mental stress" may be difficult to judge from the overt emotional state in depressed patients as well. Finally, there may be constitutional differences in stress response (see Bridges et al., 1970) which tend to obscure the influence of state variables such as particular symptoms of depression.
3. The relationship of the diurnal variation in mood and the increase of cortisol production of patients during depression, compared with the

normal state, has not yet been clearly elucidated. It cannot be ruled out that a flattening of the diurnal curve of cortisol production does occur mainly in the absence of diurnal variation in mood. On the other hand, it is well established that circadian inhibition of the HPAS can be overridden by stress; indeed, the stress response is even more pronounced around the nadir than around the zenith of the circadian curve of cortisol production (Jones et al., 1981).

4. Sleep may not be a sufficient inhibitor of enduring emotional stress such as that induced by severe depression. After all, mental processes are not extinguished during sleep. Otherwise, the relationship of the content of dreams to actual experience and the induction of sleep disturbances by emotional conflict could hardly be explained.
5. Diazepam may prevent the stress response in rats by the reduction of, e.g., anxiety induced by the stressful situation. As the drug does not alleviate depressive mood, particularly in endogenous depression, it may be an inadequate means of preventing a stress response in patients with endogenous depression. After all, drugs that are capable of diminishing the depressive state, e.g., the classical antidepressants, are also capable of normalizing the increased activity of the HPAS during depression.
6. Stress responses may differ with respect to the stimuli by which they are evoked. The spectrum of hormonal changes in "silent sufferers" with an increased activity of the HPAS has not yet been investigated. Moreover, for ethical reasons, research on emotional stress in human subjects has been limited to the acute stress response. It might, indeed, be difficult to induce chronic emotional stress since the organism tends to adapt to such stressors as usually occur in everyday life or are ethically justified in being artificially produced in experimental situations (Rose, 1980). It may well be a characteristic of some emotional disorders that it is not possible to become adapted to their associated stresses.

These arguments lead to the conclusion that stress as a causative factor of neuroendocrine dysfunctions in depression cannot be ruled out with good confidence. Therefore, it seems premature to relate these dysfunctions to disturbances in the balance of transmitter systems, assumed to underlie symptom formation in depression (Checkley, 1980; Sachar, 1975). Even if a reduced urinary excretion of 3-methoxy-4-hydroxyphenylglycol (MHPG), indeed, were to occur in a certain proportion of patients with endogenous depression, particularly bipolar depressives (Schildkraut et al., 1978), and even if this were due to a decreased norepinephrine turnover in the brain, it might not reflect the neurochemical basis of depression but rather a secondary disturbance, e.g., related to a reduced caloric intake (see Pirke and Spyra, 1982). Similarly, an overactivity of cholinergic neurons in the brains of depressed patients as postulated by some authors might occur as a consequence and not as the origin of emotional arousal and stress in the afflicted subjects. Such secondary changes in central neurotransmitters, therefore, have to be considered as possible causes of neuroendocrine dysfunctions in subtypes of depression.

## 2. REVIEW OF AUTHORS' RESEARCH ON NEUROENDOCRINOLOGY OF DEPRESSION

### 2.1. Chronobiology of Endogenous Depression

#### 2.1.1. Methodology

Our own research on neuroendocrine dysfunction in depression began with a chronobiological investigation of the endogenous subtype of the disorder. This study which has been performed within an interdisciplinary research group at the Max-Planck-Institut für Psychiatrie (MPIP) included subjects demonstrating a wide spectrum of depressive disorders ranging from mild nonpsychotic, to marked delusional depression and the unipolar as well as the bipolar variate of the disorder. Altogether 20 patients with a pure depressive syndrome of no demonstrable organic origin, 17 of whom exhibited a clear-cut endogenous symptom pattern according to thorough clinical investigation, were studied during a drug-free period of 1 to 3 weeks during depression and in 13 cases also during a drug-free period of 2 weeks after remission. Ten of these patients diagnosed as endogenously depressed and investigated completely during at least 2 weeks in depression and full remission could be compared with 10 sex- and age-matched controls who were studied for 2 weeks under the same external conditions.

According to Research Diagnostic Criteria (RDC; Spitzer et al., 1978), all 20 patients suffered from a major depression, endogenous subtype, and (with the exception of the three clinically ambiguous cases and one other case) received a score within the endogenous range ($\geq 6$) on the 10-item version of the Newcastle Scale (Carney et al., 1965). Scores of clinical and self-rating depression scales indicated depressive symptomatology of varying degrees from mild to very severe. Psychological, physiological, and biochemical measurements were taken every 3 hr during the day (from 7 a.m. to 10 p.m.) and once during the night (between 2 and 3 a.m.). Sleep was recorded polygraphically (Schulz, 1981; Schulz et al., 1979).

One additional patient with 48-hr cycles of unipolar depression, which had persisted for more than 12 years, was investigated according to the same schedule not only on a psychiatric ward (Emrich et al., 1979) but also, with kind support from Professor Aschoff, Professor Wever, and Dr. Zulley, in a special research unit completely devoid of external time cues (Dirlich et al., 1981; Doerr et al., 1979; von Zerssen et al., in press).

The excretion of urinary free cortisol (UFC; Carroll et al., 1976a) was assessed using a protein-binding method. Data analyses included the calculation of daily means, averages per time interval (with seven intervals per 24 hr), the combination of which rendered the induced waveform for diurnal variation, and furthermore, autocorrelations, power spectra, and periodograms. Among the additional analyses were correlations of the average excretion of UFC per 24 hr with other variables, e.g., the degree of severity of depression at admission according to clinical and self-rating scales.

### 2.1.2. Results

The main findings concerning UFC were (see also Doerr and von Zerssen, 1983; von Zerssen and Doerr, 1980):

1. With few exceptions, the daily excretion rates of patients and controls were within normal limits but, on the whole, they were relatively higher in the patients during depression, compared with the values of control subjects, and, as far as available, the values obtained after remission. In 7 out of 10 patients reinvestigated in full remission of a definitely endogenous episode of depression, there was a drop in the excretion rate ranging from approximately 20% to 70% of the values assessed during the episode (Fig. 2).
2. The circadian pattern of UFC excretion was well preserved in almost all patients during depression. Only in one severely anxious-depressed delusional patient with an abnormally high excretion rate could a flat-

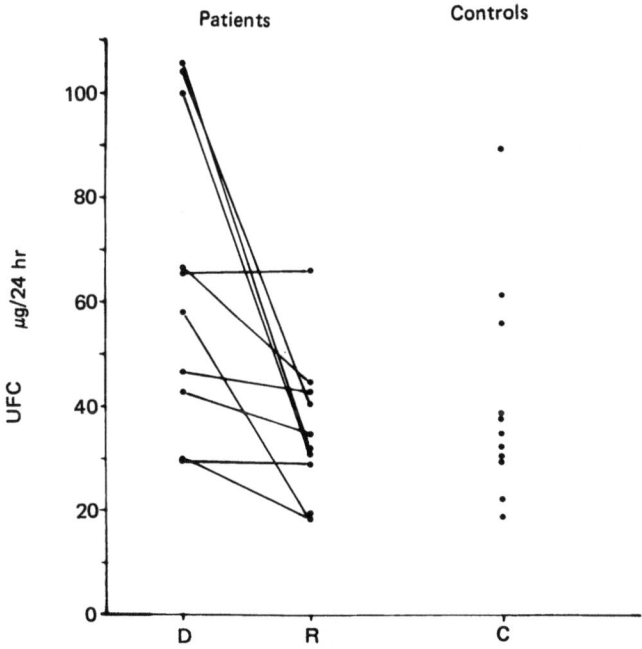

FIGURE 2. Twenty-four-hour mean of UFC excretion in endogenous depressed inpatients during depression (D) and remission (R) and in controls (C).

tening of the circadian curve be observed, but a marked variation with a drop in values from morning to night by approximately 50% was still present during the episode. The amount of UFC and the circadian variation normalized completely with the patient's recovery (von Zerssen and Doerr, 1980).
3. No indication of periods other than 24 hr could be detected in a temporal pattern of UFC excretion nor was there a consistent time shift in phase position of the circadian curve.
4. In the patient with 48-hr cycles of depression, there was a concordant cycling of UFC excretion which persisted under conditions devoid of external time cues although internal desynchronization occurred due to irregularity and shortening of the sleep–wake cycle. The UFC values, though elevated on "bad days," still remained within normal limits on the ward as well as in the isolation unit (Doerr *et al.*, 1979).
5. A significant positive correlation could be established between UFC excretion during depression and the severity of the disorder as estimated by means of a standardized rating scale, the Inpatient Multidimensional Psychiatric Scale (IMPS), at admission (von Zerssen and Doerr, 1980). This is apparently due to markedly elevated UFC values in severely depressed patients, particularly those with psychotic features (depressive delusions).

### 2.1.3. Discussion

It can be concluded from these results that enhancement of cortisol secretion is a much more typical finding in depressed patients than any changes in circadian parameters. In particular, a flattening of the circadian curve seems to occur predominantly or exclusively in a state of cortisol hypersecretion (see also Sachar, 1982, Figs. 13.3 and 13.4). This is true not only of depression but also of anorexia nervosa (Doerr *et al.*, 1980) and Cushing's syndrome (Krieger, 1982). In depression, hypersecretion with values above the upper limit of the normal range obviously occurs only in a minority of cases, especially those with a severe delusional form of the disorder.

## 2.2. Differential Diagnosis of Endogenous vs. Neurotic Depression

### 2.2.1. Methodology

Based on the findings of the chronobiological study and on reports in the literature, we expanded our research on neuroendocrine dysfunction in depression to the neurotic subtype of the disorder by comparing tests of neuroendocrine function (and sleep parameters) in patients with either neurotic or endogenous depression of comparable severity (Berger *et al.*, 1982b). Thus, only neurotic patients with a long-standing depressive symptomatology severe enough to warrant inpatient treatment, on the one hand, and patients with nonpsychotic endogenous depression who were not seriously suicidal, on the other, were included in the study. Patients for whom a differential diagnosis of endogenous or neurotic depression could not be achieved on clinical grounds were also included but treated as a separate group in data analysis. The series comprised altogether 45 patients with an RDC diagnosis of major depressive disorder, with at least moderate degree of severity according to subscales of the IMPS (Lorr and Klett, 1967), but without delusional or other psychotic symptoms.

The clinical diagnosis made independently by two experienced psychiatrists clearly indicated an endogenous type in 20 cases and a neurotic type in 19 cases and was ambiguous with respect to the endogenous-neurotic dichotomy in the remaining six cases. The clinical diagnoses were, with only very few exceptions, confirmed by RDC criteria of the endogenous vs. the nonendogenous subtype of depression and by the scores of the 10-item version of the Newcastle Scale (Carney et al., 1965). In the clinically ambiguous cases, there was a discrepancy between the operational diagnostic criteria with a trend to the diagnosis of an endogenous depression according to the RDC and to that of neurotic depression according to the Newcastle Scale. (Since in this series the scores of the Newcastle Scale were bimodally distributed, we divided them accordingly, with a score of 5 belonging to the upper range and thus indicating an endogenous depression; see Table 1).

The patients were thoroughly investigated in a drug-free state shortly after admission to the hospital. Since it had been found in the chronobiological study that a relative increase in the excretion of UFC was more typical of the patients' depressive state than any flattening of the circadian pattern, UFC was assessed in three 24-hr samples of urine without taking diurnal variation into account. Moreover, the hypoglycemic reaction to insulin (0.1 U/kg; Nathan et al., 1981) and the hGH response to the insulin-induced hypoglycemia (Gruen et al., 1975) were estimated. Suppressibility of cortisol production was ascertained outside the periods of urine collection by administering an oral dose of 1.5 mg of dexamethasone at 11 p.m. and drawing blood samples at 9 a.m., 12 noon, 3 p.m., and 4 p.m. The criterion for nonsuppression or early escape from suppression was at least one cortisol value (determined by radioimmunoassay) of $\geq 5$ µg/dl of plasma. An evening value was also assessed but not taken into account for the DST because on one occasion this assessment was part of a physostigmine-provocation test to be described later in the context of the problem of mechanisms underlying neuroendocrine dysfunction in depression.

After drug treatment of the patients with an endogenous depression and psychological treatment of the patients with neurotic depression, respectively, the excretion of UFC/24 hr was again estimated three times shortly before discharge from the hospital. Statistical evaluation consisted of group comparisons with respect to the findings at admission and the difference in the excretion of UFC between admission and discharge. Moreover, the data were compared with normal values.

TABLE 1
Clinical and Operational Classification of Major Depressive
Disorders in 45 Psychiatric Inpatients[a]

|  | ICD[b] | RDC[c] | Newcastle Scale[d] | N |
|---|---|---|---|---|
| Definitely endogenous (N = 20) | + | + | + | 19 |
|  | + | (+) | + | 1 |
|  | + | + | − | 0 |
|  | + | (+) | − | 0 |
|  | + | − | + | 0 |
|  | + | − | − | 0 |
| Ambiguous (N = 6) | ± | + | + | 1 |
|  | ± | (+) | + | 0 |
|  | ± | + | − | 1 |
|  | ± | (+) | − | 3 |
|  | ± | − | + | 0 |
|  | ± | − | − | 1 |
| Definitely neurotic (N = 19) | − | + | + | 0 |
|  | − | (+) | + | 0 |
|  | − | + | − | 0 |
|  | − | (+) | − | 3 |
|  | − | − | + | 0 |
|  | − | − | − | 16 |

[a] Key: ICD: + definitely endogenous, ± ambiguous, − definitely neurotic
RDC: + definitely endogenous, (+) probably endogenous, − not endogenous
Newcastle scale: + score ≥ 5, − score < 5
[b] See World Health Organization (1974).
[c] See Spitzer et al. (1978).
[d] See Carney et al. (1965).

### 2.2.2. Results

The results can be summarized as follows (Berger et al., 1982b):

1. Compared with control values, UFC excretion of almost all patients was within the normal range (15.3 to 83.1 μg/24 hr) during depression and remission, but on average, significantly higher after admission than before discharge (Fig. 3). The decrease of UFC was particularly pronounced in the group of endogenous depressives. The difference between the values assessed on both occasions turned out to be significantly higher in these patients than in the other two groups whereas at admission there had been practically no difference between the values of the endogenous and those of the neurotic group, and only the values of the patients with an ambiguous diagnosis had been comparatively lower.
2. The DST was abnormal in nine (20%) of the patients, with no abnormal value in the ambiguous group and practically no difference between the other two groups (five patients with an abnormal DST among 20 endogenous depressives and four among 19 neurotic depressives). Ab-

normal test results occurred predominantly in patients who, outside the test, excreted more than the average amount of UFC.

3. In most patients, blood glucose decreased by more than 50% of the initial level to values below 50 mg/dl. There were no significant group differences in initial values or in the decrease induced by insulin.

4. The hGH response to insulin-induced hypoglycemia was blunted in only three patients with endogenous depression, i.e., in less than 20%, and in one patient with neurotic depression, with no significant difference between the group means (see also Koslow et al., 1982). Notably, a blunted response was observed exclusively in dexamethasone nonsuppressors.

5. When comparing patients with either primary ($N = 36$) or secondary depression ($N = 9$) and, within the group of endogenous depressives, those with either unipolar ($N = 12$) or bipolar depression ($N = 8$), no differences in neuroendocrine function could be revealed nor did biological markers prove helpful in arriving at a definite diagnosis in clinically ambiguous cases. Curiously enough, the discrepancy between the operational diagnoses of these patients was also reflected in the biological findings: UFC excretion and the DST were normal in all of them whereas REM latency was shortened in four of the six patients.

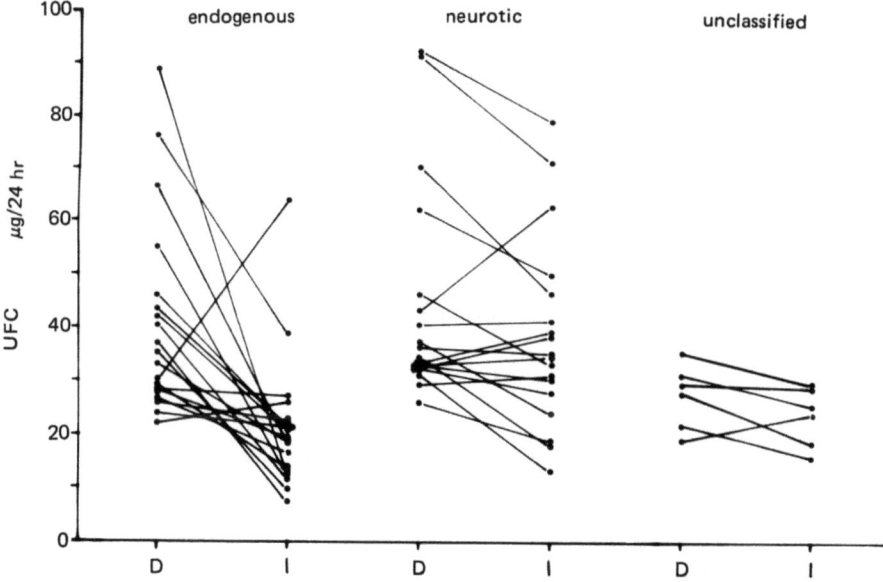

FIGURE 3. Twenty-four-hour mean of UFC excretion in endogenous, neurotic, and unclassified depressed patients during depression (D) and after improvement (I). [From Berger et al. (1982b). One case each from the endogenous and the neurotic group had to be omitted because of missing data at discharge.]

6. A composite score of three IMPS subscales (ANX + RTD + IMP), reflecting a superfactor of depressive symptomatology (von Zerssen and Cording, 1978), yielded no significant correlation with either the amount of UFC excreted in 24 hr or postdexamethasone plasma levels of cortisol, a finding which is in contrast to the significant correlation of the respective score with UFC in the chronobiological study.
7. In 13 patients with either endogenous or neurotic depression, a weight loss varying from 0.2 to 3.1 kg was observed after admission during the week preceding the DST, and seven of the nine dexamethasone nonsuppressors were found within this group.

### 2.2.3. Discussion

The results referred to so far are disappointingly negative from the viewpoint of the diagnostic validity of neuroendocrine dysfunctions in differentiating clinically relevant subtypes of depression. Only when combining the decrease in the excretion of UFC from depression to improvement (by at least 40%) with a shortening of the REM latency (to below 50 min after the beginning of sleep stage 2) can a significant biological difference be established between the two groups of definitely endogenous or neurotic depressives (see Table 2). Neither the DST nor the insulin tolerance test nor the assessment of the hGH response to insulin-induced hypoglycemia could have increased the discriminatory power of these two biological markers. However, one of them, the reduction of REM latency, is unrelated to neuroendocrine dysfunction and the other one, the decrease of UFC with remission, may reflect rather the patients' tendency to improve than basic biological differences between nosological subtypes. After all, in our investigation clinical improvement was more pronounced in patients with endogenous depression than in those with neurotic depression

TABLE 2
Classification of Endogenous and Neurotic Depressives According to Biological Markers[a]

| | Biological criterion[b] | | |
|---|---|---|---|
| | + | − | Total |
| Clinical diagnosis | | | |
| Endogenous depressive | 18 | 1 | 19 |
| Neurotic depressive | 9 | 9 | 18 |
| Total | 27 | 10 | 37 |
| $p < 0.005$ (Fisher exact test, one-tailed) | | | |

[a] Two cases (one endogenous and one neurotic depressive) omitted from the comparison because of missing data concerning UFC at discharge (see Fig. 3).
[b] Biological criterion: REM latency < 50 min and/or UFC after clinical improvement reduced by ≥ 40%.

who exhibited a more chronic, therapy-resistant course of the disorder. The results might be markedly different if one were to compare endogenous depressives with patients exhibiting a marked depressive reaction to a severely distressing life event. In the latter group, there might be dramatic changes in the functional state of the HPAS within a short time span after the traumatic situation.

Prima facie, it may be surprising that we could not reproduce our own findings with respect to a correlation between the excretion of UFC and the severity of depression in spite of using the same methods of assessing UFC and the degree of psychopathology. However, the selection of patients was different in each series. In the chronobiological study, patients with all degrees of severity were included whereas patients with only very mild degrees and, above all, those with very severe degrees of endogenous depression were excluded from the differential diagnostic investigation. Thus, in the latter project, the range of severity was markedly reduced, predominantly in the upper range where, according to the literature and our own findings in the chronobiological study, the probability of an increased activity of the HPAS seems to be particularly high.

It is to be assumed that discrepancies between our findings and reports in the literature regarding neuroendocrine differences between endogenous and neurotic depression can also be partially explained by differences in the selection of patients. The inclusion of very severely disturbed psychotic inpatients in the group of endogenous depressives and of mildly to moderately disturbed outpatients in the neurotic group might yield group differences more similar to those reported by several other investigators (see Carroll, 1982; Johnson, 1982). According to our findings, the possibility that such group differences in neuroendocrine dysfunction may emerge because of related differences in weight loss has to be seriously considered as it is well recognized that weight loss occurs predominantly in patients with the endogenous subtype of depression (Hopkinson, 1981).

## 2.3. Dexamethasone Nonsuppression in Various Psychiatric Disorders: I

### 2.3.1. Methodology

In an attempt to further elucidate factors that contribute to the variation in findings concerning neuroendocrine dysfunction in subtypes of depression, we have applied the DST in a series of well over one hundred psychiatric inpatients with various psychiatric disorders. The study was restricted to the activation of the HPAS as the central neuroendocrine phenomenon in depression. Among others, the following hypotheses were tested:

1. The incidence of abnormal test results is higher in the endogenous subtype of depression if severely depressed psychotic and suicidal patients are not excluded from testing.
2. During treatment, increased activity of the HPAS will normalize more

gradually in chronically depressed patients, i.e., mainly in patients with a severe neurotic depression, than in endogenous depressives and, in particular, patients with acute depressive reactions to stressful life events.
3. Abnormal test results can also be found in patients with psychiatric disorders other than primary depression particularly when depressive features (secondary depression), marked anxiety, or other kinds of severe negative emotions (e.g., dysphoria) are present.
4. The stress of hospital admission (Mason et al., 1965) will considerably increase positive test results in all groups of patients.

As in the differential diagnostic project, 1.5 mg of dexamethasone was administered orally at 11 p.m. Blood samples were drawn at 9 a.m., 4 p.m., and 11 p.m. Data were collected from 103 psychiatric inpatients. After excluding from the data analysis five cases with missing 11 p.m. data, 11 cases in whom the test was applied more than 3 weeks after admission, and four cases with diagnoses not occurring in patients studied during days 3 to 21, a comparison was made between test results obtained immediately (1 to 2 days) after admission ($N = 50$) and those obtained thereafter ($N = 33$), and, furthermore, between diagnostic groups. The diagnostic distribution within the sample of 83 cases thus analyzed was 42 endogenous depressives (including three schizodepressives and five patients with an additional or differential diagnosis of neurotic depression), 19 nonendogenous depressives (14 neurotic depressives and five patients with acute depressive reactions), 18 patients with various kinds of neuroses other than neurotic depression, and four psychotics other than depressives. The proportional frequency of these diagnoses was similar in both subsamples of $N = 50$ and $N = 33$.

### 2.3.2. Results

The results largely confirm the hypotheses under investigation:

1. Altogether, 38.1% of the endogenous depressives had an abnormal test result, as opposed to only 21.1% of the patients with nonendogenous depression.
2. These hypotheses could not be tested because of the relatively small number of patients with neurotic or acutely reactive forms of psychogenic depression.
3. Abnormal test results occurred in 13.6% of patients with either neuroses other than neurotic depression or psychosis other than pure depression or schizodepression. According to the case reports, however, the dexamethasone nonsuppressors had exhibited marked features of depression, anxiety, and/or dysphoria in their mental state as assessed on admission to the hospital.
4. The DST was abnormal in 36.0% of the patients assessed immediately after admission, as compared to only 15.2% of the patients investigated

at a later stage of inpatient treatment. Since the percentage was even lower (11.1%) in those studied on days 3 to 7 than in those who were tested after day 7 (20.0%), the reduced frequency of positive tests from day 3 onward was apparently due to the alleviation from the stress of hospital admission rather than to the degree of clinical improvement. The trend was similar in all diagnostic subgroups, but it was most pronounced in patients with endogenous depression, with a frequency of 52.0% in 25 patients investigated during the first 2 days after admission and of only 17.7% in those ($N = 17$) investigated thereafter. If this finding were confirmed in longitudinal studies, it might be interpreted as reflecting a particularly high sensitivity of the HPAS to environmental stressors in patients with endogenous depression. An alternative explanation would be that improvement in these patients occurred more rapidly than in other patients, as observed in the differential diagnostic study.

A comparison of all patients with an unambiguous diagnosis of endogenous depression (including the three schizodepressives) and all other patients (nonendogenous depressives, other neurotics, and psychotics except depressives) revealed a sensitivity of 43.2% and a specificity of 82.9%.

### 2.3.3. Discussion

The figures concerning sensitivity and specificity of the DST for the diagnosis of endogenous depression are not encouraging to the practicing psychiatrist: A negative result is hardly informative, and a positive result implies an almost 30% probability of a false diagnostic decision if the proportion of patients with endogenous depression and of those with other disorders could be assumed to be equal in the population under investigation (e.g., newly admitted inpatients of a psychiatric hospital). Since the percentage of endogenous depressives in most psychiatric settings is much lower than 50%, the probability of a false diagnostic decision would usually be much higher than 30%. Therefore, the test has a higher value as a research tool than as a means of diagnostic decision-making in clinical practice. This conclusion is also in accord with the many divergent findings regarding test results in subtypes of depression (e.g., Stokes *et al.*, 1976; Schlesser *et al.*, 1980; Carroll *et al.*, 1981; Carroll, 1982; Rudorfer *et al.*, 1982; Coryell *et al.*, 1982b; Berger *et al.*, 1982b) and with the fact that in patients for whom we could not achieve a clinical decision in the differential diagnosis of endogenous vs. neurotic depression (six patients in the differential diagnostic study and five patients in the DST study reported here) the DST was always normal whereas in several cases the RDC diagnosis was that of endogenous depression and in some of them (investigated in the differential diagnostic study) an abnormally low REM latency ($< 50$ min) was found.

### 2.4. Dexamethasone Nonsuppression in Various Psychiatric Disorders: II

Considering the results of our DST study and reports in the literature according to which changes in postdexamethasone cortisol levels may precede changes in the psychopathology of patients with endogenous depression (e.g., Carroll, 1980; Greden et al., 1982; Holsboer et al., 1982), we have meanwhile begun a longitudinal investigation of all new admissions to the psychiatric wards of the MPIP, using the 1.0 mg DST (Carroll et al., 1981) with blood sampling at 9 a.m., 4 p.m., and 11 p.m. The test is first applied immediately after admission and then repeated several times. A preliminary data analysis of test results obtained from 44 patients immediately after admission yielded abnormal findings in 44.4%. The frequency of a positive DST was 52.4% in all kinds of depression ($N = 21$) and 39.2% in other disorders ($N = 22$). In all diagnostic subgroups with more than three subjects, there was at least one positive test. The rather high frequency of positive tests even in patients with disorders other than depression is probably due to the "novelty stress" of hospital admission. If this were true, the DST could be regarded as a very sensitive stress indicator (see also Kalin et al., 1981), an assumption contrasting the currently held point of view (Carroll, 1980).

### 2.5. Dexamethasone Nonsuppression and Weight Loss

Since weight loss due to fasting has to be taken into account as another factor which may increase the activity of the HPAS (see Sections 1.2 and 2.2), we have performed an experimental investigation of healthy volunteers who were restricted to a diet of about 1000 cal/day for 2 weeks, thereby aiming at a weight loss of no more than 1 to 2 kg/week (Berger et al., 1982b). Twenty-nine probands (17 males and 12 females, age range 18 to 44 years) took part in the study. On the average, the subjects, all of whom had entered the trial with an almost ideal body weight, lost $1.53 \pm 0.64$ kg/week during the experimental period. Before dieting, the 1.0 mg DST revealed at least one abnormally high cortisol level in five of the probands (17.2%) who were therefore excluded from further study. During the dieting period, nine (37.5%) of the 24 remaining subjects had at least one abnormally high cortisol value. The result clearly supports the hypothesis that weight loss is to be considered a causative factor of an abnormal cortisol response to dexamethasone.

### 2.6. Dexamethasone Nonsuppression and Cholinergic Activity

It has already been mentioned that the neuronal basis of HPAS overactivity during fasting might be due to a reduced norepinephrine turnover in the brain and that stress-induced HPAS overactivity might be mediated, at least in part, by an overactivity of cholinergic neurons. Whereas the first assumption is based on experimental evidence in rats under starvation (Pirke and Spyra, 1982), the latter assumption can merely be regarded as a working hypothesis. So far, it has been demonstrated only that physostigmine is capable of inducing an early escape phenomenon from dexamethasone suppression of cortisol in the plasma of healthy volunteers (Carroll et al., 1980c). We have tried to replicate this finding and to expand it to patients with either endogenous or neurotic depression (Berger et al., 1982a) who were investigated within the differential diagnostic project as outlined above (see Section 2.2).

In three out of 21 normal subjects (Doerr and Berger, 1983), an injection of 1.0 mg of physostigmine at 4 p.m. increased the blood levels of cortisol during dexamethasone-induced suppression above the critical value of 5 µg/dl. The percentage of physostigmine-induced early escape from suppression was almost equal to these findings in the 45 depressives and did not differ between the diagnostic subgroups: The phenomenon occurred in two of the 20 endogenous depressives, two of the 19 neurotic depressives, and one of the six patients with an ambiguous diagnosis.

Since there was no difference in the results of the provocation test between patients and controls, it seems unlikely that a hypersensitivity of cholinergic receptors should exist in depressives as hypothesized by some authors (Janowsky et al., 1980; Sitaram et al., 1982). This conclusion is supported also by analogous findings of a shortened REM latency following physostigmine infusion after sleep onset in the same group of patients and in healthy controls (Berger et al., 1983a) and by related findings of another group (Nurnberger et al., 1982) concerning autonomic, hormonal, and mood changes under the influence of a cholinergic drug as trait markers in euthymic bipolar patients and controls. Nonetheless, the results are in accord with the assumption of acetylcholine as a mediator of the neuroendocrine stress response: If the cholinergic system were activated during stress and this phenomenon occurred spontaneously in a certain percentage of patients during the depressive episode, the additional cholinergic stimulus during the experiment might not be more powerful than in individuals who are not in a state of central cholinergic overactivity. On the other hand, physostigmine infusion may be a stressor which elicits a stress response by means of adverse side effects (see also Davis and Davis, 1980) as does the *anti*cholinergic drug biperiden (Benkert et al., 1981). This topic deserves further attention in psychobiological research on depression.

With respect to neuroendocrine dysfunctions in the various subtypes of depression, such factors as emotional stress, weight loss due to fasting, and others [motor activity and muscular tension, sleep disturbances, particularly early awakening, pretreatment with drugs, e.g., lithium (see Brown et al., this volume) and the like] should be ruled out as causative factors until these dysfunctions are hypothetically related to a primary disturbance in transmitter balance underlying depressive symptom formation.

## 3. SUMMARY AND CONCLUSIONS

In the literature, several neuroendocrine dysfunctions have been described for various subtypes of depression. The most consistent finding is an increased activity of the HPAS in severe forms of endogenous depression, particularly in the presence of psychotic symptomatology. Similar findings, e.g., increased cortisol production with a flattening of the circadian rhythm, were also reported to occur, even more regularly, in anorexia nervosa. Here, starvation could be identified as the causative factor. A reduced norepinephrine turnover in the brain seems to play a central role in mediating hormonal changes during starvation. Therefore, weight loss due to fasting has to be taken into account in the analysis of changes in transmitter balance and of neuroendocrine dysfunction in depression.

The whole spectrum of hormonal dysfunctions in severe endogenous depression (melancholia) is even more similar to that observed in early stages of Cushing's disease than to that observed during starvation, but is found also in other forms of Cushing's syndrome and under pharmacotherapy with ACTH or glucocorticoids. It can be inferred that the crucial neuroendocrine abnormality in melancholia is the hyperactivity of the HPAS.

In the pathogenesis of the increased HPAS activity in depression, the mental stress induced by the disorder has to be considered as an important causative factor. Objections to this concept are not well founded empirically. Hypothetically, an overactivity within the central cholinergic system may play a role in mediating the neuroendocrine stress response.

The hypotheses outlined above have been investigated by the authors in collaboration with other researchers at the Max-Planck-Institut für Psychiatrie in Munich. In a chronobiological study of patients with various degrees of severity of endogenous depression, UFC was found to be excreted in higher amounts during the episode than after remission in the majority of cases, particularly those with a severe form of the disorder. All patients exhibiting psychotic features were among the cortisol hypersecretors. In only one of them was a flattening of the circadian curve present during the episode. The average amount of UFC excreted in 24 hr correlated positively with the severity of depression rated at admission.

In an investigation of the differential diagnosis of endogenous vs. neurotic depression by means of biological markers, the UFC excretion during depression was compared with that measured at discharge when the patients were in, at least, partial remission. Both groups showed a significant decrease of UFC excretion; the values, however, exceeded the upper limit of the normal range in only very few cases. The DST with 1.5 mg was abnormal in 20% of all patients, with no significant difference between the diagnostic subgroups. A more pronounced decrease of UFC with remission in the endogenous depressives was probably due to a more marked degree of clinical improvement of these patients. UFC at admission was not clearly correlated with the severity of depression as in the chronobiological investigation, probably due to the exclusion of patients with severe, suicidal, and/or psychotic depression from the differential diagnostic study. Findings regarding the insulin tolerance test (ITT) and the hGH response during the ITT were disappointingly negative in both diagnostic subgroups. Discrepancies between these findings and those reported in the literature may partially be caused by the selection of patients rather than by basic differences in diagnostic procedures, since clinical and operational diagnoses yielded excellent agreement in the present study.

There was a correlation between the activity of the HPAS and weight loss, indicated by dexamethasone nonsuppression in most of the patients who had lost weight during the week preceding the DST. In healthy volunteers, the frequency of dexamethasone nonsuppression in the 1.0 mg DST increased markedly from baseline during a 2-week diet of approximately 1000 cal, a finding which supports the hypothesis that weight loss due to reduced caloric intake plays a role in the activation of the HPAS during episodes of depression.

In a comparison of abnormal DST results in several diagnostic subgroups of psychiatric inpatients including also endogenous depressives with all degrees of severity of the illness, the 1.5 mg DST yielded more positive results immediately after admission than during the following 2–3 weeks. The higher rate of nonsuppression was apparently due to the novelty stress of hospital admission and not to clinical improvement. The sensitivity of the test for endogenous vs. nonendogenous depression and other psychiatric disorders was 43.2% with a specificity of 82.9%, which is insufficient for diagnostic purposes.

In a longitudinal investigation of psychiatric inpatients by means of the 1.0 mg DST, there was a frequency of abnormal test results in approximately 45% immediately after admission, with not very marked differences between diagnostic groups. According to these findings, the test may be regarded as a very sensitive stress indicator rather than as a diagnostic tool for psychiatric practice.

The possibility that the central cholinergic system may play a role in DST nonsuppression was tested by administration of physostigmine to healthy volunteers and depressed patients during the DST. The results obtained so far do not support the assumption of a cholinergic hypersensitivity in depression. However, an increased activity of this system during depression cannot be ruled out.

The findings outlined here are in accord with the conclusions the authors have drawn from the literature. Contradictions to the results obtained by other authors and to varying interpretations remain to be clarified in further investigations.

## 4. REFERENCES

Amsterdam, J. D., Winokur, A., Caroff, S. N., and Conn, J., 1982, The dexamethasone suppression test in outpatients with primary affective disorder and healthy control subjects, *Am. J. Psychiatry* **139**:287.

Aro, A., Lamberg, B.-A., and Pelkonen, R., 1978, Endocrine changes in anorexia nervosa, *Acta Endocrinol. (Kbh.) Suppl.* **220**:(89):16.

Benkert, O., Klein, H. E., Hofschuster, E., and Seibold, C., 1981, Effect of the anticholinergic drug biperiden on pituitary hormones and cortisol, *Psychoneuroendocrinology* **6**:231.

Berger, M., Doerr, P., Lund, R., Bronisch, T., and von Zerssen, D., 1982a, Neuroendokrinologische Befunde und polygraphische Schlafuntersuchungen bei Patienten mit depressiven Syndromen, in: *Biologische Psychiatrie* (H. Beckmann, ed.), pp. 205–210, Thieme, Stuttgart.

Berger, M., Doerr, P., Lund, R., Bronisch, T., and von Zerssen, D., 1982b, Neuroendocrinological and neurophysiological studies in major depressive disorders: Are there biological markers for the endogenous subtype?, *Biol. Psychiatry* **17**:1217–1242.

Berger, M., Krieg, C., and Pirke, K. M., 1982c, Is the positive dexamethasone test in depressed patients a consequence of weight loss?, *Neuroendocrinol. Lett.* **4**:177.

Berger, M., Lund, R., Bronisch, T., and von Zerssen, D., 1983a, REM latency in neurotic and endogenous depression and the cholinergic REM induction test, *Psychiatry Res.*, in press.

Berger, M., Pirke, K. M., Doerr, P., Krieg, C., and von Zerssen, D., 1983b, Influence of weight loss on the dexamethasone suppression test, *Arch. Gen. Psychiatry* **40**:585.

Bleuler, M., 1964, Endokrinologische psychiatrie, in: *Psychiatrie der Gegenwart* (H. W. Gruhle, R. Jung, W. Mayer-Gross, and M. Mueller, eds.), Vol. I/1, pp. 161–252, Springer, Berlin.

Brambilla, F., Smeraldi, E., Sacchetti, E., Negri, F., Cocchi, D., and Müller, E. E., 1978, Deranged anterior pituitary responsiveness to hypothalamic hormones in depressed patients, *Arch. Gen. Psychiatry* **35**:1231.

Bridges, P. K., Jones, M. T., and Leak, D., 1970, A taxonomic study of physiological responses to a psychological stress, *J. Neurol. Neurosurg. Psychiatry* **33**:180.

Carney, M. W. P., Roth, M., and Garside, R. F., 1965, The diagnosis of depressive syndromes and the prediction of E.C.T. response, *Br. J. Psychiatry* **111**:659.

Carroll, B. J., 1979, Neuroendocrine function in mania, in: *Manic Illness* (B. Shopsin, ed.), pp. 163–176, Raven Press, New York.

Carroll, B. J., 1980, Neuroendocrine aspects of depression: Theoretical and practical significance, in: *The Psychobiology of Affective Disorders* (J. Mendels and J. D. Amsterdam, eds.), pp. 99–110, Karger, Basel.

Carroll, B. J., 1982, The dexamethasone suppression test for melancholia, *Br. J. Psychiatry* **140**:292.

Carroll, B. J., and Mendels, J., 1976, Neuroendocrine regulation in affective disorders, in: *Hormones, Behavior, and Psychopathology* (E. J. Sachar, ed.), pp. 193–224, Raven Press, New York.

Carroll, B. J., Curtis, G. C., Davies, B. M., Mendels, J., and Sugerman, A. A., 1976a, Urinary free cortisol excretion in depression, *Psychol. Med.* **6**:43.

Carroll, B. J., Curtis, G. C., and Mendels, J., 1976b, Neuroendocrine regulation in depression. I. Limbic system–adrenocortical dysfunction, *Arch. Gen. Psychiatry* **33**:1039.

Carroll, B. J., Curtis, G. C., and Mendels, J., 1976c, Neuroendocrine regulation in depression. II. Discrimination of depressed from nondepressed patients, *Arch. Gen. Psychiatry* **33**:1051.

Carroll, B. J., Feinberg, M., Greden, J. F., Haskett, R. F., James, N. M., Steiner, M., and Tarika, J., 1980a, Diagnosis of endogenous depression, *J. Affect. Dis.* **2**:177.

Carroll, B. J., Greden, J. F., and Feinberg, M., 1980b, Neuroendocrine disturbances and the diagnosis and aetiology of endogenous depression, *Lancet* **I**:321.

Carroll, B. J., Greden, J. F., Haskett, R., Feinberg, M., Albala, A. A., Martin, F. I. R., Rubin, R. T., Heath, B., Sharp, P. T., McLeod, W. L., and McLeod, M. F., 1980c, Neurotransmitter studies of neuroendocrine pathology in depression, *Acta Psychiatr. Scand. Suppl.* **280**(61):183.

Carroll, B. J., Feinberg, M., Greden, J. F., Tarika, J., Albala, A. A., Haskett, R. F., James, N. M., Kronfol, Z., Lohr, N., Steiner, M., de Vigne, J.-P., and Young, E., 1981, A specific laboratory test for the diagnosis of melancholia, *Arch. Gen. Psychiatry* **38**:15.

Checkley, S. A., 1979, Corticosteroid and growth hormone responses to methylamphetamine in depressive illness, *Psychol. Med.* **9**:107.

Checkley, S. A., 1980, Neuroendocrine tests of monoamine function in man: A review of basic theory and its application to the study of depressive illness, *Psychol. Med.* **10**:35.

Cohen, S. I., 1980, Cushing's syndrome: A psychiatric study of 29 patients, *Br. J. Psychiatry* **136**:120.

Coryell, W., Gaffney, G., and Burkhardt, P. E., 1982a, The dexamethasone suppression test and familial subtypes of depression—A naturalistic replication, *Biol. Psychiatry* **17**:33.

Coryell, W., Gaffney, G., and Burkhardt, P. E., 1982b, DSM-III melancholia and the primary-secondary distinction: A comparison of concurrent validity by means of the dexamethasone suppression test, *Am. J. Psychiatry* **139**:120.

Czernik, A., 1982, *Zur Psychophysiologie und Neuroendokrinologie von Depressionen*, Springer, Berlin.

Davis, B. M., Davis, K. L., 1980, Cholinergic mechanisms and anterior pituitary hormone secretion, *Biol. Psychiatry* **15**:303.

Dirlich, G., Kammerloher, A., Schulz, H., Lund, R., Doerr, P., and von Zerssen, D., 1981, Temporal coordination of rest–activity cycle, body temperature, urinary free cortisol, and mood in a patient with 48-hour unipolar-depressive cycles in clinical and time-cue-free environments, *Biol. Psychiatry* **16**:163.

Doerr, P., and Berger, M., 1983, Physostigmine-induced escape from dexamethasone suppression in normal adults, *Biol. Psychiatry* **18**:261.

Doerr, P., and von Zerssen, D., 1983, Die circadiane Ausscheidung an freiem Harncortisol waehrend der depressiven Phase und im freien Intervall bei Patienten mit einer endogenen Depression, in: *Depressionen* (V. Faust and G. Hole, eds.), pp. 142–150, Hippokrates, Stuttgart.

Doerr, P., von Zerssen, D., Fischler, M., and Schulz, H., 1979, Relationship between mood changes and adrenal cortical activity in a patient with 48-hour unipolar-depressive cycles, *J. Affect. Disorders* **1**:93.

Doerr, P., Fichter, M., Pirke, K. M., and Lund, R., 1980, Relationship between weight gain and hypothalamic pituitary adrenal function in patients with anorexia nervosa, *J. Steroid Biochem.* **13**:529.

Emrich, H. M., Lund, R., and von Zerssen, D., 1979, Vegetative Funktionen und körperliche Aktivität in der endogenen Depression: Verlaufsuntersuchungen von Speichelsekretion, Temperatur und Motorik bei einem Patienten mit 48-Stunden-Zyklus, *Arch. Psychiatr. Nervenkr.* **227**:227.

Fehm, H. L., Steiner, K., and Voigt, K. H., Methamphetamine-induced cortisol secretion in man is not mediated by ACTH, in: *Integrative Neurohumoral Mechanisms* (L. Angelucci, D. de Wied, E. Endroeczi, and U. Scapagnini, eds.), Elsevier, Amsterdam, in press.

Fichter, M. M., and Pirke, K. M., 1982, Somatische Befunde bei Anorexia nervosa und ihre differentialdiagnostische Wertigkeit, *Nervenarzt* **53**:635.

Fichter, M. M., Doerr, P., Pirke, K. M., and Lund, R., 1982 Behavior, attitude, nutrition and endocrinology in anorexia nervosa: A longitudinal study in 24 patients, *Acta Psychiatr. Scand.* **66**:429.

Greden, J. F., de Vigne, J.-P., Albala, A. A., Tarika, J., Buttenheim, M., Eiser, A., and Carroll, B. J., 1982, Serial dexamethasone suppression tests among rapidly cycling bipolar patients, *Biol. Psychiatry* **17**:455.

Gruen, P. H., Sachar, E. J., Altman, N., and Sassin, J., 1975, Growth hormone responses to hypoglycemia in postmenopausal depressed women, *Arch. Gen. Psychiatry* **32**:31.

Holsboer, F., Bender, W., Benkert, O., Klein, H. E., and Schmauss, M., 1980, Diagnostic value of dexamethasone suppression test in depression, *Lancet* **II**:706.

Holsboer, F., Liebl, R., and Hofschuster, E., 1982, Repeated dexamethasone suppression test during depressive illness—Normalization of test result with clinical improvement, *J. Affect. Disorders* **4**:93.

Hopkinson, G., 1981, A neurochemical theory of appetite and weight changes in depressive states, *Acta Psychiatr. Scand.* **64**:217.

Janowsky, D. S., Risch, C., Parker, D., Huey, L., and Judd, L., 1980, Increased vulnerability to cholinergic stimulation in affective-disorder patients, *Psychopharmacol. Bull.* **16**:29.

Jeffcoate, W., J., Silverstone, J. T., Edwards, C. R. W., and Besser, G. M., 1979, Psychiatric manifestations of Cushing's syndrome: Response to lowering of plasma cortisol, *Q. J. Med.* **191**:465.

Johnson, G. F. S., 1982, Endocrine dysfunction in depression, in: *Handbook of Psychiatry and*

*Endocrinology* (P. J. V. Beumont and G. D. Burrows, eds.), pp. 239–266, Elsevier, Amsterdam.

Jones, M. T., Gillham, B., Di Renzo, G., Beckford, U., and Holmes, M. C., 1981, Neural control of corticotrophin secretion, in: *Frontiers of Hormone Research*, Vol. 8, *ACTH and LPH in Health and Disease* (T. B. van Wimersma Greidanus and L. H. Rees, eds.), pp. 12–43, Karger, Basel.

Kalin, N. H., Cohen, R. M., Kraemer, G. W., Risch, S. C., Shelton, S., Cohen, M., McKinney, W. T., and Murphy, D. L., 1981, The dexamethasone suppression test as a measure of hypothalamic–pituitary feedback sensitivity and its relationship to behavioral arousal, *Neuroendocrinology* **32**:92.

Katz, J. L., Boyar, R. M., Weiner, H., Gorzynski, G., Roffwarg, H., and Hellman, L., 1976, Toward an elucidation of the psychoendocrinology of anorexia nervosa, in: *Hormones, Behavior, and Psychopathology* (E. J. Sachar, ed.), pp. 263–283, Raven Press, New York.

Kendler, K. S., and Davis, K. L., 1977, Elevated corticosteroids as a possible cause of abnormal neuroendocrine function in depressive illness, *Commun. Psychopharmacol.* **1**:183.

Koslow, S. H., Stokes, P. E., Mendels, J., Ramsey, A., and Casper, R., 1982, Insulin tolerance test: Human growth hormone response and insulin resistance in primary unipolar depressed, bipolar depressed and control subjects, *Psychol. Med.* **12**:45.

Krieger, D. T., 1982, *Cushing's Syndrome*, Springer, Berlin.

Ling, M. H. M., Perry, P. J., and Tsuang, M. T., 1981, Side effects of corticosteroid therapy, *Arch. Gen. Psychiatry* **38**:471.

Kendler, K. S., and Davis, K. L., 1977, Elevated corticosteroids as a possible cause of abnormal neuroendocrine function in depressive illness, *Commun. Psychopharmacol.* **1**:183–194.

Lorr, M., and Klett, C. J., 1967, Inpatient Multidimensional Psychiatric Scales (IMPS), revised manual, Consulting Psychologists Press, Palo Alto, Calif.

Marton, K. I., Sox, H. C., Jr., and Krupp, J. R., 1981, Involuntary weight loss: Diagnostic and prognostic significance, *Ann. Intern. Med.* **95**:568.

Mason, J. W., 1975, Psychologic stress and endocrine function, in: *Topics in Psychoneuroendocrinology* (E. J. Sachar, ed.), pp. 1–18, Grune & Stratton, New York.

Mason, J. W., Sachar, E. J., Fishman, J. R., Hamburg, D. A., and Handlon, J. H., 1965, Corticosteroid responses to hospital admission, *Arch. Gen. Psychiatry* **13**:1.

Matussek, N. M., Ackenheil, M., Hippius, H., Müller, F., Schröder, H. T., Schultes, H., and Wasilewski, B., 1980, Effect of clonidine on growth hormone release in psychiatric patients and controls, *Psychiatry Res.* **2**:25.

Nathan, R. S., Sachar, E. J., Asnis, G. M., Halbreich, U., and Halpern, F. S., 1981, Relative insulin insensitivity and cortisol secretion in depressed patients, *Psychiatry Res.* **4**:291.

Nurnberger, J. I., Jimerson, D. C., Simmons, S., Tamminga, C., Suzan Nadi, N., and Gershon, E. S., 1982, Responses to arecoline in normal twins and "well state" patients with affective disorder, Presented at the Society of Biological Psychiatry, 37th Annual Convention, Toronto, p. 91.

Pirke, K. M., and Spyra, B., 1982, Catecholamine turnover in the brain and the regulation of luteinizing hormone and corticosterone in starved male rats, *Acta Endocrinol. (Kbh)* **100**:168.

Pirke, K. M., Fichter, M., Lund, R., and Doerr, P., 1980, Die Ausschuettung des luteinisierenden Hormons waehrend des Schlafens und Wachens bei Patienten mit Anorexia nervosa, *Aktuel. Endokrinol. Stoffwechsel* **1**:147.

Pirke, K. M., Fichter, M. M., Holsboer, F., Kempin, W., Weiss, W., and Wolfram, g., 1982, 24-hour sleep–wake cycle of LH and cortisol during starvation in healthy subjects and in anorexia nervosa, *Acta Endocrinol. (Kbh.) Suppl.* **246(99)**:83.

Rose, R. M., 1980, Endocrine responses to stressful psychological events, in: *Psychiatric Clinics of North America: Advances in Psychoneuroendocrinology*, Vol. 3/2 (E. J. Sachar, ed.), pp. 251–276, Saunders, Philadelphia.

Rudorfer, M. V., Hwu, H.-G., and Clayton, P. J., 1982, Dexamethasone suppression test in primary depression: Significance of family history and psychosis, *Biol. Psychiatry* **17**:41.

Russell, G., 1977, The present status of anorexia nervosa, *Psychol. Med.* **7**:363.

Sachar, E. J., 1967, Corticosteroids in depressive illness. I. A reevaluation of control issues and the literature, *Arch. Gen. Psychiatry* **17**:544.

Sachar, E. J., 1975, Neuroendocrine abnormalities in depressive illness, in: *Topics in Psychoendocrinology* (E. J. Sachar, ed.), pp. 135–156, Grune & Stratton, New York.

Sachar, E. J., 1982, Endocrine abnormalities in depression, in: *Handbook of Affective Disorders* (E. S. Paykel, ed.), pp. 191–201, Churchill Livingstone, Edinburgh.

Sachar, E. J., Asnis, G., Nathan, R. S., Halbreich, U., Tabrizi, M. A., and Halpern, F. S., 1980, Dextroamphetamine and cortisol in depression, *Arch. Gen. Psychiatry* **37**:755.

Schildkraut, J. J., Orsulak, P. J., Schatzberg, A. F., Gudeman, J. E., Cole, J. O., Rohde, W. A., and LaBrie, R. A., 1978, Toward a biochemical classification of depressive disorders. I. Differences in urinary excretion of MHPG and other catecholamine metabolites in clinically defined subtypes of depressions, *Arch. Gen. Psychiatry* **35**:1427.

Schlesser, M. A., Winokur, G., and Sherman, B. M., 1980, Hypothalamic–pituitary–adrenal axis activity in depressive illness: Its relationship to classification, *Arch. Gen. Psychiatry* **37**:737.

Schulz, H., 1981, Sleep onset REM episodes in depression, in: *Sleep 1980* (W. P. Koella, ed.), pp. 72–84, Karger, Basel.

Schulz, H., Lund, R., Cording, C., and Dirlich, G., 1979, Bimodal distribution of REM sleep latencies in depression, *Biol. Psychiatry* **14**:595.

Shulman, R., and Diewold, P., 1977, A two-dose dexamethasone suppression test in patients with psychiatric illness, *Can. Psychiatr. Assoc. J.* **22**:417.

Sitaram, N., Nurnberger, J. I., Gershon, E. S., and Gillin, J. C., 1982, Cholinergic regulation of mood and REM sleep: Potential model and marker of vulnerability to affective disorder, *Am. J. Psychiat.* **139**:571.

Spitzer, R. L., Endicott, J., and Robins, E., 1978, Research diagnostic criteria: Rationale and reliability, *Arch. Gen. Psychiatry* **35**:773.

Stokes, P. E., Stoll, P. M., Mattson, M. R., and Sollod, R. N., 1976, Diagnosis and psychopathology in psychiatric patients resistant to dexamethasone, in: *Hormones, Behavior, and Psychopathology* (E. J. Sachar, ed.), pp. 225–229, Raven Press, New York.

Targum, S. D., Byrnes, S. M., and Sullivan, A. C., 1982, Subtypes of unipolar depression distinguished by the dexamethasone suppression test, *J. Affect. Disorders* **4**:21.

Träskman, L., Tybring, G., Åsberg, M., Bertilsson, L., Lantto, O., and Schalling, D., 1980, Cortisol in the CSF of depressed and suicidal patients, *Arch. Gen. Psychiatry* **37**:761.

Vale, W., Rivier, C., Yang, L., Minick, S., and Guilleman, R., 1978, Effects of purified hypothalamic corticotropin-releasing factor and other substances on the secretion of adrenocorticotropin and β-endorphin-like immunoactivities in vitro, *Endocrinology* **103**:1910.

Voigt, K. H., Weber, E., Fehm, H. L., and Martin, R., 1980, The concomitant storage and simultaneous release of ACTH and β-endorphin, in: *Brain and Pituitary Peptides* (W. Wuttke, A. Weindl, K. H. Voigt, and R.-R. Dries, eds.), pp. 54–64, Karger, Basel.

Voigt, K. H., Bossert, S., Bretschneider, S., Rockstroh, B., and Fehm, H. L., Disturbance of cortisol secretion in depressive patients: Lack of correlation with plasma ACTH, in: *Integrative Neurohumoral Mechanisms* (L. Angelucci, D. de Wied, E. Endroeczi, and U. Scapagnini, eds.), Elsevier, Amsterdam, in press.

von Zerssen, D., 1976, Mood and behavioral changes under corticosteroid therapy, in: *Psychotropic Action of Hormones* (T. M. Itil, G. Laudahn, and W. M. Herrmann, eds.), pp. 195–222, Spectrum, New York.

von Zerssen, D., and Cording, C., 1978, The measurement of change in endogenous affective disorders, *Arch. Psychiatr. Nervenkr.* **226**:95.

von Zerssen, D., and Doerr, P., 1980, The role of the hypothalamo–pituitary–adrenocortical system in psychiatric disorders, *Adv. Biol. Psychiatry* **5**:85.

von Zerssen, D., Dirlich, G., and Fischler, M., The influence of an abnormal time routine and therapeutic measures on 48-hour cycles of affective disorders: Chronobiological considerations, in: *Circadian Rhythms in Psychiatry* (T. A. Wehr and F. K. Goodwin, eds.), Boxwood Press, Los Angeles, in press.

Walsh, B. T., Katz, J. L., Levin, J., Kream, J., Fukushima, D. K., Hellman, L. D., Weiner, H., and Zumoff, B., 1978, Adrenal activity in anorexia nervosa, *Psychosom. Med.* **40**:499.

Whybrow, P. C., and Hurwitz, T., 1976, Psychological disturbances associated with endocrine disease and hormone therapy, in: *Hormones, Behavior, and Psychopathology* (E. J. Sachar, ed.), pp. 125–143, Raven Press, New York.

Winokur, A., Amsterdam, J., Caroff, S., Snyder, P. J., and Brunswick, D., 1982, Variability of hormonal responses to a series of neuroendocrine challenges in depressed patients, *Am. J. Psychiatry* **139**:39.

World Health Organization, 1974, *Glossary of Mental Disorders and Guide to Their Classification for Use in Conjunction with the Internal Classification of Diseases*, 8th Rev., Geneva.

CHAPTER 20

# Biological Tests in the Diagnosis and Treatment of Affective Disorders

## HARVEY A. STERNBACH, HARRY E. GWIRTSMAN, and ROBERT H. GERNER

### 1. INTRODUCTION

Psychiatry has traditionally been distanced by other branches of medicine because it lacked means whereby diagnoses and treatments could be arrived at objectively. Other specialties of medicine can turn to diagnostic aids through which differential diagnoses can be narrowed down and specific treatments instituted. Over the last 10 years psychiatry has been seeking and discovering its own diagnostic aids and classification schemes to similarly enable psychiatrists to narrow their differential diagnoses and initiate specific treatments. The affective disorders have received a large share of attention, and rightfully so, since a large segment of the population in this country will experience a major depressive episode at some time in their lives (Katz, 1980). This chapter will focus on several currently existing biological tests used to assess affective disorders and the relationships of these tests to one another and response to treatment. Our intention is to provide the reader with a brief review of each test followed by a discussion of findings from the authors' own investigations examining the relationships of these tests and treatment response. The implications of our findings, as well as those of others, will be discussed in light of current theories of the etiology of affective disorders and mechanisms underlying response to treatment.

---

HARVEY A. STERNBACH • Department of Psychiatry, University of California at Los Angeles, Neuropsychiatric Institute, Los Angeles, California 90024, and Department of Psychiatry, Brentwood Veterans Administration Hospital, Los Angeles, California 90073.   HARRY E. GWIRTSMAN • National Institutes of Health, Bethesda, Maryland 20205.   ROBERT H. GERNER • University of California, Irvine, California, and Long Beach Veterans Administration Hospital, Long Beach, California 90822.

## 2. DEXAMETHASONE SUPPRESSION TEST

The overnight dexamethasone suppression test (DST) is the most widely studied neuroendocrine test in psychiatric disorders. It arose from the observation that patients with Cushing's disease and patients on steroids often exhibited psychiatric symptoms of either depression or mania. This test involves the administration of a synthetic glucocorticoid, dexamethasone, to "turn-off corticotropin-releasing factor (CRF) by feedback inhibition" (Carroll et al., 1976). The normal response to the DST is suppression of cortisol secretion for over 24 hr after receiving dexamethasone. Although control of CRF is influenced by several neurotransmitters (Sachar, 1975; Sachar et al., 1980a,b), norepinephrine (NE) plays a major role in its regulation. NE has been found to exert a tonic inhibition upon the hypothalamic–pituitary–adrenal axis, probably by inhibiting CRF (Carroll, 1978; Schlesser et al., 1980; Sachar et al., 1980a). An abnormal DST in a depressed patient may therefore imply an NE deficiency consistent with the monoamine hypothesis of depression (Bunney and Davis, 1965; Schildkraut, 1965). The standard DST for psychiatry is to give 1 mg of dexamethasone p.o. between 11:00 and 11:30 p.m., and to obtain serum cortisol levels at 4:00 p.m. (and if possible at 11:00 p.m.) the following day. A positive (abnormal, nonsuppression) test is present if *either* cortisol is $\geq 5$ µg/dl. The DST has been confirmed to be a valid diagnostic aid in the evaluation of mental illness to the extent that a patient with an abnormal DST in the absence of pituitary–adrenal disease, anorexia nervosa, diabetes, or hepatic microsomal enzyme induction is very likely to have true "melancholia" (Carroll et al., 1981). When the DST is administered and both 4:00 p.m. and 11:00 p.m. plasma cortisol levels obtained, the sensitivity of this test for diagnosis of melancholia is as high as 67%, while the specificity is 96% (Carroll et al., 1981). Abnormal DST results have been found not to be related to sex, age, or use of common psychotropic medications. Additionally, Brown et al. (1981) have found DST nonsuppression to be significantly correlated with positive treatment response to the noradrenergic antidepressants imipramine and desipramine in unipolar depressives.

We examined a group of 37 patients with major depression (Table 1), who were given the DST with cortisol level obtained at 4:00 p.m. Twenty-three (62%) of these patients had an abnormal 4:00 p.m. cortisol level ($\geq 5$ µg/dl). Twenty-six of these patients were given open trials of methylphenidate (10 mg of methylphenidate was given p.o. in the morning and repeated in 2 hr if no robust behavioral change had occurred) and behavioral response was assessed by change of Beck Depression Inventory from predose to 2 hr after the patient's last dose. A decrease in Beck rating of $> 5$ was considered to be a positive antidepressant response. We found a significant negative association between abnormal DST and response to methylphenidate, i.e., DST nonsuppressors tend not to have an antidepressant response to methylphenidate whereas suppressors do ($p < 0.03$) (Table 2). There are several possible explanations for this finding: (1) It is possible that cortisol hypersecretion, as measured by DST

## TABLE 1
### The Dexamethasone Suppression Test (DST), TRH Stimulation Test (TST), 24-hr Urine MHPG, and Methylphenidate Stimulation Test (MST) according to Diagnostic Subtype, Sex, and Age[a]

|  | DST | | TST | | MHPG | | MST | |
|---|---|---|---|---|---|---|---|---|
|  | Positive[b] | Normal | Blunted[c] | Normal | Low[d] | Normal | Positive[e] | Normal |
| Total: | 23 | 14 | 15 | 7 | 13 | 14 | 14 | 14 |
| Unipolar | 13 | 7 | 5 | 4 | 4 | 8 | 8 | 9 |
| Bipolar | 4 | 3 | 5 | 1 | 3 | 4 | 3 | 2 |
| Schizoaffective | 3 | 1 | 2 | 1 | 3 | 1 | 1 | 3 |
| Atypical | 3 | 3 | 3 | 1 | 3 | 1 | 2 | 0 |
| Male | 9 | 7 | 6 | 4 | 6 | 4 | 5 | 5 |
| Female | 14 | 7 | 9 | 3 | 7 | 10 | 9 | 9 |
| Mean age | 43 | 38 | 40 | 41 | 37 | 43 | 36 | 41 |

[a] There were no significant differences in age or sex for any test.
[b] Cortisol $\geq 5$ μg/dl.
[c] $\Delta$TSH $< 5$ μU/ml.
[d] Females = $< 1027$ μg/24 hr, males = $< 1164$ μg/24 hr.
[e] Beck score $\Delta > 5$.

nonsuppression, may not accurately reflect a disease in central NE inhibition of CRF, but alteration of another system. However, Sachar et al. (1980b) demonstrated that the hypercorticoid state found in depression could be acutely reversed by i.v. amphetamine [which releases NE and dopamine (DA)] and that this response was not altered in monkeys by pretreatment with the DA receptor blocker, pimozide. (2) We used methylphenidate in our stimulation

## TABLE 2
### The Dexamethasone Suppression Test and Response to Methylphenidate[a]

|  | Methylphenidate response | |
|---|---|---|
|  | Positive[b] | Negative |
| DST positive[c] |  |  |
| Unipolar | 3 | 8 |
| Bipolar | 2 | 1 |
| Schizoaffective | 0 | 2 |
| Atypical | 0 | 0 |
| DST negative |  |  |
| Unipolar | 4 | 1 |
| Bipolar | 1 | 1 |
| Schizoaffective | 1 | 0 |
| Atypical | 2 | 0 |

[a] There is a significant negative association between the DST and response to methylphenidate (Fisher exact test, two-tail, $p < 0.03$).
[b] Beck score $\Delta > 5$.
[c] Cortisol $\geq 5$ μg/dl.

test whereas other studies used amphetamine. Amphetamine has been hypothesized to have different specificity for NE and DA compared with methylphenidate (Brown, 1977); however, a recent work (Ferris and Tang, 1979) has demonstrated that both agents have similar NE and DA effects, with amphetamine being slightly more potent.

In their study, Sachar et al. (1980b) noted that only their normal controls, and not the depressed patients with cortisol hypersecretion, had a mild positive behavioral response to i.v. amphetamine. Our results indicate that the depressed patients with a normal DST (suppressors) also have a positive antidepressant response to the CNS stimulant, methylphenidate. We hypothesize that these DST suppressors had adequate presynaptic NE and/or DA stores which were released by methylphenidate and functionally normal pre- and postsynaptic receptors which would react to the acute neurotransmitter release and produce the observed antidepressant effect. It is possible that their depressions were due to non-NE or non-DA mechanisms but were still acutely responsive to an increase in catecholamine stimulation or they might have a disordered mechanism for releasing neurotransmitter stores.

In contrast, we hypothesize that depression secondary to a relative NE deficiency, however, might be expected to produce supersensitive pre- and postsynaptic NE receptors (Cohen et al., 1980) and the acute administration of amphetamine-like agents to such depressed patients may produce abrupt stimulation of these supersensitive receptors and a dysphoric response. These same patients would be expected to have DST nonsuppression due to their functionally decreased central NE, and yet still show a favorable treatment response to noradrenergic antidepressants due to a sustained, time-dependent reversal of receptor supersensitivity seen with these agents (Sulser, 1978; Frazer and Mendels, 1980; Cohen et al., 1980).

Twenty of thirty-seven depressives given the DST in our study were treated with one of the noradrenergic antidepressants imipramine, desipramine, or maprotiline. The first two drugs were adjusted to optimal blood levels, and for the latter, dosage was increased to patient tolerance. Response was determined using either the Hamilton Depression Scale ($N = 12$) or retrospectively using physician, nurse, and patient reports on the chart ($N = 8$). The raters were blind to neuroendocrine test results and drug choice. We did not find any significant association between DST response and response to noradrenergic antidepressant (Table 3) in contrast to the findings of Brown et al. (1981) discussed earlier. Since Brown et al. (1981) studied only unipolar depressives, we examined a subgroup of unipolar depressives who had the DST, methylphenidate trial and treatment with a noradrenergic antidepressant ($N = 11$). Four of five patients with an abnormal DST and negative methylphenidate response had a favorable response to noradrenergic agent, while three of four with a normal DST and positive methylphenidate response also responded to noradrenergic drug (two patients had both an abnormal DST and positive response to methylphenidate and a favorable response to noradrenergic agent). We could not, therefore, explain the discrepancy in our findings with that of Brown and

TABLE 3
The Dexamethasone Suppression Test
and Response to Noradrenergic
Antidepressant[a]

| DST | Treatment response noradrenergic agent[b] | |
|---|---|---|
| | Positive[c] | Negative |
| Positive[d] | 9 | 3 |
| Negative | 4 | 4 |

[a] There was no significant association between the DST and treatment response to a noradrenergic antidepressant (Fisher exact test, two-tail).
[b] Imipramine, desipramine, or maprotiline.
[c] Hamilton Scale score $\Delta > 10$, or chart review.
[d] Cortisol $\geq 5$ µg/dl.

co-workers by the fact that we examined a more heterogenous group of depressives. The relationship between DST and treatment response might have been different had we obtained both 4:00 p.m. and 11:00 p.m. cortisol levels since this might have increased the number of DST nonsuppressors with a positive treatment response (Carroll et al., 1981). It is unlikely, however, that obtaining cortisol levels at both times would significantly have changed our results because of the very substantial proportion of nonsuppressors with negative treatment response (40%). It is possible that the relationship between DST and response to noradrenergic agents differs because they may be based upon alterations of different neurotransmitters, or that an abnormal DST predicts response to somatic therapy and not a specific neurotransmitter-sensitive condition. This is supported by findings that changes in receptor sensitivity have been reported not only with noradrenergic tricycylic antidepressants, but also with other tricyclics, MAO inhibitors, and ECT (Cohen et al., 1980) so that nonsuppressors might show treatment responses to any of these agents.

We measured 24-hr urine 3-methoxy-4-hydroxyphenylglycol (MHPG) in 24 of the 37 patients who had been given the DST. The urine MHPG was collected, preserved (Dekirmenjian and Maas, 1970), and measured by methods previously described (Dekirmenjian, 1981). MPHG is felt to be a major metabolite of CNS NE (see Section 4). Specimens were only used for MHPG analysis if total creatinine was greater than 1400 mg (or 20 mg/kg) for males or 1000 mg (or 15 mg/kg) for females (Edwards et al., 1980). Low MHPG is less than 1027 µg/24 hr for females and less than 1164 µg/24 hr for males (Dekirmenjian, 1981). We did not find any significant association between DST response and MHPG level (Table 4) although we had hypothesized that an abnormal DST would be seen in low MHPG excretors since both tests may reflect a decrease in brain NE. There are several possible explanations for this: (1) We used only one 24-hr urine sample for MHPG in each patient, while the excretion of MHPG can vary considerably on a day-to-day basis (Hollister et

TABLE 4
The Dexamethasone Suppression Test
and Urinary MHPG[a]

|  | MHPG | |
| --- | --- | --- |
|  | Low[b] | Normal |
| DST positive[c] |  |  |
| Unipolar | 2 | 6 |
| Bipolar | 2 | 2 |
| Schizoaffective | 2 | 0 |
| Atypical | 0 | 0 |
| DST negative |  |  |
| Unipolar | 2 | 3 |
| Bipolar | 1 | 1 |
| Schizoaffective | 1 | 0 |
| Atypical | 1 | 1 |

[a] There was no significant association between the DST and 24-hr urine MHPG (Fisher exact test, two-tail).
[b] Low < 1027 μg/24 hr for females, < 1164 μg/24 hr for males.
[c] Cortisol ≥ 5 μg/dl.

al., 1978, 1980); (2) urine MHPG may not be a valid measure of only central NE state. Blombery et al. (1980) have recently reported that only about 20% of urinary MHPG is derived from the brain. In addition, as one group of investigators has pointed out (Hollister et al., 1980), there is considerable overlap between the MHPG values of depressed low and high excretors and normal controls. Of the two laboratory tests employed in this study, urine MHPG was the one most subject to error because of difficulties in collection and normal variation. As a result, any possible relationship between urine MHPG and DST could be obscured by these difficulties.

At this point in time it is clear that the DST is a valuable aid in confirming the diagnosis of depression when it is abnormal. More work is needed to assess whether it can predict response to specific antidepressants. Further, although we could not differentiate DST response among subgroups of depression (possibly because of the small sample size), continuing work with the DST will be needed to determine whether this test can distinguish between schizoaffective, unipolar, bipolar, or atypical depressive illness, or genetic vs. nongenetic types as Schlesser et al. (1980) have suggested in their reports on familial pure depressive disease (FPDD), depression spectrum disease (DSD), and sporadic depressive disease (SDD). Finally, it appears probable from our experience that an abnormal DST does indicate a very high likelihood of response to one of the somatic therapies. A negative (normal) DST, on the other hand, appears to be nonpredictive of diagnosis or response and therefore is an inconclusive test.

## 3. TRH STIMULATION TEST

The TRH stimulation test measures the short-term responsiveness of the anterior pituitary to secrete TSH when stimulated with the synthetic hypothalamic tripeptide TRH. The basic component of the hypothalamic–pituitary–thyroid axis (Fig. 1) is the hypothalamic tripeptide TRH, which is carried via the hypophyseal portal venous system to the anterior pituitary where it stimulates the release of TSH. TSH, in turn, stimulates secretion of thyroxine ($T_4$) and 1-triiodothyronine ($T_3$) by the thyroid gland (Refetoff et al., 1979; Hershman, 1980). In this system, $T_4$ and $T_3$ regulate their own secretion via feedback inhibition at the level of the anterior pituitary. When synthetic TRH is administered i.v. to euthyroid controls, serum TSH rises abruptly reaching a peak between 15 and 45 min, and gradually returns to baseline by 2–4 hr (Refetoff et al., 1979). A change from baseline TSH levels of less than 5 $\mu$U/ml is considered abnormal (blunted) by most sources (Hershman, 1980). In normal controls under the age of 40, the average increase from baseline TSH after TRH administration reaches a value of 15 $\mu$U/ml (Refetoff et al., 1979).

There are a number of physiological, pathological, hormonal, and drug-related factors which can affect the outcome of this test: (1) Age may play a factor in TSH response. One report (Snyder and Utiger, 1972) suggests a diminished TSH response with increasing age; however, this is not a consistent finding (Hershman, 1980); (2) sex has been found to affect TSH response to TRH, with women showing a greater TSH response than men (Refetoff et al., 1979); (3) high cortisol levels have been found to have a suppressing effect on TSH release in man (Otsuki et al., 1973); (4) lithium has been shown to block the release of thyroid hormone from the thyroid gland (Berens et al., 1970); this results in the same effect as low levels of serum $T_3$ and $T_4$, which leads to an augmented TSH response to TRH (Lauridsen et al., 1974). A more complete list of factors that affect basal levels of serum TSH and its response to TRH can be found in Refetoff et al. (1979).

The TRH stimulation test can be administered to in- or outpatients. After an overnight fast, and with the patient recumbent and at rest for 30 min, a

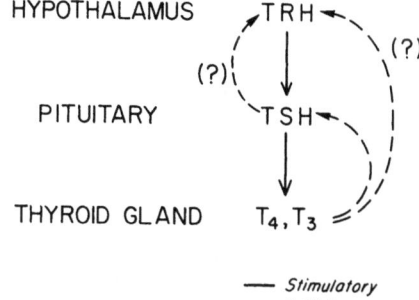

FIGURE 1. The regulatory mechanisms of the hypothalamic–pituitary–thyroid axis. $T_3$ and $T_4$ exert an inhibitory effect at the level of the pituitary. The influence of $T_3$, $T_4$, and TSH at the hypothalamic level is still unclear.

baseline TSH is obtained. Then, 500 µg of TRH is given by i.v. push and blood samples are taken at approximately 15-min intervals after injection for up to 60 min. The greatest change of TSH from baseline is used to assess TSH response. Side effects from i.v. administration of TRH may include a transient sensation of nausea, headache, warmth, desire to urinate, metallic taste, dry mouth, chest tightness, or pleasant genital sensation (Refetoff et al., 1979). When present, these effects are generally short-lived (seconds to minutes), and mild.

The use of TRH stimulation test in psychiatry evolved out of investigations into possible antidepressant effects of TRH. The earliest studies (Kastin et al., 1972; Prange et al., 1972) found a brief, but partial, antidepressant effect in depressed patients given TRH i.v. Both groups also noted a blunted TSH response to TRH in these patients. Since these early reports, there has been a proliferation of studies in psychiatric patients which have shown various abnormalities in the TRH stimulation test. A number of these studies have now shown that approximately 25–40% of depressed patients have a blunted TSH response in the absence of thyroid disease (Loosen et al., 1976; Gold et al., 1981). In addition, this blunted response may be more likely to occur in unipolar depressives and manics, rather than bipolar depressives (Extein et al., 1980a; Gold et al., 1981), and is not found in schizophrenics (Extein et al., 1980b) or personality disorders (Gold et al., 1981). A possible explanation for the blunted TSH responses in depression could be that they are secondary to elevated cortisol levels commonly found in depression (Carroll, 1978) since increased cortisol suppresses TSH (Refetoff et al., 1979). However, no correlation between TSH response to TRH and baseline cortisol levels has been found (Gold et al., 1980; Kirkegaard and Carroll, 1980).

While the underlying control of TRH and TSH is not fully elucidated, there are animal data (Reichlin, 1975; Hershman, 1980) supporting a stimulatory role for NE, and an inhibitory role for DA in release of TRH. Dopamine and L-dopa have been found to lower serum TSH and blunt the TSH response to TRH in man (Besses et al., 1975). A negative correlation was found between the TSH response to TRH and the serotonergic metabolite CSF 5-hydroxyindoleacetic acid in humans (Gold et al., 1977). Several studies in man (Kirkegaard et al., 1977; Woolf et al., 1972; Rogol et al., 1979) have shown, however, that manipulations of both alpha and beta NE receptors did not alter either basal levels of TSH or the TSH response to TRH.

Another clinical application of the TRH test is in the prediction of relapse of depressive illness. Kirkegaard and Smith (1978) found that 14 of 21 patients with "endogenous" depression who did not normalize their TSH response (an increase from baseline of $> 2$ µU/ml) after treatment relapsed within 4 months. This was in contrast to 5 of 7 patients who remained well at follow-up and those whose responses after treatment were $> 2$ µU/ml compared with their pretreatment TSH response. These results imply that the TRH stimulation test may aid in the decision of when to discontinue treatment in depressed patients who appear to be in clinical remission.

We have examined a group of depressed patients (Table 1) who were given the TRH stimulation test, DST, urine MHPG, and methylphenidate stimulation test. Twenty-two patients were given the TRH stimulation test (TST), of whom 15 (68%) had blunted responses. There were no age or sex differences between the normal and blunted responders. Nineteen patients had both TST and DST and 10 (53%) had abnormalities on both tests, while only 4 (21%) had normal responses in both tests (Table 5). There was no consistency between diagnostic subtype and test response. Gold et al., (1981) recently reported that in a group of unipolar depressives having both DST and TST, 30% had abnormalities on both tests, while only 16% had normal responses to both tests. In agreement with Extein et al. (1981), we too did not find any significant correlation between DST and TSH response to TRH. The results of all three studies indicate that abnormalities on the DST and TST can occur independently of one another, and that doing both tests can considerably increase the yield of diagnosing a major depressive episode in patients to 79–84%. Further, this strongly suggests that the abnormal blunting to TSH in depression is not merely secondary to hypersecretion of cortisol as measured by the DST, since 26% of patients were discordant for abnormalities on these two tests.

Seventeen patients in our study had the TST, methylphenidate trial, and urine MHPG levels. We did not find any association between the TST and either 24-hr urine MHPG or response to methylphenidate. Additionally, we did not find any significant relationship between the TST and general treatment response, i.e., to any antidepressant or response to a noradrenergic antidepressant (Table 6). There have been no previous studies of the correlations

TABLE 5
The Dexamethasone Suppression Test and TRH Stimulation Test (TST)[a]

|  | TST | |
|---|---|---|
|  | Blunted[b] | Normal |
| DST positive[c] |  |  |
| Unipolar | 4 | 0 |
| Bipolar | 3 | 1 |
| Schizoaffective | 1 | 1 |
| Atypical | 2 | 0 |
| DST negative |  |  |
| Unipolar | 1 | 3 |
| Bipolar | 1 | 0 |
| Schizoaffective | 0 | 0 |
| Atypical | 1 | 1 |

[a] No significant association was found between the DST and TSH response to TRH (Fisher exact test, two-tail). Ten of nineteen patients (53%) had abnormalities on both tests, while only 4 of 19 (21%) had normal responses to both tests.
[b] $\Delta TSH < 5\ \mu U/ml$.
[c] Cortisol $> 5\ \mu g/dl$.

TABLE 6
Treatment Response to Antidepressant and the TRH Test[a]

| TST | Treatment response, any antidepressant | | Treatment response, noradrenergic antidepressant | |
|---|---|---|---|---|
| | Positive[b] | Negative | Positive[b] | Negative |
| Blunted[c] | 6 | 3 | 5 | 0 |
| Normal | 4 | 2 | 2 | 1 |

[a] There was no significant relation between TSH response to TRH and general treatment response, or response to a noradrenergic antidepressant (Fisher exact test, two-tail).
[b] Hamilton Scale score $\Delta > 10$, or chart review.
[c] $\Delta$TSH $< 5$ µU/ml.

between the TST and either urine MHPG, response to short-acting CNS stimulants, or long-term treatment response to noradrenergic antidepressants. More work is needed to elucidate the mechanisms of control of TRH and TSH release since studies that have examined neurotransmitter mechanisms (discussed earlier) and CSF TRH levels in depressives (Kirkegaard et al., 1979) have been inconclusive. It can be said, however, that in the absence of the diseases and medications discussed above, i.e., thyroid disease or lithium in the past 6 months, an abnormal TSH response to TRH can significantly aid in the diagnosis of a major depressive episode in a patient. This test would be especially useful in helping to differentiate catatonias secondary to depression rather than schizophrenia. More work is needed to help elucidate whether preliminary findings of a blunted response in mania can be used to differentiate it from other irritable, excited states such as those occasionally seen in schizophrenia.

## 4. 24-HR URINE MHPG

The 24-hr urine MHPG was introduced into clinical psychiatry almost a decade ago as a measure of central catecholaminergic activity. Initial studies had shown that low MHPG was associated, in unipolar depression, with positive response to D-amphetamine and noradrenergic antidepressants such as imipramine or desipramine (Beckmann and Goodwin, 1975), and more recently maprotiline (Rosenbaum et al., 1980), and that normal MHPG was associated to treatment response with amitriptyline (Fawcett et al., 1972; Beckmann and Goodwin, 1975). The test has not been found to be uniformly useful in subcategorizing depressed states such as bipolar II or unipolar depression, or primary vs. secondary depression, although bipolar I patients tend to have lower pretreatment MHPG (Edwards et al., 1980). Additionally, MHPG has been found to vary considerably from day to day in normals and depressives (Hollister et al., 1978, 1980), and with state anxiety (Sweeney et al., 1978). Despite these shortcomings, the test has shown some ability to predict treatment response to noradrenergic agents in several studies (Beckmann and Goodwin, 1975; Rosenbaum et al., 1980; Schatzberg et al., 1980; Hollister et al., 1980) and may be clinically useful. The authors studied urine MHPG in a group of 27 patients with depression with methods described above. Thirteen (48%) of these patients had low 24-hr urine MHPG levels. There were no significant differences in either age or sex between low and normal/high MHPG excretors. Thirteen of the twenty-seven patients had baseline Hamilton Depression Rating Scale (HDRS) scores; there was no significant difference in severity of depression measured by baseline HDRS score between five low MHPG excretors (22 ± 6.25) and eight normal/high excretors (17 ± 1.77).

As mentioned earlier, there have been reports of a relationship between low urine MHPG and positive behavioral response to amphetamine. We examined 20 depressed patients who had 24-hr urine MHPG and methylphenidate trials. We hypothesized that the low MHPG excretors would tend to response more favorably to methylphenidate than the normal/high excretors. Nine of these twenty patients had low MHPG levels; however, no significant relationship was found between urine MHPG and behavioral response to methylphenidate (Table 7). We cannot explain our negative findings on the fact that we used methylphenidate while other studies used amphetamine because both have been found to have similar effects on NE and DA (Ferris and Tang, 1979). Further, when we examined 15 depressed patients who had pretreatment urine MHPG measured and were treated with either imipramine, desipramine, or maprotiline, there was no significant difference in response to these noradrenergic agents between the eight low excretors and seven normal/high excretors. In fact, six of the seven normal/high excretors had a positive treatment response to a noradrenergic agent. The results of our study have led us to conclude that urine MHPG is not an accurate predictor of treatment response to neurotrans-

TABLE 7
Urinary MHPG and Response to Methylphenidate[a]

| | Methylphenidate response | |
|---|---|---|
| | Positive[b] | Negative |
| Low MHPG[c] | | |
|   Unipolar | 2 | 1 |
|   Bipolar | 1 | 2 |
|   Schizoaffective | 1 | 1 |
|   Atypical | 1 | 0 |
| Normal/high MHPG | | |
|   Unipolar | 3 | 5 |
|   Bipolar | 1 | 0 |
|   Schizoaffective | 0 | 1 |
|   Atypical | 1 | 0 |

[a] There was no significant association between urine MHPG and behavioral response to methylphenidate (Fisher exact test, two-tail).
[b] Beck score $\Delta > 5$.
[c] Low $< 1027$ μg/24 hr for females, $< 1164$ μg/24 hr for males.

mitter-specific antidepressants. Hollister *et al.* (1980) have pointed out that there is much difficulty in collecting adequate urine volumes for reliable measurement of MHPG and that there is considerable overlap between MHPG values for depressives and normal controls. Additionally, as discussed in an earlier section, Blombery *et al.* (1980) have reported that only 20% of urine MHPG is derived from the brain, hence casting doubt on the validity of this test to measure central NE state. It is clear that more work is needed to define the role and clinical utility of the 24-hr urine MHPG in psychiatry. At present, this test does not lend itself to routine use in either in- or outpatients because of the difficulties described.

## 5. METHYLPHENIDATE STIMULATION TEST

The methylphenidate stimulation test (MST), a variant of the amphetamine trial, involves the administration of an oral, amphetamine-like compound, methylphenidate, which can have a rapid, though time-limited mood-elevating effect on depressed patients (Fawcett and Siomopoulos, 1971; Van Kammen and Murphy, 1978; Fawcett *et al.*, 1972; Weinstein, 1978). It is felt that this effect may be secondary to release and/or decrease reuptake of the catecholamines NE and DA (Fawcett *et al.*, 1972; Fawcett and Siomopoulos, 1971; Ferris and Tang, 1979). This has led to the hypothesis that some depressions are hypo-catecholaminergic and would respond to treatments that increase NE or DA. Several studies (Fawcett and Siomopoulos, 1971; Van Kammen and Murphy, 1978; Weinstein, 1978) have reported a positive association between response to D-amphetamine, which has similar actions to methylphenidate (Ferris and Tang, 1979), and subsequent amelioration of depressive symptoms with the noradrenergic antidepressants imipramine or desipramine, but not with the relatively more serotoninergic agent amitriptyline (Weinstein, 1978).

A group of 28 depressed inpatients were given the MST in open trial (Table 1). Fourteen (50%) had a positive behavioral response to methylphenidate. There were no significant age or sex differences between responders and nonresponders. Because of the small sample size in some subgroups of depressives (atypical and schizoaffective-depressed), we could not determine whether there were any significant differences among the diagnostic subgroups. We then examined a subset of 15 patients who had both the MST and treatment with a noradrenergic antidepressant (imipramine, desipramine, or maprotiline). Nine of these fifteen patients were positive responders to methylphenidate; however, there were no significant differences in response to the noradrenergic antidepressant between the positive and negative MST responders (Table 8). This challenges the currently accepted notion that negative responses to short-acting stimulants are predictive of poor treatment to noradrenergic antidepressants (Fawcett *et al.*, 1972; Maas, 1975). As discussed in a previous section, a negative response to a short-acting CNS stimulant could be the result of abrupt stimulation of supersensitive NE receptors, although such a patient may still

## TABLE 8
### Response to Methylphenidate and Noradrenergic Antidepressant[a]

| Methylphenidate response | Treatment response, noradrenergic agent[b] | |
|---|---|---|
| | Positive[c] | Negative |
| Positive[d] | 6 | 3 |
| Negative | 5 | 1 |

[a] There was no significant relationship between behavioral response to methylphenidate and treatment response to a noradrenergic antidepressant. Note that 5 of 6 negative methylphenidate responders had positive treatment responses to noradrenergic antidepressants.
[b] Imipramine, desipramine, or maprotiline.
[c] Hamilton Scale score $\Delta > 10$, or chart review.
[d] Beck score $\Delta$ 75.

have a positive treatment response to noradrenergic antidepressants secondary to more sustained reversal of NE receptor sensitivity. Although we administered the methylphenidate in open trials, we do not believe that this materially affected the outcome of our study as it was the expectation of staff and patients that only some individuals would respond to the methylphenidate, and the two psychiatrists that rated treatment response were blind to the results of MST. This qualification holds true for the relationships presented in previous sections between the MST and DST, TST and MHPG.

The administration of CNS stimulants in an effort to predict treatment response to neurotransmitter-specific antidepressants has suffered from a lack of standardization. Use of this type of test has varied with respect to doses of stimulants used, number of days used, type of trial (open vs. placebo), and method of rating response. We believe more work is needed to standardize this test so there is less confusion in comparing results of different studies that have used this test to predict treatment response. Finally, a comparison of response to both D-amphetamine and methylphenidate in the same depressed patient and use of blood levels would allow more accurate assessment of possible behavioral differences between these two CNS stimulants.

## 6. CONCLUSION

The failure of these indirect tests of central amine function to correlate with one another argues against a single neurotransmitter hypothesis of depression especially one solely involving NE. The failure of these tests to predict treatment response may mean that functional NE does not materially relate to treatment response. Rather treatment response may be determined by other factors such as decreased regulation of receptor sensitivity and may relate only indirectly to pretreatment noradrenergic function as measured by these tests

(Checkley et al., 1981; Cohen et al., 1980; Matussek et al., 1980). We suggest that future investigations should focus upon these receptor sensitivities, using pharmacologic probes such as clonidine ($\alpha_2$-receptor agonist) to determine prediction of treatment response. These tests should be related to neuroendocrine abnormalities, of which the TST and DST may be only the first of many, and behavioral responses to short-acting stimulants in order to provide a more comprehensive picture of neurotransmitter and neuroreceptor function, neuroendocrine state, and treatment response.

## 7. REFERENCES

Beckmann, H., and Goodwin, F. K., 1975, Antidepressant response to tricyclics and urinary MHPG in unipolar patients: Clinical response to imipramine or amitriptyline, *Arch. Gen. Psychiatry* **32**:17.

Berens, S. C., Bernstein, R. S., Robbins, J., and Wolff, J., 1970, Antithyroid effects of lithium, *J. Clin. Invest.* **49**:1357.

Besses, G. S., Burrow, G. N., Spaulding, S. W., and Donebedian, R. K., 1975, Dopamine infusion acutely inhibits the TSH and prolactin response to TRH, *J. Clin. Endocrinol. Metab.* **41**:985.

Blombery, P. A., Kopin, I. J., Gordon, E. K., Markey, S. P., and Ebert, M. H., 1980, Conversion of MHPG to vanillylmandelic acid: Implications for the importance of urinary MHPG, *Arch. Gen. Psychiatry* **37**:1095.

Brown, W. A., 1977, Psychologic and neuroendocrine response to methylphenidate, *Arch. Gen. Psychiatry* **34**:1103.

Brown, W. A., Haier, R. J., and Qualls, C. B., 1981, The dexamethasone suppression test in the identification of subtypes of depression differentially responsive to antidepressants, *Psychopharmacol. Bull.* **17**:88.

Bunney, W. E., Jr., and Davis, J. M., 1965, Norepinephrine in depressive reactions, *Arch. Gen. Psychiatry* **13**:483.

Carroll, B. J., 1978, Neuroendocrine function in psychiatric disorders, in: *Psychopharmacology: A Generation of Progress* (M. A. Lipton, A. DiMascio, and K. F. Killam, eds.), pp. 487–498, Raven Press, New York.

Carroll, B. J., Curtis, G. C., and Mendels, J., 1976, Neuroendocrine regulation in depression. I. Limbic system-adrenocortical dysfunction, *Arch. Gen. Psychiatry* **33**:1039.

Carroll, B. J., Feinberg, M., Greden, J. F., Tarika, J., Albala, A. A., Haskett, R. F., James, N. M., Kronfol, Z., Lohr, N., Steiner, M., de Vigne, J. P., and Young, E., 1981, A specific laboratory test for diagnosis of melancholia, *Arch. Gen. Psychiatry* **38**:15.

Checkley, S. A., Slade, A. P., Shur, E., and Dawling, S., 1981, A pilot study of the mechanism of action of desipramine, *Br. J. Psychiatry* **138**:248.

Cohen, R. M., Campbell, I. C., Cohen, M. R., Torda, T., Pickar, D., Siever, L. J., and Murphy, D. L., 1980, Presynaptic noradrenergic regulation during depression and antidepressant drug treatment, *Psychiatry Res.* **3**:93.

Dekirmenjian, H., 1981, National Psychopharmacology Laboratory, Knoxville, Tenn.

Dekirmenjian, H., and Maas, J. W., 1970, An improves procedure of 3-methoxy-4-hydroxyphenylethylene glycol determined by gas–liquid chromatography, *Anal. Biochem.* **35**:113.

Edwards, D. J., Spiker, D. G., Neil, J. F., Kopfer, D. J., and Rizk, M., 1980, MHPG excretion in depression, *Psychiatry Res.* **2**:295.

Extein, I., Pottash, A. L., Gold, M. S., Cadet, J., Sweeney, D. R., Davies, R. K., and Martin, D. M., 1980a, The thyroid-stimulating hormone response to thyrotropin-releasing hormone in mania and bipolar depression, *Psychiatry Res.* **2**:199.

Extein, I., Pottash, A. L. C., Gold, M. S., and Martin, D. M., 1980b, Differentiating mania from schizophrenia by the TRH test, *Am. J. Psychiatry* **137**:981.

Extein, I., Pottash, A. L. C., and Gold, M. S., 1981, Relationship of the thyrotropin-releasing hormone test and dexamethasone suppression test abnormalities in unipolar depression, *Psychiatry Res.* **4**:49.

Fawcett, J., and Siomopoulos, V., 1971, Dextroamphetamine response as a possible predictor of improvement with tricyclic therapy in depression, *Arch. Gen. Psychiatry* **25**:247.

Fawcett, J., Maas, J. W., and Dekirmenjian, H., 1972, Depression and MHPG excretion: Response to dextroamphetamine and tricyclic antidepressants, *Arch. Gen. Psychiatry* **26**:246.

Ferris, R. M., and Tang, F. L. M., 1979, Comparison of the effects of the isomers of amphetamine, methylphenidate and deoxypipradrol on the uptake of 1-($^3$H)norepinephrine and ($^3$H)dopamine by synaptic vesicles from rat whole brain, striatum, and hypothalamus, *J. Pharmacol. Exp. Ther.* **210**:422.

Frazer, A., and Mendels, J., 1980, Effects of antidepressant drugs on adrenergic responsiveness and receptors, in: *The Psychobiology of Affective Disorders* (J. Mendels and J. Amsterdam, eds.), pp. 72–82, Karger, Basel.

Gold, M. S., Pottash, A. L. C., Ryan, N., Sweeney, D. R., Davies, R. K., and Martin, D. M., 1980, TRH-induced TSH response in unipolar, bipolar, and secondary depressions: Possible utility in clinical assessment and differential diagnosis, *Psychoneuroendocrinology* **5**:147.

Gold, M. S., Pottash, A. L. C., Extein, I., and Sweeney, D. R., 1981, Diagnosis of depression in the 1980's, *J. Am. Med. Assoc.* **245**:1562.

Gold, P. W., Goodwin, F. K., Wehr, T., and Rebar, R., 1977, Pituitary thyrotropin response to thyrotropin-releasing hormone in affective illness: Relationship to spinal fluid amine metabolites, *Am. J. Psychiatry* **134**:1028.

Hershman, J. M., 1980, Control of thyrotropin secretion, in: *Radioimmunoassay of Hormones, Proteins and Enzymes* (A. Albertini, ed.), pp. 13–22, Excerpta Medica, Amsterdam.

Hollister, L. E., Davis, K. L., Overall, J. E., and Anderson, T., 1978, Excretion of MHPG in normal subjects: Implications for biological classification of affective disorders, *Arch. Gen. Psychiatry* **35**:1410.

Hollister, L. E., Davis, K. L., and Berger, P. A., 1980, Subtypes of depression based on excretion of MHPG and response to nortriptyline, *Arch. Gen. Psychiatry* **37**:1107.

Kastin, A. J., Ehrensing, R. H., Schalch, D. S., and Anderson, M. S., 1972, Improvement in mental depression with decreased thyrotropin response after administration of thyrotropin-releasing hormone, *Lancet* **II**:740.

Katz, M. M., 1980, Depression: A national health problem, in: *The Psychobiology of Affective Disorders* (J. Mendels and J. D. Amsterdam, eds.), pp. 1–10, Karger, Basel.

Kirkegaard, C., and Carroll, B. J., 1980, Dissociation of TSH and adrenocortical disturbances in endogenous depression, *Psychiatry Res.* **3**:253.

Kirkegaard, C., and Smith, E., 1978, Continuing therapy in endogenous depression controlled by changes in the TRH stimulation test, *Psychol. Med.* **8**:501.

Kirkegaard, C., Bjørum, N., Cohn, D., Faber, J., Lauridsen, U. B., and Nekup, J., 1977, Studies on the influence of biogenic amines and psychoactive drugs on the prognostic value of the TRH stimulation test in endogenous depression, *Psychoneuroendocrinology* **2**:131.

Kirkegaard, C., Faber, J., Hummer, L., and Rogowski, P., 1979, Increased levels of TRH in cerebrospinal fluid from patients with endogenous depression, *Psychoneuroendocrinology* **4**:227.

Lauridsen, U. B., Kirkegaard, C., and Nerup, J., 1974, Lithium and the pituitary–thyroid axis in normal subjects, *J. Clin. Endocrinol. Metab.* **39**:383.

Loosen, P. T., Prange, A. J., Jr., Wilson, I. C., and Lara, P. O., 1976, Pituitary responses of thyrotropin releasing hormone in depressed patients: A review, *Pharmacol. Biochem. Behav.* **5**:95.

Maas, J. W., 1975, Biogenic amines and depression: Biochemical and pharmacological separation of two types of depression, *Arch. Gen. Psychiatry* **32**:1357.

Matussek, N., Ackenheil, M., Hippius, H., Müller, R., Schröder, H. T., Schultes, H., and Wasilewski, B., 1980, Effect of clonidine on growth-hormone release in psychiatric patients and controls, *Psychiatry Res.* **2**:25.

Otsuki, M., Dakoda, M., and Baba, S., 1973, Influence of glucocorticoids on TRH-induced TSH response in man, *J. Clin. Endocrinol. Metab.* **36**:95.

Prange, A. J. Jr., Wilson, I. C., Lara, P. P., and Alltop, L. B., 1972, Effects on thyrotropin releasing hormone in depression, *Lancet* **II**:999.

Refetoff, S., Frank, P. H., Roubebush, C. P., and DeGroot, L. J., 1979, Clinical endocrine disorders of hypothalamus and pituitary: Evaluation of pituitary function, in: Endocrinology, Vol. 1 (L. J. Degroot, G. F. Cahill, Jr., L. Martini, D. H. Nelson, W. D. Odell, J. T. Potts, Jr., E. Steinberger, and A. I. Winegrod, eds.), pp. 196–199, Grune & Stratton, New York.

Reichlin, S., 1975, Regulations of the hypophysiotropic secretions of the brain, *Arch. Intern. Med.* **135**:1350.

Rogol, A. D., Reeves, G. D., Varma, M. M., and Blizzard, R. M., 1979, Thyroid-stimulating hormone and prolactin responses to thyrotropin-releasing hormone during infusion of epinephrine and propranolol in man, *Neuroendocrinology* **29**:413.

Rosenbaum, A. H., Schatzberg, A. F., Maruta, T., Orsulak, P. J., Cole, J. O., Grab, E. L., and Schildkraut, J. J., 1980, MHPG as a predictor of antidepressant response to imipramine and maprotiline, *Am. J. Psychiatry* **137**:1090.

Sachar, E. J., 1975, Twenty-four-hour cortisol secretory patterns in depressed and manic patients, *Prog. Brain Res.* **42**:81.

Sachar, E. J., Asnis, G., Halbreich, U., Nathan, S., and Halpern, F. S., 1980a, Recent studies in the neuroendocrinology of major depressive disorders, in: *The Psychiatric Clinics of North America*, Vol. 3, No. 2 (E. J. Sachar, ed.), pp. 313–326, Saunders, Philadelphia.

Sachar, E. J., Asnis, G., Nathan, S., Halbreich, U., Tabrizi, M. A., and Halpern, F. S., 1980b, Dextroamphetamine and cortisol in depression: Morning plasma cortisol levels suppressed, *Arch. Gen. Psychiatry* **37**:755.

Schatzberg, A. F., Orsulak, P. J., Rosenbaum, A. H., Kruger, E. R., Schildkraut, J. J., and Cole, J. O., 1980, Catecholamine measures for diagnosis and treatment of patients with depressive disorders, *J. Clin. Psychiatry* **41**:35.

Schildkraut, J. J., 1965, The catecholamine hypothesis of affective disorders: A review of supporting evidence, *Am. J. Psychiatry* **122**:509.

Schlesser, M. A., Winokur, G., and Sherman, B. M., 1980, Hypothalamic–pituitary–adrenal axis activity in depressive illness: Its relationship to classification, *Arch. Gen. Psychiatry* **37**:737.

Snyder, P. J., and Utiger, R. D., 1972, Thyrotropin response to thyrotropin releasing hormone in normal females over forty, *J. Clin. Endocrinol. Metab.* **34**:1096.

Sulser, F., 1978, Functional aspects of the norepinephrine receptor-coupled adenylate cyclase system in the limbic forebrain and its modification by drugs which precipitate or alleviate depression: Molecular approaches to an understanding of affective disorders, *Pharmakopsychiatr. Neuro-Psychopharmakol.* **11**:43.

Sweeney, D. R., Maas, J. W., and Heninger, G. R., 1978, State anxiety, physical activity, and urinary 3-methoxy-4-hydroxyphenethylene glycol excretion, *Arch. Gen. Psychiatry* **35**:1418.

Van Kammen, D. P., and Murphy, D. L., 1978, Prediction of imipramine antidepressant response by a one day *d*-amphetamine trial, *Am. J. Psychiatry* **135**:1179.

Weinstein, R., 1978, Deciding which antidepressant to use, *Am. J. Psychiatry* **135**:620.

Woolf, P. D., Lee, L. A., and Schalch, D. S., 1972, Adrenergic manipulations and TRH-induced thyrotropin (TSH) release, *J. Clin. Endocrinol. Metab.* **35**:616.

CHAPTER 21

# The Dexamethasone Suppression Test in Clinical Psychiatry

N. M. KURTZ and JEFFREY L. RAUSCH

## 1. INTRODUCTION

Over the past three decades, changes in serum cortisol have been associated with psychological and physical stress. Early studies suggested the possibility of an alteration in the feedback sensitivity of the hypothalamic–pituitary–adrenal (HPA) axis in certain psychiatric patients. As early as 1949, Pincus and associates reported a resistance in psychotic men to the normal urinary sodium change seen after administration of adrenal cortex extract. In 1954, Rizzo and colleagues noted the failure to detect the expected fall in 17-ketosteroid excretion when hydrocortisone was administered to a cyclothymic patient. By 1966, resistance of serum cortisol to dexamethasone suppression had been reported in a subgroup of depressed psychiatric patients by Stokes (1966).

Although subsequent findings have been at times discrepant (Carpenter and Bunney, 1971; Shopsin and Gershon, 1971), most studies have confirmed (Stokes *et al.*, 1975; Carroll *et al.*, 1976a,b; Carroll, 1976; Shulman and Diewold, 1977; Schlesser *et al.*, 1979; Brown *et al.*, 1979; Carroll *et al.*, 1980a; Schlesser *et al.*, 1980; Brown and Shuey, 1980; Nuller and Ostroumova, 1980) that up to 50% of patients with depressive illness have what appears to be abnormal HPA function as manifested by: (1) an early rebound from suppression of serum cortisol following an overnight dose of dexamethasone, (2) elevated basal plasma cortisol levels, (3) disregulation of the normal plasma cortisol circadian rhythm, resulting in increased secretion of cortisol at night.

This chapter will briefly review the physiology of the HPA axis and the application of the dexamethasone suppression test (DST). It will be primarily aimed at the clinician, who with proper understanding of the DST can have a

---

N. M. KURTZ • Clinical Investigation Unit, Sepulveda Veterans Administration Hospital, and Department of Psychiatry, University of California at Los Angeles, Sepulveda, California 91343. JEFFREY L. RAUSCH • Ensor Foundation Research Laboratory, William S. Hall Psychiatric Institute, Columbia, South Carolina 29202, and Department of Psychiatry, University of California at San Diego, La Jolla, California 92093.

useful clinical tool at his side for the diagnosis and treatment of depressive illness.

## 2. THE HPA AXIS

Control of plasma cortisol is reflected by hypothalamic, pituitary, limbic, and adrenal interactions. It is believed that limbic pathways acting via neurotransmitter systems play a role in modulating the secretion of corticotropin-releasing factor (CRF) by the hypothalamus. CRF is believed to be transported to the anterior lobe of the pituitary gland by the hypothalamo-pituitary portal vasculature, stimulating, in turn, the release of ACTH from the anterior lobe of the pituitary. From the pituitary, ACTH enters the peripheral bloodstream where it acts as a messenger to stimulate the secretion of cortisol from the adrenal gland. Through a negative feedback system (Jones, 1979), nonprotein bound plasma cortisol is thought to bind at both hypothalamic and pituitary receptor sites to regulate the activity of the HPA axis. Nonprotein bound cortisol makes up approximately 20% of the total plasma cortisol concentration.

It is well established that cortisol levels in plasma exhibit a daily circadian rhythm. The serum cortisol level may range from a peak of 25 $\mu$g/100 ml as it occurs in the early morning hours from about 4:00 to 6:00 a.m., to a nadir occurring around 11:00 p.m. to midnight, when the serum values are usually less than 8 $\mu$g/100 ml (Martin et al., 1977). This rhythm is responsive to acute stressful stimuli which can produce transient changes in plasma cortisol concentration.

## 3. DEXAMETHASONE SUPPRESSION TEST

Liddle, in 1960, described a method for evaluating HPA axis abnormalities which has gained wide acceptance in clinical medicine. It involves the use of dexamethasone which is a synthetic glucocorticoid. Dexamethasone in a 1.0- or 2.0-mg dose given at 11:00 p.m. is known to suppress endogenous ACTH and cortisol production in normal humans for up to 24 hr. Failure to suppress endogenous ACTH and cortisol production constitutes an abnormal DST. The test is thought to be a functional indicator of glucocorticoid feedback receptor sensitivity in the hypothalamic–pituitary system. An abnormal DST is felt to be indicative of HPA overactivity, activity unresponsive in relation to normal feedback receptor sensitivity. Classically, this test is used to screen patients suspected of having Cushing's syndrome and other neuroendocrine abnormalities, where cortisol secretion is unresponsive to feedback inhibition.

The DST is simple to perform. It involves obtaining samples of serum for cortisol on two consecutive days. On the first day at approximately 11:00 p.m., a dose of dexamethasone is given orally or i.m. Variations in the number of blood sampling points as well as the dose of dexamethasone have been used.

The usual blood sampling times have been at 8:00 a.m., 4:00 p.m., and 11:00 p.m. The dose of dexamethasone used experimentally has been 0.5, 1.0 or 2.0 mg, with 1.0 mg being the most widely accepted dose. Schlesser et al. (1980) recently reported that 93 of 179 (52%) of their patients with primary depressive illness failed to suppress using a 1.0-mg dose of dexamethasone with only an 8:00 a.m. blood sample. Stokes et al. (1975) report an abnormal DST for 9 of 11 (82%) of their medication-free depressed patients using the same 1.0-mg dose of dexamethasone with only a 9:00 a.m. blood sample. Carroll (1976) using a 2.0-mg dose and obtaining postdexamethasone samples at 8:00 a.m., 4:00 p.m., and 11:00 p.m., demonstrated resistance to dexamethasone suppression in 17 of 42 (40%) of his depressed patients. Brown et al. (1979), using the same dose and sampling times as Carroll et al. (1976b), found similar results in that 8 of 20 (40%) of their depressive population had an abnormal DST. Sachar et al. (1980), using 2.0-mg doses of dexamethasone, found that the 8:00 a.m. postdexamethasone values were not very informative in studying 23 "severely depressed patients." In that article, Sachar et al. (1980) go on to state: "This tendency to resist suppression during illness occurred primarily in the afternoon and late evening . . . ." Carroll et al. (1981) have recently compared the sensitivity of the DST to the 1.0- vs. 2.0-mg doses of dexamethasone. They concluded that the dose of dexamethasone had a strong effect on the sensitivity of the test. They showed that in both inpatients and outpatients, the 1.0-mg dose gives the physician a more sensitive test without significant loss of specificity. They also found a combination of the 4:00 p.m. and 11:00 p.m. postdexamethasone time points to be more sensitive than the 8:00 a.m. sample alone. Their findings, contrary to those of Schlesser et al. (1980), suggest that the 8:00 a.m. time point alone is not a sensitive enough indicator of an abnormal response.

Failure to adequately suppress plasma cortisol is reflected by a postdexamethasone cortisol concentration of greater than 5.0 μg/100 ml. Recently, there has begun to appear support for nonsuppression to be defined as plasma cortisol values greater than 4.0 μg/100 ml, but this awaits further validation (personal communication from Dr. A. John Rush). The studies reviewed in this chapter rely on each author's own DST methodology and criteria for an abnormal DST.

The clinician must be aware that depressive illness alone is not always responsible for an abnormal DST. A number of other disorders besides depressive illness and Cushing's syndrome show an abnormal response to dexamethasone; these include alcoholism, either during or after withdrawal (Carroll et al., 1981; Shulman and Diewold, 1977; Rees et al., 1977), anorexia nervosa (Oxenkrug, 1978), protein-calorie malnutrition (Carroll, 1978), obesity (Smith et al., 1975), malignancies with ectopic ACTH secretion (Martin et al., 1977), renovascular hypertension (Cade et al., 1967), and chronic hemodialysis patients (McDonald et al, 1979; Wallace et al., 1980).

The physician must also be aware that certain commonly used drugs may interfere with the DST. Certain drugs interfere with the metabolism of dexamethasone and thereby influence its half-life in plasma. Dexamethasone, if administered orally to drug-free normal subjects, is reported to reach peak plasma concentrations in 1–2 hr. It has a plasma half-life of approximately 4.3 hr (Duggan et al., 1975). Carroll et al. (1980b) have recently demonstrated that these pharmacokinetic properties of dexamethasone are similar in depressed patients: A depressed person does not metabolize the dexamethasone any differently than do normal subjects. This clearly suggests that the abnormal DST response seen in depressed patients is not related to difference in the metabolism of dexamethasone. Diphenylhydantoin, when administered chronically, has been shown to stimulate hepatic drug metabolizing enzymes. Haque et al. (1972) have demonstrated that chronic diphenylhydantoin administration can decrease the plasma half-life of dexamethasone by 51% which can result in a failure to suppress plasma cortisol with a single low dose of oral dexamethasone (Jubiz et al., 1970). Chronic barbiturate treatment has also been shown to reduce the half-life of dexamethasone and can result in a falsely abnormal DST (Brooks et al., 1972; Elias and Gwinup, 1980). The anticonvulsive agent carbamazepine has just recently been shown to also be able to give a falsely abnormal DST when administered chronically (personal communication from Dr. A. John Rush). Presumably this drug also acts by stimulating hepatic drug metabolizing enzymes which results in a decrease in the half-life of dexamethasone. As noted above, failure to suppress dexamethasone occurs in some alcoholics during and after drug withdrawal. Edwards et al. (1974) have demonstrated that rifampicin, an antibiotic used in the treatment of tuberculosis, alters cortisol metabolism and thus may affect DSTs. Steroids can obviously affect the HPA axis. Clinicians are well aware that steroids are commonly prescribed drugs for various conditions. Cortisol suppression will occur when patients are receiving chronic corticosteroids (Michels et al., 1967). It must be kept in mind that topically and nasally applied steroids may also result in plasma cortisol suppression and therefore may give a false reading of the DST (Michels et al., 1967). The DST in these patients may appear normal when in fact it is not.

It appears from recent studies that many psychotropic drugs do not have a significant effect on the DST (Carroll, 1976; Brown et al., 1979; Schlesser et al., 1980). These drugs include: neuroleptics, lithium carbonate, tricyclic antidepressants, and monoamine oxidase inhibitors. According to Stokes (1972), short-acting barbiturates, however, may make depressed patients even more resistant to dexamethasone suppression. In studying 31 "depressed-hospitalized" patients, he found 24 hr after a 2.0-mg oral dose of dexamethasone that nonmedicated patients suppressed to a mean level of 7.9 μg/100 ml, whereas the patients receiving from 100 to 350 mg of short-acting barbiturate at bedtime, suppressed to a mean level of 16.3 μg/100 ml. More recently, Stokes et al. (1975) have confirmed and extended these findings. They administered pentobarbital to depressed patients in doses that produced mild sedation. These patients were shown to be normally responsive to dexamethasone when drug-free. After "several days" of pentobarbital administration, these same patients became resistant to dexamethasone suppression and exhibited an abnormal DST. The most likely explanation for this finding would be the reduction in dexamethasone plasma half-life which is known to occur with barbiturate administration (Brooks et al., 1972). Nuller and Ostroumova (1980) present data suggesting that the use of benzodiazepines in unipolar depressives may produce falsely normal DSTs. They have shown that the use of benzodiazepines in unipolar depressives, who initially show an abnormal DST, may make these patients markedly more sensitive to dexamethasone suppression of their plasma cortisol concentrations. Because of the popularity of benzodiazepines, this finding can have important implications for clinicians in their ability to interpret the DST. From the Nuller and Ostroumova (1980) study, it is not clear how long their patients were taking benzodiazepines and whether the change in dexamethasone responsiveness is related to chronic remission. More research is needed in this area. A dose–response curve and a time course for the effect of benzodiazepines on dexamethasone suppression need to be investigated.

## 4. DST IN DEPRESSIVE ILLNESS

As noted above, numerous studies in psychiatric patients (Stokes et al., 1975; Carroll et al., 1976a,b; Carroll, 1976; Shulman and Diewold, 1977; Schlesser et al., 1979; Brown et al., 1979; Carroll et al., 1980a; Schlesser et al., 1980; Brown and Shuey, 1980; Nuller and Ostroumova, 1980) have confirmed that some patients with depressive illness have an abnormality of their HPA axis. The finding of an altered circadian rhythm of cortisol, increased 24-hr cortisol secretion, abnormal nocturnal cortisol release, and failure to normally suppress plasma cortisol levels with dexamethasone may occur separately or together in these patients. Recently, it has become clear that up to 40 to 50% of patients with depressive illness demonstrate some abnormality of their HPA axis (Carroll, 1976; Brown and Shuey, 1980), and that when they become euthymic their HPA dysfunction also remits (Carroll, 1976; Shulman and Diewold, 1977; Brown and Shuey, 1980). This indicates that the DST is a state and not a trait phenomenon. The patients must be in a depressive state in order to exhibit an abnormal DST. That this HPA abnormality is more than just a nonspecific response to stress and is linked to depressive illness has been repeatedly demonstrated by Carroll (1976), Brown et al. (1979), and Schlesser et al. (1980). While most investigators can agree that an abnormal DST occurs in depressed patients, there is disagreement as to whom these patients are. Part of this confusion seems to be due to the lack of uniform diagnostic criteria used between studies. Schlesser et al. (1980) have performed one of the largest studies examining DST and depression. They found that 65 of 146 (45%) of patients with primary unipolar depression and 28 of 33 (85%) of patients with bipolar depression failed to suppress all their plasma cortisol levels to less than $5.0\mu g/$100 ml after a 1.0-mg dose of dexamethasone given at 11:30 p.m. the night before. On the other hand, 42 patients with secondary unipolar depression, 61 patients with mania, and 48 patients with schizophrenia all demonstrated a normal responsiveness to dexamethasone suppression. Nuller and Ostroumova (1980) reported findings that were discrepant with those of Schlesser et al. (1980). They found that patients with endogenous depression and "depressive syndrome in patients with schizophrenia" failed to suppress dexamethasone. Although depressive syndrome in patients with schizophrenia can be construed as secondary unipolar depression, this clearly is not the case in reading their paper. In fact, it is likely that these patients are psychotically depressed unipolar depressives or are schizoaffective. In a report soon to be published, A. Rush (personal communication) has found that the primary–secondary distinction of depression did not distinguish patients who were resistant to dexamethasone suppression. This was a study conducted on over 200 patients where it seems that it was the careful evaluation of the endogenous vs. nonendogenous distinction which identified nonsuppressors from suppressors, respectively.

Schlesser et al. (1980) linked DST abnormality with the different subtypes of primary unipolar depression as described by Winokur (1979). Winokur has described three familial and possibly genetic subtypes of primary unipolar depression. These subtypes are: (1) familial pure depressive disease, (2) sporadic depressive disease, and (3) depressive spectrum disorder. Using these criteria to divide their patient population. Schlesser et al. (1980) found that DST significantly discriminated patients with familial pure depressive disease from those with depressive spectrum disorder but not from sporadic depressive disease. Familial pure depressive disease is defined as depression in a person who has a first-degree relative with depression but no first-degree relative with mania, alcohol, or antisocial personality. Sporadic depressive disease is defined as depression in a person who may or may not have a first-degree relative with depression but does have a first-degree relative with mania, alcoholism, or antisocial personality. Depressive spectrum disorder is defined as depression in a person who has no first-degree relative with a history of depression, mania, alcoholism, or antisocial personality. In the Schlesser et al. (1980) study, 76% or 38 of 50 patients with familial pure depressive disease failed to suppress with dexamethasone compared to 44% or 24 of 55 patients with sporadic depressive disorder, and only 7% or 3 of 41 patients with depressive spectrum disorder were nonsuppressors. Carroll et al. (1980a) failed to replicate Schlesser and colleagues' findings. They found that 83% or 5 of 6 of their patients with depressive spectrum disorder were resistant to suppression with dexamethasone in contrast to the 7% found in Schlesser and colleagues' study. What appears as disagreement may, upon closer examination of the investigator's methodologies, indicate an agreement. In Schlesser and colleagues' study a high percentage of patients with familial pure depressive disease were suffering from a psychotic unipolar depression whereas this was not the case with the great majority of cases included in the sporadic depressive disease group. All the patients in Carroll and colleagues' study had delusional symptoms; the authors state: "The highly selective nature of our population (and its small size) must be kept in mind when our results are compared with those of the Iowa group." It appears that both groups of investigators have found that delusional unipolar depressives have a high incidence of resistance to suppression with dexamethasone. Additional studies at other centers are needed to clarify that this statement about delusional unipolar depressives is indeed true.

An important question is: Does the DST give the clinician any indication about treatment? In a preliminary report, Brown et al. (1980) suggest that an abnormal DST identified a subgroup of primary depressive illnesses that respond preferentially to imipramine or desipramine, whereas primary depressives who are normally dexamethasone responsive respond preferentially to amitriptyline and chlorimipramine. The implications of this study are quite exciting. Imipramine and desipramine are relatively selective noradrenergic reuptake blockers. The differential response to treatment regime may have implications regarding the neurophysiology of resistances to suppression with dexamethasone in depression. The Brown et al. (1980) study does have some methodological drawbacks which should caution clinicians and questions the strength of their findings. First, the sample size was small with only four nonsuppressors receiving chlorimipramine or amitriptyline compared with four nonsuppressors receiving imipramine or desipramine. In the suppressor group, five patients taking amitriptyline or chlorimipramine were compared with six taking imipramine or desipramine. Secondly, the investigators assessed clinical response after a 2-week treatment period. The more standard time frame for assessment of treatment response to a tricyclic antidepressant is 4–6 weeks. It is unclear why the authors chose a 2-week time frame. Finally, significant changes among the treatment groups were found only in the global rating scale and there were no significant differences found when the rating from the Beck Depression Inventory and modified Hamilton Depression Rating Scale were used. In addition, neither the raters nor the patients were blind to the treatment conditions. Still, it is this kind of study that needs to be continued so that the relevance of resistance to dexamethasone suppression to treatment approach can be clarified. Dr. M. Schlesser, who has had extensive experience with the DST, in a personal communication has commented favorably on the efficacy of desipramine in treating those patients with endogenous depression who also have an abnormal DST. Although anecdotal information is important, clearly more well-controlled, double-blind studies need to be done before any definitive statements can be made regarding this point.

The studies reviewed so far in this chapter generally involved severely depressed inpatients. The question can be asked: Are these findings equally applicable to less severely depressed outpatient population? Carroll et al. (1980c) in studying 89 depressed outpatients were able to correctly identify 40% of the patients diagnosed clinically as "endogenous depression" by an abnormal DST. This demonstration of usefulness and effectiveness of the DST for outpatients who were diagnosed clinically as "endogenously depressed" has important implications for the clinician who treats depressed outpatients. It is clear that regardless of the type of practice, today's psychiatrist should be familiar with the DST and its application to depressed patients. More will be said below regarding the clinician utility of the DST.

It is clear now that the abnormal DST is a state rather than a trait phenomenon. This means that the failure to suppress with DST reverts to a normal dexamethasone response when the patients are no longer depressed. Because of this, the DST has been proposed as a possible indicator of clinical response in depressed patients. Goldberg (1980a,b) evaluated 10 patients who were dexamethasone nonsuppressors when depressed. All of the patients were treated with antidepressant medication. The antidepressant medication was stopped after the patients had been "depression-free" for 1 month. The DST was repeated immediately prior to the discontinuation of drug therapy. Five of the ten patients now showed a normal response to dexamethasone. None of these five patients relapsed in a 7-month follow-up period. On the other hand, all of the five patients who continued to show an abnormal DST while on antidepressant medication and in apparent clinical remission, had significant relapses within 2 months. To add further support to this, Greden et al. (1980) looked at 14 patients with diagnosis of major depressive disorder and abnormal DST. After clinical improvement the DST was repeated. At that time, 10 of the patients had a normal response to dexamethasone, and 8 of these 10 patients were asymptomatic for the follow-up period ranging from 7 to 36 months. All four of the patients who still had an abnormal DST at discharge had severe relapses within a couple of months. If these dramatic findings continue to be substantiated in larger numbers of patients, then the DST might be useful to clinicians to determine when to stop treatment without risk of a rapid relapse. At this point, it appears that the clinical indications seem to be that in the follow-up of patients who are known to have had a dexamethasone-resistant test during the period of their depression, that even in the absence of symptoms, the DST should be repeated periodically. An abnormal DST may then give the clinician early evidence of an impending relapse.

The DST can also be useful in helping the clinician diagnose patients presenting with confusing or mixed clinical pictures. It appears that 40 to 50% of patients who fulfill rigorous criteria for major depressive disorder will have an abnormal DST, and if DSM III criteria for melancholia are carefully applied, the incidence may be as high as 70 to 80% (A. Rush, personal communication). Clinicians must be reminded, however, that a normal response to dexamethasone will not discriminate patients with major depressive disorders from those with other psychiatric illnesses. A normal DST gives the psychiatrist little diagnostic information. It is only when the DST is abnormal that the test will be of significant value to the clinician. Case studies in the literature are continuously appearing where an abnormal DST helped clarify a confusing clinical picture. Greden and Carroll (1979) present an example of a woman who presented with catatonia. This patient presented with a clinical picture which many clinicians would have diagnosed as schizophrenia. In this case, the DST was abnormal. Because of this finding, and the patients severely debilitated state, electroconvulsive therapy was begun immediately. The patient rapidly responded and the diagnosis of primary unipolar depression was made. In this case, DST may have spared the patient a lifelong exposure to neuroleptics with its potential for serious side effects.

The status of the efficacy of the DST in children is still under investigation. Puig-Antich *et al.* (1979) have reported cortisol hypersecretion in two prepubertal children. Although they do not report these children's response to dexamethasone, it seems plausible that they, too, may show an abnormal DST. De La Fuente and Rosenbaum (1980) describe how an abnormal DST helped to clarify the diagnosis and treatment in eight atypically depressed adolescent boys. These reports suggest that the DST might prove valuable in helping the clinician in making the often difficult and at times controversial diagnosis of childhood depression.

Another area of previously suggested utility for the DST is in the evaluation of depression in the geriatric population. Very frequently, the clinician is asked to differentiate dementia from a major depression presenting with a pseudodementia. As every clinician faced with this problem already knows, this can be an extremely difficult distinction to make. Carroll *et al.* (1981) have stated that patients with dementia do not have abnormal DST results. However, a recent study reports abnormal DST results in 9 of 17 elderly patients who did not have major depressive illness, although all patients in that study did demonstrate exaggerated affects including outbursts of sadness (Spar and Gerner, 1982). More studies are necessary in the area since many patients referred to and treated as irreversible dementia may in fact have a treatable cause for their dementia, i.e., major depressive disorder.

As noted above, another practical value of the DST may be as a possible assessor of clinical state and as a predictor of relapse. Three recent studies have shown that the DST normalizes after successful electroconvulsive treatment (Greden *et al.*, 1980; De La Fuente and Rosenbaum, 1980; Gold *et al.*, 1980). One frequent clinical situation the psychiatrist faces is assessing the patient's affective status toward the end of a course of electroconvulsive therapy. If these patients have developed acute transient organic brain syndromes, this task can be very difficult. An abnormal DST at this point might support further treatments, whereas a normal DST would provide an objective indicator of response. Another major question regarding treatment is how long should a patient in remission remain on antidepressives to protect himself from relapse? From Goldberg's work (1980a,b), it appears that an abnormal DST in patient free of symptoms may predict an acute relapse. This patient should probably continue taking antidepressant medication. A patient who is symptom-free and who has converted his DST from abnormal to normal may in fact be able to stop the medication as long as there is adequate follow-up. Another potentially valuable use of the DST may be in assessing the bipolar patient who presents with mania. When the patient is in the manic state of his illness, the DST will be negative. However, as is well known, many patients when recovering from mania go on to a depressive state. It is quite possible that repeated DSTs while the patient is still manic may show that there is conversion of the DST from normal to abnormal. The abnormal DST at this point may predict that this patient will go on to a depressive phase of his illness and that appropriate antidepressant treatment should be initiated. Studies confirming this will need to be done but after reviewing the literature it is certainly suggestive that this may well be the case.

## 5. SUMMARY

This chapter has attempted to critically review the expanding literature of the DST in psychiatry. Although there are some negative reports, the majority of studies indicate that up to 50% of patients with primary depressive illness will fail to suppress plasma cortisol concentrations to less than $5\mu g/100$ ml 24 hr following an overnight 1.0-mg dose of dexamethasone.

An abnormal DST seems to be specific for depressive illness, assuming that the subject has no other medical illness nor is taking any of the medications that are known to confound the results. Clinically, the test may be particularly useful for difficult diagnostic questions relating to major depressive disorder. Numerous studies have validated the finding that an abnormal DST reverts to normal when the depressed patient responds to therapy. Reports are beginning to appear that in those patients who have successfully attempted suicide the incidence of abnormal DSTs which did not convert are quite high. Clearly, this has important implications for both outpatient and inpatient management of depressed patients with abnormal DSTs. Preliminary results suggest that the demonstration of an abnormal DST in recently depressed patients who are in remission will predict an early relapse if their antidepressant medication is stopped.

Finally, the significance of the DST as a research tool deserves mention in that it may provide valuable perspectives into the mechanisms of affective illness. An abnormal DST indicates a decreased functional feedback sensitivity in the HPA axis. There are several other emerging examples of decreased neuroendocrine responsiveness in the HPA axis in subgroups of depressed patients, such as blunted TSH response to TRH (Loosen and Prange, 1980), blunted GH response to a variety of agonists (Risch *et al.*, 1981), and blunted prolactin response to opiate receptor stimulation (Extein *et al.*, 1980; Judd *et al.*, 1982). These blunted hormone responses might parallel common changes in neurotransmitter activity or receptor sensitivity, possibly in conjunction with altered circadian rhythmicity.

One neurotransmitter hypothesized to be involved is acetylcholine, since physostigmine has been shown to induced escape from dexamethasone suppression in normal subjects (Carroll *et al.*, 1980d; Berger *et al.*, 1980). However, it is not yet possible to delineate the exact neurotransmitter mechanisms involved, and several neurotransmitters are proposed to be involved in the release of cortisol.

CRF has been associated with increased behavioral activation in rats (Sutton *et al.*, 1982). Thus, in humans, one might expect increased adrenocortical activity in mania and decreased activity in depression, yet, in fact, the opposite is true. This may constitute a sort of temporal asynchrony between adrenocortical activity and behavior in affective illness. Within the limbic system, evidence suggests that both the amygdala and hippocampus are reciprocally involved in steroid regulation (Mandell *et al.*, 1963), areas where the phenomenon of kindling may occur. Kindling has been suggested as a theoretical model of psychosis (Post and Kopanda, 1976). Perhaps this theoretical model might explain DST abnormalities in psychiatric patients as a "pathological phase coherence" (Mandell *et al.*, 1982) of neuronal activity, superseding normal feedback inhibitory signals in the case of depression. Although the mechanisms are not yet completely known, research is under way aimed at elucidating the complex series of pathways converging on the hypothalamus, and the DST may be but one initial step in the direction of understanding the neuroendocrinology of affective illness.

## 6. REFERENCES

Berger, M., Doerr, P., Lund, R., and von Zerssen, D., 1980, Modification of the dexamethasone suppression test and the REM-latency by physostigmine in normal and depressed patients, 12th CINP Congress, Progress in Neuro-Psychopharmacology Supplement Abstract 58, p. 78.

Brooks, S. M., Werk, E. E., Ackerman, S. J., Sullivan, I., and Thrasher, K., 1972, Adverse effects of phenobarbital on corticosteroid metabolism in patients with bronchial asthma, *N. Engl. J. Med.* **286**:1125.

Brown, W. A., and Shuey, I., 1980, Response to dexamethasone and subtypes of depression, *Arch. Gen. Psychiatry* **37**:747.

Brown, W. A., Johnston, R., and Mayfield, D., 1979, The 24-hour dexamethasone suppression test in a clinical setting: Relationship to diagnosis, symptoms, and response to treatment, *Am. J. Psychiatry* **136**:543.

Brown, W. A., Haier, R. J., and Qualls, C. B., 1980, Dexamethasone suppression test identified subtypes of depression which respond to different antidepressants, *Lancet* **I**:928.

Cade, R., Shires, D. L., Barrow, M. V., and Thomas, W. C., Jr., 1967, Abnormal diurnal variation of plasma cortisol in patients with renovascular hypertension, *J. Clin. Endocrinol. Metab.* **27**:800.

Carpenter, W. T., and Bunney, W. E., Jr., 1971, Adrenal cortical activity in depressive illness, *Am. J. Psychiatry* **128**:31.

Carroll, B. J., 1976, Limbic system–adrenal cortex regulation in depression and schizophrenia, *Psychosom. Med.* **38**:106.

Carroll, B. J., 1978, Neuroendocrine function in psychiatric disorders, in: *Psychopharmacology: A Generation of Progress* (M. A. Lipton, A. DiMascio, and K. F. Killam, eds.), pp. 487–496, Raven Press, New York.

Carroll, B. J., Curtis, G. C., and Mendels, J., 1976a, Neuroendocrine regulation in depression. I. Limbic system–adrenocortical dysfunction, *Arch. Gen. Psychiatry* **33**:1039.

Carroll, B. J., Curtis, G. C., and Mendels, J., 1976b, Neuroendocrine regulation in depression. II. Discrimination of depressed from non-depressed patients, *Arch. Gen. Psychiatry* **33**:1051.

Carroll, B. J., Greden, J. F., Feinberg, M., James, N. M., Haskett, R. F., Steiner, M., and Tarika, J., 1980a, Neuroendocrine dysfunction in genetic subtypes of primary unipolar depression, *Psychiatry Res.* **2**:251.

Carroll, B. J., Schroeder, K., Mukhopadhyay, S., Greden, J. F., Feinberg, M., Ritchie, J., and

Tarika, J., 1980b, Plasma dexamethasone concentrations and cortisol suppression in patients with endogenous depression, *J. Clin. Endocrinol. Metab.* **51**:433.

Carroll, B. J., Feinberg, M., Greden, J. F., Haskett, R. F., James, N. M., Steiner, M., and Tarika, J., 1980c, Diagnosis of endogenous depression: Comparison of clinical, research and neuroendocrine criteria, *J. Affect. Disorders* **2**:177.

Carroll, B. J., Greden, J. F., Haskett, R., Feinberg, M., Albala, A. A., Martin, F. I. R., Rubin, R. T., Heath, B., Sharp, P. T., McLeod, W. L., and McLeod, M. F., 1980d, Neurotransmitter studies of neuroendocrine pathology in depression, *Acta Psychiatr. Scand.* **61**(Suppl. 280):183.

Carroll, B. J., Feinberg, M., Greden, J. F., Tarika, J., Albala, A. A., Haskett, R. F., James, N. M., Kronfol, Z., Lohr, N., Steiner, M., de Vigne, J. P., and Young, E., 1981, A specific laboratory test for the diagnosis of melancholia: Standardization, validation, and clinical utility, *Arch. Gen. Psychiatry* **38**:15.

De La Fuente, J. R., and Rosenbaum, A. H., 1980, Neuroendocrine dysfunction and blood levels of tricyclic antidepressants, *Am. J. Psychiatry* **137**:1260.

Duggan, D. E., Yeh, K. C., Matalia, M., Ditzler, C. A., and McMahon, F. G., 1975, Bioavailability of oral dexamethasone, *Clin. Pharmacol. Ther.* **18**:205.

Edwards, O. M., Courtenay-Evans, R. J., Galley, J. M., Hunter, J., and Tait, A. D., 1974, Changes in cortisol metabolism following rifampicin therapy, *Lancet* **II**:549.

Elias, A. N., and Gwinup, G., 1980, Effects of some clinically encountered drugs on steroid synthesis and degradation, *Metabolism* **29**:582.

Extein, I., Pottash, A. L. C., Gold, M. S., Sweeney, D. R., Martin, D., and Goodwin, F. K., 1980, Deficient prolactin response to morphine in depressed patients, *Am. J. Psychiatry* **137**:845.

Gold, M. S., Pottash, A. L. C., Extein, I. R. L., and Sweeney, D. R., 1980, Dexamethasone suppression tests in depression and response to treatment, *Lancet* **I**:1190.

Goldberg, I. K., 1980a, Dexamethasone suppression tests as an indicator of safe withdrawal of antidepressant therapy, *Lancet* **I**:376.

Goldberg, I. K., 1980b, Dexamethasone suppression tests in depression and response to treatment, *Lancet* **II**:92.

Greden, J. F., and Carroll, B. J., 1979, The dexamethasone suppression test as a diagnostic aid in catatonia, *Am. J. Psychiatry* **136**:1199.

Greden, J. F., Albala, A. A., Haskett, R. F., James, N. M., Goodman, L., Steiner, M., and Carroll, B. J., 1980, Normalization of DST: A laboratory index of recovery from endogenous depression, *Biol. Psychiatry* **15**:449.

Haque, N., Thrasher, K., Werk, E. E., Jr., Knowles, H. C., Jr., and Sholiton, L. J., 1972, Studies on dexamethasone metabolism in man: Effect of diphenylhydantoin, *J. Clin. Endocrinol. Metab.* **34**:44.

Jones, M. T., 1979, Control of adrenocortical hormone secretion, in: *The Adrenal Gland* (V. H. T. James, ed.), pp. 99–130, Raven Press, New York.

Jubiz, W., Meikle, A. W., Levinson, R. A., Mizutani, S., West, C. D., and Tyler, F. H., 1970, Effect of diphenylhydantoin on the metabolism of dexamethasone: Mechanism of the abnormal dexamethasone suppression in humans, *N. Engl. J. Med.* **283**:11.

Judd, L. L., Risch, S. C., Parker, D. C., Janowsky, D. S., Segal, D. S., and Huey, L. Y., 1982, The effect of a methadone challenge on the prolactin and growth hormone responses of psychiatric patients and normal controls, *Psychopharmacol. Bull.* **18**:204.

Liddle, G. W., 1960, Tests of pituitary–adrenal suppressibility in the diagnosis of Cushing's syndrome, *J. Clin. Endocrinol. Metab.* **20**:1539.

Loosen, P. T., and Prange, A. J., 1980, TRH: A useful tool for psychoneuroendocrine investigation, *Psychoneuroendocrinology* **5**:63.

McDonald, W. J., Golper, T. A., Mass, R. D., Kendall, J. W., Porter, G. A., Girard, D. E., and Fischer, M. D., 1979, Adrenocorticotropin–cortisol axis abnormalities in hemodialysis patients, *J. Clin. Endocrinol. Metab.* **48**:92.

Mandell, A. J., Chapman, L. F., Rand, R. W., and Walter, R. D., 1963, Plasma corticosteroids: Changes in concentration after stimulation of hippocampus and amygdala in man, *Science* **139**:1212.

Mandell, A. J., Knapp, S., Ehlers, C. L., and Russo, P. V., 1982, The stability of constrained randomness: Lithium prophylaxis at several neurobiological levels, in: *The Neurobiology of the Mood Disorders* (R. M. Post and J. C. Ballenger, eds.), Williams & Wilkins, Baltimore.

Martin, J. B., Reichlin, S., and Brown, G. M., 1977, Regulation of ACTH secretion and its disorders, in: *Clinical Neuroendocrinology* (J. B. Martin, ed.), pp. 179–200, Davis, Philadelphia.

Michels, M. I., Smith, R. E., and Heimlich, E. M., 1967, Adrenal suppression and intranasally applied steroids, *Ann. Allergy* **25**:569.

Nuller, J. L., and Ostroumova, M. N., 1980, Resistance to inhibiting effect of dexamethasone in patients with endogenous depression, *Acta. Psychiatr. Scand.* **61**:169.

Oxenkrug, G. F., 1978, Dexamethasone test in alcoholics, *Lancet* **II**:795.

Pincus, G., Hoagland, H., Freeman, H., Elmadjian, F., and Romanoff, L., 1949, A study of pituitary–adrenocortical function in normal and psychotic men, *Psychosom. Med.* **11**:74.

Post, R. M., and Kopanda, R. T., 1976, Cocaine, kindling, and psychosis, *Am. J. Psychiatry* **133**:627.

Puig-Antich, J., Chambers, W., Halpern, F., Hanlon, C., and Sachar, E. J., 1979, Cortisol hypersecretion in prepubertal depressive illness: A preliminary report, *Psychoneuroendocrinology* **4**:191.

Rees, L. H., Besser, G. M., Jeffcoate, W. J., Goldie, D. J., and Marks, V., 1977, Alcohol-induced pseudo-Cushing's syndrome, *Lancet* **I**:726.

Risch, S. C., Kalin, N. H., and Murphy, D. L., 1981, Pharmacological challenge strategies: Implications for neurobehavioral mechanisms in affective disorders and treatment approaches, *J. Clin. Pharmacol.* **1**:238.

Rizzo, N. D., Fox, H. M., Laidlaw, S. C., and Thorn, G. W., 1954, Concurrent observations of behavior changes and of adrenocortical variations in a cyclothymic patient during a period of 12 months, *Ann. Intern. Med.* **41**:798.

Sachar, E. J., Asnis, G., Holbreich, U., Nathan, S., and Halpern, F., 1980, Recent studies in the neuroendocrinology of major depressive disorders, in: *Psychiatric Clinics of North America: Advances in Psychoendocrinology* (E. J. Sachar, ed.), pp. 313–326, Saunders, Philadelphia.

Schlesser, M. A., Winokur, G., and Sherman, B. M., 1979, Genetic subtypes of unipolar primary depressive illness distinguished by hypothalamic–pituitary–adrenal axis activity, *Lancet* **I**:739.

Schlesser, M. A., Winokur, G., and Sherman, B. M., 1980, Hypothalamic–pituitary–adrenal axis activity in depressive illness: Its relationship to classification, *Arch. Gen. Psychiatry* **37**:737.

Shopsin, B., and Gershon, S., 1971, Plasma cortisol response to dexamethasone suppression in depressed and control patients, *Arch. Gen. Psychiatry* **24**:320.

Shulman, R., and Diewold, P., 1977, A two-dose dexamethasone suppression test in patients with psychiatric illness, *Can. Psychiatr. Assoc. J.* **22**:417.

Smith, S. R., Bledsoe, T., and Chhetri, M. K., 1975, Cortisol metabolism and the pituitary–adrenal axis in adults with protein-calorie malnutrition, *J. Clin. Endocrinol. Metab.* **40**:43.

Spar, J. E., and Gerner, R., 1982, Does the dexamethasone suppression test distinguish dementia from depression?, *Am. J. Psychiatry* **139**:238.

Stokes, P. E., 1966, Pituitary suppression in psychiatric patients, The Endocrine Society, Program of the 48th meeting.

Stokes, P. E., 1972, Studies on the control of adrenocortical function in depression, in: *Recent Advances in Psychobiology of the Depressive Illness* (T. A. Williams, M. M. Katz, and J. A. Shields, eds.), pp. 199–220, DHEW Publ. No. (ASM) 70-9053, Superintendent of Documents, Washington, D.C.

Stokes, P. E., Pick, G. R., Stoll, P. M., and Nunn, W. D., 1975, Pituitary–adrenal function in depressed patients: Resistance to dexamethasone suppression, *J. Psychiatr. Res.* **12**:271.

Sutton, R. E., Koob, G. f., Le Moal, M., Rivier, J., and Vale, W., 1982, Corticotropin-releasing factor (CRF) produces behavioral activation in rats, *Nature (London)* **297**:331.

Wallace, E. Z., Rosman, P., Toshav, N., Sacerdote, A., and Balthazar, A., 1980, Pituitary–adrenocortical function in chronic renal failure: Studies in episodic secretion of cortisol and dexamethasone suppressibility, *J. Clin. Endocrinol. Metab.* **50**:46.

Winokur, G., 1979, Familial (genetic) subtypes of pure depressive disease, *Am. J. Psychiatry* **136**:911.

CHAPTER 22

# The TRH Test in the Diagnosis of Affective Disorders and Schizophrenia

MARK S. GOLD, A. CARTER POTTASH, and IRL EXTEIN

## 1. INTRODUCTION

An impressive body of genetic (Winokur *et al.*, 1969; Rosenthal and Kety, 1968), biological, phenomenological, and pharmacological response (Spitzer *et al.*, 1978; Schou *et al.*, 1954; Casey *et al.*, 1960; Davis, 1975, 1976; Taylor *et al.*, 1974; Pope and Lipinski, 1978) data support the concept of affective disorders and schizophrenia as distinct entities. These data further support the division of depressive disorders into primary "major depressive disorders" and secondary "minor" and "depressive spectrum" disorders, as well as the division of major depressions into unipolar and bipolar subtypes (Spitzer *et al.*, 1978; Schlesser *et al.*, 1979; Goodwin and Extein, 1979). The development of specific and effective pharmacotherapies for affective disorders and schizophrenia has underlined the need for accurate diagnosis of these disorders. One strategy for augmenting clinical diagnosis is to measure biological correlates that, in addition to elucidating the biological mechanisms of psychiatric disorders, may more clearly define homogeneous groups of psychiatric patients and guide the choice of pharmacotherapy.

Neuroendocrine testing has been used to study the biological substrate of psychiatric illness, particularly the affective disorders (Ettigi and Brown, 1977; Sachar, 1975; Carroll, 1978). With the development of reliable diagnostic nosology, psychiatrists have reported significant abnormalities in neuroendocrine function and response to provocative stimuli in patients with primary major

---

MARK S. GOLD and A. CARTER POTTASH • Research Facilities, and Psychiatric Diagnostic Laboratories of America, Fair Oaks Hospital, Summit, New Jersey 07901, and Department of Psychiatry, Yale University, New Haven, Connecticut 06520.   IRL EXTEIN • Falkirk Hospital, Central Valley, New York 10917.

depressive disorders. Significant and reproducible neuroendocrine abnormalities have been reported in such patients for cortisol secretion (Sachar *et al.*, 1973), cortisol suppression after dexamethasone (Sachar *et al.*, 1973; Carroll *et al.*, 1976), LH (Altman *et al.*, 1975), GH response to provocative stimuli (Sachar *et al.*, 1971), TSH response to TRH (Prange *et al.*, 1972; Kastin *et al.*, 1972; Takahashi *et al.*, 1973, 1974; Coppen *et al.*, 1974; Ehrensing *et al.*, 1974; Maeda *et al.*, 1975; Kirkegaard *et al.*, 1975, 1978; Hollister *et al.*, 1976; Loosen *et al.*, 1977; Gold *et al.*, 1977; Brambilla *et al.*, 1978; Gold *et al.*, 1979; Gold *et al.*, 1981; Mendlewicz *et al.*, 1979; Asnis *et al.*, 1980), and GH response to TRH (Gold *et al.*, 1979). Although dexamethasone suppression test abnormalities are found in patients with major depression and not other patient groups (Sachar *et al.*, 1973; Carroll *et al.*, 1976), and have been reported to correlate with family history (Schlesser *et al.*, 1979), neuroendocrine abnormalities have not been otherwise found reliably to separate important subgroups of depressed patients who appeared similar. For example, both bipolar and unipolar depressed patients have cortisol hypersecretion which is not adequately suppressed by dexamethasone administration (Sachar *et al.*, 1973; Carroll *et al.*, 1976).

The differential diagnosis of depression, and hence choice of treatment, can be difficult for the clinician in many cases. Current evidence suggests that major depressive disorders diagnosed by strict clinical criteria are entities distinct from other dysphoric syndromes (Spitzer *et al.*, 1978; Goodwin and Extein, 1979; Klein, 1974). Minor depressive disorder (Spitzer *et al.*, 1978) is commonly seen in the course of adjustment reactions and personality disorders (Goodwin and Extein, 1979; Spitzer, 1980). This distinction between major and minor depression is important for many reasons (Spitzer *et al.*, 1978; Klein, 1974), including the finding that major depressions tend to respond better than minor depressions to antidepressant medications (Goodwin and Extein, 1979; Klein, 1974; Goodwin, 1977). Distinguishing major from minor depression in patients on the basis of symptoms and history alone can be extremely difficult, making choice of antidepressant medication, psychotherapy, or a combination difficult as well (Goodwin and Extein, 1979; Klein, 1974; Goodwin, 1977). Within the group of patients with major depression, unipolar and bipolar patients have genetic and biological characteristics distinct from each other (Winokur *et al.*, 1969; Goodwin and Extein, 1979). In addition, bipolar patients are more likely than unipolars to become manic on antidepressants (Goodwin, 1977; Bunney, 1978; Wehr and Goodwin, 1979), and are more likely to have an antidepressant response to lithium (Goodwin, 1977, Goodwin *et al.*, 1972). Bipolar and unipolar depression can be clinically identical, difficult to reliably distinguish cross-sectionally, and are distinguished primarily by the history of mania (Spitzer *et al.*, 1978). However, this differential can be difficult in patients with histories of mild hypomanias or no family history and misleading in depressed patients evaluated during their first few episodes of affective illness (Winokur *et al.*, 1969). This latter group may contain many "false unipolars" who are eventually correctly given the diagnosis of bipolar disorder (Winokur *et al.*, 1969; Goodwin, 1977). Thus, treatment decisions such as whether to use a tricyclic antidepressant or a monoamine oxidase inhibitor (MAOI) or lithium may have to be based on frequently inadequate ambiguous clinical information, and would be aided by meaningful laboratory tests which might separate unipolar from bipolar depression.

In addition, the differential diagnosis of mania versus schizophrenic psychosis is often difficult. Recognition of the specific efficacy of lithium and neuroleptics in the acute (Schou *et al.*, 1954; Casey *et al.*, 1960) and prophylactic (Davis, 1975, 1976) treatment of bipolar affective disorder and schizophrenia, respectively, has resulted in appropriately increased emphasis on accurately distinguishing these two disorders (Taylor *et al.*, 1974; Pope and Lipinski, 1978). However, patients with classic bipolar disorder can, in the manic state, develop paranoia and Schneiderian "first rank" symptoms thought to be diagnostic of schizophrenia (Pope and Lipinski, 1978; Carlson and Goodwin, 1973). Likewise, patients with schizophrenic disorder can, in the acutely psychotic state, manifest hyperactivity, irritability, and grandiosity suggestive of a manic state (Pope and Lipinski, 1978). These patients are sometimes referred to as having "good prognosis schizophrenia" or schizoaffective disorders (Spitzer *et al.*, 1978; Robins and Guze, 1970). Knowledge of the longitudinal course can improve the clinician's ability to distinguish mania from schizophrenia but, in many cases, does not eliminate the need to make treatment decisions such as whether to use lithium, based on ambiguous cross-sectional clinical presentations. Again, a laboratory test with diagnostic implications would be a significant advance.

The TSH response to TRH ("the TRH test") is an endocrinological test used in the diagnosis of disease of the hypothalamic–pituitary–thyroid axis (Hershman and Pittman, 1971; Haigler *et al.*, 1971; Lamberg and Gordin, 1978; Martin *et al.*, 1977). The TSH response to TRH has been reported to be decreased in major unipolar depression; these data have been confirmed by a large number of independent investigators (Prange *et al.*, 1972; Kastin *et al.*, 1972; Takahashi *et al.*, 1973, 1974; Coppen *et al.*, 1974; Ehrensing *et al.*, 1974; Maeda *et al.*, 1975; Kirkegaard *et al.*, 1975, 1978; Hollister *et al.*, 1976; Loosen *et al.*, 1977; Brambilla *et al.*, 1978; Gold *et al.*, 1977, 1979, 1981; Mendlewicz *et al.*, 1979; Asnis *et al.*, 1980), but due to a lack of a clear understanding or working hypothesis which might explain this abnormality, as well as questions about whether the abnormality is primary or secondary, and other factors, no consensus has emerged as to the relevance of this test to the diagnosis of depression or the practice of clinical psychiatry (Carroll, 1978). More recently, it has been reported that TSH response to TRH is increased in bipolar depression (Gold *et al.*, 1977, 1979, 1981. This group finding has not been confirmed by other investigators (Takahashi *et al.*, 1974; Kirkegaard *et al.*, 1978; Mendlewicz *et al.*, 1979). Preliminary investigation has also demonstrated a decreased TSH reponse to TRH in mania compared to schizophrenic psychosis (Extein *et al.*, 1980a,b). In order to further evaluate the possible utility of the TRH test in psychiatric differential diagnosis, we have studied the TSH response to TRH in patients with affective, schizophrenic, and schizoaffective disorders.

## 2. METHODS

Seventy inpatients admitted to the Fair Oaks Hospital Neuropsychiatric Evaluation Unit were included in this study of the TSH response to TRH. These 70 patients consisted of 10 consecutive admissions who met Research Diagnostic Criteria for each of seven diagnostic groups: major depressive disorder, primary bipolar subtype; major depressive disorder, primary unipolar subtype; minor depressive disorder; manic disorder; schizophrenia, undifferentiated subtype; schizoaffective disorder, manic type; and schizoaffective disorder, depressed type. The patients with minor depressive disorder all met DSM III criteria (Wehr and Goodwin, 1979) for personality disorder or adjustment reaction. The schizophrenic patients were all actively psychotic. Patients with clinical evidence of thyroid disease (White and Walsmsley, 1979) or other endocrine diseases were excluded from the study. Patients with alcohol or drug addiction or organic brain syndrome were also excluded. None of the patients had been on lithium carbonate for at least 6 months prior to the TRH test. Serum thyroxine ($T_4$), triiodothyronine ($T_3$) uptake, free thyroxine index [FTI: $(T_4 \times T_3-$ uptake$)/30$], and TSH levels were determined prior to the TRH infusion and were within normal limit ($T_4$, 5–13 mg/100 ml; $T_3$ uptake, 24–36%; TSH, 0–10 $\mu$IU/ml) for all patients (Wallach, 1978). With the exception of five manic patients who received neuroleptics (maximum 4 days), all patients were free of all medication except flurazepam for at least 1 week prior to theTRH infusion.

An indwelling venous catheter was placed in patients who were at bedrest by 8:00 a.m. after overnight fast. At 9:00 a.m., 500 $\mu$g of synthetic TRH was administered over 30 sec through the catheter. Before and 15, 30, 60, and 90 min after TRH administration, small samples of blood were obtained for measurement of serum TSH in duplicate by radioimmunoassay (Gold et al., 1981). The maximum TSH response, or $\Delta$TSH, was determined for each patient by subtracting the basline TSH from the peak TSH level after TRH infusion (Martin et al., 1977). TSH data are expressed as the mean ± standard error of the mean ($\mu$IU/ml) for 10 patients per group. In the unipolar, bipolar, manic, and schizophrenic patients, serum cortisol levels ($\mu$g/100 ml) were determined in duplicate by radioimmunoassay at 4:00 p.m., 12 midnight, and 8:00 a.m. prior to the TRH infusion. These values were plotted against time and a 24-hr area-under-the-curve for cortisol secretion was determined for each patient in units of $\mu$g/100 ml.

Since TSH and $\Delta$TSH values may not be normally distributed, and since the patients per diagnostic group provided an inadequate indication of true population distribution in each group, the conservative nonparametric Mann–Whitney U test was used to analyze group differences. Since there were directional hypotheses for the major comparisons (among depressive subgroups and between schizophrenia and mania), one-tailed direct comparisons were justifiable. Since midnight and 24-hr cortisol levels were not normally distributed, the Mann–Whitney U test was used for these values also. One-way analysis of variance was used for other baseline characteristics, with the least significant difference test used for post-hoc between-group comparisons. In order to determine the possible influence of baseline characteristics on the observed diagnostic group differences in $\Delta$TSH, one-way and two-way analyses of covariance were employed, covarying out all different combinations of baseline variables (age, sex, $T_4$, $T_3$ uptake, FTI, and baseline TSH). Although the $\Delta$TSH values were not normally distributed, this parametric statistical approach produced such robust statistical differences in $\Delta$TSH among diagnostic groups that we have chosen to present the results as support for the conclusion that the diagnostic group differences are not artifacts of differing baseline group characteristics.

## 3. RESULTS

TSH values peaked at 15 or 30 min for all patients except one unipolar, one bipolar, two schizophrenic, and one schizoaffective depressed patient, who peaked at 60 min (Figs. 1–3). The mean $\Delta$TSH for each group was: unipolar depression, 7.0 ± 0.9; minor depression, 11.2 ± 1.2; bipolar depression, 14.7 ± 1.4; mania, 5.5 ± 0.9; schizophrenia, 9.5 ± 1.1; schizoaffective mania, 9.8 ± 1.5; schizoaffective depression, 8.0 ± 1.1 (Figs. 4–5). Unipolar patients had significantly lower $\Delta$TSH values than bipolar depressed patients ($p < 0.001$, Mann–Whitney U test). In comparison to patients with minor depression, patients with unipolar depression had significantly lower $\Delta$TSH values ($p < 0.01$), whereas bipolar depressed patients had significantly higher $\Delta$TSH values ($p < 0.04$). Manic patients had significantly lower $\Delta$TSH values than patients with schizophrenia, undifferentiated subtype, and patients with schizoaffective disorder, manic type ($p < 0.01$, Mann–Whitney U test). There were no significant differences in $\Delta$TSH values between patients with schizophrenia undifferentiated subtype, schizoaffective disorder manic type, or schizoaffective disorder depressed type. Statistical significance of the differences in TSH levels between depressive subgroups and between mania and schizophrenia at each of the 15-, 30-, 60-, and 90-min time points after TRH infusion is presented in Figs. 1 and 2.

Several factors that have been reported to influence the TRH test are summarized in Table 1. These include age, sex $T_4$, $T_3$ uptake, FTI, and baseline TSH. In addition, midnight serum cortisols and 24-hr area-under-the-curve cortisol values for patients with unipolar depression, bipolar depression, mania, and schizophrenia are also presented in Table 1. Analysis of variance showed significant differences among groups only for the FTI (see Table 1). It is noteworthy that both unipolar and bipolar depressed patients had somewhat

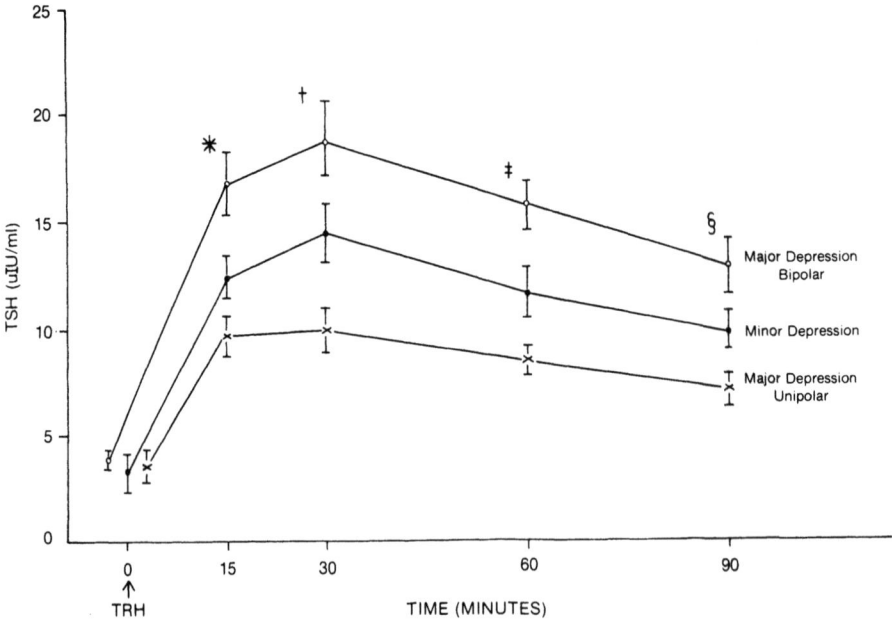

FIGURE 1. Serum TSH levels at 0, 15, 30, 60, and 90 min after infusion of 500 µg of TRH in groups of depressed patients. Each value represents the mean ± S.E.M. for 10 patients. *, unipolar < bipolar ($p < 0.001$); unipolar < minor ($p < 0.05$); minor < bipolar ($p < 0.05$). †, unipolar < bipolar ($p < 0.001$); unipolar < minor ($p < 0.01$). ‡, unipolar < bipolar ($p < 0.001$); unipolar < minor ($p < 0.05$); minor < bipolar ($p < 0.05$). §, unipolar < bipolar ($p < 0.001$); unipolar < minor ($p < 0.01$); minor < bipolar ($p < 0.05$).

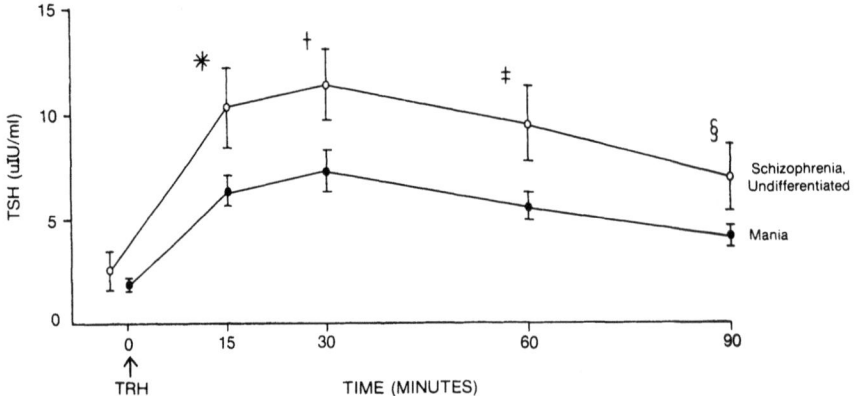

FIGURE 2. Serum TSH levels at 0, 15, 30, 60, and 90 min after i.v. infusion of 500 µg of TRH in schizophrenic and manic patients. Each value represents the mean ± S.E.M. for 10 patients. *, mania < schizophrenia ($p < 0.05$). †, mania < schizophrenia ($p < 0.05$). ‡, NS. §, NS.

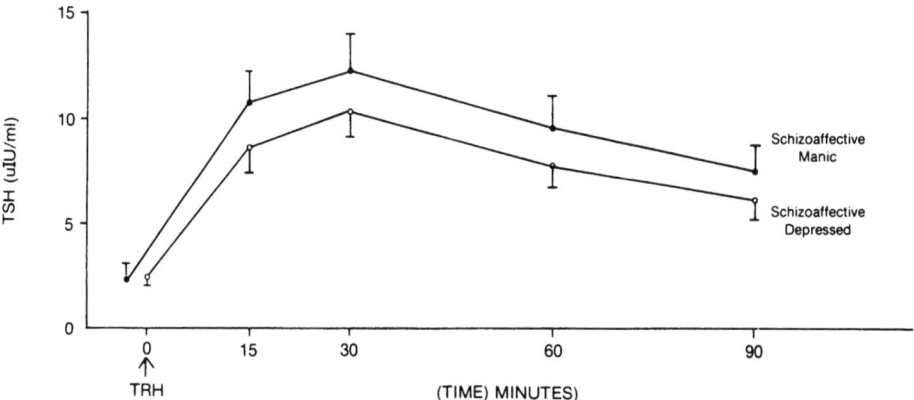

FIGURE 3. Serum TSH levels at 0, 15, 30, 60, and 90 min after i.v. infusion of 500 μg of TRH in schizoaffective patients. Each value represents the mean ± S.E.M. for 10 patients.

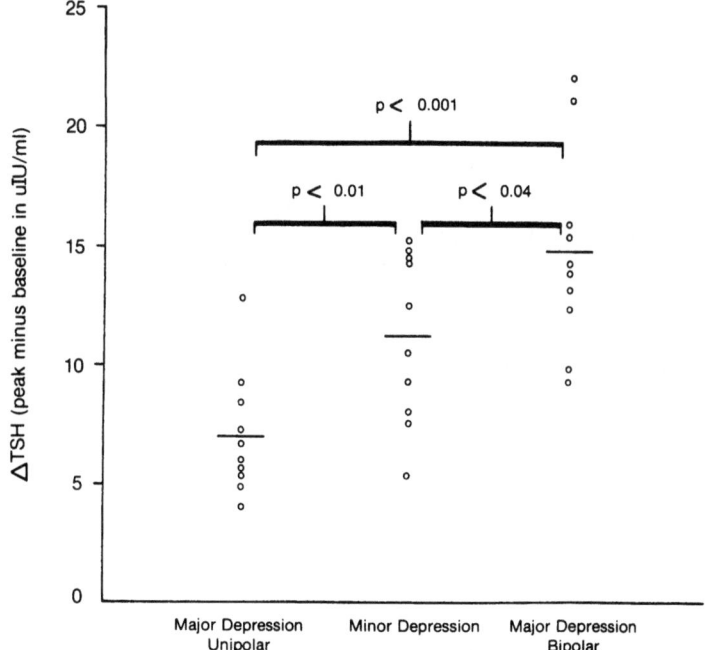

FIGURE 4. Maximum change in serum TSH level (peak TSH − baseline), or ΔTSH, after infusion of 500 μg of TRH in groups of depressed patients. Each circle represents the ΔTSH for an individual patient. Each line represents the mean for 10 patients.

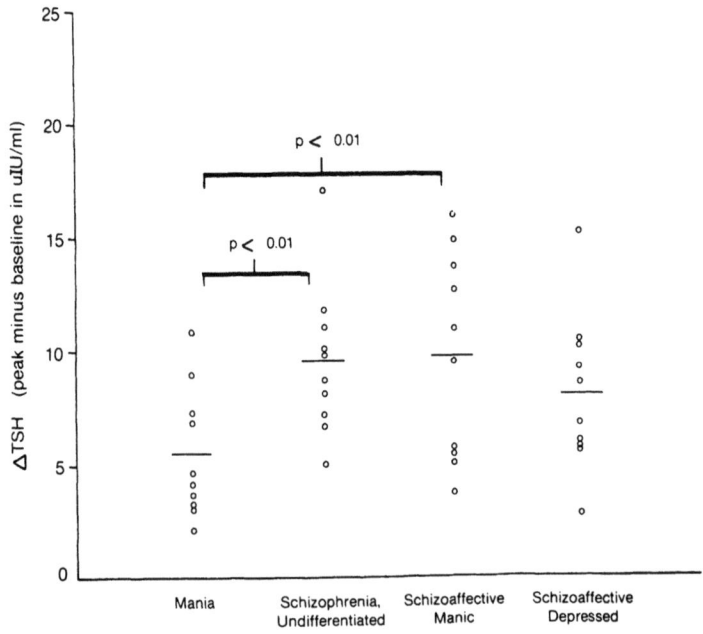

FIGURE 5. Maximum change in serum TSH level (peak TSH − baseline), or $\Delta$TSH, after infusion of 500 μg of TRH in manic, schizophrenic, and schizoaffective patients. Each circle represents the $\Delta$TSH for a individual patient. Each line represents the mean for 10 patients.

higher baseline TSH values. Thus, although there are subgroup differences in baseline TSH, these differences are not related to subgroup differences in $\Delta$TSH, which was highest in bipolar depressed patients but lowest in unipolar depressed patients. In fact, across all groups, the correlation coefficient between $\Delta$TSH and baseline TSH was only $r = 0.33$, thus accounting for only 10% of the variance. In addition, across all groups there was no correlation greater than 0.17 between $\Delta$TSH and either age, sex, $T_4$, $T_3$ uptake, FTI, midnight cortisol, or 24-hr cortisol area-under-the-curve. Although the $\Delta$TSH values were not entirely normally distributed, we applied analysis of covariance to the values to control for the influence of baseline TSH on $\Delta$TSH, and to factor out such influences as age, sex $T_4$, $T_3$ uptake, and FTI. No matter what combination of factors was removed, the diagnosis factor remained significant at the $p < 0.001$ level, suggesting robust diagnostic group differences that cannot be adequately accounted for by any other variable measured. The mean $\Delta$TSH of the five manic patients who had received neuroleptics in the week preceding the TRH test was $6.2 \pm 1.7$, not significantly different from the mean $\Delta$TSH of $4.8 \pm 1.1$ in the five neuroleptic free-manic patients.

## 4. DISCUSSION

The use of 500 μg of TR to provide maximum stimulation (Haigler et al., 1971) of pituitary TSH response has been widely used in medicine in the di-

TABLE 1
Baseline Characteristics of Diagnostic Groups

| | Age | Sex (M/F) | $T_4$ (μg/100 ml) | $T_3$ uptake (% of control) | Free $T_4$ index (μg/100 ml) | TSH (μIU/ml) | Midnight serum cortisol (μg/100 ml) | 24-hr cortisol area-under-the-curve (μg/100 ml-hr) |
|---|---|---|---|---|---|---|---|---|
| Unipolar depression | 41.6 ± 4.4 | 6/4 | 10.7 ± 0.9 | 32.0 ± 1.2 | 11.2 ± 0.1[b] | 3.6 ± 0.8 | 12.1 ± 3.0 | 440.6 ± 48.7 |
| Minor depression | 32.4 ± 4.9 | 3/7 | 8.7 ± 0.6 | 30.2 ± 0.9 | 8.8 ± 0.5 | 3.3 ± 0.9 | | |
| Bipolar depression | 38.1 ± 4.8 | 2/8 | 10.2 ± 1.1 | 30.1 ± 1.0 | 10.0 ± 0.9[c] | 3.9 ± 0.4 | 10.9 ± 2.8 | 460.2 ± 90.3 |
| Mania | 38.0 ± 5.1 | 4/6 | 9.2 ± 0.5 | 29.2 ± 1.0 | 9.0 ± 0.4 | 2.0 ± 0.3 | 9.9 ± 2.5 | 391.0 ± 43.2 |
| Schizophrenia, undifferentiated | 24.8 ± 2.2 | 5/5 | 7.9 ± 0.5 | 29.6 ± 0.6 | 7.8 ± 0.4 | 2.6 ± 0.9 | 13.2 ± 3.2 | 431.6 ± 46.4 |
| Schizoaffective, manic | 33.5 ± 3.9 | 3/7 | 9.4 ± 0.7 | 31.3 ± 1.3 | 9.8 ± 0.5 | 2.5 ± 0.6 | | |
| Schizoaffective, depressed | 29.3 ± 2.9 | 4/6 | 8.0 ± 0.6 | 29.8 ± 0.8 | 8.6 ± 0.5 | 2.5 ± 0.4 | —[e] | —[e] |
| F | 1.8 | —[d] | 1.8 | 1.0 | 2.8 | 1.4 | | |
| p | NS | | NS | NS | 0.02 | NS | NS | NS |

[a] Each value is the mean ± S.E.M. for 10 patients per group.
[b] Significantly higher than schizophrenia ($p < 0.01$); minor depression, mania, and schizoaffective depressed ($p < 0.05$).
[c] Significantly higher than schizophrenia ($p < 0.05$).
[d] Not applicable.
[e] Mann–Whitney U test applied because of distributions.

agnosis of disease of the hypothalamic–pituitary–thyroid axis (Hershman and Pittman, 1971; Haigler et al., 1971; Lamberg and Gordin, 1978; Martin et al., 1977). The TSH response to TRH in patients with minor depressions reported here is in the same range as that reported for normal subjects (Martin et al., 1977; Lamberg and Gordin, 1978; Haigler et al., 1971; Hershman and Pittman, 1971).

The results reported here are consistent with the widely recognized and reported finding that the TRH-induced TSH elevation is blunted in major depressive disorder, specifically in primary unipolar major depressive disorder, but not in minor depressive disorder (Prange et al., 1972; Kastin et al., 1972; Takahashi et al., 1973, 1974; Coppen et al., 1974; Ehrensing et al., 1974; Maeda et al., 1975; Kirkegaard et al., 1975, 1978; Hollister et al., 1976; Loosen et al., 1977; Brambilla et al., 1978; Gold et al., 1977, 1979, 1981; Mendlewicz et al., 1979; Asnis et al., 1980). Data reported here also suggest that bipolar depressed patients have an increased response to TRH, consistent with other reports in the literature (Gold et al., 1977, 1979, 1981; Extein et al., 1980d). However, some other authors have failed to find decreased responses in unipolar patients (Takahashi et al., 1974; Amsterdam et al., 1979) while finding normal or augmented responses in bipolar depressed patients. Similarly, some authors have found decreased rather than increased responses in patients with a bipolar depression (Takahashi et al., 1974; Kirkegaard et al., 1978), or have found responses which differ in the basis of menstrual status (Mendlewicz et al., 1979). These discrepancies may be due in part to the low dose of TRH used in several of the other studies, or due to the chronicity or treatment-resistant status of patients who are generally referred to research settings, or other factors. This study also finds a significant decrease in TSH response to TRH in mania, consistent with several reports of both nonsignificant trends to blunted response (Kirkegaard et al., 1978; Takahashi et al., 1975) and significant decreases in response (Extein et al., 1980a,b). In addition, it appears that the TSH response to TRH switches with switch in mood in manic depressive illness (Gold et al., 1981; Extein et al., 1980a). The results here are consistent with previous reports of normal TSH response to TRH in secondary and characterological depression, and in schizophrenia (Loosen et al., 1977; Kirkegaard et al., 1978; Gold et al., in press; Asnis et al., 1980). Previous reports have suggested possible factors which may be artifactually related to the observed diagnostic differences in TSH response to TRH (Hershman and Pittman, 1971; Haigler et al., 1971; Lamberg and Gordin, 1978; Martin et al., 1977). Blunted TSH response to TRH can be produced by hyperthyroidism, elevate glucocorticoids, alcohol, some drugs, and advanced age in males (Hershman and Pittman, 1971; Haigler et al., 1971; Lamberg and Gordin, 1978; Martin et al., 1977; Duick and Wahner, 1979; Loosen and Prange, 1980). An augmented TSH response to TRH can be produced by hypothyroidism (Hershman and Pittman, 1971; Haigler et al., 1971; Lamberg and Gordin, 1978; Martin et al., 1977). However, all the patients in this study were euthyroid and the significant intergroup differences noted above are not attributable to differences in age, sex, baseline thyroid functioning, or cortisol secretion. Alcoholic patients were excluded and except for five manic patients receiving neuroleptics, all were free of all medication except for flurazepam. Benzodiazepines have been shown to have little or no effect on the TRH test (Kirkegaard et al., 1977). The five manic patients on neuroleptics, which have been shown to slightly augment theTSH response to TRH (Kirkegaard et al., 1977), did not differ significantly from the five manic patients who did not receive neuroleptics. Thus, the blunted TSH response in manics compared to schizophrenics cannot be explained by the neuroleptic use in the subset of the manic patients. Lithium can cause clinical or subclinical hypothyroidism and hence might augment TSH response to TRH in bipolar depressed patients who are taking or have recently taken lithium. However, none of our bipolar patients had taken lithium for at least 6 months prior to the study and most had never taken lithium. Also, unipolar and bipolar depressed patients did not differ significantly on any measure of baseline thyroid functioning. Blunting of the TSH response to TRH in unipolar compared to bipolar depressed patients and in manic patients compared to schizophrenic patients cannot be attributed to differences in glucocorticoid status as assessed by diurnal cortisol secretion in this study (Loosen and Prange, 1980). Thus, the differences in the TSH response to TRH between unipolar, bipolar, and minor depression, and between mania and schizophrenic psychosis do not seem to be artifacts of any of the factors known to influence the TRH test, and may reflect basic biological differences in these psychiatric disorders.

These findings demonstrate agreement between clinical diagnosis using the Research Diagnostic Criteria (RDC) (Spitzer et al., 1978), and the magnitude of the TSH response to TRH. This agreement suggests that in cases in which it is difficult clinically to distinguish unipolar, bipolar, and minor depression, or to distinguish mania from schizophrenic psychosis, the TRH test may assist the clinician in diagnosis and the choice of treatment.

In the differential diagnosis of depression, either a blunting or an augmenting of the TSH response to TRH may help distinguish patients with a major depressive disorder from those with minor depressions (Kirstein et al., 1980). This distinction is important because patients with major depressions are more likely than patients with minor depressions to require and benefit from antidepressant medications (Goodwin and Extein, 1979; Klein, 1974; Goodwin, 1977). Further research is needed to determine if alterations in the TRH test in depressed patients predict response to antidepressant medications. This would be particularly interesting in patients whose TRH test results do not "agree" with the clinical diagnosis—for example, patients with minor depression and a blunted TSH response. Among patients with major depressions, the TSH response to TRH may help in the differential diagnosis of unipolar versus bipolar depression, and hence may have implication for prognosis and pharmacotherapy. A bipolar depressed patient would be more likely to have an antidepressant response to lithium (Goodwin, 1977; Goodwin et al., 1972) or need lithium as an adjunct to a tricyclic or MAOI, both of which produce manias (Bunney, 1978), or rapid mood cycling (Wehr and Goodwin,

1979) in a significant portion of bipolar but not unipolar depressed patients. Again, investigation of the pharmacological response and course of patients whose TRH test "disagrees" with the clinical diagnosis, such as unipolars with an augmented TSH response, would be particularly important. These patients may be "false unipolars" (Winokur et al., 1969; Goodwin, 1977) who could perhaps be identified before they develop manic episodes. TRH testing has also been shown to predict high relapse potential in depressed patients in one large study when these patients were retested at the time of clinical recovery (Kirkegaard et al., 1977). These data also suggest the potential utility of the TRH test in psychiatric treatment decision making.

In the differential diagnosis of mania versus schizophrenic psychosis, a blunted TSH response to TRH may help identify the manic patient. The ability to differentiate mania from schizophrenic psychosis is crucial for several reasons, including the more optimistic prognosis of affective illness (Pope and Lipinski, 1978; Robins and Guze, 1970), greater therapeutic efficacy of lithium in bipolar illness (Schou et al., 1954; Davis, 1976; Pope and Lipinski, 1978; Goodwin, 1977; Hershman and Pittman, 1971) and neuroleptics in schizophrenia (Casey et al., 1960; David, 1975; Pope and Lipinski, 1978), and the risk of tardive dyskinesia in improperly diagnosed bipolar patients treated with long-term neuroleptics (Davis, 1975). It would be important to document whether the difference in the TRH test between patients with manic disorder and patients with schizophrenic disorder undifferentiated subtype holds in differentiating mania versus paranoid schizophrenia as well. It would also be important in further studies to control for differences in activity. Further research is needed to determine if the results of the test predict outcome or response to lithium versus neuroleptics in manic and schizophrenic patients. There has been much interest in recent years in the response to lithium of subgroups of schizophrenic patients (Pope and Lipinski, 1978). Of particular interest again may be the pharmacological response of patients whose TRH test results are not consistent with clinical diagnosis, e.g., lithium response in schizophrenic patients with blunted TSH response. The TRH test in patients with schizoaffective disorders needs to be explored as well. Although the TRH test in the study reported here did not distinguish schizoaffective disorder manic type from schizophrenic disorder undifferentiated type, it would be important to investigate the differences in clinical course and response to lithium of patients with schizoaffective disorder manic type with and without a blunted TSH response to TRH.

Finally, given these observed diagnostic group differences, we might speculate about possible underlying mechanisms. First, neuroendocrine abnormalities may serve as biological markers for changes in the central monoamine neurotransmitters, norepinephrine (NE), dopamine (DA), and serotonin (5-hydroxytryptamine; 5 HT), which are implicated in the etiology and pharmacotherapy of these disorders (Goodwin and Extein, 1979; Sachar, 1975; Carroll, 1978). Because many hypothalamic releasing factors are under monoaminergic control, changes in the functional activity of one or more monoamine neurotransmitter systems may be expected to produce changes in a number of endocrine systems. The release of TSH from the anterior pituitary is mediated by the hypothalamic tripeptide TRH. The release of TRH is regulated by NE, 5HT, and DA (Grimm and Reichlin, 1973; Reichlin et al., 1974; Reichlin, 1975; Besser et al., 1975; Chen and Meites, 1975). NE and DA stimulate and 5HT inhibits TRH release. Thus, changes in the hypothalamic–pituitary–thyroid axis may be an index of changes in central monoaminergic transmission in psychiatric disorders. Though the neurochemical abnormalities which result in abnormal TSH response to TRH are still unclear, there are several possible abnormalities in monoamine systems which could account for the TRH abnormalities reported here. Our findings are consistent with the hypothesis that mania is characterized by increased central noradrenergic and possibly dopaminergic transmission (Bunney, 1978; Schildkraut, 1978). This would lead to increased TRH release and subsequent homeostatic adjustment of the pituitary TRH receptors to become less sensitive or are "down regulated." An inverse relationship between CSF 5-hydroxyindoleacetic acid and TSH response to TRH in unipolar depressed patients has been reported (Gold et al., 1977). This suggests that unipolar depressed patients with blunted TSH response to TRH have increased serotonergic activity. Since depletion of NE can result in TRH reduction (Grimm and Reichlin, 1973; Reichlin et al., 1974), the blunted TSH response to TRH in unipolar depressed patients may be consistent with decreased functional NE activity in unipolar depression (Schildkraut, 1978; Bunney and Davis, 1965). These data suggest that patients with unipolar depression and blunted TSH response to TRH may have NE hypoactivity and 5HT hyperactivity. Many other viable explanations of the TRH test abnormalities exist, involving other neurotransmitters and neuromodulators. An understanding of why unipolar depressed and manic patients manifest similar TRH test abnormalities also requires further study.

Second, the diagnostic group differences may reflect primary dysfunction of the hypothalamic–pituitary–thyroid axis or endogenous TRH system (Prange et al., 1972). It has been suggested that a flat or absent TSH response to TRH may reflect deficiency in endogenous TRH with subsequent deficient pituitary TSH reserve. This hypothesis is supported by data which suggest that TSH responses return to normal after depressed patients are chronically administered TRH (Van den Burg et al., 1976), but not by other recent data on TRH levels in CSF (Kirkegaard et al., 1979). This hypothesis has been difficult to assess due to the lack of a reliable TRH radioimmunoassay. There is evidence for TRH receptors throughout the brain and especially the limbic system (Burt and Snyder, 1975), evidence that TRH itself may outside of the hypothalamic–pituitary–thyroid axis be a centrally active neuromodulator or "endogenous arousal factor" (Renaud and Martin, 1975; Plotnikoff et al., 1974; Hökfelt et al., 1975), and may have antidepressant effects (Prange et al., 1972; Loosen and Prange, 1980). It is possible that alterations in TRH release are not just an index of changes in brain monoamines but are etiologically related to affective illness.

In conclusion, the TRH test is a neuroendocrine test that can be performed clinically and may be useful in psychiatric diagnosis and treatment planning (Extein *et al.*, 1980c; Gold *et al.*, 1979, 1980a,b, 1981). The TRH test appears to be useful in differentiating similarly appearing unipolar, bipolar, and minor depressive disorders (Gold *et al.*, 1979, 1980a,b, 1981), and in differentiating mania from schizophrenic psychosis (Extein *et al.*, 1980b). The TRH test (Prange *et al.*, 1972; Gold *et al.*, 1979, in press; Extein *et al.*, 1980b,c), dexamethasone suppression test (Sachar, 1975; Carroll, 1978; Carroll *et al.*, 1976), and other biological tests (Sachar *et al.*, 1971, 1973; Maas *et al.*, 1968; Extein *et al.*, 1980d; Gold *et al.*, 1978; Pickar *et al.*, 1981) may at some time be added to the present standard diagnostic criteria (Spitzer *et al.*, 1978; Spitzer, 1980) for affective disorders. Researchers need to explore the possibility that the TRH test may, by identifying biologically homogeneous subgroups of patients, help the clinician make treatment decisions, such as whether to use antidepressant medications and choice of tricyclics versus lithium in depression, and choice of lithium versus neuroleptics in acute psychosis.

## 5. SUMMARY

TSH was measured in serum before and 15, 30, 60, and 90 min after i.v. administration of 500 μg of TRH to 10 consecutive euthyroid patients in each of seven groups determined by Research Diagnostic Criteria. $\Delta$TSH (peak minus baseline) was calculated for each patient. The mean $\Delta$TSH ($\mu$IU/ml $\pm$ SEM) of 7.0 $\pm$ 0.9 in major unipolar depression and 14.7 $\pm$ 1.4 in major bipolar depression were different from each other ($p < 0.001$) and respectively lower ($p < 0.01$) and higher ($p < 0.04$) than that of 11.2 $\pm$ 1.2 in minor depression. The mean $\Delta$TSH of 5.5 $\pm$ 0.9 in mania was lower ($p < 0.01$) than that of 9.5 $\pm$ 1.1 in undifferentiated schizophrenia. Differences were unexplainable by differences in baseline thyroid function. Thus, the TRH test may be useful in differentiating unipolar, bipolar, and minor depressions, and in differentiating mania from schizophrenia.

ACKNOWLEDGMENTS. The authors would like to give special thanks to the staff of the Fair Oaks Hospital Neuropsychiatric Evaluation and Research Unit and D. Martin and E. Howard for their skillful and diligent help in carrying out this study.

## 6. REFERENCES

Altman, N., Sachar, E. J., Gruen, P. H., Halpern, F. S., and Eto, S., 1975, Reduced plasma LH concentration in postmenopausal depressed women, *Psychosom. Med.* **37**:274.

Amsterdam, J. D., Winokur, A., Mendels, J., and Snyder, P., 1979, Distinguishing depression subtypes of thyrotropin response to TRH testing, *Lancet* **II**:904.

Asnis, G. M., Nathan, R. S., Halbreich, U., Halpern, F. S., and Sachar, E. J., 1980, TRH tests in depression, *Lancet* **I**:424.
Besser, G. S., Burrow, G. N., Spaulding, S. W., and Donabedian, R. K., 1975, Dopamine infusion acutely inhibits the TSH and prolactin response to TRH, *J. Clin. Endocrinol. Metab.* **41**:985.
Brambilla, F., Smeraldi, E., Sacchetti, E., Negri, F., Cocchi, D., and Müller, E. E., 1978, Deranged anterior pituitary responsiveness to hypothalamic hormones in depressed patients, *Arch. Gen. Psychiatry* **35**:1231.
Bunney, W. E., Jr., 1978, Psychopharmacology of the switch process in affective illness, in: *Psychopharmacology: A Generation of Progress* (M. A. Lipton, A. DiMascio, and K. F. Killam, eds.), pp. 1249–1259, Raven Press, New York.
Bunney, W. E., Jr., and Davis, J. M., 1965, Norepinephrine in depressive reactions, *Arch. Gen. Psychiatry* **13**:483.
Burt, D. R., and Snyder, S. H., 1975, Thyrotropin-releasing hormone (TRH): Apparent receptor binding in rat brain membrane, *Brain Res.* **92**:309.
Carlson, G. A., and Goodwin, F. K., 1973, The stages of mania: A longitudinal analysis of the manic episode, *Arch. Gen. Psychiatry* **28**:221.
Carroll, B. J., 1978, Neuroendocrine function in psychiatric disorders, in: *Psychopharmacology: A Generation of Progress* (M. A. Lipton, A. DiMascio, and K. F. Killam, eds.), pp. 487–497, Raven Press, New York.
Carroll, B. J., Curtis, G. C., and Mendels, J., 1976, Neuroendocrine regulation in depression, *Arch. Gen. Psychiatry* **33**:1039.
Casey, J. F., Bennett, I. F., Lindley, C. J., Hollister, L. E., Gordon, M. H., and Springer, N. N., 1960, Druge therapy in schizophrenia, *Arch. Gen. Psychiatry* **2**:210.
Chen, H. J., and Meites, J., 1975, Effects of biogenic amines and TRH on release of prolactin and TSH in the rat, *Endocrinology* **96**:10.
Coppen, A., Peet, M., Montgomery, S., Bailey, J., Marks, V., and Woods, P., 1974, Thyrotropin-releasing hormone in the treatment of depression, *Lancet* **II**:433.
Davis, J. M., 1975, Overview: Maintenance therapy in psychiatry. I. Schizophrenia, *Am. J. Psychiatry* **132**:1237.
Davis, J. M., 1976, Overview: Maintenance therapy in psychiatry. II. Affective disorders, *Am. J. Psychiatry* **133**:1.
Duick, D. S., and Wahner, H. W., 1979, Thyroid axis in patients with Cushing's syndrome, *Arch. Intern. Med.* **139**:767.
Ehrensing, R. H., Kastin, A. J., Schalch, D. S., Friesen, H. G., Vargas, J. R., and Schally, A. V., 1974, Affective state and thyrotropin and prolactin responses after repeated injections of thyrotropin-releasing hormone in depressed patients, *Am. J. Psychiatry* **131**:714.
Ettigi, P. G., and Brown, G. M., 1977, Psychoneuroendocrinology of affective disorders: An overview, *Am. J. Psychiatry* **134**:493.
Extein, I., Pottash, A. L. C., Gold, M. S., Cadet, J., Sweeney, D. R., Davies, R. K., and Martin, D. M., 1980a, The thyroid-stimulating hormone response to thyrotropin-releasing hormone in mania and bipolar depression, *Psychiatry Res.* **2**:199.
Extein, I., Pottash, A. L. C., Gold, M. S., and Martin, D. M., 1980b, Differentiating mania from schizophrenia by the TRH test, *Am. J. Psychiatry* **137**:981.
Extein, I., Pottash, A. L. C., Gold, M. S., Sweeney, D. R., Kirstein, L., and Martin, D. M., 1980c, The TRH test in the differential diagnosis of depression, mania, and schizophrenia, in: *Abstract Book, XI Congress of the International Society of Psychoneuroendocrinology*, p. 65, Florence, Italy.
Extein, I., Pottash, A. L. C., Gold, M. S. Sweeney, D. R., Martin, D. M., and Goodwin, F. K., 1980d, Deficient prolactin response to morphine in depressed patients, *Am. J. Psychiatry* **137**:845.
Gold, M. S., Pottash, A. L. C., Davies, R. K., Ryan, N., Sweeney, D. R., and Martin, D. M., 1979, Distinguishing unipolar and bipolar depression by thyrotropin release test, *Lancet* **II**:411.
Gold, M. S., Pottash, A. L. C., Kleber, H. D., Extein, I., Kirstein, L., Sweeney, D. R., and

Martin, D. M., 1980a, TRH test in diagnosis of unipolar depression, in: *Abstract Book, XI Congress of the International Society of Psychoneuroendocrinology*, p. 75, Florence, Italy.

Gold, M. S., Pottash, A. L. C., Ryan, N., Sweeney, D. R., Davies, R. K., and Martin, D. M., 1980b, TRH-induced TSH response in unipolar, bipolar, and secondary depressions: Possible utility in clinical assessment and differential diagnosis, *Psychoneuroendocrinology* **5**:147.

Gold, M. S., Pottash, A. L. C., Extein, I., Martin, D. M., Howard, E., and Mueller, E. A., 1981, The TRH test in the diagnosis of major and minor depression, *Psychoneuroendocrinology* **6**(2):159–169.

Gold, P. W., Goodwin, F. K., Wehr, T., and Rebar, R., 1977, Pituitary thyrotropin response to thyrotropin-releasing hormone in affective illness: Relationship to spinal fluid amine metabolites, *Am. J. Psychiatry* **134**:1028.

Gold, P. W., Reus, V. K., and Goodwin, F. K., 1978, Vasopressin in affective illness—Hypothesis, *Lancet* **I**:1233.

Goodwin, F. K., 1977, Drug treatment of affective disorders: General principles, in: *Psychopharmacology in the Practice of Medicine* (M. E. Jarvik ed.), pp. 241–253, Appleton–Century–Crofts, New York.

Goodwin, F. K., and Extein, I., 1979, The biological basis of affective disorders, in: *Progress in the Functional Psychoses* (R. Cancro, L. Shapiro, and M. Kesselman, eds.), pp. 129–152, Spectrum Publications, New York.

Goodwin, F. K., Murphy, D. L., Dunner, D. L., and Bunney, W. E., Jr., 1972, Lithium response in unipolar versus bipolar depression, *Am. J. Psychiatry* **129**:44.

Grimm, Y., and Reichlin, S., 1973, Thyrotropin-releasing hormone (TRH): Neurotransmitter regulation of secretion by mouse hypothalamic tissue *in vitro*, *Endocrinology* **93**:626.

Haigler, E. D., Jr., Pittman, J. A., Jr., and Hershman, J. M., 1971, Direct evaluation of pituitary thyrotropin reserve utilizing synthetic thyrotropin-releasing hormone, *J. Clin. Endocrinol. Metab.* **33**:573.

Hershman, J. M., and Pittman, J. A., Jr., 1971, Control of thyrotropin secretion in man, *N. Engl. J. Med.* **285**:997.

Hökfelt, T., Fuxe, K., Johansson, O., Jeffcoate, S., and White, N., 1975, Thyrotropin-releasing hormone containing nerve terminals in certain brain stem nuclei and the spinal cord, *Neurosci. Lett.* **1**:133.

Hollister, L. E., Davis, K. L., and Berger, P. A., 1976, Pituitary response to thyrotropin-releasing hormone in depression, *Arch. Gen. Psychiatry* **33**:1393.

Kastin, A. J., Schalch, D. S., Ehrensing, R. H., and Anderson, M. S., 1972, Improvement in mental depression with decreased thyrotropin response after administration of thyrotropin-releasing hormone, *Lancet* **II**:740.

Kirkegaard, C., Nørlem, N., Lauridsen, U. B., Bjørum, N., and Christiansen, C., 1975, Protirelin stimulation test and thyroid function during treatment of depression, *Arch. Gen. Psychiatry* **32**:1115.

Kirkegaard, C., Bjørum, N., Cohn, D., Lauridsen, U. B., and Nekup, J., 1977, Studies on the influence of biogenic amines and psychoactive drugs on the prognostic value of the TRH stimulation test in endogenous depression, *Psychoneuroendocrinology* **2**:131.

Kirkegaard, C., Bjørum, N., Cohn, D., and Lauridsen, U. B., 1978, Thyrotropin-releasing hormone (TRH) stimulation test in manic-depressive illness, *Arch. Gen. Psychiatry* **35**:1017.

Kirkegaard, C., Faber, J., Hummer, L., and Rogowski, P., 1979, Increased levels of TRH in cerebrospinal fluid from patients with endogenous depression, *Psychoneuroendocrinology* **4**:227.

Kirstein, L., Gold, M. S., Pottash, A. L. C., Extein, I., Sweeney, D. R., and Davies, R., 1980, Clinical correlates of the TRH infusion test, in: *Abstract Book, XI Congress of the International Society of Psychoneuroendocrinology*, p. 71, Florence, Italy.

Klein, D. F., 1974, Endogenomorphic depression: A conceptual and terminological revision, *Arch. Gen. Psychiatry* **31**:447.

Lamberg, B. A., and Gordin, A., 1978, Abnormalities of thyrotropin secretion and clinical implications of the thyrotropin-releasing hormone stimulation test, *Ann. Clin. Res.* **10**:171.

Loosen, P. T., and Prange, A. J., Jr., 1980, Thyrotropin-releasing hormone (TRH): A useful tool for psychoneuroendocrine investigation, *Psychoneuroendocrinology* **5**:63.

Loosen, P. T., Prange, A. J., Jr., Wilson, I. C., Lara, P. P., and Pettus, C., 1977, Thyroid stimulating hormone response after thyrotropin-releasing hormone in depressed, schizophrenic and normal women, *Psychoneuroendocrinology* **2**:137.

Maas, J. W., Fawcett, J., and Dekirmenjian, H., 1968, 3-Methoxy-4-hydroxy-phenylglycol (MHPG) excretion in depressive states, *Arch. Gen. Psychiatry* **19**:129.

Maeda, K., Kato, Y., Ohgo, S., Chihara, K., Yoshimoto, Y., Yamaguchi, N., Kuromaru, S., and Imura, H., 1975, Growth hormone and prolactin release after injection of thyrotropin-releasing hormone in patients with depression, *J. Clin. Endocrinol. Metab.* **40**:501.

Martin, J. P., Reichlin, S., and Brown, G. M. (eds.), 1977, Regulation of TSH secretion and its disorders, in: *Clinical Neuroendocrinology*, pp. 201–228, Davis, Philadelphia,.

Mendlewicz, J., Linowski, P., and Brauman, H., 1979, TSH responses to TRH in women with unipolar and bipolar depression, *Lancet* **II**:1079.

Pickar, D., Davis, G. C., Schulz, C., Extein, I., Wagner, R., Naber, D., Gold, P. W., Van Kammen, D. P., Goodwin, F. K., Wyatt, R. J., Li, C. H., and Bunney, W. E., Jr., 1981, Behavioral and biological effects of acute beta-endorphin injection in schizophrenic and depressed patients, *Am. J. Psychiatry* **138**:160.

Plotnikoff, N. P., Prange, A. J., Jr., Breese, G. R., and Wilson, I. C., 1974, Thyrotropin-releasing hormone: Enhancement of DOPA activity in thyroidectomized rats, *Life Sci.* **14**:1271.

Pope, H. G., and Lipinski, J. F., Jr., 1978, Diagnosis in schizophrenia and manic-depressive illness, *Arch. Gen. Psychiatry* **35**:811.

Prange, A. J., Jr., Wilson, I. C., Lara, P. P., Alltop, L. B., and Breese, G. R., 1972, Effects of thyrotropin-releasing hormone in depression, *Lancet* **II**:999.

Reichlin, S., 1975, Regulation of the hypophysiotropic secretions of the brain, *Arch. Intern. Med.* **135**:1350.

Reichlin, S., Jackson, I., and Seyler, L. E., 1974, Regulation of the secretion of TRH and LHRH, in: *Frontiers in Neurology and Neuroscience Research* (P. Seeman and G. M. Brown, eds.), pp. 48–59, University of Toronto Press, Toronto.

Renaud, L. P., and Martin, J. B., 1975, Thyrotropin-releasing hormone (TRH): Depressant action on central neuronal activity, *Brain Res.* **86**:150.

Robins, E., and Guze, S. B., 1970, Establishment of diagnostic validity in psychiatric illness: Its application to schizophrenia, *Am. J. Psychiatry* **126**:983.

Rosenthal, D., and Kety, S. S. (eds.), 1968 *The Transmission of Schizophrenia*, Pergamon Press, Elmsford, N.Y.

Sachar, E. J., 1975, Evidence for neuroendocrine abnormalities in the major mental illnesses, in: *Biology of the Major Psychoses* (D. X. Freedman, ed.), pp. 347–358, Raven Press, New York.

Sachar, E. J., Finkelstein, J., and Hellman, L., 1971, Growth hormone responses in depressive illness. I. Response to insulin tolerance test, *Arch. Gen. Psychiatry* **25**:263.

Sachar, E. J., Hellman, L., Roffwarg, H. P., Halpern, F. S., Fukushima, D. K., and Gallagher, T. F., 1973, Disrupted 24-hour patterns of cortisol secretion in psychotic depression, *Arch. Gen. Psychiatry* **28**:19.

Schildkraut, J. J., 1978, Current status of the catecholamine hypothesis of affective disorders, in: *Psychopharmacology: A Generation of Progress* (M. A. Lipton, A. DiMascio, and K. F. Killam, eds.), pp. 1223–1234, Raven Press, New York.

Schlesser, M. A. Winokur, G., and Sherman, B. M., 1979, Genetic subtypes of unipolar primary depressive illness distinguished by hypothalamic–pituitary–adrenal axis activity, *Lancet* **I**:739.

Schou, M., Juel-Nielsen, N., Stromgren, E., and Voldy, H., 1954, The treatment of manic psychoses by the administration of lithium salts, *J. Neurol. Neurosurg. Psychiatry* **17**:250.

Spitzer, R. L. (Chairperson), 1980, American Psychiatric Association: Diagnostic and Statistical Manual of Mental Disorders, Task Force, Nomenclature and Statistics, 3rd ed., Washington, D.C.

Spitzer, R. L., Endicott, J., and Robins, E., 1978, Research diagnostic criteria: Rationale and reliability, *Arch. Gen. Psychiatry* **35**:773.

Takahashi, S., Kondo, H., Yoshimura, M., and Ochi, Y., 1973, Anti-depressant effect of thyrotropin-releasing hormone (TRH) and the plasma thyrotropin levels in depression, *Folia Psychiatr. Neurol. Jpn.* **27**:305.

Takahashi, S., Kondo, H., Yoshimura, M., and Ochi, Y., 1974, Thyrotropin responses to TRH in depressive illness: Relation to clinical subtypes and prolonged duration of depressive episode, *Folia Psychiatr. Neurol. Jpn.* **28**:335.

Takahashi, S., Kondo, H., Yoshimura, M., and Ochi, Y., 1975, Thyroid function levels and thyrotropin responses to TRH administration in manic patients receiving lithium carbonate, *Folia Psychiatr. Neurol. Jpn.* **29**:231.

Taylor, M. A. Gaztanaga, P., and Abrams, R., 1974, Manic-depressive illness and acute schizophrenia: A clinical, family history and treatment-response study, *Am. J. Psychiatry* **131**:678.

Van de Burg, W., von Praag, H. M., Bos, E. R. H., Piers, D. A., VanZanten, A. K., and Doorenbos, H., 1976, TRH by slow, continuous infusion: An antidepressant?, *Psychol. Med.* **6**:393.

Wallach, J. (ed.), 1978, in: *Interpretation of Diagnostic Test*, 3rd ed., p. 27, Little, Brown, Boston.

Wehr, T. A., and Goodwin, F. K., 1979, Rapid cycling in manic depressives induced by tricyclic antidepressants, *Arch. Gen. Psychiatry* **36**:555.

White, S. H., and Walsmsley, R. N., 1979, Can the initial assessment of thyroid function be improved?, *Lancet* **I**:933.

Winokur, G., Clayton, P. J., and Reicht, T., 1969, *Manic-Depressive Illness*, Mosby, St. Louis.

CHAPTER 23

# Aspects of Thyroid Axis Function in Depression
## A Review

ARTHUR J. PRANGE, JR., and PETER T. LOOSEN

## 1. INTRODUCTION

Venerable interest in the possible participation of the hypothalamic–pituitary–thyroid (HPT) axis in affective disorders has been rekindled by advances in physiological understanding of the axis, by technological progress, and by specific findings. It is now known that thyrotropin-releasing hormone (TRH), a tripeptide, is the hypothalamic substance that prompts the secretion of the anterior pituitary substance, thyroid-stimulating hormone (TSH, thyrotropin); it is now possible, by radioimmunoassay, to measure all the hormones of the HPT axis. It is now recognized that there is indeed a fault in HPT function in at least some patients with depression. A certain fraction of such patients, a quarter or more depending on criteria, show a blunted TSH response to TRH injection, in the absence of a usual endocrine explanation. In this brief review we will concern ourselves not only with this finding, which may be regarded as an aspect of the diagnostic use of HPT axis physiology, but also with the therapeutic use of hormones of the axis.

## 2. THE TSH RESPONSE TO TRH

We reported our initial experience in giving TRH in 1972. Two of ten women with unipolar depression showed a virtually absent TSH response while the response of a third woman was abnormally low. Hyperthyroidism, which

---

ARTHUR J. PRANGE, Jr. • Department of Psychiatry, Biological Sciences Research Center, University of North Carolina at Chapel Hill, School of Medicine, Chapel Hill, North Carolina 27514. PETER T. LOOSEN • Clinical Research Unit, Dorothea Dix Hospital, Raleigh, North Carolina 27611.

is the usual explanation for such findings, was ruled out by usual means (Prange *et al.*, 1972). At present 41 studies pertaining to a total of 917 patients show that about 25% of patients with primary depression show a blunted TSH response to TRH stimulation. Four studies involving 20 patients have been negative. We have reviewed these data in detail elsewhere (Loosen and Prange, 1982).

To be classified as blunted, we have required that a TSH response be less than 5 μU/ml over baseline, a rise less than that seen in our series of age- and sex-matched controls. In the TRH test the hormone is usually given intravenously and blood is taken at regular times before and after injection for later assay of TSH (and prolactin and growth hormone, if desired). Only TSH responses will be considered in this review.

The TSH response deficit seems not to be a function of increasing age (Takahashi *et al.*, 1974; Loosen *et al.*, 1977; Brambilla *et al.*, 1980), previous drug intake (Loosen *et al.*, 1977; Burger and Patel, 1977; Kirkegaard *et al.*, 1978), or severity of illness (Maeda *et al.*, 1975; Hollister *et al.*, 1976; Karlberg *et al.*, 1978). Furthermore, elevations of thyroid hormones, which would produce enhanced negative feedback on the pituitary gland, do not provide an explanation. Depressed patients who show TSH blunting are nearly always euthyroid. Indeed, some may show blunting and *low* thyroid hormone levels (Takahashi *et al.*, 1974; Hatotani *et al.*, 1977), which suggests a derangement of feedback within the axis. The notion of disturbed feedback is reinforced by our preliminary observation (Loosen *et al.*, 1980) that pretreatment with thyroid hormones attenuates the TSH response in normal subjects but not in depressed patients who show a normal, i.e., nonblunted, response.

The observation of a blunted TSH response to TRH in depressed patients can be viewed in different ways. While no data are available as regards the usefulness of the TRH test for treatment choice, the test has been used to predict both outcome to standard treatments (antidepressant drugs, ECT) and prognosis for early relapse after remission induced by these treatments. TRH has been given during depression and again in recovery, the change in maximum TSH levels between the two tests being used as the variable. In the two studies in which this was done, the magnitude of the increase in the TSH peak response correlated with a favorable clinical response (Brambilla *et al.*, 1980; Langer *et al.*, 1980). In addition, in another study the persistence of a low TSH response predicted, as it were, an early relapse (Kirkegaard and Bjørum, 1980), though this was not confirmed (Langer *et al.*, 1980).

Is a blunted TSH response a trait marker or a state marker? The TRH test has been done during depression and again after remission in several studies. In most studies the TSH response, being reduced in depression, normalized upon recovery (Hatotani *et al.*, 1977; Gregoire *et al.*, 1977; Tsutsui *et al.*, 1979; Mendlewicz *et al.*, 1979; Brambilla *et al.*, 1980). In two studies (Maeda *et al.*, 1975; Kirkegaard *et al.*, 1975), some blunted responses persisted into remission; in five other studies (Coppen *et al.*, 1974; Loosen *et al.*, 1977; Mendlewicz *et al.*, 1979; Asnis *et al.*, 1980; Papakostas *et al.*, 1981), persistence was the rule. Thus, a blunted TSH response in depression is apparently sometimes a trait marker and sometimes a state marker. As a trait marker the sensitivity of blunting must be less than 25%; this is its approximate frequency in depressive attacks and it does not always persist. In a similar way, specificity is also imperfect, as is the case with other apparent trait markers for mental illness. The exceptions to specificity, however, are interesting. TSH blunting occurs in anorexia nervosa, mania, and alcoholism (Loosen and Prange, 1982), as well as in depression, and all these conditions bear some relationship to depression. The absence of TSH blunting in schizophrenic patients lends some specificity to the finding within mental disorders (Loosen and Prange, 1982).

The question also arises as to the distribution of TSH blunting in diagnostic subtypes of depression. Some studies have reported that the fault occurs only in primary depression (Kirkegaard et al., 1978; Gold et al., 1980; Asnis et al., 1980; Brambilla et al., 1980; Papakostas et al., 1981) while others have found it in secondary depression as well (Takahashi et al., 1974; Ettigi et al., 1979; Koenig et al., 1980). In a similar way, differential occurrence in unipolar vs. bipolar patients is unclear. In most reports (Coppen et al., 1974; Maeda et al., 1975; Loosen et al., 1977; Kirkegaard et al., 1978; Amsterdam et al., 1979; Bjørum and Kirkegaard, 1979; Mendlewicz et al., 1979),no significant difference emerged between the two conditions; patients in both categories showed blunting about equally.

Whatever its clinical utility may prove to be, the mechanism of the blunted TSH response is presently not well understood. A detailed discussion of this matter is given elsewhere (Loosen and Prange, 1982). One possibility is that the fault is the result of chronic TRH hypersecretion, causing downregulation of the pituitary thyrotroph cells and lowered sensitivity to TRH administration. A second hypothesis is that the fault results from excessive inhibition of the thyrotroph by substances from the brain such as somatostatin, neurotensin, and dopamine.

It is well known that cortisol inhibits the TSH response (Re et al., 1976; Otsuki et al., 1973) and it is well known that cortisol is often elevated in depression (Carroll and Mendels, 1976). For these reasons, relationships between serum cortisol elevation and TSH blunting have been investigated. In a preliminary study we reported that cortisol elevations were, indeed, correlated with TSH blunting in a small sample of unipolar depressed women (Loosen et al., 1978a,b). However, several investigators subsequently addressed this point and did not find such a relationship (Langer et al., 1980; Gold et al., 1980; Kirkegaard and Carroll, 1980; Sachar et al., 1980; Papakostas et al., 1981). At present it appears extremely doubtful that adrenal activation accounts for more than a few, if any, instances of TSH blunting in depression.

## 3. THERAPEUTIC USES OF HORMONES OF THE HPT AXIS

### 3.1. TRH

In our first clinical assessment of TRH, we undertook a double-blind, placebo-controlled, crossover study of 10 women with unipolar depression

(Prange and Wilson, 1972; Prange et al., 1972). These patients showed a prompt, though partial and short-lived, improvement. Improvement was maximum 24 hr after i.v. injection of 0.6 mg of the hormone. A 50% reduction of scores on the Hamilton Rating Scale for Depression was noted; at the end of a week, mean scores had returned to baseline levels. Many investigators then quickly began to perform trials of TRH in one or another dose, orally or i.v., by any of several schedules, with various dependent variables in heterogeneous groups of patients. These factors have made it difficult to summarize the findings. Table 1 displays the results of both oral administration and i.v. administration of TRH in depressed patients.

Only a few researchers have investigated the possibility that TRH may accelerate or enhance the antidepressant effect of tricyclic antidepressants (TCAs). In a double-blind study of 29 patients, Mountjoy et al. (1974) gave TRH (40 mg/day orally), or placebo, with TCAs and found no significant difference between groups. Sørensen et al. (1974) found that i.v. TRH did not enhance standard treatments in a single-blind study of four patients.

Ruiz Ruiz et al. (1980) performed an extensive double-blind study of 90 patients, 52 men and 38 women, with various forms of depression. They used TRH alone and then in combination with imipramine (IMI) or amitriptyline (AMI). Medications were given i.v. for a few days and then orally. At the end of 30 days there were no reliable differences in the improvement rates of the three groups of 30 patients: TRH alone; IMI alone; AMI alone. Patients with unsatisfactory responses ($N = 13$, 15, and 18, respectively) were brought for-

TABLE 1
Trials of TRH as an Antidepressant Agent

|  | Oral administration | | Intravenous administration | |
| --- | --- | --- | --- | --- |
|  | Positive | Negative | Positive | Negative |
| Single-blind or open | 1 study (1 patient)[a] | 0 studies | 4 studies (188 patients)[d] | 5 studies (61 patients)[f] |
| Double-blind | 1 study (4 patients)[b] | 5 studies (72 patients)[c] | 7 studies (133 patients)[e] | 9 studies (129 patients)[g] |
| Totals | 2 studies (5 patients) | 5 studies (72 patients) | 11 studies (321 patients) | 14 studies (190 patients) |

[a] Van der Vis-Melsen et al. (1972).
[b] Itil et al. (1975).
[c] Huey et al. (1975), Kieley et al. (1976), Mountjoy et al. (1974), Sugarman et al. (1975), and Turek and Rocha (1974).
[d] Chazot et al. (1974), Itil et al. (1975), Maggini et al. (1974), and Obiols et al. (1974). The report of Chazot et al. (1974) contains two studies. Results were positive in patients with unipolar depression, negative in patients with neurotic, postpartum, and schizoid depression.
[e] Kastin et al. (1972), Lipton and Goodwin (1975), Pecknold and Ban (1977), Prange et al. (1972), Ruiz Ruiz et al. (1980), and Van den Burg et al. (1975, 1976). The report of Lipton and Goodwin (1975) is classified as both positive and negative. The physician rater found clear TRH advantages while the nurse raters did not.
[f] Chazot et al. (1974), Deniker et al (1974), Dimitrikoudi et al. (1974), Sørensen et al. (1974), and Takahashi et al. (1973).
[g] Benkert et al. (1974), Coppen et al (1974), Ehrensing et al. (1974), Evans et al. (1975), Hall et al. (1975), Hollister et al. (1974), Lipton and Goodwin (1975), Vogel et al. (1977), and Widerlov and Sjostrom (1975).

ward to a second trial. Patients who had failed on TRH alone were given larger oral doses (16–24 mg daily); patients who had failed on IMI or on AMI were given in addition to the TCA, oral TRH, 8 mg/day. At the end of a second 30-day period the rates of satisfactory outcome in the three groups were 15, 47, and 44%, respectively. Thus, the addition of TRH appeared to produce about a 45% improvement rate in TCA nonresponders, though the improvement rate that may have resulted from an additional 30 days of TCA alone is uncertain. It appeared that, in the patients who had not responded to TRH alone in the initial period, an increase in the dose of TRH was not useful.

### 3.2. TSH

This hormone, like other hormones of the anterior pituitary, is a large peptide. Its exact amino acid sequence varies from species to species (Wilbur, 1979).

Studies of the possible behavioral effects of TSH are rare both in man and in animals. We performed one such study in 20 women with primary depression, who were randomly assigned to either of two treatment groups (Prange et al., 1970). Patients in the experimental group were given an i.m. injection of 10 IU of bovine TSH on day 1 and again on day 8, while patients in the comparison group received saline injections. Patients in both groups received IMI, 150 mg orally, beginning on day 1 and daily thereafter for 28 days. Observers were ignorant of treatments. Their assessments showed that TSH injection greatly accelerated the antidepressant action of IMI. While TSH may have exerted independent behavioral effects, it surely released thyroid hormones from the thyroid gland. These hormones are known to accelerate the antidepressant action of TCAs, and this is a sufficient explanation for what was observed.

### 3.3. L-Triiodothyronine

We have performed four studies, comprising 40 women and 20 men, to determine whether L-triiodothyronine ($T_3$) would accelerate the antidepressant response to IMI (Prange et al., 1969; Wilson et al., 1970; Prange, 1971; Prange et al., 1976). We added $T_3$, 25 μg/day (vs. placebo), to a regimen of IMI. In two studies of women with primary depression (unipolar or bipolar), remission was obtained about twice as fast in $T_3$–IMI groups as in the placebo–IMI groups. In a series of depressed men $T_3$ failed to accelerate their responses but they responded faster to IMI alone than women had. Women who had received $T_3$–IMI achieved the rapidity of response demonstrated by IMI-treated men. Findings concerning $T_3$ were confirmed by Wheatley (1972) and Coppen et al. (1972). In both studies women benefited more than men from the adjunctive use of the thyroid hormone. Feighner et al. (1972) used larger doses of IMI and a shorter course of $T_3$. They found only a trend for $T_3$-treated patients to improve more rapidly than controls.

We have treated depressed patients by using $T_3$ alone in rapidly increasing doses. The hormone appeared early in the course of treatment to be as effective as standard doses of IMI, but its effect later in the study was offset by toxicity. While $T_3$ may exert a partial antidepressant effect, a point of considerable theoretical interest, the hormone given alone, appears to be impractical as a clinical remedy (Prange *et al.*, 1976).

The acceleration of TCA response by $T_3$, described above, and the remedy by $T_3$ of patients who have failed to respond to a usual TCA trial should be regarded as separate, though related, matters. Earle (1970), Hatotani *et al.* (1974), Ogura *et al.* (1974), and Goodwin *et al.* (1981) have studied the latter phenomenon. All found positive results from adding the hormone. About two-thirds to three-fourths of failed patients were converted to successes in a few days.

## 4. DISCUSSION

The definitive value of measuring the TSH response to TRH in mental patients remains to be determined. The response may be useful as a diagnostic tool, as a trait marker, or in predicting the adequacy of treatment, but more data are needed to evaluate these possibilities. What is quite evident, however, is that a defect in the HPT axis does exist in some depressed patients. Until recently there was no basis for such an assertion, for studies of other hormones of the HPT axis have not revealed clear abnormalities (Prange and Loosen, 1981).

The above consideration, the existence of an HPT fault in some depressed patients, renders more appealing and plausible the previous empirical findings that hormones of the HPT axis have some therapeutic value in depressed patients. Indeed, it is hormones only of the HPT axis that seem, at this time, clearly to have therapeutic potential. One study found that dexamethasone, a synthetic adrenal steroid, can be combined with IMI with therapeutic benefit (McClure and Cleghorn, 1968), but another study did not confirm this (Feighner *et al.*, 1972). A group of investigators has claimed that a carefully selected group of treatment-resistant depressed women is responsive to large doses of estrogen (Klaiber *et al.*, 1979), but confirmation is lacking. We administered a potent estrogen with IMI to unselected depressed women and found that the hormone, ethinyl estradiol, increased toxicity (Prange, 1971; Prange *et al.*, 1976). We gave testosterone, with IMI, to depressed men, but found it converted depression to a paranoid syndrome (Wilson *et al.*, 1974; Prange *et al.*, 1976). One group of investigators has suggested that MSH release-inhibiting factor (MIF I), Pro-Leu-Gly amide, given orally is useful in depressed patients (Ehrensing and Kastin, 1974, 1978), but this finding has not been studied by others. β-Endorphin has been used by many workers, but the results appear equivocal both in depression and in schizophrenia (Verebey *et al.*, 1978).

At the present time TRH appears less promising as a therapeutic agent than as a diagnostic tool. As noted above, design of therapeutic trials has varied so extensively that to summarize them is difficult. The very large doses used in some trials may also have confounded results. In this connection it should be noted that, in the two trials of MIF I in depression, small doses were more effective than large ones (Ehrensing and Kastin, 1974, 1978). Route of administration also appears to be important. A perusal of Table 1 suggests that TRH is more effective i.v. than orally. A long-acting congener with increased penetration of the blood–brain barrier, we think, would deserve a clinical trial. Finally it should be noted that TRH appears to exert behavioral activity not only in depression but also in a number of other conditions—mania, schizophrenia, hyperkinetic syndrome of children, Parkinson's disease, and normal volunteers (Prange et al., 1979). Its effects are surely not disease-specific, but nonspecificity in itself need not detract from therapeutic potential.

TSH has been only superficially examined as a behaviorally active agent in animals and man. It would be easy to study such possible effects in animals by ablating the thyroid gland. Since this cannot be done ethically in humans, it is difficult to know whether behavioral effects evoked by TSH in man are due to TSH itself or to the thyroid secretions it inevitably evokes. We noted enhanced antidepressant response when IMI treatment was prefaced by TSH injection, but the enhanced secretion of thyroid hormones is an adequate explanation. Whatever its behavioral effects may be, clearly TSH has won an important diagnostic place in psychiatry as a dependent variable after TRH administration.

Of the various hormones of the HPT axis, $T_3$ presently shows the most therapeutic potential. $T_3$ by itself appears not to be useful; apparently relatively large doses must be used and these are apt to cause systemic toxicity. When $T_3$ is used as an adjunct to TCAs, however, smaller doses can be used and enhanced toxicity is not noted. Indeed, some authors have reported diminished toxicity, though this may be an artifact of enhanced therapeutic effect.

As an adjunct to TCAs, $T_3$ finds use both to accelerate the response of drug-naive patients and to convert TCA nonresponders to responders. In neither case does $T_3$ seem to affect blood levels of TCA. Garbutt et al. (1979) showed that $T_3$ has no effect on blood levels of IMI or its chief metabolite, desmethylimipramine, and no clear effect on the actions of these agents on electrocardiographic findings. These findings support the observation, noted above, that $T_3$, in doses up to 50 μg/day, does not increase overall toxicity of the therapeutic regimen. They also bear upon the question whether $T_3$, when added to a TCA regimen, accomplished something that could not be accomplished by increasing TCA dosage. Bánki (1975) examined this question in a study of 96 TCA failures. For some he added $T_3$ to the TCA regimen; 39 of 52 promptly improved. For others he increased the dosage of TCA; 10 of 44 improved. In a comparative study, Bánki (1977) found respective improvement rates of 23/33 and 4/16. Goodwin et al. (1981) found that, in some TCA failures who later responded to $T_3$, TCA serum levels had earlier been in the range that is usually therapeutic.

The above considerations beg the question of how $T_3$, used adjunctively, exerts its therapeutic effects. It is presently not possible to decide whether, with adjunctive $T_3$, one is treating a hidden disorder of the HPT axis or, on the other hand, is exerting some pharmacologic effect on, as one possible example, amine receptor sensitivity. Some points bear on this issue. When $T_3$ is used from the outset, its benefits pertain much more, if not exclusively, to women. Just as women respond better to $T_3$, so it is the women, compared to men, have slower reflex times, whether normal or depressed (Ziegler and Levine, 1925; Prange et al., 1969; Whybrow et al., 1972), and more often experience frank hypothyroidism (Ingbar and Woeber, 1974). The small doses of $T_3$ employed, which have only minor metabolic effects, also suggest that a fault is being treated. A study by Tsutsui et al. (1979) bears on this theme. They identified 11 patients with long-standing depression, a poor response to TCAs, and a blunted TSH response to TRH, i.e., a demonstrated fault in the HPT axis. Ten improved promptly when given $T_3$, suggesting that $T_3$ may be most effective when given to patients with a thyroid axis fault. Whybrow et al. (1972) found that the greater the spontaneous thyroid activation of depressed patients, the prompter the subsequent response to IMI. Perhaps giving $T_3$ to depressed patients only accomplishes promptly what the organism would later usually accomplish spontaneously, i.e., a tendency toward remission and a readiness to respond to treatment. The resolution of this issue and related ones awaits further study.

## 5. SUMMARY

There are both diagnostic and therapeutic aspects to the relationships between the HPT axis and depression. The use of TRH to stimulate the release of TSH has revealed that some depressed patients show a subnormal response. All hormones of the HPT axis—TRH, TSH, $T_3$, and thyroxine—have been used experimentally in one or another way for therapeutic purposes. Among these, $T_3$ has been shown to exert two useful effects. It accelerates the therapeutic response to tricyclic drugs in previously untreated depressed women. Among men as well as women, it converts to therapeutic successes patients who have previously been tricyclic failures.

## 6. REFERENCES

Amsterdam, J. D., Winokur, A., Mendels, J., and Snyder, P., 1979, Distinguishing depression subtypes by thyrotropin response to TRH testing, Lancet **II**:904.

Asnis, G. M., Nathan, R. S., Halbreich, U., Halpern, F. S., and Sachar, E. J., 1980, TRH tests in depression, Lancet **1**:424.

Bánki, C. M., 1975, Triiodothyronin alkalmazasa a depressio kezelesében [Triiodothyronine in the treatment of depression], Orv. Hetil. **116**:2543.

Bánki, C. M., 1977, Cerebrospinal fluid amine metabolites after combined amitriptyline–triiodothyronine treatment of depressed women, Eur. J. Clin. Pharmacol. **11**:311.

Benkert, O., Gordon, A., and Martschke, D., 1974, The comparison of thyrotropin-releasing hormone, luteinizing hormone-releasing hormone and placebo in depressive patients using a double-blind cross-over technique, *Psychopharmacologia* **40**:191.

Bjørum, N., and Kirkegaard, C., 1979, Thyrotropin-releasing hormone test in unipolar and bipolar depression, *Lancet* **II**:694.

Brambilla, F., Smeraldi, E., Bellodi, L., Sachetti, E., and Mueller, E., 1980, Neuroendocrine correlates and monoaminergic hypothesis in primary affective disorder, in: *Progress in Psychoneuroendocrinology* (F. Brambilla, G. Racagni, and D. de Wied, eds.), pp. 235-245, Elsevier/North-Holland, Amsterdam.

Burger, H. G., and Patel, Y. C., 1977, TSH and TRH: Their physiological regulation and the clinical application of TRH, in: *Clinical Neuroendocrinology* (L. Martini and G. M. Besser, eds.), Academic Press, New York.

Carroll, B. J., and Mendels, J., 1976, Neuroendocrine regulation in affective disorders, in: *Hormones, Behavior and Psychopathology* (E. J. Sachar, ed.), pp. 193-224, Raven Press, New York.

Chazot, G., Chalumeau, A., Aimard, G., Mornex, R., Garde, A., Schott, B., and Girard, P. F., 1974, Thyrotropin releasing hormone and depressive states: From agroagonines to TRH, *Lyon Med.* **231**:831.

Coppen, A., Whybrow, P. C., Noguera, R., Maggs, R., and Prange, A. J., Jr., 1972, The comparative antidepressant value of L-tryptophan and imipramine with and without attempted potentiation by liothyronine, *Arch. Gen. Psychiatry* **26**:234.

Coppen, A., Montgomery, S., Peet, M., and Bailey, J., 1974, Thyrotropin-releasing hormone in the treatment of depression, *Lancet* **II**:433.

Deniker, P., Ginestet, D., Loo, H., Zarifian, E., and Cottereau, M. J., 1974, Preliminary study of the action of hypothalamic thyrostimulin (TRH) in depressive states, *Ann. Med. Psychol.* **1**:249.

Dimitrikoudi, M., Hanson-Norty, E., and Jenner, F. A., 1974, T.R.H. in psychoses, *Lancet* **I**:456.

Earle, B. V., 1970, Thyroid hormone and tricyclic antidepressants in resistant depressions, *Am. J. Psychiatry* **126**:1667.

Ehrensing, R. H., and Kastin, A. J., 1974, Melanocyte-stimulating hormone release inhibiting hormone as an antidepressant: A pilot study, *Arch. Gen. Psychiatry* **30**:63.

Ehrensing, R. H., and Kastin, A. J., 1978, Dose-related biphasic effect of prolyl-leucyl-glycinamide (MIF-I) in depression, *Am. J. Psychiatry* **135**:562.

Ehrensing, R. H., Kastin, A. J., Schalch, D. S., Friesen, H. G., Vargas, J. R., and Shally, A. V., 1974, Affective state and thyrotropin and prolactin responses after repeated injections of thyrotropin-releasing hormone in depressed patients, *Am. J. Psychiatry* **131**:714.

Ettigi, P. G., Brown, G. M., and Seggie, J. A., 1979, TSH and LH responses in subtypes of depression, *Psychosom. Med.* **41**:203.

Evans, L. E. J., Hunter, P., Hall, R., Johnston, M., and Roy, V. M., 1975, A double-blind trial of intravenous thyrotropin-releasing hormone in the treatment of reactive depression, *Br. J. Psychiatry* **127**:227.

Feighner, J. P., King, L. J., Schuckit, M. A., Croughan, J., and Briscoe, W., 1972, Hormonal potentiation of impipramine and ECT in primary depression, *Am. J. Psychiatry* **128**:1230.

Garbutt, J., Malekpour, B., Brunswick, D., Jonnalagadda, M. R., Jolliff, L., Podolak, R., Wilson, I., and Prange, A. J., Jr., 1979, Effects of triiodothyronine on drug levels and cardiac function in depressed patients treated with imipramine, *Am. J. Psychiatry* **136**:980.

Gold, M. S., Pottash, A. L. C., Ryan, N., Sweeney, D. R., Davies, R. K., and Martin, D. M., 1980, TRH induced TSH response in unipolar, bipolar, and secondary depressions: Possible utility in clinical assessment and differential diagnosis, *Psychoneuroendocrinology* **5**:147.

Goodwin, F. K., Prange, A. J., Jr., Post, R. M., Muscettola, G., and Lipton, M. A., 1981, L-Triiodothyronine converts tricycle antidepressant non-responders to responders. *Am. J. Psychiatry* **139**:334-338.

Gregoire, F., Brauman, H., De Buck, R., and Corvilain, J., 1977, Hormone release in depressed patients before and after recovery, *Psychoneuroendocrinology* **2**:303.

Hall, R., Hunter, P. R., Price, J. S., and Mountjoy, C. Q., 1975, Thyrotropin-releasing hormone in depression, *Lancet* **I**:162.

Hatotani, N., Tsujimura, R., Nishikubo, M., Yamaguchi, T., Endo, M., and Endo, J., 1974, Endocrinological studies of depressive states, with special reference to hypothalamo-pituitary function, First World Congress of Biological Psychiatry, Buenos Aires, Abstract 101.

Hatotani, N., Nomura, J., Yamaguchi, T., and Kitayama, I., 1977, Clinical and experimental studies on the pathogenesis of depression, *Psychoneuroendocrinology* **2**:115.

Hollister, L. E., Berger, P., Ogle, F. L., Arnold, R. C., and Johnson, A., 1974, Protirelin (TRH) in depression, *Arch. Gen. Psychiatry* **31**:468.

Hollister, L. E., Davis, K. L., and Berger, P. A., 1976, Pituitary response to thyrotropin-releasing hormone in depression, *Arch. Gen. Psychiatry* **33**:1393.

Huey, L. Y., Janowsky, D. S., Mandell, A. J., Judd, L. L., and Pendery, M., 1975, Preliminary studies on the use of thyrotropin-releasing hormone in manic states, depression, and the dysphoria of alcohol withdrawal, *Psychopharmacol. Bull.* **11**:24.

Ingbar, S. H., and Woeber, K. A., 1974, The thyroid gland, in: *Textbook of Endocrinology* (R. H. Williams, ed.), pp. 95–232, Saunders, Philadelphia.

Itil, T. M., Patterson, C. D., Polvan, N., Bigelow, A., and Bergey, B., 1975, Clinical and CNS effects of oral and I.V. thyrotropin-releasing hormone in depressed patients, *Dis. Nerv. Syst.* **36**:529.

Karlberg, B. E., Kjellman, B. F., and Kägedal, B., 1978, Treatment of endogenous depression with oral thyrotropin-releasing hormone and amitriptyline, *Acta Psychiatr. Scand.* **58**:389.

Kastin, A. J., Ehrensing, R. H., Schalch, D. S., and Anderson, M. S., 1972, Improvement in mental depression with decreased thyrotropin response after administration of thyrotropin-releasing hormone, *Lancet* **II**:740.

Kieley, W. F., Adrian, A. D., Lee, J. H., and Nicoloff, J. T., 1976, Therapeutic failure of oral thyrotropin-releasing hormone in depression, *Psychosom. Med.* **38**:233.

Kirkegaard, C., and Bjørum, N., 1980, TSH responses to TRH in endogenous depression, *Lancet* **I**:152.

Kirkegaard, C., and Carroll, B. J., 1980, Dissociation of TSH and adrenocortical disturbances in endogenous depression, *Psychiatry Res.* **3**:253.

Kirkegaard, C., Norlem, N., Lauridsen, U. B., Bjørum, N., and Christiansen, C., 1975, Protirelin stimulation test and thyroid function during treatment of depression, *Arch. Gen. Psychiatry* **32**:1115.

Kirkegaard, C., Bjørum, N., Cohn, D., and Lauridsen, U. B., 1978, Thyrotropin-releasing hormone (TRH) stimulation test in manic depressive illness, *Arch. Gen. Psychiatry* **35**:1017.

Klaiber, E. L., Broverman, D. M., Vogel, W., and Kobayashi, Y., 1979, Estrogen therapy for severe persistent depressions in women, *Arch. Gen. Psychiatry* **36**:550.

Koenig, G., Langer, G., Schoenbeck, G., Schuessler, M., and Reiter, H., 1980, Neuroendocrine status of depressed patients at admission and during clomipramine treatment, Proc. XI Int. Congr. Soc. Psychoendocrinol., Florence, Italy, p. 74.

Langer, G., Schoenbeck, G., Koenig, G., Reiter, H., Schuessler, M., Aschauer, H., and Lesch, O., 1980, Evidence for neuroendocrine involvement in the therapeutic effects of antidepressant drugs, in: *Progress in Psychoneuroendocrinology* (F. Brambilla, G. Racagni, and D. de Wied, eds.), pp. 197–208, Elsevier/North-Holland, Amsterdam.

Lipton, M. A., and Goodwin, F. K., 1975, A controlled study of thyrotropin releasing hormone in hospitalized depressed patients, *Psychopharmacol. Bull.* **11**:28.

Loosen, P. T., and Prange, A. J., Jr., 1982, Serum thyrotropin response to thyrotropin-releasing hormone in psychiatric patients: A review, *Am. J. Psychiatry* **139**:405.

Loosen, P. T., Prange, A. J., Jr., Wilson, I. C., Lara, P. P., and Pettus, C., 1977, Thyroid stimulating hormone response after thyrotropin releasing hormone in depressed, schizophrenic and normal women, *Psychoneuroendocrinology* **2**:137.

Loosen, P. T., Prange, A. J., Jr., and Wilson, I. C., 1978a, Influence of cortisol on TRH-induced TSH response in depression, *Am. J. Psychiatry* **135**:244.

Loosen, P. T., Prange, A. J., Jr., and Wilson, I. C., 1978b, The thyrotropin response to thyrotropin-releasing hormone in psychiatric patients: Relation to serum cortisol, *Prog. Neuro-Psychopharmacol.* **2**:479.

Loosen, P. T., Wilson, I. C., and Prange, A. J., Jr., 1980, Endocrine and behavioral changes in depression after thyrotropin-releasing hormone (TRH): Alteration by pretreatment with thyroid hormones, *J. Affect. Disorders* **2**:267.

McClure, D. J., and Cleghorn, R. A., 1968, Suppression studies in affective disorders, *Can. Psychiatry. Assoc. J.* **13**:477.

Maeda, K., Kato, Y., Ohgo, S., Chihara, K., Yoshimoto, Y., Yamaguchi, N., Kuromaru, S., and Imura, H., 1975, Growth hormone and prolactin release after injection of thyrotropin-releasing hormone in patients with depression. *J. Clin. Endocrinol. Metab.* **40**:501.

Maggini, C., Guazzelli, M., Mauri, M., Carrara, S., Formaro, P., Martino, E., Macchina, E., and Baschieri, L., 1974, Sleep, clinical and endocrine studies in depressive patients treated with thyrotropin releasing hormone, in: *Second European Congress of Sleep* (W. P. Koella, P. Levin, and M. Bertini, eds.), Karger, Basel.

Mendlewicz, J., Linkowski, P., and Brauman, H., 1979, TSH responses to TRH in women with unipolar and bipolar depression, *Lancet* **II**:1079.

Mountjoy, C. Q., Weller, M., Hall, R., Price, J. S., Hunter, P., and Dewar, J. H., 1974, A double-blind crossover sequential trial of oral thyrotropin-releasing hormone in depression, *Lancet* **I**:958.

Obiols, J., Pujol, J., and Obiols-Llandrich, J., 1974, Hormonas hipotalamicasy function tiroidea en los sindromes de presivos, 1st World Congress of Biological Psychiatry, Buenos Aires.

Ogura, C., Okuma, T., Uchida, Y., Imai, S., Yogi, H., and Sunami, Y., 1974, Combined thyroid (triiodothyronine)–tricyclic antidepressant treatment in depressive states, *Folia Psychiatr. Neurol. Jpn.* **28**:179.

Otsuki, M., Dakoda, M., and Baba, S., 1973, Influence of glucocorticoids on TRF induced TSH response in man, *J. Clin. Endocrinol. Metab.* **36**:95.

Papakostas, Y., Fink, M., Lee, J., Irwin, P., and Johnson, L., 1981, Neuroendocrine measures in psychiatric patients: Course and outcome with ECT, *Psychiatry Res.* **4**:55.

Pecknold, J. C., and Ban, T. A., 1977, TRH in depressive illness, *Int. Pharmacopsychiatry* **12**:166.

Prange, A. J., Jr., 1971, Therapeutic and theoretical implications of imipramine–hormone interactions in depressive disorders, in: *Proceedings of the V World Congress of Psychiatry*, Excerpta Medica, Amsterdam.

Prange, A. J., Jr., and Loosen, P. T., 1981, Findings in affective disorders relevant to the thyroid axis, melanotropin, oxytocin, and vasopressin, Psychoneuroendocrinology Symposium, Hamilton, Ontario.

Prange, A. J., Jr., and Wilson, I. C., 1972, Thyrotropin releasing hormone (TRH) for the immediate relief of depression: A preliminary report, *Psychopharmacologia* **26**:82.

Prange, A. J., Jr., Wilson, I. C., Rabon, A. M., and Lipton, M. A., 1969, Enhancement of imipramine antidepressant activity by thyroid hormone, *Am. J. Psychiatry* **126**:457.

Prange, A. J., Jr., Wilson, I. C., Knox, A., McClane, T. K., and Lipton, M. A., 1970, Enhancement of imipramine by thyroid stimulating hormone: Clinical and theoretical implications, *Am. J. Psychiatry* **127**:191.

Prange, A. J., Jr., Wilson, I. C., Lara, P. P., Alltop, L. B., and Breese, G. R., 1972, Effects of thyrotropin releasing hormone in depression, *Lancet* **II**:999.

Prange, A. J., Jr., Wilson, I. C., Breese, G. R., and Lipton, M. A., 1976, Hormonal alteration of imipramine response: A review, in: *Hormones, Behavior, and Psychopathology* (E. J. Sachar, ed.), pp. 41–67, Raven Press, New York.

Prange, A. J., Jr., Loosen, P. T., and Nemeroff, C. B., 1979, Peptides: Application to research in nervous and mental disorders, in: *New Frontiers of Psychotropic Drug Research* (S. Fielding, ed.), pp. 117–189, Futura Publishing Co., Mt. Kisko, N.Y.

Re, R. N., Kourides, I. A., Ridgway, E. C., Weintraub, B. D., and Maloof, F., 1976, The effect of glucocorticoid administration on human pituitary secretion of thyrotropin and prolactin, *J. Clin. Endocrinol. Metab.* **43**:338.

Ruiz Ruiz, M., Zardon Perez, J. M., and Brotat Ester, M., 1980, Investigacion controlada con

T.R.H. en las afecciones depresivas [Controlled investigation with TRH in depression], in: *Progresos en Psicofarmacologia*, Vol. 2 (M. Ruiz Ruiz, ed.), pp. 25–34, Bepya, Barcelona.

Sacher, E. J., Halbreich, U., Asnis, G., Nathan, S., and Halpern, F. S., 1980, Neuroendocrine disturbance in depression, in: *Progress in Psychoneuroendocrinology* (F. Brambilla, G. Racagni, and D. de Wied, eds.), pp. 265–272, Elsevier/North-Holland, New York.

Sørensen, R., Svendsen, K., and Schou, M., 1974, T.R.H. in depression, *Lancet* **II**:865.

Sugarman, A. A., Mueller, P. S., Swartzburg, M., and Rochford, J., 1975, Abbott-38579 (synthetic TRH) in the treatment of depression: A controlled study of oral administration, *Psychopharmacol. Bull.* **11**:30.

Takahashi, S., Kondo, H., Yoshimura, M., and Ochi, Y., 1973, Antidepressant effect of thyrotropin-releasing hormone (TRH) and the plasma thyrotropin levels in depression, *Folia Psychiatr. Neurol. Jpn.* **27**:305.

Takahashi, S., Kondo, H., Yoshimura, M., and Ochi, Y., 1974, Thyrotropin responses to TRH in depressive illness: Relation to clinical subtypes and prolonged duration of depressive episode, *Folia Psychiatr. Neurol. Jpn.* **28**:355.

Tsutsui, S., Yamazaki, Y., Namba, T., and Tsushima, M., 1979, Combined therapy of T3 and antidepressants in depression, *J. Int. Med. Res.* **7**:138.

Turek, I. S., and Rocha, J., 1974, Oral thyrotropin-releasing hormone (TRH) in depressive illness, *J. Clin. Pharmacol.* **14**:612.

Van den Burg, W., van Praag, H. M., Bos, E. R. H., Piers, D. A., van Zanten, A. K., and Doorenbos, H., 1975, Thyrotropin-releasing hormone (TRH) as a possible quick-acting but short-lasting antidepressant, *Psychol. Med.* **5**:404.

Van den Burg, W., van Praag, H. M., Bos, E. R. H., Piers, D. A., van Zanten, A. K., and Doorenbos, H., 1976, TRH by slow, continuous infusion: An antidepressant?, *Psychol. Med.* **6**:393.

Van der Vis-Melsen, M. J. E., and Wiener, J. D., 1972, Improvement in mental depression with decreased thyrotropin response after administration of thyrotropin-releasing hormone, *Lancet* **II**:1415.

Verebey, K., Volavka, J., and Clouet, D., 1978, Endorphins in psychiatry: An overview and a hypothesis, *Arch. Gen. Psychiatry* **35**:877.

Vogel, H. P., Benkert, O., Illig, R., Mueller-Oerlinghausen, B., and Poppenberg, A., 1977, Psychoendocrinological and therapeutic effects of TRH in depression, *Acta Psychiatr. Scand.* **56**:223.

Wheatley, D., 1972, Potentiation of amitriplyline by thyroid hormone, *Arch. Gen. Psychiatry* **26**:229.

Whybrow, P. C., Coppen, A., Prange, A. J., Jr., Noguera, R., and Bailey, J. E., 1972, Thyroid function and the response to liothyronine in depression, *Arch. Gen. Psychiatry* **26**:242.

Widerlov, E., and Sjostrom, R., 1975, Effects of thyrotropin releasing hormone on endogenous depression, *Nord Psykiatr. Tidsskr.* **29**:503.

Wilbur, J. G., 1979, Human pituitary thyrotropin, in: *Endocrinology*, Vol. 1 (L. J. DeGroot and G. H. Cahill, Jr., eds.), Grune & Stratton, New York.

Wilson, I. C., Prange, A. J., Jr., McClane, T. K., Rabon, A. M., and Lipton, M. A., 1970, Thyroid-hormone enhancement of imipramine in nonretarded depressions, *N. Engl. J. Med.* **282**:1063.

Wilson, I. C., Prange, A. J., Jr., and Lara, P. P., 1974, Methyltestosterone with imipramine in men: Conversion of depression to paranoid reaction, *Am. J. Psychiatry* **131**:21.

Ziegler, L. H., and Levine, B. S., 1925. The influence of emotional reactions on basal metabolism, *Am. J. Med. Sci.* **169**:68.

CHAPTER 24

# Chrononeuroendocrinology of Depression
## An Interpretation of Neuroendocrine Rhythms on the Basis of Pharmacological Studies

ALESSANDRO AGNOLI,
MASSIMO BALDASSARRE, FABRIZIO STOCCHI,
NICOLA MARTUCCI, GIOVANNI MURIALDO,
CLAUDIO ZAULI, ROSARIA D'URSO, and
PAOLO FALASCHI

## 1. INTRODUCTION

The recent progress in neurobiology has redefined most of the classical concepts in the neurosciences. Not even the role of the neurotransmitters, from the study of which many data about the pathophysiology of the nervous system were drawn, has remained unchanged by the new considerations. In fact, their functions, previously believed to be transmission of biological messages from neuron to neuron or from neuron to a peripheral receptor, have been newly defined (Calne, 1979).

According to the circumstances, the neurotransmitter is considered as a complex factor able to carry out many different functions: neurotransmission, neuromodulation, and neurohormonal control. This new approach, made possible by actual knowledge of the systems of hypothalamic–pituitary hormonal

---

ALESSANDRO AGNOLI • I Clinica Neurologica, University of Rome, 00161 Rome, Italy. MASSIMO BALDASSARRE, FABRIZIO STOCCHI, and NICOLA MARTUCCI • Clinica Neurologica, University of L'Aquila, L'Aquila, Italy. GIOVANNI MURIALDO and CLAUDIO ZAULI • Semeiotica Medica, I.S.M.I., University of Genoa, Genoa, Italy. ROSARIA D'URSO and PAOLO FALASCHI • V Clinica Medica, University of Rome, Rome, Italy.

control and of the neuronal influences on the hypothalamus (Schally and Arimura, 1977), has made it possible to trace a bridge between neurotransmission and hormonal function.

The monoaminergic pathways terminating in the hypothalamic areas, wherein the synthesis of peptides having excitatory or inhibitory influences on the production of the pituitary hormones occurs, are known (Martin, 1973; de Wied and de Jong, 1974). Well known is the role dopamine (DA) and the DA tuberoinfundibular pathway play in the inhibitory control of prolactin (PRL) (Del Pozo and Brownell, 1979); analogously, dopaminergic, noradrenergic, and serotoninergic influences act on the control of secretion of GH (Müller, 1973; Woolf and Lee, 1977).

Less certain data were obtained from the study of the neuronal control of TSH even if it seems possible to hypothesize an inhibitory role of DA (Scanlon et al., 1979). Similarly, a catecholaminergic mechanism would be implicated in the control of LH (Ojeda and McCann, 1974; Knobil and Plant, 1978) while both serotonin (5HT) and DA would modulate ACTH regulation (Cavagnini et al., 1976; Lamberts and Birkenhäger, 1976).

Relationships between neurotransmitter amines and mood control are now accepted; the correlation between the modifications of cerebral 5HT and some forms of depression as well as between reduction of DA levels and psychomotor aspects of the inhibited depression has been known (van Praag, 1977; Agnoli et al., 1978; Agnoli and Ruggieri, 1979).

Therefore, in order to postulate a relationship between hormones and monoamines, it appears evident that the alterations in the amount of the former must necessarily induce modifications in the latter and vice versa (Agnoli et al., 1980a). To exemplify in the endocrine pathology a hormonal alteration should translate into an alteration of mood while a neurotransmitter imbalance, e.g., in depression, should induce hormonal modifications. Regarding the first point, it is, as yet, difficult to provide a simple answer: the variables which make the results unclear appear to be too many, beginning with the definition of "depression" which in some studies is not always correctly used. Eliminating the less controlled results, it is, however, possible to point out several meaningful correlations between endocrine pathology and depression (Prange et al., 1977, for a review). In hypothyroidism, it has long been known that depressive correlates exist and pathologies associated with high levels of ACTH, e.g., Cushing's and Addison's disease, frequently exhibit depressive symptoms.

Conversely, notwithstanding the greater occurrence of depression in females, no significant correlation seems to exist between depression and changes in sexual hormones as is evidenced by studies on effects of menopause and oral contraceptives (Prange et al., 1977, for a review).

More information can be deduced from the examination of the results of the studies concerning the second point, e.g., hormonal modifications that appear in the course of depression.

## 1.1. Cortisol

Studies conducted with basal blood levels of cortisol brought results that, even though not always in agreement, seem to point out an increased activity of the ACTH–cortisol axis in primary depression. The cortisol hyperfunction does not seem to be responsible to nonspecific factors, like stress (Carroll *et al.*, 1976b), as it was first held, but appears to characterize such patients in a specific way (Sachar, 1976), particularly in view of high levels of cortisol not only in the blood but also in the cerebral spinal fluid (CSF) of depressed subjects (Träskman *et al.*, 1980).

An alteration of cortisol appears confirmed by studies with the dexamethasone suppression test: numerous authors (Carroll *et al.*, 1976b; Brown *et al.*, 1979; Brown and Shuey, 1980; Schlesser *et al.*, 1979) agree to individualize 40 to 50% of depressed subjects that do not show a response to dexamethasone. This test tends, in patients affected by primary depression, to normalize itself at the end of the treatment and to the resolution of the clinical picture. The circadian secretory profile of cortisol is also similarly altered in depression.

It is characterized by quantitative and numerical increase in the cortisol secretive bursts, which appear in the period of the day (8:00 p.m.–2:00 a.m.) during which they are absent in normal subjects (Sachar *et al.*, 1973b; Carroll *et al.*, 1976a).

## 1.2. PRL

Contradictory indications emerge from the study of the PRL basal level in depression; several authors (Sachar *et al.*, 1973a; Maeda *et al.*, 1975) reported higher levels of the hormone in depressed patients compared to controls, a fact not confirmed by other authors (Ehrensing *et al.*, 1974; Arana *et al.*, 1977).

Results of pharmacological activation are also contradictory. While Sachar *et al.* (1973a) reported normal suppression of PRL secretion after L-dopa, other workers have observed both an increase (Mendlewicz *et al.*, 1977) and an accentuated reduction (Gold *et al.*, 1976) in PRL levels compared to controls in subjects with bipolar depression, following the administration of L-dopa. Similarly, both an increase (Maeda *et al.*, 1975) and a reduction (Ehrensing *et al.*, 1974; Linnoila *et al.*, 1979) in PRL secretion, compared to controls, has been reported in response to TRH administration.

Such a disparity of results can be explained if one considers the multiple factors, environmental and pharmacological, that modify the hormone levels (Noel *et al.*, 1972) and the existence of circadian variations in the rhythm of PRL secretion (Sassin *et al.*, 1972). The study of secretory patterns of the hormone during 24 hr offers interesting information which would lead one to believe that a qualitative perturbation in the secretion of PRL exists; in fact, an evening increase and a greater presence of hormonal fluctuations (Halbreich *et al.*, 1979) and a modification of the hormonal characteristics of the circadian rhythm (Polleri *et al.*, 1979, 1980; Martucci *et al.*, 1980) in depressed patients as well as in controls have been described.

### 1.3. GH

The evaluation of the basal levels of GH in depression has been inconclusive, particularly in view of the pattern of GH secretion, which occurs in "bursts" and thus does not present a basic stable level and also due to an enormous number of variable factors which influence GH secretion (Roth *et al.*, 1963; Quabbe *et al.*, 1966).

The study of hormonal modifications following pharmacological stimulus has had concordant results; in fact, the observations of a limited response to GH to L-dopa appear uncertain after observation of the inhibitory influence exercised by estrogens on GH (Sachar, 1976). The observation of a reduced response of the hormone to the insulin-induced hypoglycemia in depression seems unequivocal (Endo *et al.*, 1974; Gruen *et al.*, 1975) as is reduced response to 5-HTP (Takahashi *et al.*, 1974) and an increased response to TRH (Maeda *et al.*, 1975) compared to controls.

Interesting observations come from the study of the circadian rhythm of GH in depressed patients, where an absence of secretory bursts of GH in relation to sleep with reduction in the quantity of slow-wave sleep (Roffwarg *et al.*, 1970; Skilkrut *et al.*, 1975) has been noted.

Considering the influence of sleep patterns on the secretion of GH and the frequent occurrence of sleep disturbances in depressed patients, this finding supposes a qualitative disorganization in the hormonal patterns in depression.

### 1.4. TSH

The observation of an association between hypothyroidism and depressive symptoms has focused attention on a possible imbalance of the pituitary–thyroid axis in depression.

The majority of studies have failed to conclusively show alterations in the basal levels of thyroid hormone in depressive patients, although some authors emphasize the existence of values at the upper limit of normal range (Whybrow *et al.*, 1972).

Despite the above findings, there seems to be sufficient agreement to believe that in depressed patients, stimulus tests such as TRH induce a limited TSH response (Hollister *et al.*, 1976; Kirkegaard *et al.*, 1978; Asnis *et al.*, 1980a) which tends to normalize itself after clinical improvement (Langer *et al.*, 1980; Kirkegaard and Bjørum, 1980).

### 1.5. LH

Modifications in the functional activity of gonadotropins have been reported in depression; some authors (Altman *et al.*, 1975) find the levels of LH reduced in menopausal depressive women compared to age-matched controls. This finding is confirmed by others (Brambilla *et al.*, 1978, 1980) who have shown an accentuated response of FSH and LH after TRH stimulus.

On the basis of these early data the question may be asked whether the described neuroendocrine alterations are merely concomitant to the affective disorders, or are correlated to the etiology and pathophysiology of the disease, being the expression of the same neurotransmitter imbalance that caused it, along with the involvement of the hypothalamic–pituitary system.

If the latter assumption is true, it may be further asked whether endocrine indices might be helpful in the diagnosis and monitoring of the treatment. A definitive answer to these questions is, at the moment, quite difficult: first, because of the discrepancies that emerge when reported data are considered, and then because of the importance of questions arising out of methodological considerations.

The problem arises since, in many instances, the changes in the investigated variable, namely the hormone level, are quite small, and differentiation from "baseline value" may not be possible. Hence, dynamic evaluations such as changes in response to procedures which either stimulate or inhibit the hormonal secretion and chronobiological studies became necessary. The necessity of chronobiological approach is further emphasized when one considers the existence of a specific rhythm in the depressive pathology and, in general, in psychopathology. Moreover, the fluctuations of a variable, like hormonal level, are an expression of the spontaneous variations in the neuroregulators and transmitters and indicate the primary effects of imbalanced neurotransmitter relationships, which are eventually altered by use of stimulating or inhibiting procedures; the latter thus add an external variable. It is, therefore, evident that a chronobiological evaluation can provide more direct and "natural" information about eventual hormonal modifications in depression.

## 2. PRL

### 2.1. Materials and Methods

Twenty subjects affected by primary depression were studied, 3 males and 17 females, of whom 11 were postmenopausal and 6 fertile. None were undergoing any pharmacological treatment. Before the test, all patients were accustomed to the ward schedule for 4 days, during which time they were subjected to pharmacological washout. They were allowed to sleep from 10:00 p.m. to 7:00 a.m., and breakfast, lunch, and dinner were served at 8:00 a.m., 12:30 p.m., and 6:00 p.m., respectively. Sleep was clinically monitored.

Subjects were bled by means of a catheter in the forearm vein, so that movements were not restricted and easy bleeding was possible even during sleep. Blood was drawn every 2 hr starting at 8:00 a.m. and sera were rapidly obtained and stored at $-20°C$ until assayed. Measurement of PRL was made by a double-antibody RIA method using Biodata (Milano) reagents. The intraassay and interassay variations were 4.5 and 10.5%, respectively. The clinical evaluation of the patients was performed using the Hamilton Rating Scale for Depression and the investigators administering this test were unaware of the design of the study.

Twenty-seven healthy volunteers were used as controls, 10 males and 17 females. The hormonal test in the 6 fertile depressive patients and in the 12 fertile controls was performed during the early follicular phase of the menstrual cycle.

In 15 depressive patients, the test was repeated after the resolution of the clinical picture which was obtained in 5 cases following treatment with chlorimipramine, in 4 with nomifensine, and in 6 with CF 253997, an ergolinic derivative with dopaminergic and serotoninergic activity (Polleri et al., 1979).

In 4 female patients, 2 in fertile age and 2 in menopause, PRL samples were taken every hour and a nocturnal EEG was recorded.

### 2.2. Results

The examination of basal PRL values in the depressed patients and in the controls (Table 1) shows a substantial uniformity of the data. In fact, even if differences exist between mean basal values of the hormone in the male depressed subjects compared to controls, the depressed patients showing higher values, the finding is obtained in an exiguous number of patients, and is not statistically significant.

Also not significant in the statistical analysis are the observed differences in basal PRL values between fertile and menopausal depressed women when compared with controls of the same age.

The clinical type, evaluated in women and differentiated with prevalent anxiety or inhibition, according to the subgroups of the Hamilton Rating Scale, does not seem to be associated with any difference in hormonal levels. However, while differentiating hormonal values in the mean of the diurnal (8:00 a.m. to 10:00 p.m.) and nocturnal (12:00 p.m. to 6:00 a.m.) levels, it was ob-

TABLE 1
Mean PRL Values in Depressive Patients and Controls[a]

|  | Males | Females | | | |
|---|---|---|---|---|---|
|  |  | Fertile women | Menopausal women | Anxious type | Inhibited type |
| Depressive patients | 9.37 ± 1.10 ($N = 3$)[b] | 16.2 ± 4.93 ($N = 6$) | 12.4 ± 4.48 ($N = 2$) | 13.4 ± 4.79 ($N = 2$) | 13.7 ± 5.56 ($N = 6$) |
| Controls | 7.75 ± 1.56 ($N = 10$) | 18.2 ± 5.9 ($N = 12$) | 16.9 ± 7.73 ($N = 5$) |  |  |

[a] PRL values (ng/ml) are the mean (± S.D.) of the 24-hr values.
[b] $N$ = number of cases.

FIGURE 1. Mean ± S.D. (vertical lines) of diurnal (d) and nocturnal (n) levels of PRL (ng/ml) in depressive women, subdivided in fertile (6 cases), menopausal (11 cases), anxious (11 cases), and inhibited (6 cases) patients. Twelve fertile and five menopausal normal women formed the control group. The statistical analysis (Student's $t$ test) was performed, for each studied datum, between the diurnal and nocturnal value and between the control vs. depressive value.
* = $p < 0.05$; *** = $p < 0.001$.

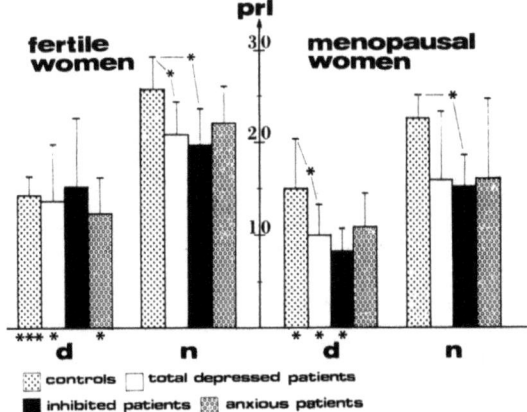

served in the group of depressive women, subdivided into fertile and menopausal (Fig. 1), that in fertile women there is no significant difference in diurnal PRL levels compared to controls and compared to all the parameters considered (controls, depressive patients, anxious or inhibited type). Conversely, in the same group, the mean nocturnal values show significant differences in the correlation between controls and total group of depressive patients and between controls and inhibited type.

While comparing the mean nocturnal and diurnal PRL levels in fertile women, it is observed that depressive women show less nocturnal elevation compared to controls (51.8% vs. 79.9%), and this difference is particularly marked in the inhibited type (3 cases) (22.55% vs. 44.11% in the anxious type—3 cases).

Observing the results obtained in the group of depressed women in menopause, it is noted that the mean diurnal PRL values are significantly lower compared to controls whereas no significant difference is noted in the hormonal values in the inhibited and anxious types (3 and 8 cases, respectively), even if the former group showed lower values in general. Similarly, even though the mean nocturnal PRL values in the menopausal group also appear overall reduced compared to controls, the only statistically significant finding was the difference between controls and inhibited forms of depression.

As opposed to the observations made in fertile women, the nocturnal elevation of PRL levels in menopausal patients is of much smaller magnitude and is comparable in both groups, depressed patients and controls (36.62% vs. 32.94%). In addition, the elevation in the anxious type is even smaller than that in the inhibited form (33.0% vs. 46.27%), in this group.

The results obtained from the 24-hr secretory pattern in 4 female cases, 2 in fertile age and 2 in menopause, are shown in Fig. 2 compared to 6 controls in menopause. It can be noted that while in the control group the PRL shows a progressive rise in the course of the day with maximum values in the night, in the depressed women the profile is different and shows a phase of hormonal secretion in the afternoon.

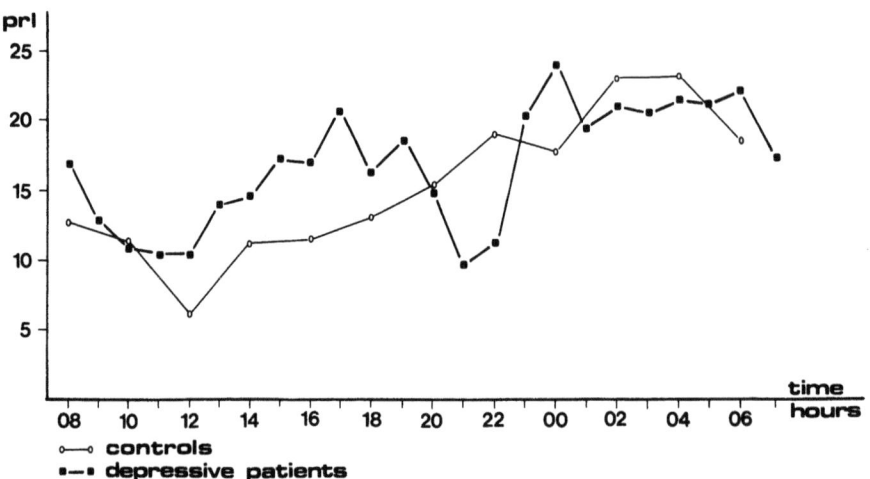

FIGURE 2. A 24-hr secretory pattern of PRL (ng/ml) in four depressive and in six normal women.

After treatment with chlorimipramine and resolution of the clinical picture in 5 cases, an overall increase in the total as well as mean diurnal and nocturnal PRL levels is observed (Fig. 3). The increase after chlorimipramine treatment were in the vicinity of 35% for all parameters (total PRL 39.15%, nocturnal PRL 37.75%, diurnal PRL 36.78%); the levels were almost identical to those observed in the normal menopausal women. The treatment in 4 cases with nomifensine does not seem to produce any modification in either nocturnal or

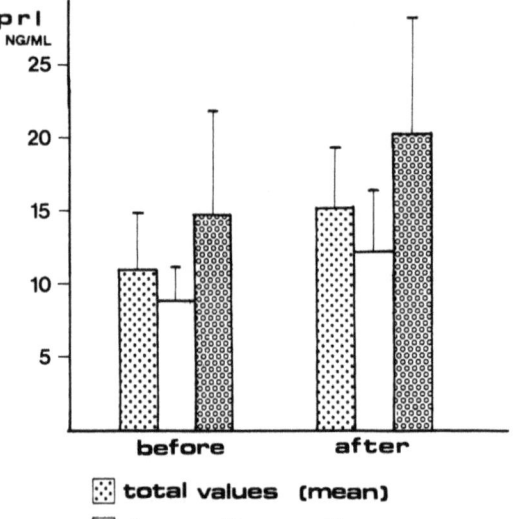

FIGURE 3. Mean diurnal, nocturnal, and total PRL values in five depressive patients before and after treatment with chlorimipramine.

diurnal PRL levels (mean ± S.D. PRL: basal, 10.07 ± 3.43; nocturnal, 12.44 ± 5.05; diurnal, 8.77 ± 3.21 vs. 9.87 ± 1.94, 11.68 ± 2.07 and 8.96 ± 2.2, respectively).

In the same way, CF 253997, in 6 cases studied, causes no appreciable modification of hormonal levels at the end of treatment compared to the pretreatment values.

## 3. GH

### 3.1. Materials and Methods

Nineteen subjects affected by primary depression, 4 males and 15 females (4 of fertile age), were studied. The criteria for clinical evaluation, the modalities of environmental adaptation, and the technique of sampling were analogous to those described above.

The blood samples were taken every 2 hr and processed for GH determination by RIA using Biodata (Milano) reagent. The intraassay and interassay variations were 5.5 and 11.2%, respectively, and the normal values at 8:00 a.m. were below 6 ng/ml.

In 5 cases, all females, the hormonal samples were taken every 30 min, and an EEG was recorded during the night.

In 10 subjects, all females, treatment with chlorimipramine (4 cases) and CF 253997 (6 cases) was performed, at the end of which circadian hormonal levels were reexamined in a similar manner.

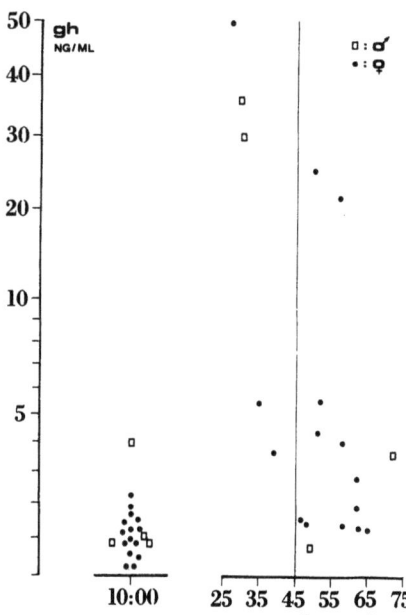

FIGURE 4. GH values at 10:00 a.m. and correlation between hormonal nocturnal peaks and age in depressive patients.

All the subjects being studied were free from metabolic pathology that could alter the secretion of hormone being examined.

## 3.2. Results

The basal GH levels, evaluated at 10:00 a.m., in our depressive patients are found in the normal range oscillating between 0.25 and 4 ng/ml (Fig. 4). However, contrary to normal GH secretory pattern, the depressive patients studied showed nocturnal secretory peaks even in subjects above 45 years of age.

From the examination of the characteristics present in patients with anomalous secretory peaks, no significant differences were noted either in the clinical type or in the response to the treatment.

The circadian profile of the hormone in the 5 cases in which samples were taken every 30 min confirmed the existence of secretory nocturnal episodes in at least 2 cases of depressed menopausal females (Fig. 5).

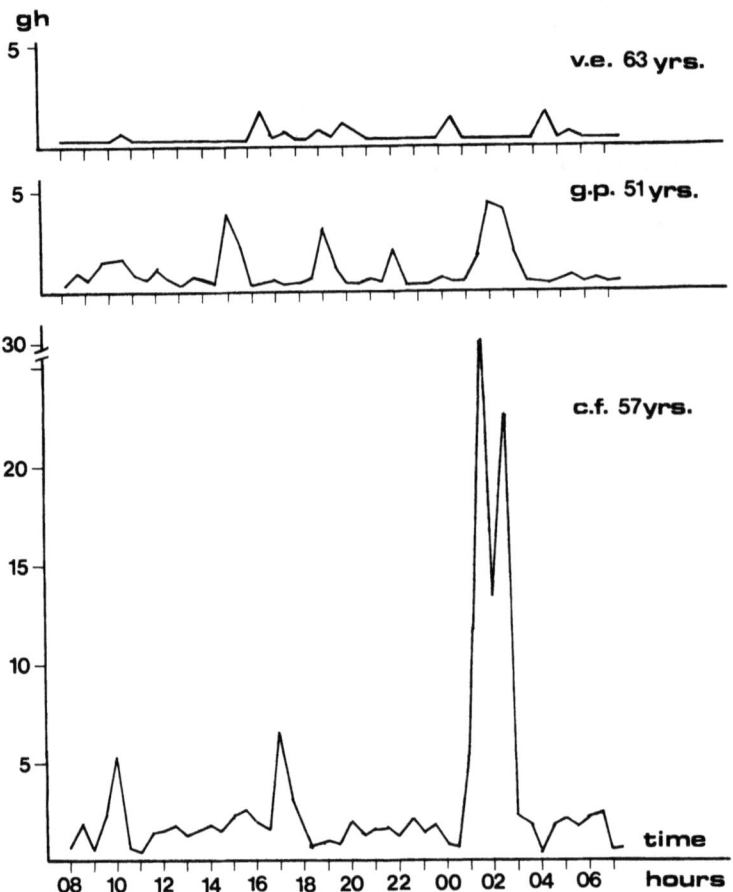

FIGURE 5. A 24-hr rhythm of GH (ng/ml) in three depressed menopausal women.

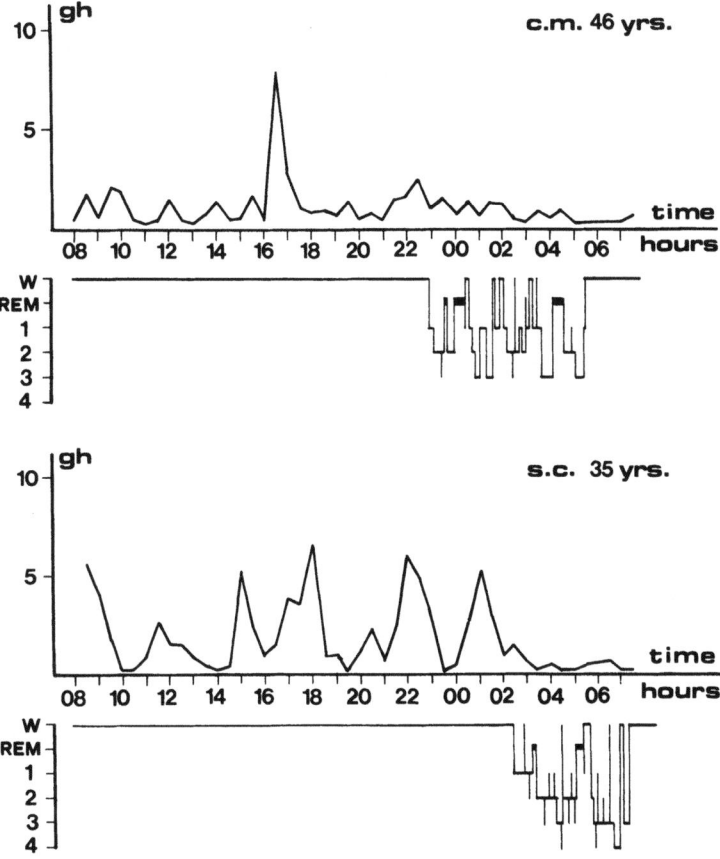

FIGURE 6. A 24-hr rhythm of GH (ng/ml) and sleep pattern in two depressive fertile women.

In the fertile depressive females (Fig. 6), comparing the GH secretory profile with the sleep patterns, a dissociation between stages of slow-wave sleep and secretory hormonal peaks is seen, with the presence of numerous secretory diurnal episodes only partially correlated to afternoon naps.

In the 4 cases studied, the treatment with chlorimipramine does not produce significant variation in the mean level of GH (1.09 ± 0.9: pretreatment; 0.96 ± 0.34: posttreatment). On the contrary, CF 253997 caused an increase in GH (43.2%) which was not statistically significant (1.62 ± 0.8 vs. 2.32 ± 1.31).

## 4. CORTISOL

### 4.1. Materials and Methods

Five patients affected by primary depression, 3 in menopause and 2 in fertile age, ranging in age from 35 to 65 years, were studied. The evaluation of the clinical status was carried out with the Hamilton Rating Scale and according to the procedure described above the blood samples were taken every hour for 24 hr for the determination of the cortisol level, using a Radim (Roma) [$^{125}$I]cortisol kit. The intraassay and interassay variations were 6.3 and 11.5%, respectively. The normal values were 5–20 μg/100 ml at 8:00 a.m.

In 2 patients of the group studied, the test was repeated with the same characteristics at the resolution of the clinical picture after treatment with chlorimipramine.

### 4.2. Results

The examination of the 24-hr cortisol pattern shows that while in 3 cases (Fig. 7) the normal circadian profile appears substantially preserved, characterized by a hormonal increase in the last hours of sleep and in the first hours

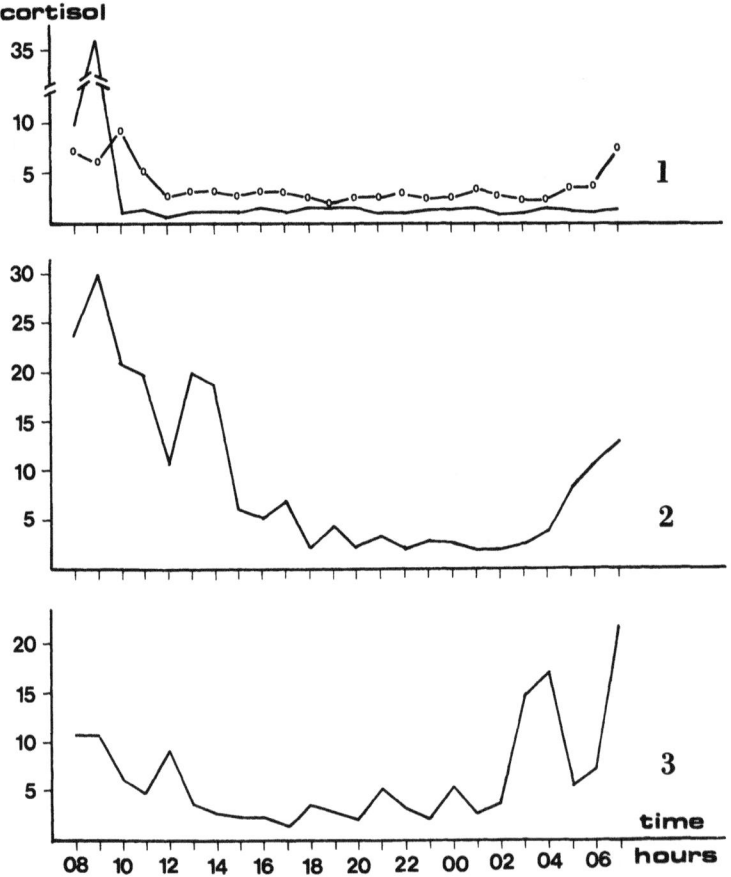

FIGURE 7. Secretory pattern of cortisol (μg/100 ml) in three depressive women. In case 1, the line with open circles indicates the hormonal profile after treatment with chlorimipramine.

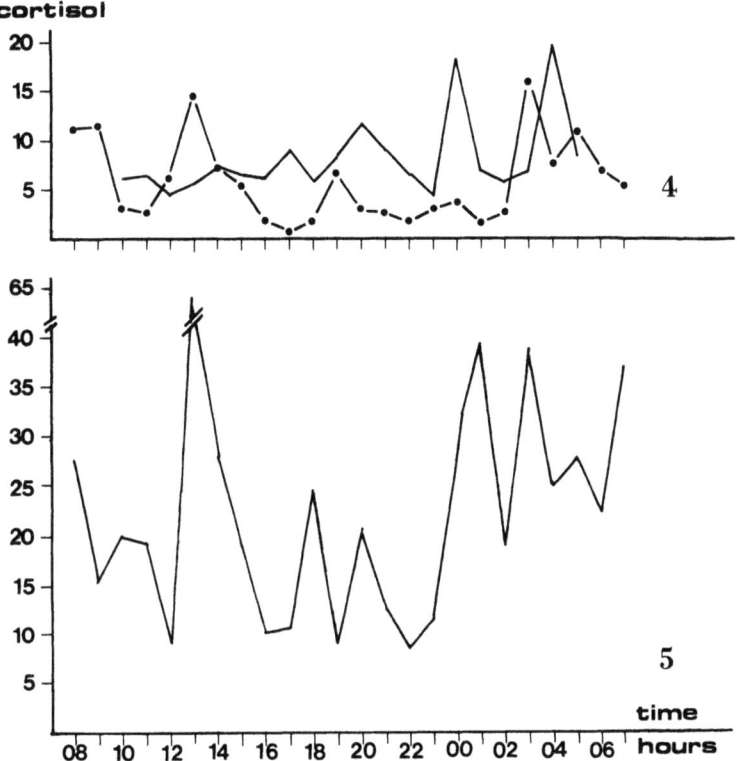

FIGURE 8. A 24-hr pattern of cortisol (μg/100 ml) in two depressive women. The line with the solid circles (case 4) represents the hormonal profile after treatment with chlorimipramine.

of awakening (Bessett *et al.*, 1979), in 2 cases (Fig. 8) the secretion appears to be characterized by numerous, elevated, and irregular secretory bursts. It is interesting to note, even if these data are inconclusive because of the small number of cases, that the subjects with an irregular cortisol profile had anxious depression.

In 2 patients who underwent a chlorimipramine treatment, a modification in the hormonal pattern with a quantitative reduction and tendency to regularization of the hormonal profile is observed (Figs. 7, 8).

## 5. LH

### 5.1. Materials and Methods

In 5 female subjects, 3 menopausal and 2 fertile age, affected by primary depression, the hormonal level was monitored for LH for 24 hr with samples every hour. The LH evaluation was performed by RIA using Biodata (Milano) reagents and the intraassay and interassay variations were 5.6 and 10.5%, re-

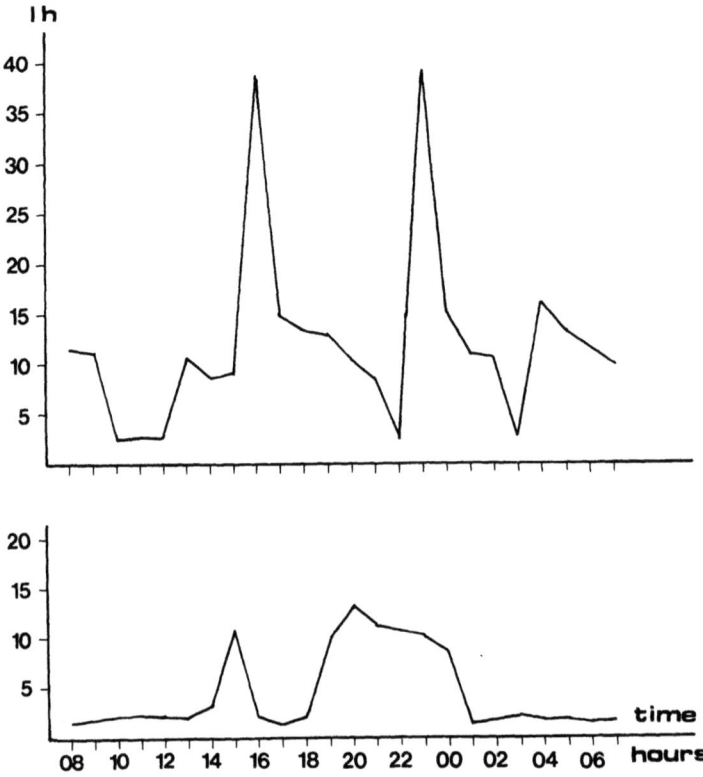

FIGURE 9. A 24-hr pattern of LH (μIU/ml) in two fertile depressive women.

spectively. The normal values in females in the early follicular phase ranged between 5 and 15 μIU/ml.

## 5.2. Results

The results obtained in the menopausal cases show hormonal levels at upper limits of the evaluation scale compatible with normal levels of women of the same age.

In 2 cases of fertile women (Fig. 9) a secretory pattern characterized by frequent irregular peaks prevalently found in the diurnal hours is observed.

## 6. DISCUSSION

### 6.1. PRL

The results obtained in our study led us to believe that in depression the basal PRL levels are to be found in a normal range.

The conclusion drawn by Carroll (1978) indicating the increase of basal hormonal levels in depression cannot be confirmed in our studies. Similarly, we did not observe any differences between the two clinical types, anxious and inhibited. Concerning this conclusion, which could place us in contradiction with the supposed DA reduction in depression (van Praag, 1977), we believe that it does not invalidate the DA hypothesis but limits itself to indicate the absence of a direct correlation between eventual DA deficit and PRL secretion.

Such an observation is confirmed by our previous work on Parkinson's disease (Ruggieri *et al.*, 1979; Agnoli *et al.*, 1980b), a condition in which a well-known DA nigrostriatal reduction does not induce any significant modification in the basal PRL levels, which led us to suppose a relative independence between the DA nigrostriatal and tuberoinfundibular systems.

In the same way in depression, at least considering the basal levels of PRL, there seems to be an absence of direct correlation between the DA tuberoinfundibular system and the DA mesolimbic and mesocortical systems.

A substantial uniformity in the hormonal data, compared to the normal, is to be found in the diurnal values of fertile women.

On the contrary, considering a dynamic parameter like the nocturnal elevation of the hormone, a minor rise of PRL in depressed women is especially evident in the inhibited type (22.5% vs. 44.1% of the anxious type).

This result not only confirms the utility and sensitivity of evaluating dynamic hormonal parameters in the depressive pathology, but also leads us to believe that a different hormonal pattern exists in the symptomatological groups studied according to age.

In fact, in depressed menopausal women, in addition to a reduction of the PRL levels both diurnal and nocturnal, an index of a reduction of the DA-inhibitory role of the estrogens (Labrie *et al.*, 1980), a minor nocturnal elevation of the hormone in the anxious depressed women is verified.

In keeping with this hypothesis of a prevalent DA deficit in inhibited depression, the observation of a slightly higher basal PRL level in fertile inhibited women with respect to anxious women can be explained as an expression of a reduction of DA-inhibitory control in the presence of a DA-inhibitory action of the circulating estrogens.

Such an interpretation cannot explain the findings in the depressed menopausal women in whom, contrary to the anxious depressed women, are seen much higher basal levels than the inhibited patients.

This leads us to believe that such mechanical transposition is too coarse, excluding other factors which, other than DA and 5HT, regulate the secretion of PRL, or that the estrogens, whose role is only partially known, play a much more complex role in the modulation of the hormone.

The examination of the 24-hr secretory pattern (Fig. 2) indicates a disorganization of the rhythm in depressed patients compared to controls; even if the nocturnal increase of the hormone is maintained, numerous PRL secretory episodes during the afternoon period are observed in depressed patients but not in the control group in whom normal episodes were reported in the literature (Sassin *et al.*, 1972).

Although the above-stated observations agree with others (Halbreich *et al.*, 1979), it should be emphasized that no matter how interesting, the conclusion is as yet speculative because of the few cases examined (4 in our study, 7 in Halbreich and co-workers' study) and of the lack of distribution in the depressed patients on the basis of sex and menopausal condition or none.

After treatment with chlorimipramine and resolution of the clinical picture, an overall increase in PRL levels is noted, a conclusion in agreement with that published by other authors (Asnis *et al.*, 1980b). This conclusion, associated with the observation that the treatment with nomifensine and CF 253997 does not induce appreciable modifications of the hormone, although it improves the clinical picture, leads us to believe that the increase of PRL is not secondary to the disappearance of the symptoms, but is a function of the treatment with chlorimipramine, a substance whose serotoninergic action is known (Francis *et al.*, 1976).

## 6.2. GH

The observation of the presence of nocturnal peaks in subjects above 45 years, a finding not observed in normals (Finkelstein *et al.*, 1972; Quabbe, 1977), seems to indicate the existence of specific alterations of GH in depression.

The disorganization of the secretory patterns is mostly evident (Fig. 6), indicating that the hormonal secretory peaks do not appear to correlate with the stages of slow-wave sleep, during which there is sometimes a decrease of hormonal values. This appears to be in agreement with the observations of other authors (Roffwarg *et al.*, 1970; Skilkrut *et al.*, 1975) who found an absence of secretory peaks during the stages of slow-wave sleep in depressed patients.

The findings of normal basal value and the presence of nocturnal secretory bursts, abnormal for the age, led us to believe that in depression, a dynamic alteration of GH secretory rhythms exists, which is a probable function of imbalance of the neurotransmitters modulating the hormonal secretion.

To validate such a hypothesis, ulterior confirmation is necessary, considering variables other than the transmitters which are able to influence the secretory rhythm of the hormone such as estrogens (Quabbe *et al.*, 1966) and glucocorticoids (Krieger and Glick, 1974).

## 6.3. Cortisol

The basal cortisol values do not differ significantly from the normal, confirming the substantial regularity of the basal level in depression, already discussed in the literature.

On the contrary, the results obtained in the cases of anxious depression appear interesting, and are in agreement with the profile reported in the literature (Sachar *et al.*, 1973b; Carroll *et al.*, 1976a) on the presence in the depressed patients of a qualitative and quantitative alteration of the secretory bursts. However, it should be emphasized that such a result, even considering the exiguousness of the sample studied, is to be observed only in the forms in which anxious symptomatology prevails. This fact leads us to hypothesize the existence of a correlation of the modifications obtained more with anxiety than with depression.

The pharmacological treatment seems to tend to reduce, regularizing, the cortisol secretory profile; such a result, rather than a specific pharmacological action, seems to relate more to the resolution of the symptomatology.

## 6.4. LH

The examination of the results obtained from the LH evaluation shows, in the cases studied, a substantial normality of the hormonal data in the menopausal group. In the two fertile depressed women studied, a prevalence of secretory bursts in diurnal hours is not in agreement with the normal secretory profile, which is characterized by secretory phases irregularly but uniformly distributed during the course of the day (Yen *et al.*, 1974). It should be emphasized, however, that it is difficult to interpret the LH secretory profile with samples taken every hour, considering the rapid alterations in the secretion of the hormone which tends to determine frequent and close peaks, not demonstrable with the method of hourly samples (Judd, 1979).

## 7. CONCLUSION

In conclusion, it appears evident that, in the field of neuroendocrinology, it is not possible to offer a comprehensive hypothesis that can explain precisely the role and the interactions between neurotransmitter systems and hormonal systems in psychopathology.

Notwithstanding this, all the data obtained, even if they cannot be systematized into an overall hypothesis, represent an unequivocal contribution, although partial, to the research of the biological parameters that can furnish an explanation or enable us to monitor the evolution of psychopathological forms.

However, it is necessary to develop methodological uniformity, which will fix valid limits and thus overcome the uncertainties often felt in such studies due to dissimilarities in the cases studied, and the criteria and methods used.

This appears even more important in the field of chrononeuroendocrinology, a sector which, if explored in our opinion, can establish the indices able to differentiate the forms of depression and the instruments to show the effect of pharmacological treatment on biological rhythms.

ACKNOWLEDGMENT. We acknowledge the collaboration and technical assistance furnished by Professor G. Nappi of the Neurological Clinic of the University of Pavia and the Chronobiological Research Center of L'Aquila.

## 8. REFERENCES

Agnoli, A., and Ruggieri, S., 1979, An appraisal of clinical studies on the dopaminergic hypothesis of depression, in: *Neuropsychopharmacology* (G. Saletu, ed.), pp. 185–196, Pergamon Press, Elmsford, N.Y.

Agnoli, A., Ruggieri, S., Cerone, G., Aloisi, P., Baldassarre, M., and Stramentinoli, G., 1978, The dopamine hypothesis of depression: Results of treatment with dopaminergic drugs, in: *Depressive Disorders* (S. Garattini, ed.), pp. 447–458, Schattauer Verlag, Stuttgart.

Agnoli, A., Baldassarre, M., Del Roscio, S., Palesse, N., and Martucci, N., 1980a, Neuroendocrinologia degli stati depressivi, *Riv. Psichiatria* 1:106.

Agnoli, A., Falaschi, P., Baldassarre, M., Rocco, A., Del Roscio, S., D'Urso, R., and Ruggieri, S., 1980b, Neuroendocrine approach in parkinsonian syndromes, in: *Progress in Psychoneuroendocrinology* (F. Brambilla, G. Racagni, and D. de Wied, eds.), pp. 315–324, Elsevier/North-Holland, Amsterdam.

Altman, N., Sachar, E. J., Gruen, P. H., Halpern, F. S., and Eto, S., 1975, Reduced plasma LH concentration in postmenopausal depressed women, *Psychosom. Med.* 37:274.

Arana, G., Boyd, A. E., III, Reichlin, S., and Lipsitt, D., 1977, Prolactin levels in mild depression, *Psychosom. Med.* 39:193.

Asnis, G. M., Nathan, R. S., Halbreich, U., Halpern, F. S., and Sachar, E. J., 1980a, TRH tests in depression, *Lancet* I:424.

Asnis, G. M., Nathan, R. S., Halbreich, U., Halpern, F. S., and Sachar, E. J., 1980b, Prolactin changes in major depressive disorders, *Am. J. Psychiatry* 137:1117.

Besset, A., Bonardet, A., Billiard, M., Descomps, B., De Paulet, A., and Passouant, P., 1979, Circadian patterns of growth hormone and cortisol secretions in narcoleptic patients, *Chronobiologia* 6:19.

Brambilla, F., Smeraldi, E., Sacchetti, E., Negri, F., Cocchi, D., and Müller, E. E., 1978, Deranged anterior pituitary responsiveness to hypothalamic hormones in depressed patients, *Arch. Gen. Psychiatry* 35:1231.

Brambilla, F., Smeraldi, E., Bellodi, L., Sacchetti, E., and Müller, E. E., 1980, Neuroendocrine correlates and monoaminergic hypothesis in primary affective disorders, in: *Progress in Psychoneuroendocrinology* (F. Brambilla, G. Racagni, and D. de Wied, eds.), pp. 235–245, Elsevier/North-Holland, Amsterdam.

Brown, W. A., and Shuey I., 1980, Response to dexamethasone and subtype of depression, *Arch. Gen. Psychiatry* 37:747.

Brown, W. A., Johnston, R., and Mayfield, D., 1979, The 24-hour dexamethasone suppression test in a clinical setting: Relationship to diagnosis, symptoms and response to treatment, *Am. J. Psychiatry* 136:543.

Calne, D. B., 1979, Neurotransmitters, neuromodulators and neurohormones, *Neurology* 29:1517.

Carroll, B. J., 1978, Neuroendocrine function in psychiatric disorders, in: *Psychopharmacology: A Generation of Progress* (M. A. Lipton, A. DiMascio, and K. F. Killam, eds.), pp. 487–497, Raven Press, New York.

Carroll, B. J., Curtis, G. C., and Mendels, J., 1976a, Neuroendocrine regulation in depression. I. Limbic system–adrenocortical dysfunction, *Arch. Gen. Psychiatry* 33:1039.

Carroll, B. J., Curtis, G. C., and Mendels, J., 1976b, Neuroendocrine regulation in depression. II. Discrimination of depressed from non-depressed patients, *Arch. Gen. Psychiatry* 33:1051.

Cavagnini, F., Raggi, U., Micossi, P., Di Landro, A., and Invitti, C., 1976, Effect of an antiserotoninergic drug, metergoline, on ACTH and cortisol response to insulin hypoglycemia and lysine-vasopressin in man, *J. Clin. Endocrinol. Metab.* 43:306.

Del Pozo, E., and Brownell, J., 1979, Prolactin. I. Mechanisms of control, peripheral actions and modification by drugs, *Horm. Res.* 10:143.

de Wied, D., and de Jong, W., 1974, Drug effects and hypothalamic–anterior pituitary function, *Annu. Rev. Pharmacol.* 14:389.

Ehrensing, R. H., Kastin, A. J., Schalch, D. S., Friesen, H. G., Vargas, J. R., and Schally, A.

V., 1974, Affective state and thyrotropin and prolactin responses after repeated injections of thyrotropin-releasing hormone in depressed patients, *Am. J. Psychiatry* **131**:714.

Endo, M., Endo, J., Nishikubo, M., Yamaguchi, T., and Hatotani, N., 1974, Endocrine studies in depression, in: *Psychoneuroendocrinology* (N. Hatotani and M. Tsu, eds.), pp. 22–31, Karger, Basel.

Finkelstein, J. W., Boyar, R. M., Roffward, H. P., Kream, J., and Hellman, L., 1972, Age-related change in twenty-four-hour spontaneous secretion of growth hormone, *J. Clin. Endocrinol. Metab.* **35**:665.

Francis, A. F., Williams, P., Williams, R., Link, J., Cole, E. N., and Hughes, D., 1976, The effect of clorimipramine on prolactin levels—Pilot studies, *Postgrad. Med. J.* **52**:87.

Gold, P. W., Goodwin, F. K., Wehr, T., Rebar, R., and Sack, R., 1976, Growth hormone and prolactin response to levodopa in affective illness, *Lancet* **II**:1308.

Gruen, P. H., Sachar, E. J., Altman, N., and Sassin, J., 1975, Growth hormone responses to hypoglycemia in postmenopausal depressed women, *Arch. Gen. Psychiatry* **32**:31.

Halbreich, U., Grunhaus, L., and Ben-David, M., 1979, Twenty-four-hour rhythm of prolactin in depressive patients, *Arch. Gen. Psychiatry* **36**:1183.

Hollister, L. E., Davis, K. L., and Berger, P. A., 1976, Pituitary response to thyrotropin-releasing hormone in depression, *Arch. Gen. Psychiatry* **33**:1393.

Judd, H. L., 1979, Biorhythms of gonadotropins and testicular hormone secretion, in: *Endocrine Rhythms* (D. T. Krieger, ed.), pp. 299–324, Raven Press, New York.

Kirkegaard, C., and Bjørum, N., 1980, TSH responses to TRH in endogenous depression, *Lancet* **I**:152.

Kirkegaard, C., Bjørum, N., Cohn, D., and Lauridsen, U. B., 1978, Thyrotrophin-releasing hormone (TRH) stimulation test in manic-depressive illness, *Arch. Gen. Psychiatry* **35**:1017.

Knobil, E., and Plant, T. M., 1978, Neuroendocrine control of gonadotropin secretion in female rhesus monkey, in: *Frontiers in Neuroendocrinology*, Vol. 5 (W. F. Ganong and L. Martini, eds.), pp. 249–264, Raven Press, New York.

Krieger, D. T., and Glick, S. M., 1974, Sleep EEG stages and plasma growth hormone concentration in states of endogenous and exogenous hypercortisolemia or ACTH evaluation, *J. Clin. Endocrinol. Metab.* **39**:986.

Labrie, F., Ferland, L., Di Paolo, T., and Veilleux, R., 1980, Modulation of prolactin secretion by sex steroids and thyroid hormones, in: *Central and Peripheral Regulation of Prolactin Function* (R. M. MacLeod, and V. Scapagnini, eds.), pp. 97–113, Raven Press, New York.

Lamberts, S. W., and Birkenhäger, J. C., 1976, Effect of bromocriptine in pituitary-dependent Cushing's syndrome, *J. Endocrinol.* **70**:315.

Langer, G., Schönbeck, G., Koinig, G., Lesch, O., Schüssler, M., and Waldhäusl, W., 1980, Antidepressant drugs and the hypothalamic–pituitary–thyroid axis, *Lancet* **I**:100.

Linnoila, M., Lamberg, B. A., Rosberg, G., Karonen, S. L., and Welin, M. G., 1979, Thyroid hormones and TSH, prolactin and LH responses to repeated TRH and LRH injections in depressed patients, *Acta Psychiatr. Scand.* **59**:536.

Maeda, K., Kato, Y., Ohgo, S., Chihara, K., Yoshimoto, Y., Yamaguchi, N., Kuromaru, S., and Imura, H., 1975, Growth hormone and prolactin release after injection of thyrotropin-releasing hormone in patients with depression, *J. Clin. Endocrinol. Metab.* **40**:501.

Martin, J. B., 1973, Neuronal regulation of growth hormone secretion, *N. Engl. J. Med.* **288**:1384.

Martucci, N., Palesse, N., Baldassarre, M., Agnoli, A., and Polleri, A., 1980, Studio cronobiologico della prolattinemia nella depressione ciclica monopolare, *Riv. Psichiatria* **1**:208.

Mendlewicz, J., Linkowski, P., and Brauman, H., 1977, Growth hormone and prolactin response to levodopa in affective illness, *Lancet* **I**:652.

Müller, E. E., 1973, Nervous control of growth hormone secretion, *Neuroendocrinology* **11**:338.

Noel, G. L., Suh, H. K., Stone, J. G., and Frantz, A. G., 1972, Human prolactin and growth hormone release during surgery and other conditions of stress, *J. Clin. Endocrinol. Metab.* **35**:840.

Ojeda, S. R., and McCann, S. M., 1974, Evidence for participation of a catecholaminergic mechanism in the post-castration rise in plasma gonadotropins, *Neuroendocrinology* **12**:295.

Polleri, A., Murialdo, G., Masturzo, P., Martucci, N., Palesse, N., Agnoli, A., and Gasparetto, B., 1979, Spontaneous prolactin secretory pattern in depressive patients, in: *Neuroendocrine Correlates in Neurology and Psychiatry* (E. E. Müller and A. Agnoli, eds.), pp. 255–261, Elsevier/North-Holland, Amsterdam.
Polleri, A., Masturzo, P., Murialdo, G., Baldassarre, M., Martucci, N., Nappi, G., Savoldi, F., Muratorio, A., Murri, L., and Gasparetto, B., 1980, Chronobiology of prolactin secretion: A marker in physiology and pathology, in: *Central and Peripheral Regulation of Prolactin* (R. M. MacLeod and U. Scapagnini, eds.), pp. 207–220, Raven Press, New York.
Prange, A. J., Jr., Lipton, M. A., Nemeroff, C. B., and Wilson, J. C., 1977, The role of hormones in depression, *Life Sci.* **20**:1305.
Quabbe, H. J., 1977, Chronobiology of growth hormones secretion, *Chronobiologia* **4**:217.
Quabbe, H. J., Schilling, E., and Helge, H., 1966, Pattern of growth hormone secretion during a 24-hour fast in normal adults, *J. Clin. Endocrinol.* **26**:1173.
Roffwarg, H. P., Sachar, E., Finkelstein, J., Curti, J., Ellman, S., Kream, J., Fishman, M., Gallagher, F. T., and Hellman, L., 1970, Sleep stage pattern in depression in relation to nocturnal plasma cortisol and human growth hormone, *Psychophysiology* **7**:323.
Roth, J., Glick, S. M., Yalow, R. S., and Berson, S. A., 1963, Secretion of human growth hormone: Physiologic and experimental modification, *Metabolism* **12**:577.
Ruggieri, S., Falaschi, P., Baldassarre, M., D'Urso, R., Frajese, G., and Agnoli, A., 1979, Neuroendocrine response to active drugs in Parkinson's disease, in: *Neuroendocrine Correlates in Neurology and Psychiatry* (E. E. Müller and A. Agnoli, eds.), pp. 127–137, Elsevier/North-Holland, Amsterdam.
Sachar, E. J., 1976, Neuroendocrine abnormalities in depressive illness, in: *Topics in Psychoendocrinology* (E. J. Sachar, ed.), pp. 135–156, Grune & Stratton, New York.
Sachar, E. J., Frantz, A. G., Altman, N., and Sassin, J., 1973a, Growth hormone and prolactin in unipolar and bipolar depressed patients: Responses to hypoglycemia and L-dopa, *Am. J. Psychiatry* **130**:1362.
Sachar, E. J., Hellmann, L., Roffwarg, H. P., Halpern, F. S., Fukushima, D. K., and Gallagher, T. F., 1973b, Disrupted 24-hour patterns of cortisol secretion in psychotic depression, *Arch. Gen. Psychiatry* **28**:19.
Sassin, J. F., Franz, A. G., Weitzman, E. D., and Kapen, S., 1972, Human prolactin: 24 hour pattern with increased release during sleep, *Science* **177**:1205.
Scanlon, M. F., Pourmand, M., McGregor, A. M., Rodriguez-Arnao, M. D., Hall, K., Gomez-Pan, A., and Hall, R., 1979, Some current aspects of clinical and experimental neuroendocrinology with particular reference to growth hormone, thyrotropin and prolactin, *J. Endocrinol. Invest.* **2**:307.
Schally, A. V., and Arimura, A., 1977, Physiology and nature of hypothalamic regulatory hormones, in: *Clinical Neuroendocrinology* (L. Martini and G. M. Besser, eds.), pp. 1–42, Academic Press, New York.
Schlesser, M. A., Winokur, G., and Sherman, B. M., 1979, Genetic subtypes of unipolar primary depressive illness distinguished by hypothalamic–pituitary–adrenal axis activity, *Lancet* **I**:739.
Skilkrut, R., Chandra, O., Osswald, M., Rüther, E., Baarfüsser, B., and Matussek, N., 1975, Growth hormone release during sleep and with thermal stimulation in depressed patients, *Neuropsychobiology* **1**:70.
Takahashi, S., Kondo, H., and Yoshimura, M., 1974, Growth hormone responses to adminiatration of L-5-hydroxytryptophan (L-5-HTP) in manic depressive psychosis, in: *Psychoneuroendocrinology* (N. Hatotani and M. Tsu, eds.), pp. 32–38, Karger, Basel.
Träskman, L., Tybring, G., Asberg, M., Bertilsson, L., Lantto, O., and Schalling, D., 1980, Cortisol in CSF of depressed and suicidal patients, *Arch. Gen. Psychiatry* **37**:761.
van Praag, H. M., 1977, Significance of biochemical parameters in the diagnosis, treatment and prevention of depressive disorders, *Biol. Psychiatry* **12**:101.
Whybrow, P. C., Coppen, A., Prange, A. J., Jr., Noguera, R., and Bailey, J. E., 1972, Thyroid function and the response to liothyronine in depression, *Arch. Gen. Psychiatry* **26**:242.

Woolt, P. D., and Lee, L., 1977, Effect of the serotonin precursor trytophan on pituitary hormone secretion, *J. Clin. Endocrinol. Metab.* **45:**123.

Yen, S. S. C., Tsai, C. C., Vandenberg, G., and Parker, D. C., 1974, Ultradian fluctuations of gonadotropins, in: *Biorhythms and Human Reproduction* (M. Ferin, R. M. Richart, and R. Van de Wiele, eds.), pp. 204–216, Wiley, New York.

CHAPTER 25

# Alterations in Circadian Secretion of Pituitary and Pineal Hormones in Affective Disorders

JULIEN MENDLEWICZ

## 1. INTRODUCTION

Biological rhythms are an integral part of life. The day and night 24-hr period is present in almost all biological systems, remaining constant in various environments. This observation has led to the concept of endogenous biological clocks. The exact synchronization of the 24-hr period also involves external synchronizers ("zeitgebers"). Besides circadian rhythms, living organisms also present shorter *infradian* rhythms such as the REM period of sleep (paradoxical sleep) or longer *ultradian* rhythms such as the menstrual cycle in female.

The exact nature of biological clocks remains unclear, but the hypothalamic–pituitary axis seems to play a role in the maintenance of circadian rhythms both in animals and in man (Richter, 1958). It is by now clear that the pituitary gland constitutes a major link in the neuroendocrine axis in man. Central neurotransmitters regulate the secretion of the hypothalamic neurohormones which in turn may affect brain monoamine metabolism. These neuroendocrine parameters are subjected to circadian variations both in animals and in man and they may be implicated in the pathogenesis of periodic psychoses. A new approach called *chronophysiology* permits the study of temporal changes in clinical and physiological symptoms in diseases such as cancer, endocrine disturbances, cardiovascular and psychiatric conditions. Furthermore, treatment response may also show important temporal variations probably due to circadian variations in drug metabolism and receptor sensitivity (chronopharmacology).

Some affective disorders are characterized by cycles of depressive and manic episodes and by periodic as well as diurnal disturbances in mood and

---

JULIEN MENDLEWICZ • Department of Psychiatry, University Clinics of Brussels, Erasme Hospital, Free University of Brussels, 1070 Brussels, Belgium.

biological functions such as sleep, energy, appetite, and sex (evening improvements and morning worsening). These diurnal changes may be related to desynchronization of day–night variation of the mood and drive system. According to this concept, desynchronization may be an important pathophysiological aspect of depression with some biological rhythms following its nonendogenous, or free-running, circadian period, deviated from the normal 24-hr period (in general phase advanced). Furthermore, experimental studies have shown day and night shift (i.e., desynchronization phenomenon) to modify energy level and concentration abilities; while sleep deprivation has been reported to temporarily alleviate depression. It may be paradoxical to improve depression with sleep deprivation since this condition is usually associated with a reduction in sleep. The mechanism of sleep deprivation is unknown but it may be understood in light of the desynchronization theory. During sleep deprivation, the previously desynchronized rhythm may come into play with the free-running cycle, and when this resynchronization takes place, the depression may be relieved. As soon as the patient resumes his sleep pattern, desynchronization occurs again, and the depression may return (chronotherapy).

The study of circadian and ultradian rhythms of biological functions is thus of great importance in psychopathology, in particular, manic-depression. According to this desynchronization hypothesis, manic-depression may be conceptualized as a "biological clock" disorder.

Over several centuries, observations of remarkably predictable recurrences of periodic psychoses have stimulated interest and raised the hope that studies of these patients might help us understand important aspects of the pathophysiology of affective psychoses (Gjessing, 1960; Jenner, 1968). Despite the fact that precise periodicities of psychosis are rare, there is, nevertheless, a marked statistical tendency to approach a specific timetable in a large number of patients. Such phenomena are highly relevant to the structural changes in mood observed in affectively ill patients.

A minority of manic-depressive patients, called "rapid cyclers," show an unusually rapid shift from depression to mania and vice versa. These patients switch rapidly from the zenith of mania to the nadir of depression with little free normal intervals. This switch to mania offers a unique opportunity to monitor some biological variables in relation to sudden mood changes. Phase shifts in biochemical circadian rhythms in manic-depressive patients have also been suggested for steroids, electrolyte rhythms, neurochemical metabolites, and body temperature. While these studies based on brief observations suggest there are circadian disturbances in manic-depressives, longitudinal observations are essentially to demonstrate consistent circadian patterns in affectively ill patients.

The existence of circadian variations in the release of several pituitary hormones, such as ACTH, TSH, prolactin, and adrenal hormones such as cortisol, in man has been well documented. Moreover, for several of these hormones, such as ACTH, TSH, cortisol, and melatonin, it has been shown that higher frequency nonperiodic oscillations, corresponding to secretory episodes, were superimposed to the basal circadian rhythm. As a consequence of this hormonal variability, consistent studies of hormonal secretion imply repeated blood sampling over long periods of time (i.e., 24-hr periods). This approach has been used by Sachar et al. (1970) who found that the normal 24-hr pattern of cortisol secretion was disrupted in some depressed patients. There was an increase in the number of secretory episodes, with active secretion during the normal nonsecretory period, and with elevation of all peaks of plasma cortisol throughout the 24 hr. The pattern almost returned to normal after the patient recovered. Previous reports have shown that the effects of dexamethasone and of insulin-induced hypoglycemia on cortisol secretion were reduced in depressed patients (Carroll, 1969), an observation relating to the same endocrine dysfunction. They postulated the existence of an "abnormal drive from limbic areas," i.e., there was a central limbic dysfunction in depression.

Other studies suggest that these disturbances cannot be explained entirely as a simple stress response, since these abnormalities are present in nonanxious patients, during sleep, and are not corrected after the administration of large doses of sedative medications (Stokes, 1972).

In light of the biogenic amine hypothesis of affective illness, recent studies have shown alterations in circadian and seasonal rhythms of various neurotransmitter substances in the plasma of depressed patients (Riederer et al., 1974). The above arguments led us to examine circadian variations of pituitary-pineal hormone levels in manic-depression.

## 2. METHOD

We have investigated pituitary activity in patients suffering from primary affective disorders. In this chapter, we are now reporting preliminary results on the 24-hr secretion of prolactin, TSH, and melatonin during the depressed phase of manic-depressive illness in subjects diagnosed as bipolar manic-depressives, i.e., patients experiencing both manic and depressive episodes, and unipolar depressives, suffering from depression alone.

Serum TSH ($\mu$U/ml) and prolactin ($\mu$U/ml) were measured by a double-antibody radioimmunoassay (Golstein and Vanhaelst, 1973; Aubert et al., 1974; Badawi et al., 1974). Plasma concentration of melatonin was determined by radioimmunoassay (Weinberg et al., 1979). Estimated amplitudes and phases (day and night) of TSH, prolactin, and melatonin patterns observed in depressed patients were compared to the estimations obtained for control patterns recorded in healthy volunteers. All patients studied were free of medications for at least 1 week prior to the investigation and were hospitalized in an inpatient unit for a primary depressive episode severe enough to warrant hospitalization. Patients were diagnosed as bipolar and unipolar depressive (Mendlewicz and Fleiss, 1974). Severity of the depressive illness was assessed by the Hamilton Rating Scale. Blood samples were drawn for 24 hr through a plastic indwelling catheter. Blood samples were collected every hour during the daytime and every 30 min during the night. All patients were confined to bed, had normal breakfast, lunch, supper, and their nocturnal sleep was not interrupted. Daytime sleep was prevented and sleep times were recorded by trained nurses. All patients and controls were investigated throughout the calendar year.

## 3. RESULTS

The prolactin patterns of all depressive patients as a group ($N = 18$) showed no significant difference with patterns observed in healthy subjects ($N = 6$ males). Nevertheless, the mean prolactin level over 24 hr was significantly lower in bipolar patients as compared to unipolars. This was mainly due to the absence of sleep-related elevation of prolactin in 6 out of 8 bipolar patients (75%) in whom maximum prolactin secretion occurred during wakefulness (phase advanced), whereas maximum prolactin concentrations were observed during sleep in all unipolar patients as in normal controls.

The diurnal patterns of TSH levels studied in unmedicated depressed female patients ($N = 13$) differed greatly from those exhibited by normal subjects previously investigated ($N = 6$ males, 10 females). The mean 24-hr TSH level was lower in all depressed patients when compared to normals. In these patients, the rhythm appears to be desynchronized, the early morning peak being absent. In some cases, a maximum secretion occurred before midnight. Higher frequency variation of plasma TSH could also be observed between unipolar and bipolar patients. Furthermore, thyroid function was found to be normal in all patients and controls.

One interesting aspect of the desynchronization hypothesis could be related to the increased incidence of affective (manic and depressive) relapses in spring and autumn when daylight is either longer or shorter. It is thus possible that in genetically predisposed subjects, some circadian physiological clock parameters may be desynchronized during these periods. In those susceptible individuals, circadian desynchronization may be triggered by ultradian seasonal variations and then continue into a free-running cycle. Melatonin is a pineal hormone particularly sensitive to day–night changes in exposure to light. It is also under central noradrenergic control. Melatonin is thus of great interest in studying cyclic manic-depressive patients. Preliminary data on 24-hr plasma melatonin concentrations are available for 4 female depressed patients before and after treatment and 5 normal controls (males). Secretory episodes are observed during wakefulness in all subjects (phase advanced), but they are of higher magnitude in depressed patients and seem to appear at abnormal times (late afternoon or evening before onset of sleep). Furthermore, the circadian rhythm of melatonin is less apparent in depressed patients. The night/day ratio for melatonin is 1.38 in depressed patients and 2.8 in normals. Nocturnal rise of melatonin is almost absent in 3 of 4 depressed patients who show an elevation of the pituitary hormone during daytime. Finally, no significant changes in 24-hr melatonin patterns can be seen in depressed patients after antidepressant treatment and following remission. It is thus clear that the altered circadian secretion of melatonin in depression is not state dependent.

We have previously demonstrated striking alterations of 24-hr plasma dopamine-β-hydroxylase activity in the same depressed patients, with more episodic variations during daytime and the absence of dopamine-β-hydroxylase circadian rhythms in some depressed patients. These observations may provide an objective and quantitative biological indicator of the alteration of circadian peripheral dopaminergic activity in affective illness (Van Cauter and Mendlewicz, 1978). There are several parameters which could not be rigorously controlled in our studies. Among those are sex, age, and ovarian status. Furthermore, the absence of sleep EEG records does not permit disregarding the possible influence of sleep disturbances on circadian hormonal rhythms, although the abnormalities described in our depressed patients do not seem to disappear after clinical remission (including normalization of sleep) following antidepressant treatment.

Alterations in circadian rhythms for plasma pituitary and pineal hormones such as prolactin, TSH, and melatonin are present in some depressed patients and it is tempting to hypothesize that these circadian disturbances may be related to desynchronization phenomenon in manic-depressive illness. This desynchronization may induce primary modifications of circadian rhythms of central catecholaminergic and serotoninergic activities in affective illness, as suggested by the typical alterations in 24-hr plasma dopamine-β-hydroxylase activity which we have previously described in manic-depressive patients. Moreover, it is possible that cholinergic–adrenergic interactions are also of significance. Finally, there is no reason to assume that groups of patients labeled as "depressive" are necessarily similar on genetic and biochemical grounds. As a matter of fact, we have previously shown that it is possible to differentiate between several genetic subroups in depressive illness (Mendlewicz, 1974). It is thus conceivable that some form of depressive illness may be associated with an abnormality in serotonin metabolism while catecholaminergic deficiencies may be present in other forms of depression, and that there is a complex imbalance between several neurotransmitters. The rapid changes in mood and the abruptness of the desynchronization observed in manic-depressive patients make it unlikely that these phenomena are related to primary oscillations in central neurotransmitter synthesis or turnover rate. More fruitful hypotheses may be formulated in terms of infradian, ultradian, or circadian variations in specific brain receptor sensitivity or more complex behavioral modulation through endogenous neuropeptide substances. This may be more consistent with the concept of internal desynchronization of biological rhythms in some manic-depressive patients. As circadian clock frequency may be transmitted on an X-chromosome gene as has been shown in animal studies (Konopka and Benzer, 1971) (*Drosophila melanogaster*) and may increase with age, a circadian etiology is consistent with the genetics and age distribution of manic-depressive illness as indicated by Kripke *et al.* (1978). The brain distribution of other releasing factors and peptides has not yet been reported and when the brain effects of hypothalamic hormones and peptides will be further elucidated, the clinical conditions in which their actions are investigated may be better understood.

Nevertheless, the chronobiological studies described above combining enzymatic and neuroendocrine evaluation over long periods of time in the study of abnormal behavior are most promising and may enable us to better understand cyclical alterations of hypothalamic and pituitary functions in man, although it is still premature to draw firm conclusions from the neuroendocrine abnormalities as to the specific nature of the underlying neurotransmitter or neuropeptide disturbances in psychopathology.

## 4. SUMMARY

Affective disorders have been characterized by circadian alterations in physiological functions indicating that these illnesses could be considered as biological clock disorders. The involvement of the hypothalamic–pituitary axis has also been implicated in the pathophysiology of depressive disorders. Recent advances in radioimmunological methods have made it possible to measure and monitor throughout 24 hr several hormones which may be of importance in mood disorders. This chapter reviews some of the work on biological rhythms in relation to neuroendocrine variables, and summarizes recent work showing striking alterations in the 24-hr secretory patterns of such hormones as prolactin, TSH, melatonin in depressed patients diagnosed as unipolar and bipolar. These hormonal circadian data are consistent with a model of biological rhythm disturbances in the major affective disorders.

## 5. REFERENCES

Aubert, M. L., Becker, R. L., Saxena, B. B., and Raiti, S., 1974, Report of the National Pituitary Agency: Collaborative study of the radioimmunoassay of human prolactin, *J. Clin. Endocrinol. Metab.* **38**:1115.

Badawi, M., Bila, S., L'Hermite, M., Perez-Lopez, R. F., and Robyn, C., 1974, Comparative evaluation of radioimmunoassay for human prolactin using anti-ovine and anti-human prolactin sera, in: *Radioimmunoassays and Related Procedures in Medicine* (C. Robyn, ed.), Vol. 1, pp. 411–422, International Atomic Energy Agency Press, Vienna.

Carroll, B. J., 1969, Hypothalamic–pituitary function in depressive illness: Insensitivity to hypoglycemia, *Br. Med. J.* **3**:27.

Gjessing, R., 1960, Beiträg zur Kenntniss der Pathophysiologie des Katatonen Stupors. *Arch. Psychiatr. Nervenkr.* **200**:350.

Golstein, J., and Vanhaelst, L., 1973, Influence of thyrotrophin-free serum on the radioimmunoassay of human thyrotropin, *Clin. Chim. Acta.* **49**:141.

Jenner, F. A., 1968, Periodic psychoses in the light of biological rhythm research, *Int. Rev. Neurobiol.* **11**:129.

Konopka, R. J., and Benzer, S., 1971, Clock mutants of *Drosophila melanogaster*, *Proc. Natl. Acad. Sci. USA* **68**:2112.

Kripke, D. F., Mullaney, D. J., Atkinson, M., and Wolf, S., 1978, Circadian rhythm disorders in manic-depressives, *Biol. Psychiatry* **13**:335.

Mendlewicz, J., 1974, Le concept d'hétérogénéité dans la psychose maniaco-dépressive, *L'Inf. Psychiatr.* **2**:1044.

Mendlewicz, J., and Fleiss, J. L., 1974, Linkage studies with X-chromosome markers in bipolar (manic-depressive) and unipolar (depressive) illnesses, *Biol. Psychiatry* **9**:261.

Richter, C. P., 1958, Abnormal but regular cycles in behavior and metabolism in rats and catatonic-schizophrenics, in: *Psychoneuroendocrinology* (M. Reiss, ed.), pp. 168–181, Grune & Stratton, New York.

Riederer, P., Birkmayer, W., Neumayer, E., Ambrozi, L., and Lineauer, W., 1974, The daily rhythm of HVA, VMA (VA), and 5HIAA in depressive syndrome, *J. Neural Transm.* **35**:23.

Sachar, E. J., Hellman, L., and Fukushima, D. K., 1970, Cortisol production in depressive illness, *Arch. Gen. Psychiatry* **23**:289.

Stokes, P. E., 1972, Studies on the control of adrenocortical function in depression, in: *Recent Advances in the Psychobiology of Depressive Illnesses* (T. A. williams, M. M. Katz, and J. A. Shield, eds.), pp. 199–220, U. S. Department of Health, Education and Welfare, Washington, D.C.

Van Cauter, E., and Mendlewicz, J., 1978, 24-hour dopamine–beta-hydroxylase pattern: A possible biological index of manic-depression, *Life Sci.* **22**:147.

Weinberg, U., D'Eletto, R. D., Weitzman, E. D., Erlich, S., and Hollander, C. S., 1979, Circulating melatonin in man: Episodic secretion throughout the light–dark cycle, *J. Clin. Endocrinol. Metab.* **48**:114.

CHAPTER 26

# A Study of Circadian Variation of Platelet Serotonin Uptake and Serum Cortisol in Patients with Major Depression

JEFFREY L. RAUSCH, NANDKUMAR S. SHAH, EARL A. BURCH, ALEXANDER G. DONALD, and ETHAN V. MUNSON

1. INTRODUCTION

The blood platelet shares many characteristics in common with synaptosomes, and has been suggested as a model for central serotoninergic neuronal functioning (Murphy et al., 1970; Stahl, 1977). Serotonin (5HT) has been indirectly implicated in the pathophysiology of depression (van Praag and Korf, 1971) and, as such, the uptake of 5HT into the platelet assumes potential importance as a possible indicator of central 5HT uptake activity, as both platelets and synaptosomes share similar Michaelis constants ($K_m$) for 5HT uptake to the nearest order of magnitude (Stahl and Meltzer, 1978; Smith et al., 1978).

Previous studies have reported a decreased platelet 5HT uptake in depressed patients (Hallstrom et al., 1976; Coppen et al., 1978; Tuomisto et al., 1979). Generally, these studies have investigated platelet 5HT uptake kinetics, by incubating platelets with varied concentrations of 5HT for a fixed time period, and using calculated kinetic constants with which to compare patients and controls.

---

JEFFREY L. RAUSCH • Ensor Foundation Research Laboratory, William S. Hall Psychiatric Institute, Columbia, South Carolina 29202, and Department of Psychiatry, University of California at San Diego, La Jolla, California 92093.    NANDKUMAR S. SHAH • Ensor Research Foundation    EARL A. BURCH, and ALEXANDER G. DONALD • William S. Hall Psychiatric Institute, and Department of Neuropsychiatry and Behavioral Sciences, University of South Carolina School of Medicine, Columbia, South Carolina 29208.    ETHAN V. MUNSON • Department of Psychiatry, University of California at San Diego, La Jolla, California 92093.

For our study, platelet 5HT uptake was examined from a time course obtained from each subject, i.e., by incubating platelets for varied time intervals at a fixed concentration ($10^{-6}$ M) of 5HT. In this way, accumulation of 5HT into the platelets could be plotted for each subject over a 10-min incubation period.

Another dimension of our study was to investigate how the time of day would affect the platelet 5HT uptake in normals and depressed patients. Little data were available from previous studies. From twice-daily blood sampling, Oxenkrug et al. (1978) suggested a circadian rhythm for platelet 5HT uptake, reporting disturbed rhythms in five psychotic depressives with diurnal mood variations. Wirz-Justice and Pühringer (1978) reported altered diurnal rhythms of platelet 5HT content in depressed patients, measuring platelet 5HT content, but not uptake, at four times during the day. In our study, platelet 5HT uptake was determined at four times of day, in conjunction with serum cortisol.

Many studies have reported serum cortisol elevations in depressed patients, these elevations being often resistant to suppression with dexamethasone (Carroll et al., 1976). Disrupted patterns of circadian cortisol secretion have been reported in psychotic depressives (Sachar et al., 1973), and an increased incidence of depression has been reported in patients with Cushing's syndrome (Kelly et al., 1980). From a review of the literature, Krieger (1978) concluded that regulation of circadian cortisol secretion involved serotoninergic mechanisms. The present study examined serum cortisol in conjunction with platelet 5HT uptake in order to assess for temporal correlations between the circadian patterns of these two parameters.

Finally, in our study we had an opportunity to repeat measures of platelet 5HT uptake and serum cortisol in a subgroup of depressed patients who completed 4 weeks' treatment with a tricyclic antidepressant, either imipramine or clomipramine.

## 2. METHODS AND MATERIALS

Seven patients, one male and six females, age 21–58, newly hospitalized in the Institute's wards between January and May 1980, were selected for the study. All were euthyroid, and free of other medical illnesses. Each patient had prominent vegetative signs of depression, and met DSM III criteria for major depressive episode. Five patients had not taken tricyclic antidepressants prior to admission, and all patients underwent a 1-week drug-free period prior to blood sampling.

Each patient was studied in conjunction with an age ($\pm$ 5 years) and sex-matched control, selected from personnel who worked in the same institute. Mean POMS* depression scores were 2.1 for controls and 36.8 for patients. All subjects gave informed consent before entry into the study.

---

* *Profile of Mood States*, Educational and Industrial Testing Service, San Diego, California 92107.

All subjects were placed on a low-monoamine diet with no aspirin during the 3 days prior to blood donation. Each patient and matched control simultaneously donated 20 ml of blood via plastic syringe anticoagulated with 1 ml of acid citrate dextrose (NIH formula A; Wyatt et al., 1973). Blood collection was performed promptly at 7:00 a.m., 10:00 a.m., 1:00 p.m., and 4:00 p.m. under identical conditions for each patient and matched control.

Blood was centrifuged at $175g$ for 10 min at room temperature to prepare platelet-rich plasma (PRP) (Shah et al., 1980). Platelet counts were obtained microscopically for each subject's PRP (Shah et al., 1980). Mean platelet counts did not significantly differ between patients and controls. One-milliliter aliquots of PRP were added to polyallomer test tubes and preincubated in a Dubnoff metabolic shaking-water bath at 37°C for 10 min. To each tube, 0.1 ml of $[^{14}C]$-5HT ($10^{-6}$ M) was added and incubations were carried out for 0.5, 1, 3, 5, and 10 min. The incubation periods ended by removal of the tubes from the bath and placement in ice, where uptake is minimal, according to our experience and that of Stahl and Meltzer (1978). The tubes were then centrifuged at $5000g$ for 5 min at 4°C, in order to separate platelets from plasma. Plasma was removed as completely as possible and the remaining platelet plug resuspended in 1.0 ml of distilled water by sonication. Ten milliliters of scintillation fluid (Scintiverse, Fisher Scientific Co.) was then added to each platelet suspension and the radioactivity was counted in a Nuclear Chicago Mark II liquid scintillation counter using $^{133}Ba$ as an external standard. The uptake was calculated as nanomoles of $[^{14}C]$-5HT per $10^{11}$ platelets.

Serum cortisol determinations were performed by radioimmunoassay (Abraham, 1973; Manlimos and Abraham, 1975). This was done from an aliquot of PRP, saved for this purpose, from each blood sample. In this way, it was possible to determine the serum cortisol and platelet 5HT uptake values for each time of day from each subject. Data from each of these determinations were then subjected to an analysis of variance.

## 3. RESULTS

It can be seen from Fig. 1 that the active uptake was generally complete after 5-min incubation at a 5HT concentration of $10^{-6}$ M; hence, this time interval was chosen for comparison of patients with controls.

Time of day was found to be an important variable in the measurement of platelet 5HT uptake within subjects. Individual subjects showed significant changes across times of day accounting for 30.6% of the variance, $F(39, 78) = 7.246$, $p < 0.001$. This was almost as large an effect as that of the difference between subjects, which accounted for 51.9% of the variance.

This time of day effect was different for different subjects, and analysis of variance found no significant group × time interaction which could distinguish depressives from controls.

The depressed patients had significantly lower platelet 5HT uptake ($p <$

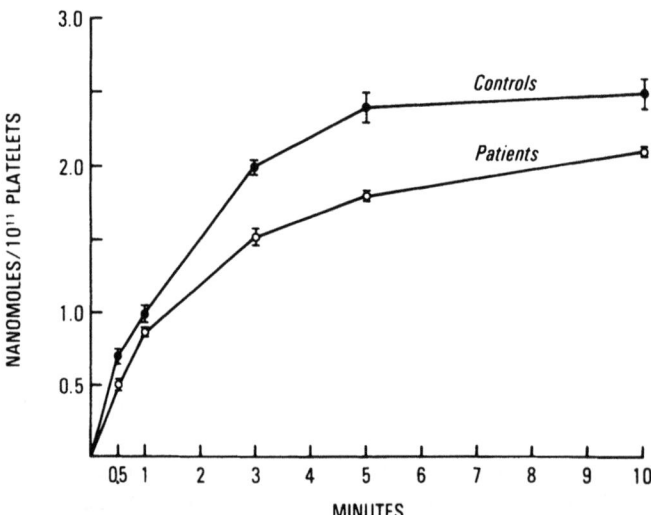

FIGURE 1. Time course of platelet 5HT uptake for controls and patients, averaged for different times of day (mean ± S.E.M.).

0.01), as compared to the control group. This difference was present for all times of day measured (Fig. 2). Five patients in this study had no history of tricyclic antidepressant therapy prior to admission, and their platelets, before the treatment, demonstrated the decreased uptake in comparison to controls (Table 1). This is important because previous findings of reduced platelet 5HT

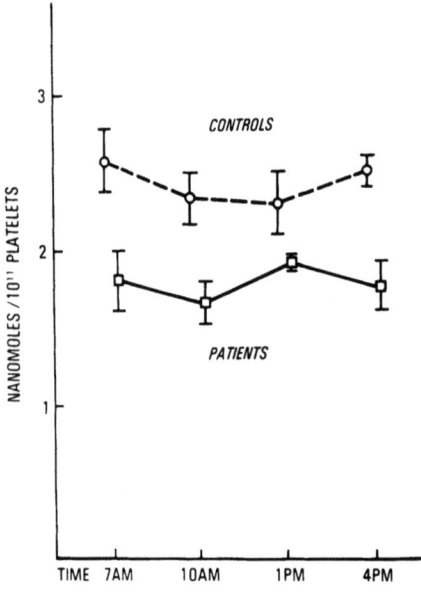

FIGURE 2. Circadian variation of [$^{14}$C]-5HT platelet uptake in depressed patients and controls, 5-min uptake values (mean ± S.E.M.) plotted for each time of day.

## TABLE 1
### Platelet Serotonin Uptake[a]

| | N | 7 a.m. | 10 a.m. | 1 p.m. | 4 p.m. | Average range | Group main effect | Group × time interaction |
|---|---|---|---|---|---|---|---|---|
| Patients | 7 | 1.81 ± 0.19 | 1.66 ± 0.14 | 1.91 ± 0.05 | 1.76 ± 0.16 | 0.67 | 17.017[b] | 0.877[c] |
| Controls | 7 | 2.57 ± 0.20 | 2.33 ± 0.16 | 2.30 ± 0.20 | 2.50 ± 0.10 | 0.74 | $F(1, 12)$ | $F(3, 36)$ |
| Patients (no tricyclics prior to admission) | 5 | 1.66 ± 0.13 | 1.70 ± 0.16 | 1.96 ± 0.06 | 1.76 ± 0.23 | 0.56 | 12.65[b] | 1.16[c] |
| Controls | 5 | 2.32 ± 0.17 | 2.32 ± 0.16 | 2.22 ± 0.2 | 2.38 ± 0.10 | 0.68 | $F(1, 8)$ | $F(3, 24)$ |
| Patients (posttreatment) | 4 | 0.53 ± 0.21 | 0.48 ± 0.18 | 0.50 ± 0.20 | 0.55 ± 0.22 | 0.085 | | |

[a] Values are mean ± S.E.M.
[b] $p < 0.01$.
[c] NS.

uptake in depression have been criticized as possibly reflecting prior tricyclic treatment (Murphy et al., 1978).

As can be seen from Fig. 3, treatment with a tricyclic antidepressant markedly inhibited 5HT uptake in depressed patients' platelets, as measured after 4 week's successful treatment with either imipramine ($N = 2$) or clomipramine ($N = 2$). These tricyclics also markedly inhibited the daily fluctuations of platelet 5HT uptake, although it is possible that the assay's sensitivity to fluctuations is less at the lower levels of uptake which are seen after inhibition from tricyclic antidepressants.

Serum cortisol determinations revealed significantly higher levels in the depressed patient group (Fig. 4), as is consistent with previous reports (Carroll et al., 1976). The difference was least marked at 7:00 a.m. when both groups demonstrated highest cortisol levels. The depressed group's cortisol levels tended to remain at relatively high levels throughout the day, whereas the control group's cortisol levels fell more markedly after 7:00 a.m. These findings are consistent with the observations of Sachar et al. (1973) who found that psychotic depressives secreted large quantities of cortisol at times when secretion is otherwise normally minimal.

Sachar et al. (1973) also reported that patients' elevated cortisol levels tended toward normalization after treatment. In our study, cortisol data (Table 2) were available for three patients who completed the treatment. With one exception (one patient, one time of day), there was a tendency toward normalization of cortisol values posttreatment. However, the group size is not large enough here to allow for any possible statistical significance, and it is not known whether the changes would be due to the tricyclic antidepressant treatment per se, or due to nonspecific factors such as length of hospitalization.

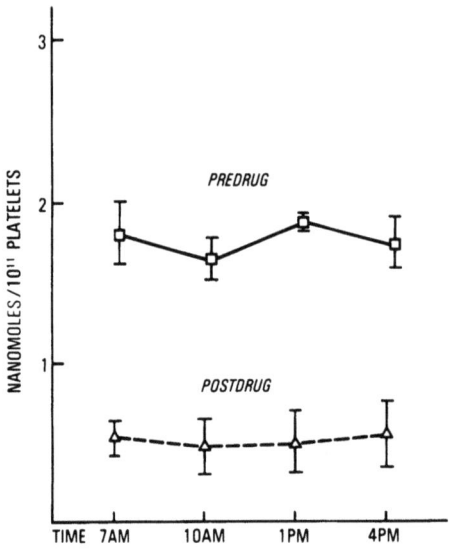

FIGURE 3. [$^{14}$C]-5HT platelet uptake in patients before and after treatment with a tricyclic antidepressant.

TABLE 2
Serum Cortisol[a]

| | N | 7 a.m. | 10 a.m. | 1 p.m. | 4 p.m. | Average range | Group main effect | Group × time interaction |
|---|---|---|---|---|---|---|---|---|
| Patients | 7 | 180.2 ± 12.2 | 127.3 ± 14.0 | 127.2 ± 16.1 | 113.9 ± 21.1 | 85.0 | 16.53[b] | 2.20[c] |
| Controls | 7 | 152.6 ± 13.6 | 48.0 ± 6.1 | 61.5 ± 11.7 | 53.8 ± 8.1 | 101.6 | $F(1, 12)$ | $F(3, 36)$ |
| Patients (no tricyclics prior to admission) | 5 | 184.2 ± 15.1 | 124.6 ± 20.0 | 113.4 ± 8.5 | 99.9 ± 26.9 | 75.8 | 9.08[d] | 0.67[c] |
| Controls | 5 | 155.5 ± 17.4 | 47.4 ± 8.5 | 76.5 ± 6.7 | 48.3 ± 10.0 | 103.0 | $F(1, 8)$ | $F(3, 24)$ |
| Patients (posttreatment) | 3 | 201.7 ± 70.4 | 98.9 ± 29.4 | 90.2 ± 35.0 | 78.7 ± 33.5 | 111.9 | | |

[a] ng/ml ± S.E.M.
[b] $p < 0.05$.
[c] NS.
[d] $p < 0.01$.

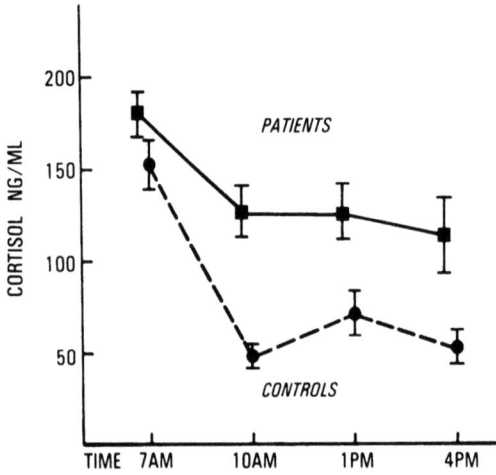

FIGURE 4. Circadian variation of serum cortisol levels in depressed patients and controls.

Some evidence exists, however, that 5HT reuptake inhibitors can alter serum cortisol secretion (Syvälahti et al., 1979).

Intraclass correlations between platelet 5HT uptake and serum cortisol revealed no temporal relationship between the circadian variations of these two parameters, for either controls or patients. Four time points during the day, however, may not be frequent enough blood sampling to rule out any temporal relationship between platelet 5HT uptake and serum cortisol variations.

## 4. DISCUSSION

This study found a reduction in the platelet serotonin uptake of depressed patients, as measured by a time course for [$^{14}$C]-5HT. Recent kinetic studies have reported a decreased maximum velocity ($V_{max}$) of platelet 5HT uptake in unipolar depression (Coppen et al., 1978; Tuomisto et al., 1979). One study has reported a decreased $V_{max}$ in bipolar depressed patients compared to unipolar depressed patients or bipolar manic patients and controls (Meltzer et al., 1980). Another study has reported decreased platelet uptake in endogenous depression, and not in reactive depression (Hallstrom et al., 1976). In our study, the subjects were unipolar depressed patients with clear vegetative symptoms characteristic of endogenous depression.

Some reports are at variance with those above. Shaw et al. (1971) reported no changes with depression; however, it has been suggested (Tuomisto et al., 1979) that the failure to detect differences may be due to technical factors such as long incubation times or high 5HT concentrations. Despite the long incubation times (10–90 min) employed by Shaw et al. (1971), it is interesting that the difference means, although nonsignificant, is in the direction of lower values in depressed patients than those in controls.

Other studies, employing spectrofluorometry to compare platelet 5HT content to platelet 5HT uptake, have also been at variance. Oxenkrug (1979) reported higher platelet 5HT uptake in depression, using a spectrofluorometric methodology. Also, Wirz-Justice and Pühringer (1978) reported no difference in uptake in depression, but demonstrated a diurnal rhythm of platelet 5HT content which was altered in depression.

The present study found a daily variation in platelet 5HT uptake, suggestive of a circadian rhythm. We were fortunate to have studied the same normal volunteer at identical times of day on two separate occasions (Fig. 5), and it is interesting that the pattern of variation for both days was almost identical for this one subject, whereas the pattern between subjects varied much more. Not enough data are present to know if different individuals have uniquely consistent rhythms for platelet 5HT uptake. Analysis of variance could not distinguish a specific pattern for circadian variation of platelet 5HT uptake which could differentiate patients from controls, other than the fact that patients had significantly lower uptake per se.

In this study, there was a very marked inhibition of uptake after treatment with either clomipramine or imipramine. This is as was expected, since previous studies have indicated that tricyclic antidepressants lower the reuptake of 5HT in synaptosomes and platelets (Lingjaerde, 1976; Tuomisto et al., 1980; Ross et al., 1980).

It has been suggested that previous tricyclic treatment would not likely be responsible for platelet 5HT reductions seen in depressed patients, since these depression-related changes are reflected as changes in $V_{max}$ and not $K_m$, whereas tricyclic antidepressants affect $K_m$ and not $V_{max}$ (Tuomisto et al., 1979). However, one report indicates that the tricyclic effect may be more complicated than previously thought, with evidence that clomipramine can af-

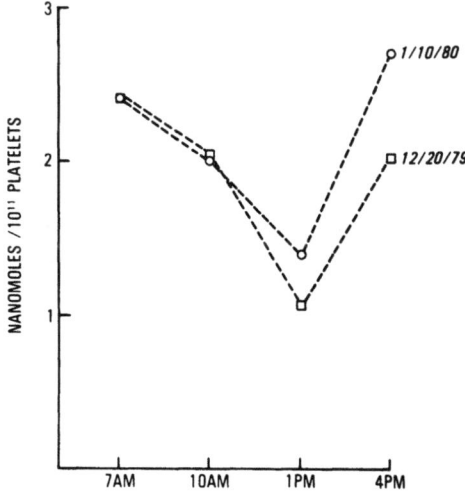

FIGURE 5. Circadian variation of platelet 5HT uptake, as studied in one normal volunteer on two separate days.

fect both $K_m$ and $V_{max}$ (Lingjaerde, 1979). As mentioned above, five patients in the present study had not taken tricyclic antidepressants prior to admission, and all patients underwent a 1-week washout period prior to testing.

It has been reported that depressed patients have a decreased density of platelet imipramine binding sites (Briley et al., 1980; Paul and Rehavi, 1981). Furthermore, it has been suggested that, in fact, imipramine selectivity labels 5HT uptake sites (Langer et al., 1980; Paul and Rehavi, 1980). Thus, our findings of decreased 5HT uptake in these patients may be consistent with the notion of a decreased density of 5HT uptake sites in the platelets of depressed patients. However, the exact pathophysiological significance of these findings awaits further clarification.

In a strain of rats with genetic 5HT storage pool deficiency (Fawn-Hooded rats), there is a reported absence of imipramine binding sites on both platelets and synaptosomes (Dumbrille-Ross and Wa Tang, 1981). This finding strengthens the possibility that platelets may reflect pathophysiological changes in 5HT neurons, with respect to 5HT uptake.

Since both the changes seen in depression and those seen in tricyclic antidepressant treatment are in the same direction of reduced uptake, the reduction of platelet 5HT uptake seen in depression may either reflect a more generalized serotoninergic inactivity associated with depression, or may reflect a compensatory cell membrane change related to factors such as cholinergic functioning, which may be regulated by serotoninergic input (Vizi et al., 1981).

The postulate that platelet 5HT uptake reflects central serotoninergic activity was addressed indirectly by our examination of platelet 5HT uptake in conjunction with serum cortisol.

Previous investigation (Scapagnini et al., 1971) had indicated that circadian variation of brain 5HT content generally paralleled the curve of plasma corticosterone in the rat. More recent evidence suggests that 5HT neurons may stimulate the release of CRF from the hypothalamus (Fuller, 1981). Krieger (1973) has suggested that "cholinergic and serotonergic mechanisms are involved in the circadian release of adrenocorticotropic hormone, whereas stress-induced release of the hormone may have cholinergic and adrenergic or dopaminergic components."

Our study found no correlation between the daily variations of plasma cortisol and platelet 5HT uptake, although it is possible that blood sampling was too infrequent for any relationship to become apparent. More work will be necessary both to understand how circadian variations of 5HT and serum cortisol may interact, and what role any interaction may play in depressive illness.

## 5. SUMMARY

Platelet 5HT uptake and serum cortisol were measured four times during the day (7:00 a.m., 10:00 a.m., 1:00 p.m., 4:00 p.m.) in drug-free depressed patients and compared to age- and sex-matched controls. In both patients and controls, there were variations in the platelet 5HT uptake during different times of the day. In comparison with controls, patients had significantly reduced platelet 5HT uptake, and increased serum cortisol levels. After treatment with a tricyclic antidepressant, the platelet 5HT uptake was markedly inhibited. These findings are discussed from the perspective of the platelet as a model for the central serotoninergic neuron.

ACKNOWLEDGMENTS. The financial support from the Health Resources Foundation of the William S. Hall Psychiatric Institute is gratefully acknowledged. The authors are thankful to Ms. Jane Yates and Mr. William McAmis for technical assistance and to Ms. Anita Cole for secretarial help.

## 6. REFERENCES

Abraham, G. E., 1973, Radioimmunoassay of plasma steroid hormones, in: *Modern Method of Steroid Analysis* (E. Heftman, ed.), pp. 451–470, Academic Press, New York.
Briley, M. S., Langer, S. Z., Raisman, R., Sechter, D., and Zarifian, E., 1980, Tritiated imipramine binding sites are decreased in platelets of untreated depressed patients, *Science* **209**:303.
Carroll, B. J., Curtis, G. C., and Mendels, J., 1976, Neuroendocrine regulation in depression. I. Limbic system–adrenocortical dysfunction, *Arch. Gen. Psychiatry* **33**:1039.
Coppen, A., Swade, C., and Wood, K., 1978, Platelet 5-hydroxytryptamine accumulation in depressive illness, *Clin. Chim. Acta.* **87**:165.
Dumbrille-Ross, A., and Wa Tang, S., 1981, Absence of high affinity [$^3$H] imipramine binding in platelets and cerebral cortex of fawn-hooded rats, *Eur. J. Pharmacol.* **72**:137.
Fuller, R. W., 1981, Serotonergic stimulation of pituitary–adrenocortical function in rats, *Prog. Neuroendocrinol.* **32**:118.
Hallstrom, C. O., Rees, W. L., Pare, C. M., Trenchard, A., and Turner, P., 1976, Platelet uptake of 5-hydroxytryptamine and dopamine in depression, *Postgrad. Med. J.* **52**(Suppl. 3):40.
Kelly, W. F., Checkley, S. A., and Bender, D. A., 1980, Cushing's syndrome, tryptophan and depression, *Br. J. Psychiatry* **136**:125.
Krieger, D. T., 1973, Neurotransmitter regulation of ACTH release, *Mt. Sinai J. Med. N.Y.* **40**:302.
Krieger, D. T., 1978, Endocrine processes and serotonin, in: *Serotonin in Health and Disease: The Central Nervous System* (W. E. Essman, ed.), Vol. III, pp. 51–67, Spectrum Publications, New York.
Langer, S. Z., Moret, C., Raisman, R., Dubocovich, M. L., and Briley, M., 1980, High affinity [$^3$H] imipramine binding in rat hypothalamus: Association with uptake of serotonin but not of norepinephrine, *Science* **210**:1133.
Lingjaerde, O., 1976, Effect of doxepin on uptake and efflux of serotonin in human blood platelets in vitro, *Psychopharmacologia* **47**:183.
Lingjaerde, O., 1979, Inhibitory effect of clomipramine and related drugs on serotonin uptake in platelets: More complicated than previously thought, *Psychopharmacology* **61**:245.
Manlimos, F. S., and Abraham, G. E., 1975, Chromatographic purification of tritiated steroids: Prior to use in radioimmunoassay, *Anal. Lett.* **8**:403.
Meltzer, H. Y., Arora, R., Baber, R., Busch, D., Kaskey, G., Nasr, S., Piyakalamala, S., and Tricou, B. J., 1980, Serotonin uptake in blood platelets of depressed patients, Society of Biological Psychiatry, 35th Annual Convention and Scientific Program, Boston.
Murphy, D. L., Colburn, R. W., Davis, J. M., and Bunney, W. E., Jr., 1970, Imipramine and lithium effects on biogenic amine transport in depressed and manic-depressed patients, *Am. J. Psychiatry* **127**:339.

Murphy, D. L., Campbell, I. C., and Costa, J. L., 1978, The brain serotonergic system in affective disorders, *Prog. Neuropsychopharmacol.* **2**:5.

Oxenkrug, G. H., 1979, The content and uptake of 5HT by blood platelets in depressive patients, *J. Neural Transm.* **45**:285.

Oxenkrug, G. H., Prakhje, I., and Mikhalenke, I. N., 1978, Disturbed circadian rhythm of 5HT uptake by blood platelets in depressive psychosis, *Act. Nerv. Super.* **20**:66.

Paul, S. M., and Rehavi, M., 1981, Imipramine binding in man and depressive illness, Presented at the *American Psychiatric Association* annual meeting, New Orleans.

Ross, S. B., Aperia, B., Beck-Friis, J., Jansa, S., Wetterberg, L., and Aberg, A., 1980, Inhibition of 5-hydroxytryptamine uptake in human platelets by antidepressant agents *in vivo*, *Psychopharmacology* **67**:1.

Sachar, E. J., Hellman, L., Roffwarg, H. P., Halpern, F. S., Fukushima, D. K., and Gallagher, T. F., 1973, Disrupted 24-hour patterns of cortisol secretion in psychotic depression, *Arch. Gen. Psychiatry* **28**:19.

Scapagnini, U., Moberg, G. P., Van Loon, G. R., De Groot, J., and Ganong, W. F., 1971, Relation of brain 5-hydroxytryptamine content to the diurnal variation plasma corticosterone in the rat, *Neuroendocrinology* **7**:90.

Shah, N. S., Burch, E. A., Yates, J. D., May, D. A., Donald, A. G., Freed, J. E., and Pressley, L. C., 1980, Dopamine uptake by human blood platelets, *Res. Commun. Psychol. Psychiatr. Behav.* **5**:25.

Shaw, D. M., MacSweeney, D. A., Woolcock, N., and Bevan-Jones, A. B., 1971, Uptake and release of $^{14}$C-5-hydroxytryptamine by platelets in affective illness, *J. Neurol. Neurosurg. Psychiatry* **34**:224.

Smith, L. T., Hanson, D. R., and Omenn, G. S., 1978, Comparisons of serotonin uptake by blood platelets and brain synaptosomes, *Brain. Res.* **146**:400.

Stahl, S. M., 1977, The human platelet: A diagnostic and research tool for the study of biogenic amines in psychiatric and neurologic disorders, *Arch. Gen. Psychiatry* **34**:509.

Stahl, S. M., and Meltzer, H. Y., 1978, A kinetic and pharmacologic analysis of 5-hydroxytryptamine transport by human platelets and platelet storage granules: Comparison with central serotonergic neurons, *J. Pharmacol. Exp. Ther.* **205**:118.

Syvälahti, E., Eneroth, P., and Ross, S. B., 1979, Acute effects of zimelidine and alaproclate, two inhibitors of serotonin uptake, on neuroendocrine function, *Psychiatry Res.* **1**:111.

Tuomisto, J., Tukiainen, E., and Ahlfors, V. G., 1979, Decreased uptake of 5-hydroxytryptamine in blood platelets from patients with endogenous depression, *Psychopharmacology* **65**:141.

Tuomisto, J., Tukiainen, E., Voutilainen, R., and Tuomainen, P., 1980, Inhibition of 5-hydroxytryptamine and noradrenaline uptake in platelets and synaptosomes incubated in plasma from human subjects treated with amitriptyline or nortriptyline: Utilization of the principle for a bioassay method, *Psychopharmacology* **69**:137.

van Praag, H. M., and Korf, J., 1971, Endogenous depressions with and without disturbances in the 5-hydroxytryptamine metabolism: A biochemical classification, *Psychopharmacologia* **19**:148.

Vizi, E. S., Hàrsing, L. G., Jr., and Zsilla, G., 1981, Evidence of the modulatory role of serotonin in acetylcholine release from striatal interneurons, *Brain. Res.* **212**:89.

Wirz-Justice, A., and Pühringer, W., 1978, Seasonal incidence of an altered diurnal rhythm of platelet serotonin in unipolar depression, *J. Neural Transm.* **42**:45.

Wyatt, R. J., Murphy, D. L., Belmaker, R., Cohen, S., Donnelly, C. H., and Pollin, W., 1973, Reduced monoamine oxidase activity in platelets: A possible genetic marker for vulnerability to schizophrenia, *Science* **179**:916.

CHAPTER 27

# Neuroendocrine Evaluation of Catecholaminergic Function in Man

## Application to Research in Psychiatry and Neurology

S. LAL and N. P. V. NAIR

## 1. INTRODUCTION

Apomorphine (Apo), a dopamine (DA) receptor agonist in animals and man (Sourkes and Lal, 1975; Tsang and Lal, 1977; Lee *et al.*, 1978), has been widely used in clinical research (Lal, 1981; Neumeyer *et al.*, 1981) and has been the subject of several recent reviews (Sourkes and Lal, 1975; Colpaert *et al.*, 1976; Di Chiara and Gessa, 1978). Apo and clonidine, an α-adrenergic receptor agonist (Andén *et al.*, 1970; Kobinger, 1978; Lovinger *et al.*, 1976; Autret *et al.*, 1977), increase growth hormone (GH) secretion (Lal *et al.*, 1972a, 1973, 1975a; Brown *et al.*, 1973). The GH response to these two agents provides a simple clinical approach to the study of DA and norepinephrine (NE) function in psychiatric and neurological disorders and to the investigation of the mode of action of psychoactive drugs and various treatment modalities in psychiatry (Lal and Nair, 1979; 1980; Lal *et al.*, 1980a). The present chapter reviews some of the variables that affect Apo- and clonidine-induced GH secretion in normal subjects and points to the limitations as well as the application of this neuroendocrine approach in psychiatric and neurological research.

---

S. LAL • Douglas Hospital Research Centre, Department of Psychiatry, Montreal General Hospital, and Department of Psychiatry, McGill University, Montreal, Quebec, Canada.  N. P. V. NAIR • Douglas Hospital Research Centre, and Department of Psychiatry, McGill University, Montreal, Quebec, Canada.

## 2. APOMORPHINE-INDUCED GH SECRETION

GH secretion in man is governed by many factors (Lal and Martin, 1980). In certain pathological states such as acromegaly (Chiodini et al., 1974) Apo decreases GH secretion in contrast to a stimulatory effect in normal subjects.

The selective action of Apo on DA receptors (Andén et al., 1967; Lal et al., 1972b) together with its short half-life (Van Tyle and Burkman, 1971; Symes et al., 1976) and prompt effect in stimulating GH secretion following s.c. administration make this aporphine alkaloid an ideal agent for clinical investigation. The individual peak GH concentration usually occurs within 30–60 min inclusive (Lal et al., 1975b). In contrast, the GH response to oral L-dopa, which is also used to assess central DA function, is much more variable in terms of number of subjects who show an adequate GH response, the promptness of response, and the time of onset of the individual peak values (Lal et al., 1975b). Further, unlike Apo, L-dopa affects the turnover of NE and serotonin (5HT), which also modulate GH secretion in man (Lal and Martin, 1980). These data point to important advantages of Apo over L-dopa as a pharmacological tool in neuroendocrine research. The GH response to Apo is highly reproducible (Rotrosen et al., 1979). The response is independent of stress. In keeping with the latter, the increase in GH is independent of changes in cortisol (Brown et al., 1974; Lal et al., 1975b).

Several factors modify the response to Apo. Whereas the GH response to ambulation (Frantz and Rabkin, 1965), arginine (Merimee et al., 1966), insulin hypoglycemia (Merimee and Fineberg, 1971), and synthetic 1–24 ACTH (Stjernholm et al., 1975) is greater in women than in men, following Apo the GH response is significantly lower in women (Ettigi et al., 1975; Lal et al., 1980a). Further, many normal women fail to respond to Apo in contrast to a consistent response in normal men (Ettigi et al., 1975; Lal et al., 1980a). A lower GH response in women compared with men has also been reported after L-dopa (Sachar et al., 1975). The reason for this gender difference in response to dopaminergic drugs is unclear. In women on the birth control pill (estrogen–progesterone combination) and in postmenopausal women who are administered estrogens, the GH response to Apo is significantly higher than in women on no medication (Ettigi et al., 1975; Lal et al., 1980a).

No difference in Apo-induced GH response is noted between pre- and postmenopausal women (Lal et al., 1980a). This suggests that in contrast to exogenously administered estrogens, physiological levels of estrogens have little effect on GH secretion provoked by Apo. This is in keeping with the preliminary observation of the absence of variation in GH response in the follicular and luteal phase of the menstrual cycle (Ettigi et al., 1975). However, in the latter study plasma steroid levels were not measured.

Testosterone enhances GH secretion following arginine or insulin hypoglycemia in males with delayed sexual development (Martin et al., 1968; Illig and Prader, 1970) but not in normal men (Merimee and Fineberg, 1971). The effect of androgens on Apo-induced GH secretion has not been studied in normal individuals. However, in patients with liver cirrhosis who have abnormally low total plasma testosterone concentrations, the GH response to Apo is significantly lower than in cirrhotics with normal testosterone values (Fig. 1). The magnitude of the GH response to Apo in cirrhotics significantly correlates with the concentration of plasma testosterone (Lal and Nair, 1979; Lal et al., 1981c).

A diminished GH response to Apo with age has been noted by Maany et al. (1975). In a larger group of male subjects there was no significant inverse correlation between age and peak GH response. However, there was some indication of a diminished response in men aged 45 to 66 years compared with men aged 20 to 42 (Lal et al., 1980a).

The GH response to a variety of stimuli, including L-dopa, is diminished in obese subjects (Fingerhut and Krieger, 1974). A similar impaired GH response is noted with Apo (Tolis et al., 1975). In the study of Tolis et al. (1975) in which a similar dose (μg/kg) of Apo was administered to tall nonobese men and short obese subjects, the obese subjects were 35–115% of ideal body weight. At what point deviations from ideal body weight affect the GH response is unknown. The percentage of adiposity increases with age even in the nonobese (Dudl et al., 1973) so that purported changes in Apo-induced GH response with age may reflect an effect of increased adiposity.

Glucose loading blunts the GH response to a variety of stimuli including L-dopa (Mims et al., 1973). A similar inhibitory effect on the Apo response was noted by Ettigi et al. (1975) but not by Nilsson (1975).

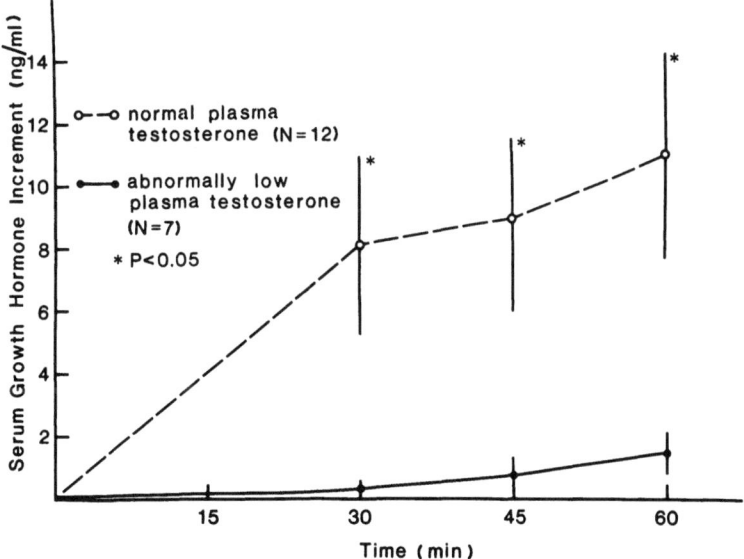

FIGURE 1. Apomorphine-induced GH response in patients with liver cirrhosis with normal and abnormally low plasma testosterone concentrations. Lal, S., Oravec, M., Aronoff, A., Kiely, M. E., Guyda, H., Solomon, S., and Nair, N. P. V. (unpublished data). Male patients were injected with apomorphine HCl (0.75 mg s.c.) at 0 min.

In view of the variables that affect the GH response to Apo, any studies undertaken to evaluate DA function in man should take into consideration the effects of gender, obesity, hormone therapy, and age and should be conducted in the fasting state. Also consideration must be given to the occurrence of spontaneous episodes of GH secretion, stress effects, and other physiological factors that affect GH secretion independently of any selective drug action on hypothalamic–pituitary DA receptors.

## 3. EFFECT OF PSYCHOACTIVE DRUGS ON APOMORPHINE-INDUCED GH SECRETION

Neuroleptics, except reserpine, block DA receptors and this effect is believed to subserve their antischizophrenic properties as well as their ability to induce parkinsonism (Snyder et al., 1974) and increase prolactin (PRL) secretion (Gruen et al., 1978; Lal and Rastogi, 1981). In keeping with this mode of action of neuroleptics, chlorpromazine (Lal et al., 1973), haloperidol (Rotrosen et al., 1979), as well as the allegedly selective DA blocker pimozide (Lal et al., 1977) antagonize Apo-induced GH secretion (Table 1). Clozapine, an effective therapeutic agent in schizophrenia, however, has little or no tendency to induce parkinsonism (Nair et al., 1977) or increase PRL secretion (Nair et al., 1979; Lal and Nair, 1980). Also, behavioral, biochemical, and endocrinological indices of DA receptor blockade in animals have been reported as either absent or only weakly present (Bürki et al., 1975). These observations

TABLE 1
Effect of Drugs on Apomorphine-Induced GH Secretion

| Drug | Class | Effect | Reference |
|---|---|---|---|
| Chlorpromazine | Neuroleptic | Antagonizes | Lal et al. (1973) |
| Haloperidol | Neuroleptic | Antagonizes | Rotrosen et al. (1979) |
| Pimozide | Neuroleptic | Antagonizes | Lal et al. (1977) |
| Clozapine | Neuroleptic | Antagonizes | Nair et al. (1979) |
| Benztropine | Anticholinergic | No effect | Lal et al. (1975b) |
| Choline | Acetylcholine precursor | Enhances | Lal et al. (1981a) |
| Methysergide | 5HT antagonist | No effect | Lal et al. (1977) |
| Cyproheptadine | 5HT antagonist | No effect | Rotrosen et al. (1979) |
| L-Tryptophan | 5HT precursor | No effect | Lal et al. (1980b) |
| Melatonin | Pineal hormone | No effect | Koulu and Lammintausta (1979) |
| Baclofen | GABAergic agent | No effect | Nair et al. (1980) |
|  |  | No effect | Koulu et al. (1980) |
| Sodium valproate | GABAergic agent | No effect | Nair et al. (1980) |
|  |  | No effect | Koulu et al. (1980) |
| Diazepam | GABAergic agent | Antagonizes | Koulu et al. (1980) |
| Lithium | Antimanic agent | No effect | Lal et al. (1978) |
| Naloxone | Opiate antagonist | No effect | Lal et al. (1979a) |
| Levallorphan | Opiate antagonist | No effect | Lal et al. (1979a) |

led to the suggestion that the therapeutic effect of clozapine is not mediated by DA receptor blockade (Bürki et al., 1975). However, in chronic schizophrenics withdrawn from neuroleptics for at least 2 weeks, clozapine significantly inhibited Apo-induced GH secretion (Nair et al., 1979). The weak PRL stimulatory effect of clozapine cannot be explained on the basis of its intrinsic antimuscarinic properties (Lal et al., 1979b). Also, the antimuscarinic property of clozapine cannot account for the inhibitory effect on Apo-induced GH secretion as the anticholinergic agent benztropine has no effect on the Apo-induced GH response (Lal et al., 1975b). It would appear then that clozapine has a differential effect on functionally different DA systems in man. Observations with clozapine suggest that there are pharmacological differences in DA receptors that modulate PRL secretion and extrapyramidal function on the one hand and DA receptors that modulate GH secretion and mediate the antischizophrenic effects of tranquilizers on the other.

The capacity of a drug to stimulate PRL secretion has been proposed as a screening test for potential antischizophrenic agents (Gruen et al., 1978). In light of findings with clozapine, inhibition of Apo-induced GH secretion in conjunction with an absent PRL stimulating effect may provide a better screening test for selection of antischizophrenic agents, especially those which are devoid of the usual side effects of neuroleptics, namely, those associated with hyperprolactinemia and extrapyramidal side effects, both parkinsonism and, perhaps, tardive dyskinesia. In regard to the latter, tardive dyskinesia has never been associated with clozapine (Cole and Gardos, 1980). Because of hematological side effects, clozapine itself has been withdrawn from use in clinical trials in North America.

Apo decreases PRL in man (Lal et al., 1973; Martin et al., 1974) and this effect has been used to evaluate DA receptor function in schizophrenia (Ettigi et al., 1976; Rotrosen et al., 1979; Tamminga et al., 1977; Meltzer et al., 1980). However, as observations with clozapine suggest that DA receptors modulating PRL secretion may be pharmacologically distinct from those mediating antischizophrenic effects, the usefulness of this measure is questionable.

Anticholinergic agents improve Parkinson's disease. The failure of benztropine to enhance Apo-induced GH secretion (Lal et al., 1975b) points to different modulatory mechanisms of DA function in basal ganglia and the hypothalamic–pituitary axis. The acetylcholine precursor choline worsens Parkinson's disease (Papavasiliou and Rosal, 1979) and improves patients with tardive dyskinesia and Huntington's chorea (Davis et al., 1976). In contrast to this inhibitory effect of choline on basal ganglia DA function, choline enhances DA function in the hypothalamic–pituitary axis as reflected in the augmentation of Apo-induced GH secretion (Lal et al., 1981a). Regional differences in the organization of DA systems point to difficulties in extrapolating studies on the hypothalamic–pituitary axis to other regions of brain.

5HT mechanisms modulate striatal DA function in animals (Sahakian et al., 1979). In man, the putative 5HT antagonist methysergide and the peripheral 5HT antagonist cyproheptadine have no effect on Apo-induced GH secretion

(Lal et al., 1977; Rotrosen et al., 1979). The 5HT precursor L-tryptophan, which is used in psychiatric and neurological research to increase selectively central indoleamine function, in a dose sufficient to increase plasma free tryptophan more than 12-fold has no effect on Apo-induced GH secretion (Lal et al., 1980b). In man, changes in free serum tryptophan parallel changes in the CNS as reflected by ventricular CSF concentrations (Young et al., 1976). Thus, the failure of tryptophan to alter Apo-induced GH secretion (Lal et al., 1980b) suggests that central changes in 5HT function induced by tryptophan do not alter DA function, at least in the hypothalamic–pituitary axis.

There are reports that the GABAergic agents baclofen (Frederiksen, 1975), sodium valproate (Linnoila et al., 1976), and diazepam (Rao, 1964) exert antischizophrenic effects. GABAergic agents inhibit the release of DA (Stock et al., 1973), at least in the striatum. In man, both short-term (Koulu et al., 1980) and long-term administration (Nair et al., 1980) of sodium valproate and baclofen have no effect on Apo-induced GH secretion. Baclofen (but not sodium valproate) inhibits L-dopa-induced GH secretion (Nair et al., 1980; Koulu et al., 1980) and diazepam inhibits the GH release induced by L-dopa as well as Apo (Koulu et al., 1980). These observations indicate that in the hypothalamic-pituitary axis, as in the striatum in man (Nair et al., 1978), baclofen inhibits DA function by a presynaptic effect. Diazepam, on the other hand, exerts an inhibitory effect on both pre- and postsynaptic DA function. The inhibitory effect of baclofen and diazepam on DA function may be relevant to their purported antischizophrenic effects.

There is some evidence that narcotic antagonists exert clinical antischizophrenic effects (Buchsbaum et al., 1980). Also, naloxone reverses β-endorphin-induced catatonia in the rat, a behavioral state that has been likened to catatonia in schizophrenia (Bloom et al., 1976) and, in some studies, inhibits Apo-induced stereotyped behavior in rodents (Cox et al., 1976; Malick et al., 1977). The failure of naloxone or levallorphan to antagonize Apo-induced GH secretion (Lal et al., 1979a) suggests that if narcotic antagonists exert an antischizophrenic action, then this effect is not mediated by DA receptor blockade.

There is some evidence that enhancement of DA mechanisms may play a role in mania associated with manic-depressive psychosis (Gerner et al., 1976). In animals, lithium prevents neuroleptic-induced DA supersensitivity (Pert et al., 1978) and inhibits DA-sensitive adenylate cyclase (Schorderet, 1977). In man, lithium attenuates the euphoriant effect of amphetamine (Van Kammen and Murphy, 1975). These observations raise the possibility that lithium may induce its therapeutic effect by modulating DA mechanisms. However, lithium has no effect on Apo-induced GH secretion, at least in subjects without evidence of an affective disorder (Lal et al., 1978).

It should be mentioned that the failure of some drugs to modify the GH response to Apo may be related to the use of inadequate doses of experimental drug or the use of too large a dose of Apo.

## 4. EFFECT OF SLEEP DEPRIVATION AND ELECTROCONVULSIVE THERAPY ON APOMORPHINE-INDUCED GH SECRETION

Sleep deprivation (Gerner et al., 1979) and electroconvulsive therapy (ECT) exert an antidepressant effect in patients with a primary affective disorder. The mechanism by which these procedures induce amelioration is unknown. In view of the putative role of altered DA function in the pathophysiology of manic-depressive psychosis and changes in DA function that occur in animals following sleep deprivation (Tufik et al., 1978) or ECT (Papeschi et al., 1974), various investigators have looked at the effect of these treatment procedures on Apo-induced GH secretion in man. In normal subjects, 24-hr sleep deprivation decreases the GH response to Apo (Lal et al., 1981b). The reversal of the effects of sleep deprivation on vigilance and waking EEG by amphetamine in man (Hartmann et al., 1977) may indicate that sleep deprivation also decreases DA function in brain regions outside the hypothalamic–pituitary axis. Whether similar changes occur in manic-depressive patients is unknown. In contrast to clinical studies in animals, sleep deprivation enhances DA function as indicated by the increase in behavioral response to Apo (Tufik et al., 1978).

ECT improves parkinsonian symptoms (Balldin et al., 1980). In depressed patients ECT has no effect on Apo-induced GH secretion (Modigh et al., 1981) but enhances the PRL lowering effect of Apo (Balldin, 1981). These observations in parkinsonian and depressed patients are compatible with an increase in sensitivity of striatal DA receptors and of DA receptors modulating PRL secretion and no effect on DA receptors modulating GH secretion. This differential effect of ECT on responsiveness of GH and PRL secretion to Apo again emphasizes the heterogeneity of DA receptors in brain and difficulties in generalizing conclusions drawn from one DA system to another.

## 5. APOMORPHINE-INDUCED GH SECRETION IN PSYCHIATRIC AND NEUROLOGICAL DISORDERS

Changes in Apo-induced GH secretion have been observed in certain extrapyramidal disorders (Table 2). In Parkinson's disease, the response is decreased (Brown et al., 1973). As degenerative changes are known to occur in the hypothalamus of parkinsonian patients (Langston and Forno, 1978), the decreased response to Apo may reflect destruction of DA receptor sites. In Huntington's chorea, neuronal changes are also noted in the hypothalamus (Bruyn, 1973). In this disorder, both decreases (Levy et al., 1979) and increases (Müller et al., 1979) in Apo-induced GH secretion have been described. Spasmodic torticollis may be associated with a variety of extranuchal neurological symptoms (Baxter and Lal, 1979) though the site and extent of the lesion are unknown (Lal, 1979). In a single case described by Brown et al., (1973), there

## TABLE 2
### Apomorphine-Induced GH Secretion in Psychiatric and Neurological Disorders

| Diagnosis | GH response | Reference |
|---|---|---|
| Unipolar depression | No change | Maany et al. (1979) |
| | No change | Garver et al. (1977) |
| Bipolar depression | No change | Maany et al. (1979) |
| Reactive depression | No change | Maany et al. (1979) |
| Mania | Trend for decrease | Garver et al. (1977) |
| Acute schizophrenia | Increase | Pandey et al. (1977) |
| | No change | Meltzer et al. (1980) |
| Chronic schizophrenia | Decrease | Ettigi et al. (1976) |
| | No change | Pandey et al. (1977) |
| | Bimodal response | Rotrosen et al. (1979) |
| | No change | Meltzer et al. (1980) |
| Schizophrenic catalepsy | No change | Cervantes et al. (1977) |
| Anorexia nervosa | Decrease | Sherman and Halmi (1977) |
| | Decrease | Casper et al. (1977) |
| Pseudocyesis | Decrease | Tulandi et al. (1981) |
| Tardive dyskinesia | Decrease | Ettigi et al. (1976) |
| | No change | Tamminga et al. (1977) |
| Huntington's chorea | Increase | Müller et al. (1979) |
| | Decrease | Levy et al. (1979) |
| Parkinson's disease | Decrease | Brown et al. (1973) |
| Dystonia musculorum deformans | No change | Brown et al. (1973) |
| Spasmodic torticollis | Decrease | Brown et al. (1973) |
| | No change | Lal et al. (1979c) |
| Hepatic encephalopathy | Decrease | Lal et al. (1981c) |

was no response to Apo. In contrast, in a series of 10 patients the GH response was similar to that of well-matched controls (Lal et al., 1979c). The failure of the patient of Brown et al. (1973) to respond may have been related to the smaller dose of Apo used and to the fact that normal women may show no response to Apo. The normal response to Apo suggests that the lesion in spasmodic torticollis does not extend to involve hypothalamic–pituitary DA function.

Chronic neuroleptic therapy increases DA receptor sensitivity in animals (Tarsy and Baldessarini, 1977; Symes et al., 1977). Such a change in receptor function has been implicated in the pathophysiology of tardive dyskinesia (Tarsy and Baldessarini, 1977). In chronic male schizophrenics who had been on neuroleptics for 5 years or more and who were withdrawn from all drugs for 2 to 15 weeks, the GH response to Apo was significantly lower than in controls (Ettigi et al., 1976). Further, in schizophrenic patients with tardive dyskinesia, the response was lower than in schizophrenic patients without tardive dyskinesia. The diminished GH response could not be explained on the basis of persistent neuroleptic blockade of DA receptors. Interestingly, there was a significant inverse correlation between peak GH concentrations and duration of neuroleptic treatment but no correlation between peak GH concentration and duration of schizophrenic psychosis. These observations suggest that the diminished GH response was a consequence of neuroleptic therapy rather than a consequence of schizophrenia (Ettigi et al., 1976) and that in the hypothalamic–pituitary axis, prior neuroleptic therapy induces a subsensitivity of DA receptors. In contrast to the findings of Ettigi et al. (1976), Tamminga et al. (1977) found no difference between patients with tardive dyskinesia and controls. The discrepancy may be related to differences in duration of prior neuroleptic treatment.

The DA hypothesis of schizophrenia implicates an enhancement of DA mechanisms in brain (Meltzer and Stahl, 1976). In none of four studies of chronic schizophrenia was the GH response to Apo increased (Ettigi et al., 1976; Tamminga et al., 1977; Rotrosen et al., 1979; Meltzer et al., 1980). In the study of Rotrosen et al. (1979), a greater variability in GH response was evident among the schizophrenic subjects compared with controls. These findings to date suggest that if supersensitivity of DA receptors is a pathophysiological factor in the etiology of schizophrenia, this does not extend to the tuberoinfundibular DA system. In contrast to the findings in chronic schizophrenia, Pandey et al. (1977) found an increase in GH response in acute schizophrenic patients of both sexes compared with chronic patients or a younger group of controls. Meltzer et al. (1980), however, found no difference between controls, acute schizophrenics, and chronic schizophrenics. Schizophrenics who were ill for less than 4 years, however, had significantly greater values than those ill for 4 years or more. Further, they found a significant correlation between GH response and duration of illness which is in contrast to the findings of Ettigi et al. (1976). Meltzer et al. (1980) did not present data on the relationship of duration of neuroleptic therapy and GH response.

In animals, drugs that impair striatal DA function induce catalepsy (Papeschi, 1972). However, the relevance of catalepsy in animals to schizophrenic catalepsy though often implied is questionable (Lal et al., 1980c). Changes in hypothalamic DA function have been postulated to occur in the clinical disorder (Powers et al., 1976). In a single patient with schizophrenic catalepsy, the GH response to Apo was normal (Cervantes et al., 1977).

In depressed patients (unipolar and bipolar manic-depressives), the GH response to Apo (Maany et al., 1979) or L-dopa (Sachar et al., 1975) is unchanged. In preliminary data a trend was noticed for a decreased response to Apo in mania (Garver et al., 1977).

The etiology of anorexia nervosa is unknown though a disruption of hypothalamic function has been proposed. Two studies reported inadequate GH responses to Apo in patients with anorexia nervosa (Sherman and Halmi, 1977; Casper et al., 1977). Unfortunately, the interpretation by these authors of their data is questionable because basal GH concentrations are elevated in anorexia nervosa (as was noted in their own patients), and the effect of Apo in normal subjects with elevated baseline concentrations is unknown. In view of autoregulation of GH secretion in man (Abrams et al., 1971), a modifying effect of elevated baseline GH concentrations cannot be excluded. Also, neither study employed controls which is particularly important as many normal women fail to show an adequate GH response to Apo. The attempt of Casper *et al.* (1977) to compare their patients with the controls of Ettigi *et al.* (1975) is methodologically unsound. Also the data provided by Ettigi and associates do not lend themselves to the statistical methods used by Casper *et al.* (1977).

In a single case of pseudocyesis, the GH response to Apo was absent but returned to normal following resolution (Tulandi *et al.*, 1981).

There is evidence that DA neurotransmission is impaired in hepatic encephalopathy (Morgan *et al.*, 1980). The turnover of DA as assessed by the probenecid technique, however, is normal (Lal *et al.*, 1974). This raises the possibility that DA receptor function may be altered in this condition. Though the GH response to Apo is decreased in hepatic encephalopathy, the impairment is a function of the concentration of circulating testosterone rather than hepatic encephalopathy per se. Thus, a similar impairment is noted in patients with cirrhosis who are without hepatic encephalopathy but who have a decrease in testosterone concentration (Lal and Nair, 1979; Lal *et al.*, 1981c).

## 6. CLONIDINE-INDUCED GH SECRETION

Clonidine has been less extensively studied than Apo. The response to clonidine is more variable and less consistent than with Apo (Lal *et al.*, 1975a). Clonidine induces a transient hyperglycemia which precedes the elevation of GH. L-Dopa, the precursor of DA and NE, stimulates glucagon secretion whereas Apo is without effect (Lorenzi *et al.*, 1977). This raises the possibility that the increase in GH following clonidine may be mediated indirectly by an effect on glucagon secretion rather than a direct effect on the hypothalamic–pituitary axis. In this regard, glucagon is known to stimulate GH secretion in man. However, recent studies have shown that clonidine-induced GH secretion is independent of glucagon secretion (Lal *et al.*, 1981d). Also the GH response is independent of cortisol secretion (Lal *et al.*, 1975a).

There is a trend for a lower response to clonidine in women than in men (Lal *et al.*, 1980a; Matussek, 1978). Premenopausal women have a greater response than postmenopausal women (Matussek *et al.*, 1980). The response is antagonized by glucose loading (Lal *et al.*, 1980a). An age effect has not been observed (Matussek *et al.*, 1980). In regular consumers of alcohol the GH response to clonidine is diminished.

The GH response to clonidine is decreased in patients with an endogenous depression (Matussek, 1978; Matussek *et al.*, 1980). Similar decreases have been observed after amphetamine (Langer *et al.*, 1976) and insulin hypoglycemia (Sachar *et al.*, 1975). These data are compatible with the catecholamine hypothesis which implicates an impairment of noradrenergic function in the pathophysiology of endogenous depression.

Recent findings have shown an increase in CSF NE in chronic schizo-

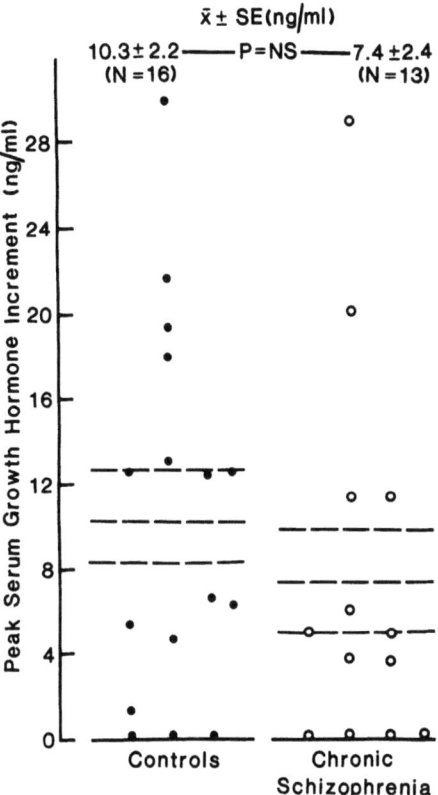

FIGURE 2. Clonidine-induced GH secretion in chronic schizophrenia. Male subjects were injected with clonidine (0.15 mg i.v.) over a 10-min period commencing at 0 min and samples taken at −30, 0, 15, 30, 45, 60, and 90 min from commencement of clonidine injection. Each point represents the peak GH increase of one subject. Lal, S., Nair, N. P. V., Thavundayil, J., and Guyda, H. (unpublished data). Interrupted lines represent $\bar{X} \pm$ S.E.

phrenia (Lake *et al.*, 1980). The GH response to clonidine in chronic schizophrenia is unchanged (Matussek *et al.*, 1980) (Fig. 2).

## 7. SUMMARY

The GH response to the DA receptor agonist Apo and the α-adrenergic receptor agonist clonidine provide a simple clinical approach to the evaluation of DA and NE function, respectively, in psychiatric and neurological disorders and the investigation of the mode of action of psychotropic drugs and other treatment modalities. Interpretation of results, however, requires careful consideration of the physiology of GH regulation and variables that modify the GH response to provocative agents in normal subjects. This neuroendocrine approach only evaluates catecholaminergic function in the hypothalamic–pituitary axis so that results may not reflect catecholaminergic function in other brain regions.

ACKNOWLEDGMENTS. This work was supported in part by the Medical Research Council (Canada). We thank Nadia Zajac for her excellent secretarial assistance.

## 8. REFERENCES

Abrams, R. L., Grumbach, M. M., and Kaplan, S. L., 1971, The effect of administration of human growth hormone on the plasma growth hormone, cortisol, glucose, and free fatty acid response to insulin: Evidence for growth hormone autoregulation in man, *J. Clin. Invest.* **50**:940.

Andén, N.-E., Rubenson, A., Fuxe, K., and Hökfelt, T., 1967, Evidence for dopamine receptor stimulation by apomorphine, *J. Pharm. Pharmacol.* **19**:627.

Andén, N.-E., Corrodi, H., Fuxe, K., Hökfelt, B., Hökfelt, T., Rydin, C., and Svensson, T., 1970, Evidence for a central noradrenaline receptor stimulation by clonidine, *Life Sci.* **9**:513.

Autret, A., Minz, M., Beillevaire, T., Cathala, H. P., and Schmitt, H., 1977, Effect of clonidine on sleep patterns in man, *Eur. J. Clin. Pharmacol.* **12**:319.

Balldin, J., 1981, Experimental and clinical studies on neuroendocrine and behavioural effects of electroconvulsive therapy, reports from the Department of Psychiatry and Neurochemistry, St. Jorgen's Hospital, University of Göteborg, No. 14, Gotab, Kungälv, Sweden.

Balldin, J., Edén, S., Granérus, A.-K., Modigh, K., Svanborg, A., Walinder, J., and Wallin, L., 1980, Electroconvulsive therapy in Parkinson's syndrome with 'on–off' phenomenon, *J. Neural Transm.* **47**:11.

Baxter, D. W., and Lal, S., 1979, Essential tremor and dystonic syndromes, *Adv. Neurol.* **24**:373.

Bloom, F. A., Segal, D., Ling, N., and Guillemin, R., 1976, Endorphins: Profound behavioral effects in rats suggest new etiological factors in mental illness, *Science* **194**:630.

Brown, W. A., Van Woert, M. H., and Ambani, L. M., 1973, Effect of apomorphine on growth hormone release in humans, *J. Clin. Endocrinol. Metab.* **37**:463.

Brown, W. A., Krieger, D. T., and Van Woert, M. H., 1974, Dissociation of growth hormone and cortisol release following apomorphine, *J. Clin. Endocrinol. Metab.* **38**:1127.

Bruyn, G. W., 1973, Neuropathological changes in Huntington's chorea, *Adv. Neurol.* **1**:399.

Buchsbaum, M. S., Davis, G. C., and van Kammen, D. P., 1980, Diagnostic classification and the endorphin hypothesis of schizophrenia: Individual differences and psychopharmacological strategies, in: *Perspectives in Schizophrenia Research* (C. Baxter and T. Melnechuk, eds.), pp. 177–194, Raven Press, New York.

Bürki, H. R., Eichenberger, E., Sayers, A. C., and White, T. G., 1975, Clozapine and the dopamine hypothesis of schizophrenia, a critical appraisal, *Pharmakopsychiatr. Neuro-psychopharmakol.* **8**:115.

Casper, R. C., Davis, J. M., and Pandey, G. N., 1977, The effect of nutritional status and weight changes on hypothalamic function tests in anorexia nervosa, in: *Anorexia Nervosa* (R. A. Vigersky, ed.), pp. 137–147, Raven Press, New York.

Cervantes, P., Lal, S., Smith, F., and Guyda, H., 1977, Dopaminergic function in two patients with catalepsy, *Acta Psychiatr. Scand.* **55**:214.

Chiodini, P. G., Liuzzi, A., Botalla, G., Cremascoli, G., and Silvestrini, F., 1974, Inhibitory effect of dopaminergic stimulation on GH release in acromegaly, *J. Clin. Endocrinol. Metab.* **38**:200.

Cole, J. O., and Gardos, G., 1980, Tardive dyskinesia, in: *Psychopharmacology Update* (J. O. Cole, ed.), pp. 151–164, Collamore Press, Lexington, Mass.

Colpaert, F. C., Van Bever, W. F. M., and Leysen, J. E. M. F., 1976, Apomorphine: Chemistry, pharmacology, biochemistry, *Int. Rev. Neurobiol.* **19**:225.

Cox, B., Ary, M., and Lomax, P., 1976, Changes in sensitivity to apomorphine during morphine dependence and withdrawal in rats, *J. Pharmacol. Exp. Ther.* **196**:637.

Davis, K. L., Hollister, L. E., Barchas, J. D., and Berger, P. A., 1976, Choline in tardive dyskinesia and Huntington's chorea, *Life Sci.* **19**:1507.

Di Chiara, G., and Gessa, G. L., 1978, Pharmacology and neurochemistry of apomorphine, *Adv. Pharmacol. Chemother.* **15**:87.

Dudl, R. J., Ensinck, J. W., Palmer, H. E., and William, R. H., 1973, Effect of age on growth hormone secretion in man, *J. Clin. Endocrinol. Metab.* **37**:11.

Ettigi, P., Lal, S., Martin, J. B., and Friesen, H. G., 1975, Effect of sex, oral contraceptives, and glucose loading on apomorphine-induced growth hormone secretion, *J. Clin. Endocrinol. Metab.* **40**:1094.

Ettigi, P., Nair, N. P. V., Lal, S., Cervantes, P., and Guyda, H., 1976, Effect of apomorphine on growth hormone and prolactin secretion in schizophrenic patients, with and without oral dyskinesia, withdrawn from chronic neuroleptic therapy, *J. Neurol. Neurosurg. Psychiatry* **39**:870.

Fingerhut, M., and Krieger, D. T., 1974, Plasma growth hormone response to L-dopa in obese subjects, *Metabolism* **23**:267.

Frantz, A. G., and Rabkin, M. T., 1965, Effects of estrogen and sex differences on secretion of human growth hormone, *J. Clin. Endocrinol. Metab.* **25**:1470.

Frederiksen, P. K., 1975, Baclofen in the treatment of schizophrenia, *Lancet* **I**:702.

Garver, D. L., Pandey, G. N., Hengeveld, C., and Davis, J. M., 1977, Growth hormone response and central aminergic systems in affective diseases, *Psychopharmacol. Bull.* **13**:61.

Gerner, R. H., Post, R. M., and Bunney, W. E., Jr., 1976, A dopaminergic mechanism in mania, *Am. J. Psychiatry* **133**:1177.

Gerner, R. H., Post, R. M., Gillin, J. C., and Bunney, W. E., Jr., 1979, Biological and behavioral effects of one night's sleep deprivation in depressed patients and normals, *J. Psychiatr. Res.* **15**:21.

Gruen, P. H., Sachar, E. J., Langer, G., Altman, N., Leifer, M., Frantz, A., and Halpern, F. S., 1978, Prolactin responses to neuroleptics in normal and schizophrenic subjects, *Arch. Gen. Psychiatry* **35**:108.

Hartmann, E., Orzack, M. H., and Branconnier, P., 1977, Sleep deprivation deficits and their reversal by d- and l-amphetamine, *Psychopharmacology* **53**:185.

Illig, R., and Prader, A., 1970, Effect of testosterone on growth hormone secretion in patients with anorchia and delayed puberty, *J. Clin. Endocrinol. Metab.* **30**:615.

Kobinger, W., 1978, Central alpha-adrenergic systems as targets for hypotensive drugs, *Rev. Physiol. Biochem. Pharmacol.* **81**:39.

Koulu, M., and Lammintausta, R., 1979, Effect of melatonin on L-tryptophan and apomorphine-stimulated growth hormone secretion in man, *J. Clin. Endocrinol. Metab.* **49**:70.

Koulu, M., Lammintausta, R., and Dahlström, S., 1980, Effects of some gamma-aminobutyric acid (GABA)-ergic drugs on the dopaminergic control of human growth hormone secretion, *J. Clin. Endocrinol. Metab.* **51**:124.

Lake, C. R., Sternberg, D. E., van Kammen, D. P., Ballenger, J. C., Ziegler, M. G., Post, R. M., Kopin, I. J., and Bunney, W. E., Jr., 1980, Schizophrenia: Elevated cerebrospinal fluid norepinephrine, *Science* **207**:331.

Lal, S., 1979, Pathophysiology and pharmacotherapy of spasmodic torticollis: A review, *Can. J. Neurol. Sci.* **6**:427.

Lal, S., 1981, Clinical studies with apomorphine, in: *Clinical Pharmacology of Apomorphine and Other Dopaminomimetics*, Vol. 2 (G. U. Corsini and G. L. Gessa, eds.), pp. 1-11, Raven Press, New York.

Lal, S., and Martin, J. B., 1980, Neuroanatomy and neuropharmacological regulation of neuroendocrine function, in: *Handbook of Biological Psychiatry*, Vol. 3 (H. M. van Praag, M. H. Lader, O. J. Rafaelsen, and E. J. Sachar, eds.), pp. 101-167, Dekker, New York.

Lal, S., and Nair, N. P. V., 1979, Growth hormone and prolactin responses in neuropsychiatric research, *Dev. Neurol.* **2**:179.

Lal, S., and Nair, N.P.V., 1980, Effect of neuroleptics on prolactin and growth hormone secretion in man, *Dev. Endocrinol.* **9**:223.

Lal, S., and Rastogi, R. B., 1981, Effect of neuroleptics on pituitary function in man, in: *Neuroendocrine Regulation and Altered Behavior* (P. D. Hrdina and R. L. Singhal, eds.), pp. 169–183, Croom Helm, London.

Lal, S., de la Vega, C. E., Sourkes, T. L., and Friesen, H. G., 1972a, Effect of apomorphine on human growth hormone secretion, *Lancet* **II**:661.

Lal, S., Sourkes, T. L., Missala, K., and Belendiuk, G., 1972b, Effects of aporphine and emetine alkaloids on central dopaminergic mechanisms in rats, *Eurp. J. Pharmacol.* **20**:71.

Lal, S., de la Vega, C. E., Sourkes, T. L. and Friesen, H. G., 1973, Effect of apomorphine on growth hormone, prolactin, luteinizing hormone and follicle-stimulating hormone levels in human serum, *J. Clin. Endocrinol. Metab.* **37**:719.

Lal, S., Aronoff, A., Garelis, E., Sourkes, T. L., Young, S. N., and de la Vega, C. E., 1974, Cerebrospinal fluid homovanillic acid, 5-hydroxyindoleacetic acid, lactic acid, and pH before and after probenecid in hepatic coma, *Clin. Neurol. Neurosurg.* **77**:142.

Lal, S., Tolis, G., Martin, J. B., Brown, G. M., and Guyda, H., 1975a, Effect of clonidine on growth hormone, prolactin, luteinizing hormone, follicle-stimulating hormone and thyroid stimulating hormone in the serum of normal men, *J. Clin.Endocrinol. Metab.* **41**:827.

Lal, S., Martin, J. B., de la Vega, C. E., and Friesen, H. G., 1975b, Comparison of the effect of apomorphine and L-dopa on serum growth hormone levels in normal men, *Clin. Endocrinol.* **4**:277.

Lal, S., Guyda, H., and Bikadoroff, S., 1977, Effect of methysergide and pimozide on apomorphine-induced growth hormone secretion in men, *J. Clin. Endocrinol. Metab.* **44**:766.

Lal, S., Nair, N. P. V., and Guyda, H., 1978, The effect of lithium on hypothalamic–pituitary dopaminergic function, *Acta Psychiatr. Scand.* **57**:91.

Lal, S., Nair, N. P. V., Cervantes, P., Pulman, J., and Guyda, H., 1979a, Effect of naloxone or levallorphan on serum prolactin concentrations and apomorphine-induced growth hormone secretion, *Acta Psychiatr. Scand.* **59**:173.

Lal, S., Mendis, T., Cervantes, P., Guyda, H., and De Rivera, J. L., 1979b, Effect of benztropine on haloperidol-induced prolactin secretion, *Neuropsychobiology* **5**:327.

Lal, S., Hoyte, K., Kiely, M. E., Sourkes, T. L., Baxter, D. W., Missala, K., and Andermann, F., 1979c, Neuropharmacological investigation and treatment of spasmodic torticollis, *Adv. Neurol.* **24**:335.

Lal, S., Nair, N. P. V., Cervantes, P., and Guyda, H., 1980a, Use of drug-induced growth hormone and prolactin responses in psychiatric research, in: *Progress in Psychoneuroendocrinology* (F. Brambilla, G. Racagni, and D. de Wied, eds.), pp. 295–307, Elsevier/North-Holland, Amsterdam.

Lal, S., Young, S. N., Cervantes, P., and Guyda, H., 1980b, Effect of L-tryptophan on apomorphine-induced growth hormone secretion in normal subjects, *Pharmakopsychiatr. Neuro-Psychopharmakol.* **13**:331.

Lal, S., Cervantes, P., and Nair, N. P. V., 1980c, Failure of naloxone to reverse catalepsy associated with schizophrenia: A case report, *Psychiatr. J. Univ. Ottawa* **5**:160.

Lal, S., Etienne, P., Thavundayil, J., Nair, N. P. V., Collier, B., Rastogi, R., Guyda, H., and Schwartz, G., 1981a, Effect of choline on central dopaminergic function in normal subjects, *J. Neural Transm.* **50**:29.

Lal, S., Thavundayil, J., Nair, N. P. V., Etienne, P., Rastogi, R., Schwartz, G., Pulman, J., and Guyda, H., 1981b, Effect of sleep deprivation on dopamine receptor function in normal subjects, *J. Neural Transm.* **50**:39.

Lal, S., Oravec, M., Aronoff, A., Kiely, M. E., Guyda, H., Solomon, S., and Nair, N. P. V., 1981c, Hypothalamic–pituitary dopaminergic function in hepatic failure in man, *J. Neural Transm.* **53**:7.

Lal, S., Tolis, G., McDonald, T. J., Cervantes, P., and Dupré, J., 1981d, Effect of clonidine on growth hormone and glucagon secretion, *Horm. Metab. Res.* **13**:648.

Langer, G., Heinze, B., Reim, B., and Matussek, N., 1976, Reduced growth hormone responses to amphetamine in "endogenous" depressive patients: Studies in normal, "reactive" and "endogenous" depressive, schizophrenic, and chronic alcoholic subjects, *Arch. Gen. Psychiatry* **33**:1471.

Langston, J. W., and Forno, L. S., 1978, The hypothalamus in Parkinson disease, *Ann. Neurol.* **3**:129.

Lee, T., Seeman, P., Tourtellotte, W. W., Farley, I. J., and Hornykiewicz, O., 1978, Binding of $^3$H-neuroleptics and $^3$H-apomorphine in schizophrenic brains, *Nature (London)* **274**:897.

Levy, C. H., Carlson, H. E., Sowers, J. R., Goodlett, R. E., Tourtellotte, W. W., and Hershman, J. M., 1979, Growth hormone and prolactin secretion in Huntington's disease, *Life Sci.* **24**:743.

Linnoila, M., Viukari, M., and Hietala, O., 1976, Effect of sodium valproate on tardive dyskinesia, *Br. J. Psychiatry* **129**:114.

Lorenzi, M., Tsalikian, E., Bohannon, N. V., Gerich, J. E., Karam, J. H., and Forsham, P. H., 1977, Differential effects of L-dopa and apomorphine on glucagon secretion in man: Evidence against central dopaminergic stimulation of glucagon, *J. Clin. Endocrinol. Metab.* **45**:1154.

Lovinger, R., Holland, J., Kaplan, S., Grumbach, M., Boryczka, A. T., Shackleford, R., Salmon, J., Reid, I. A., and Ganong, W. F., 1976, Pharmacological evidence for stimulation of growth hormone secretion by a central noradrenergic system in dogs, *Neuroscience* **1**:443.

Maany, I., Frazer, A., and Mendels, J., 1975, Apomorphine: Effect on growth hormone, *J. Clin. Endocrinol. Metab.* **40**:162.

Maany, I., Mendels, J., Frazer, A., and Brunswick, D., 1979, A study of growth hormone release in depression, *Neuropsychobiology* **5**:282.

Malick, J. B., Millingsley, M. L., Kubena, R. K., and Goldstein, J. M., 1977, Evaluation of naloxone in laboratory tests predictive of clinical antipsychotic activity, *Commun. Psychopharmacol.* **1**:475.

Martin, J. B., Lal, S., Tolis, G., and Friesen, H. G., 1974, Inhibition by apomorphine of prolactin secretion in patients with elevated serum prolactin, *J. Clin. Endocrinol. Metab.* **39**:180.

Martin, L. G., Clark, J. W., and Connor, T. B., 1968, Growth hormone secretion enhanced by androgens, *J. Clin. Endocrinol. Metab.* **28**:425.

Matussek, N., 1978, Effect of amphetamine and clonidine on human growth hormone release in psychiatric patients and controls, in: *Depressive Disorders* (S. Garattini, ed.), pp. 431–446, Schattauer Verlag, Stuttgart.

Matussek, N., Ackenheil, M., Hippius, H., Muller, F., Schröeder, H. T., Schultes, H., and Wasilewski, B., 1980, Effect of clonidine on growth hormone release in psychiatric patients and controls, *Psychiatry Res.* **2**:25.

Meltzer, H. Y., and Stahl, S. M., 1976, The dopamine hypothesis of schizophrenia: A review, *Schizophr. Bull.* **2**:19.

Meltzer, H. Y., Busch, D., So, R., Holcomb, H., and Fang, V. S., 1980, Neuroleptic-induced elevations in serum prolactin levels: Etiology and significance, in: *Perspectives in Schizophrenia Research* (C. Baxter and T. Melnechuk, eds.), pp. 149–176, Raven Press, New York.

Merimee, T. J., and Fineberg, S. E., 1971, Studies of the sex based variation of human growth hormone secretion, *J. Clin. Endocrinol. Metab.* **33**:896.

Merimee, T. J., Burgess, J. A., and Rabinowitz, D., 1966, Sex determined variation in serum insulin and growth hormone response to amino acid stimulation, *J. Clin. Endocrinol. Metab.* **26**:791.

Mims, R. B., Scott, C. L., Modebe, O. M., and Bethune, J. E., 1973, Prevention of L-dopa-induced growth hormone stimulation by hyperglycemia, *J. Clin. Endocrinol. Metab.* **37**:660.

Modigh, K., Balldin, J., Edén, S., Granérus, A.-K., and Walinder, J., 1981, Electroconvulsive therapy and receptor sensitivity, *Acta Psychiatr. Scand.* (Suppl. 290) **63**:91.

Morgan, M. Y., Jakobovits, A. W., James, I. M., and Sherlock, S., 1980, Successful use of bromocriptine in the treatment of chronic hepatic encephalopathy, *Gastroenterology* **78**:663.

Müller, E. E., Parati, E. A., Cocchi, D., Zanardi, P., and Caraceni, T., 1979, Dopaminergic drugs on growth hormone and prolactin secretion in Huntington's disease, *Adv. Neurol.* **23**:319.

Nair, N. P. V., Zickerman, V., and Schwartz, G., 1977, Dopamine and schizophrenia: A reappraisal in the light of clinical studies with clozapine, *Can. Psychiatr. Assoc. J.* **22**:285.

Nair, N. P. V., Yassa, R., Ruiz-Navarro, J., and Schwartz, G., 1978, Baclofen in the treatment of tardive dyskinesia, *Am. J. Psychiatry* **135**:1562.

Nair, N. P. V., Lal, S., Cervantes, P., Yassa, R., and Guyda, H., 1979, Effect of clozapine on apomorphine-induced growth hormone secretion and serum prolactin concentrations in schizophrenia, *Neuropsychobiology* **5**:136.

Nair, N. P. V., Lal, S., and Guyda, H., 1980, Effects of GABA-ergic drugs on nigrostriatal and hypothalamic–pituitary dopaminergic systems, *Brain Res. Bull.* **5**(Suppl. 2):427.

Neumeyer, J. L., Lal, S., and Baldessarini, R. J., 1981, Historical highlights of the chemistry, pharmacology and early clinical uses of apomorphine, in: *Clinical Pharmacology of Apomorphine and other Dopaminomimetics* Vol. 1 (G. U. Corsini and G. L. Gessa, eds.), pp. 1–18, Raven Press, New York.

Nilsson, K. O., 1975, Lack of effect of hyperglycaemia on apomorphine-induced growth hormone release in normal man, *Acta Endocrinol. (Kbh.)* **80**:230.

Pandey, G. N., Garver, D. L., Tamminga, C., Erickson, S., Ali, S. I., and Davis, J. M., 1977, Postsynaptic supersensitivity in schizophrenia, *Am. J. Psychiatry* **134**:518.

Papavasiliou, P. S., and Rosal, V., 1979, Effects of choline in patients with levodopa-induced dyskinesia, *Nutr. Brain* **5**:335.

Papeschi, R., 1972, Dopamine, extrapyramidal system and psychomotor function, *Psychiatr. Neurol. Neurochir.* **75**:13.

Papeschi, R., Randrup, A., and Lal, S., 1974, Effect of ECT on dopaminergic and noradrenergic mechanisms. I. Effect on the behavioral changes induced by reserpine, alpha-methyl-*p*-tyrosine or amphetamine, *Psychopharmacologia* **35**:149.

Pert, A., Rosenblatt, J. E., Sivit, C., Pert, C. B., and Bunney, W. E., Jr., 1978, Long-term treatment with lithium prevents the development of dopamine receptor supersensitivity, *Science* **201**:171.

Powers, P., Douglass, T. S., and Waziri, R., 1976, Hyperpyrexia in catatonic states, *Dis. Nerv. Syst.* **37**:359.

Rao, A., 1964, A controlled trial with "valium" in some psychiatric disorders, *Indian J. Psychiatry* **6**:188.

Rotrosen, J., Angrist, B., Gershon, S., Paquin, J., Branchey, L., Oleshansky, M., Halpern, F., and Sachar, E. J., 1979, Neuroendocrine effects of apomorphine: Characterization of response patterns and application to schizophrenia research, *Br. J. Psychiatry* **135**:444.

Sachar, E. J., Altman, N., Gruen, P. H., Glassman, A., Halpern, F. S., and Sassin, J., 1975, Human growth hormone response to levodopa: Relation to menopause, depression, and plasma dopa concentration, *Arch. Gen. Psychiatry* **32**:502.

Sahakian, B. J., Wurtman, R. J., Barr, J. K., Millington, W. R., and Chiel, H. J., 1979, Low tryptophan diet decreases brain serotonin and alters response to apomorphine, *Nature (London)* **279**:731.

Schorderet, M., 1977, Lithium-inhibition of cyclic AMP accumulation induced by dopamine in isolated retinae of the rabbit, *Biochem. Pharmacol.* **26**:167.

Sherman, B. M., and Halmi, K. A., 1977, Effect of nutritional rehabilitation on hypothalamic–pituitary function in anorexia nervosa, in: *Anorexia Nervosa* (R. A. Vigersky, ed.), pp. 211–224, Raven Press, New York.

Snyder, S. H., Banerjee, S. P., Yamamura, H. I., and Greenberg, D., 1974, Drugs, neurotransmitters, and schizophrenia: Phenothiazines, amphetamines, and enzymes synthesizing psychomimetic drugs aid schizophrenia research, *Science* **184**:1243.

Sourkes, T. L., and Lal, S., 1975, Apomorphine and its relation to dopamine in the nervous system, *Adv. Neurochem.* **1**:247.

Stjernholm, M. R., Alsever, R. N., and Beck, P., 1975, Growth hormone release after synthetic 1–24 ACTH: Effects of estrogen and sex, *J. Clin. Endocrinol. Metab.* **40**:516.

Stock, G., Magnusson, T., and Andén, N.-E, 1973, Increase in brain dopamine after axotomy or treatment with gammahydroxybutyric acid due to elimination of the nerve impulse flow, *Naunyn-Schmiedebergs Arch. Pharmakol.* **278**:347.

Symes, A. L., Lal, S., and Sourkes, T. L., 1976, Time-course of apomorphine in the brain of the immature rat after apomorphine injection, *Arch. Int. Pharmacodyn. Ther.* **223**:260.

Symes, A. L., Lal, S., Young, S. N., Tsang, D., and Sourkes, T. L., 1977, Effect of chronic chlorpromazine administration or prior treatment with reserpine on brain apomorphine concentrations and apomorphine-induced stereotyped behaviour in the rat, *Eur. J. Pharmacol.* **43:**173.

Tamminga, C. A., Smith, R. C., Pandey, G., Frohman, L. A., and Davis, J. M., 1977, A neuroendocrine study of supersensitivity in tardive dyskinesia, *Arch. Gen. Psychiatry* **34:**1119.

Tarsy, D., and Baldessarini, R. J., 1977, The pathophysiologic basis of tardive dyskinesia, *Biol. Psychiatry* **12:**431.

Tolis, G., Lal, S., and Pinter, E., 1975, Reduced growth hormone responses to apomorphine and L-dopa in obesity, in: *Symposium of the International Society of Psychoendocrinology*, pp. 335–339, Publishing House of the Hungarian Academy of Sciences, Viségrad.

Tsang, D., and Lal, S., 1977, Effect of monoamine receptor agonists and antagonists on cyclic AMP accumulation in human cerebral cortex slices, *Can. J. Physiol. Pharmacol.* **55:**1263.

Tufik, S., Lindsey, C. J., and Carlini, E. A., 1978, Does REM sleep deprivation induce a supersensitivity of dopaminergic receptors in the rat brain?, *Pharmacology* **16:**98.

Tulandi, T., McInnes, R. A., Mehta, A., and Tolis, G., 1981, Pseudocyesis: Pituitary function before and after resolution of symptoms, *Obstet. Gynecol.* **59:**119.

van Kammen, D. P., and Murphy, D. L., 1975, Attenuation of the euphoriant and activating effects of *d*- and *l*-amphetamine by lithium carbonate treatment, *Psychopharmacologia* **44:**215.

Van Tyle, W. K., and Burkman, A. M., 1971, Spectrofluorometric assay of apomorphine in brain tissue, *J. Pharm. Sci.* **60:**1736.

Young, S. N., Lal, S., Feldmuller, F., Sourkes, T. L., Ford, R. M., Kiely, M., and Martin, J. B., 1976, Parallel variation of ventricular CSF tryptophan and free serum tryptophan in man, *J. Neurol. Neurosurg. Psychiatry* **39:**61.

CHAPTER 28

# Neuroendocrine Studies in Schizophrenia

GHANSHYAM N. PANDEY, SYED I. ALI,
REGINA CASPER, and JOHN M. DAVIS

## 1. INTRODUCTION

An alteration in the function of dopamine (DA) has been related to the pathophysiology of schizophrenic illness and has led to the so-called "dopamine hypothesis of schizophrenia." Although there has been only indirect evidence to support this hypothesis (see Meltzer and Stahl, 1976), this has stimulated extensive clinical research for investigating the role of DA in schizophrenic illness. The initial research was directed either toward studying the levels of DA and its metabolites in urine, plasma, and CSF or examining the enzymes associated with the biosynthesis and metabolism of DA in schizophrenic patients. In particular, these studies attempted to elucidate the role of presynaptic dopaminergic mechanisms in schizophrenia. However, in recent years the role of postsynaptic mechanisms such as DA receptor response or DA receptor density has been studied in patients with schizophrenic illness. Such studies have been primarily stimulated by the observations that the potency of antipsychotic drugs is related to their ability to block DA receptors in the brain.

In order to evaluate the role of postsynaptic dopaminergic mechanisms in schizophrenic illness, several strategies have been employed. In one set of studies, the sensitivity of postsynaptic DA receptors has been determined by quantitating the number of DA receptors in the postmortem brain of schizophrenic patients, using radiolabeled ligand-binding techniques. The other strategy primarily uses neuroendocrine studies for determining CNS DA receptor response under *in vivo* situations, in schizophrenic patients and normal controls. Both strategies have some limitations and, therefore, can provide only

---

GHANSHYAM N. PANDEY, REGINA CASPER, and JOHN M. DAVIS • Department of Research, Illinois State Psychiatric Institute, Chicago, Illinois, and Department of Psychiatry, University of Illinois College of Medicine, Chicago, Illinois 60612.   SYED I. ALI • Department of Research, Illinois State Psychiatric Institute, Chicago, Illinois 60612.

limited information. This chapter will discuss the results obtained by using these strategies in studying the role of dopaminergic mechanisms in schizophrenic illness.

## 2. DA HYPOTHESIS OF SCHIZOPHRENIA

The evidence for excessive dopaminergic activity in schizophrenia is generally derived from pharmacologic and behavioral studies with DA agonists and antagonists. A variety of stimulant drugs, such as amphetamine and methylphenidate or apomorphine which presumably increase dopaminergic activity either by releasing DA or by stimulating DA receptors, when administered to animals cause a typical behavioral syndrome known as stereotypy (Randrup and Munkvard, 1970). This behavior is inhibited by drugs such as phenothiazines, which are potent blockers of DA receptors. Furthermore, it is observed that amphetamine worsens schizophrenic behavior when administered to schizophrenic patients (Angrist et al., 1974a,b). Similarly, L-dopa caused an exacerbation of their psychosis, including increased paranoia and hallucinations when administered to schizophrenic patients (Yaryura-Tobias et al., 1970; Angrist et al., 1973).

However, major evidence (Yaryura-Tobias et al., 1970; Angrist et al., 1974a,b) supporting the DA hypothesis is derived from the fact that neuroleptic drugs which are beneficial in schizophrenic illness are potent blockers of DA receptors. Furthermore, the observation that the clinical potency of the neuroleptic drugs is directly related to their ability to block DA receptors provides additional evidence that the beneficial effect of these drugs may be related to the blockade of DA receptors (Seeman et al., 1976).

Although indirect support for the DA hypothesis has been compelling, the clinical evidence suggesting a role for DA in the etiology of schizophrenic illness has been lacking. One of the earlier strategies for studying dopaminergic function in schizophrenia involves the measurement of DA and its metabolites in the urine and CSF of schizophrenic patients. Homovanillic acid (HVA) is the major metabolite of DA, formed as a result of both O-methylation and deamination by the enzymes catechol-$o$-methyltransferase (COMT) and monoamine oxidase (MAO). If schizophrenic illness were associated with increased levels of DA, an alteration in HVA levels would be expected. Chase et al. (1970) found normal levels of HVA in CSF obtained from schizophrenic patients. Rimon et al. (1971) and Bowers (1973) measured HVA levels in the CSF of schizophrenic patients after the administration of probenecid which blocks the transport of HVA to blood via the choroid plexus, thus increasing the levels of HVA in CSF. Although Rimon et al. (1971) found no differences in the HVA levels in the CSF between schizophrenic patients and nonschizophrenic psychiatric control group, they observed significantly higher CSF HVA levels in patients with paranoid schizophrenia. Similarly, Bowers (1973) and Post et al. (1975) did not observe significant differences in CSF HVA levels between acute schizophrenic and control groups. Several other studies have provided inconclusive results.

As studies with presynaptic dopaminergic mechanisms have provided uncertain findings (Crow et al., 1979), recent studies have examined the possibilities of alterations in postsynaptic dopaminergic mechanisms in schizophrenia. Since those drugs which are beneficial in the treatment of schizophrenics are potent blockers of postsynaptic DA receptors, the possibility exists that schizophrenic illness may be associated with hyperactivity of the DA receptors. However, studies providing direct evidence for DA receptor alterations in schizophrenic patients have been difficult because of lack of a suitable peripheral marker for central DA function. Therefore, such strategies as neuroendocrine studies, quantitation of DA receptors using radiolabeled ligand-binding technqiue in the postmortem brain have been employed. Although the results from such studies are often complicated and difficult to interpret, nonetheless, these studies have been useful in studying the role of DA in schizophrenic illness. The results obtained by using neuroendocrine strategies by several groups of investigators, including ourselves, will be discussed in the following sections.

## 3. NEUROENDOCRINE STUDIES

The neuroendocrine systems provide a very useful tool for studying central monoamine function in man. This is because it has been shown that not only the release of some of the hormones is regulated by neural mechanisms, but also that several neurotransmitters control the release of these hormones. Of particular interest in studies of dopaminergic function in schizophrenic illness are the pituitary hormones, prolactin (PRL) and growth hormone (GH). We will restrict ourselves to the studies of GH in schizophrenia.

## 4. GH RELEASE IN SCHIZOPHRENIA

The release of GH from the pituitary is regulated by two hypothalamic regulating hormones. One of these is excitatory and the other is inhibitory in nature (Brown et al., 1978; Martin, 1973). The cell bodies that are considered to secrete the releasing factors are located in the arcuate nuclei of the medial basal hypothalamus. While the identity of the GH-releasing factor has not been determined, an inhibitory factor was isolated (Brazeau et al., 1973) in the course of attempts to isolate and purify the GH-releasing factor. This inhibitory factor has been named somatostatin.

A variety of stimuli cause the release of GH in man (see Brown et al., 1978). These include physiologic, psychologic, and pharmacologic stimuli. GH secretion is increased by several metabolic stimuli, such as insulin-induced hypoglycemia, and by administration of certain amino acids. Emotional arousal, stress and onset of sleep cause an increase in GH secretion. Administration of L-dopa, apomorphine, amphetamine, etc., also causes increases in GH secretion.

In relation to schizophrenia, the regulation of GH release by monoamines is of particular interest. It has been observed that α-adrenergic blockade causes decrease in GH secretion (Blackard and Heidingsfelder, 1968). Similarly, administration of L-dopa and of the dopamine agonist apomorphine causes rapid increase in GH secretion (Boyd et al., 1970; Lal et al., 1973).

Taken together these observations indicate that the release of GH is regulated by certain monoamines in the brain. Whereas DA appears to play a greater role in the secretion of GH, norepinephrine and serotonin may also be involved in its regulation. The GH system, therefore, provides a unique opportunity to test monoamine functions in psychiatric disorders and DA function, in particular, in schizophrenic illness.

## 5. GH RELEASE AS A TEST FOR CENTRAL DA RECEPTOR SENSITIVITY

Neuroendocrine tests may provide valid methods for studying central monoamine functions in psychiatric disorders. However, as suggested by Checkley (1980), certain pharmacologic criteria must be satisfied before assuming that a hormone response to a particular drug results from the stimulation of a particular neurotransmitter receptor. Some of the important criteria listed by Checkley (1980) are:

1. The same hormone response should result from the administration of all drugs which stimulate the specified receptor.
2. In each case the hormone response to receptor stimulation should be inhibited by drugs which selectively block that receptor but should not be inhibited by drugs which selectively block other receptor sites.
3. The response can only be considered to be a measure of receptor sensitivity if the drug in question acts directly at a receptor. If the drug indirectly acts upon transmitter release or reuptake, then changes in hormone response may be caused by changes in these phenomena or by changes in receptor sensitivity.
4. Other minor criteria may involve issues relating to the blood–brain barrier.

The secretion of GH after stimulation with selective DA agonists such as apomorphine (Lal et al., 1973), bromocriptine (Camanni et al., 1975), etc., generally appears to meet the above criteria for testing DA receptor sensitivity in mental disorders. The GH response to these agonists is blocked by DA agonists (Lal et al., 1977) but not by adrenergic or serotonin antagonists. GH release by other less selective DA agonists such as L-dopa, methylphenidate (Brown et al., 1978), and amphetamine (Besser et al., 1969) could also be used for studying DA function, but are less specific tests for studying DA receptor sensitivity.

Since the release of GH after stimulation with DA agonists appears to be an adequate index of central DA receptor sensitivity, it provides a useful tool for testing the DA hypothesis of schizophrenia. It is, therefore, not surprising that in recent years several investigators have used this approach for studying the role of DA function and DA receptor sensitivity in schizophrenic illness.

## 6. GH STUDIES IN SCHIZOPHRENIA

GH release following administration of DA agonists has been used to assess dopaminergic function in patients with schizophrenic illness. Generally three types of agonists have been used: (1) the direct and more selective DA agonist, apomorphine, (2) the indirect DA agonist, amphetamine, and (3) the precursor of DA, L-dopa. Whereas apomorphine-induced GH release may provide an *in vivo* assessment of central DA receptor sensitivity, amphetamine- and L-dopa-induced GH release may be considered as a dopaminergic or a mixed dopaminergic–adrenergic response.

Lal *et al.* (1973) reported the effect of apomorphine administration in normal human subjects. They administered 0.25 to 1.5 mg of apomorphine s.c. to normal healthy subjects and determined serum GH levels up to 150 min following apomorphine administration. They observed that the GH response peak occurred between 30 and 60 min and that the levels of GH released were related to the administered dose of apomorphine.

In a preliminary report, we (Pandey *et al.*, 1977) studied apomorphine-induced GH release in a group of hospitalized schizophrenic patients and non-hospitalized normal controls. The patients were diagnosed according to the Research Diagnostic Criteria (RDC) specified by Spitzer and Endicott. Apomorphine challenge was performed at the end of a 2-week baseline drug-free period. In order to evaluate GH response to apomorphine, blood was drawn from the fasting subjects in the morning by an indwelling catheter at $-60$, $-30$, 0, 20, 40, 60, 80, 120, and 160 min surrounding the administration of 0.75 mg of apomorphine s.c. levels of GH were determined in the plasma obtained at the above periods.

GH levels following apomorphine administration increased in normal subjects, reaching peak level at 60–80 min after apomorphine administration and then reaching toward baseline levels. A typical GH response curve in a normal subject is shown in Fig. 1.

In this study (Pandey *et al.*, 1977) we performed apomorphine tests on a total of 32 subjects: 9 acute schizophrenic patients, 8 chronic schizophrenic patients, 7 additional chronic schizophrenic patients with tardive dyskinesia (TD), and 8 normal control volunteers. In order to compare GH response, peak levels of GH were compared between groups following apomorphine administration.

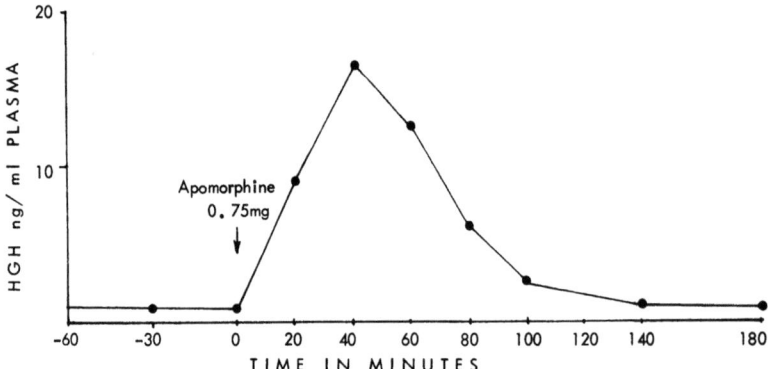

FIGURE 1. Apomorphine-induced GH secretion in a normal control. Blood was drawn from a fasting subject (normal control) in the morning at −60, −30, 0, 20, 40, 60, 80, 120, and 160 min surrounding the administration of 0.75 mg of apomorphine s.c. The blood was immediately spun and GH levels in the plasma were determined by radioimmunoassay.

The mean GH responses in the various groups studied are shown in Table 1. The combined schizophrenic population without TD had a slightly higher mean GH peak (23.5 ± 4.4 ng/ml) as compared to normal control subjects. However, the distribution of GH peaks in the schizophrenic population appeared to be bimodal. When the schizophrenic population was subgrouped as acute and chronic, it was observed that the mean GH value for patients with acute schizophrenia was significantly higher than that of the chronic schizophrenic group. The mean GH level in the acute schizophrenic group was also significantly higher as compared to the normal control group. However, the mean GH peak in chronic schizophrenics was slightly lower, but not significantly, than the normal control group. The mean GH level in the chronic schizophrenic patients with TD was very similar to that of the chronic schizophrenic patients and not significantly different from normal control groups.

Contrary to our expectations, we did not observe increased apomorphine-stimulated GH release in schizophrenic illness. However, a subgroup of pa-

TABLE 1
Apomorphine-Induced GH Response in Schizophrenic Patients and Normal Control Subjects

| Diagnostic group | No. of men | No. of women | Age (mean ± S.E.) | GH peak (ng/ml plasma, mean ± S.E.) |
|---|---|---|---|---|
| Normal subjects ($N = 8$) | 4 | 4 | 30.1 ± 3.4 | 18.3 ± 3.1 |
| Acute schizophrenic patients ($N = 9$) | 5 | 4 | 21.7 ± 1.2 | 30.7 ± 4.4[a] |
| Chronic schizophrenic patients ($N = 8$) | 5 | 3 | 21.2 ± 0.9 | 12.3 ± 3.7 |
| Chronic schizophrenic patients with TD ($N = 7$) | 3 | 4 | 48.7 ± 13.2 | 12.0 ± 2.4 |

[a] This value was significantly higher than that for normal volunteers ($p < 0.05$) and that for chronic schizophrenic patients without TD ($p < 0.01$).

tients diagnosed as "acute schizophrenia" did show significantly increased peak GH levels as compared to a normal control group. Does this suggest hyporesponsivity of DA receptors in a subgroups of schizophrenic patients?

Our observations do raise several questions: (1) Is the schizophrenic group heterogeneous in terms of a biochemical (physiological) response? (2) Are there several types of DA receptors involved in schizophrenic illness? (3) Is increased GH response to apomorphine a manifestation of a defect associated with one type of DA receptor? (4) Is GH response related to chronicity of the illness?

Several lines of evidence suggest heterogeneity of schizophrenic illness. Studies with several biologic markers indicate that not all schizophrenic patients share an abnormality of common biologic markers whereas some subtypes of schizophrenic groups have a low platelet MAO activity that others do not. Similarly a subgroup of the schizophrenic population responds to lithium therapy (Garver, 1981). Therefore, it is possible that the patients in this study diagnosed according to RDC as having acute schizophrenia may have had a different biologic disorder than those diagnosed as having chronic schizophrenia. The persistence of symptoms in the chronic schizophrenic patients, as opposed to the acute patients who showed recovery with periods of good functioning, may indicate a different course in these two putative diagnostic entities. Therefore, it is not unreasonable to suggest a different etiology for acute schizophrenia.

A second possible explanation is that the patients diagnosed as chronic might at one time have had an increased DA receptor sensitivity like that observed among acute patients, but that some adaptive process associated with the chronicity of the illness or with the pharmacotherapy for the illness interfered with this sensitivity. Chronic patients had had greater exposure to neuroleptic drugs: six of eight of the chronic patients were taking neuroleptics at the time of admission, in contrast with only one of nine acute patients. Even though in the present study the subjects had been drug-free for 2-3 weeks before the neuroendocrine testing, the possibility that prolonged residual effects of previous exposure to neuroleptics could cause reductions of GH responses should not be ruled out. Eight of nine acute schizophrenic patients were drug-free for at least 3 months, so it is unlikely that previous drug exposure could account for their observed elevations in the two systems.

GH response following administration of DA agonists has been studied in schizophrenic patients by several groups of investigators. Ettigi *et al.* (1976) measured GH levels following apomorphine administration in 17 chronic schizophrenic patients who were withdrawn from chronic neuroleptic therapy for up to 2 weeks and in 21 normal control volunteers. In general, they observed a decreased GH response to apomorphine in chronic schizophrenic patients withdrawn from neuroleptics as compared to normal controls.

GH response to apomorphine and L-dopa administration was studied by Rotrosen *et al.* (1976, 1978) in a group of recently hospitalized chronic schizophrenic patients. They observed no differences between the apomorphine-induced GH peak responses between 22 chronic schizophrenic subjects and 9 normal controls. Similarly, GH response to L-dopa was not significantly different between schizophrenic patients and normal controls.

Meltzer et al. (1981) studied GH and PRL levels following apomorphine administration in 13 chronic, 5 acute, and 15 normal control subjects. They did not find significant differences in the GH response between the groups studied. However, they observed that GH response in the schizophrenic patients who were ill for less than 4 years had significantly higher GH responses as compared to the schizophrenic group ill for more than 4 years. They speculated that GH response to apomorphine may be related to the duration of illness.

Although there appears to be some disagreement between the results reported by several groups, it generally appears that chronic schizophrenic patients have in general decreased GH response to apomorphine and acute schizophrenics or the schizophrenic patients who have been sick for a shorter duration have increased GH response to apomorphine.

In a recent study (Pandey et al., 1979), we have replicated our preliminary report on GH secretion in acute schizophrenics. In this study, apomorphine-induced GH secretion was examined in 16 normal subjects (8 males, 8 females), 21 patients with acute schizophrenia (12 males, 9 females) and 14 patients with chronic schizophrenia (8 males, 6 females). The mean peak GH levels following apomorphine administration are shown in Table 2 and the distribution of peak GH levels in the subgroups is shown in Fig. 2. Similar to our earlier reports, the mean peak GH levels were significantly higher in patients with acute schizophrenia as compared with normal controls and chronic schizophrenics. Chronic schizophrenic groups did not differ significantly from the normal control group.

Although we have replicated our own preliminary study using a larger sample size, the finding that the peak GH levels in acute schizophrenia are elevated needs to be replicated by other groups.

There appear to be at least two other explanations for the failure to observe elevated GH responses in chronic schizophrenic patients. The GH strategy is a measure of DA receptor response in the tuberoinfundibular region of the brain, and either an abnormal or a normal GH response may not be related to

TABLE 2
Apomorphine-Induced GH Response in Schizophrenic Patients and Normal Controls

| Diagnostic group | N | GH peak (ng/ml plasma, mean ± S.E.) |
|---|---|---|
| Normal subjects | 16 | 18.66 ± 2.13 |
| Acute schizophrenics | 21 | 31.54 ± 3.91[a] |
| Chronic schizophrenics | 14 | 16.39 ± 3.53 |

[a] $p < 0.007$, as compared to normal subjects and chronic schizophrenics.

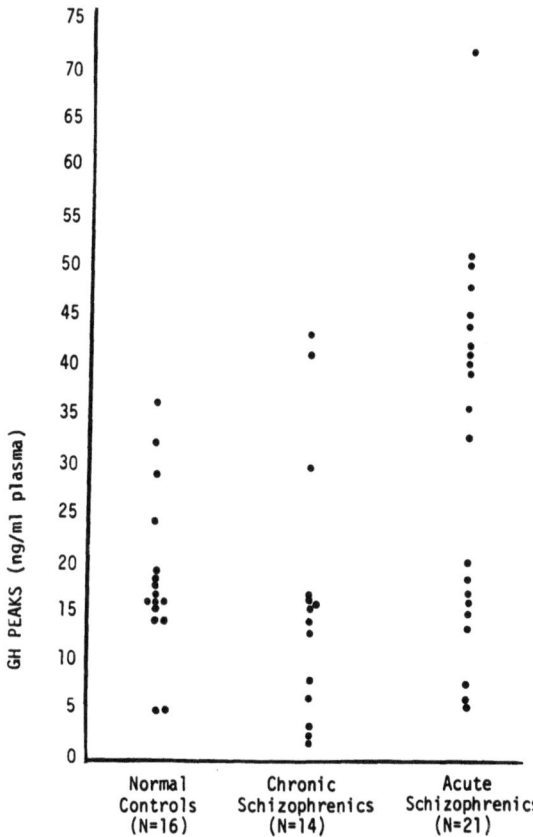

FIGURE 2. Apomorphine-induced GH response in schizophrenic patients and normal controls. GH levels following apomorphine administration were determined in each subject according to the procedure described in Fig. 1. The figure shows the distribution of peak GH level in the subgroups.

the DA receptor response in other areas of the brain. For example, acute schizophrenia may be associated with altered DA receptor responsiveness in the tuberoinfundibular region, whereas chronic schizophrenia may involve altered DA receptor responsiveness in some other areas of the brain, such as the caudate.

Second, acute and chronic schizophrenic illness may be associated with altered responsiveness of DA receptors of different subtypes. In recent years, the existence of at least three types of DA receptors has been suggested. The recent reports of an increased number of DA receptors in the postmortem brain of schizophrenic patients may appear to be consistent with this explanation. These studies and their implications for schizophrenic illness are discussed below.

## 7. DA RECEPTORS IN SCHIZOPHRENIA

DA receptor function in schizophrenia has been studied by both direct and indirect methods. The indirect methods include the use of neuroendocrine strategies, such as measurement of GH and PRL and the determination of DA-sensitive adenylate cyclase in postmortem brains. In recent years, the development of radioligand-binding techniques for the quantitation and characterization of DA receptors has provided a tool for the studies of DA receptor function in schizophrenia. Using this technique several groups of investigators have reported increased density of DA receptors in the brains of schizophrenic patients. Owen *et al.* (1978) observed that spiroperidol binding in caudate, putamen, and nucleus accumbens in the postmortem brains of the schizophrenic patients was significantly higher as compared with control population. Lee *et al.* (1978) and Lee and Seeman (1980) also reported increased [$^3$H]haloperidol and [$^3$H] spiperone binding in the caudate, putamen, and nucleus accumbens from the postmortem schizophrenic brain, as compared to controls. However, they did not observe any significant differences in [$^3$H]apomorphine binding between schizophrenic patients and controls. MacKay *et al.* (1978, 1980) reported increased [$^3$H]spiperone binding in the caudate and nucleus accumbens from schizophrenic patients who had been on neuroleptic medication up until death. They did not observe differences in those patients who had been off medication for more than a month before death.

Different subtypes of DA receptors have been reported in the brain (Seeman, 1981). Of these subtypes, the DA receptor which is labeled by antagonists such as haloperidol and spiperone appears to be increased in the schizophrenic brain, whereas those DA receptors which are labeled by agonists such as apomorphine or those DA receptors which are linked to adenylate cyclase appear unchanged in the postmortem schizophrenic brain. The results reported by us and other groups of investigators to the effect that apomorphine-induced DA responses in chronic schizophrenics are similar to those of normal controls appear to be consistent with the findings of Lee *et al.* (1978) and Carenzi *et al.* (1975) concerning normal [$^3$H]apomorphine binding and normal apomorphine-stimulated adenylate cyclase activity in the postmortem schizophrenic brain.

## 8. CONCLUSION

Several strategies have been employed to validate the DA hypothesis of schizophrenia. The evidence to support this hypothesis has been primarily pharmacologic. Initial studies of the role of DA in the pathophysiology of schizophrenia centered on presynaptic mechanisms. Such studies which included determinations of the levels of DA and its metabolites, and the enzymes associated with its degradation such as MAO, had provided equivocal results. In recent years, the focus appears to have shifted on the postsynaptic mechanisms.

The evidence that schizophrenic illness may be associated with hyperactivity of an increased number of DA receptors is primarily derived from the ligand binding studies using the postmortem schizophrenic brain. The observation that the density of one subtype of DA receptor ($D_2$ receptor) is increased in the postmortem schizophrenic brain is consistent and is reported by several groups of investigators. No differences in the density of other subtypes of DA receptors have been observed in the postmortem brain of schizophrenic patients and normal controls. Whether or not the increase in DA $D_2$ receptor density is related to prolonged treatment with neuroleptics or chronicity is not clear.

The neuroendocrine strategy appears to be very useful for studying DA receptor function in man, since the release of GH and PRL can be regulated by DA agonists and antagonists through their actions on DA receptors in the hypothalamus and the pituitary. The release of GH as a result of stimulation of DA receptors by apomorphine, a DA agonist, provides a useful index of central DA receptor function in schizophrenic illness. Several investigators have studied the release of GH following apomorphine administration in schizophrenic patients and normal controls. Most investigators either do not observe any significant differences or report a slight decrease in GH secretion between chronic schizophrenic patients and normal controls.

In our preliminary report we did not observe significant differences in the mean peak GH levels between chronic schizophrenic patients and normal controls. However, we observed that acute schizophrenic patients had significantly higher GH levels as compared to normal controls and chronic schizophrenic patients. We have replicated our earlier findings using a larger patient population.

A global evaluation of the results of GH studies and the DA receptor binding studies indicates that the schizophrenic population may be divided into at least two groups, each having a different etiology. Whereas chronic schizophrenics may have increased DA $D_2$ receptor density but normal density of other DA recognition sites, acute schizophrenics may have elevated DA receptor response to apomorphine. This response probably is not mediated by $D_2$ receptors.

It therefore appears that more than one strategy should be employed for studying the role of dopaminergic function in schizophrenia. It is in this context that the neuroendocrine strategy may prove most useful.

## 9. REFERENCES

Angrist, B., Sathananthan, G., and Gershon, S., 1973, Behavioral effects of L-DOPA in schizophrenic patients, *Psychopharmacologia* **31**:1.

Angrist, B., Sathananthan, G., Wilk, S., and Gershon, S., 1974a, Amphetamine psychosis: Behavioral and biochemical aspects, *J. Psychiatr. Res.* **11**:13.

Angrist, B., Lee, H. K., and Gershon, S., 1974b, The antagonism of amphetamine-induced symptomatology by a neuroleptic, *Am. J. Psychiatry* **131**:817.

Besser, G. M., Butler, P. W. P., Landon, J., and Rees, L., 1969, Influence of amphetamines on plasma corticosteroid and growth hormone levels in man, *Br. Med. J.* **IV**:528.

Blackard, W. G., and Heidingsfelder, S. A., 1968, Adrenergic receptor control mechanism for growth hormone secretion, *J. Clin. Invest.* **47**:1407.

Bowers, M. B., Jr., 1973, 5-Hydroxyindoleacetic acid (5-HIAA) and homovanillic acid (HVA) following probenecid in acute psychotic patients treated with phenothiazines, *Psychopharmacologia* **28:**309.

Boyd, A. E., III. Lebovitz, H. E., and Pfeiffer, J. B., 1970, Stimulation of human growth hormone secretion by L-DOPA, *N. Engl. J. Med.* **283:**1425.

Brazeau, P., Vale, W., Burgus, R., Ling, N., Butcher, M., Rivier, J., and Guillemin, R., 1973, Hypothalamic polypeptide that inhibits the secretion of immunoreactive pituitary growth hormone, *Science* **179:**77.

Brown, W. A., Corriveau, D. P., and Ebert, M. H., 1978, Acute psychologic and neuroendocrine effects of dextroamphetamine and methylphenidate, *Psychopharmacology* **58:**189.

Camanni, F., Massara, F., Belforte, L., and Molinatti, G. M., 1975, Changes in plasma growth hormone levels in normal and acromegalic subjects following administration of 2-bromo-alpha-ergocryptine, *J. Clin. Endocrinol. Metab.* **40:**363.

Carenzi, A., Gillin, J. C., Guidotti, A., Schwartz, M. A., Trabucchi, M., and Wyatt, R. J., 1975, Dopamine-sensitive adenyl cyclase in human caudate nucleus: A study in control subjects and schizophrenic patients, *Arch. Gen. Psychiatry* **32:**1056.

Chase, T. N., Schnur, J. A., and Gordon, E. K., 1970, Cerebrospinal fluid monoamine catabolites in drug-induced extrapyramidal disorders, *Neuropharmacology* **9:**265.

Checkley, S. A., 1980, Neuroendocrine tests of monoamine function in man: A review of basic theory and its application to the study of depressive illness, *Psychol. Med.* **10:**35.

Crow, T. J., Baker, H. F., Cross, A. J., Joseph, M. H., Lofthouse, R., Longden, A., Owen, F., Riley, G. J., Glover, V., and Killpack, W. S., 1979, Monoamine mechanisms in chronic schizophrenia: Post-mortem neurochemical findings, *Br. J. Psychiatry* **134:**249.

Ettigi, P., Nair, N. P. V., Lal, S., Cervantes, P., and Guyda, H., 1976, Effect of apomorphine on growth hormone and prolactin secretion in schizophrenic patients, with or without oral dyskinesia, withdrawn from chronic neuroleptic therapy, *J. Neurol. Neurosurg. Psychiatry* **39:**870.

Garver, D. L., 1981, Patterns of illness in lithium responsive patients, Scientific Proceedings of the American Psychiatric Association.

Lal, S., De la Vega, C. E., Sourkes, T. L., and Friesen, H. G., 1973, Effect of apomorphine on growth hormone, prolactin, luteinizing hormone and follicle-stimulating hormone levels in human serum, *J. Clin. Endocrinol. Metab.* **37:**719.

Lal, S., Guyda, H., and Bikadoroff, S., 1977, Effect of methylsergide and pimozide on apomorphine-induced growth hormone secretion in men, *J. Clin. Endocrinol. Metab.* **44:**766.

Lee, T., and Seeman, P., 1980, Elevation of brain neuroleptic/dopamine receptors in schizophrenia, *Am. J. Psychiatry* **137:**191.

Lee, T., Seeman, P., Tourtellotte, W. W., Farley, I. J., and Hornykiewicz, O., 1978, Binding of $^3$H-neuroleptics and $^3$H-apormorphine in schizophrenic brains, *Nature (London)* **274:**897.

MacKay, A. V. P., Dobie, A., Bird, E. D., Spokes, E. G., Quik, M., and Iversen, L. L., 1978, $^3$H-spiperone binding in normal and schizophrenic postmortem human brain, *Life Sci.* **23:**527.

MacKay, A. V. P., Bird, E. D., Spokes, E. G., Rosser, M., Iversen, L. L., Creese, I., and Snyder, S. H., 1980, Dopamine receptors and schizophrenia: Drug effect or illness?, *Lancet* **II:**915.

Martin, J. B., 1973, Neural regulation of growth hormone secretion, medical progress, *N. Engl. J. Med.* **288:**1384.

Meltzer, H. Y., and Stahl, S. M., 1976, The dopamine hypothesis of schizophrenia: A review, *Schizophr. Bull.* **2:**19.

Meltzer, H. Y., Busch, D., and Fang, V. S., 1981, Hormones, dopamine receptors and schizophrenia, *Psychoneuroendocrinology* **6:**17.

Owen, F., Crow, T. J., Poulter, M., Cross, A. J., Longden, A., and Riley, G. J., 1978, Increased dopamine receptor sensitivity in schizophrenia, *Lancet* **II:**223.

Pandey, G. N., Garvey, D. L., Tamminga, C., Ericksen, S., Ali, S. I., and Davis, J. M., 1977, Post-synaptic supersensitivity in schizophrenia, *Am. J. Psychiatry* **134:**518.

Pandey, G. N., Casper, R. C., Davis, J. M., Ali, S. I., and Garver, D. L., 1979, Elevated Growth Hormone Response in Schizophrenia: Syllabus and Scientific Proceedings, American Psychiatric Association, p. 345.

Post, R. M., Fink, E., Carpenter, W. T., Jr., and Goodwin, F. K., 1975, Cerebrospinal fluid amine metabolites in acute schizophrenia, *Arch. Gen. Psychiatry* **32**:1063.

Randrup, A., and Munkvard, I., 1970, Biochemical, anatomical and psychological investigations of stereotyped behavior induced by amphetamines, in: *Amphetamines and Related Compounds* (E. Costa and S. Garattini, eds.), pp. 695–713, Raven Press, New York.

Rimon, R., Roos, B., Rakkolainen, V., and Alanen, Y., 1971, The content of 5-hydroxyindoleacetic acid and homovanillic acid in the cerebrospinal fluid of patients with acute schizophrenia, *J. Psychosom. Res.* **15**:375.

Rotrosen, J., Angrist, B. M., Gershon, S., Sachar, E. J., and Halpern, F. S., 1976, Dopamine receptor alteration in schizophrenia: Neuroendocrine evidence, *Psychopharmacology* **51**:1.

Rotrosen, J., Angrist, B., and Paquin, J., 1978, Neuroendocrine studies with dopamine agonists in schizophrenia, *Psychopharmacol. Bull.* **14**:14.

Seeman, P., 1981, Brain dopamine receptors, *Pharmacol. Rev.* **32**:229.

Seeman, P., Lee, T., Chau-Wong, M., and Wong, K., 1976, Antipsychotic drug doses and neuroleptic/dopamine receptors, *Nature (London)* **261**:717.

Yaryura-Tobias, J. A., Wolpert, A., Dana, L., and Merlis, S., 1970, Action of L-DOPA in drug induced extrapyramidalism, *Dis. Nerv. Syst.* **31**:60.

CHAPTER 29

# Exogenous Peptides and Schizophrenia

MAN MOHAN SINGH and STANLEY R. KAY

1. INTRODUCTION

The discovery that a group of systemic peptide hormones belonging to the hypothalamic–pituitary system is distributed in other parts of the central nervous system (CNS) and have neuronal functions in addition to their hormonal duties, and that another group of neuroactive peptides is distributed both in the brain and in the digestive tract and related structures, has been a development of major importance in neuroendocrinology and the neurosciences (Krieger and Martin, 1981a,b). The neuronal and hormonal activities of the neurohumoral peptides, though coordinated in some cases, are often served by different components of the peptide molecules (Marx, 1975a,b; de Kloet and de Wied, 1980). The brain and gut functions of the brain–gut peptides may also be independent, although the observation of prompt increase in gastric secretion by intrahypothalamic injection of gastrin in rats suggests that the two may, in some cases, be connected (Tepperman and Evered, 1980). The possibility of the latter is also suggested by the finding that cholecystokinins (CCKs) in the gut may come from the brain along the vagus (Dockray, 1979). However, CCKs are unusual among neuropeptides with both CNS and peripheral distributions, in that their CNS concentrations are much greater than those in the periphery (Emson *et al.*, 1980; Krieger and Martin, 1981a).

The neuropeptide physiology may prove to be of great significance in biological psychiatry. One reason is the rapidly growing evidence for the psychophysiological and behavioral actions of many of the neuropeptides. Thus, various pituitary hormones, including adrenocorticotropin, various components of β-lipotropin (endorphins, enkephalins, and melanotropins), vasopressin, and oxytocin, have actions on active and passive avoidance behavior,

---

MAN MOHAN SINGH • Creedmoor Psychiatric Center, Queens Village, New York 11427.   STANLEY R. KAY • Bronx Psychiatric Center and Albert Einstein College of Medicine, New York, New York 10461.

learning, memory, attention, and arousal (de Kloet and de Wied, 1980; Kovacs et al., 1979; Bohus, 1979). Sensory processes of various kind, including pain, seem to be affected by endorphins and substance P, and possibly by vasoactive intestinal peptide (VIP) and CCKs (Henry, 1977; Akil et al., 1978; Oyama et al., 1980; Hökfelt et al., 1980a; Krieger and Martin, 1981a,b). Another reason is the finding that certain neuropeptides are found alongside classical neurotransmitters, such as dopamine, norepinephrine, and acetylcholine, in neuronal systems suspected to be important in psychiatric disorders (Hökfelt et al., 1980a,b). Particularly interesting in this regard is the coexistence of CCKs and dopamine in the neurons of the mesolimbic dopaminergic system (Hökfelt et al., 1980a,b,c), which has been implicated in schizophrenia. Based on the observation that CCKs inhibit dopamine release in this system (Hökfelt et al., 1980a), it has been suggested that CCKs may be inhibitory modulators of mesolimbic dopamine neurons and that their deficiency may be a factor in the dopaminergic hyperactivity of this system suspected in certain forms of schizophrenia. It has, therefore, been proposed that administered CCKs may ameliorate schizophrenia by reducing excessive mesolimbic dopaminergic activity. In support of this idea, neuroleptic-like activity of CCK-8 has been observed in mice (Zetler, 1980b,c, 1981).

Of the various neuropeptides, endorphins have so far attracted the greatest investigative attention in psychiatry. Endorphin excess and endorphin deficiency have both been proposed and tested in schizophrenia, but neither hypothesis has received impressive support, and indications are that at least the opioid components of the brain endorphin mechanisms are probably not of major importance for the pathogenesis of schizophrenia (Mackay, 1979; Deakin, 1980; Emrich, 1981). The possibility that the behavioral effects of endorphins not dependent upon opioid receptors may be more relevant to schizophrenia has just begun to be explored (*infra*).

Some of the other neuropeptides that have not yet attracted much attention in psychiatry may well prove to be of greater significance for the pathogenesis of schizophrenia than endorphins. CCKs acting as neuromodulators in the mesolimbic dopaminergic system were mentioned earlier. VIP may also be of interest because of its suspected role as a vasodilator in the regulation of brain circulation (Heistad et al., 1980; Emson et al., 1980) and its coexistence with acetylcholine (Ach) (Larsson et al., 1976). There is evidence to suggest that in chronic nonparanoid schizophrenics, prefrontal brain circulation is relatively reduced and that it fails to react to tasks of the type that produce increased frontal circulation in normal individuals (Ingvar and Franzén, 1974; Franzén and Ingvar, 1975). Also, these types of patients seem to be unusually resistant to the central as well as peripheral effects of anticholinesterases and muscarinic blockers (Singh and Lal, 1982). These defects could conceivably be related to impaired VIP-acetylcholine regulation of the frontal circulation. Major concentrations of VIP are in areas of focal interest in schizophrenia research, i.e., the limbic frontal areas (Brodman areas, 6, 10, 11, and 13) and limbic nuclei (Emson et al., 1979, 1980). These areas, along with the basal ganglia, also account for most of the ACh concentrations in the brain (Robinson, 1983). Processes of VIP and cholinergic neurons, disposed in similar fashion, have been demonstrated in relation to pial and other cerebral blood vessels (Owman *et al.*, 1974; Larsson *et al.*, 1976), and *in vitro* as well as *in vivo* studies (through intracarotid injections) have shown that both VIP and ACh increase cerebral blood flow, especially in the gray matter, and that whereas ACh seems to have a direct vasodilatory action, VIP may have more of a modulatory role (Larsson *et al.*, 1976; Heistad *et al.*, 1980; Krieger and Martin, 1981a,b). Furthermore, there is strong evidence for an ACh role in the activation of arousal mechanisms (Vanderwolf and Robinson, 1981); preliminary data suggest a similar role for VIP (Said, 1979). Disturbance of arousal and attention is an important aspect of the pathophysiology of schizophrenia (for references see Kay and Singh, 1979; Singh and Lal, 1982).

A relatively recent development in the neuropeptide field has been the discovery of psychoactive food-derived or synthetic exogenous peptides (*infra*) and the emergence of the possibility that such peptides, entering the body through oral or systemic routes, may play a pathological or beneficial role in brain functions. Such a possibility is increased by the findings suggesting that biologically active peptides or peptide–immune complexes may cross the gut barrier as well as the blood–brain barrier. This chapter reviews some data on this issue as well as observations implicating exogenous psychoactive peptides in the pathogenesis and treatment of schizophrenia.

## 2. ACCESS OF BIOLOGICALLY ACTIVE EXOGENOUS PEPTIDES TO THE BRAIN

For any exogenous peptide to affect brain functions, it has either to cross various barriers to reach, and directly act on, the target neurons or to produce extraneuronal changes, such as alteration of the blood–brain barrier, interference with chemical or neural feedback to the hypothalamic–pituitary system and/or the brain, formation of immune complexes with access to the brain, or inhibition of the transport of neurotransmitter precursors or other necessary materials across the blood–brain barrier, that can indirectly influence neuronal functioning. In this context, it should be noted that there may be two different blood–brain barriers, i.e., the main or "classical" vascular barrier and the choroid plexus–cerebrospinal fluid system, which differ in structure and function, and have been postulated to have different implications for brain disorders (Rudin, 1980). In a series of papers, Rudin (1980, 1981a,b) has argued that the choroid plexus or the second blood–brain barrier, which has close association with the periventricular limbic or primary personality brain, may be particularly important in the development of mental illnesses such as schizophrenia, in which, in contrast to the focal brain defects associated with attacks through the classical blood–brain barrier, there is a general and fundamental change in the whole personality. He has proposed that, acting through this route, exogenous peptides, immune complexes, or viruses may be responsible for the production of various forms of schizophrenia.

One type of evidence for the access of peptides to the brain has come from studies of the behavioral and neurophysiological effects of the systemically administered neuropeptides. In reviews of such studies (Marx, 1975a,b; Kastin et al., 1979a,b; Klee and Zioudrou, 1980), it has been noted that, when systemically administered, neuropeptides such as MSH, somatostatin, TRH, LHRH, ACTH, and endorphins have CNS actions in animals and man that can be demonstrated through tests such as dopa potentiation, serotonin potentiation, oxotremorine reversal, conditioned avoidance, complex-maze learning, analgesia, memory, and attention. Electrophysiological studies have led to the same conclusion (e.g., Koranyi et al., 1977). The peptides, however, seem to have a short life in the systemic circulation and may be undetectable in blood by the time CNS effects appear (Kastin et al., 1979a,b). The CNS-related actions may not only occur sometime after the peptides are injected into the circulation but also may be long-lasting (Plotnikoff et al., 1975; Koranyi et al., 1977).

More direct evidence for the ability of biologically active peptides to cross the blood-brain barrier has also been reported (Greenberg et al., 1976; Hemmings, 1978a,b; Rapoport et al., 1980). In a series of studies in adult rats, Hemmings (1978a,b) has shown that the feeding of $^{125}$I- or $^{131}$I-labeled proteins, including ferritin, IgG, hemoglobin, and gliadins, leads to the appearance of peptides and large-molecule breakdown products in the brain and other body tissues, and that these products can not only be demonstrated within the brain cells but are usually, if not always, antigenically intact, i.e., they are recognized by labeled antisera raised against the ingested proteins. Of particular interest is the apparent permeability of the blood-brain barrier to $\alpha$-gliadin fragments, since this protein, which forms part of cereal-grain glutens, has been implicated in the pathogenesis of celiac disease (gluten enteropathy) as well as schizophrenia (infra).

These studies of Hemmings, corroborated by other investigations (Williams and Hemmings, 1978; Williams, 1979b), also show that intact molecules or macromolecular fragments of isotopically labeled ferritin, $\gamma$-globulin, hemoglobin, and gliadin can be absorbed into circulation from the gut in rats. This absorption, which is believed to occur through some form of assisted diffusion, is demonstrable both in young rodents, where it is quite marked, and in adult rodents, where it is significant (e.g., 0.5 to 1.0% of $\alpha$-gliadin) but much less marked (Hemmings, 1978a; Williams, 1979b). However, Hemmings (1978a) noted that the proteins studied were relatively resistant to proteolysis in rats and that observations with such materials may not be applicable to proteins subject to full hydrolysis in the gut. This means that his observations on rodents may be generalizable to humans only to the extent that certain proteins are partially or fully resistant to proteolysis or in cases where, due to genetic factors or intestinal pathology, proteins are not fully hydrolyzed in the gut or the gut barrier becomes defective. However, there is evidence to suggest that many whole proteins and large protein fragments can be absorbed through the gut in humans—more readily in the neonatal period and less so in adult life (see Matthews and Payne, 1975; Matthews and Adibi, 1976, for reviews and references). Lipid-soluble peptides and those resistant to proteolytic enzymes are most likely to be absorbed. Small hydrophobic peptides may enter through diffusion or the peptide transport mechanisms, and there is good evidence to suggest that several small peptides appear intact in the peripheral blood during absorption. At least some neuroactive peptides, e.g., TRH, which is extremely resistant to proteolytic enzymes, have been shown to be absorbed through the gut and the possibility exists that there may be active transport mechanisms for the absorption of certain biologically active peptides. Finally, in relation to the issue of permeability of various tissue barriers to psychoactive peptides, it should be noted that many of the polypeptide hormones have been reported to be able to enter cells (Kolata, 1978). Despite being large charged molecules, they seem to be able to cross the cell membranes and may, therefore, produce long-term intracellular changes in addition to short-term effects related to actions on specific surface receptors. Hemmings (1978a), in his studies on absorption of a α-gliadin, found that some of the labeled breakdown products of this food protein may have entered into brain cells. This seems significant in that not only may biologically active exogenous peptides be able to cross various barriers to reach the brain but, by entering the neurons, they may be able to produce long-term changes.

## 3. CEREAL GRAIN PROTEINS AND SCHIZOPHRENIA

In a series of papers over more than a decade, Dohan (1966a,b, 1969a,b, 1970, 1978, 1980) has elaborated the hypothesis that polypeptides derived from the proteins in cereal grains (wheat, rye, barley, and oats) may be an exogenous factor that interacts with a genetic predisposition to evoke one or more forms of schizophrenia. Several types of indirect evidence have gone into the making of this hypothesis, and in recent years there has been some direct suggestive evidence in schizophrenic patients.

Bender noted in 1953 that the schizophrenic child was unusually subject to celiac disease—a relatively uncommon, genetically determined condition now known to be due to intolerance to cereal grain proteins. Graff and Handford (1961) reported that 4 of 37 adult male schizophrenics admitted over a period of 1 year had a history of celiac disease in childhood. Dohan (1969a,b, 1970) later made inquiries of a number of clinicians and noted that schizophrenia and celiac disease seemed to occur some time in the life of the same person far more frequently than expected by chance alone. The lifetime probability for celiac disease in the general population is about 1 in 3000 (Black, 1964), while that for schizophrenia is about 1 in 100. The chance probability for the two to occur in the same person would, therefore, be about 0.03%. The actual figures from the responses to Dohan's inquiries were between 1 and 10%, i.e., 30 to 300 times greater.

Dohan also carried out an epidemiological survey of admissions to mental hospitals during World War II in Norway, Finland, Sweden, Canada, and the USA, in relation to changes in wheat and rye consumption and a number of other parameters such as unemployment, alcohol consumption, availability of mental hospital beds and doctors, and the type of involvement, or the lack of it, in the war (Dohan, 1966a,b). The results showed a highly significant correlation between changes from prewar levels in the consumption of wheat and rye and the admissions for schizophrenia, with reduced consumption due to wartime shortages being associated with a considerable reduction in admissions. Such a correlation was not obtained for the other psychoses, nor did changes in schizophrenia admissions correlate with the other factors studied.

This led Dohan to suggest that schizophrenia and celiac disease, both probably polygenic, may be genetically related and have perhaps one or more genes in common, while their phenotypic differences are determined by genes peculiar to each condition. He also suspected that constituents of cereal grains may in some way be implicated in schizophrenia, as are the cereal grain glutens in the pathogenesis of celiac disease. However, at this stage, his suggestion of a genetic relationship between the two conditions rests on only rather anecdotal clinical information, for no systematic investigation of this question has been carried out.

Some recent epidemiological data seem to further support this thesis. Dale (1981) has reported marked differences in the prevalence rates for schizophrenia in various Pacific Island populations of Micronesia. Starting with almost total absence of schizophrenia in the Marshall Islands and the Polynesian atolls to the east, the rates became progressively higher in moving toward the west and closer to the continental mass of Asia, and presumably an increasingly closer contact with modern civilization and greater exposure to cereal grains, and substances such as beer which contain grain products. Dohan (personal communication) has calculated a correlation of 0.8 ($p < 0.04$) between the prevalence rates in various parts of Micronesia and the dollar value of wheat and barley beer imports. Dohan has also obtained information suggesting that in a health survey between 1947 and 1955 in the Yap Island in Micronesia, when inhabitants ate some rice but no other grains, no overt schizophrenics were found. In Dale's survey (*supra*), which came after over 30 years of Americanization and increasing consumption of cereal grains and barley beer, this western Micronesian island has a prevalence rate for schizophrenia of 9.7 per thousand—the highest for any part of Micronesia. Further epidemiological information has come from Papua, New Guinea, and the Solomon Islands (Dohan *et al.*, 1982) suggesting that in tribal villages and remote areas where diet has little or no cereal grains, schizophrenia is virtually nonexistent, while in the more Westernized coastal regions, where cereal grains have become part of the native diet, schizophrenia is much more common.

Both the association between celiac disease and schizophrenia and the possible role of wheat gluten-type proteins in the latter, are also suggested by the psychiatric and psychosocial characteristics of celiacs and the often dramatic improvement in these characteristics with a gluten-free diet.

Many authors have described certain personality changes as characteristic of celiac patients and also noted that such changes are not observed in cases with comparable degrees of malabsorption and malnutrition from other causes, such as ulcerative colitis (Käser, 1961; Asperger, 1961; Townley and Anderson, 1967; Paulley, 1959; Challacombe et al., 1972; Anderson et al., 1972; Cooke, 1976). Definite psychic symptoms, sometimes reaching psychotic proportions, are believed to occur in all celiacs. The patients are said to be quiet, inward, disinterested, difficult, weepy, repetitious, avoidant and anxious when approached, lacking in exploratory behavior, uncommunicative, paranoid, and at times delusional and psychotic. In other words, they show many schizoid characteristics. These features may be present in the so-called "preceliac" syndrome (Daynes, 1956) and are among the earliest signs and symptoms to improve with a gluten-free diet. Conversely, feeding a relatively heavy dose (30 g) of gluten to a celiac patient can result in an acute psychotic reaction (Dohan, 1970, 1978). Normal humans fed up to 150 g a day of gluten for many weeks do not develop psychiatric symptoms (Levine et al., 1966). Emotional stress and conflict have been found to be important in the precipitation of symptoms in celiacs (Prugh, 1951; Paulley, 1959; Grant, 1972), as they are in the development of symptoms and relapses in schizophrenics (Brown and Birley, 1968; Brown et al., 1972). Also, as with schizophrenics (Bateson et al., 1956; Lidz, 1973), the relationships between celiacs and their parents, especially mothers, are often markedly disturbed.

A possible link between schizophrenia and celiac disease is also suggested by the occurrence of motor end-plate alterations in schizophrenics, similar to those seen in celiac disease (Crayton and Meltzer, 1976). Further support is provided by the similarity in the fasting plasma levels of amino acids in schizophrenics and celiacs and the difference between these levels and those in healthy controls (Manowitz, 1978).

Prompted by such observations, Dohan et al. (1969) and Dohan and Grasberger (1973) attempted to test the effect of a cereal grain-free diet, and then of secretly adding wheat gluten to this diet, on the response of schizophrenics to routine treatment. The time it took to transfer newly admitted patients from a closed ward to an open ward was used as the criterion of response to treatment.

During a 175-day period, through random assignment, 47 schizophrenics received a cereal grain-free, milk-free diet, while 55 schizophrenics received a high-cereal-grain diet. The median length of stay on the closed ward for the two groups combined was 7 days. However, 62% of the schizophrenics in the cereal-free group were released *before* this day, whereas only 36% of the high-cereal group were released in the same time ($p = 0.009$). The differences in proportions were also significant at the end of days 5, 6, 8, and 9. The average length of stay on the closed ward was 17.5 days for the cereal-free group and

30.6 for the high-cereal group. The nonschizophrenics admitted during the study period showed no apparent dietary effect. Fifty such patients were divided between the two dietary groups; half of each group were released on or before the median day.

However, neither the patients nor the staff were blind to the diets used in this part of the study. Therefore, during a second period of 143 days, patients were again randomly assigned either to a high-cereal diet or to the cereal-free diet, but this time wheat gluten was secretly added to the latter so that both groups were actually on a high-gluten diet. The schizophrenics in the two groups now did not differ significantly in response, suggesting that staff bias or a nonspecific psychological effect of the diet had probably not accounted for the more rapid release of the cereal-free-group patients in the first period. The nonschizophrenic dietary groups did not differ in releases.

A subsequent study examined the effect of test diets, given during stay on the locked ward, on the total length of stay until discharge from the hospital (Dohan and Grasberger, 1973). Only patients admitted for the first time to that particular hospital were included. The patient samples came from the above two study periods plus those from a third period of 148 days during which subjects were randomly assigned to a cereal grain-free, milk-free diet or a high-cereal diet. The experimental diets were given only during the patients' stay in the locked ward, which in most cases was less than 15 days. Thereafter, they received the ordinary hospital diets until discharge. The results showed that during the first 90 days after admission to the hospital, the schizophrenics assigned to the cereal grain-free, milk-free diet were discharged more than twice as fast as those in the high-cereal group.

From these and the previous findings, Dohan and his associates concluded that the ingestion of cereal grains may be pathogenic in those with the genotype for schizophrenia and that, as in celiac disease, the harmful factor is contained in the gluten fraction of wheat and other cereal grains.

A more sophisticated and detailed clinical test of this hypothesis was conducted by the authors (Singh and Kay, 1976a,b,c; Singh, 1978). In view of the fact that schizophrenia is both cross-sectionally and longitudinally heterogenous, we studied a group of carefully selected schizophrenics according to an A–B–A' design (Fig. 1) in which each patient served as his own control and the cross-sectional as well as longitudinal aspects of schizophrenia could be considered. Fourteen patients, six with a progressive course of illness (all hebephrenics) and eight with an episodic course of illness (four catatonics, one possible hebephrenic, three paranoids), were involved. There were 11 females and 4 males of single status and 9 to 15 years of education. Five patients were white and the rest black. The age at onset varied from 15 to 23 years, while the duration of illness was 0 to 1 year for three cases, 2 to 5 years for five cases, and 6 to 13 years for six cases. The patients were maintained on a strict cereal grain-free, milk-free diet on a locked ward and observed drug-free for 2 weeks, and then on neuroleptic medication for 12 weeks. Along with medication, the patients were given, "double-blind," a flavored drink that contained soy flour as placebo control in the first and last 4 weeks (i.e., periods A and A') and 30 g a day of wheat gluten in the middle 4 weeks (period B) of the 12-week period on medication (Fig. 1). The neuroleptic dosages were titrated against clinical response in the first 2 weeks of period A and thereafter held constant. It was hypothesized that if wheat gluten were pathogenic to schizophrenics, the gluten challenge would interrupt or reverse therapeutic progress, which would be reinstated after the challenge ended. Without a significant effect of the test intervention, one would expect progressive clinical improvement in patients because neuroleptics are proven therapeutic agents in schizophrenia.

In analyzing the data, it was important to recognize this progression of events or order effect, that is, the three periods were not independent but were related to each other as successive components of a continuous therapeutic process. Therefore, both periods A and A' needed to be considered simultaneously in relation to test period B. Since the unaltered course of therapeutic changes in these patients was unknown, we calculated the deviation in course produced by the test intervention by subtracting B from $(A + A')/2$ using ratings at the end of each 4-week period. As illustrated in Fig. 1, this gave a very conservative estimate of the gluten effect as the deviation was determined from an estimated unaltered course much less favorable than would normally be expected.

The patient-by-patient differences values were obtained for each of the parameters and then analyzed parametrically with correlated $t$ tests, and also nonparametrically with $\chi^2$ tests. A particular reason for the latter, which is considered a less powerful test, was that such a test is valid in scales lacking

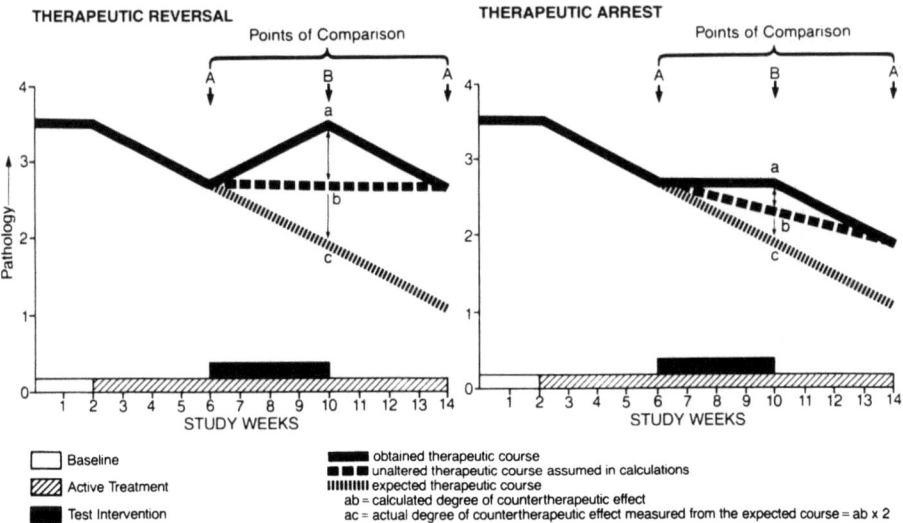

FIGURE 1. Longitudinal A–B–A' research design and statistical evaluation of the effect on therapeutic course of an intervention in period B.

true interval properties, as is probably the case with psychiatric scales, and requires no assumptions about distribution.

The main measures were 33 parameters on a Psychopathology Rating Schedule (Singh and Kay, 1975a) rated every 2 weeks by a psychiatrist and a pyschologist after in-depth interviews. In the parametric analyses, the hypothesized gluten effects were apparent in 26 psychopathology parameters and reached statistical significance in 13 (Table 1). Similarly, on $\chi^2$ analysis, in 24 of the parameters, the patients who showed the hypothesized gluten effects outnumbered those who showed the opposite effects; these differences were statistically significant for 7 measures (Table 1). Considering the patients by parameters, in 11 cases (79%) a majority of the parameter-by-parameter changes were in the hypothesized direction, while the opposite was true in only three cases ($\chi^2 = 4.57$, $df = 1$, $p < 0.05$).

Theoretically, in a univariate analysis of the kind used, at $p < 0.10$ one might get significant effects by chance alone in 3 of the 33 parameters. The actual number of significant effects was much greater and all the parameters shown in Table 1 changed in the hypothesized direction. In fact, there was only one parameter (Passive/Apathetic Withdrawal) in which changes in the opposite direction were significant at $p < 0.10$. Furthermore, many of the parameters involved were those characteristic for schizophrenia. Also, most of the patients, overall, changed in the hypothesized direction.

The study also had a built-in opportunity for cross-validation of the test of the hypothesis discussed above. Completely independently of the psycho-

TABLE 1
Psychopathology Ratings: Significant Results of Nonparametric and Parametric Tests (One-Tailed) of Hypothesized Wheat Gluten Effects, i.e., $B > (A + A')/2$

| Parameter | Chi square | | Correlated $t$ test | |
|---|---|---|---|---|
| | $\chi^2$ | $p$ | $t$ | $p$ |
| Poor Rapport | 1.34 | — | 2.25 | <0.02 |
| Preoccupied Behavior | 6.24 | <0.02 | 2.72 | <0.01 |
| Hostile/Fearful Social Avoidance | 3.00 | <0.10 | 2.22 | <0.02 |
| Uncooperativeness | 0.34 | — | 1.38 | <0.10 |
| Stereotyped Thinking | 3.00 | <0.10 | 1.35 | <0.10 |
| Difficulty in Abstract Thinking | 1.60 | — | 2.63 | <0.02 |
| Poor Judgment and Insight | 0.40 | — | 1.90 | <0.05 |
| Bizarre and Unusual Thought Content | 0.70 | — | 1.45 | <0.10 |
| Grandiosity | 3.60 | <0.10 | 0.98 | — |
| Elation | 3.58 | <0.10 | 1.59 | <0.10 |
| Depression | 1.60 | — | 1.40 | <0.10 |
| Anxiety | 1.00 | — | 1.89 | <0.05 |
| Tension State | 3.00 | <0.10 | 2.19 | <0.02 |
| Poor Attention | 3.00 | <0.10 | 1.16 | — |
| Altered State of Awareness | 0.00 | — | 1.46 | <0.10 |

TABLE 2
Social Avoidance Behavior and Social Participation Scales

|  | Daytime | | | Evening | |
| --- | --- | --- | --- | --- | --- |
|  | Soy flour Weeks 3–6 | Wheat gluten Weeks 7–10 | Soy flour Weeks 11–14 | Soy flour Weeks 3–6 | Wheat gluten Weeks 7–10 | Soy flour Weeks 11–14 |
| Withdrawal | $+p < 0.001$ | ns | ns | $+p < 0.01$ | ns | ns |
| Nonverbal Communication | $+p < 0.001$ | ns | ns | $+p < 0.05$ | ns | $+p < 0.05$ |
| Affective Responsiveness | $+p < 0.001$ | ns | ns | ns | ns | $+p < 0.05$ |
| Overactive Social Avoidance | ns | ns | $-p < 0.05$ | $+p < 0.01$ | ns | $+p < 0.05$ |
| Hostile/Fearful Social Avoidance | ns | $-p < 0.01$ | ns | $+p < 0.02$ | ns | $+p < 0.05$ |
| Social Participation | $+p < 0.01$ | ns | ns | $+p < 0.01$ | ns | $+p < 0.05$ |

[a] Results of Cox-Stuart tests of trend based on daily group mean ratings by daytime and evening nursing staff. Probability levels and directions of change (+, clinical improvement, −, clinical worsening) are given for significant trends.

pathology ratings, the ward staff gave twice-daily ratings on six different scales for social behavior (Table 2). Cox–Stuart tests of trend that could be applied to these showed significant therapeutic trends both before and after, but none during, the gluten challenge period.

At the time of discharge from the research unit, each patient received a therapeutic outcome rating that allowed him to be classified in a good- or poor-outcome group. As shown in Fig. 2, the hypothesized gluten effect predominated in the poor-outcome patients. Thus, in eight out of nine poor-outcome cases, a majority of the psychopathology and social parameters changed in the hypothesized direction with the gluten challenge ($\chi^2 = 5.44$, $df = 1$, $p < 0.02$), while in only two of the five good-outcome cases was this so. The poor-outcome group was comprised of six hebephrenics (all nuclear schizophrenics with progressive course), two catatonics, and only one paranoid. This suggested the possibility that the pathological gluten effect may be a particular characteristic of the poor-prognosis, nuclear, nonparanoid (mostly hebephrenic) schizophrenics. The possible significance of this was further highlighted by the fact that

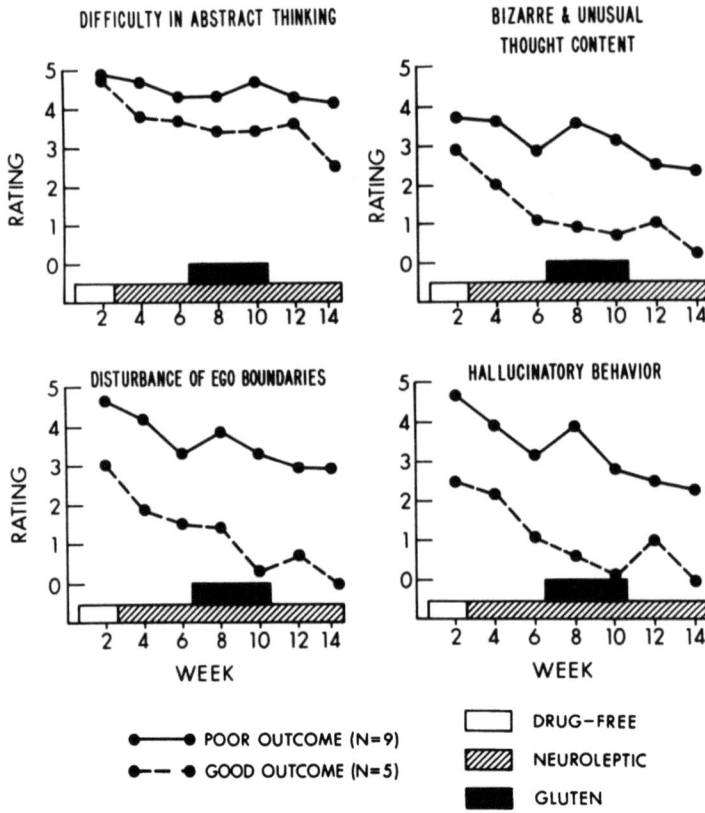

FIGURE 2. The effect of wheat gluten challenge on the therapeutic course of eight characteristic aspects of schizophrenia in the poor- and good-outcome cases.

the countertherapeutic effects of anticholinergic agents observed in previous studies of similar research design (Singh and Smith, 1973; Singh and Kay, 1975a,b,c) were found, on subgroup analyses, to be predominantly a characteristic of the good-prognosis, good-therapeutic-outcome patients (Singh and Kay, 1978a; Singh, 1978). This suggested that the schizophrenic process may consist of a neuroleptic-responsive component, which is accentuated by anticholinergics, and a neuroleptic-resistant component, which is differently enhanced by wheat gluten in patients receiving neuroleptics that continue to control the drug-responsive component (Singh, 1978). This may mean one of two things: (1) there are two types of schizophrenics, those who are gluten sensitive and those who are not, or (2) the neuroleptics are able to block a certain part of the disease process even in the presence of wheat gluten so that what we saw in our study was just a part of the pathogenic effect of wheat gluten in schizophrenics.

The contrast between wheat gluten effects and anticholinergic effects was also apparent in the patterns of clinical parameters most involved (Singh and

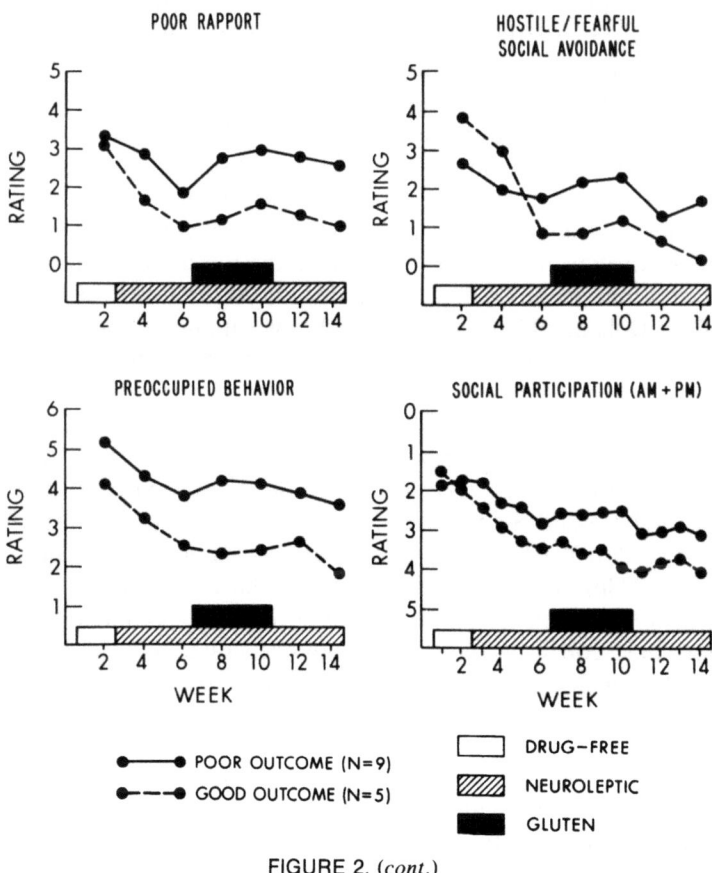

FIGURE 2. (cont.)

Kay, 1978a; Singh, 1978). The gluten effect was seen particularly in clinical features such as Preoccupied Behavior (Autism), Poor Rapport, Difficulty in Abstract Thinking, Poor Judgement and Insight, and Stereotyped Thinking, which would be expected especially to characterize the chronicity-prone, poor-prognosis forms of schizophrenia. However, it was much less apparent in the "acute" symptoms such as Sleeplessness, Disorientation, Disorganized Thinking, Hostility and Uncooperativeness, which were all significantly increased by the anticholinergics. This not only emphasized the differentiation noted above but highlighted the fact that, contrary to popular practice, especially in the so-called multivariate approaches to data analysis, various psychopathological parameters cannot be regarded as being clinically equivalent in significance or as having a homogeneity of covariance. It would be inadvisable, therefore, to lump them all together.

Analyses of neuroleptic dosages showed that the gluten-period changes were not due to patients receiving smaller amounts of medication. If anything the dosages were somewhat higher in this period than in the other periods. However, our study did not unequivocally exclude the possibility that gluten effects were due to interference with neuroleptic absorption, or to any interference with neuroleptic metabolism, bioavailability, or pharmacological activity. The predominance of gluten effects in the treatment-resistant patients seemed to argue against this. This conclusion was further supported by the observation that the extrapyramidal reactions reflecting pharmacological effects of neuroleptic medication were as frequent in the gluten period as in the gluten-free periods. A study published since then in which the effect of wheat gluten feeding on haloperidol absorption was investigated in rats showed no significant absorption interference but raised the possibility that wheat gluten may affect the pharmacokinetics of the drug in the brain (Freed et al., 1978; Singh, 1979b; Dohan, 1979b). More recently, wheat gluten has been shown not to affect haloperidol blood concentrations in schizophrenics (Luchins et al., 1980).

Our work has been contested on technical grounds for not employing a placebo control group and for not applying a multivariate statistical test or a more "robust" nonparametric test such as Wilcoxon matched pairs signed-ranks test (Meltzer and Stahl, 1976; Ziemba et al., 1978; Levy and Weinreb, 1976; Smith, 1976). We have discussed these issues in some detail (Singh and Kay, 1976c, 1978b) and have maintained that the criticisms are not only unwarranted but show a poor understanding of the basic technical problems our studies have been concerned with. Thus, our decision to use multiple univariate tests was determined by the fact that schizophrenia is a heterogeneous disorder and that the measures used were not highly correlated as evident from the observation that attempts to find one or two highly discriminative, robust, and reliable psychopathological factors for schizophrenia have been unsuccessful and, indeed, some factor-analytical studies have generated as many as 25 different factors in this condition (Fleiss et al., 1971).

Unfortunately, no large-scale convincing investigation has been reported so far to settle issues raised by Dohan's and our work and to unequivocally prove or reject the role of cereal grain glutens in schizophrenia. Only two relatively minor clinical studies have been published since our report in 1976.

Rice et al. (1978) studied 16 extremely chronic schizophrenics with an average of 13 years of almost continuous hospitalization. It was an uncontrolled investigation in which for the first 4 weeks extra wheat gluten was added to the normal hospital diet, while in the next 8 weeks cereal grains and milk were substantially, but not completely, eliminated from the diet (see Dohan, 1979a). Medications were continued unchanged throughout. One patient worsened in the first period. This patient and one other patient improved considerably in the second period; the other 14 cases showed no apparent effect. In commenting on this study, Dohan (1979a) and Singh (1979a) pointed out that not only was the choice of a highly chronic population, in which the disease process had probably become irreversible, inadvisable for the test of any etiological hypothesis, but also that the exclusion diet was not strict and was given for a period that may be too short for very chronic cases to show improvement. It was also noted (Singh, 1979a) that the patient sample of this study was very much more chronic and intractable than that investigated by Singh and Kay (1976a,c) and that it had very few nonparanoid schizophrenics, who accounted for 80% of the total sample and most of the "gluten responders" in the Singh and Kay study.

Recently, Potkin et al. (1981) reported a negative study involving only eight cases. The patients, though not as chronic as those in Rice et al. study, were much more chronic and intractable than those in Singh and Kay study. Three cases were of paranoid type and five of chronic undifferentiated type, and all cases were "well-stabilized" on traditional treatment. Thus, many of the criticisms discussed above are also applicable to this study. It suffers further from having a very small sample of a population known to be heterogeneous; the chances of a type II error are large in a study of this kind.

So, an unequivocal clinical confirmation of the role of cereal grain proteins in one or more types of schizophrenia must wait further work that takes into account the shortcomings of the previous studies. In the meantime, a number of biochemical and experimental reports of relevance to this hypothesis have appeared.

In view of the suspected etiological link between celiac disease and schizophrenia, a number of studies have tested schizophrenics with immunological tests of the type known to be positive in a high proportion of celiacs. Dohan et al. (1972) examined a large number of sera from hospitalized patients with serious psychiatric and neurological problems, as well as those from normal controls, for the presence of gliadin antibodies measured by the ability of sera to agglutinate gliadin-sensitized latex particles. The test was positive in only 3.1% of controls but in 20.1% of schizophrenics, suggesting that schizophrenia is associated with increased permeability to ingested gliadin and increased frequency of antigliadin antibodies. However, the finding proved to be nonspecific since other seriously ill hospitalized psychiatric patients had the same frequency (20.6%) of antigliadin antibodies as did the schizophrenics. Hospitalized neurological and other medical-surgical patients also had an increased frequency of antigliadin antibodies (13.7%) though not to the extent that the psychiatric patients did. The "positive" patients in the nonpsychiatric group were generally those with "hopeless" crippling diseases. The suggestion, therefore, was that the gut-barrier function may be impaired in states of sustained emotional distress and turmoil leading to absorption of antigenic gliadin polypeptides and the formation of antigliadin antibodies in an abnormal proportion of such cases. Essentially similar observations have been reported by Hekkens et al. (1980) who used the more sophisticated ELISA technique that detected abnormal, antigliadin antibody titers in the sera of 98% celiacs. Two other studies compared schizophrenics with healthy controls in terms of antibodies against a dozen or so vegetable "seed" foodstuffs (Mascord et al., 1978; Hekkens et al., 1980). The schizophrenics significantly differed from controls only in showing greater propensity for producing antibody to wheat. Rye showed the same tendency but not to the same extent. A recent small-sample study by McGuffin et al. (1981) failed to find a significant difference between the sera from schizophrenics and controls in terms of wheat gluten antibodies. On the other hand, Ashkenazi et al. (1979) have reported evidence suggesting much higher than normal frequency of abnormal cellular immunoreactivity to wheat gluten fractions in schizophrenic and psychotic individuals. In a semiquantitative test that measured the production of a leukocyte migration inhibition factor by peripheral blood lymphocytes in response to challenge with gluten fractions, they found that the test was positive (i.e., lymphocytes were sensitized to gluten fractions) in 16 out of 31 patients with schizophrenia ($N = 21$) or schizophrenia-like psychoses ($N = 10$); the proportions of positive tests were the same for schizophrenics and other psychotics. The schizophrenics significantly differed from both healthy controls and celiacs. In a subsequent communication, Ashkenazi et al. (1981) reported that in healthy controls, this test was positive in only 3% of cases while in celiacs it was positive in 82% of cases; 19% of cases with nonceliac gastrointestinal diseases also gave positive tests. Thus, the schizophrenics and psychotics had much higher percentages of abnormal tests than controls and even patients with nonceliac bowel disturbances but somewhat lower percentages than the celiacs. However, unlike the celiacs, schizophrenics and psychotics showed no evidence of malabsorption on testing, suggesting a rather selective gut-barrier defect for polypeptides derived from gluten.

These studies clearly suggest that schizophrenics do absorb gluten peptides and that this happens in a much higher proportion—perhaps 25 to 50% of cases—than in healthy individuals and even in those with nonceliac gastrointestinal disturbance. However, the gluten components recognized by the immunological tests may not be of specific importance to schizophrenia, for they seem to occur with abnormal frequency in other psychiatric as well as in hopeless nonpsychiatric disorders. It is possible that gluten peptides pathogenic in schizophrenia are small peptides that do not produce immune responses. On the other hand, it may be that the absorbed peptides interact with a genetically programmed process specific to schizophrenia to produce brain effects (Dohan, 1980). One possible mechanism may be through the formation of immune complexes which are deposited in specific brain areas (Rudin, 1980, 1981a,b), while another may be through the development of autoimmunity. A significant body of literature suggesting increased levels of immunoglobulins in schizophrenic sera, and some suggesting the presence of antibrain antibodies, may be relevant in this contest (Durell and Archer, 1976; Pulkkinen, 1980; Gowdy, 1980, Pandey et al., 1981), especially when one considers recent literature indicating significant cortical atrophy and functional brain damage in chronic schizophrenics (Ingvar and Franzén, 1974; Franzén and Ingvar, 1975; Weinberger et al., 1979a,b).

The presence of psychoactive, and possibly psychotoxic, peptides in cereal grain glutens, which on being released by normal enzymatic digestion are available for absorption, has been suggested by other workers (Zioudrou and Klee, 1979; Zioudrou et al., 1979; Klee and Zioudrou, 1980). Using a number of biochemical and biological assays, they found that pepsin digestate, prepared under conditions resembling those in the stomach, of wheat gluten and milk protein casein contained significant concentrations of peptides with naloxone-reversible opioid activity and opiate receptor binding in the brain. Termed exorphins, such peptides from gluten were generally more potent than those from casein, and in some tests, they were many times more potent than morphine (Klee and Zioudrou, 1980). Digestates of proteins from oats, rye, and soy, as well as ovalbumin, lactalbumin, and γ-globulin, contained no exorphins. However, digestates of proteins from wheat, barley, maize, oats, and rye contained other peptides which acted as opiate antagonists, although they did not seem to interact directly with opiate receptors (Zioudrou and Klee, 1979). These peptides stimulated the assay systems which are inhibited by opiates and were, therefore, termed the "stimulatory fraction." On later analyses, this fraction proved to be adenosine, which has been shown to have vasomotor, autonomic, hormonal, and immunological actions in mammalial tissues and has been suspected of being a neurotransmitter (Klee and Zioudrou, 1980). This fraction, as well as the exorphins, were noted to be resistant to further proteolysis in the intestine.

Thus, it would seem that wheat gluten contains peptides which act like a mixture of opiate agonists and antagonists. This may be of relevance to the pathogenesis of schizophrenia, since drugs with mixed opiate agonist–antagonist activity such as nalorphine and cyclazocine may produce marked dysphoria and psychotic symptoms (Jaffé and Martin, 1975; Martin, 1967). On the other hand, since the available evidence seems to suggest that opioid actions of endorphin-like peptides may be unimportant for schizophrenia (Emrich,

1981; Deakin, 1980), the so-called stimulatory peptides may prove to be of greater relevance to schizophrenia because of its neurotoxic actions. Some support for this idea comes from a report of high whole blood and erythrocyte levels of adenosine in schizophrenics (Hansen, 1972). Also, in relation to Horrobin's (1977) hypothesis of a prostaglandin deficiency in schizophrenia, it has been reported that adenosine diphosphate, which strongly stimulates $PGE_1$ synthesis in platelets of normal individuals, fails to do so in the platelets of schizophrenics. It may be, of course, that neither of the endorphin-related peptides is of importance for schizophrenia, and that there are other as yet undiscovered psychotoxic peptides in wheat gluten digestates which are pathogenic in schizophrenia. These peptides may act on the brain after absorption or indirectly interfere with brain functioning through, for example, interference with gut–brain peptides such as CCK or VIP, or sending false feedback messages to the brain concerning peptides in the periphery.

In this context, it may be mentioned that a Norwegian group has reported abnormal urinary excretion of biologically active peptides or peptide–protein complexes in schizophrenics (Trygstad *et al.*, 1980; Reichelt *et al.*, 1981; Nygaard *et al.*, 1981). Using Sephadex gel chromatography, they found that urine from adult and child schizophrenics contained greater than normal amounts of peptide–protein complexes which produced somewhat diagnosis-specific chromatography patterns. Hebephrenic schizophrenics tended to have over three times the average normal peak activity at 900- to 1500-ml elution fraction, while the paranoids seemed to show at least twice the average normal activity at 600- to 900-ml elution fraction. Purified peptides from 900- to 1500-ml fraction produced catatonia in animals, while those from the 600- to 900-ml fraction produced dopamine uptake inhibition, dopamine hyperactivity, analgesia, GABA and glutamic acid release, cholinergic blockade, loss of righting reflex, and hypothermia. Thus, the peptides involved were clearly active in assay systems of known relevance to schizophrenia. These peptides, which have not yet been identified, may be of endogenous origin, or they may be derived from absorbed exogenous peptides.

Behavioral changes of the kind pertinent to schizophrenia have been observed in rats following wheat gluten or gliadin feeding and, presumably, absorption of psychoactive peptides from these materials. Amphetamine-induced changes in behavior at present provide the best experimental "model" of schizophrenic psychosis, and the ability to block amphetamine-caused excitation and stereotypy of behavior is the best known empirical feature for predicting antipsychotic activity in any drug (Randrup and Munkvad, 1965, 1967; Ellinwood *et al.*, 1973; Singh and Lal, 1982). It is, therefore, of considerable interest that in two separate investigations, wheat gluten feeding was found to increase amphetamine behaviors (Taylor, 1978; Williams, 1979a). Taylor (1978) tested adult rats for amphetamine effects after the animals had been on a normal diet alone or a normal diet plus intraperitoneal injections of gliadin for 2 days; the gliadin-pretreated animals showed significantly greater levels of stereotypy but lesser excitation. L-Methionine pretreatment also had a similar effect, whereas L-tryptophan pretreatment had a similar but less marked effect. The latency of onset was shortest in the gliadin-treated rats. Pimozide, a neuroleptic, was able to prevent amphetamine stereotypy in all groups but had a differential effect on activity level. Gliadin pretreatment significantly potentiated activity reduction by pimozide leading to a near-cataleptic state, methionine had no effect, while tryptophan actually produced a significant increase in activity level. Changes with gliadin occurred at the very low dose of 0.1 mg/kg. Williams (1979a) studied wheat gluten effect on amphetamine behaviors in two groups of rats, one maintained on a normal diet and the other on a "wheat-free" diet; in a control group, amphetamine effect was tested after normal diet but no gluten supplement. Immunological assays for wheat proteins in body tissues were also carried out. The results indicated an enhancement of amphetamine-induced activation and stereotypy in both test groups. The latency of activation was shorter but that for stereotypy longer in the "wheat-free" test group than in the "normal diet" test group. However, the degree of activation and stereotypy in the end were most marked in the "wheat-free" test group. Corresponding to this, much greater amounts of antigenic wheat protein products appeared in the blood and tissues of the "wheat-free" test group than in the "normal diet" test group. Thus, it seemed clear that absorbed wheat gluten products had a potentiating effect on the amphetamine-induced behaviors. In a subsequent study, Williams and Wood (1981) observed that wheat protein feeding on its own increased activity level in rats and that this acceleration of activity was cancelled by noloxone given parenterally, suggesting the involvement of endorphin-like peptides in wheat gluten. In a different type of study, Woodley et al. (1980) noted the effect of gliadin feeding in neonatal rats on the development of complex motor skills and avoidance behavior. Two litters from two pregnant rats on a gliadin-free diet were studied. For 28 days after birth, a third of each litter were fed a daily supplement of gliadin, another third bovine serum albumin, and the remaining third, dilute sucrose. The gliadin-fed animals significantly differed from others in terms of complex motor skills such as clinging and rope climbing. Also, gliadin animals showed greater avoidance behavior in response to bright light. This recognizably preliminary work has been followed up by another small study by the same group (J. F. Woodley, personal communication). Two female rats were put on diets, one gliadin, one casein, the day after impregnation by the same male. Respective diets were continued in the litters produced by the two females during the postnatal study period. The gliadin animals were faster to turn on a slope until the 24th day after birth, and they showed greater ambulation in the open field from the 10th day onward. Thus, the suggestion is that gliadin may affect behavior during development and lead to hyperactivity and possibly increased avoidance and exploratory behavior. These observations bear some kinship to those by Williams and Wood (1981) suggesting an increase in spontaneous activity of rats by wheat gluten. In this context, a further point of interest is that the first of the Woodley group studies, the gliadin-fed animals had twice the brain acetylcholinesterase activity of all the other study animals, suggesting a possibly decreased brain cholinergic function due to greater acetylcholine destruction. Elsewhere, we had reviewed evidence suggesting the important role of central cholinergic mechanisms in the inhibitory control of exploratory behavior, stereotyped repetitious behaviors, and actively emitted behaviors in general (Singh and Lal, 1982).

## 4. NONOPIOID ENDORPHIN-RELATED PEPTIDES AND SCHIZOPHRENIA

If the putative schizopathic factors in cereal grain proteins prove to be the endorphin-related peptides, then the information in this section represents the other side of the coin, i.e., the possibility that certain endorphin-related peptides may have antischizophrenic properties. This need not be considered a contradiction because certain endorphin peptides have been shown to have opposite behavioral effects. Thus, in terms of conditioned avoidance and brain self-stimulation behaviors, α-endorphin seems to act like amphetamine and increases these behaviors, while γ-endorphin (which differs from α-endorphin only in having leucine at the end of the peptide molecule) behaves like neuroleptics and suppresses these behaviors (van Ree et al., 1981). Furthermore, there may be a dissociation between the opioid actions and other behavioral actions of these molecules, so that similarity of endorphins such as α- and γ-endorphins in terms of opioid activity, which can be blocked by naloxone, may have no bearing on their nonopioid actions, which are not blocked by naloxone, and in which they may be opposed to each other (van Ree et al., 1981). Studies by de Wied's group (de Wied et al., 1978a, b, 1980; Dorsa et al., 1979) have indeed revealed that the removal of the first amino acid (tyrosine) in the γ-endorphin peptide chain produces a loss of opioid activity but leaves the other behavioral activity intact. The neuropeptide destyrosine γ-endorphin (DTγE) thus produced was in fact even more potent in this respect and was hypothesized to be the brain's own antischizophrenic agent. Further studies suggested that even after the first five amino acids, comprising the enkephalin molecule, were removed from the γ-endorphin chain, the neuroleptic-like activity remained intact and that this activity was, therefore, enclosed in the desenkephalin γ-endorphin (DEγE) so produced. The C-terminal amino acid leucine in this molecule seemed to be essential for this activity (de Wied et al., 1980). Both DTγE and DEγE showed relatively selective dopamine antagonism in the mesolimbic system, and their normal presence could be demonstrated in the rat brain as well as in human cerebrospinal fluid and brain tissue (van Ree et al., 1981). In vitro studies suggested that the brain tissue could synthesize both these γ-endorphins, thus raising the possibility that γ-endorphins may be normally produced by the brain to protect against the type of psychotic dysfunction that characterizes schizophrenia (van Ree et al., 1981). If this were to be the case, it would lead one to suspect that the schizopathic effect of cereal grain proteins might be due to endorphin-related peptides absorbed from the digestates of these proteins. Such peptides, likely to be responsible for the amphetamine-potentiating effects of gliadin feeding (*supra*), might act by blocking the supposed neuroleptic-like actions of γ-endorphins, or potentiating the suggested amphetamine-like actions of α-endorphins.

The possibility of antischizophrenic actions of DTγE and DEγE has been tested in a number of small preliminary studies of various designs in which the peptides were administered by intramuscular injections in daily doses of 1 to 5 mg for 5 to 15 days. van Ree *et al.* (1981) and van Praag *et al.* (1982) have reviewed the four studies conducted by their group in the Netherlands, while Emrich *et al.* (1981) have reviewed the data from six published and unpublished clinical studies by their and other groups.

The four Dutch studies (one open and three blind) involved a total of 40 schizophrenics; in two of these, neuroleptics were used concurrently, and in two others, the neuroleptics were stopped before the study. Twenty-three patients received DTγE: 7 showed no response, 6 showed mild to moderate improvement, while 10 improved from a moderate to marked degree. DEγE treatment was used in the remaining 17 patients with the following results: no response, 2; slight–moderate improvement, 5; moderate–marked improvement, 10. Some patients had an exacerbation of symptoms after initial improvement. No significant adverse effects were noted with either peptide, and the absence of extrapyramidal reactions was particularly noteworthy. Most of the good responders belonged to the hebephrenic and paranoid subtypes, while the poor responders generally belonged to the catatonic, residual, and schizoaffective types. There were also negative correlations between response to treatment and the duration of the psychosis, the duration of the last psychotic episode, and the duration of neuroleptic medication, suggesting that the more acute the disease process, the better the response to DTγE and DEγE treatment.

The six studies reviewed by Emrich *et al.* (1981) involved a total of 54 schizophrenics. Fourteen patients, mostly with acute psychosis, were reported as improved to varying degrees, while 40 patients, largely of chronic type, were considered unresponsive. No significant adverse effects were observed. Thus, basically, the results of these studies seemed to be in agreement with the findings of the Dutch studies in that the relatively acute or recent-onset patients improved significantly with the neuropeptide treatment, while the chronic, intractable patients proved to be refractory. This may mean that with chronicity, the disease process becomes irreversible and, therefore, refractory to all types of treatment, or that different types of disease processes are involved, of which only some are responsive to the γ-endorphin peptides. It should be emphasized, however, that all the studies were rather brief investigations, so that it would be difficult to be certain the nonresponders would remain so in a longer course of treatment, or that the improvement would persist with continued treatment in the responders.

## 5. CCK-LIKE PEPTIDES AND SCHIZOPHRENIA

Possibly related to the endorphins, at least in function, are the octa- and tetrapeptide terminal fragments of the 33-peptide molecule of the gut peptide hormone CCK. These fragments (CCK-8 and CCK-4) have been demonstrated in the brain of all mammalian species and, along with vasoactive intestinal peptides, are unique among neuropeptides in being concentrated in the frontal and limbic cortical areas and the deep nuclei of the limbic system (Emson *et al.*, 1980). The CCKs are also the most abundant of the known brain peptides, and there is evidence to suggest that some of the peripheral CCKs may travel from the brain down vagal nerves (Dockray, 1979; Emson *et al.*, 1980; Krieger and Martin, 1981a). Their access to the brain from peripheral circulation is suggested by a large number of behavioral effects produced by quite small amounts of systemically administered CCKs (Zetler, 1980a,b,c, 1981). These effects include endorphin-like analgesia which is reversed by naloxone (Zetler, 1980b,c); indeed, CCK-8 is many times more potent as an analgesic than is morphine. A number of other behavioral effects are, however, unrelated to this opioid activity and are naloxone-resistant.

A synthetic decapeptide analog of CCK-8, caerulein (ceruletide, CCK-10), which is normally used in the study of the biliary system and in the treatment of intestinal hypotonia and paralytic ileus (Ersparmer, 1970; Ganzina and Santamaria, 1976), has been found not only to show all the behavioral actions of CCK-8 but to have much greater potency than the endogenous peptide in many of the tests. Both peptides are more potent than diazepam in prolonging hexobarbital sleeping time and in blocking the convulsant actions of picrotoxin, but seem to act through a different mechanism as suggested by the fact that, unlike diazepam, they are ineffective against the convulsant actions of bicuculline and pentetrazol (Zetler, 1980a, 1981). In raising the convulsive threshold, these peptides contrast with neuroleptics but in many other respects they have a behavioral pharmacological profile similar to that of a potent neuroleptic. Thus, in an experimental comparison with haloperidol, these peptides, especially caerulein, were equally or more potent in producing typical neuroleptic effects such as atropine-reversible catalepsy, ptosis, inhibition of rearing, and attenuation of methylphenidate-induced stereotyped behaviors (Zetler, 1981). All the drugs were given by systemic injections. In many instances, the peptide effects were quite prolonged. The peptides differed from haloperidol not only in being anticonvulsant but also in suppressing appetite. On the basis of such data, it was suggested that CCKs may be effective antipsychotic agents while lacking the undesirable neuroleptic actions on seizure threshold and appetite (Zetler, 1981). This hypothesis is further strengthened by the observation that CCKs coexist with dopamine in the mesolimbic system (Hökfelt *et al.*, 1980b), which is implicated in the antipsychotic actions of neuroleptics (Mathysse and Kety, 1975), and that they seem to inhibit dopamine release in the nucleus accumbens–tuberculum olfactorum region (Hökfelt *et al.*, 1980a). It has indeed been suggested that an imbalance between CCK peptides and dopamine in the mesolimbic system, whereby a loss or decrease in the peptides leads to an "overactive" dopamine system, may be etiological in schizophrenia (Hökfelt et al., 1980a).

Based on these observations, Moroji et al. (1982) have carried out the first clinical trial of caerulein in chronic schizophrenics. Twenty treatment-resistant schizophrenics were studied nonblind to determine the efficacy of intramuscularly administered caerulein. The neuroleptic medication was continued unchanged and caerulein was used in two doses of 0.3 and 0.6 µg/kg. A single dose of 0.3 µg/kg in 12 patients produced significant improvement in mood and typical features of schizophrenia such as emotional withdrawal, mannerisms and posturing, uncooperativeness, and blunted affect, which lasted for the whole 3-week period of observation. A single injection of 0.6 µg/kg in 20 patients produced a somewhat more pronounced improvement that was still evident 3 weeks later. In both instances, the improvement was much more marked in the first week after the injection and began to wear off thereafter. The only adverse effects were slight abdominal pain and borborygmus in a few cases; these effects appeared within 2 hr after the injection and disappeared spontaneously a few minutes later. Thus, observations in this very preliminary open-design study were suggestive of a beneficial effect of CCK peptides in neuroleptic-resistant schizophrenics, an effect that seemed to be long-lasting and was accompanied by few side effects. Apart from the possibility of observer bias, it has to be considered that the effect may have been due to some interaction between the peptide and neuroleptic medication. Therefore, the investigators are now in the process of conducting further investigations using a double-blind research design and drug-free patients (personal communication).

## 6. CONCLUSIONS AND PERSPECTIVES

The amphetamine "model" psychosis, apparently related to amphetamine-induced increase in catecholaminergic activity (Randrup and Munkvad, 1966) and the close correlation observed between dopaminergic blockade and neuroleptic potency (Creese et al., 1978) led to the formulation of the dopamine overactivity hypothesis of schizophrenia (Randrup and Munkvad, 1972; Mathysse, 1973; Snyder, 1973). A refinement of the hypothesis postulated the site of antipsychotic action of neuroleptics and of the dopamine dysfunction in schizophrenia to be in the mesolimbic system (Andén, 1972; Stevens, 1973). A great deal of research has been generated by this hypothesis, but no direct evidence of dopamine hyperactivity has been forthcoming in schizophrenia (Crow, 1978; Sandler, 1978). Does this mean that the hyperactivity of the dopamine systems is not a factor in schizophrenia?

In retrospect, one wonders why the question was not asked as to how one could really consider and talk about a chronically developing and highly protracted illness such as schizophrenia in terms of extremely brief chemical phenomena involved in dopaminergic, and other forms, of neurotransmission. Furthermore, even if dopaminergic dysfunction in schizophrenia were to remain a viable concept, there is the problem of heterogeneity of schizophrenia; not only does the amphetamine "model" psychosis seem to relate to only a subtype of this syndrome (Connell, 1958), but it is clear that neuroleptics are only partially effective in schizophrenia and cannot, therefore, lead to a complete understanding of the pathogenesis of this syndrome (Singh and Smith, 1973; Singh and Kay, 1975a,b; Kay and Singh, 1979). It has been suggested, for example, that the dopaminergic defect may not be hyperactivity per se, but supersensitivity of the dopamine receptors (Carlsson, 1978; Crow, 1978; Langer et al., 1981). This not only fails to provide an insight into the etiology of the neuroleptic-resistant schizophrenia but serves merely to raise the question as to why there should be supersensitivity in the dopamine receptors in schizophrenia, for evidence of neuronal damage of the sort that can cause denervation supersensitivity is lacking, and the chronic neuroleptic treatment which seems to cause such supersensitivity apparently leads to dyskinesia and not to schizophrenic exacerbation (Klawans, 1973; Gianutsos et al., 1974; Crane, 1978).

We believe the data reviewed here suggest that the study of endogenous and exogenous neuropeptides offers considerable promise for finding answers to these questions. One reason for this belief is that, unlike classical neurotransmitters, neuropeptides often seem to have slower but long-lasting neuronal effects, so that through their molecular and or modulatory actions they could more reasonably be expected to produce chronic changes, including possibly receptor hyper- and hyposensitivities, of the type leading to schizophrenias. A second reason is that the nonpeptide transmitters are estimated to account for only about 40% of the synapses in the brain (Krieger and Martin, 1981a,b), so that a significant role of the neuropeptides in the normal and abnormal brain functions is highly likely. A third reason relates to the data reviewed in this chapter suggesting that exogenous neuroactive peptides can pass through the gut and the blood–brain barriers so that a long-term exposure to psychoactive peptides derived from common foods such as cereal grains would be just the kind of condition that, in susceptible individuals, could produce a chronic disease such as schizophrenia.

From all the neuropeptides, biological psychiatry's selection of opioid peptides as the first choice for study in schizophrenia may have been unfortunate, because it was based on the rather weak foundation of an observation of immobility with β-endorphin and the conjecture of a possible antipsychotic property through comparison of this commonplace pharmacological phenomenon with the neuroleptic-induced catalepsy (Jaquet and Marks, 1976). Not only has a great deal of research effort failed to suggest a significant role for opioid mechanisms in schizophrenia (Emrich, 1981) but basic evidence, supported at least to some extent by the clinical data, that we have reviewed here suggests that nonopioid endorphin-like peptides may have much greater significance for the pathogenesis and treatment of schizophrenia. For the pathogenesis, the amphetamine-like actions of the endogenous α-endorphins, which appear to have no relation to their opioid actions, may be of interest, while from the treatment point-of-view, the nonopioid γ-endorphin fragments may be considered as the brain's own antipsychotic agents because they are quite similar to neuroleptics in the behavioral effects, such as brain self-stimulation suppression, that predict antipsychotic activity. Their suggestive effectiveness in clinical studies, given in just one dose a day, and the absence of any extrapyramidal reactions, make them of particular practical and theoretical interest.

The CCK peptides in the brain may prove to be of even greater interest because not only are they present in high concentrations in the limbic brain, and also appear to coexist with dopamine in the mesolimbic system, but they have biochemical and behavioral profiles predictive of potent antipsychotic activity. Interestingly, these peptides are also many times more potent than morphine as an analgesic, but their other behavioral actions seem independent of this opioid action. A very preliminary clinical study supports the prediction of antipsychotic activity of CCK peptides. More importantly, it suggests a very-long-lasting action which may be demonstrable even in neuroleptic-resistant patients and which is accompanied by no extrapyramidal reactions and few other side effects.

Both the nonopioid endorphin peptides and the CCK peptides in the brain may also prove to be of relevance to the apparent pathogenic effects of cereal grain proteins in schizophrenics. The linkage might lie in the nonopioid psychoactive peptides in these proteins, which may be responsible for amphetamine potentiation in animals, and which, when absorbed in schizophrenics, may activate or exacerbate the disease process by interfering with γ-endorphins and CCKs and/or promoting the amphetamine-like actions of α-endorphins.

Finally, it seems to us that the VIPs, which appear to be in the same areas as the CCKs, may be of importance for schizophrenia. We suspect this because of their apparent coexistence and functional collaboration with acetylcholine in the cerebral blood vessels, and possibly within the limbic system. Poor frontal circulation and refractoriness to cholinergic stimulation in chronic nonparanoid schizophrenics, and reversible cholinergic hypoactivity in acute, neuroleptic-responsive, nonparanoid schizophrenics (Singh and Lal, 1982) may well prove to be due to disturbance of VIP–acetylcholine relationships in the brain. Also, the preliminary observations of Woodley et al. (1980) suggesting much higher than control brain acetylcholinesterase activity in gliadin-fed animals seem to be of considerable interest in this context. It suggests a possible link between the schizopathic effect of food-derived peptides, cholinergic dysfunctions, and poor frontal circulation in schizophrenics.

ACKNOWLEDGMENT. We thank Barbara Mason for her assistance in the preparation of the manuscript.

## 7. REFERENCES

Akil, H., Richardson, D. E., Barchas, J. D., and Li, C. H., 1978, Appearance of β-endorphin-like immunoreactivity in human ventricular cerebrospinal fluid upon analgesic electrical stimulation, *Proc. Natl. Acad. Sci. USA* **75**:5170.
Andén, N. E., 1972, Dopamine turnover in the corpus striatum and the limbic system after treatment with neuroleptic and anti-acetylcholine drugs, *J. Pharm. Pharmacol.* **24**:905.
Anderson, C. M., Gracey, M., and Burke, V., 1972, Coeliac disease: Some still controversial aspects, *Arch. Dis. Child.* **47**:292.
Ashkenazi, A., Krasilowsky, D., Levin, S., Idar, D., Kalian, M., Or, A., Ginat, Y., and Halperin, B., 1979, Immunologic reaction of psychotic patients to fractions of gluten, *Am. J. Psychiatry* **136**:1306.
Ashkenazi, A., Levin, S., and Idar, D., 1981, Immunological assays for coeliac disease, *Lancet* **II**:687.
Asperger, H., 1961, Die psychopathologie des coeliakiekranken kindes, *Ann. Paediatr.* **197**:346.
Bateson, G., Jackson, D. D., Haley, J., and Weakland, J., 1956, Toward a theory of schizophrenia, *Behav. Sci.* **1**:251.
Bender, L., 1953, Childhood schizophrenia, *Psychiatr. Q.* **27**:663.
Black, J. A., 1964, Possible factors in the incidence of coeliac disease, *Acta Paediatr.* **53**:109.
Bohus, B., 1979, Effects of ACTH-like neuropeptides on animal behavior and man, *Pharmacology* **18**:113.
Brown, G. W., and Birley, J. L. T., 1968, Crises and life changes and the onset of schizophrenia, *J. Health Soc. Behav.* **9**:203.
Brown, G. W., Birley, J. L. T., and Wing, J. K., 1972, Influence of family life on the course of schizophrenic disorders: A replication, *Br. J. Psychiatry* **121**:241.
Carlsson, A., 1978, Antipsychotic drugs, neurotransmitters, and schizophrenia, *Am. J. Psychiatry* **135**:164.
Challacombe, D. N., MacCulloch, M. J., and Birtles, C. J., 1972, Controlled measures of exploratory movement in a coeliac child during gluten withdrawal, *Arch. Dis. Child.* **47**:823.
Connell, P. H., 1958, *Amphetamine Psychosis*, Maudsley Monograph No. 5, Chapman & Hall, London.
Cooke, W. T., 1976, Neurological manifestations of malabsorption, in: *Handbook of Clinical Neurology*, Vol. 28, *Metabolic and Deficiency Diseases of the Nervous System* (P. J. Vinken and G. W. Bruyn, eds., in collaboration with H. L. Klawans), pp. 225–239, Elsevier/North-Holland, Amsterdam.
Crane, G. E., 1978, Tardive dyskinesia and related neurologic disorders, in: *Handbook of Psychopharmacology*, Vol. 10, *Neuroleptics and Schizophrenia* (L. L. Iversen, S. D. Iversen, and S. H. Snyder, eds.), pp. 165–196, Plenum Press, New York.
Crayton, J. W., and Meltzer, H. Y., 1976, Motor endplate alterations in schizophrenic patients, *Nature (London)* **264**:658.
Creese, I., Burt, D. R., and Snyder, S. H., 1978, Biochemical actions of neuroleptic drugs: Focus on dopamine receptors, in: *Handbook of Psychopharmacology*, Vol. 10, *Neuroleptics and Schizophrenia* (L. L. Iversen, S. D. Iversen, and S. H. Snyder, eds.), pp. 37–89, Plenum Press, New York.
Crow, T. J., 1978, An evaluation of the dopamine hypotheses of schizophrenia, in: *The Biological Basis of Schizophrenia* (G. Hemmings and W. A. Hemmings, eds.), pp. 63–78, MTP Press, Lancaster, U.K.
Dale, P. W., 1981, Prevalence of schizophrenia in the Pacific Island populations of Micronesia, *J. Psychiatr. Res.* **16**:103.
Daynes, G., 1956, Bread and tears—Naughtiness, depression and fits due to wheat sensitivity, *Proc. R. Soc. Med.* **49**:391.
Deakin, J. F. W., 1980, Opiates, opioid peptides and their possible relevance to schizophrenia, in:

*Biochemistry of Schizophrenia and Addiction* (G. Hemmings, ed.), pp. 39–51, University Park Press, Baltimore.

de Kloet, R., and de Wied, D., 1980, The brain as target tissue for hormones of pituitary origin: Behavioral and biochemical studies, in: *Frontiers in Neuroendocrinology*, Vol. 6 (L. Martini and W. F. Ganong, eds.), pp. 157–201, Raven Press, New York.

de Wied, D., Bohus, B., van Ree, J. M., and Urban, I., 1978a, Behavioral and electrophysiological effects of peptides related to lipotropin (β-LPH), *J. Pharmacol. Exp. Ther.* **204**:570.

de Wied, D., Kovács, G. L., Bohus, B., van Ree, J. M., and Greven, H. M., 1978b, Neuroleptic activity of the neuropeptide β-LPH62–77 ([des-Tyr$^1$]-γ-endorphin; DTγE), *Eur. J. Pharmacol.* **49**:427.

de Wied, D., van Ree, J. M., and Greven, H. M., 1980, Neuroleptic-like activity of peptides related to (des-Tyr$^1$)-γ-endorphin: Structure activity studies, *Life Sci.* **26**:1575.

Dockray, G. J., 1979, Cholecystokinin in brain and gut: Origins, evolution and identity, in: *Gut Peptides, Secretion, Function and Clinical Aspects* (A. Miyoshi, ed.), pp. 237–244, Elsevier/North-Holland, Amsterdam.

Dohan, F. C., 1966a, Wartime changes in hospital admissions for schizophrenia: A comparison of admission for schizophrenia and other psychoses in six countries during WWII, *Acta Psychiatr. Scand.* **42**:1.

Dohan, F. C., 1966b, Cereals and schizophrenia: Data and hypothesis, *Acta Psychiatr. Scand.* **42**:125.

Dohan, F. C., 1969a, Schizophrenia: Possible relationship to cereal grains and celiac disease, in: *Schizophrenia: Current Concepts and Research* (D. V. D. Siva-Sankar, ed.), pp. 539–551, P. J. D. Publications, Hicksville, N.Y.

Dohan, F. C., 1969b, Is celiac disease a clue to the pathogenesis of schizophrenia?, *Ment. Hyg.* **53**:525.

Dohan, F. C., 1970, Coeliac disease and schizophrenia, *Lancet* **I**:897.

Dohan, F. C., 1978, Schizophrenia: Are some food-derived polypeptides pathogenic? Coeliac disease as a model, in: *The Biological Basis of Schizophrenia* (G. Hemmings and W. A. Hemmings, eds.), pp. 167–177, MTP Press, Lancaster, U.K.

Dohan, F. C., 1979a, Celiac-type diets in schizophrenia [letter], *Am. J. Psychiatry* **136**:733.

Dohan, F. C., 1979b, Schizophrenia: Glutens and neuroleptics, *Biol. Psychiatry* **14**:851.

Dohan, F. C., 1980, Hypothesis: Genes and neuroactive peptides from food as cause of schizophrenia, in: *Neural Peptides and Neuronal Communication* (E. Costa and M. Trabucchi, eds.), pp. 535–548, Raven Press, New York.

Dohan, F. C., and Grasberger, J. C., 1973, Relapsed schizophrenics: Earlier discharge from the hospital after cereal-free, milk-free diet, *Am. J. Psychiatry* **130**:685.

Dohan, F. C., Grasberger, J. C., Lowell, F. M., Johnston, H. T., Jr., and Arbegast, A. W., 1969, Relapsed schizophrenics: More rapid improvement on milk- and cereal-free diet, *Br. J. Psychiatry* **115**:595.

Dohan, F. C., Martin, L., Grasberger, J. C., Boehme, D., and Cottrell, J. C., 1972, Antibodies to wheat gliadin in blood of psychiatric patients: Possible role of emotional factors, *Biol. Psychiatry* **5**:127.

Dohan, F. C., Harper, E. H., Clark, M. H., Rodriguez, R., and Zigas, V., 1982, Where is schizophrenia rare? Paper presented at 38th Annual Convention of Society of Biological Psychiatry, New York City.

Dorsa, D. M., van Ree, J. M., and de Wied, D., 1979, Effects of (des-Tyr$^1$)-γ-endorphin and α-endorphin on substantia nigra self-stimulation, *Pharmacol. Biochem. Behav.* **10**:899.

Durell, J., and Archer, E. G., 1976, Plasma proteins in schizophrenia: A review, *Schizophr. Bull.* **2**:147.

Ellinwood, E. H., Jr., Sudilovsky, A., and Nelson, L. M., 1973, Evolving behavior in the clinical and experimental amphetamine (model) psychoses, *Am. J. Psychiatry* **130**:1088.

Emrich, H. M. (ed.), 1981, *The Role of Endorphins in Neuropsychiatry*, Karger, Basel.

Emrich, H. M., Zaudig, M., von Zerssen, D., Kissling, W., Dirlich, G., and Herz, A., 1981, Action of [des-Tyr$^1$)-γ-endorphin in schizophrenia, *Mod. Probl. Pharmacopsychiatry* **17**:279.

Emson, P. C., Fahrenkrug, J., and Spokes, E. G. S., 1979, Vasoactive intestinal polypeptide (VIP): Distribution in normal human brain and in Huntington's disease, *Brain Res.* **173**:174.

Emson, P. C., Hunt, S. P., Rehfeld, J. F., Golterman, N., and Fahrenkrug, J., 1980, Cholecystokinin and vasoactive intestinal polypeptide in the mammalian CNS: Distribution and possible physiological roles, in: *Neural Peptides and Neuronal Communication* (E. Costa and M. Trabucchi, eds.), pp. 63–74, Raven Press, New York.

Ersparmer, V., 1970, Progress report: Caerulein, *Gut* **11**:79.

Fleiss, J. L., Gurland, B. J., and Cooper, J. E., 1971, Some contributions to the measurement of psychopathology, *Br. J. Psychiatry* **119**:647.

Franzén, G., and Ingvar, D. H., 1975, Abnormal distribution of cerebral activity in chronic schizophrenia, *J. Psychiatr. Res.* **12**:199.

Freed, W. J., Luchins, D. J., Gillin, J. C., and Wyatt, R. J., 1978, Wheat gluten impedes absorption of haloperidol, *Biol. Psychiatry* **13**:769.

Ganzina, F., and Santamaria, A., 1976, Caerulein (ceruletide): A review, *Acta Gastroenterol. Belg.* **39**:169.

Gianutsos, G., Drawbaugh, R. B., Hynes, M. D., and Lal, H., 1974, Behavioral evidence for dopaminergic supersensitivity after chronic haloperidol, *Life Sci.* **14**:887.

Gowdy, J. M., 1980, Immunoglobulin levels in psychotic patients, *Psychosomatics* **21**:751.

Graff, H., and Handford, A., 1961, Celiac syndrome in the case history of five schizophrenics, *Psychiatr. Q.* **35**:306.

Grant, E. C., 1972, Nonverbal communication in the mentally ill, in: *Non-Verbal Communication* (R. A. Hinde, ed.), pp. 349–358, Cambridge University Press, London.

Greenberg, R., Whalley, C. E., Jourdikian, F., Mendelson, I. S., Walter, R., Nikolics, K., Coy, D. H., Schally, A. V., Kastin, A. J., 1976, Peptides readily penetrate the blood–brain barrier: Uptake of peptides by synaptosomes is passive, *Pharmacol. Biochem. Behav. (Suppl.)* **5**:151.

Hansen, O., 1972, Blood nucleoside and nucleotide studies in mental disease, *Br. J. Psychiatry* **121**:341.

Heistad, D. D., Marcus, M. L., Said, S. I., and Gross, P. M., 1980, Effect of acetylcholine and vasoactive intestinal peptide on cerebral blood flow, *Am. J. Physiol.* **239**:H73.

Hekkens, W. T. J. M., Schipperijn, A. J. M., and Freed, D. L. J., 1980, Antibodies to wheat proteins in schizophrenia: Relationship or coincidence, in: *The Biochemistry of Schizophrenia and Addiction* (G. Hemmings, ed.), pp. 125–133, University Park Press, Baltimore.

Hemmings, W. A., 1978a, The absorption of large breakdown products of dietary proteins into the body tissues including brain, in: *The Biological Basis of Schizophrenia* (G. Hemmings and W. A. Hemmings, eds.), pp. 239–257, MTP Press, Lancaster, U.K.

Hemmings, W. A., 1978b, The entry into the brain of large molecules derived from dietary protein, *Proc. R. Soc. London (Biol.)* **200**:175.

Henry, J. L., 1977, Substance P and pain: A possible relation in afferent transmission, in: *Substance P* (U. S. Von Euler and B. Pernow, eds.), pp. 231–240, Raven Press, New York.

Hökfelt, T., Johansson, O., Ljungdahl, Å., Lundberg, J. M., and Schultzberg, M., 1980a, Peptidergic neurones, *Nature (London)* **284**:515.

Hökfelt, T., Rehfeld, J. F., Skirboll, L., Ivemark, B., Goldstein, M., and Markey, K., 1980b, Evidence for coexistence of dopamine and CCK in meso-limbic neurones, *Nature (London)* **285**:476.

Hökfelt, T., Lundberg, J. M., Schultzberg, M., Johansson, O., Ljungdahl, Å., and Rehfeld, J., 1980c, Coexistence of peptides and putative transmitters in neurons, in: *Neural Peptides and Neural Communication* (E. Costa and M. Trabucchi, eds.), pp. 1–23, Raven Press, New York.

Horrobin, D. F., 1977, Schizophrenia as a prostaglandin deficiency disease, *Lancet* **I**:936.

Ingvar, D. H., and Franzén, G., 1974, Distribution of cerebral activity in chronic schizophrenia, *Lancet* **II**:1484.

Jaffé, J. H., and Martin, W. R., 1975, Narcotic analgesics and antagonists, in: *The Pharmacological Basis of Therapeutics* (L. S. Goodman and A. Gilman, eds.), pp. 245–283, Macmillan Co., New York.

Jacquet, Y. F., and Marks, N., 1976, The C-fragment of β-lipotropin: An endogenous neuroleptic or antipsychotogen?, *Science* **194**:632.
Käser, H., 1961, Diagnose and klinik der coeliakie, *Ann. Paediatr.* **197**:320.
Kastin, A. J., Olson, R. D., Schally, A. V., and Coy, D. H., 1979a, CNS effects of peripherally administered brain peptides, *Life Sci.* **25**:401.
Kastin, A. J., Ehrensing, R. H., Coy, D. H., Schally, A. V., and Kostrzewa, R. M., 1979b, Behavioral effects of brain peptides, including LH-RH, in: *Psychoneuroendocrinology in Reproduction: An Interdisciplinary Approach* (L. Zichella and P. Pancheri, eds.), pp. 69–80, Elsevier/North-Holland, Amsterdam.
Kay, S. R., and Singh, M. M., 1979, Cognitive abnormality in schizophrenia: A dual-process model, *Biol. Psychiatry* **14**:155.
Klawans, H. L., 1973, The pharmacology of tardive dyskinesia, *Am. J. Psychiatry* **118**:509.
Klee, W. A., and Zioudrou, C., 1980, The possible actions of peptides with opioid activity derived from pepsin hydrolysates of wheat gluten and of other constituents of gluten in the function of the central nervous system, in: *The Biochemistry of Schizophrenia and Addiction* (G. Hemmings, ed.), pp. 53–76, University Park Press, Baltimore.
Kolata, G. B., 1978, Polypeptide hormones: What are they doing in cells?, *Science* **201**:895.
Koranyi, L., Whitmoyer, D. I., and Sawyer, C. H., 1977, Effect of thyrotropin-releasing hormone, luteinizing hormone-releasing hormone, and somatostatin on neuronal activity of brain stem reticular formation and hippocampus in the female rat, *Exp. Neurol.* **57**:807.
Kovács, G. L., Bohus, B., and Versteeg, D. H. G., 1979, The effects of vasopressin on memory processes: The role of noradrenergic neurotransmission, *Neuroscience* **4**:1529.
Krieger, D. T., and Martin, J. B., 1981a, Brain peptides, *N. Engl. J. Med.* **304**:876.
Krieger, D. T., and Martin, J. B., 1981b, Brain peptides, *N. Engl. J. Med.* **304**:944.
Langer, D. H., Brown, G. L., and Docherty, J. P., 1981, Dopamine receptor supersensitivity and schizophrenia: A review, *Schizophr. Bull.* **7**:208.
Larsson, L. I., Edvinsson, L., Fahrenkrug, J., Håkanson, R., Owman, C., Schaffalitzky de Muckadell, O., and Sundler, F., 1976, Immunohistochemical localization of a vasodilatory polypeptide (VIP) in cerebrovascular nerves, *Brain. Res.* **113**:400.
Levine, R. A., Briggs, G. W., Harding, R. S., and Nolte, L. B., 1966, Prolonged gluten administration in normal subjects, *N. Engl. J. Med.* **274**:1109.
Levy, D. L., and Weinreb, H. J., 1976, Wheat gluten—Schizophrenia findings, *Science* **194**:448.
Lidz, T., 1973, *Origin and Treatment of Schizophrenic Disorders*, Basic Books, New York.
Luchins, D. J., Freed, W. J., Potkin, S., Rosenblatt, J. E., Gillin, J. C., and Wyatt, R. J., 1980, Wheat gluten and haloperidol [letter], *Biol. Psychiatry* **15**:819.
McGuffin, P., Gardiner, P., and Swinburne, L. M., 1981, Schizophrenia, celiac disease and antibodies to food, *Biol. Psychiatry* **16**:281.
Mackay, A. V. P., 1979, Psychiatric implications of endorphin research, *Br. J. Psychiatry* **135**:470.
Manowitz, P., 1978, Amino acid levels in schizophrenia: A clue to etiology, *Biol. Psychiatry* **13**:489.
Martin, W. R., 1967, Opioid antagonists, *Pharmacol. Rev.* **19**:463.
Marx, J. L., 1975a, Learning and behavior (I): Effects of pituitary hormones, *Science* **190**:367.
Marx, J. L., 1975b, Learning and behavior (II): The hypothalamic peptides, *Science* **190**:545.
Mascord, I., Freed, D., and Durant, B., 1978, Antibodies to foodstuffs in schizophrenia, *Br. Med. J.* **1**:1351.
Mathysse, S., 1973, Antipsychotic drug actions, a clue to the neuropathology of schizophrenia?, *Fed. Proc.* **32**:200.
Mathysse, S. W., and Kety, S. S. (eds.), 1975, *Catecholamines and Schizophrenia*, Pergamon Press, Elmsford, N.Y.
Matthews, D. M., and Adibi, S. A., 1976, Peptide absorption, *Gastroenterology* **71**:151.
Matthews, D. M., and Payne, J. W. (eds.), 1975, *Peptide Transport in Protein Nutrition*, Elsevier, Amsterdam.
Meltzer, H. Y., and Stahl, S. M., 1976, The dopamine hypothesis of schizophrenia: A review, *Schizophr. Bull.* **2**:19.

Moroji, T., Watanabe, N., Aoki, N., and Itoh, S., 1982, Antipsychotic effects of caerulein, a decapeptide chemically related to cholecystokinin octapeptide, on chronic schizophrenia, *Arch. Gen. Psychiatry* **39**:485.

Nygaard, J. A., Foss, T., and Trygstad, O., 1981, Chromatographic profiles of urinary peptide-protein complexes in patients considered to have a schizophrenic-autistic syndrome, Presented at the IIIrd World Congress of Biological Psychiatry, Stockholm.

Owman, C., Edvinsson, L., and Nielsen, K. C., 1974, Autonomic neuroreceptor mechanisms in brain vessels, *Blood Vessels* **11**:2.

Oyama, T., Jin, T., Yamaya, R., Ling, N., and Guillemin, R., 1980, Profound analgesic effects of β-endorphin in man, *Lancet* **I**:122.

Pandey, R. S., Gupta, A. K., and Chaturvedi, U. C., 1981, Autoimmune model of schizophrenia with special reference to antibrain antibodies, *Biol. Psychiatry* **16**:1123.

Paulley, J. W., 1959, Emotion and personality in the etiology of steatorrhea, *Am. J. Dig. Dis.* **4**:352.

Plotnikoff, N. P., White, W. F., Kastin, A. J., and Schally, A. V., 1975, Gonadotropin releasing hormone (GnRH): Neuropharmacological studies, *Life Sci.* **17**:1685.

Potkin, S. G., Weinberger, D., Kleinman, J., Nasrallah, H., Luchins, D., Bigelow, L., Linnoila, M., Fischer, S. H., Bjornsson, T. D., Carman, J., Gillin, J. C., and Wyatt, R. J., 1981, Wheat gluten challenge in schizophrenic patients, *Am. J. Psychiatry* **138**:1208.

Prugh, D. G., 1951, A preliminary report on the role of emotional factors in idiopathic celiac disease, *Psychosom. Med.* **13**:220.

Pulkkinen, E., 1980, Some connections between immunoglobulins and schizophrenia, in: *Biochemistry of Schizophrenia and Addiction* (G. Hemmings, ed.), pp. 111–124, University Park Press, Baltimore.

Randrup, A., and Munkvad, I., 1965, Special antagonism of amphetamine-induced abnormal behavior: Inhibition of stereotyped activity with increase of some normal activities, *Psychopharmacologia* **7**:416.

Randrup, A., and Munkvad, I., 1966, Role of catecholamines in the amphetamine excitatory response, *Nature (London)* **211**:540.

Randrup, A., and Munkvad, I., 1967, Stereotyped activities produced by amphetamine in several animal species and man, *Psychopharmacologia* **11**:300.

Randrup, A., and Munkvad, I., 1972, Evidence indicating an association between schizophrenia and dopaminergic hyperactivity in the brain, *J. Orthomol Psychiatry* **1**:2.

Rapoport, S. I., Klee, W. A., Pettigrew, K. D., and Ohno, K., 1980, Entry of opioid peptides into the central nervous system, *Science* **207**:84.

Reichelt, K. L., Hole, K., Hamberger, A., Saelid, G., Edminson, P. D., Braestrup, C. B., Lingjaerde, O., Ledaal, P., and Orbeck, H., 1981, Biologically active peptide-containing fractions in schizophrenia and childhood autism, in: *Neurosecretion and Brain Peptides* (J. B. Martin, S. Reichlin, and K. L. Bick, eds.), pp. 627–643, Raven Press, New York.

Rice, J. R., Ham, C. H., and Gore, W. E., 1978, Another look at gluten in schizophrenia, *Am. J. Psychiatry* **135**:1417.

Robinson, S. E., 1983, Cholinergic pathways in the brain, in: *Central Cholinergic Mechanisms and Adaptive Dysfunctions* (M. M. Singh, D. M. Warburton, and H. Lal, eds.), Plenum Press, New York.

Rudin, D. O., 1980, The choroid plexus and system disease in mental illness. I. A new brain attack mechanism via the second blood–brain barrier, *Biol. Psychiatry* **15**:517.

Rudin, D. O., 1981a, The choroid plexus and system disease in mental illness. III. The exogenous peptide hypothesis of mental illness, *Biol. Psychiatry* **16**:489.

Rudin, D. O., 1981b, The major psychoses and neuroses as omega-3 essential fatty acid deficiency syndrome: Substrate pellagra, *Biol. Psychiatry* **16**:837.

Said, S. I., 1979, Vasoactive intestinal polypeptide (VIP) as a neural peptide, in: *Gut Peptides, Secretion, Function and Clinical Aspects* (A. Moyoshi, ed.), pp. 268–274, Elsevier/North-Holland, Amsterdam.

Sandler, M., 1978, The dopamine hypothesis revisited, in: *The Biological Basis of Schizophrenia* (G. Hemmings and W. A. Hemmings, eds.), pp. 79–85, MTP Press, Lancaster, U.K.

Singh, M. M., 1978, Some insights into the pathogenesis of schizophrenia, in: *The Biological Basis of Schizophrenia* (G. Hemmings and W. A. Hemmings, eds.), pp. 179–195, MTP Press, Lancaster, U.K.

Singh, M. M., 1979a, Celiac-type diets in schizophrenia [letter], *Am. J. Psychiatry* **136**:733.

Singh, M. M., 1979b, Schizophrenia: Glutens and neuroleptics [letter], *Biol. Psychiatry* **14**:853.

Singh, M. M., and Kay, S. R., 1975a, A comparative study of haloperidol and chlorpromazine in terms of clinical effects and therapeutic reversal with benztropine in schizophrenia: Theoretical implications for potency differences among neuroleptics, *Psychopharmacologia* **43**:103.

Singh, M. M., and Kay, S. R., 1975b, A longitudinal therapeutic comparison between two prototypic neuroleptics (haloperidol and chlorpromazine) in matched groups of schizophrenics: Nontherapeutic interactions with trihexyphenidyl. Theoretical implications for potency differences, *Psychopharmacologia* **43**:115.

Singh, M. M., and Kay, S. R., 1975c, Therapeutic reversal with benztropine in schizophrenics: Practical and theoretical significance, *J. Nerv. Ment. dis.* **160**:258.

Singh, M. M., and Kay, S. R., 1976a, Wheat gluten as a pathogenic factor in schizophrenia, *Science* **191**:401.

Singh, M. M., and Kay, S. R., 1975b, Gluten and schizophrenia, technical comment letter, *Lancet* **II**:689.

Singh, M. M., and Kay, S. R., 1976c, Wheat gluten—Schizophrenia findings, *Science* **194**:448.

Singh, M. M., and Kay, S. R., 1978a, Nosological and prognostic distinctions in schizophrenia: Pharmacological validation in terms of therapeutic antagonism between anticholinergic anti-Parkinsonism drugs and neuroleptics, *Neuropsychobiology* **4**:288.

Singh, M. M., and Kay, S. R., 1978b, Therapeutic antagonism between anticholinergics and neuroleptics: Possible involvement of cholinergic mechanisms in schizophrenia, *Schizophr. Bull.* **4**:3.

Singh, M. M., and Lal, H., 1982, Central cholinergic mechanisms, neuroleptic action and schizophrenia, in: *Clinical Applications of Neuropharmacology* (W. Essman and L. Valzelli, eds.), pp. 337–389, Spectrum, New York.

Singh, M. M., and Smith, J. M., 1973, Reversal of some therapeutic effects of an antipsychotic agent by an anti-Parkinsonism drug, *J. Nerv. Ment. Dis.* **157**:50.

Smith, J. M., 1976, Wheat gluten—Schizophrenia findings [letter], *Science* **194**:448.

Snyder, S. H., 1973, Amphetamine psychosis: A "model" schizophrenia mediated by catecholamines, *Am. J. Psychiatry* **130**:61.

Stevens, J. R., 1973, An anatomy of schizophrenia?, *Arch. Gen. Psychiatry* **29**:177.

Taylor, M., 1978, A preliminary investigation of dietary constituents and amphetamine-induced abnormal behavior, in: *The Biological Basis of Schizophrenia* (G. Hemmings and W. A. Hemmings, eds.), pp. 213–216, MTP Press, Lancaster, U.K.

Tepperman, B. L., and Evered, M. D., 1980, Gastrin injected into the lateral hypothalamus stimulates gastric acid in rats, *Science* **209**:1142.

Townley, R. W., and Anderson, C. M., 1967, Coeliac disease: A review, *Ergeb. Inn. Med. Kinderheilkd.* **26**:1.

Trygstad, O. E., Reichelt, K. L., Foss, I., Edminson, P. D., Saelid, G., Bremer, J., Hole, K., Ørbeck, H., Johansen, J. H., Bøler, J. B., Titlestad, K., and Opstad, P. K., 1980, Patterns of peptides and protein-associated peptide complexes in psychiatric disorders, *Br. J. Psychiatry* **136**:59.

Vanderwolf, C. H., and Robinson, T. E., 1981, Reticulo-cortical activity and behavior: A critique of the arousal theory and a new synthesis, *Behav. Brain. Sci.* **4**:459.

van Praag, H. M., Verhoeven, W. M. A., van Ree, J. M., and de Wied, D., 1982, The treatment of schizophrenic psychoses with $\gamma$-type endorphins, *Biol. Psychiatry* **17**:83.

van Ree, J. M., Verhoeven, W. M. A., van Praag, H. M., and de Wied, D., 1981, Neuroleptic-like and antipsychotic effects of $\gamma$-type endorphins, *Mod. Probl. Pharmacopsychiatry* **17**:266.

Weinberger, D. R., Torrey, E. F., Neophytides, A. N., and Wyatt, R. J., 1979a, Lateral cerebral ventricular enlargement in chronic schizophrenia, *Arch. Gen. Psychiatry* **36**:735.

Weinberger, D. R., Torrey, E. F., Neophytides, A. N., and Wyatt, R. J., 1979b, Structural abnormalities in the cerebral cortex of chronic schizophrenic patients, *Arch. Gen. Psychiatry* **36**:935.

Williams, E. W., 1979a, The effect of dietary wheat protein on rat behavior, *J. Orthomol. Psychiatry* **8**:113.

Williams, E. W., 1979b, Transmission of dietary proteins through the adult rat gut, in: *Protein Transmission through Living Membranes* (W. A. Hemmings, ed.), pp. 259–268, Elsevier/North-Holland, Amsterdam.

Williams, E. W., and Hemmings, W. A., 1978, Intestinal uptake and transport of proteins in the adult rat, *Proc. R. Soc. London (Biol.)* **203**:177.

Williams, E. W., and Wood, H. P., 1981, The effect of wheat proteins on rat behavior and the effect of naloxone hydrochloride on this response, *Nutr. Res.* **1**:187.

Woodley, J. F., Sterchi, E. E., Bridges, J. F., Forsyth, T., Faulkner, L., Lucy, J., and Makin, A., 1980, The digestion and absorption of dietary protein, in: *The Biochemistry of Schizophrenia and Addiction* (G. Hemmings, ed.), pp. 277–285, University Park Press, Baltimore.

Zetler, G., 1980a, Anticonvulsant effects of caerulein and cholecystokinin octapeptide, compared with those of diazepam, *Eur. J. Pharmacol.* **65**:297.

Zetler, G., 1980b, Effects of cholecystokinin-like peptides on rearing activity and hexobarbital-induced sleep, *Eur. J. Pharmacol.* **66**:137.

Zetler, G., 1980c, Analgesia and ptosis caused by caerulein and cholecystokinin octapeptide (CCK-8), *Neuropharmacology* **19**:415.

Zetler, G., 1981, Central depressant effects of caerulein and cholecystokinin octapeptide (CCK-8) differ from those of diazepam and haloperidol, *Neuropharmacology* **20**:277.

Ziemba, T., Meltzer, H. Y., and Davis, J. M., 1978, Do anticholinergics antagonize antipsychotic drug action?, *Schizophr. Bull.* **4**:7.

Zioudrou, C., and Klee, W. A., 1979, Possible roles of peptides derived from food proteins in brain function, in: *Nutrition and the Brain*, Vol. 4 (R. J. Wurtman and J. J. Wurtman, eds.), pp. 125–158, Raven Press, New York.

Zioudrou, C., Streaty, R. A., and Klee, W. A., 1979, Opioid peptides derived from food proteins: The exorphins, *J. Biol. Chem.* **254**:2446.

CHAPTER 30

# Hormonal Responses to D-Amphetamine in Schizophrenia

DANIEL P. VAN KAMMEN, S. CHARLES SCHULZ, and ALAN D. ROGOL

## 1. INTRODUCTION

The effects of D-amphetamine in man and particularly in schizophrenic patients have been studied because of the hypothesized disturbance in the dopamine (DA) and norepinephrine (NE) systems in the illness. In this chapter, we incorporated several of our studies of the effects of D-amphetamine on hormonal secretion. We studied PRL, which is under tonic inhibitory control of DA (MacLeod and Lamberts, 1979; Clemens and Shaar, 1980; Meites, 1973), growth hormone (GH), which can be released by both NE (Massara and Camanni, 1972) and DA (Malas et al., 1983; Meltzer et al., 1979), and cortisol, which is under partial control of NE (Sachar et al., 1981). Furthermore, we examined immunoreactive (i.r.) β-endorphin levels assuming that the euphoria that is experienced may be mediated by endorphin release as well as the hypothesized β-endorphin disturbance in schizophrenia (Berger et al., 1980; Davis et al., 1980; Bloom et al., 1976).

Animal studies have shown that D-amphetamine raises (Lu and Meites, 1971), decreases (Meltzer et al., 1979), or leaves PRL levels unchanged (Ravitz and Moore, 1977), and increases GH concentration (Marantz et al., 1976; Besser et al., 1969). In normal subjects, PRL levels rose, remained the same, or decreased (Dommissee et al., 1982; Nurnberger et al., 1982a,b; Wells et al., 1978). GH, cortisol, and β-endorphin (i.r.) levels rose in response to D-amphetamine in normals.

The hormonal responses to D-amphetamine reflect in part different aspects of the hypothalamic–pituitary–adrenal axis. Therefore, our aim was to compare

---

DANIEL P. VAN KAMMEN and S. CHARLES SCHULZ • Western Psychiatric Institute and Clinic, University of Pittsburgh, Pittsburgh, Pennsylvania 15213. ALAN D. ROGOL • University of Virginia Medical Center, Charlottesville, Virginia 22908.

the endocrine response to D-amphetamine in our schizophrenic patients (van Kammen et al., 1978; Schulz et al., 1981, 1982; van Kammen et al., in prep.) with findings of other studies in man (Table 1). Cortisol and PRL responses appear to be normal although a potentially disturbed β-endorphin (i.r.) and GH regulation in schizophrenia may require further study. These hormonal studies are part of an extensive evaluation of the DA hypothesis of schizophrenia (van Kammen, 1979) with D-amphetamine (van Kammen et al., 1978, 1980, 1981, 1982a,b,c).

## 2. METHODS

### 2.1. Subjects

Subjects in all three studies were physically healthy drug-free schizophrenic patients, diagnosed with the Research Diagnostic Criteria (RDC) (Spitzer et al., 1975, 1978). They were schizophrenic and schizoaffective (mainly schizophrenic) patients (DSM III; RDC criteria). Patients showed five or more of the 12 differential symptoms of the International Pilot Study of Schizophrenia (IPSS). All subjects were voluntary admissions to the 4-East clinical-research unit at the National Institutes of Health in Bethesda, Maryland, with the exception of two patients in the third study (β-endorphin infusion) who were studied at St. Elizabeths' Hospital, Washington, D.C., and who were not rated on the IPSS differential system. All subjects had signed informed consent forms for the procedures, and adhered to a controlled-monoamine, low-caffeine, and no-alcohol diet. The infusions took place after overnight rest and fasting from 11:00 p.m. the previous night. In studies 1 and 3, patients received the infusion at 8:15 a.m. while in a sitting position; in study 2 the infusion was given at 9:00 a.m. while lying in bed. Thirty minutes prior to the infusion, a 0.9% saline infusion was started in the left forearm through a 19-gauge butterfly needle. A heparin lock for repeated blood drawings was placed in the other arm. All blood samples were put on ice immediately, centrifuged at 4°C, and the plasma stored at −20°C.

### 2.2. Study 1: PRL Response to D-Amphetamine

Sixteen patients participated in this study. Under double-blind conditions, during chronic placebo administration of at least 3 weeks, 16 patients received i.v. infusions of 20 mg D-amphetamine ($N = 16$) or lactose placebo ($N = 14$) over 30 sec in 0.9% saline followed by a flush of saline. The two infusions were separated by 3–5 days. Blood was drawn just prior to the amphetamine or lactose placebo as well as at 5, 15, 25, 40, and 120 min postinfusion. Interassay variation was less than 15% and intraassay variation was less than 10%. Four patients had never received neuroleptics and the other 12 had been on placebo for at least 3 weeks. Of these patients, 5 men and 12 women again received an

TABLE 1
Responses to Amphetamines and Methylphenidate: Literature Survey

| Authors | Subjects | Sex | No. | Age | Dose | Time of day | PRL | GH | Cortisol | Comments |
|---|---|---|---|---|---|---|---|---|---|---|
| Slater et al. (1976) | Depressed, drug-free: 5 BPI, 4 UP | F | 9 | Not given | 30 mg D-amph p.o. | 9:00 am | ↑ | | | Rises at 120 and 180 min |
| van Kammen et al. (1978) | Drug-free, schizophrenics | M/F 10/6 | 16 | 22.5 (18–32) | 20 mg D-amph i.v. | 8:15 am | ↑ | | | Lithium doubled PRL rise following amph in 1 and decreased in 2 patients. Standard deviation too large during pimozide treatment to show statistically significant increase with D-amph |
| | Pimozide treated | | 9 | | | | (↑) | | | |
| | Lithium treated | | 11 | | | | ↑ | | | |
| Schulz et al. (1981) | Drug-free schizophrenics | M/F 6/3 | 9 | 26 | 20 mg D-amph i.v. | 9:00 am | ↑ | NS | | For comments, see text |
| Nurnberger et al. (1981a,b) | Normal volunteers; 13 pairs of monozygotic twins; 3 pairs of dizygotic twins | Not given | | (18–40) | 0.3 mg/kg D-amph i.v. | 9:00 am | ↑ | ↑ | ↑ | PRL rises variably but replicable and under genetic control. GH, PRL, cortisol changes intercorrelated |
| | 11 BPI (well state) | | | | 0.3 mg/kg D-amph i.v. | 9:00 am | ↑ | | | |
| Wells et al. (1978) | Normal volunteers | M | 6 | (21–31) | 10–20 mg D-amph placebo | 4:00 pm | ↓ | | | Nonsignificant increase with 10 mg at 3 hr. Significant dose-dependent PRL decrease at 1 hr |
| Halbreich et al. (1980, 1981) | Normal volunteers | M | 12 | (24–31) | 0.15 mg/kg, 0.1 mg/kg D-amph i.v. | 9:30 am | ↑ | ↑↑ | | |
| | Postmenopausal | F | 8 | (46–62) | 0.15 mg/kg, 0.1 mg/kg D-amph i.v. | 9:30 am | ↑ | ↑ | | Postmenopausal women show less of an increase in GH and PRL than young men. Increases higher in pm than in am but not in young men |

(cont.)

TABLE 1 (cont.)

| Authors | Subjects | Sex | No. | Age | Dose | Time of day | PRL | GH | Cortisol | Comments |
|---|---|---|---|---|---|---|---|---|---|---|
| Dommissee et al. (1982) | Normal volunteers | M/F 5/5 | 10 | (19–38) | 20 mg D-amph placebo-controlled p.o. | 9:00 am | NS | ↑ | | Dose may have been too low to have raised PRL. Peak GH at 10:30 am ($p = 0.04$) |
| Besser et al. (1969) | | M | 8 | (19–24) | 10 mg D-amph | 9:30 am | | NS | ↑ | Rise in ACTH (i.r.); highest in evening |
| | | M | 6 | (20–28) | 15 mg methamph | 6:00 pm | | ↑ | ↑↑ | |
| Butler et al. (1968) | Healthy normal volunteers | M | 32 | (18–32) | 5 mg D-amph; 20 mg chlordiazepoxide Mixture of chlordiazepoxide and D-amph p.o. | 6:15 pm | | | ↑ → NS | Compared to placebo, D-amph increased and chlordiazepoxide decreased cortisol. The mixture did not show a change from placebo (at 90 and 180 min.) 4 groups of 11 patients |
| Rees et al. (1970) | Normal volunteers | M | 6 | (19–34) | 15 mg methamph i.v.; no placebo control | 4:00 pm | | ↑ | ↑ | Propranolol (0.15 mg/kg) and thymoxamine (0.10 mg/kg) increased GH response ($p < 0.05$) at 30, 45, 60 min; GH peaked at 45 min; fluorogenic corticosteroids are blocked by thymoxamine; effect on propranolol response is increased at 45 min ($p < 0.05$) only |
| Checkley and Crammer (1977) | Depressed and recovered | M/F 2/8 | 10 | 54.6 (39–70) | 15 mg/75 kg methamph i.v. | 5:35 pm | | NS NS | ↑ NS | 3 weeks drug-free. Corticosteroid increase is not present in symptomatic phase of depression; GH response is not significantly different when pa- |

| Study | Subjects | Sex | N | Age | Drug/dose | Time | GH | Cortisol | Comments |
|---|---|---|---|---|---|---|---|---|---|
| Checkley (1979) | Endogenous depression | M/F 6/14 | 20 | (17–76) | 15 mg/75 kg methamph i.v. | 5:15 pm | NS | NS | tients are recovered; GH increase is less than in normal controls (male) and middle-aged females. Reactive depressed patients have higher GH response than endogenous depressives |
| | Reactive depression | 3/3 | 6 | | | | | | |
| | Other functional psychosis | 3/5 | 8 | | | | | | |
| | Other psychosis | 7/3 | 10 | | | | | ↑ | No group differences in GH response; endogenous depressed patients no cortisol increase |
| Langer et al. (1976) | Normal volunteers | M/F 11/10 | 21 | (21–64) | 0.1 mg/kg D-amph, placebo i.v. | 9:00 am | ↑ | | 6 weeks drug-free. Low GH response in normal volunteers (smokers) and depressed (nonsmokers). Endogenous depressives have smaller responses ($p < 0.01$); alcoholics and depressives responded similarly as normals. Acute schizophrenic group too small to be evaluated separately against chronic patients |
| | Depressed | 8/10 | 17 | (21–64) | | | (↑) | | |
| | Schizophrenics | 5/3 | 8 | (26–51) | | | ↑ | | |
| | Chronic alcoholics | 5/1 | 6 | (31–55) | | | ↑ | | |
| Langer and Matussek (1977) | Normal volunteers | M/F 7/3 | 10 | (27–46) | 0.1 mg/kg D-amph and L-amph i.v. placebo | 9:30 am | ↑ | | Number of deficient GH response equal for D-amph and L-amph placebo; placebo did not induce GH rise (not shown) |
| Janowsky et al. (1978) | Psychotics | M/F 32/12 | 44 | 31 (18–61) | 0.5 mg/kg methylphenidate i.v. | Not given | ↑ | | On neuroleptics less than off neuroleptics ($p < 0.05$); baseline similar; schizophrenics show a smaller response than nonschizophrenics ($p < 0.005$) |
| | Schizophrenics | 11/8 | 19 | | | | | | |
| | a. on neuroleptics | | 21 | | | | | | |
| | b. off neuroleptics | | 23 | | | | | | |

*(cont.)*

TABLE 1 (cont.)

| Authors | Subjects | Sex | No. | Age | Dose | Time of day | PRL | GH | Cortisol | Comments |
|---|---|---|---|---|---|---|---|---|---|---|
| Sachar et al. (1980) | Normal volunteers<br>Depressed | M/F 5/0<br>3/8 | 5<br>11 | (23–25)<br>(29–62) | 0.1 mg/kg D-amph i.v. | 9:30 am | | | ↑<br>↓ | Depressed patients have elevated cortisol prior to infusion; suppression of cortisol levels by D-amph |
| Sachar et al. (1981) | Normal volunteers<br>Depressed | M<br>F<br>M/F 8/14 | 10<br>8<br>22 | (23–32)<br>(49–61)<br>(18–64) | 0.15 mg/kg D-amph i.v. | 9:30 am ($N = 40$)<br>8:30 pm ($N = 36$) | | | ↑<br>↓<br>↓ | 30% of depressed patients failed to suppress cortisol with dexamethasone. Cortisol response in evening at 60 min |
| Parkes et al. (1977) | Normal volunteers | M/F 5/3 | 8 | (24–41) | 20 mg D- and L-amph | 9:00 am | | ↑ | | In normals, GH response is quite variable |
| | Narcoleptics | M/F 14/12 | 12<br>12<br>6 | (25–65) | 20 mg D-amph, 20 mg L-amph, 30 mg D-amph | 9:00 am | | | | Twelve narcoleptics were on D-amphetamine; one on L-amphetamine (1–35 years). In narcoleptics, no increase with 20 mg D- or L-amph; D- and L-amph equipotent. Only 2 out of 6 who received 30 mg D-amph showed an increase |
| Brown (1977) | Normal volunteers | M | 17 | (21–37) | Placebo; 10 and 20 mg methylphenidate p.o. | 10:00 am | | ↑ | NS | HGH correlates with elation (20 mg); 2 patients (of 14) were omitted because of elevated baseline HGH variable; 20 mg dose: 7 out of 12 had increased HGH; authors concluded DA raises GH |
| Brown et al. (1978) | Normal volunteers | M | 59 | (19–37) | 20 mg methylphenidate 10 and 20 mg D-amph | 9:00 am | | ↑<br>↑ | NS<br>↑ | Placebo controlled; only D-amph raised cortisol |

| Study | Diagnosis | Sex | N | (Age) | Dose | Time | Response | Comments |
|---|---|---|---|---|---|---|---|---|
| Halbreich et al. (1982) | Normal | M | 6 | (24–31) | 0.1 mg/kg | 9:30 am | ↑ | GH response decreases after menopause. Endogenous depressives may have normal response if controlled for age, menstrual cycle status |
| | Normal | F | 7 | (46–62) | 0.15 mg/kg D-amph | 7:00 pm | (↑) | |
| | Endogenous depressive (postmenopausal) | F | 6 | (52–62) | | | (↑) | |
| | Endogenous depressive (premenopausal) | F | 3 | (29–35) | | | (↑) | |
| | Endogenous depressive | M | 2 | (45, 65) | | | ↑ | |
| | Atypical depressive | Not given | 6 | (26–54) | | | ↑ | |
| Greenhill et al. (1981) | Hyperactive children | M | 13 | (6–9.5) | 10–30 mg/day D-amph (in 2 doses) | All night (every 20 min sampling) | ↓  = | Inhibition of weight velocity is correlated with PRL suppression. All-night-sleep endocrine secretion was measured before and after 6 months of chronic D-amph treatment. Chronic treatment decreased sleep-associated PRL increase ($p < 0.005$); GH was unaffected |
| Aarskog et al. (1977) | Hyperkinetic children | M/F 17/3 | 20 | (6–13) | 15 mg D-amph 20 mg methylphenidate p.o. | Morning after awakening | ↑ ↑ | Seven subjects were retested after 6–8 months of treatment with methylphenidate; at that time, delayed and paradoxical GH response to D-amph |

infusion of 20 mg D-amphetamine after 15 to 25 days of lithium carbonate treatment, when lithium levels were $0.9 \pm 0.1$ mEq/liter on the morning of the infusion. Similarly, 7 men and 4 women again received D-amphetamine infusions after chronic pimozide treatment of 15–49 days (mean 34 days) with a mean daily dose of 8.4 mg pimozide for the last 7 days prior to the infusion. The last dose of lithium or pimozide was given at 10:00 p.m. the night prior to the 8:15 a.m. infusion. Chronic lithium and pimozide treatments were given in a random sequence to those patients who received both chronic treatments. Eleven patients received placebo and amphetamine infusions when they were drug-free (van Kammen et al., 1978).

### 2.3. Study 2: PRL, GH, Cortisol, and β-Endorphin (i.r.) Response to Two Doses of D-Amphetamine

Nine drug-free (range 41–92 days, mean 50 days) patients received D-amphetamine or placebo infusion 2 hr after awakening. Patients remained in bed throughout the blood sampling. Three baseline samples were drawn (at 8:30, 8:45, and 9:00 a.m.) just prior to the infusion. All patients received saline placebo, 2.5 mg D-amphetamine, and 20 mg D-amphetamine in random order. The infusion was completed over 20 sec followed by 10 ml saline flush. Blood samples were drawn subsequently at 20, 30, 40, 60, 90, 120, and 180 min following the infusion for assays of PRL, GH, cortisol, and β-endorphin (i.r.). Neuroendocrine response to placebo or D-amphetamine was determined by repeated measures of variance. In this group of patients, duration of illness and Phillips scores for premorbid functioning and other clinical variables were obtained independently at the time of admission (Schulz et al., *1981, 1982*; van Kammen et al., in prep.).

### 2.4. Assays

PRL, GH, and cortisol levels were determined in triplicate with a homologous double-antibody radioimmunoassay. The PRL and GH values of Studies 2 and 3 were obtained in the same laboratory (ADR). The β-endorphin (i.r.) values were obtained in a similar radioimmunoassay (Naber et al., 1981). Amphetamine levels were obtained with a double-antibody radioimmunoassay with interassay variation of 15% and intraassay variation of 10% (Ebert et al., 1976). In all situations, samples were assayed without knowledge of the drug status, endocrinological or behavioral response.

## 3. RESULTS

### 3.1. Study 1: PRL Response to D-Amphetamine

Serum PRL levels were raised in 11, unchanged in 4, and decreased in 1 of the 16 patients in response to 20 mg D-amphetamine. After the paired placebo infusion, a decline was observed in PRL levels in 13 of 14 patients. This direction in change in PRL following the two infusions was significantly different at 25 min ($N = 14$, sign test, $p < 0.002$). Placebo infusion PRL values ($N = 14$) were significantly reduced (Fig. 1).

After D-amphetamine, PRL levels in the men ($N = 10$) rose significantly at 25 min ($9.7 \pm 0.7$ vs. $13.3 \pm 1.6$ ng/ml, $p < 0.01$), and at 40 min ($9.7 \pm 0.7$ vs. $14.9 \pm 2.3$ ng/ml, $p < 0.02$) compared to preinfusion baseline. A smaller rise observed in the women did not reach significance ($15.3 \pm 1.8$ vs. $18.3 \pm 1.5$ ng/ml, $p = $ NS; $15.3 \pm 1.8$ vs. $16.2 \pm 2.5$ ng/ml, $p = $ NS).

To evaluate data from both groups combined, PRL values in response to amphetamine and placebo were expressed in percent of preinfusion levels. Profile analysis of variance was used to compare the effect of D-amphetamine on PRL over time to similar data collected from the placebo infusion series. The time-related pattern of PRL response to the drug was significantly different from the nondrug condition ($f = 9.36$, $df = 10$, $p < 0.012$). The overall postdrug

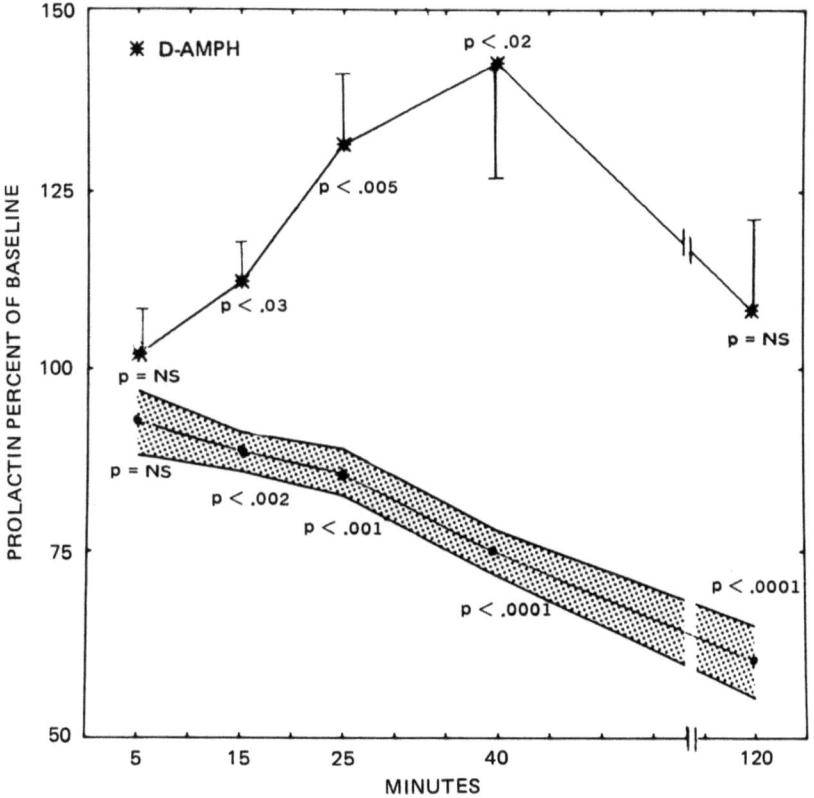

FIGURE 1. Percent changes compared to preinfusion (baseline) PRL values indicate increases following D-amphetamine and decreases following placebo (dotted area).

level of PRL was significantly higher after D-amphetamine ($f = 13.54$, $df = 10$, $p = 0.004$). This indicates that administration of D-amphetamine resulted in significantly elevated PRL levels. As expected, the women had higher preamphetamine PRL values than the men ($15.3 \pm 1.8$ vs. $9.7 \pm 0.07$ ng/ml, $p < 0.05$).

## PRL Response: Comparisons after Chronic Pimozide and Lithium Pretreatment

Pimozide raised the preinfusion baseline to levels ranging from 16 to 120 ng/ml (women: $76.6 \pm 25.3$ ng/ml; men: $44.5 \pm 12.4$ ng/ml). Lithium pretreatment did not affect preinfusion levels in the women ($16.1 \pm 1.8$ vs. $16.4 \pm 1.1$ ng/ml. $N = 4$) or in the men ($8.1 \pm 1.0$ vs. $9.5 \pm 3.0$ ng/ml, $p = $ NS, $N = 5$).

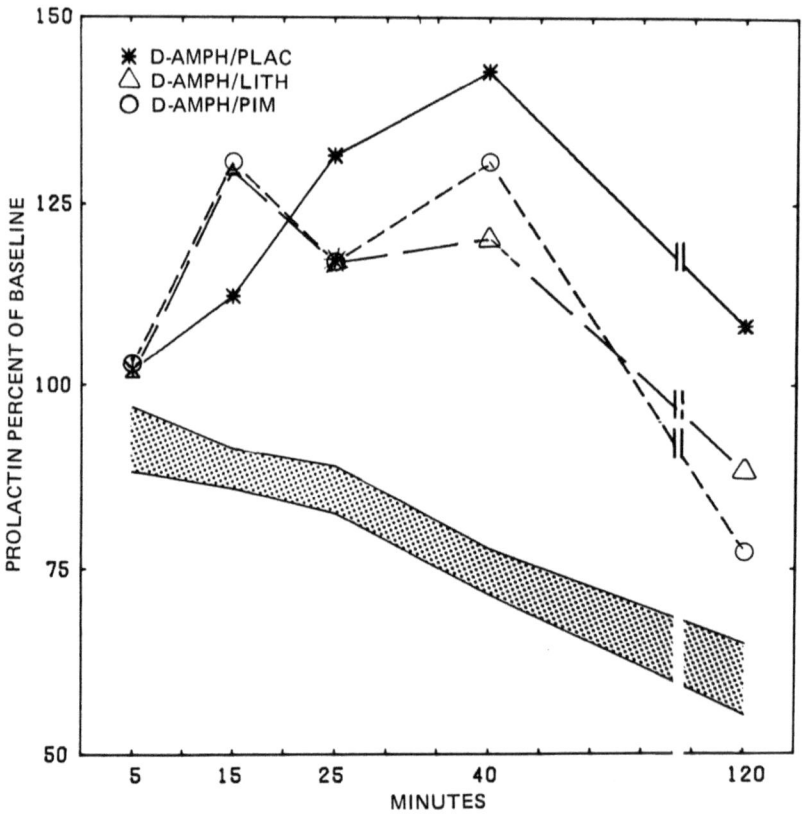

FIGURE 2. Pimozide and lithium pretreatment do not seem to affect the amphetamine-induced rise consistently. Pimozide raised the preinfusion values. Only with lithium pretreatment was a significant rise observed.

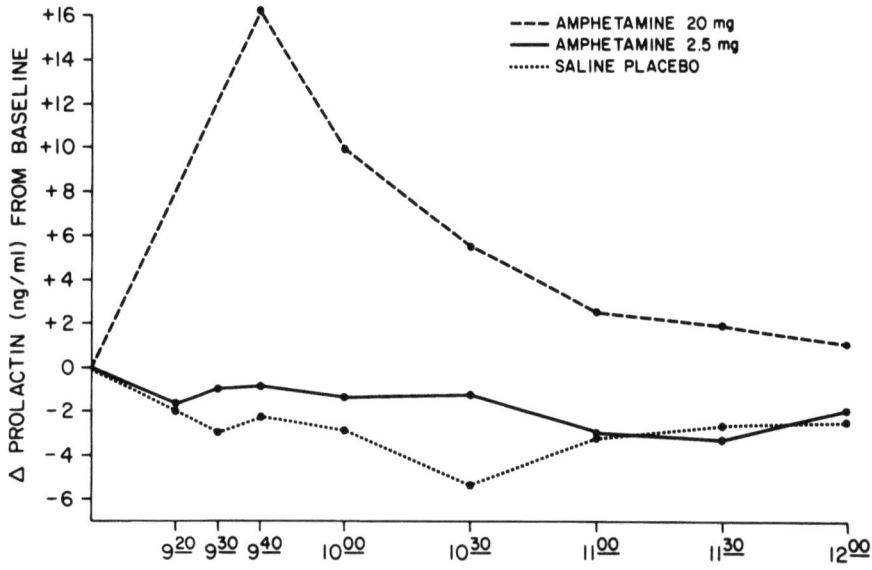

FIGURE 3. The replication study under more stringent conditions (bedrest, noninterview stress). Similar results as in study 1 were observed. 2.5 mg D-amphetamine i.v. had no effect.

The relative rise following D-amphetamine was again observed after chronic pimozide treatment, although when compared to placebo, a significant percent increase was observed at 5 min only ($p < 0.05$) but not at subsequent intervals. This failure to observe a significant PRL rise may be due to the high variability of levels during pimozide treatment. Percent PRL increase following D-amphetamine on lithium was significantly different from the placebo values at 25 min (82 ± 6% vs. 117 ± 13%, $p < 0.05$) and at 40 min (73 ± 6% vs. 120 ± 14%, $p < 0.05$). Percent changes following D-amphetamine were indistinguishable regardless of pretreatment with placebo, pimozide, or lithium (Fig. 2).

### 3.2. Study 2: Two Doses of D-Amphetamine

#### 3.2.1. PRL Response

Following the infusion of 20 mg D-amphetamine, serum PRL levels increased significantly compared to both placebo and the 2.5 mg D-amphetamine administration (ANOVA, $p < 0.05$) (Fig. 3). Three patients showed a marked increase in PRL concentration (240% over baseline) with values as high as 44 ng/ml. The other six patients showed no difference in PRL levels following the D-amphetamine infusion of both doses or placebo. There were no significant correlations between amphetamine blood levels and PRL response when controlled for dose, although in those patients with the large PRL increase, no significant rise was seen after the 2.5-mg infusion. The three patients with the larger response to 20 mg D-amphetamine could not be separated from the other six patients by sex, drug response, duration of illness, age of onset, number of days drug-free, premorbid functioning, or CT scan.

### 3.2.2. GH Response

There was no significant rise in GH following D-amphetamine infusion (ANOVA). One patient had to be excluded because of elevated baseline levels. Only three patients showed an increase over 5 ng/ml whereas six patients did not respond appreciably to either dose of D-amphetamine.

### 3.2.3. Cortisol Response

In the five patients studied, cortisol levels rose following 20 mg D-amphetamine compared to baseline or to placebo (van Kammen et al., in prep.). After placebo, cortisol levels declined significantly. Peak response occurred at 1 hr following infusion compared to placebo (11.5 ± 3.0 vs. 21.9 ± 3.5 mg/dl, $p < 0.05$, ($N = 5$), more than 90% increase). Following placebo, a decline was noted (16.6 ± 4.1 vs. 11.5 ± 3.0 mg/dl) and following 20 mg D-amphetamine, an increase (15.6 ± 2.9 vs. 21.9 ± 3.5 mg/dl).

### 3.2.4. β-Endorphin (i.r.) Response

Radioimmunoassay β-endorphin levels were significantly elevated ($p < 0.004$) by 20 mg D-amphetamine infusion but not significantly more than following saline infusion which also led to a significant increase ($p < 0.05$) (Fig. 4). Therefore, no significant amphetamine effect was noted for the group. There

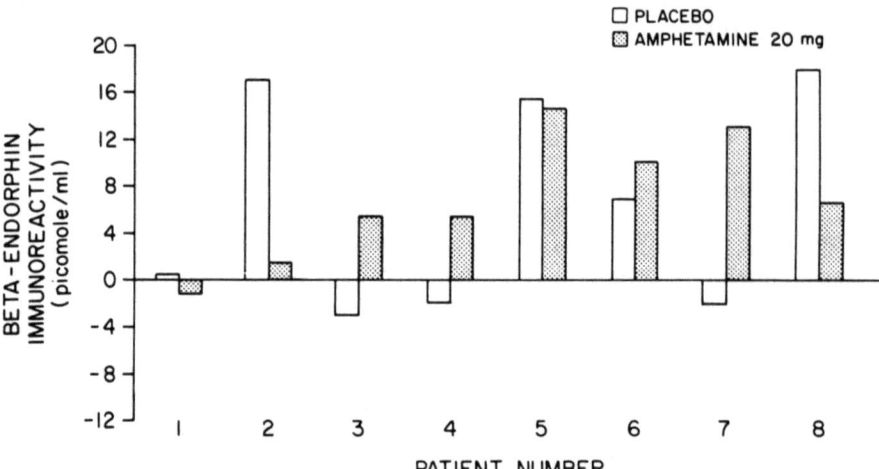

FIGURE 4. Peak response of β-endorphin (i.r.) following saline placebo and D-amphetamine i.v.

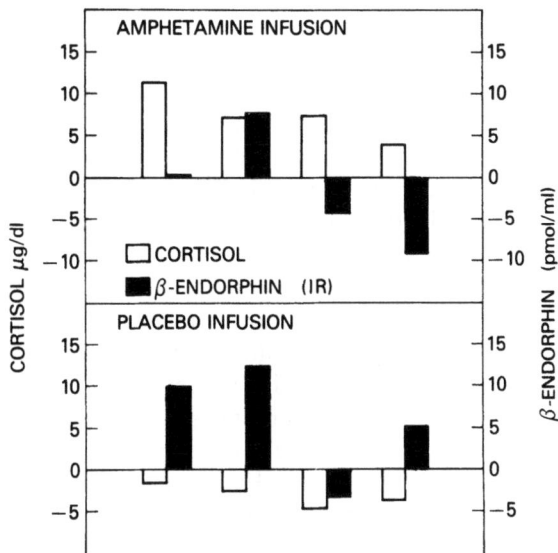

FIGURE 5. Cortisol and β-endorphin (i.r.) levels were measured in four patients following a saline placebo and 20 mg D-amphetamine. There was no relationship between the two hormone level changes.

was considerable variance in response. There were no significant correlations between the immunoreactive β-endorphin-like substance and cortisol (Fig. 5), PRL, or GH concentrations following D-amphetamine or placebo.

## 4. DISCUSSION

### 4.1. PRL Response

The increase in PRL following D-amphetamine administration is consistent with most reports of studies in humans (see Table 1), but contradict the dopaminergic hypothesis, i.e., D-amphetamine raises PRL levels contrary to DA or oral L-dopa administration which suppress PRL levels. Although β-endorphin releases PRL in men (Schulz et al., 1980), PRL levels did not correlate with β-endorphin (i.r.) levels. A better way of testing this potential relationship may be with a more specific antibody or receptor assay. Zacur et al. (1976) reported that there is some evidence that NE facilitates PRL release by inhibition of DA pathways. These findings suggest that (1) primary and secondary processes (presynaptic DA receptors) regulating PRL release are different in man than in animals (Meltzer et al., 1979); (2) mechanisms other than DA (e.g., NE, serotonin, bombesin) activated by D-amphetamine may have a direct effect on PRL release; (3) D-amphetamine conceivably has a direct effect upon the lactotrophs; or (4) that D-amphetamine affects the PRL-releasing factor. Kuczenski (1977) demonstrated a biphasic dopaminergic response to amphetamine in rats, and Groves et al. (1979) showed that increased DA release can decrease DA neuron firing activity, which might explain the "paradoxical" response.

### 4.2. GH Response

In normal controls and depressed patients, GH responses to DA agonists such as apomorphine (Meltzer *et al.*, 1981; Malas *et al.*, 1983) seem to be "stable," i.e., a similar response is observed after each repeated observation. Repeated amphetamine administration is associated with similar responses in a given individual over time (Nurnberger *et al.*, 1981, 1982). However, in the schizophrenic patients, this is not the case.

Good premorbid functioning was associated with greater increases in GH response following apomorphine (Malas *et al.*, 1983). In poor premorbid or process schizophrenic patients, no GH response was noted. This relationship with premorbid functioning suggests a hypo- or nondopaminergic development in some chronic schizophrenic patients, whereas acutely ill patients who were functioning well prior to the onset of their illness may have a hyperdopaminergic disorder (Meltzer *et al.*, 1981; Malas *et al.*, 1983). Such patients presumably do not have an enlarged ventricle brain ratio on CT scanning. The decreased GH response to D-amphetamine was not related to these clinical variables in our patients. D-Amphetamine exerts stimulating effects at several different levels in the hypothalamic–pituitary axis leading to release of GH which prevents a specific biochemical interpretation. Chronicity could also lead to a more blunted response. Due to these phenomena, GH responses to DA agonists will have only limited value in trait-related evaluating in schizophrenia but may be informative about the stability of the clinical state, predict drug response or impending relapse.

Presumably, amphetamine-induced release of DA and NE acts on both the hypothalamus and the pituitary. Both catecholamines may release somatostatin (GH-inhibiting hormone), but also may affect GRF, the GH-releasing factor. The variable responses observed across subjects suggest that the GH release to this indirectly acting DA agonist is indirectly related to the activity of, for instance, somatostatin. However, CSF somatostatin levels did not correspond with the peak response (van Kammen, Reichlin *et al.*, unpublished data). Thymoxamine, an α-receptor blocker, does not block the methylamphetamine-induced or amphetamine-induced GH release but, rather, increases it, although less so than propranolol, a β-adrenergic receptor blocker (Rees *et al.*, 1970). Propranolol increases L-dopa induced GH levels also (Camanni *et al.*, 1974). The α-receptor blocker phentolamine blocks the GH increase following amphetamine in animals (Besser *et al.*, 1969). In monkeys, pimozide does not block the amphetamine-induced GH response (Marantz *et al.*, 1976). Similarly, Nurnberger *et al.* (1981) were unable to block the GH response to D-amphetamine with acute haloperidol administration, suggesting that other (nondopaminergic) mechanisms such as NE may be involved, even though apomorphine does increase GH release. The GH response to D-amphetamine is dose-dependent and those schizophrenic patients who respond with an increase may have a more variable response over time than depressed and normal subjects (Table 1).

### 4.3. Endorphins, Cortisol, and Stress Response

We studied the cortisol responses to examine any link between this hormone and β-endorphin with respect to D-amphetamine response. ACTH and β-endorphin are supposedly released together and are derived from the same prohormone, priopiomelanocortin. D-Amphetamine raises cortisol levels, purportedly by noradrenergic mechanisms, except in depressed patients in whom cortisol levels are decreased (Sachar et al., 1980, 1981). Cohen et al. (1981) showed that D-amphetamine (0.3 mg/kg) increased β-endorphin (i.r.) in normal subjects, which correlated significantly with the cortisol responses ($r = 0.76$, $p < 0.05$). Dubois et al. (1981) reported that surgical stress produced significant increases in both β-endorphin (i.r.) and cortisol in eight patients, suggesting that these hormonal responses to D-amphetamine and to surgical stress might be comparable in nonpsychiatric patients. However, dexamethasone, which suppresses cortisol release, failed to suppress β-endorphin levels in humans and rhesus monkeys (Kalin et al., 1980). Kalin et al. (1980) suggested another pathway for β-endorphin release not directly linked to the ACTH-releasing system. In our schizophrenic patients, immunoreactive β-endorphin-like substances were not released consistently with cortisol. Different stressors may have different endocrine effects. If we consider D-amphetamine administration a stressor, the increases in PRL and GH may reflect the stress response. There was no evidence that the increase or decrease in psychosis following D-amphetamine (van Kammen et al., 1978) related to the endocrine responses. Nor did the hormonal responses to D-amphetamine in schizophrenics correlate with each other, in contrast to the report by Nurnberger et al. (1981) who observed significant correlations in normals between increases in PRL, GH, and cortisol following D-amphetamine administration. Brown et al. (1978) did not observe a correlation between cortisol and GH in response to methylphenidate, an agent that does not release cortisol as does D-amphetamine.

The β-endorphin (i.r.) increases in schizophrenic patients following placebo may be a reflection of their inability to modulate external stimulation as has been suggested by psychophysiological studies (Zahn, 1980), but the β-endorphin (i.r.) response study requires replication before further conclusions can be drawn.

### 5. SUMMARY

Twenty milligrams of D-amphetamine (i.v.) induced increases in plasma PRL, GH, cortisol, and β-endorphin (i.r.) levels in man. The dose response was such that patients receiving 20 mg D-amphetamine showed large increases in cortisol and PRL concentrations, but only small increments in the concentrations of GH and β-endorphin (i.r.); 2.5 mg had no effect upon PRL or GH.

In drug-free schizophrenic patients, cortisol and PRL response appeared to be similar as in normal subjects. However, GH release was relatively unresponsive to D-amphetamine, and β-endorphin (i.r.) role also following placebo.

Comparison with previously published studies suggests that only the GH and β-endorphin (i.r.) but not the PRL and cortisol responses to i.v. D-amphetamine may be disturbed in schizophrenia. The β-endorphin (i.r.) response to placebo, in particular, requires further study.

## 6. REFERENCES

Aarskog, D., Ferang, F. Ø., Kløve, H., Støa, K. F., and Thorsen, T., 1977, The effect of stimulant drugs, dextroamphetamine and methylphenidate, on secretion of growth hormone in hyperactive children, *J. Pediatr.* **90**:136.

Berger, P. A., Watson, S. J., Akil, H., Elliot, G. R., Rubin, R. T., Pfefferbaum, A., Davis, K. L., Barchas, J. D., and Li, C. H., 1980, Beta-endorphin and schizophrenia, *Arch. Gen. Psychiatry* **37**:635.

Besser, G. M., Butler, P. W. P., Landon, J., and Rees, L., 1969, Influence of amphetamines on plasma corticosteroid and growth hormone levels in man, *Br. Med. J.* **4**:528.

Bloom, F., Segal, D., Ling, N., and Guillemin, R., 1976, Endorphins: Profound behavioral effects in rats suggest new etiological factors in mental illness, *Science* **194**:630.

Brown, W. A., 1977, Psychologic and neuroendocrine response to methylphenidate, *Arch. Gen. Psychiatry* **34**:1103.

Brown, W. A., Corriveau, D. P., and Ebert, M. H., 1978, Acute psychologic and neuroendocrine effects of dextroamphetamine and methylphenidate, *Psychopharmacology* **58**:189.

Butler, P. W. P., Besser, G. M., and Steinberg, H., 1968, Changes in plasma cortisol induced by dexamphetamine and chlordiazepoxide given alone and in combination in man, *J. Endocrinol.* **40**:391.

Camanni, F., Massara, F., and Molinatti, G. M., 1974, Propranolol enhancement of L-DOPA induced growth hormone stimulation, *Biomedicine* **21**:241.

Checkley, S. A., 1979, Corticosteroid and growth hormone responses to methylamphetamine in depressive illness, *Psychol. Med.* **9**:107.

Checkley, S. A., and Crammer, J. L., 1977, Hormone responses to methylamphetamine in depression: A new approach to the noradrenaline depletion hypothesis, *Br. J. Psychiatry* **131**:582.

Clemens, J. A., and Shaar, C. J., 1980, Control of prolactin secretion in mammals, *Fed. Proc.* **39**:2588.

Cohen, M. R., Nurnberger, J. I., Pickar, D., Gershon, E., and Bunney, W. E., Jr., 1981, Dextroamphetamine infusions in normals result in correlated increases of plasma β-endorphin and cortisol immunoreactivity, *Life Sci.* **29**:1243.

Davis, G. C., Buchsbaum, M. S., Naber, D., and van Kammen, D. P., 1980, Effects of opiates and opiate antagonists on somatosensory evoked potentials in patients with schizophrenia and normal adults, in: *Recent Advances in Biological Psychiatry* (C. Perris and L. von Knorring, eds.), pp. 73–80, Karger, Basel.

Dommissee, P., Schulz, S. C., Narasimhachari, N., and Hamer, R., 1982, The neuroendocrine and behavioral response to dextro-amphetamine in normal individuals, Presented at the Society of Biological Psychiatry, Toronto.

Dubois, M., Pickar, D., Cohen, M. R., Roth, Y. F., MacNamara, T., and Bunney, W. E., Jr., 1981, Surgical stress in humans is accompanied by an increase in plasma beta-endorphin immunoreactivity, *Life Sci.* **29**:1249.

Ebert, M. H., van Kammen, D. P., and Murphy, D. L., 1976, Plasma levels of amphetamine and behavioral response, in: *Pharmacokinetics of Psychoactive Drugs* (L. A. Gottschalk and S. Merlis, eds.), pp. 157–169, Spectrum, New York.

Greenhill, L. L., Puig-Antich, J., Chambers, W., Rubinstein, B., Halpern, F., and Schar, E. J.,

1981, Growth hormone, prolactin, and growth responses in hyperkinetic males treated with d-amphetamine, *J. Am. Acad. Child Psychiatry* **20**:84.

Groves, P. M., Staunton, D. A., Wilson, C. J., and Young, S. J., 1979, Sites of action of amphetamine intrinsic to catecholaminergic nuclei: Catecholaminergic presynaptic dendrites and axons, *Prog. Neuro-Psychopharmacol.* **3**:315.

Halbreich, U., Asnis, G. M., Halpern, F., Tabrizi, M. A., and Sachar, E. J., 1980, Diurnal growth hormone responses to dextroamphetamine in normal young men and post-menopausal women, *Psychoneuroendocrinology* **5**:339.

Halbreich, U., Sachar, E. J., Asnis, G. M., Nathan, R. S., and Halpern, F. S., 1981, The prolactin response to intravenous dextroamphetamine in normal young men and postmenopausal women, *Life Sci.* **28**:2337.

Halbreich, U., Sachar, E. J., Asnis, G. M., Quitkin, F., Nathan, R. S., Halpern, F. S., and Klein, D. F., 1982, Growth hormone response to dextroamphetamine in depressed patients and normal subjects, *Arch. Gen. Psychiatry* **39**:189.

Janowsky, D. S., Leichner, P., Parker, D., Judd, L. L., Huey, L., and Clopton, P., 1978, The effect of methylphenidate on serum growth hormone: Influence of antipsychotic drugs and diagnosis, *Arch. Gen. Psychiatry* **35**:1384.

Kalin, N. H., Risch, S. C., Cohen, R. M., Insel, T., and Murphy, D. L., 1980, Dexamethasone fails to suppress β-endorphin plasma concentrations in humans and rhesus monkeys, *Science* **209**:827.

Kuczenski, R., 1977, Biphasic effects of amphetamine on striatal dopamine dynamics, *Eur. J. Pharmacol.* **46**:249.

Langer, G., and Matussek, N., 1977, Dextro- and L-amphetamine are equipotent in releasing human growth hormone, *Psychoneuroendocrinology* **2**:379.

Langer, G., Heinze, G., Reim, B., and Matussek, N., 1976, Reduced growth hormone responses to amphetamine in "endogenous" depressive patients: Studies in normal, "reactive" and "endogenous" depressive, schizophrenic, and chronic alcoholic subjects, *Arch. Gen. Psychiatry* **33**:1471.

Lu, K. H., and Meites, J., 1971, Inhibition by L-Dopa and monoamine oxidase inhibitors of pituitary prolactin release; stimulation by methyl-dopa and d-amphetamine, *Proc. Soc. Exp. Biol. Med.* **137**:480.

MacLeod, R. M., and Lamberts, S. W. J., 1979, Clinical and fundamental correlates in dopaminergic control of prolactin secretion, in: *Neuroendocrine Correlates in Neurology and Psychiatry* (E. E. Müller and A. Agnoli, eds.), Elsevier/North-Holland, Amsterdam.

Malas, K. L., van Kammen, D. P., DeFraites, E. J., Brown, G., and Gold, P. W., 1983, Reduced growth hormone response to apomorphine in schizophrenic patients with poor premorbid functioning, *Biol. Psychiatry* **18**:255.

Marantz, R., Sachar, E. J., Weitzman, E., and Sassin, J., 1976, Cortisol and GH responses to d- and l-amphetamine in monkeys, *Endocrinology* **99**:459.

Massara, F., and Camanni, F., 1972, Effects of various adrenergic receptor stimulating and blocking agents on human growth hormone secretion, *J. Endocrinol.* **54**:195.

Meites, J., 1973, *Human Prolactin*, American Elsevier, New York.

Meltzer, H. Y., Fessler, R. G., Simonovic, M., Docherty, J., and Fang, V. S., 1979, Effect of d- and l-amphetamine on rat plasma prolactin levels, *Psychopharmacology* **61**:63.

Meltzer, H. Y., Bush, D., and Fang, V. S., 1981, Hormones, dopamine receptors, and schizophrenia, *Psychoneuroendocrinology* **6**:17.

Naber, D., Pickar, D., Post, R. M., van Kammen, D. P., Waters, R. N., Ballenger, J. C., and Goodwin, F. K., 1981, Endogenous opioid activity and β-endorphin immunoreactivity in CSF of psychiatric patients and normal volunteers, *Am. J. Psychiatry* **138**:1457.

Nurnberger, J. I., Jr., Gershon, E. S., Simmons, S., Ebert, M., Kessler, L. R., Dibble, E. D, Jimerson, S. S., Brown, G. M., Gold, P. W., Jimerson, D. C., Guroff, J. J., and Storch, F. I., 1982, Behavioral, biochemical and neuroendocrine responses to amphetamine in normal twins and "well state" bipolar patients *Psychoneuroendocrinology*, **7**:163–176.

Nurnberger, J. I., Jr., Gershon, E. S., Jimerson, D. C., Buchsbaum, M. S., Gold, P. W., Brown,

G., and Ebert, M., 1981, Pharmacogenetics of *d*-amphetamine response in man, in: *Genetic Research Strategies for Psychobiology and Psychiatry* (E. S. Gershon, S. Matthysse, X. O. Breakefield, and R. D., Ciaranello, eds.), pp. 257–268, Boxwood Press, New York.

Parkes, J. D., Debono, A. G., Jenner, P., and Walters, J., 1977, Amphetamines, growth hormone and narcolepsy, *Br. J. Clin. Pharmacol.* **4**:343.

Ravitz, A. J., and Moore, K. E., 1977, Effects of amphetamine, methylphenidate and cocaine on serum prolactin concentrations in the male rat, *Life Sci.* **21**:267.

Rees, L., Butler, P. W. P., Gosling, C., and Besser, G. M., 1970, Adrenergic blockade and the corticosteroid and growth hormone responses to methylamphetamine, *Nature (London)* **228**:565.

Sachar, E. J., Asnis, G., Nathan, R. S., Halbreich, U., Tabrizi, M. A., and Halpern, F. S., 1980, Dextroamphetamine and cortisol in depression: Morning plasma cortisol levels suppressed, *Arch. Gen. Psychiatry* **37**:755.

Sachar, E. J., Halbreich, U., Asnis, G. M., Nathan, R. S., Halpern, F. S., and Ostrow, L., 1981, Paradoxical cortisol responses to dextroamphetamine in endogenous depression, *Arch. Gen. Psychiatry* **38**:1113.

Schulz, S. C., Wagner, R., van Kammen, D. P., Rogol, A. D., Davis, G. C., Wyatt, R. J., Pickar, D., Bunney, W. E., Jr., and Li, C. H., 1980, Prolactin response to beta-endorphin in man, *Life Sci.* **27**:1735.

Schulz, S. C., van Kammen, D. P., Pickar, D., Cohen, M., and Naber, D., 1982, Response of plasma beta-endorphin immunoreactivity: *d*-Amphetamine and placebo in schizophrenic patients, *Psychiatry Res.* **7**:171.

Schulz, S. C., van Kammen, D. P., Rogol, A. D., Ebert, M., Pickar, D., Cohen, M. R., and Naber, D., 1981b, Amphetamine increases prolactin but not growth hormone or endorphins in schizophrenic patients, *Psychopharmacol. Bull.* **17**:193.

Slater, S., de la Vega, C. E., Skyler, J., and Murphy, D. L., 1976, Plasma prolactin stimulation by fenfluramine and amphetamine, *Psychopharmacol. Bull.* **12**:26.

Spitzer, R. L., Endicott, J. E., and Robbins, E., 1975, 1978, *Research Diagnostic Criteria (RDC) for a Selected Group of Functional Disorders*, New York Psychiatric Institute, New York.

van Kammen, D. P., 1979, The dopamine hypothesis of schizophrenia revisited, *Psychoneuroendocrinology* **4**:37.

van Kammen, D. P., Docherty, J. P., Marder, S. R., Siris, S. G., and Bunney, W. E., Jr., 1978, *d*-Amphetamine raises serum prolactin in man: Evaluations after chronic placebo, lithium and pimozide treatment, *Life Sci.* **23**:1487.

van Kammen, D. P., Docherty, J. P., Marder, S. R., Schulz, S. C., and Bunney, W. E., Jr., 1980, Lack of behavioral supersensitivity to *d*-amphetamine after pimozide withdrawal, *Arch. Gen. Psychiatry* **37**:287.

van Kammen, D. P., Docherty, J. P., Marder, S. R., and Bunney, W. E., Jr., 1981, Acute amphetamine response predicts antidepressant and antipsychotic responses to lithium carbonate in schizophrenic patients, *Psychiatry Res.* **4**:313.

van Kammen, D. P., Docherty, J. P., and Bunney, W. E., Jr., 1982a, Prediction of early relapse after pimozide discontinuation by response to *d*-amphetamine during pimozide treatment, *Biol. Psychiatry* **17**:233.

van Kammen, D. P., Docherty, J. P., Marder, S. R., Schulz, S. C., Dalton, L., and Bunney, W. E., Jr., 1982b, Antipsychotic effects of pimozide in schizophrenia: Treatment response prediction with acute dextroamphetamine administration, *Arch. Gen. Psychiatry* **39**:261.

van Kammen, D. P., Bunney, W. E., Jr., Docherty, J. P., Marder, S. R., Ebert, M. H., Rosenblatt, J. E., and Rayner, J. N., 1982c, *d*-Amphetamine induced heterogeneous changes in psychotic behavior in schizophrenia, *Am. J. Psychiatry* **139**:991.

van Kammen, D. P., Schulz, S. C., and Davis, G. C., *d*-Amphetamine raises cortisol in schizophrenic patients with and without naltrexone pretreatment, in preparation.

Wells, B., Silverstone, T., and Rees, L., 1978, The effect of oral dextroamphetamine on prolactin secretion in man, *Neuropharmacology* **17**:1060.

Zacur, H. A., Foster, G. V., and Tyson, J. E., 1976, Multifactorial regulation of prolactin secretion, *Lancet* **I**:410.

Zahn, T., 1980, Predicting outcome from measures of attention and autonomic functioning, in: *Perspectives in Schizophrenia Research* (C. Baxter and T. Melneche, eds.), pp. 81–106, Raven Press, New York.

CHAPTER 31

# Neuroleptics and Prolactin
## A Second Look

WALTER ARMIN BROWN

## 1. INTRODUCTION

In 1955, shortly after its introduction as an antipsychotic, clinicians noted that chlorpromazine occasionally induced galactorrhea (Hooper *et al.*, 1961). It was not until 1971, however, that chlorpromazine, and other neuroleptics as well, were found to regularly elevate serum prolactin (PRL) (Beumont *et al.*, 1974). This observation and our current understanding of its mechanism awaited a series of conceptual and methodological advances.

In the early 1960s, rat experiments measuring PRL with the laborious pigeon crop assay showed that the brain's long-known inhibitory effect on PRL release could be accounted for by a PRL-inhibiting substance contained in the hypothalamus (Pasteels, 1961). Soon a number of pharmacologic agents, neuroleptics among them, were found to block the inhibitory effect of this hypothalamic material (Danon *et al.*, 1963). The discovery and development of the radioimmunoassay technique made it possible by the late 1960s to measure serum concentrations of PRL in both animals and humans with a degree of sensitivity and specificity.

The 1970s saw a flurry of studies replicating, extending, and elaborating the observation that neuroleptics induce PRL release. Interest in this phenomenon was generated and sustained by its predictability and robustness and by the hope that the PRL response might bear some relation to a neuroleptic's clinical effects.

But enthusiasm lagged when it appeared that the PRL-elevating and antipsychotic effects of neuroleptics differed in their time course and dose–response characteristics (Gruen *et al.*, 1978a). Even more disheartening were inconsistencies among the studies concurrently examining clinical response and

---

WALTER ARMIN BROWN • Neuroendocrine Research Program, Providence Veterans Administration Medical Center, and Department of Psychiatry, Brown University, Providence, Rhode Island 02912.

PRL levels during acute neuroleptic treatment (Gruen *et al.*, 1978b; Meltzer and Fang, 1976); and studies of patients with tardive dyskinesia consistently failed to show a relationship between either baseline or pharmacologically provoked PRL levels and the presence or absence of this disorder (Asnis *et al.*, 1979).

By 1980 the initial enthusiasm over the potential relevance of PRL to clinical psychiatry had been replaced by the consensus that the PRL response was of no use in either guiding or monitoring neuroleptic treatment. Reexamination of the early studies, however, along with recent observations obtained under controlled conditions and, most compelling, preliminary reports examining neuroleptic plasma levels in relation to PRL suggest that a second look is warranted. But before scrutinizing the relationships among the PRL-elevating, extrapyramidal, and antipsychotic effects of neuroleptics, we first look at how the different brain systems underlying these effects respond to neuroleptics.

## 2. ANATOMICAL AND PHYSIOLOGIC CONSIDERATIONS

Neuroleptic-induced PRL release is mediated by the tuberoinfundibular dopamine (DA) system, a small neural pathway with cell bodies in the arcuate nucleus of the hypothalamus and nerve terminals adjacent to portal vessels in the median eminence. DA released from these nerve terminals travels through the portal vasculature and acts directly upon the lactotropic pituitary cells to inhibit the synthesis and release of PRL. The hypothalamic DA neurons may also stimulate the release of a PRL-inhibiting peptide from the median eminence. Although it is clear that the pituitary contains DA receptors and that DA can act directly on the lactotropic cells, the relative physiologic roles of DA and a PRL-inhibiting peptide in PRL regulation are yet to be established. Neuroleptics block the inhibiting effect of DA on PRL release at the median eminence, pituitary, or both sites.

The extrapyramidal side effects of neuroleptics including acute dystonia, parkinsonian symptoms, and tardive dyskinesia are mediated, at least in part, by the nigrostriatal DA system. The brain systems mediating the antipsychotic effects of neuroleptics are as yet unknown; but the neuropharmacologic effects of these drugs and specifically the high correlation between antipsychotic potency and affinity for DA receptors suggest involvement of a DA system. By exclusion of dopaminergic tracts unlikely to be involved in cognition and emotion, the mesolimbic and mesocortical DA systems are considered the most likely sites for the antipsychotic effects.

Clearly the PRL-elevating, extrapyramidal, and antipsychotic effects of neuroleptics involve different brain sites. Yet at a macroscopic and physiologic level these DA systems share a number of characteristics. The pituitary DA receptors involved in PRL regulation, striatal DA receptors involved in extrapyramidal symptoms, and the DA receptors involved in the antipsychotic effects of neuroleptics and putatively, as suggested by preliminary postmortem studies, in the pathogenesis of schizophrenia are all of the D2 type; they are not linked to adenylate cyclase and have a higher affinity for neuroleptics than for DA agonists (Seeman, 1980). The similarities among these DA receptors in their ultrastructure and affinities for pharmacologic agents are reflected in similarities in the dose–response characteristics for the neuroleptic effects which they mediate.

## 3. POTENCY AND DOSE–RESPONSE

The potencies of neuroleptics in alleviating psychotic symptoms, inducing extrapyramidal side effects and elevating serum PRL, are highly correlated and all of these effects are strongly correlated with a neuroleptic's affinity for D2 receptors (Seeman, 1980; Creese et al., 1976). Alleviation of psychotic symptoms, extrapyramidal side effects, and maximum PRL elevation occur with relatively small amounts of haloperidol and fluphenazine, requiring two to three times more trifluoperazine and perphenazine and 50 times more chlorpromazine and thioridazine for all three effects. These drugs have the same relative potency with respect to their affinity for DA (D2) receptors. These observations are the strongest and most direct evidence to date that binding to DA receptors is a central neuropharmacological event in the antipsychotic, extrapyramidal, and PRL-elevating effects of neuroleptics.

A critical issue with respect to the clinical relevance of PRL measurement during neuroleptic treatment is the relationship among the dose–response curves for the PRL-elevating and clinical effects. More specifically, what are the neuroleptic dose ranges over which these responses occur?

Although 25 years of clinical experience has generated considerable information and led to fairly standard practices regarding the neuroleptic treatment of schizophrenia, the relationship between dose and antipsychotic response is not entirely clear. People vary widely in their neuroleptic dose requirements and among a group of patients treated with more than the very lowest doses there is little if any relationship between neuroleptic dose and clinical response. Nevertheless, clinical experience, including controlled drug trials, has provided a broad and widely accepted antipsychotic dose range. For the treatment of acute schizophrenia the dose range is from 200 to 1600 mg/day of chlorpromazine or its equivalent and for maintenance treatment 50 to 400 mg/day (Hollister, 1977).

During the initial weeks and months of neuroleptic treatment a substantial proportion of patients develop extrapyramidal symptoms. These symptoms are not clearly dose related (Rao et al., 1980; Wiles et al., 1976) and patients appear to vary quite widely in their propensity to develop them. Clinical experience indicates that for a patient with this propensity extrapyramidal symptoms require some minimum neuroleptic dose and can be alleviated by lowering the dose.

Both early and recent studies have examined PRL elevation during the hours after a single neuroleptic dose. Although people vary widely in their PRL response to a given neuroleptic dose, serum PRL predictably peaks 1 to 2 hr after i.m. injection and gradually falls over the next 6 hr. Maximum PRL elevation in these circumstances is achieved with single neuroleptic doses far below the daily antipsychotic dose (Gruen et al., 1978b; Rubin et al., 1976; Meltzer et al., 1981) and the sensitive range of the neuroleptic dose–PRL response curve is from 0.5 to 1.5 mg of haloperidol or 25 to 75 mg of chlorpromazine. Serum PRL concentration does not rise further with higher doses. These findings have been interpreted to mean that the PRL response to neuroleptics occurs at a much lower dose than the antipsychotic response, that maximum PRL levels are reached at neuroleptic doses far below the therapeutic range, and, therefore, that serum PRL cannot reflect a neuroleptic's availability for therapeutic purposes.

But the antipsychotic response to neuroleptics is usually assessed under conditions so different from the acute PRL response that the doses required cannot be meaningfully compared. In both clinical practice and research the antipsychotic response is usually assessed over days and weeks, not hours of neuroleptic treatment; and, although rapid neuroleptization studies suggest that low neuroleptic doses can reduce psychotic symptoms within hours, we really do not know the immediate effect on schizophrenic symptoms of a low (within the PRL-elevating range) neuroleptic dose. Thus, whether or not the PRL-elevating and antipsychotic dose ranges are comparable when both responses are assessed within hours after a single dose remains an open question.

In contrast, when measured during neuroleptic treatment under steady-state conditions (12 hr after an oral dose of neuroleptic), serum PRL is clearly sensitive to changes in neuroleptic dose within the therapeutic range. Under these conditions serum PRL continues to rise until the upper end of the antipsychotic dose range is reached (Ohman and Axelsson, 1978; Rubin et al., 1980; Bjorndal et al., 1980). Maximum PRL elevation does not occur until patients are treated with a daily neuroleptic dose greater than 80 mg of haloperidol or 600 mg of thioridazine. In schizophrenic patients responsive to neuroleptics, it is questionable whether doses above these provide additional antipsychotic effect.

TABLE 1
Serum PRL Concentration during Withdrawal from Long-Term Neuroleptic Treatment ($N = 26$)

| | Weeks | | | |
|---|---|---|---|---|
| | 1 | 2 | 3 | 4 |
| Prolactin[a] (mean ± S.E.M., ng/ml) | 10.7 ± 1.6 | 8.4 ± 1.3 | 6.2 ± 0.9 | 4.2 ± 0.4 |
| Neuroleptic dose (mean mg/day, CPZ equiv.) | 335.6 | 214.6 | 105.6 | 0 |

[a] $F = 16.9, p < 0.001$.

We have systematically withdrawn 26 schizophrenic patients from long-term neuroleptic treatment and measured PRL in a single morning sample during maintenance treatment, dosage reduction, and the drug-free state (Table 1). As neuroleptic dose was reduced by one-third each week, we found a stepwise reduction in serum PRL concentration with significant differences in PRL among the maintenance, two-thirds dose, one-third dose, and drug-free conditions. These data as well indicate that serum PRL concentration measured under steady-state conditions is sensitive to and reflects changes in neuroleptic dose within the therapeutic range.

## 4. TOLERANCE

Brain DA systems may differ in their response to long-term neuroleptic treatment. Clinical observation suggests that some degree of tolerance develops to the acute extrapyramidal side effects of neuroleptics. The increase in human CSF homovanillic acid noted during the first weeks of neuroleptic treatment, which may be primarily an index of striatal DA turnover, attenuates several weeks after (Bowers and Heninger, 1981; Post and Goodwin, 1975) and rat studies assessing both biochemical and behavioral dimensions of nigrostriatal DA system activity consistently show adaptive changes with long-term neuroleptic treatment. Tolerance develops to the increase in striatal DA turnover seen at the onset of neuroleptic treatment, the number of striatal DA receptors increases, and, as indicated by biochemical and behavioral responses to DA agonists, there is, consequent to long-term neuroleptic treatment, an enhancement of DA receptor sensitivity (Burt et al., 1977; Lal et al., 1977).

According to the prevailing clinical wisdom, tolerance does not develop to the antipsychotic effect of neuroleptics. The data base underlying this widely held belief is unclear. Although many schizophrenic patients remain in remission on constant-dose long-term neuroleptic treatment, over 20% relapse within a given year. Nonadherence does not in itself account for the high rate of relapse; one study comparing patients maintained on oral and i.m. fluphenazine showed that by 1 year after hospital discharge both groups had a relapse rate approaching 40% (Hogarty et al., 1979). Surely a number of social, psychological, and pharmacological factors other than tolerance could contribute to relapse in neuroleptic-maintained patients. As to whether or not tolerance does develop to the antipsychotic effects the clinical and biochemical data needed for a definitive answer are not yet available.

Several lines of evidence, however, suggest that tolerance develops to the PRL-elevating effect of neuroleptics. A number of investigators have observed that serum PRL concentrations in patients on long-term neuroleptic treatment are within the normal range (DeRivera et al., 1976; Naber et al., 1979) and we have shown in two studies that when patients who have been on long-term neuroleptic treatment are withdrawn from such treatment, kept drug-free, and then restarted on their previous neuroleptic dose, they show a significant increase in PRL over that during chronic treatment (Brown and Laughren, 1981b). We have also shown that in patients maintained on a constant neuroleptic dose, serum PRL falls during the first 4 weeks of neuroleptic treatment and after 4 months reaches a stable level higher than the drug-free level but substantially lower than that at the onset of neuroleptic treatment (Brown and Laughren, 1981b). These shifts in PRL appear to occur in the face of stable serum neuroleptic levels. This apparent tolerance to the PRL-elevating effect of neuroleptics is consistent with the changes observed in rat PRL responses after both surgical destruction of tuberoinfundibular DA neurons and long-term neuroleptic treatment. Under these conditions baseline PRL is decreased and the sensitivity of the PRL-inhibiting response to DA agonists is enhanced, suggesting an increase in the sensitivity of pituitary DA receptors (Cheung and Weiner, 1978; Lal et al., 1977).

## 5. NEUROLEPTIC SERUM LEVELS

Although within an individual the clinical response to neuroleptics is apparently dose related and the PRL response clearly dose related, among a group of neuroleptic-treated patients the neuroleptic dose bears no consistent relationship to either clinical response or serum PRL. Measurement of neuroleptic serum levels points to at least part of the reason; individuals on the same neuroleptic dose vary widely in serum neuroleptic levels (Davis et al., 1972). Consequently, there is little correlation among a group of patients between neuroleptic dose and neuroleptic serum level. Recent studies suggest that the clinical response to neuroleptics is more closely related to neuroleptic serum level than it is to neuroleptic dose.

Until recently, complex and expensive chemical methods were required to measure serum neuroleptic levels. Among the few studies measuring chlorpromazine or thioridazine by these methods and comparing their serum levels to prolactin, one showed a positive correlation between serum PRL concentration and chlorpromazine levels (Wiles et al., 1976), whereas others failed to show a consistent relationship (Kolakowska et al., 1976; Nikitopoulou et al., 1977). A confounding factor in these studies is the presence of active metabolites so that measurement of the parent compound alone is not an accurate reflection of the drug's biological activity.

Recent serum level studies have used the newly developed radioreceptor assay for neuroleptics, which measures total serum neuroleptic activity, that contributed by both the parent drug and its active metabolites, or have applied this or other assay techniques to the measurement of neuroleptics, such as haloperidol, which have weakly active or inactive metabolites. The results of these studies have been quite consistent with respect to the relationship between neuroleptic serum level and PRL. Among neuroleptic-treated patients under both steady-state conditions and after an acute neuroleptic dose, there is a high correlation between neuroleptic levels and serum PRL which is sustained throughout the therapeutic dose range. Rubin et al. (1980) found a correlation of 0.83 between serum haloperidol and serum PRL levels; Rao et al. (1980) report a correlation of 0.83 between plasma haloperidol and PRL levels, and Meltzer et al. (1981) found in patients treated with chlorpromazine a correlation of 0.92 between serum PRL and serum neuroleptic activity measured by radioreceptor assay. Consistent with these data is the observation that over the 24 weeks after fluphenazine decanoate injections are discontinued, the gradual decline of plasma fluphenazine levels is accompanied by a gradual fall in PRL levels (Wistedt et al., 1981).

We have examined in 61 patients taking a variety of neuroleptics the relationship between serum PRL and serum neuroleptic activity measured by radioreceptor assay. Since in this assay serum neuroleptic levels for one drug can be quite different from those for a clinically equivalent dose of another, we converted the neuroleptic levels for each drug to normal scores in order to compare serum PRL concentration to neuroleptic levels among the various drugs. We found a significant positive correlation ($r = 0.55, p < 0.001$) between serum PRL and neuroleptic levels across drug groups. Thus, irrespective of specific neuroleptic medication, patients with relatively low neuroleptic levels for a given drug have relatively low PRL and those with relatively high levels have relatively high PRL. We have also found, and not surprisingly so in view of the foregoing data, a high correlation within individuals ($r = 0.85$ to $0.95$) between serum neuroleptic levels and serum PRL.

Although a few studies have failed to show a relationship between neuroleptic serum levels and PRL (Smith et al., 1979; Linnolia et al., 1980) the majority of studies are consistent in showing a linear relationship between serum PRL and neuroleptic levels throughout the therapeutic dose range. In view of the multiple variables affecting an individual's PRL regulation, the significant positive correlations observed between absolute PRL concentration and neuroleptic levels are quite compelling.

## 6. SERUM PRL AND ANTIPSYCHOTIC RESPONSE

A handful of studies have examined the relationship between serum PRL and the antipsychotic response to neuroleptic treatment. These studies have compared scores on a symptom rating scale to the increment in or absolute level of PRL during the initial weeks of neuroleptic treatment or have compared serum PRL in patients classified as treatment responders and nonresponders. With a few exceptions (Ohman and Axelsson, 1978; Meltzer and Fang, 1976), these studies have failed to show a relationship between the antipsychotic and PRL responses to neuroleptic treatment (Gruen et al., 1978b; Bjorndal et al., 1980; Dotti et al., 1981; Rao et al., 1980). Patients who show a poor or absent clinical response during several weeks of neuroleptic treatment usually have PRL levels as high as those who show a good response. Although, as these data would seem to suggest, there may be no relationship between the antipsychotic and PRL response to neuroleptics, the studies generating these data have almost invariably used research designs which render the data difficult to interpret and the conclusions tenuous.

A substantial subgroup of schizophrenic patients, perhaps as recent data suggest those who have structural brain abnormalities, are resistant to neuroleptic treatment. Such patients do not improve irrespective of neuroleptic dose or serum level. There is a tendency to treat these drug-resistant patients with relatively high doses of neuroleptic and we found them to have relatively high serum neuroleptic levels (Brown et al., 1982). In such patients, PRL levels, as a reflection of neuroleptic activity, would not be expected to co-vary with clinical response. If such patients are included in a study, and usually no attempt is made to exclude them, they are likely to obscure any relationship that might exist between PRL or any other correlate of neuroleptic activity and antipsychotic response.

Also likely to obscure a relationship between antipsychotic response and PRL is the clinical practice, common to most of these studies, of changing the neuroleptic dose in response to the patient's symptomatology. Treatment at a constant dose for some period of time and a range of dose levels among patients is necessary to assess the relationship between clinical response and serum PRL.

In a recent study (Evans et al., 1980), patients were maintained on a constant dose of haloperidol (10 mg/day) for 8 weeks and clinical response examined in relationship to serum haloperidol and PRL levels. PRL rose significantly over baseline in the treatment responders, the authors report, but in the nonresponders did not change over the drug-free baseline level. Most of the nonresponders, they say, subsequently showed a satisfactory clinical response when the dose of haloperidol was raised. Although mean serum haloperidol levels were somewhat higher in the responders (4.8 ng/ml) than in the nonresponders (2.9 ng/ml), there was considerable overlap in these levels between the groups, and serum haloperidol was not as strong a correlate of clinical response as serum PRL. These data, in suggesting that an antipsychotic response to neuroleptics requires a degree of neuroleptic bioavailability sufficient to elevate serum PRL, are not inconsistent with previous findings. Although this information is not necessarily of immediate clinical usefulness, it points to the potential value of scrutinizing in greater detail the relationships between the clinical and PRL response during the treatment of acute schizophrenia.

Two groups of investigators have examined the relationship between serum PRL and psychotic relapse in schizophrenic patients maintained on neuroleptic treatment (Brown and Laughren, 1981a; Wistedt, 1981). Both groups have found that patients with relatively low serum PRL during maintenance neuroleptic treatment are more vulnerable to relapse than those with relatively high PRL.

Laughren et al. (1979) evaluated 11 patients during maintenance neuroleptic treatment, a 4-week drug-free trial, and upon return to neuroleptic treatment. The three patients who showed psychotic relapse during the drug-free trial were the three with the lowest serum PRL concentrations (all < 6.0 ng/ml) during maintenance neuroleptic treatment. This group went on to study in 28 schizophrenic patients the relationship between serum PRL during maintenance treatment and the latency to relapse after neuroleptic withdrawal. Patients with low PRL (< 6.0 ng/ml) relapsed significantly earlier than those with higher PRL (6.5 vs. 18.0 drug-free weeks) (Brown and Laughren, 1981a). These data are consistent with Wistedt's (1981) observation that patients who relapse after withdrawal of neuroleptic medication show a greater fall in PRL levels following drug withdrawal than those who remain stable. These data suggest that low serum PRL during and following maintenance neuroleptic treatment may identify a subgroup of schizophrenic patients with a high propensity to relapse.

We have observed a similar phenomenon in remitted patients continued on constant-dose neuroleptic treatment (Brown et al., 1982). Serum PRL concentration was significantly lower in those who relapsed during the 6 months after PRL was measured than in those who remained stable (5.0 vs. 12.6 ng/ml, $p < 0.05$). Consistent with this, Wistedt (1981) found among schizophrenic patients on long-acting neuroleptics that those who relapsed had lower PRL levels than those who remained stable. The high correlation between serum neuroleptic levels and PRL during chronic neuroleptic treatment (Brown et al., 1982) and the fact that low serum neuroleptic levels are an equal if not more powerful predictor of relapse than low PRL levels (Brown et al., 1982; Wistedt, 1981) suggest that in patients on continuous neuroleptic treatment the relationship between low serum PRL and relapse may be accounted for by the relatively low neuroleptic levels in patients with low PRL.

## 7. SERUM PRL AND EXTRAPYRAMIDAL SIDE EFFECTS

Blockade of striatal DA receptors is among the critical pathophysiologic events underlying the early appearing extrapyramidal side effects of neuroleptics. One might hypothesize that PRL levels would be higher in patients who develop such symptoms than in those who do not. Yet, the aggregated data from the few studies examining this issue are inconclusive. Wiles et al. (1976) found an association between parkinsonian side effects and relatively high serum PRL, and Rao et al. (1980) found a significant positive correlation between scores on an extrapyramidal side effects scale and serum PRL. Others, however, find no relationship between the degree of neuroleptic-induced PRL elevation and extrapyramidal side effects (Dotti et al., 1981). These inconsistencies are not surprising in view of the lack of a consistent relationship between either neuroleptic dose or serum levels and the development of extrapyramidal symptoms (Kolakowska et al., 1976; Rao et al., 1980). Clearly there are marked individual differences in the propensity to develop extrapyramidal symptoms and such symptoms are determined by factors other than the biological activity of neuroleptics alone. Yet the value of serum PRL as a predictor or correlate of early extrapyramidal symptoms may be worthy of further study. Such study, to provide definitive information, would have to control carefully for the sex of the patients, dose of neuroleptic, and timing of PRL and extrapyramidal symptom measurement.

Tardive dyskinesia, a late-appearing extrapyramidal side effect, is now recognized as a serious and not infrequent complication of long-term neuroleptic treatment. The most compelling hypothesis regarding its pathophysiology is that of striatal DA receptor sensitivity resulting from chronic DA receptor blockade. Because DA receptor activity in the striatum cannot be directly assessed in humans, researchers have pursued the possibility that DA activity in the tuberoinfundibular tract as assessed by serum concentrations of PRL may reflect or parallel the activity of the striatal DA system.

Although there is some evidence in rats for enhanced tuberoinfundibular tract DA receptor sensitivity following neuroleptic treatment (Lal et al., 1977), attempts to demonstrate such enhanced sensitivity in patients with chronic tardive dyskinesia have so far been unsuccessful. Schizophrenic patients with and without tardive dyskinesia do not appear to differ in their serum PRL concentrations during neuroleptic treatment, after abrupt neuroleptic withdrawal, and in the drug-free condition (Asnis et al., 1979). Furthermore, several studies have shown that schizophrenic patients with and without tardive dyskinesia do not differ in their GH and PRL responses to DA agonists and antagonists (Tamminga et al., 1977; Asnis et al., 1979). In these studies, however, the attempt to demonstrate enhanced sensitivity has involved assessing the degree of neuroendocrine response to a given pharmacologic challenge dose. Supersensitivity in fact involves a lowered threshold for response and is demonstrated when a response occurs at a lower than usual challenge dose. The research design appropriate for demonstrating supersensitivity is difficult to apply in clinical studies.

Another methodological problem in the studies to date has been their focus on patients with chronic persistent tardive dyskinesia. There is considerable evidence for heterogeneity in the pathophysiology underlying these tardive dyskinesia syndromes (Tarsy and Baldessarini, 1977; Casey and Denney, 1977). Patients who develop transient tardive dyskinesia in the context of neuroleptic withdrawal may be most likely to have an underlying enhancement of DA receptor sensitivity. A preliminary study suggests, in fact, that these withdrawal dyskinesias are accompanied by elevated serum GH levels and lowered serum PRL levels, endocrine findings consistent with enhanced tuberoinfundibular DA receptor sensitivity (Brown and Laughren, 1980).

Thus, the information that changes in serum PRL can provide about the pathophysiology or clinical course of tardive dyskinesia is not yet entirely clear. It is likely, in view of the known modulating effects of estrogen on neuronal activity, that any relationship between serum PRL and tardive dyskinesia will be complex. A preliminary study, for example, shows that serum PRL concentrations are significantly higher in women with severe tardive dyskinesia than in those with mild tardive dyskinesia; but that men with severe, mild, and no tardive dyskinesia do not differ in serum PRL levels (Glazer et al., 1981).

## 8. PRL AND NEUROLEPTIC BIOAVAILABILITY

The high correlations being reported between neuroleptic plasma levels and serum PRL suggest that PRL may be a measure of neuroleptic bioavailability. Some intriguing observations suggest in fact that serum PRL may be superior to serum neuroleptic levels for this purpose. Methods for measuring serum neuroleptic levels usually assess total serum neuroleptic, both protein-bound and free. The free neuroleptic fraction, less than 10% of the total neuroleptic, is the only biologically active component. Although the correlations between total and free neuroleptic are usually high (Rubin et al., 1980), neuroleptic serum level is nonetheless an indirect measure of the drug's availability for therapeutic purposes.

Several lines of evidence suggest that serum PRL may be a better correlate of free (biologically active) neuroleptic than of total serum neuroleptic. Rubin et al. (1980) report a slightly higher correlation between free haloperidol concentration and serum PRL than between total haloperidol concentration and PRL. We find (Brown et al., 1982) as have others (Cohen et al., 1980; Calil et al., 1979) that total serum neuroleptic activity as measured by radioreceptor assay varies widely among groups of patients on clinically equivalent doses of different neuroleptics. Patients taking thioridazine, for example, have serum neuroleptic levels 10 to 20 times higher than those taking other neuroleptics, and patients taking fluphenazine have relatively low neuroleptic levels. These differences in serum level among clinically equivalent doses of different neuroleptics may be related to differences in their blood brain ratios and affinities for plasma proteins. Yet, despite these marked differences in total serum neuroleptic activity, we find that patients on different neuroleptics do not differ in serum PRL levels (Brown et al., 1982). Serum PRL levels among patients taking different neuroleptics are related not to absolute neuroleptic levels as measured by radioreceptor assay, but rather to the relative neuroleptic level for a given drug. These data suggest that serum PRL may reflect, more so than absolute serum neuroleptic levels, a neuroleptic's biological activity.

Nikitopoulou et al. (1977) studied the physiological and psychological effects of thioridazine in six healthy men. They found that plasma thioridazine and its metabolites were not as strong a correlate as plasma PRL of thioridazine's central effects (drowsiness and EEG changes). They conclude that monitoring of plasma PRL during antipsychotic treatment may be a more useful guide to management than drug plasma levels.

## 9. MONITORING THE PRL RESPONSE

Because of individual differences in serum PRL profiles over the hours following a single neuroleptic dose, PRL must be measured at several points over the first few hours in order to define the extent of the PRL response (Rubin and Forster, 1980; Busch et al., 1979). Yet when PRL is measured under steady-state conditions (12 hr after the last dose of neuroleptic) a single morning sample appears to provide meaningful data.

We have observed considerable individual stability in serum PRL concentration assessed in this manner over weeks and months of long-term neuroleptic treatment and equal stability during the drug-free condition in men (Brown and Laughren, 1981a,b). The average coefficient of variation for serum PRL within subjects ranges from 20 to 25% during constant-dose neuroleptic treatment and correlation coefficients for serum PRL in samples taken at both weekly and monthly intervals range from 0.78 to 0.97 (Brown and Laughren, 1981b). Furthermore, serum PRL concentration in a single morning sample is sensitive to changes in neuroleptic dose and correlates with serum neuroleptic levels (Brown et al., 1982; Rubin et al., 1980). These data in the aggregate suggest that serum PRL concentration in a single morning sample can reflect stable individual differences in PRL secretion. Such a sampling procedure when carried out under conditions suitably controlled with respect to time of day, menstrual cycle phase, and time from drug ingestion may be sufficient for monitoring PRL in a clinical context.

## 10. CONCLUSIONS

Information to date bearing on the relationships among neuroleptic level, serum PRL, and clinical response suggests that serum PRL reflects neuroleptic bioavailability and that the PRL response to neuroleptics parallels and reflects CNS responses. The observed relationships between serum PRL and clinical response warrant continued scrutiny.

## 11. REFERENCES

Asnis, G. M., Sachar, E. J., Langer, G., Halpern, F. S., and Fink, M., 1979, Normal prolactin responses in tardive dyskinesia, *Psychopharmacology* **66**:247.

Beumont, P. J. V., Gelder, M. G., Friesen, H. G., Harris, G. W., MacKinnon, P. C. B., Mandelbrote, B. M., and Wiles, D. H., 1974, The effects of phenothiazines on endocrine function. I. Patients with inappropriate lactation and amenorrhoea, *Br. J. Psychiatry* **124**:413.

Bjorndal, N., Bjerre, M., Gerlach, J., Kristjansen, P., Magelund, G., Oestrich, I. H., and Waehrens, J., 1980, High dosage haloperidol therapy in chronic schizophrenic patients: A double-blind study of clinical response, side effects, serum haloperidol, and serum prolactin, *Psychopharmacology* **67**:17.

Bowers, M. B., Jr., and Heninger, G. R., 1981, Cerebrospinal fluid homovanillic acid patterns during neuroleptic treatment, *Psychiatry Res.* **4**:285.

Brown, W. A., and Laughren, T. P., 1980, Growth hormone release and the tardive dyskinesia of neuroleptic withdrawal, *Lancet* **I**:259.

Brown, W. A., and Laughren, T. P., 1981a, Low serum prolactin and early relapse following neuroleptic withdrawal, *Am. J. Psychiatry* **183**:237.

Brown, W. A., and Laughren, T. P., 1981b, Tolerance to the prolactin elevating effect of neuroleptics, *Psychiatry Res.* **5**:317.

Brown, W. A., Laughren, T. P., Chisholm, E., and Williams, B. W., 1982, Low serum neuroleptic levels predict relapse in schizophrenic patients, *Arch. Gen. Psych.* **39**:998.

Burt, D. R., Creese, I., and Snyder, S. H., 1977, Antischizophrenic drugs: Chronic treatment elevates dopamine receptor binding in brain, *Science* **196**:326.

Busch, D. A., Fang, V. S., and Meltzer, H. Y., 1979, Serum prolactin levels following intramuscular chlorpromazine: Two- and three-hour response as predictors of six-hour response, *Psychiatry Res.* **1**:153.

Calil, H. M., Avery, D. H., Hollister, L. E., Creese, I., and Snyder, S. H., 1979, Serum levels of neuroleptics measured by dopamine radioreceptor assay and some clinical observations, *Psychiatry Res.* **1**:39.

Casey, D. E., and Denney, D., 1977, Pharmacological characterization of tardive dyskinesia, *Psychopharmacology* **54**:1.

Cheung, C. Y., and Weiner, R. I., 1978, In vitro supersensitivity of the anterior pituitary to dopamine inhibition of prolactin secretion, *Endocrinology* **102**:1614.

Cohen, B. M., Lipinski, J. F., Harris, P. Q., Pope, H. G., Jr., and Friedman, M., 1980, Clinical use of the radioreceptor assay for neuroleptics, *Psychiatry Res.* **2**:173.

Creese, I., Burt, D. R., and Snyder, S. H., 1976, Dopamine receptor binding predicts clinical and pharmacological potencies of antischizophrenic drugs, *Science* **192**:481.

Danon, A., Dikstein, S., and Sulman, F. G., 1963, Stimulation of prolactin secretion by perphenazine in pituitary–hypothalamus organ culture, *Proc. Soc. Exp. Biol. Med.* **114**:366.

Davis, J. M., Erickson, S., and Dekirmenjian, H., 1972, Plasma levels of anti-psychotic drugs and clinical response, in: *Psychopharmacology: A Generation of Progress* (M. A. Lipton, A. DiMascio, and K. F. Killman, eds.), pp. 905–915, Raven Press, New York.

DeRivera, J. L., Lal, S., Ettigi, P., Hontela, S., Muller, H. F., and Friesen, H. G., 1976, Effect of acute and chronic neuroleptic therapy on serum prolactin levels in men and women of different age groups, *Clin. Endocrinol. (Oxford)* **5**:273.

Dotti, A., Lostia, O., Rubino, I. A., Bersant, G., Carilli, L., and Zorretta, D., 1981, The prolactin response in patients receiving neuroleptic therapy: The effect of fluphenazine decanoate, *Prog. Neuro-Psychopharmacol.* **5**:69.

Evans, L. E. J., Beumont, P. J. V., Luttrell, B., Henniker, A. J., and Penny, J., 1980, Prolactin response as a guide to neuroleptic treatment of schizophrenia, *Prog. Reprod. Biol.* **6**:260.

Glazer, W. M., Moore, D. C., Bowers, M. B., and Brown, W. A., 1981, Serum prolactin and tradive dyskinesia, *Am. J. Psychiatry* **138**:1493.

Gruen, P. H., Sachar, E. J., Langer, G., Altman, N., Leifer, M., Frantz, A., and Halpern, F. S., 1978a, Prolactin responses to neuroleptics in normal and schizophrenic subjects, *Arch. Gen. Psychiatry* **35**:108.

Gruen, P. H., Sachar, E. J., Altman, N., Langer, G., Tabrizi, M. A., and Halpern, F. S., 1978b, Relation of plasma prolactin to clinical response in schizophrenic patients, *Arch. Gen. Psychiatry* **35**:1222.

Hogarty, G. E., Schooler, N. R., Ulrich, R., Mussare, F., Ferro, P., and Herron, E., 1979, Fluphenazine and social therapy in the aftercare of schizophrenic patients: Relapse analysis of a two-year controlled study of fluphenazine decanoate and fluphenazine hydrochloride, *Arch. Gen. Psychiatry* **36**:1283.

Hollister, L. E., 1977, Antipsychotic medications and the treatment of schizophrenia, in: *Psychopharmacology: From Theory to Practice* (J. Barchas, P. Berger, R. Ciaranello, and G. Elliott, eds.), pp. 121–150, Oxford University Press, London.

Hooper, J. H., Jr., Welch, V. C., Point, P., and Schackelford, R. T., 1961, Abnormal lactation associated with tranquilizing drug therapy. *J. Am. Med. Assoc.* **178**:506.

Kolakowska, T., Wiles, D. H., Gelder, M. G., and McNeilly, A. S., 1976, Clinical significance of plasma chlorpromazine levels. II. Plasma levels of the drug, some of its metabolites and prolactin in patients receiving long-term phenothiazine treatment, *Psychopharmacology* **49**:101.

Lal, H., Brown, W. A., Drawbaugh, R., Hynes, M., and Brown, G., 1977, Enhanced prolactin inhibition following chronic treatment with haloperidol and morphine, *Life Sci.* **20**:101.

Laughren, T. P., Brown, W. A., and Williams, B. W., 1979, Serum prolactin and clinical state during neuroleptic treatment and withdrawal, *Am. J. Psychiatry* **136**:108.

Linnolia, M., Viukari, M., Vaisanen, K., and Auvinen, J., 1980, Plasma neuroleptic and prolactin levels in mentally retarded patients, *Acta Pharmacol. Toxicol.* **46**:159.

Meltzer, H. Y., and Fang, V. S., 1976, The effect of neuroleptics on serum prolactin in schizophrenic patients, *Arch. Gen. Psychiatry* **33**:279.

Meltzer, H. Y., Busch, D. A., Creese, I. R., Snyder, S. H., and Fang, V. S., 1981, Effect of intramuscular chlorpromazine on serum prolactin levels in schizophrenic patients and normal controls, *Psychiatry Res.* **5**:95.

Naber, D., Fischer, H., and Ackenheil, M., 1979, Effect of long-term neuroleptic treatment on dopamine tuberoinfundibular system: Development of tolerance?, *Commun. Psychopharmacol.* **3**:59.

Nikitopoulou, G., Thorner, M., Crammer, J., and Lader, M., 1977, Prolactin and psychophysiologic measures after single doses of thioridazine, *Clin. Pharmacol. Ther.* **21**:422.

Ohman, R., and Axelsson, R., 1978, Relationship between prolactin response and antipsychotic effect of thioridazine in psychiatric patients, *Eur. J. Clin. Pharmacol.* **14**:111.

Pasteels, J. L., 1961, Secretion of prolactin by the pituitary in tissue culture, *C. R. Acad. Sci.* **253**:2140.

Post, R. M., and Goodwin, F. K., 1975, Time dependent effects of phenothiazines on dopamine turnover in psychiatric patients, *Science* **190**:488.

Rao, V. A. R., Bishop, M., and Coppen, A., 1980, Clinical state, plasma levels of haloperidol and prolactin: A correlation study in chronic schizophrenia, *Br. J. Psychiatry* **137**:518.

Rubin, R. T., and Forster, B., 1980, Haloperidol stimulation of prolactin secretion: How many blood samples are needed to define the hormone response?, *Commun. Psychopharmacol.* **4**:41.

Rubin, R. T., Poland, R. E., O'Conner, D., Gouin, P. R., and Tower, B. B., 1976, Selective neuroendocrine effects of low-dose haloperidol in normal adult men, *Psychopharmacology* **47**:135.

Rubin, R. T., Forsman, A., Heykants, J., Ohman, R., Tower, B., and Michiels, M., 1980, Serum haloperidol determinations in psychiatric patients: Comparison of methods and correlation with serum prolactin level, *Arch. Gen. Psychiatry* **37**:1069.

Seeman, P., 1980, Brain dopamine receptors, *Pharmacol. Rev.* **32**:229.

Smith, R. C., Tamminga, C. A., Crayton, J. W., Dekirmenjian, H., and Davis, J. M., 1979, Relationship of butaperazine blood levels to plasma prolactin in chronic schizophrenic patients, *Psychopharmacology* **66**:29.

Tamminga, C. A., Smith, R. C., Pandey, G., Frohman, L. A., and Davis, J. M., 1977, A neuroendocrine study of supersensitivity in tardive dyskinesia, *Arch. Gen. Psychiatry* **34**:1199.

Tarsy, D., and Baldessarini, R. J., 1977, The pathophysiologic basis of tardive dyskinesia, *Biol. Psychiatry* **12**:431.

Wiles, D. H., Kolakowska, T., McNeilly, A. S., Mandelbrote, B. M., and Gelder, M. G., 1976, Clinical significance of plasma chlorpromazine levels. I. Plasma levels of the drug, some of its metabolites and prolactin during acute treatment, *Psychol. Med.* **6**:407.

Wistedt, B., 1981, Withdrawal of long-acting neuroleptics in schizophrenic outpatients: Clinical and biological findings, *Acta Univ. Ups.* **397**:1.

Wistedt, B., Wiles, D., and Kolakowska, T., 1981, Slow decline of plasma drug and prolactin levels after discontinuation of chronic treatment with depot neuroleptics, *Lancet* **I**:1163.

CHAPTER 32

# Neuroendocrine Changes during the Course of Neuroleptic Treatment of Schizophrenic Patients

NORBERT NEDOPIL,
JOHANNES WEISS-BRUMMER, and
ECKART RÜTHER

## 1. INTRODUCTION

The blockade of postsynaptic dopaminergic receptors is considered to be the most important biological factor in the action of neuroleptic drugs for the treatment of schizophrenia (Carlsson, 1978; Johnstone et al., 1978). A wide variety of indirect evidence (Crow et al., 1976; Van Kammen, 1979; Snyder, 1976; Meltzer and Stahl, 1976) and some direct evidence in postmortem brains of schizophrenic patients (Lee and Seeman, 1977) supported the dopamine hypothesis of schizophrenia, i.e., that the productive symptoms of schizophrenia are associated with a relative excess of dopaminergic activity in certain subcortical brain structures.

Meltzer (1979) lists six major dopaminergic pathways: (1) retinal, (2) incertohypothalamic, (3) nigrostriatal, (4) mesocortical, (5) mesolimbic, and (6) tuberoinfundibular. Only two of these are believed to have relevance for the pathogenesis of schizophrenia: the mesolimbic tract (Snyder, 1972) and the mesocortical fibers (Meltzer, 1979). The antipsychotic activity of neuroleptic drugs is attributed to the inhibition of dopaminergic transmission in these neurons (Snyder, 1972).

The tuberoinfundibular tract plays an important role in the control of the release of hormones from the anterior lobe of the pituitary. It exerts a tonic inhibitory effect on the secretion of prolactin (PRL) (MacLeod, 1976) and a

---

NORBERT NEDOPIL, JOHANNES WEISS-BRUMMER, and ECKART RÜTHER • Psychiatric Hospital of the University of Munich, 8000 Munich 2, FRG.

stimulating effect on the secretion of human growth hormone (HGH) (Martin, 1976). Blockade of dopaminergic receptors in these neurons by neuroleptics leads to an increase of PRL secretion into blood. Langer et al. (1977) found that the PRL response after single i.m. doses of neuroleptics corresponded to their antipsychotic potency. Stimulation of dopaminergic receptors in this area leads to an increase of HGH serum levels. The injection of apomorphine, a directly stimulating agent of dopaminergic receptors, leads to an increase of HGH in the serum of healthy controls (Brown et al., 1973; Lal et al., 1973), an effect which could be reversed by a treatment with neuroleptics like pimozide (Lal et al., 1977); clozapine also reversed the HGH-stimulating effect of apomorphine in schizophrenic patients (Nair et al., 1979).

The relevance of these findings for clinical practice depends on the answers to the following questions:

1. Does the action of neuroleptics on one dopaminergic system (the tuberoinfundibular tract, measured by the increase of PRL and the suppression of the apomorphine-induced HGH stimulation) reflect their action on other dopaminergic systems, which are believed to play a role in the psychotic process?
2. Does the measurement of the neuroendocrinological parameters (PRL, HGH) indicate the amount of neuroleptics needed to block central dopaminergic receptors, thus helping to find the optimal doses for the antipsychotic efficacy in an individual patient?
3. Are the time courses of the effects of neuroleptics similar in the different dopaminergic systems, or can their effect on one system predict their effect on another, e.g., does the amount of the PRL stimulation after the onset of neuroleptic treatment predict the later antipsychotic effect of the treatment?

In order to answer these questions the neuroendocrine parameters, PRL and HGH, were monitored during the course of both a constant-dose and an adapted-dose treatment, and before and after apomorphine challenge. Data were also obtained before the onset of treatment and after withdrawal from neuroleptic drugs.

## 2. METHODS

For the apomorphine challenge, blood was taken from an indwelling venous catheter 60, 30, and 0 min before and for 2 hr every 15 min after s.c. injection of 0.5 mg of apomorphine. Blood was centrifuged instantly and serum samples were analyzed for PRL and HGH. In the other patients, PRL was determined from serum samples taken in the morning.

Psychopathological assessment was documented for all patients on the AMDP system (Scharfetter, 1972), for most patients on the BPRS (Overall and Gorham, 1962) and on the GAS (Endicott et al., 1976). Extrapyramidal side effects were documented according to the scale developed by Simpson and Angus (1970).

Patients selected for these studies met the ICD-8 criteria for schizophrenia (ICD Nos. 295.1 to 295.3) (WHO, 1974).

## 3. RESULTS

### 3.1. PRL

#### 3.1.1. Neuroleptic Treatment Adjusted to Clinical Needs

In a preliminary study 37 patients (14 males and 23 females, mean age 33 years) were treated with different neuroleptics in adjusted daily doses. Before treatment and after a treatment period of at least 4 weeks (range 4 to 10 weeks), psychopathological ratings, examinations for extrapyramidal symptoms, and blood sampling for PRL determination were made on the same day. Daily doses of neuroleptics were converted to chlorpromazine equivalents (Davis, 1974; Haase, 1978). The average daily dose was equivalent to $1.48 \pm 2.1$ g ($\pm$ S.D.) of chlorpromazine. PRL levels of women were about twice as high ($39.2 \pm 31.2$ ng/ml) as of men ($21.0 \pm 21.1$ ng/ml). No difference in PRL levels was found between patients, who received additional biperiden medication for extrapyramidal symptoms ($30.4 \pm 21.0$ ng/ml) and those who did not ($31.1 \pm 20.8$ ng/ml). Neither the daily dose nor the total amount of neuroleptics applied influenced the level of PRL as calculated by correlation statistics. The clinical parameters, psychopathological state, and improvement of psychosis, as well as extrapyramidal symptoms did not correlate with the PRL serum levels either. The data were calculated for the whole group and for male and female patients separately without yielding new results.

#### 3.1.2. Time Course of PRL Levels during Neuroleptic Treatment

In a second group of 16 patients (13 men, 3 women) the time course of PRL serum levels was monitored during a 30-day treatment with haloperidol in a free dosage. Before the beginning of the treatment and at intervals of 4 to 6 days, blood was taken for PRL analysis. The haloperidol medication was averaged over these 4 to 6 days for comparison with PRL levels. Average haloperidol doses remained almost constant at about 20 mg/day during the first 15 days of treatment and then fell gradually to 12 mg/day (day 30). Average PRL levels rose from $15.0 \pm 14.8$ ng/ml before treatment to a maximal average value of $42.0 \pm 27.8$ ng/ml ($p < 0.001$) on day 10. It then decreased slowly, but insignificantly. In seven patients remaining in the study until day 30, the mean value of PRL was $24.0 \pm 16.6$ ng/ml.

During the first 5 days of treatment, patients with a higher daily dose of haloperidol had higher PRL levels; however, this difference was not significant. During the following treatment periods, no difference either in haloperidol dos-

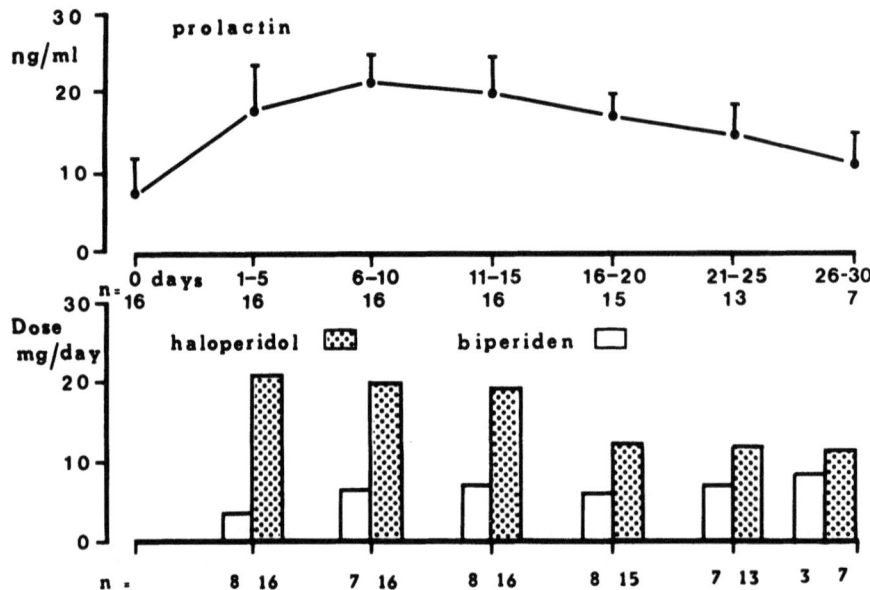

FIGURE 1. Time course of PRL levels during haloperidol treatment in free dosage. Average haloperidol and biperiden doses are given at the bottom.

age or in PRL levels was observed. Apparently additional antiparkinson medication did not influence PRL levels significantly (Fig. 1). No significant correlation between the individual haloperidol doses over the 4- to 6-day period and the subsequent PRL level could be calculated. Again no relationship between hormonal and clinical changes could be established.

### 3.1.3. Fixed-Dose Treatment

In a third study, two groups of patients were treated double-blindly with different, but fixed doses of benperidol. Six men and seven women received 3 mg of benperidol, 5 men and 8 women 12 mg of benperidol each day for 20 days of treatment. In the 3-mg group, PRL levels rose from a value of 10.0 ± 5.0 ng/ml before treatment to 50.0 ± 27.0 ng/ml on day 10 and then fell again to 40.0 ± 19.0 ng/ml on day 20. In the 12-mg group, the respective values were 6.4 ± 3.5 ng/ml before treatment, 90.0 ± 7.4 ng/ml on day 10, and 100.0 ± 34.5 ng/ml on day 20. The average height of the PRL levels of the 12-mg group was found to be approximately double the PRL levels of the 3-mg population. However, interindividual variability was great and the differences became significant only on day 20 ($p < 0.05$).

In this study, too, there was a significant difference in PRL levels between men and women, the latter again doubling the former.

|  | Men | Women |  |
| --- | --- | --- | --- |
| 3-mg group |  |  |  |
| Day 10 | 35 ± 12 ng/ml | 27 ± 26 ng/ml | $p < 0.05$ |
| Day 20 | 32 ± 4.5 ng/ml | 66 ± 40 ng/ml | ns |
| 12-mg group |  |  |  |
| Day 10 | 40 ± 14.5 ng/ml | 104 ± 34 ng/ml | $p < 0.01$ |
| Day 20 | 42 ± 24 ng/ml | 126 ± 51 ng/ml | $p < 0.05$ |

From these data it can be concluded that the difference in PRL levels between the high- and low-dose groups could only be seen in the female patients. The same observation could be made regarding the increase of PRL levels from the pretreatment value to the value on day 10:

|  | Men | Women |  |
| --- | --- | --- | --- |
| 3 mg | 28 ± 12 ng/ml | 60 ± 31 ng/ml | ns |
| 12 mg | 36 ± 8 ng/ml | 90 ± 36 ng/ml | $p < 0.01$ |

Again women had a larger increase than men and this increase was more pronounced in the 12-mg group. However, because of the large interindividual variance, no significant difference could be calculated between the two dosage groups.

### 3.1.4. Discontinuation of Medication

Five days after withdrawal of benperidol, the PRL levels of four patients were determined. It had considerably decreased by then, but not reached the respective values obtained before the beginning of treatment (Fig. 2).

### 3.1.5. Clinical Changes and Prolactin

In 13 patients benperidol plasma levels were determined. Correlations between PRL levels and benperidol plasma levels were found not to be significant on day 10 or day 20. This holds true for the group as a whole as well as for men and women calculated separately. Correlations were also calculated to find relations between extrapyramidal side effects and PRL plasma levels, between psychopathological status or psychopathological improvement and PRL serum levels or increase in PRL. However, no conclusive results could be obtained, although there was a significantly negative correlation between the presence of hallucinations and delusions and PRL levels in the high-dosage group, but not in the low-dosage group. In the low-dosage group extrapyramidal side effects were significantly correlated with PRL plasma levels on day 10 for women only. However, these correlations could be merely chance products, since there was no difference between the dosage groups as far as psycho-

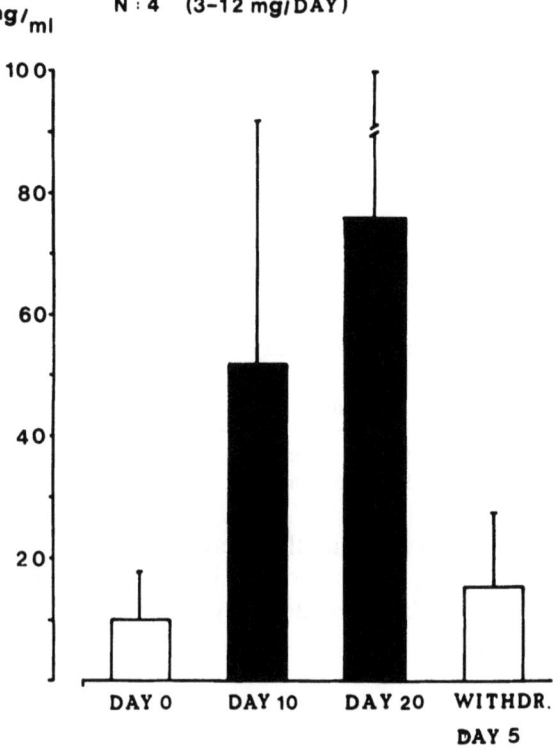

FIGURE 2. PRL levels before, during, and 5 days after discontinuation of benperidol treatment.

TABLE 1
Correlations between PRL Levels and Extrapyramidal Symptoms (Simpson), and PRL Levels and Psychopathology (Hallucinations and Delusions) during Neuroleptic Treatment with Benperidol

| Correlations | Day 10 | | Day 20 | |
|---|---|---|---|---|
| | r | p | r | p |
| PRL: Simpson | 0.20 | ns | 0.30 | ns |
| Women,  3 mg | 0.72 | $p < 0.05$ | | ns |
|    12 mg | 0.73 | $p < 0.05$ | | ns |
| PRL: Psychopathology | | | | |
|   PRL-AMDP (hallucin) | −0.34 | ns | −0.50 | $p < 0.05$ |
|   PRL-AMDP (paranoid) | −0.25 | ns | −0.54 | $p < 0.05$ |
|   PRL-$\Delta$ AMDP (hallucin) | −0.02 | ns | 0.03 | ns |
|   PRL-$\Delta$ AMDP (paranoid) | −0.27 | ns | 0.32 | ns |

pathological status or improvement were concerned, nor was there a significant difference in Simpson scores on day 10.

### 3.2. Apomorphine-Induced HGH Response

Apomorphine challenge were performed on 22 patients: 12 of them received haloperidol in a free dosage, and 10 received penfluridol in a fixed-dosage schedule.

#### 3.2.1. Haloperidol in an Adjusted Dosage

Patients receiving haloperidol were tested before the beginning of treatment and after 3 weeks of treatment; seven of them were retested 5 days after withdrawal of all medication. Unfortunately, not all patients were tested at all times. The average daily dose of haloperidol was $31.9 \pm 15.4$ mg.

The data of the HGH response at the different points of time are shown in Fig. 3. Before treatment there is a significant increase of growth hormone 60 min after s.c. injection of 0.5 mg of apomorphine. This increase is almost completely suppressed after 3 weeks of haloperidol treatment. Although there is again a small increase 5 days after withdrawal, this increase is still not significant.

In seven patients, who were tested after 3 weeks of treatment and 5 days after withdrawal of haloperidol, the increase of the apomorphine-induced HGH response remained insignificant. The PRL levels, however, which were determined in all nine of the withdrawn patients and could be compared intraindividually to their respective PRL levels after 3 weeks of treatment, were significantly reduced 5 days after discontinuation of treatment (Fig. 4). This result confirms the data obtained from the patients withdrawn from benperidol.

Apomorphine had little influence on the PRL levels elevated by neuroleptic treatment. Sixty minutes after the application of apomorphine, the mean values decreased from $47.2 \pm 38.3$ ng/ml to $43.1 \pm 34.3$ ng/ml. Five days after withdrawal, however, the slightly elevated PRL levels fell significantly from a mean of $12.6 \pm 16.3$ ng/ml to $9.7 \pm 16.2$ ng/ml ($p < 0.05$).

Again it was calculated whether the hormonal changes correlated to the psychopathological changes. No significant correlation could be obtained between any of these parameters. In eight patients the HGH response was completely blocked by neuroleptic treatment; two of these patients, however, had an increase of psychotic symptom scores. In two other patients the HGH response was not completely suppressed after 3 weeks of haloperidol treatment; their psychopathology scores, however, had improved, and one of them was almost free of psychotic symptoms.

A relationship between HGH responses and PRL serum levels could not be established. Correlations were insignificant before and during neuroleptic therapy and after its discontinuation.

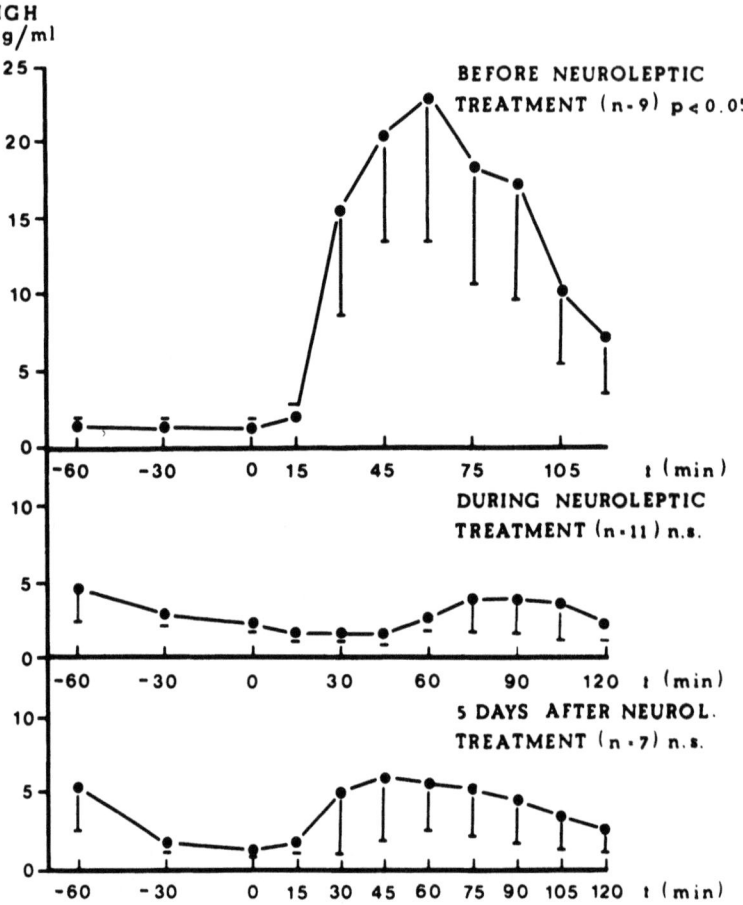

FIGURE 3. Apomorphine-induced HGH response before, during, and 5 days after neuroleptic treatment.

TABLE 2
Penfluridol Treatment: Correlations between Apomorphine-Induced HGH Response and Psychopathology and between HGH Response and PRL Levels

| Correlation | | N | r | p |
|---|---|---|---|---|
| Psychopathology | | | | |
| ΔHGH 3 weeks Th. | AMPD tot. | 7 | −0.54 | ns |
| PRL | | | | |
| ΔHGH 1 week Th. | PRL | 6 | 0.04 | ns |
| 3 weeks Th. | PRL | 6 | 0.33 | ns |

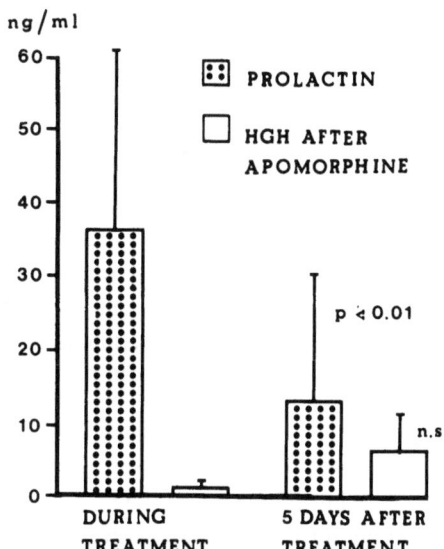

FIGURE 4. PRL levels and apomorphine-induced HGH response during and 5 days after discontinuation of neuroleptic treatment.

### 3.2.2. Penfluridol in a Fixed Dose

Ten patients received 40 mg of penfluridol the first 2 days of treatment; the dose was then reduced to 20 mg/day for the remaining week and to 15 mg/day for the next 2 weeks. In seven patients an apomorphine challenge was performed after 1 and 3 weeks. Three patients had only one apomorphine test after 1 week of treatment.

The HGH response after apomorphine was completely suppressed in all patients at both times. After 1 week the maximum HGH level, 75 min after the injection of apomorphine, was $2.0 \pm 3.2$ ng/ml, and after 3 weeks it was $1.0 \pm 1.4$ ng/ml. PRL levels rose to $53.4 \pm 29.9$ ng/ml in the first and to $35.9 \pm 19.8$ ng/ml in the third week of treatment. An influence of apomorphine on the PRL serum levels could not be observed at any of the two test times.

In this study, also, no significant correlation could be calculated between HGH response and PRL serum levels (Table 2). Whereas the HGH response was completely blocked in all patients, there was a considerable variation of PRL plasma levels. Again no correlation could be calculated between the hormonal parameters and the psychopathological parameters.

## 4. DISCUSSION

In accordance with the findings of most authors (Gruen *et al.*, 1978; Meltzer and Fang, 1976; Wode-Helgodt *et al.*, 1978) these studies revealed an increase of serum PRL during neuroleptic treatment. This increase is attributed to the blockade of dopaminergic receptors in the pituitary and on the tuberoinfundibular tract. From *in vitro* experiments, animal studies and from human pharmacological experiments, one can deduce a similar rank order of neuroleptics in their potency to either displace each other from dopaminergic receptors (Snyder and Bennett, 1976), to produce extrapyramidal side effects (Haase, 1978), to increase serum PRL levels (Langer *et al.*, 1977), or to work clinically as antipsychotic drugs (Davis, 1974). Thus, it could be concluded that dopaminergic receptors should work quite similarly throughout the brain and even from species to species. The response of the dopaminergic receptor in one system should reflect the response of the receptor of another system.

In fact, Meltzer and Fang (1976) found a weak correlation between the increase of serum PRL and clinical improvement in male patients and in another study both in male and female patients (Meltzer *et al.*, 1978). Siris *et al.* (1978) reported similar correlations, and Sedvall *et al.* (1975) found a correlation between PRL levels in cerebrospinal fluid and clinical improvement in male patients treated with chlorpromazine.

Similar to the findings of Gruen *et al.* (1978), we did not find any correlation between serum PRL and clinical improvement and only in a few subgroups significant correlations between clinical outcome and PRL serum levels. We could, however, reproduce a few other results: In our studies female patients had higher PRL levels than men under neuroleptic treatment. The same finding has been reported by many authors (DeRivera *et al.*, 1976; Meltzer and Fang, 1976; Meltzer *et al.*, 1978; Wode-Helgodt *et al.*, 1978; Gruen *et al.*, 1978).

In our studies patients with higher doses of neuroleptics had higher levels of PRL than those with lower doses. In the study comparing two different doses of benperidol, this difference could be attributed almost exclusively to the female patients. This might be explained by the greater capacity of women to secrete PRL, and a greater variability of female patients in their PRL response to neuroleptics. Gruen *et al.* (1978) had similar results using 60 mg butaperazine or 70 to 100 mg of loxapine. However, neither in their nor in our studies was there any direct correlation between doses of neuroleptics and PRL serum levels. Such correlations have been reported by Kolakowska *et al.* (1975) and Wode-Helgodt *et al.* (1978), using lower doses of chlorpromazine. Therefore, direct correlations between dose of neuroleptic applied and PRL plasma level seem to exist only up to certain relatively low doses. Further evidence for such a supposition is, that maximal stimulation of PRL is achieved with much lower doses of neuroleptics than clinically administered in single-dose experiments (Hays and Rubin, 1979; Langer *et al.*, 1977).

Thus, the monitoring of PRL plasma levels in clinical situations is only of very limited use for the study of the pharmacokinetics of neuroleptic drugs.

The effect of neuroleptics on PRL plasma levels is very rapid, peak levels being reached 30 min after i.m. injection of 50 mg of chlorpromazine (Meltzer, 1979). In chronic treatment the maximal effect seems to occur 6 to 10 days after the onset of neuroleptics (Fig. 1). From the literature it has not yet been decided how long the elevated PRL levels can be observed in chronic neuroleptic treatment. Meltzer and Fang (1976) claim that tolerance to neuroleptic-induced PRL secretion did not develop in a longitudinal study. Naber et al. (1980) reported, on the other hand, that PRL levels returned to normal after 5 to 15 years of neuroleptic treatment. Similar findings were reported by Beumont et al. (1974) and Kolakowska et al. (1975).

Another open question is, which mechanisms lead to tolerance in these patients, since even after very prolonged treatment, patients exhibiting low PRL serum levels respond to a challenge of sulpiride with an increase of PRL (Müller et al., 1981). This would indicate that PRL is still adequately secreted and that dopaminergic receptors do not change significantly in their sensitivity.

In our studies, PRL levels returned to normal 5 days after the discontinuation of neuroleptic therapy. Beumont et al. (1974) reported that it lasted 7 days, Wode-Helgodt et al. (1978) 30 hr, and Meltzer and Fang (1976) 22 hr until normalization of PRL levels was observed.

The rapid response of PRL levels after the beginning of neuroleptic treatment and the rapid normalization after its discontinuation would allow to, at least, monitor the compliance of patients under acute treatment, since this appears to be the parameter most rapidly influenced by neuroleptics.

The effect of apomorphine on PRL and HGH levels is the result of the stimulation of postsynaptic dopaminergic receptors. Although some authors claim that low doses of apomorphine stimulate preferentially inhibitory, presynaptic, dopaminergic receptors (Corsini et al., 1978; Feinberg and Carroll, 1979), this is apparently not the case with dopaminergic neurons ending in the pituitary. The stimulating effect of apomorphine on growth hormone secretion is antagonized by neuroleptics like chlorpromazine (Lal et al., 1973), pimozide (Lal et al., 1977), haloperidol (Rotrosen et al., 1979), and clozapin (Nair et al., 1979). In our study it was also completely blocked by relatively high doses of penfluridol and in 8 of 10 patients by average doses of haloperidol. With the applied dose of 0.5 mg apomorphine no changes in the HGH levels or in the PRL levels elevated by neuroleptic drugs could be observed during medical treatment. Neither the growth hormone responses of the unmedicated patients, nor the differences of growth hormone response before and during neuroleptic treatment was an indicator of the outcome of neuroleptic treatment. Rotrosen et al. (1976) reported that patients with high growth hormone responses had a poorer outcome to neuroleptic therapy than patients with comparatively low HGH responses to apomorphine. Pandey et al. (1977), however, reported that acutely ill schizophrenic patients had higher HGH responses, whereas chronic schizophrenic patients were reported to have lower HGH responses according to Tamminga et al. (1977) and Ettigi et al. (1976). Chronic patients do not respond as well clinically to neuroleptic therapy as acute patients.

In the acute schizophrenic patients of our study, suppression of apomorphine-induced HGH secretion was complete 7 days after the beginning of treatment, indicating that the blockade of dopaminergic receptors stimulating the secretion of HGH and of those responsible for the suppression of PRL secretion run similar time courses. Five days after withdrawal of neuroleptic drugs the two parameters behaved quite differently. Whereas PRL had almost returned to normal, the HGH response remained suppressed, indicating that the influence of neuroleptic drugs is different on the two dopaminergic receptor populations.

After a chronic treatment with neuroleptics for more than 12 years the HGH response was still blunted with a mean increase of 5 mg (Müller et al., 1981), while PRL serum levels had returned to normal. The authors reported that 12 days after withdrawal from neuroleptic drugs the HGH response had increased to a mean of 12 ng/ml. This value did not change significantly until 30 days after withdrawal, when neuroleptic therapy was begun again. This was attributed to the fact that chronic schizophrenic patients have similar HGH responses as acute patients and healthy controls, and that these patients do not develop supersensitivity in the dopaminergic neurons concerned.

The results from chronic schizophrenic patients, on the other hand, strengthen the hypothesis that the two receptor populations, one responsible for the suppression of PRL secretion and the other one for the stimulation of growth hormone secretion, behave differently. The different reactions of two receptor populations, situated anatomically so closely together, make it doubtful whether the parameters obtained from the testing of one dopaminergic system can reflect the activity of another dopaminergic system.

The clinical relevance of these parameters as tests for antipsychotic activity of drugs in individual patients appears to be very limited. On the other hand, there are several questions that still deserve attention: It is still unclear as to how many years of treatment elapse before the PRL levels normalize and what consequences this may have for further treatment. Brown and Laughren (1981) concluded that low serum PRL levels in chronically treated, schizophrenic patients may augure early relapse after withdrawal from neuroleptics. A second question would be whether there are different dopaminergic receptors with different time courses of reaction in the other dopaminergic tracts. The sequence of onset of extrapyramidal side effects, early dyskinesia, parkinsonism, akathisia might be a hint toward such a possibility.

## 5. SUMMARY

The blockade of postsynaptic dopaminergic receptors is considered to be the most important biological factor of the action of neuroleptics. Neuroendocrine studies in the psychopharmacology of neuroleptics have two major advantages:

1. They allow to link preclinical results—even from animal studies—to clinical practice.
2. They are a useful tool to examine quantitatively receptor functions altered by psychopharmacological agents in a clinical situation.

These receptor functions are either studied by the direct effect of the drugs on the levels of hormones that are under the control of dopaminergic input, or by the stimulation of a hormone by a dopaminergic agonist.

PRL is under tonic inhibitory control of dopamine. There is no controversy about the increase of PRL plasma levels during neuroleptic therapy. Controversial results have, however, been obtained concerning the relevance of these results for the clinical improvement of individual patients.

Data from own studies, obtained from 79 patients under different treatment schedules (adjusted dose, fixed dose) with different neuroleptics (haloperidol, benperidol, penfluridol, phenothiazines), demonstrated that all neuroleptics led to a significant increase of PRL plasma levels. This increase was twice as high in women than in men. Correlations between PRL plasma levels and clinical outcome could not be calculated in any of the studied subgroups during the treatment, which lasted between 20 and 40 days, respectively.

Apomorphine, a postsynaptic dopaminergic agonist, stimulates growth hormone secretion, an effect also blocked by neuroleptic drugs. In a study on 22 schizophrenic patients treated with either haloperidol or penfluridol, the growth hormone (HGH) response after apomorphine was suppressed in all but 3 subjects. Suppression was complete in the remaining 19 patients. However, no significant correlation between HGH response and the psychopathological parameters, nor between HGH response and PRL increase could be established. Five days after discontinuation of neuroleptic therapy HGH stimulation to apomorphine remained still suppressed in 7 patients, whereas PRL plasma levels were already decreased, suggesting a different time course of adaptation in the two receptor populations.

These data suggest that the two receptor populations react differently to neuroleptic influence, although they are anatomically in close proximity. In the light of these results the clinical relevance of peripherally obtained parameters as tests for the antipsychotic action of neuroleptic drugs in individual patients appears to be doubtful.

Both tests may, however, have relevance in examining the changes of sensitivity of dopaminergic receptors during long-term treatment with, and after withdrawal from neuroleptic drugs. Further studies should aim in that direction.

## 6. REFERENCES

Beumont, P. J. V., Gelder, M. G., Friesen, H. G., Harris, G. W., MacKinnon, P. C. B., Mandelbrote, B. M., and Wiles, D. H., 1974, The effects of phenothiazines on endocrine function. I. Patients with inappropriate lactation and amenorrhea, *Br. J. Psychiatry* **124**:413.

Brown, W. A., and Laughren, T., 1981, Low serum prolactin and early relapse following neuroleptic withdrawal, *Am. J. Psychiatry* **138**:237.

Brown, W. A., van Woert, M. H., and Ambani, L. M., 1973, Effect of apomorphine on growth hormone release in humans, *J. Clin. Endocrinol. Metab.* **37**:463.

Carlsson, A., 1978, Antipsychotic drugs, neurotransmitters and schizophrenia, *Am. J. Psychiatry* **135**:164.

Corsini, G. U., Onali, P., Masala, C., Cianchetti, C., Mangoni, A., and Gessa, G., 1978, Apomorphine hydrochloride induced improvement in Huntington's chorea: Stimulation of dopamine receptor, *Arch. Neurol.* **35**:27.

Crow, T. J., Deakin, J. F. W., Johnstone, E. C., and Longden, A., 1976, Dopamine and schizophrenia, *Lancet* **II**:563.

Davis, J. M., 1974, Dose equivalence of the antipsychotic drugs, *J. Psychiatr. Res.* **11**:65.
De Rivera, J. L., Lal, S., Ettigi, P., Hontela, S., Muller, H. F., and Friesen, H. G., 1976, Effects of acute and chronic neuroleptic therapy on serum prolactin levels in men and women of different age groups, *Clin. Endocrinol.* **5**:273.
Endicott, J., Spitzer, R. L., Fleiss, J. V., and Cohen, J., 1976, The global assessment scale: A procedure for measuring overall severity of psychiatric disturbances, *Arch. Gen Psychiatry* **33**:766.
Ettigi, P., Nair, N. P. V., Lal, S., Cervantes, P., and Guyda, H., 1976, Effect of apomorphine on growth hormone and prolactin secretion in schizophrenic patients, with or without oral dyskinesia, withdrawn from chronic neuroleptic therapy, *J. Neurol. Neurosurg. Psychiatry* **39**:870.
Feinberg, M., and Carroll, B. J., 1979, Effects of dopamine agonists and antagonists in Tourette's disease, *Arch. Gen. Psychiatry* **36**:979.
Gruen, P. H., Sachar, E. J., Altman, N., Langer, G., Tabrizi, M. A., and Halpern, F. S., 1978, Relation of plasma prolactin to clinical response in schizophrenic patients, *Arch. Gen. Psychiatry* **35**:1222.
Haase, H. J., 1978, *Therapie mit Psychopharmaka und anderen seelisches Befinden beeinflussenden Medikamenten*, pp. 121–136, Schattauer, Munich.
Hays, S. E., and Rubin, R. T., 1979, Variability of prolactin response to intravenous and intramuscular haloperidol in normal adult men, *Psychopharmacology* **61**:17.
Johnstone, E. C., Crow, T. J., Frith, C. D., Carney, M. W. P., and Price, J. S., 1978, Mechanism of the antipsychotic effect in the treatment of acute schizophrenia, *Lancet* **I**:848.
Kolakowska, T., Wiles, D. H., McNeilly, A. S., and Gelder, M. G., 1975, Correlation between plasma levels of prolactin and chlorpromazine in psychiatric patients, *Psychol. Med.* **5**:214.
Lal, S., De la Vega, C. E., Sourkes, T. L., and Friesen, H. G., 1973, Effect of apomorphine on growth hormone, prolactin, luteinizing hormone and follicle-stimulating hormone levels in human serum, *J. Clin. Endocrinol. Metab.* **37**:719.
Lal, S., Guyda, H., and Bikadoroff, S., 1977, Effect of methysergide and pimozide on apomorphine-induced growth hormone secretion in men, *J. Clin. Endocrinol. Metab.* **44**:766.
Langer, G., Sachar, E. J., Halpern, F. S., Gruen, P. H., and Solomon, M., 1977, The prolactin response to neuroleptic drugs. A test of dopaminergic blockade: Neuroendocrine studies in normal men, *J. Clin. Endocrinol. Metab.* **45**:996.
Lee, T., and Seeman, P., 1977, Dopamine receptors in normal and schizophrenic brains, *Neurosci. Abstr.* **3**:443.
MacLeod, R. M., 1976, Regulation of prolactin secretion, in: *Frontiers in Neuroendocrinology* (L. Martini and W. F. Ganong, eds.), Vol. 4, pp. 189–194, Raven Press, New York.
Martin, J. B., 1976, Brain regulation of growth hormone secretion, in: *Frontiers in Neuroendocrinology* (L. Martini and W. F. Ganong, eds.), Vol. 4, pp. 129–168, Raven Press, New York.
Meltzer, H. Y., 1979, Biochemical studies in schizophrenia, in: *Disorders of the Schizophrenic Syndrome* (L. Bellak, ed.), pp. 45–135, Basic Books, New York.
Meltzer, H. Y., and Fang, V. S., 1976, The effect of neuroleptics on serum prolactin in schizophrenic patients, *Arch. Gen. Psychiatry* **33**:279.
Meltzer, H. Y., and Stahl, S. M., 1976, The dopamine hypothesis of schizophrenia: A review, *Schizophr. Bull.* **2**:19.
Meltzer, H. Y., Fang, V. S., and Goode, D. J., 1978, The effect of neuroleptics and alpha-methylparatyrosine and serum prolactin levels in laboratory animals and man, *Psychopharmacol. Bull.* **14**:5.
Müller, F., Ackenheil, M., Albus, M., May, G., and Zander, K., 1981, Changes in dopamine receptor sensitivity of chronic schizophrenics after long-term neuroleptic treatment and after drug withdrawal. Third World Congress of Psychiatry (Abstract).
Naber, D., Ackenheil, M., Laakmann, G., Fischer, H., and von Werder, K., 1980, Basal and stimulated levels of prolactin, TSH and LH in serum of chronic schizophrenic patients, long term treated with Neuroleptics. *Pharmakopsychiatr. Neuro-Psychopharmakol.* **13**:325.
Nair, N. P., Lal, S., Cervantes, P., Yassa, R., and Guyda, H., 1979, Effect of clozapine on apo-

morphine-induced growth hormone secretion and serum prolactin concentrations in schizophrenia, *Neuropsychobiology* **5**:136.

Overall, J. E., and Gorham, D. R., 1962. The Brief Psychiatric Rating Scale, *Psychol. Rep.* **10**:799.

Pandey, G. N., Garver, D. L., Tamminga, C., Ericksen, S., Ali, S. I., and Davis, J. M., 1977, Postsynaptic supersensitivity in schizophrenia, *Am. J. Psychiatry* **134**:518.

Rotrosen, J., Angrist, B. M., Gershon, S., Sachar, E. J., and Halpern, F. S., 1976, Dopamine receptor alteration in schizophrenia: Neuroendocrine evidence, *Psychopharmacology* **51**:1.

Rotrosen, J., Angrist, B. M., Gershon, S., Paquin, J., Branchey, L., Olenshansky, M., Halpern, F., and Sachar, E. J., 1979, Neuroendocrine effects of apomorphine: Characterization of response patterns and application to schizophrenia research, *Br. J. Psychiatry* **135**:444.

Scharfetter, C. H., 1972, Das AMD-P-System, Springer, Berlin.

Sedvall, G., Alfredsson, G., Bjerkenste, L., Eneroth, P., Fryö, B., Härnryd, C., and Wode-Helgodt, B., 1975, Biochemical correlates to antipsychotic drug action in man, in: *The Impact of Biology in Modern Psychiatry* (E. S. Gershon, R. E. Belmaker, S. S. Kety, and M. Rosenbaum, eds.), pp. 153–177, Plenum Press, New York.

Simpson, G. M., and Angus, J. W., 1970, A rating scale for extrapyramidal side effects, *Acta Psychiatr. Scand.* **212**:11.

Siris, S. G., van Kammen, D. P., and De Fraites, E. G., 1978, Serum prolactin and antipsychotic response to pimozide in schizophrenia, *Psychopharmacol. Bull.* **14**:11.

Snyder, S. H., 1972, Catecholamines in the brain as mediators of amphetamine psychosis, *Prog. Neurobiol.* **1**:151.

Snyder, S. H., 1976, The dopamine hypothesis of schizophrenia: Focus on the dopamine receptor, *Am. J. Psychiatry* **133**:197.

Snyder, S. H., and Bennett, J. P., 1976, Neurotransmitter receptors in the brain: Biochemical identification, in: *Biogenic Amine Receptors* (L. L. Iversen, S. D. Iversen, and S. H. Snyder, eds.), pp. 153–177, Plenum Press, New York.

Tamminga, C. A., Smith, R. C., Pandey, G., Frohman, L. A., and Davis, J. M., 1977, A neuroendocrine study of supersensitivity in tardive dyskinesia, *Arch. Gen. Psychiatry* **34**:1199.

Van Kammen, D. P., 1979, The dopamine hypothesis of schizophrenia revisited, *Psychoneuroendocrinology* **4**:37.

WHO (World Health Organization), 1974, International classification of diseases, WHO, Geneva.

Wode-Helgodt, B., Borg, S., Fryö, B., and Sedvall, G., 1978, Clinical effects and drug concentrations in plasma and cerebrospinal fluid in psychotic patients treated with fixed doses of chlorpromazine, *Acta Psychiatr. Scand* **58**:149.

CHAPTER 33

# Effects of Long-Term Neuroleptic Treatment on Serum Levels of Prolactin, TSH, LH, and Norepinephrine and on α-Adrenergic and Dopaminergic Receptor Sensitivity

Relations to Tardive Dyskinesia

DIETER NABER and FRANZ MÜLLER

## 1. INTRODUCTION

Neuroendocrine effects of catecholamine receptor-blocking neuroleptic drugs have been investigated in numerous studies (Ackenheil, 1981; Brambilla *et al.*, 1979a; Burt *et al.*, 1977; Carlsson, 1978, Collu *et al.*, 1977; Ettigi *et al.*, 1976; Meltzer *et al.*, 1978; Pandey *et al.*, 1977; Sachar, 1978); in particular, the increase of serum prolactin (PRL) after acute neuroleptic treatment is well documented (Gruen *et al.*, 1978; Langer *et al.*, 1977; Rubin and Hays, 1980, Wode-Helgodt, *et al.*, 1977). There are, however, only a few studies dealing with the neuroendocrine effects of chronic neuroleptic treatment. As both differ distinctly, inferring long-term effects from short-term effects is inappropriate (Lerner *et al.*, 1977). Moreover, there are striking differences in the reactivity of the tuberoinfundibular, nigrostrial, mesolimbic, and mesocortical dopaminergic systems to long-term treatment with neuroleptics (Bowers and Rozitis, 1974; Julou *et al.*, 1977).

Since neuroleptic therapy mostly extends over months and years, neuroendocrine effects occurring with chronic treatment need further attention,

---

DIETER NABER and FRANZ MÜLLER • Psychiatric Hospital of the University of Munich, Munich, FRG.

especially with regard to the variety of the neuroendocrine side effects: menstrual irregularities, galactorrhea, altered libido, oligospermia, gynecomastia, edema, weight gain, and both hypo- and hyperglycemia (Shader and DiMascio, 1970). These disorders are thought to result mainly from the blockade of catecholamine receptors in the hypothalamus, thus influencing the hypothalamic regulation of anterior pituitary hormones (de Wied and de Jong, 1974).

In this chapter, we report of studies on the effect of long-term neuroleptic treatment on serum levels of PRL, TSH, LH, and norepinephrine (NE) as well as on α-adrenergic and dopaminergic receptor sensitivity. Finally, relations between drug treatment, hormonal levels and receptor sensitivity with regard to tardive dyskinesia were investigated.

## 2. METHODS AND MATERIALS

Our own studies were carried out on chronic schizophrenic inpatients in a state hospital. They were physically well, without clinical evidence of endocrinopathy or organic brain disease, and did not receive any hormonal medication. The men, aged from 35 to 65 years (48 ± 10), had been treated with neuroleptics for 5–21 (13 ± 3) years. The average daily dosage, converted to chlorpromazine equivalents (CPZe) according to the clinical potency, was 410 ± 215 mg CPZe, and the lifetime neuroleptic dosage 1690 ± 890 g CPZe. The women, aged from 24 to 72 (51 ± 8) years, were under neuroleptics for 5–20 (13 ± 6) years, their average daily dosage was 590 ± 330 mg CPZe, and the total dosage was 2390 ± 1585 g CPZe.

Neuroendocrine stimulation tests and blood sampling were performed on patients under basal metabolic conditions at 8 a.m., about 60 min after getting up and 12 hr after the last dose of neuroleptics.

## 3. RESULTS AND DISCUSSION

### 3.1. Effect of Long-Term Neuroleptic Treatment on PRL Serum Levels

Among the effects of neuroleptic drugs on the secretion of pituitary hormones, the elevation of serum PRL is best documented (Gruen *et al.*, 1978; Langer *et al.*, 1977; Meltzer and Fang, 1976; Rubin and Hays, 1980; Wode-Helgodt *et al.*, 1977). As serum PRL levels have been found to remain elevated even after 1–3 months of neuroleptic treatment, it was suggested that tolerance to the PRL-increasing effect of neuroleptics does not develop (Gruen *et al.*, 1978; Meltzer and Fang, 1976). However, long-term studies do not support this hypothesis: DeRivera *et al.* (1976) investigated schizophrenic patients who had received neuroleptics for 9–16 years; about 20% of the men and 40% of the women had normal levels. Wilson *et al.* (1975) found normal PRL levels in about 25% of schizophrenic women treated with neuroleptics for 8 years. Beumont et al.(1974) reported that all male and 20% of female patients had normal PRL levels; all were treated with neuroleptics for more than 10 years. Two other studies (Aratö et al., 1979; Martin-DuPan and Baumann, 1979) performed on male patients after neuroleptic therapy for about 2 years, showed normal values in about 50% of patients in each case. In another group of patients, treated for at least 5 years, two out of four men and three out of seven women had PRL levels within the normal range (Kolakowska et al., 1976).

The results of these long-term studies indicate that within 3–24 months of neuroleptic treatment, certain patients, in particular males, may develop tolerance to the PRL-increasing effect. This hypothesis has been confirmed by our own studies, in which all 23 male and 29 of the 35 female chronic schizophrenic patients had normal PRL levels (Naber et al., 1979). In further studies similar results were obtained. Only 1 out of 37 male and 6 out of 37 female patients had levels higher than the normal range (Fig. 1). A possible depletion of the lactotrophic cells, by which the pituitary could no longer secrete large amounts of PRL, was excluded by the results of TRH administration: This potent stimulus of PRL secretion induced a marked increase of PRL levels (Fig. 2). As mentioned above, these studies implicate the development of tolerance in the dopamine tuberoinfundibular system. This tolerance seems to be at least partially reversible, as 5 days after discontinuation of chronic neuroleptic therapy in a study of 16 schizophrenic patients, serum PRL fell from 7.0 ± 5.0 ng/ml to 4.6 ± 4.4 ng/ml (Müller et al., 1980). Both mean levels are

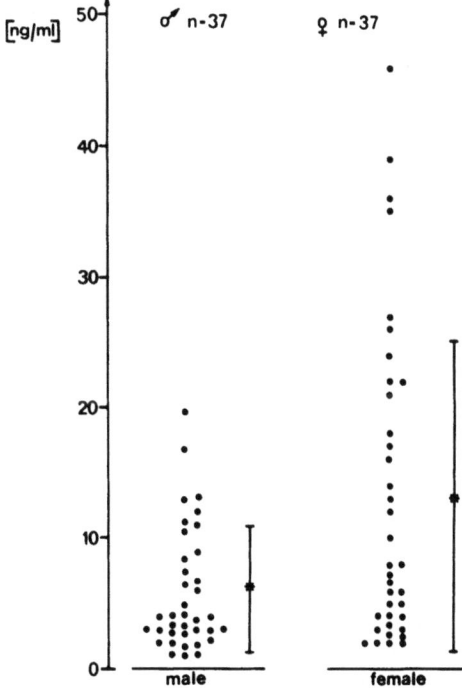

FIGURE 1. PRL plasma levels of chronic schizophrenic patients treated with neuroleptics for 6–21 years. Male patients: mean ± S.D. 6.1 ± 4.8 ng/ml. Female patients: 13.1 ± 11.9 ng/ml. Normal range: males 0–20 ng/ml, females 0–25 ng/ml.

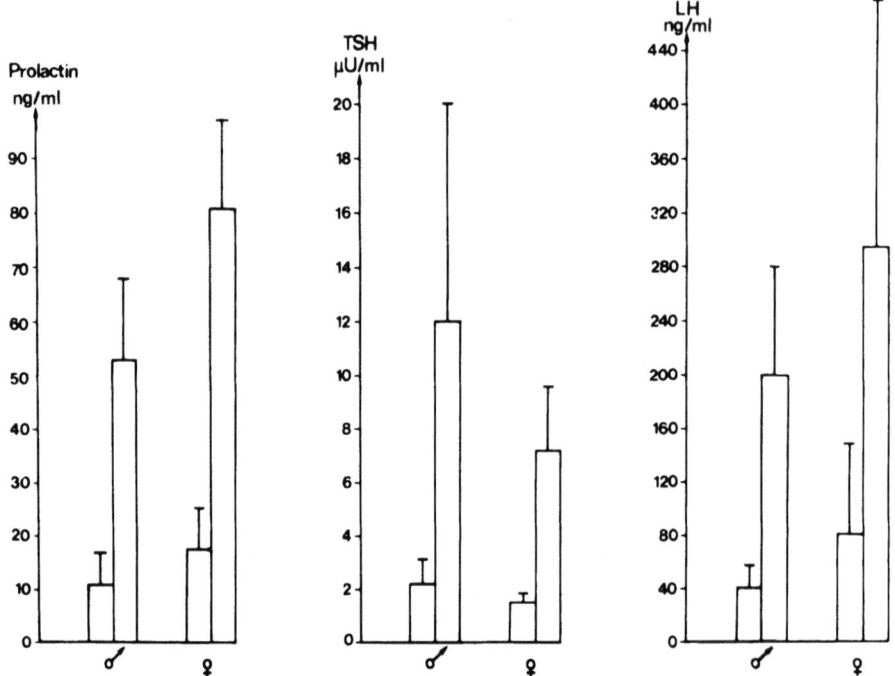

FIGURE 2. Basal and stimulated (0.2 mg TRH, 0.025 mg LHRH) plasma levels of PRL, TSH, and LH in chronic schizophrenic patients treated with neuroleptics for 6–21 years (12 males, 10 females). Male patients' PRL: basal 10.9 ± 5.8 ng/ml; stimulated 53 ± 15 ng/ml. TSH: basal 2.2 ± 0.9 µU/ml; stimulated 12.0 ± 8.0 µU/ml. LH: basal 39.7 ± 15.2 ng/ml; stimulated 199 ± 81 ng/ml. Female patients' PRL: basal 17.2 ± 8.1 ng/ml; stimulated 81 ± 16 ng/ml. TSH basal: 1.5 ± 0.3 µU/ml; stimulated 7.2 ± 2.4 µU/ml. LH basal: 80.5 ± 68.4 ng/ml; stimulated: 295 ± 186 ng/ml. Normal ranges: TSH basal 0.35 µU/ml; stimulated 3–18 µU/ml. LH males: basal 25–75 ng/ml; stimulated 75–360 ng/ml. Females: basal 30–240 ng/ml; stimulated 80–750 ng/ml.

within the normal range, but the significant decrease seems to reflect a reversal of adaptation of the tuberoinfundibular system. A similar decrease of normal PRL levels in serum (Laughren et al., 1979; Zander et al., 1981) as well as in cerebrospinal fluid (Zander et al., 1981) was observed after the discontinuation of neuroleptic drugs for 10 and 30 days.

The significant correlation between serum PRL and neuroleptic dosage found acute treatment (Gruen et al., 1978; Langer et al., 1977; Wiles et al., 1976; Wode-Helgodt et al., 1977) was not observed after long-term therapy (Müller et al., 1980; Naber et al., 1980b).

### 3.2. Effect of Long-Term Neuroleptic Treatment on TSH Serum Levels

Basal TSH secretion or TSH response to TRH has been found to be decreased after acute administration of pimozide (Collu et al., 1977); other groups did not observe any change after application of chlorpromazine (Sekso et al.,

1977; Wode-Helgodt et al., 1977), haloperidol (Naber et al., 1980c), or pimozide (Lamberg et al., 1977). Kirkegaard et al. (1977) found an increase by chlorpromazine, haloperidol, and biperiden. Similar conflicting results have been observed after chronic neuroleptic treatment; Lambert et al. (1977) found a subnormal TSH response to TRH after treatment with chlorpromazine or thioridazine, while Czernik and Kleesiek (1979) presented normal values of basal and stimulated TSH in patients treated with depot neuroleptics.

After neuroleptic treatment for 9–21 years, mean levels of basal as well as of stimulated TSH serum levels in 12 male and 10 female schizophrenic patients were within the normal range (Naber et al., 1980a). In contrast to the normal levels in all females, basal TSH levels in one man and stimulated TSH in two men were abnormally high (Fig. 2). In males, the neuroleptic dosage correlated significantly with the stimulated TSH levels.

These results are similar to those of Czernik and Kleesiek (1979) suggesting that long-term neuroleptic treatment does not markedly affect TSH secretion.

### 3.3. Effect of Long-Term Neuroleptic Treatment on LH Serum Levels

Acute studies concerning the effect of neuroleptics on LH secretion have reported conflicting results: While most groups did not find any change (Brambilla et al., 1979b; Murray et al., 1975; Rubin et al., 1976; Wode-Helgodt et al., 1977), Collu et al. (1977) observed a decrease, and Brambilla et al. (1975) an increase of LH serum levels. Studies on long-term effects did not show any significant alterations of LH values during neuroleptic treatment or after withdrawal (Beumont et al., 1974; Czernik and Kleesiek, 1979). In another group of patients on long-term treatment with depot neuroleptics, only those with sexual dysfunction had elevated LH levels. Patients with normal sexual function had normal levels despite their neuroleptic treatment (Arató et al., 1979).

Our study confirmed these results: Basal and after stimulation with LHRH, the mean levels of both sexes were within the normal range (Fig. 2). There was a significant correlation between duration of neuroleptic treatment and stimulated LH levels in males, but no difference in duration or dosage of neuroleptic therapy was found between 4 patients with abnormally low basal levels and the other 18 patients with normal LH secretion. These low levels may be caused by long-term hospitalization, as Brambilla et al. (1975) as well as Johnstone et al. (1977) found a low LH secretion among unmedicated, chronic schizophrenic inpatients. To summarize, LH secretion does not appear to be definitely affected by chronic neuroleptic administration.

### 3.4. Effect of Long-Term Neuroleptic Treatment on NE Serum Levels

Neuroleptics, not only blocking the dopaminergic but also the adrenergic receptors (Carlsson, 1978; Keller et al., 1973), cause an activation of presynaptic noradrenergic neurons. This feedback mechanism leads to an increase of the NE turnover (Andén et al., 1970; Bartholini et al., 1973; Bürki et al., 1974).

In humans, acute clozapine treatment has been found to increase NE serum levels (Sarafoff et al., 1979). After chronic therapy with different neuroleptics for 6–21 years, all 23 male and 35 female schizophrenic patients had markedly elevated levels [mean ± S.D., male patients, 0.75 ± 0.44 ng/ml; female patients, 0.85 ± 0.42 ng/ml; normal range 0.21 ± 0.08 ng/ml (Naber et al., 1980b)] (Fig. 3). These results are in agreement with other studies on patients who had been on long-term treatment with neuroleptics; Müller et al. (1980) found a four-fold elevation of serum NE in 16 chronic schizophrenic patients, while 13 patients of Zander et al. (1981) showed a three-fold increase. Five days after drug withdrawal, serum NE decreased from 0.84 ± 0.51 ng/ml to 0.50 ± 0.27 ng/ml (Müller et al., 1980).

The levels of catecholamines, reflecting the degree of sympathetic neuronal activity, can be altered by a variety of factors such as emotional stress, motoric activity or posture, environment, and methods of blood sampling (Lake et al., 1976). In all the studies mentioned above, no significant correlation between NE serum levels and age, blood pressure, pulse rate, or orthostatic variation was found. Under the same procedure of blood collection, acute untreated

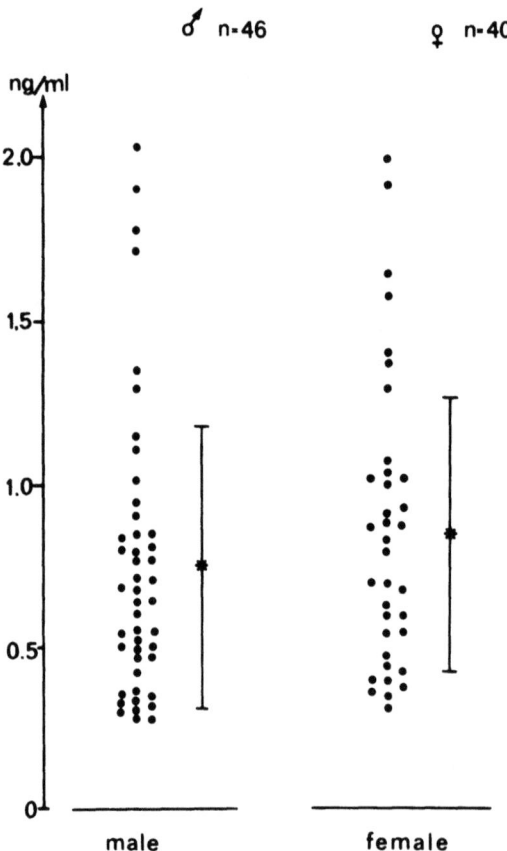

FIGURE 3. NE plasma levels (mean ± S.D.) of chronic schizophrenic patients treated with neuroleptics for 6–21 years. Male patients: 0.75 ± 0.44 ng/ml. Female patients: 0.85 ± 0.42 ng/ml. Normal range: 0.21 ± 0.08 ng/ml.

schizophrenic patients had levels of 0.30 ± 0.15 ng/ml (Ackenheil et al., 1979). These values are significantly higher than those of normal controls, but significantly less than the levels of those patients who had been treated long-term with neuroleptics. It therefore is suggested that the elevation of serum NE levels under chronic neuroleptic treatment is at least partly induced by the pharmacological treatment.

### 3.5. Changes of Receptor Sensitivity after Long-Term Neuroleptic Treatment

The blockade of dopaminergic and adrenergic receptors is a well-documented effect of neuroleptic drugs (Carlsson, 1978). Supersensitivity is suggested to develop as a compensatory mechanism to the reduced stimulation (Christensen and Møller-Nielsen, 1980). This induction of supersensitivity, measured e.g. by the increase of apomorphine-induced stereotypes, occurs already after 1–2 days of neuroleptic therapy (Christensen and Møller-Nielsen, 1980). In another study, the hypothesized DA receptor supersensitivity was confirmed by receptor binding after repeated neuroleptic administration (Burt et al., 1977). For example, in rats, chronic treatment with haloperidol, perphenazine, or reserpine elicited a 20–25% increase in striatal DA receptor binding. The development of supersensitivity is time dependent (Muller and Seeman, 1978) and reversible, since 17 days after drug withdrawal the receptor binding was no longer elevated (Burt et al., 1977).

In humans, the sensitivity of dopaminergic and α-adrenergic receptors can be studied by stimulation with the corresponding agonists apomorphine and clonidine, both of which penetrate the blood–brain barrier and increase GH secretion (Lal et al., 1973; Rotrosen et al., 1979). To test the reliability of GH response to apomorphine, Rotrosen et al. (1979) stimulated subjects several times and found a high interindividual variability, but a high intraindividual reproducibility.

### 3.6. GH Response after Stimulation with Clonidine

The stimulation of GH by clonidine has been shown to be a useful tool for measuring central α-adrenergic receptor sensitivity (Lal et al., 1975; Matussek et al., 1980). In studies on normal controls, the majority of subjects had a GH response to clonidine of 5–25 ng/ml; only some subjects showed no stimulation effects (GH < 5 ng/ml, Matussek et al., 1980).

The effect of long-term neuroleptic treatment on GH response to clonidine was investigated in 16 chronic schizophrenic patients treated with neuroleptics for 12 ± 4 years (Müller et al., 1980). Only 2 patients had a clonidine-induced increase of GH (10 vs. 20 ng/ml), whereas 14 patients had a blunted GH response. Nine of the sixteen patients were tested again 5 days after discontinuation of neuroleptic therapy (Fig. 4): Only 2 patients had a stimulated GH secretion of more than 5 ng/ml. Baseline serum GH levels of these patients

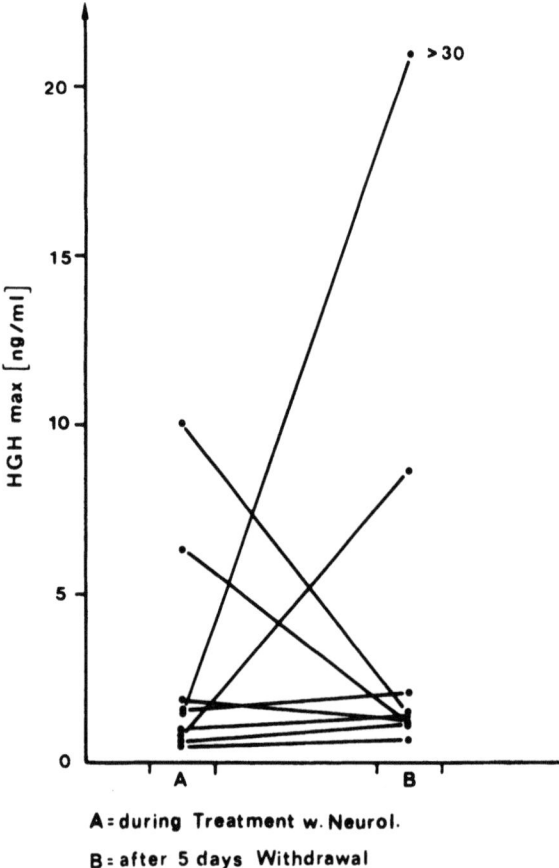

FIGURE 4. Maximal GH serum levels after clonidine (0.15 mg i.v.) stimulation of nine male, chronic schizophrenic inpatients treated with neuroleptics for 6–21 years. Mean ± S.D. during treatment 2.74 ± 3.40 ng/ml; mean ± S.D. after 5 days' drug withdrawal 5.84 ± 11.19 ng/ml.

under neuroleptic treatment and after 5 days of drug withdrawal did not differ significantly from those of normal controls.

These data on long-term-treated schizophrenic patients provide no evidence of α-adrenergic supersensitivity under neuroleptic treatment and after 5 days of drug withdrawal. However, our results tend to show an α-adrenergic subsensitivity. A cautious interpretation should consider the following: since our population was a heterogeneous sample of chronic schizophrenic patients, discriminating between drug-induced effects and different hormonal responses in subtypes of schizophrenia (Matussek et al., 1980) is difficult. Drug discontinuation of 5 days with regard to neuroleptic treatment of 10–20 years could be insufficient and therefore unable to change α-adrenergic receptor sensitivity. In further studies, the interval of drug discontinuation should be extended.

### 3.7. GH Response after Stimulation with Apomorphine

Apomorphine, a direct-acting DA agonist, stimulates the release of GH from the anterior pituitary (Christensen and Møller-Nielsen, 1980; Ettigi et al.,

1976; Lal et al., 1975; Pandey et al., 1977; Rotrosen et al., 1979). In acute untreated schizophrenic patients, apomorphine induced a marked GH stimulation of more than 20 ng/ml serum. This GH response to apomorphine, indicating the neuroleptic DA receptor blockade, was significantly reduced by treatment with haloperidol (Ackenheil, 1981). However, this short-term neuroleptic treatment did not induce DA receptor supersensitivity as it was postulated on the basis of animal experiments (Christensen and Møller-Nielsen, 1980); 5 days of drug withdrawal did not cause a significant change of apomorphine-induced GH response (Ackenheil, 1981). Other studies on the effect of drug withdrawal after long-term neuroleptic treatment did not reveal evidence for a DA receptor supersensitivity either. Pandey et al. (1977) observed in chronic schizophrenic patients, withdrawn from neuroleptics for 2–3 weeks, significantly lower GH response than in acute schizophrenic patients and slightly lower levels than in normal controls. Moreover, in comparison to normals, Ettigi et al. (1976) demonstrated a significantly lower GH response after apomorphine stimulation in chronic schizophrenic patients withdrawn from neuroleptics for 3–15 weeks. The time course of DA receptor sensitivity after drug withdrawal was investigated by Müller et al. (1981, unpublished results); under neuroleptic treatment, only 3 out of 16 patients had stimulated GH plasma levels of more than 5 ng/ml (Fig. 5). After drug withdrawal of 12 days, a significant increase of GH response to apomorphine was observed. After 30 and 90 days, however, mean levels of GH decreased; only about 50% of patients

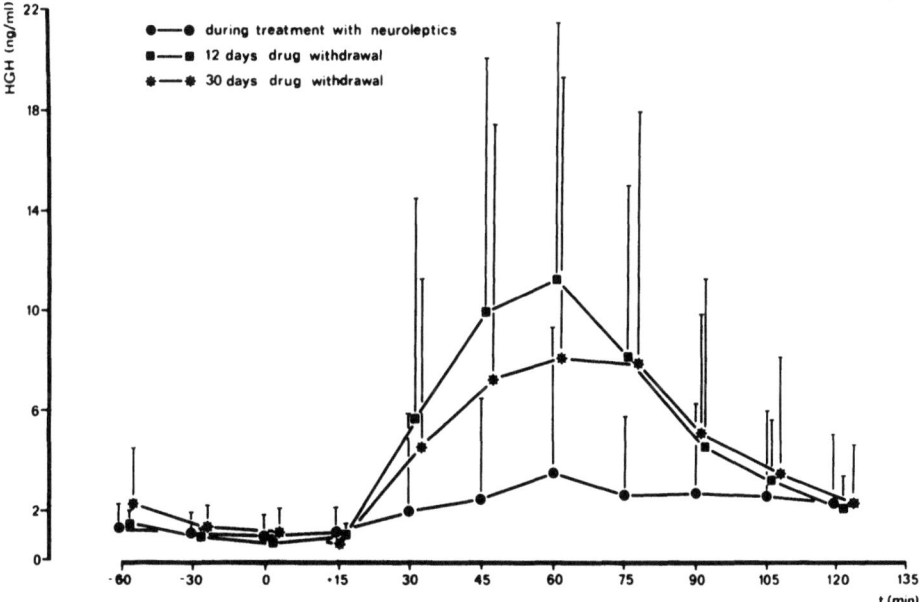

FIGURE 5. GH serum levels before and after apomorphine application (0.5 mg s.c.) of 11 chronic schizophrenic patients (9 males and 2 females) treated with neuroleptics for 6–21 years and 12 and 30 days after drug withdrawal.

had stimulated GH levels higher than 5 ng/ml. These data provide no evidence for hypothalamic DA receptor supersensitivity, because the significant increase in GH response on day 12 probably corresponds to a readjustment from mostly blunted GH responses under neuroleptic treatment back to stimulated levels similar to those of normal controls.

### 3.8. Relations of Receptor Sensitivity and Hormonal Levels to Tardive Dyskinesia

There is much evidence that long-term neuroleptic treatment may induce symptoms of tardive dyskinesia in predisposed subjects and it was concluded that the temporal schedule and daily amount of neuroleptics may be more important than the lifetime dosage (see reviews of Gerlach, 1979; Klawans *et al.*, 1980). The pathogenesis of tardive dyskinesia has been hypothesized to relate to the chronic DA receptor blockade and to the resulting receptor hypersensitivity (Klawans *et al.*, 1980). Therefore, neuroendocrine DA-mediated hypersensitivity as indicated in an increase of apomorphine-induced GH secretion might be expected. However, this theory could not be confirmed: In

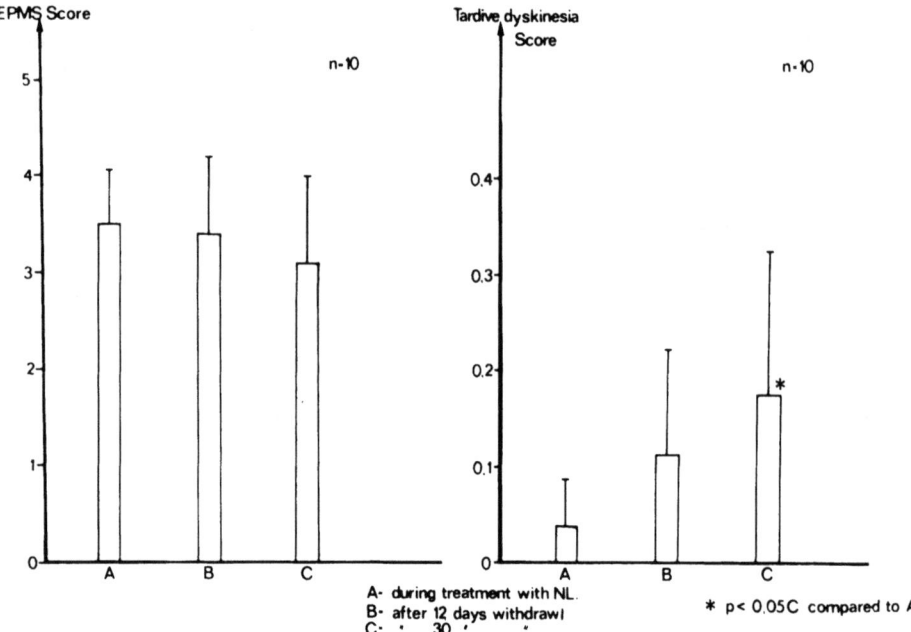

FIGURE 6. Extrapyramidal motor symptoms (EPMS) and tardive dyskinesia scores of 11 chronic schizophrenic inpatients (9 males, 2 females) treated with neuroleptics for 6–21 years. EPMS score: mean ± S.D. under neuroleptic treatment 3.45 ± 0.62; mean ± S.D. after 12 days' drug withdrawal 3.4 ± 0.8; mean ± S.D. after 30 days' drug withdrawal 3.13 ± 0.7. Tardive dyskinesia score: mean ± S.D. under neuroleptic treatment 0.03 ± 0.05; mean ± S.D. after 12 days' drug withdrawal 0.11 ± 0.11; mean ± S.D. after 30 days' drug withdrawal 0.18 ± 0.17.*, significant difference ($p < 0.05$) between the time under neuroleptic treatment and the time after 30 days' drug withdrawal. NL, neuroleptic.

patients withdrawn from neuroleptics for up to 15 weeks, Ettigi et al. (1976) observed in subjects with tardive dyskinesia a significantly lower apomorphine-induced peak GH concentration, compared to patients without tardive dyskinesia. Pandey et al. (1977) reported no difference in GH secretion after stimulation with apomorphine between chronic schizophrenic patients with and without tardive dyskinesia. Of 18 chronic schizophrenic inpatients under neuroleptic treatment, six subjects had mild orofacial symptoms of tardive dyskinesia. After 30 days of drug discontinuation, symptoms and incidence of tardive dyskinesia were slightly but significantly increased (Müller et al., 1981, unpublished results) (Fig. 6). Neither under neuroleptic treatment nor after drug withdrawal were symptoms of tardive dyskinesia significantly correlated to apomorphine-induced GH secretion. In a previous study, dosage and duration of neuroleptic treatment, NE and PRL serum levels were not correlated to tardive dyskinesia either (Naber et al., 1980b).

Neuroendocrine studies have therefore not revealed evidence of a hypothalamic hypersensitivity in patients with tardive dyskinesia. However, neuroleptic receptor blockade is a general process involving different systems; tardive dyskinesia may be associated with a selective DA supersensitivity of certain receptor populations only.

## 4. CONCLUSIONS

Neuroendocrine effects of long-term neuroleptic therapy were investigated in several studies on chronic schizophrenic patients, hospitalized and treated with neuroleptics for 6–21 years. The following results were obtained.

1. Under chronic neuroleptic treatment the majority of patients had PRL plasma levels within the normal range, which decreased significantly after drug withdrawal. Reversible adaptation of the tuberoinfundibular DA system is suggested, since an exhaustion of the lactotrophic cells was excluded.
2. TSH and LH secretion are not markedly impaired by chronic neuroleptic treatment.
3. Long-term neuroleptic treatment induces a significant reversible elevation of NE plasma levels.
4. The clonidine-induced GH secretion in schizophrenic patients was mostly blunted under chronic neuroleptic treatment and after 5 days of drug withdrawal. This suggests an α-adrenergic subsensitivity.
5. The apomorphine-induced GH secretion in schizophrenic patients was mostly blunted under neuroleptic treatment. After drug withdrawal, stimulated GH plasma levels normalized. These data do not suggest a DA receptor supersensitivity during and after long-term neuroleptic treatment.
6. Symptoms of tardive dyskinesia were not significantly correlated to apomorphine-induced GH secretion, neither to plasma levels of PRL and NE, nor to dosage and duration of neuroleptic treatment.

ACKNOWLEDGMENT. We would like to acknowledge the support for this work given by the Deutsche Forschungsgemeinschaft.

## 5. REFERENCES

Ackenheil, M., 1981, Biochemical effects of apomorphine: Contribution to schizophrenia research, in: *Apomorphine and Other Dopaminomimetics, Vol. 2, Clinical Research* (G. L. Gessa and G. U. Corsini, eds.), pp. 215–225, Raven Press, New York.

Ackenheil, M., Albus, M., Müller, F., Müller, T., Welter, D., Zander, K., and Engel, R., 1979, Catecholamine response to short-time stress in schizophrenic and depressive patients, in: *Catecholamines: Basic and Clinical Frontiers*, Vol. 2 (E. Usdin, A. Coppen, and J. Barchas, eds.), pp. 1937–1941, Pergamon Press, Elmsford, N.Y.

Andén, N. E., Butcher, S. G., Carrodi, H., Fuxe, K., and Ungerstedt, U., 1970, Receptor activity and turnover of dopamine and noradrenaline after neuroleptics, *Eur. J. Pharmacol.* **11**:303.

Aratö, M., Erdös, A., and Polgf, M., 1979, Endocrinological changes in patients with sexual dysfunction under long-term neuroleptic treatment, *Pharmakopsychiatr. Neuro-psychopharmakol.* **12**:426.

Bartholini, G., Keller, H. H., and Pletscher, A., 1973, Effect of neuroleptics on endogenous norepinephrine in rat brain, *Neuropharmacology* **12**:751.

Beaumont, P. J. V., Corker, C. S., Friesen, H. G., Kolakowska, T., Mandelbrote, B. M., Marshall, J., Murray, M. A. F., and Wiles, D. H., 1974, The effects of phenothiazines on endocrine function. II. Effects in men and post-menopausal women, *Br. J. Psychiatry* **124**:420.

Bowers, M. B., Jr., and Rozitis, A., 1974, Regional differences in homovanillic acid concentrations after acute and chronic administration of antipsychotic drugs, *J. Pharm. Pharmacol.* **26**:743.

Brambilla, F., Guerrini, A., Gaustalla, A., Rovere, C., and Riggi, F., 1975, Neuroendocrine effects of haloperidol therapy in chronic schizophrenics, *Psychopharmacologia* **44**:17.

Brambilla, F., Bellodi, L., Negri, F., Smeraldi, E., and Malagoli, G., 1979a, Dopamine receptor sensitivity in the hypothalamus of chronic schizophrenics after haloperidol therapy: Growth hormone and prolactin response to stimuli, *Psychoneuroendocrinology* **4**:329.

Brambilla, F., Riggi, F., Guerrini, A., Malagoli, G., and Guastalla, A., 1979b, The effect of prolonged sulpiride therapy on pituitary–gonadal function in male psychotic patients, *Clin. Ther.* **2**:91.

Bürki, H. R., Ruch, W., Asper, H., Baggiolini, M., and Stille, G., 1974, Effect of single and repeated administration of clozapine on the metabolism of dopamine and noradrenaline in the brain of the rat, *Eur. J. Pharmacol.* **27**:180.

Burt, D. R., Creese, I., and Snyder, S. H., 1977, Antischizophrenic drugs: Chronic treatment elevates dopamine receptor binding in brain, *Science* **196**:326.

Carlsson, A., 1978, Antipsychotic drugs, neurotransmitters, and schizophrenia, *Am. J. Psychiatry* **135**:164.

Christensen, A. V., and Møller-Nielsen, I., 1980. On the supersensitivity of DA-receptors after single and repeated administration of neuroleptics, in: *Tardive Dyskinesia* (W. E. Fann, R. C. Smith, J. M. Davis, and E. F. Domino, eds.), pp. 35–50, MTP Press, New York.

Collu, R., Jequier, J. C., Leboeuf, J., and Ducharme, J. R., 1977, Endocrine effects of pimozide, a specific dopaminergic blocker, *J. Clin. Endocrinol. Metab.* **44**:981.

Czernik, A., and Kleesiek, K., 1979, Effects of depot neuroleptics on pituitary hormones, *Nervenarzt* **50**:527.

DeRivera, J. L., Lal, S., Ettigi, P., Hontela, S., Muller, H. F., and Friesen, H. G., 1976, Effects of acute and chronic neuroleptic therapy on serum prolactin levels in men and women of different age groups, *Clin. Endocrinol.* **5**:273.

de Wied, D., and de Jong, W., 1974, Drug effects and hypothalamic–anterior pituitary function, *Annu. Rev. Pharmacol.* **14**:389.

Ettigi, P., Nair, N. P. V., Lal, S., Cervantes, P., and Guyda, H., 1976, Effect of apomorphine on growth hormone and prolactin secretion in schizophrenic patients, with or without oral dyskinesia, withdrawn from chronic neuroleptic therapy, *J. Neurol. Neurosurg. Psychiatry* **39**:870.

Gerlach, J., 1979, Tardive dyskinesia, *Dan. Med. Bull.* **26**:209.

Gruen, H. P., Sachar, E. J., Langer, G., Altman, N., Leifer, M., Frantz, A., and Halpern, F. S., 1978, Prolactin responses to neuroleptics in normals and schizophrenic subjects, *Arch. Gen. Psychiatry* **35**:108.

Johnstone, E. C., Crow, T. J., and Mashiter, K., 1977, Anterior pituitary hormone secretion in chronic schizophrenia—An approach to neurohumoral mechanisms, *Psychol. Med.* **7**:223.

Julou, L., Scatton, B., and Glowinski, J., 1977, Acute and chronic treatment with neuroleptics: Similarities and differences in their action in nigrostriatal, mesolimbic and mesocortical dopaminergic neurons, in: *Advances in Biochemical Psychopharmacology* Vol. 16 (E. Costa and G. L. Gessa, eds.), pp. 617–624, Raven Press, New York.

Keller, H. H., Bartholini, G., and Pletscher, A., 1973, Increase of 3-methoxy-4-hydroxyphenylethylene glycol in rat brain by neuroleptic drugs, *Eur. J. Pharmacol.* **23**:183.

Kirkegaard, C., Bjørum, N., Cohn, D., Faber, J., Lauridsen, U. B., and Nekup, J., 1977, Studies on the influence of biogenic amines and psychoactive drugs on the prognostic value of the TRH stimulation test in endogenous depression, *Psychoneuroendocrinology* **2**:131.

Klawans, H. L., Goetz, C. G., and Perlik, S., 1980, Tardive dyskinesia: Review and update, *Am. J. Psychiatry* **137**:900.

Kolakowska, T., Wiles, D. H., Gelder, M. G. and McNeilly, A. S., 1976, Clinical significance of plasma chlorpromazine levels, *Psychopharmacology* **49**:101.

Lake, C. R., Ziegler, M. G., and Kopin, I. J., 1976, Use of plasma norepinephrine for evaluation of sympathetic neuronal function in man, *Life Sci.* **18**:1315.

Lal, S., De la Vega, C. E., Sourkes, T. L., and Friesen, H. G., 1973, Effect of apomorphine on growth hormone, prolactin, luteinizing hormone and follicle-stimulating hormone levels in human serum, *J. Clin. Endocrinol. Metab.* **37**:719.

Lal, S., Tolis, G., Martin, J. B., Brown, G. M., and Guyda, H., 1975, Effect of clonidine on growth hormone, prolactin, luteinizing hormone, follicle stimulating hormone, and thyroid-stimulating hormone in the serum of normal men, *J. Clin. Endocrinol. Metab.* **41**:827.

Lamberg, B. A., Linnoila, M., Fogelholm, R., Olkinuora, M., Kotilainen, P., and Saarinen, P., 1977, The effect of psychotropic drugs on the TSH-response to thyroliberin (TRH), *Neuroendocrinology* **24**:90.

Langer, G., Sachar, E. J., Halpern, F. S., Gruen, P. H., and Solomon, M., 1977, The prolactin résponse to neuroleptic drugs. A test of dopaminergic blockade: Neuroendocrine studies in normal men, *J. Clin. Endocrinol. Metab.* **45**:996.

Laughren, T. P., Brown, W. A., and Williams, B. W., 1979, Serum prolactin and clinical state during neuroleptic treatment and withdrawal, *Am. J. Psychiatry* **136**:108.

Lerner, P., Nosé, P., Gordon, E. K., and Lovenberg, W., 1977, Haloperidol: Effect of long-term treatment on rat striatal dopamine synthesis and turnover, *Science* **197**:181.

Martin-Du-Pan, R., and Baumann, D., 1979, Neuroendocrine effects of chronic neuroleptic therapy in male psychiatric patients, *Psychoneuroendocrinology* **3**:245.

Matussek, N., Ackenheil, M., Hippius, H., Müller, F., Schroöder, T., Shultes, H., and Wasilewski, B., 1980, Effect of clonidine on growth hormone release in psychiatric patients and controls, *Psychiatry Res.* **2**:25.

Meltzer, H. Y., and Fang, V. S., 1976, The effect of neuroleptics on serum prolactin in schizophrenic patients, *Arch. Gen. Psychiatry* **33**:279.

Meltzer, H. Y., Goode, D. J., and Fang, V. S., 1978, The effect of psychotropic drugs on endocrine function. I. Neuroleptics, precursors and agonists, in: *Psychopharmacology: A Generation of Progress* (M. A. Lipton, A. DiMascio, and K. F. Killam, eds.), pp. 509–529, Raven Press, New York.

Müller, F., Bartl, S., Fischer, B., Wörner, J., and Ackenheil, M., 1980, Hormonal studies after long-term treatment with neuroleptics, Abstracts of the 12th CINP Congress, Göteborg, Sweden, p. 255.

Muller, P., and Seeman, P., 1978, Dopaminergic supersensitivity after neuroleptics: Time course and specificity, *Psychopharmacology* **60:**1.

Murray, M. A., Bancroft, J. H., Anderson, D. C., Tennent, T. G., and Carr, P. J., 1975, Endocrine changes in male sexual deviants after treatment with anti-androgens, oestrogens or tranquilizers, *J. Endocrinol.* **67:**179.

Naber, D., Fischer, B., and Ackenheil, M., 1979, Effect of long-term neuroleptic treatment on dopamine tuberoinfundibular system: Development of tolerance? *Commun. Psychopharmacol.* **3:**59.

Naber, D., Ackenheil, M., Laakmann, G., Fischer, B., and von Werder, K., 1980a, Basal and stimulated levels of prolactin, TSH and LH in serum of chronic schizophrenic patients, long-term treated with neuroleptics, *Pharmakopsychiatr. Neuro-Psychopharmakol.* **13:**325.

Naber, D., Finkbeiner, C., Fischer, B., Zander, K. J., and Ackenheil, M., 1980b, Effect of long-term neuroleptic treatment on prolactin and norepinephrine levels in serum of chronic schizophrenics: Relations to psychopathology and extrapyramidal symptoms, *Neuropsychobiology* **6:**181.

Naber, D., Steinböck, H., and Greil, W., 1980c, Effects of short- and long-term neuroleptic treatment on thyroid function, *Prog. Neuropsychopharmacol.* **4:**199.

Pandey, G. N., Garver, D. L., Tamminga, C., Ericksen, S., Ali, S. I., and Davis, J. M., 1977, Postsynaptic supersensitivity in schizophrenia, *Am. J. Psychiatry* **134:**518.

Rotrosen, J., Angrist, B., Gershon, S., Paquin, J., Branchey, L., Oleshansky, M., Halpern, F., and Sachar, E. J., 1979, Neuroendocrine effects of apomorphine: Characterization of response patterns and application to schizophrenia research, *Br. J. Psychiatry* **135:**444.

Rubin, R. T., and Hays, S. E., 1980, The prolactin secretory response to neuroleptic drugs: Mechanisms, applications and limitations, *Psychoneuroendocrinology* **5:**121.

Rubin, R. T., Poland, R. E., O'Connor, D., Gouin, P. R., and Tower, B. B., 1976, Selective neuroendocrine effects of low-dose haloperidol in normal adult men, *Psychopharmacologia* **47:**135.

Sachar, E. J., 1978, Neuroendocrine response to psychotropic drugs, in: *Psychopharmacology: A Generation of Progress* (M. A. Lipton, A. DiMascio, and K. F. Killam, eds.), pp. 499–507, Raven Press, New York.

Sarafoff, M., Davis, L., and Rüther, E., 1979, Clozapine induced increase of human plasma norepinephrine, *J. Neural Transm.* **46:**175.

Sekso, M., Solter, M., Banovac, K., Vizner, B., and Petek, M., 1977, The effect of dopaminergic drugs on pituitary hormone release in normal subjects, *Acta Med. Iugosl.* **31:**321.

Shader, R. I., and DiMascio, A., 1970, *Psychotropic Drug Side Effects: Clinical and Theoretical Perspectives*, Williams & Wilkins, Baltimore.

Wiles, D. H., Kolakowska, T., McNeilly, A. S., Mandelbrote, B. M., and Gelder, M. G., 1976, Clinical significance of plasma chlorpromazine levels. I. Plasma levels of the drug, some of its metabolites and prolactin during acute treatment, *Psychol. Med.* **6:**407.

Wilson, R. G., Hamilton, J. R., Boyd, W. D., Forrest, A. P. M., Cole, E. N., Boyns, A. R., and Griffiths, K., 1975, The effect of long term phenothiazine therapy on plasma prolactin, *Br. J. Psychiatry* **127:**71.

Wode-Helgodt, B., Eneroth, P., Fryö, B., Gullberg, B., and Sedvall, G., 1977, Effect of chlorpromazine treatment on prolactin levels in cerebrospinal fluid and plasma of psychotic patients, *Acta Psychiatr. Scand.* **56:**280.

Zander, K. J., Fischer, B., Zimmer, R., and Ackenheil, M., 1981, Long-term neuroleptic treatment of chronic schizophrenic patients: Clinical and biochemical effects of withdrawal, *Psychopharmacology* **73:**43.

CHAPTER 34

# Central Nervous System and Pituitary Dopaminergic Defects in Hyperprolactinemic States

CARLO FERRARI, ANNA MARIA MATTEI, and
PIER GIORGIO CROSIGNANI

## 1. INTRODUCTION

The fundamental role exerted by dopamine (DA) in the inhibitory control of prolactin (PRL) secretion at both hypothalamic and pituitary levels has been well established (Weiner and Ganong, 1978, for a review). In recent years many investigations have therefore been performed to evaluate the possible impairment of dopaminergic inhibition of PRL secretion in hyperprolactinemic states of different etiology, and much evidence in support of this possibility has indeed been obtained (Crosignani *et al.*, 1977, 1980a; Fine and Frohman, 1978; Müller *et al.*, 1978; Lim *et al.*, 1979; Ferrari *et al.*, 1980a, 1981a; Reschini *et al.*, 1980; Frohman *et al.*, 1981). Thus, many patients with idiopathic or adenomatous hyperprolactinemia do not increase serum PRL levels after DA receptor blockade by sulpiride administration, but a PRL response to this stimulus occurs in the same subjects during concurrent infusion of exogenous DA (Crosignani *et al.*, 1977; Ferrari *et al.*, 1979), indicating that lack of PRL increase after sulpiride is due to insufficient DA concentration outside the blood–brain barrier (most likely in the pituitary gland) in these conditions. Another dopaminergic defect occurring at pituitary level in some hyperprolactinemic patients is shown by the failure of PRL levels to suppress after administration of direct DA agonists like DA itself or L-dopa and bromocriptine. This defect is only uncommonly encountered in patients with idiopathic or adenomatous hyperprolactinemia (Reschini *et al.*, 1980; Crosignani *et al.*, 1980a, Ferrari *et al.*, 1980b), while it

---

CARLO FERRARI, ANNA MARIA MATTEI, and PIER GIORGIO CROSIGNANI • II Department of Medicine, Fatebenefretelli Hospital and IV Department of Obstetrics and Gynecology, University of Milan, Milan, Italy.

is uniformly present in uremic subjects (Lim et al., 1979; Frohman et al., 1981), in whom it is reverted by successful renal transplantation.

These data suggest the existence of a functional derangement of DA receptors at the pituitary level in uremia. Cirrhotic patients with hyperprolactinemia also exhibit partial or sometimes total resistance to L-dopa administration which is reversible with improvement of their disease (Borzio et al., 1981). Resistance of some hyperprolactinemic patients to DA agonists is confirmed by the results of chronic bromocriptine treatment, which in some cases fails to lower PRL levels even at high drug doses (Ferrari et al., 1981a). Interestingly, in uremic patients chronic bromocriptine therapy is able to overcome this resistance (Frohman et al., 1981). The most recently recognized dopaminergic defect in hyperprolactinemic states is resistance to the PRL-lowering effect of CNS dopaminergic activation induced either by L-dopa plus carbidopa after carbidopa pretreatment (Fine and Frohman, 1978; Crosignani et al., 1980a; Frohman et al., 1981) or by nomifensine administration (Müller et al., 1978; Ferrari et al., 1980a; Crosignani et al., 1980a). This defective PRL inhibition occurs in most patients with prolactinomas as well as with idiopathic hyperprolactinemia (Crosignani et al., 1980a), in uremic subjects (Frohman et al., 1981), and in some hyperprolactinemic patients with liver cirrhosis (Borzio et al., 1981), polycystic ovaries, or hypothyroidism (Moriondo et al., 1980).

On the other hand, another quite characteristic abnormality found in the majority of hyperprolactinemic patients, and especially in those with pituitary adenomas, is failure of PRL to rise after TRH (Kleinberg et al., 1977; Crosignani et al., 1980b), a stimulus which acts independently of the pituitary DA concentration. Lack of PRL response to TRH implies that neither adenomatous nor normal lactotrophs may increase PRL release after this challenge. While adenomatous cells might be irresponsive because of alteration of receptor sites or because of already maximal secretion (Kleinberg et al., 1977), the failure of PRL release by normal lactotrophs has been attributed to functional suppression by a short-loop feedback system involving increased activity of tuberoinfundibular DA neurons as a result of high circulating PRL levels (Hökfelt and Fuxe, 1972; Healy et al., 1977). Other lines of evidence suggesting the possible existence of increased CNS dopaminergic activity in hyperprolactinemia include the reported stimulation of LH release by metoclopramide (Quigley et al., 1979) and lack of LH inhibition by bromocriptine (Evans et al., 1980) as well as exaggerated TSH responses to metoclopramide (Quigley and Yen, 1980) and domperidone (Scanlon et al., 1981).

However, a normal LH inhibition by DA infusion or by DA agonist drugs in hyperprolactinemic women has been found by others (Ferrari et al., 1981b; Lachelin et al., 1977), and stimulation of LH release by metoclopramide has not been confirmed, while such a response has been induced by this drug after pretreatment with carbidopa plus L-dopa to increase CNS dopaminergic tone, in both hyperprolactinemic and normal women (Elli et al., 1980). Therefore, the question of whether there is an increased or decreased CNS dopaminergic activity in hyperprolactinemia is still unsettled, possibly because the situation may be different in different brain areas controlling the secretion of different pituitary hormones. In the present chapter our data on the PRL responses to various dopaminergic and antidopaminergic agents in hyperprolactinemic states of different etiology will be reviewed to gain further insight into the dopaminergic defects occurring in these conditions.

## 2. SUBJECTS

A large number of hyperprolactinemic patients (serum PRL levels, as mean of at least three samples, greater than 20 ng/ml for females and 15 ng/ml for males) has been investigated with different neuropharmacological agents.

The diagnosis of macroprolactinoma was made by standard X-rays of the skull and by CT scans; microprolactinoma was diagnosed by finding the characteristic abnormalities by sellar tomography. The diagnosis of hypothalamic hyperprolactinemia was established by CT scans and confirmed at surgery. The patients with normal sella turcica on tomography and on CT scans were considered to have idiopathic hyperprolactinemia. Some hyperprolactinemic patients with acromegaly due to large pituitary adenomas were also studied.

The modalities and the criteria of normal PRL responses to the different neuropharmacological tests performed are summarized in Table 1.

## 3. RESULTS AND DISCUSSION

### 3.1. Defective Pituitary DA Concentration in Hyperprolactinemic States

As shown in Tables 2 and 3, most hyperprolactinemic patients did not further increase PRL levels after administration of the DA antagonists sulpiride

TABLE 1
Neuropharmacological Tests Used for the Study of Dopaminergic Regulation of PRL Secretion in Hyperprolactinemia

| Test | Times of blood sampling (min) | Normal PRL response (% increase or decrease) |
|---|---|---|
| Sulpiride (100 mg i.m.) | −15, 0, 20, 30, 60 | 100% increase |
| Domperidone (2–8 mg i.v.) | −15, 0, 30, 60 | 100% increase |
| Dopamine (5 μg/kg/min for 120 min) | −15, 0, 30, 60, 90, 120 | 50% decrease |
| L-Dopa (500 mg p.o.) | −15, 0, 30, 60, 90, 120, 150, 180 | 50% decrease |
| Bromocriptine (2.5 mg p.o.) | −15, 0, 120, 180, 240 | 50% decrease |
| Carbidopa (35 mg p.o.) plus L-dopa (100 mg p.o.)[a] | −15, 0, 30, 60, 90, 120, 150, 180 | 50% decrease |
| Nomifensine (200 mg p.o.) | −15, 0, 120, 150, 180, 210, 240 | 35% decrease |

[a] After pretreatment with carbidopa 50 mg every 6 hr for four doses.

## TABLE 2
Serum PRL Response to Sulpiride (100 mg i.m.) in Patients with Hyperprolactinemia of Different Etiology. Shown are Basal and Peak PRL Levels (ng/ml)

| Patient No. | Idiopathic hyperPRL | | Microprolactinoma | | Macroprolactinoma | | Acromegaly | | Hypothalamic hyperPRL | |
|---|---|---|---|---|---|---|---|---|---|---|
| | Basal | Peak | Basal | Peak | Basal | Peak | Basal | Peak | Basal | Peak |
| 1 | 16 | 234 | 292 | 400 | 3000 | 3000 | 25 | 41 | 31 | 36 |
| after surgery | | | 27 | 28 | | | | | | |
| 2 | 16 | 290 | 116 | 170 | 28 | 36 | 37 | 57 | 20 | 21 |
| after surgery | | | 68 | 83 | | | | | | |
| 3 | 15 | 120 | 55 | 86 | 26 | 63 | 26 | 62 | 21 | 22 |
| after surgery | | | 50 | 120 | | | | | | |
| 4 | 19 | 92 | 73 | 80 | 28 | 70 | 25 | 83 | 37 | 60 |
| 5 | 48 | 86 | 125 | 220 | 54 | 83 | 25 | 78 | 18 | 30 |
| 6 | 65 | 77 | 89 | 92 | 50 | 93 | 36 | 46 | 52 | 60 |
| 7 | 22 | 99 | 239 | 284 | 1740 | 2280 | | | | |
| 8 | 42 | 45 | 53 | 56 | 314 | 374 | | | | |
| 9 | 16 | 55 | 106 | 106 | 1100 | 960 | | | | |
| 10 | 16 | 114 | 23 | 23 | | | | | | |
| 11 | 22 | 252 | 52 | 63 | | | | | | |

| | | | | | |
|---|---|---|---|---|---|
| 12 | 20 | | | | |
| 13 | 27 | | | | |
| 14 | 60 | | | | |
| 15 | 45 | | | | |
| 16 | 200 | | | | |
| 17 | 48 | 306 | 520 | 760 | |
| 18 | 16 | 109 | 181 | 218 | |
| 19 | 125 | 112 | 46 | 86 | |
| 20 | 21 | 54 | 76 | 103 | |
| 21 | 19 | 200 | 76 | 86 | |
| 22 | 34 | 375 | 137 | 143 | |
| 23 | 16 | 113 | 38 | 36 | |
| 24 | 35 | 264 | 59 | 75 | |
| 25 | 140 | 156 | 151 | 159 | |
| 26 | 17 | 71 | 217 | 253 | |
| 27 | 22 | 220 | 80 | 101 | |
| 28 | 52 | 310 | 350 | 401 | |
| 29 | 87 | 229 | 110 | 114 | |
| 30 | 61 | 127 | 176 | 135 | |
| 31 | 96 | 26 | 307 | 437 | |
| | | 24 | | | |
| | | 150 | | | |
| | | 137 | | | |
| | | 287 | | | |
| | | 107 | | | |

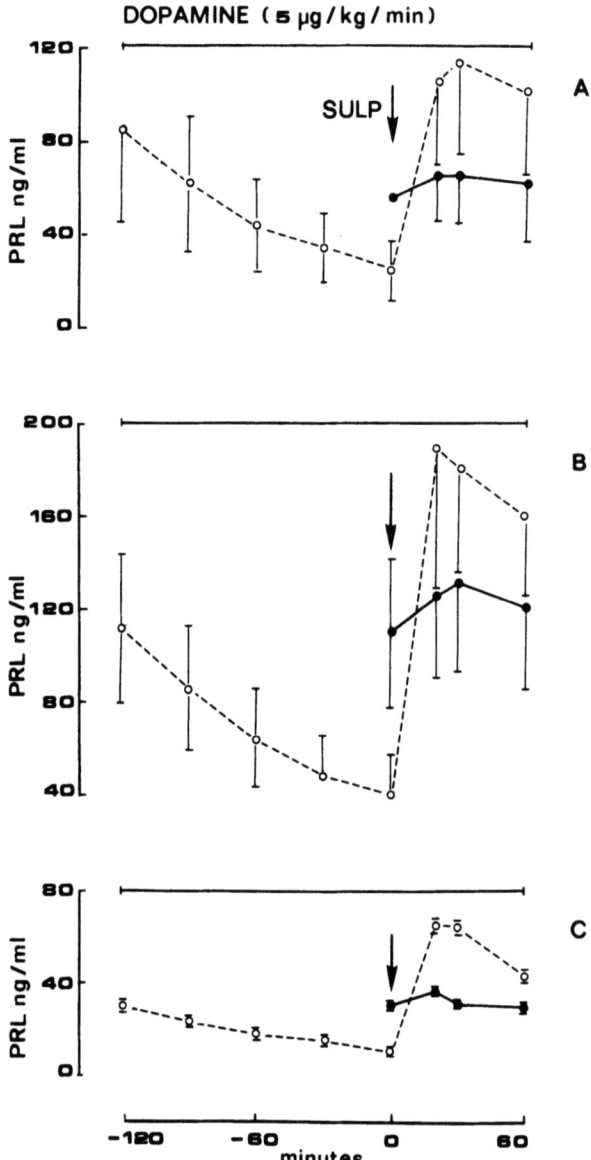

FIGURE 1. Serum PRL response to sulpiride (100 mg i.m.) without (solid lines) and with (broken lines) concurrent DA infusion in hyperprolactinemic patients unresponsive to sulpiride alone. (A) Idiopathic hyperprolactinemia ($N = 5$); (B) pituitary microadenoma ($N = 10$); (C) hypothalamic hyperprolactinemia ($N = 3$).

TABLE 3
Serum PRL Response to Domperidone (2–8 mg i.v.) in Patients with
Hyperprolactinemia of Different Etiology. Shown are Basal and Peak PRL Levels
(ng/ml)

| Patient No. | Idiopathic hyperPRL | | Microprolactinoma | | Macroprolactinoma | |
|---|---|---|---|---|---|---|
| | Basal | Peak | Basal | Peak | Basal | Peak |
| 1 | 24 | 100 | 202 | 220 | 95 | 110 |
| 2 | 25 | 104 | 89 | 110 | 875 | 1110 |
| 3 | | | 39 | 100 | | |
| 4 | | | 44 | 72 | | |
| 5 | | | 225 | 220 | | |
| 6 | | | 90 | 140 | | |
| 7 | | | 64 | 100 | | |
| 8 | | | 62 | 60 | | |
| 9 | | | 145 | 170 | | |
| 10 | | | 171 | 180 | | |

and domperidone. Nevertheless, it is noteworthy that at least a doubling of basal values occurred in the majority of subjects with idiopathic hyperprolactinemia and only rarely in patients with pituitary adenoma. No one of six patients with hyperprolactinemia due to hypothalamic disease showed a PRL response to sulpiride. However, DA infusion resulted in lowering of PRL levels and restoration of normal responsivity to both sulpiride (Fig. 1) and domperidone (Fig. 2) in patients unresponsive to DA antagonists given alone. These findings confirm and extend our previous data (Crosignani *et al.*, 1977; Ferrari *et al.*, 1979) and show that in most hyperprolactinemic patients there is a lack of DA at the pituitary level, a defect which may be corrected by administering

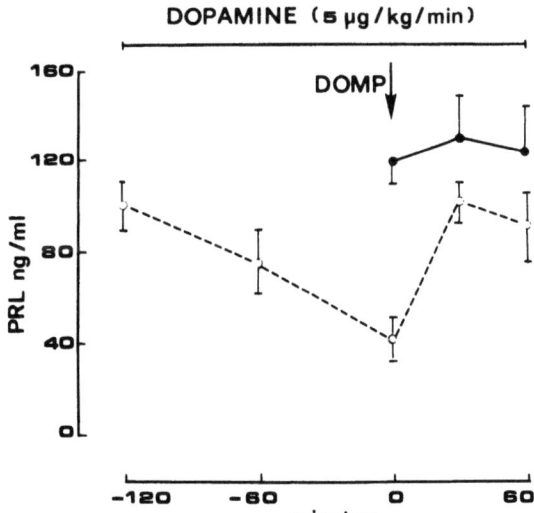

FIGURE 2. Serum PRL response to domperidone (2–8 mg i.v.) without (solid lines) and with (broken lines) concurrent DA infusion in 10 patients with pituitary microadenoma unresponsive to domperidone alone.

exogenous DA. The similarity of the responses obtained with the two DA antagonists supports the idea that the PRL-releasing activity of sulpiride is directly exerted on the lactotrophs, since domperidone does not cross the blood–brain barrier.

### 3.2. Defective Pituitary DA Receptors in Hyperprolactinemic States

As shown in Tables 4–6, only a minority of hyperprolactinemic patients both with and without evidence of pituitary adenoma did not adequately suppress PRL levels after acute administration of DA agonists such as DA itself, L-dopa, and bromocriptine, which directly act on DA receptors at the pituitary level. These findings imply the existence of normally functioning DA receptors on pituitary lactotrophs in the great majority of subjects. This is also true in patients chronically treated with bromocriptine (Fig. 3). Interestingly, most of the subjects resistant to the PRL-lowering action of the standard bromocriptine doses (5 mg/day) suppressed their PRL levels by increasing drug doses up to 20 mg/day (Table 7), implying that the DA receptor defect, when present, may be only partial.

### 3.3. CNS Dopaminergic Defect in Hyperprolactinemic States

In addition to the above-discussed DA defects, recently introduced neuropharmacologic tests which specifically activate DA neurotransmission in the CNS without direct action on the pituitary showed the existence of defective CNS DA function in most hyperprolactinemic patients either with or without

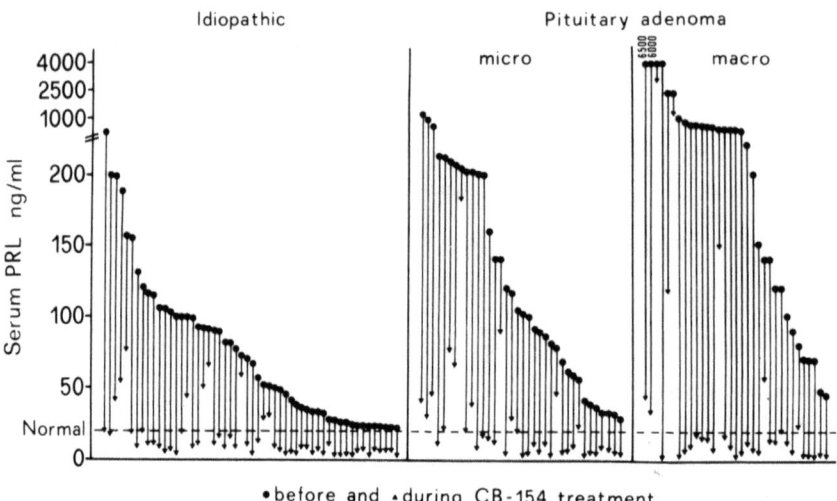

FIGURE 3. Effect of bromocriptine treatment on serum PRL levels in hyperprolactinemic patients. The levels under treatment refer to the mean of the values obtained during the last month of therapy.

TABLE 4

Serum PRL Response to Dopamine Infusion (5 μg/kg·min for 120 min) in Patients with Hyperprolactinemia of Different Etiology. Shown are Basal and Nadir PRL Levels (ng/ml)

| Patient No. | Idiopathic hyperPRL | | Microprolactinoma | | Macroprolactinoma | | Acromegaly | | Hypothalamic hyperPRL | |
|---|---|---|---|---|---|---|---|---|---|---|
| | Basal | Nadir | Basal | Nadir | Basal | Nadir | Basal | Nadir | Basal | Nadir |
| 1 | 71 | 25 | 65 | 49 | 1000 | 600 | 24 | 8 | 22 | 11 |
| 2 | 68 | 43 | 195 | 70 | 660 | 240 | 40 | 17 | 45 | 21 |
| 3 | 50 | 15 | 88 | 33 | 400 | 160 | 39 | 17 | 30 | 6 |
| 4 | 148 | 88 | 100 | 65 | 92 | 80 | | | | |
| 5 | 85 | 50 | 87 | 81 | 480 | 188 | | | | |
| 6 | 73 | 16 | 61 | 43 | 62 | 11 | | | | |
| 7 | 148 | 28 | 79 | 70 | 1630 | 1234 | | | | |
| 8 | 20 | 6 | 92 | 27 | 1525 | 650 | | | | |
| 9 | 28 | 6 | 63 | 24 | 39 | 11 | | | | |
| 10 | 57 | 16 | 130 | 23 | | | | | | |
| 11 | 77 | 31 | 56 | 16 | | | | | | |
| 12 | 33 | 15 | 200 | 70 | | | | | | |
| 13 | 240 | 64 | 98 | 36 | | | | | | |
| 14 | 28 | 6 | 200 | 96 | | | | | | |
| 15 | 42 | 8 | 40 | 10 | | | | | | |
| 16 | | | 25 | 9 | | | | | | |
| 17 | | | 120 | 28 | | | | | | |
| 18 | | | 195 | 96 | | | | | | |
| 19 | | | 91 | 18 | | | | | | |
| 20 | | | 215 | 70 | | | | | | |
| 21 | | | 90 | 26 | | | | | | |
| 22 | | | 74 | 10 | | | | | | |
| 23 | | | 32 | 14 | | | | | | |
| 24 | | | 264 | 39 | | | | | | |
| 25 | | | 77 | 17 | | | | | | |
| 26 | | | 225 | 56 | | | | | | |
| 27 | | | 32 | 10 | | | | | | |
| 28 | | | 145 | 48 | | | | | | |
| 29 | | | 146 | 28 | | | | | | |
| 30 | | | 20 | 8 | | | | | | |

TABLE 5
Serum PRL Response to L-Dopa (500 mg p.o.) in Patients with Hyperprolactinemia of Different Etiology. Shown are Basal and Nadir PRL Levels (ng/ml)

| Patient No. | Idiopathic hyperPRL | | Microprolactinoma | | Macroprolactinoma | | Acromegaly | | Hypothalamic hyperPRL | |
|---|---|---|---|---|---|---|---|---|---|---|
| | Basal | Nadir | Basal | Nadir | Basal | Nadir | Basal | Nadir | Basal | Nadir |
| 1 | 46 | 21 | 140 | 80 | 400 | 80 | 26 | 6 | 23 | 7 |
| 2 | 25 | 4 | 41 | 15 | 80 | 64 | 24 | 9 | 49 | 10 |
| 3 | 99 | 35 | 101 | 26 | 39 | 8 | 16 | 7 | 35 | 5 |
| 4 | 18 | 5 | 805 | 81 | 480 | 300 | 37 | 11 | 17 | 4 |
| 5 | 38 | 14 | 32 | 9 | 520 | 180 | | | | |
| 6 | 19 | 4 | 60 | 28 | 82 | 11 | | | | |
| 7 | 71 | 27 | 78 | 20 | 39 | 9 | | | | |
| 8 | 90 | 24 | 39 | 17 | | | | | | |
| 9 | 46 | 6 | 80 | 24 | | | | | | |
| 10 | 101 | 44 | 37 | 9 | | | | | | |
| 11 | 30 | 8 | 115 | 22 | | | | | | |
| 12 | 240 | 44 | 24 | 12 | | | | | | |
| 13 | 42 | 12 | 260 | 50 | | | | | | |
| 14 | 20 | 10 | | | | | | | | |
| 15 | 105 | 30 | | | | | | | | |
| 16 | 280 | 170 | | | | | | | | |
| 17 | 50 | 32 | | | | | | | | |
| 18 | 50 | 16 | | | | | | | | |
| 19 | 60 | 22 | | | | | | | | |
| 20 | 192 | 79 | | | | | | | | |
| 21 | 98 | 45 | | | | | | | | |

## TABLE 6
Serum PRL Response to Bromocriptine (2.5 mg p.o.) in Patients with Hyperprolactinemia of Different Etiology. Shown are Basal and Nadir PRL Levels (ng/ml)

| Patient No. | Idiopathic hyperPRL | | Microprolactinoma | | Macroprolactinoma | | Acromegaly | |
|---|---|---|---|---|---|---|---|---|
| | Basal | Nadir | Basal | Nadir | Basal | Nadir | Basal | Nadir |
| 1 | 97 | 39 | 115 | 34 | 41 | 6 | 42 | 18 |
| after surgery | | | 47 | 8 | | | | |
| 2 | 32 | 14 | 200 | 160 | 173 | 66 | 27 | 6 |
| after surgery | | | 61 | 6 | | | | |
| 3 | 24 | 4 | 85 | 30 | 1000 | 1000 | 29 | 6 |
| 4 | 26 | 3 | 104 | 90 | 2078 | 1312 | | |
| 5 | 32 | 22 | 59 | 18 | 82 | 30 | | |
| 6 | 34 | 10 | 90 | 52 | | | | |
| 7 | 43 | 10 | 245 | 63 | | | | |
| 8 | 54 | 25 | 219 | 69 | | | | |
| 9 | 65 | 32 | 95 | 36 | | | | |
| 10 | 23 | 3 | 243 | 36 | | | | |
| 11 | 34 | 12 | 56 | 12 | | | | |
| 12 | 50 | 32 | 22 | 1 | | | | |
| 13 | 20 | 3 | 45 | 14 | | | | |
| 14 | 19 | 5 | 79 | 7 | | | | |
| 15 | 53 | 8 | 110 | 36 | | | | |
| 16 | 28 | 14 | 150 | 14 | | | | |
| 17 | 140 | 91 | 300 | 140 | | | | |
| 18 | 96 | 26 | 297 | 252 | | | | |
| 19 | | | 222 | 60 | | | | |
| 20 | | | 754 | 85 | | | | |
| 21 | | | 45 | 7 | | | | |

radiological evidence of pituitary tumor (Fine and Frohman, 1978; Müller et al., 1978; Ferrari et al., 1980a; Crosignani et al., 1980a; Frohman et al., 1981).

Our current experience with carbidopa plus L-dopa and with nomifensine testing is summarized in Tables 8 and 9, and shows that in fact most hyperprolactinemic patients do not show adequate PRL suppression with both these neuropharmacological agents, irrespective of the underlying etiology. These findings suggest that the majority of patients with idiopathic hyperprolactinemia may harbor a microprolactinoma in preradiological stage.

In 41 hyperprolactinemic subjects in whom both carbidopa plus L-dopa and nomifensine tests have been performed, a positive correlation between the PRL suppression obtained with the two agents was found ($r = 0.46, p < 0.0025$), despite the fact that a dissociation between the two responses occurred in some patients. These data confirm that the two tests may give results generally similar, but not always superimposable. The occurrence of dissociated responses in individual cases (mostly PRL inhibition with carbidopa plus L-dopa but not with nomifensine) may be due to the fact that the former treatment results in increased hypothalamic concentration of DA (Porter, 1971) which may in turn

### TABLE 7
PRL Suppression by Increasing Doses of Bromocriptine in Hyperprolactinemic Patients Poorly Responsive to Chronic Therapy with 5 mg/day

| Patient No. | Diagnosis | Serum PRL (ng/ml) | | | | | |
|---|---|---|---|---|---|---|---|
| | | Basal | 5 mg | 7.5 mg | 10 mg | 15 mg | 20 mg |
| 1 | Idiopathic | 120 | 60 | — | 46 | — | — |
| 2 | Idiopathic | 92 | 57 | 100 | 68 | 95 | — |
| 3 | Idiopathic | 72 | 35 | 52 | — | 60 | — |
| 4 | Idiopathic | 160 | 54 | 38 | — | — | — |
| 5 | Micro | 61 | 50 | 6 | — | — | — |
| 6 | Micro | 158 | 60 | 6 | 2 | — | — |
| 7 | Micro | 220 | 65 | 50 | 32 | — | — |
| 8 | Micro | 101 | 31 | 41 | 35 | 31 | 20 |
| 9 | Micro | 638 | 90 | 31 | — | — | — |
| 10 | Micro | 126 | 39 | 10 | 2 | — | — |
| 11 | Micro | 89 | 58 | 24 | — | — | — |
| 12 | Micro | 60 | 34 | 16 | — | — | — |
| 13 | Macro | 300 | 190 | 144 | 76 | — | — |
| 14 | Macro | 6700 | 350 | 90 | 35 | — | — |
| 15 | Macro | 400 | 300 | 220 | 12 | — | — |
| 16 | Macro | 300 | 60 | 24 | 17 | 12 | — |
| 17 | Macro | 6000 | 590 | 130 | 27 | — | — |
| 18 | Macro | 500 | 220 | 160 | 50 | 20 | — |
| 19 | Macro | 1000 | 180 | 16 | 1 | — | — |

reach the pituitary gland via the portal circulation in sufficient amounts to inhibit PRL release. Not surprisingly, there was no correlation between the PRL inhibition by either nomifensine or carbidopa plus L-dopa and the PRL stimulation by sulpiride or domperidone, confirming that these two kinds of tests evaluate the integrity of different dopaminergic mechanisms involved in the inhibition of PRL release.

## 4. CONCLUSIONS

As discussed in the previous sections, the different neuropharmacological tests employed in this investigation probe the integrity of three different dopaminergic mechanisms inhibiting PRL secretion, and provide evidence for the existence of defective dopaminergic regulation at three different levels in hyperprolactinemic states. Individual patients may show one, two, or all of these defects, involving pituitary DA concentration, pituitary DA receptors, and CNS component of dopaminergic inhibition of PRL release; no one of them may be apparent in a few cases. The first and the third of these defects are very common in hyperprolactinemia, while the second is relatively uncommon. Although the existence of defective dopaminergic inhibition of PRL secretion has no value

## TABLE 8
Serum PRL Response to L-Dopa (100 mg p.o.) Plus Carbidopa (35 mg p.o.) after Pretreatment with Carbidopa (50 mg Every 6 hr for 4 Doses) in Patients with Hyperprolactinemia of Different Etiology. Shown are Basal and Nadir PRL Levels (ng/ml)

| Patient No. | Idiopathic hyperPRL | | Microprolactinoma | | Macroprolactinoma | | Acromegaly | | Hypothalamic hyperPRL | |
|---|---|---|---|---|---|---|---|---|---|---|
| | Basal | Nadir | Basal | Nadir | Basal | Nadir | Basal | Nadir | Basal | Nadir |
| 1 | 108 | 85 | 160 | 100 | 340 | 200 | 34 | 28 | 60 | 30 |
| 2 | 15 | 10 | 45 | 25 | 78 | 78 | 29 | 10 | 48 | 18 |
| 3 | 100 | 113 | 107 | 45 | 430 | 380 | 24 | 11 | | |
| 4 | 27 | 8 | 500 | 400 | 570 | 375 | 32 | 12 | | |
| 5 | 50 | 24 | 285 | 21 | 110 | 49 | | | | |
| 6 | 24 | 5 | 62 | 60 | 35 | 19 | | | | |
| 7 | 72 | 58 | 51 | 18 | 24 | 10 | | | | |
| 8 | 100 | 56 | 48 | 25 | | | | | | |
| 9 | 50 | 9 | 60 | 34 | | | | | | |
| 10 | 90 | 88 | 35 | 17 | | | | | | |
| 11 | 28 | 12 | 140 | 36 | | | | | | |
| 12 | 240 | 144 | 20 | 12 | | | | | | |
| 13 | 42 | 21 | 290 | 200 | | | | | | |
| 14 | 28 | 18 | | | | | | | | |
| 15 | 130 | 100 | | | | | | | | |
| 16 | 380 | 200 | | | | | | | | |
| 17 | 52 | 34 | | | | | | | | |
| 18 | 60 | 44 | | | | | | | | |
| 19 | 60 | 44 | | | | | | | | |
| 20 | 176 | 150 | | | | | | | | |

## TABLE 9
Serum PRL Response to Nomifensine (200 mg p.o.) in Patients with Hyperprolactinemia of Different Etiology. Shown are Basal and Nadir PRL Levels (ng/ml)

| Patient No. | Idiopathic hyperPRL | | Microprolactinoma | | Macroprolactinoma | | Acromegaly | | Hypothalamic hyperPRL | |
|---|---|---|---|---|---|---|---|---|---|---|
| | Basal | Nadir | Basal | Nadir | Basal | Nadir | Basal | Nadir | Basal | Nadir |
| 1 | 68 | 58 | 130 | 130 | 25 | 26 | 105 | 50 | 38 | 36 |
| 2 | 69 | 76 | 58 | 30 | 920 | 410 | 19 | 18 | 72 | 40 |
| 3 | 35 | 15 | 45 | 36 | 370 | 308 | 58 | 40 | 40 | 26 |
| 4 | 36 | 31 | 115 | 97 | 69 | 70 | | | 27 | 26 |
| 5 | 23 | 8 | 64 | 52 | 29 | 20 | | | | |
| 6 | 78 | 64 | 45 | 40 | 440 | 280 | | | | |
| 7 | 96 | 92 | 62 | 50 | 560 | 500 | | | | |
| 8 | 46 | 26 | 68 | 50 | 58 | 49 | | | | |
| 9 | 118 | 114 | 41 | 28 | 43 | 50 | | | | |
| 10 | 36 | 20 | 111 | 62 | 600 | 380 | | | | |
| 11 | 44 | 40 | 23 | 10 | | | | | | |
| 12 | 29 | 20 | 93 | 76 | | | | | | |
| 13 | 100 | 94 | 101 | 64 | | | | | | |
| 14 | 56 | 38 | 305 | 240 | | | | | | |
| 15 | 53 | 34 | 34 | 15 | | | | | | |
| 16 | 56 | 37 | 32 | 34 | | | | | | |
| 17 | 27 | 13 | 60 | 60 | | | | | | |
| 18 | 21 | 17 | 434 | 332 | | | | | | |
| 19 | 109 | 90 | | | | | | | | |
| 20 | 42 | 39 | | | | | | | | |
| 21 | 56 | 20 | | | | | | | | |
| 22 | 20 | 8 | | | | | | | | |
| 23 | 26 | 28 | | | | | | | | |
| 24 | 135 | 100 | | | | | | | | |
| 25 | 80 | 20 | | | | | | | | |
| 26 | 120 | 96 | | | | | | | | |
| 27 | 185 | 177 | | | | | | | | |
| 28 | 250 | 240 | | | | | | | | |
| 29 | 77 | 74 | | | | | | | | |

### TABLE 10
Percent Incidence of Three Different Defects in Dopaminergic Inhibition of PRL Secretion in Hyperprolactinemic States of Various Etiology as Assessed by Neuropharmacological Testing

| Dopaminergic defect | Idiopathic hyperPRL | Micro-prolactinoma | Macro-prolactinoma | Acromegaly | Hypothalamic hyperPRL |
|---|---|---|---|---|---|
| Pituitary DA concentration[a] | 33 (33)[e] | 94 (36) | 82 (11) | 50 (6) | 100 (6) |
| Pituitary DA receptors[b] | 17 (47) | 19 (53) | 37 (16) | 0 (7) | 0 (7) |
| Pituitary DA receptors[c] | 9 (54) | 8 (36) | 9 (33) | — | — |
| CNS DA inhibition[d] | 74 (31) | 84 (19) | 80 (10) | 50 (4) | 75 (4) |

[a] As assessed by impaired PRL stimulation by sulpiride or domperidone.
[b] As assessed by impaired PRL inhibition by DA, L-dopa, or bromocriptine acute testing.
[c] As assessed by failure of PRL levels to decrease below 50% of basal during chronic bromocriptine therapy.
[d] As assessed by impaired PRL inhibition by carbidopa plus L-dopa or nomifensine.
[e] Number of tested subjects is shown in parentheses.

in the differential diagnosis of hyperprolactinemic states (Crosignani et al., 1980a,b; Ferrari et al., 1980b; Reschini et al., 1980), the incidence of the three recognized dopaminergic defects may be different according to the etiology of hyperprolactinemia, as shown in Table 10. Thus, evidence of defective pituitary DA concentration is almost uniformly found in prolactinomas and in patients with hypothalamic lesions, but is relatively uncommon in idiopathic disease; defective CNS dopaminergic inhibition of PRL secretion is very common in all the categories of hyperprolactinemic patients, while evidence of defective pituitary DA receptors is only uncommonly found. It would appear that the only difference between adenomatous and idiopathic disease lies in the frequency of impaired DA concentration at the pituitary level. The different prevalence of pituitary DA and CNS defects in the same groups of patients clearly indicates the different nature of the two disturbances.

Finally, the present data indicating the high incidence of pituitary DA defect in the presence of functioning DA receptors at lactotrophs confirm that DA agonist therapy is the rational approach to the treatment of hyperprolactinemia.

ACKNOWLEDGMENTS. The collaboration of Dr. A. Liuzzi and Dr. P. G. Chiodini in the study of chronic bromocriptine treatment is gratefully acknowledged. The studies here reported were partially supported by C.N.R. (Rome) Special Projects Biology of Reproduction and Control of Neoplastic Growth.

## 5. REFERENCES

Borzio, M., Caldara, R., Ferrari, C., Barbieri, B., Borzio, F., and Romussi, M., 1981, Growth hormone and prolactin secretion in liver cirrhosis: Evidence for dopaminergic dysfunction, *Acta Endocrinol.* **97**:441.
Crosignani, P. G., Reschini, E., Peracchi, M., Lombroso, G. C., Mattei, A., and Caccamo, A.,

1977, Failure of dopamine infusion to suppress the plasma prolactin response to sulpiride in normal and hyperprolactinemic subjects, *J. Clin. Endocrinol. Metab.* **45**:841.

Crosignani, P. G., Ferrari, C., Malinverni, A., Barbieri, C., Mattei, A., Caldara, R., and Rocchetti, M., 1980a, Effect of central nervous system dopaminergic activation on prolactin secretion in man: Evidence for a common central defect in hyperprolactinemic patients with and without radiological signs of pituitary tumors, *J. Clin. Endocrinol. Metab.* **51**:1068.

Crosignani, P. G., Ferrari, C., Mattei, A., Malinverni, A. G., and Picciotti, M. C., 1980b, Functional evaluation of hyperprolactinemic states, in: *Central and Peripheral Regulation of Prolactin Function* (R. M. MacLeod and U. Scapagnini, eds.), pp. 287–291, Raven Press, New York.

Elli, R., Ballabio, M., Scaperotta, R. C., Rondena, M., Travaglini, P., and Faglia, G., 1980, On dopaminergic control of LH secretion in hyperprolactinemic states, in: *Endocrinology of Human Infertility: New Aspects*, Oxford, 1980, Abstract Book, p. 95.

Evans, W. S., Rogol, A. D., MacLeod, R. M., and Thorner, M. O., 1980, Dopaminergic mechanisms and luteinizing hormone secretion. I. Acute administration of the dopamine agonist bromocriptine does not inhibit luteinizing hormone release in hyperprolactinemic women, *J. Clin. Endocrinol. Metab.* **50**:103.

Ferrari, C., Travaglini, P., Caldara, R., Moriondo, P., Mattei, A., Crosignani, P. G., and Faglia, G., 1979, Restoration of the prolactin response to sulpiride by metergoline administration in hyperprolactinemic patients, *Neuroendocrinology* **29**:338.

Ferrari, C., Crosignani, P. G., Caldara, R., Picciotti, M. C., Malinverni, A. G., Barattini, G., Rampini, P., and Telloli, P., 1980a, Failure of nomifensine administration to discriminate between tumorous and nontumorous hyperprolactinemia, *J. Clin. Endocrinol. Metab.* **50**:23.

Ferrari, C., Travaglini, P., Mattei, A., Caldara, R., Moriondo, P., Romussi, M., and Crosignani, P. G., 1980b, Inhibition of prolactin secretion by acute and chronic metergoline treatment in hyperprolactinemic patients with pituitary microadenoma or other disorders, in: *Pituitary Microadenomas* (G. Faglia, M. A. Giovanelli, and R. M. MacLeod, eds.), pp. 399–406, Academic Press, New York.

Ferrari, C., Mattei, A., Benco, R., Barattini, G., Caldara, R., Vergadoro, F., Rampini, P., Reschini, E., and Crosignani, P. G., 1981a, Medical treatment of hyperprolactinemic states, in: *Endocrinology of Human Infertility: New Aspects* (P. G. Crosignani and B. Rubin, eds.), pp. 139–159, Academic Press, New York.

Ferrari, C., Rampini, P., Malinverni, A. G., Scarduelli, C., Benco, R., Caldara, R., Barbieri, C., Testori, G. P., and Crosignani, P. G., 1981b, Inhibition of luteinizing hormone release by dopamine infusion in healthy women and in various pathophysiological conditions, *Acta Endocrinol.* **97**:436.

Fine, S. A., and Frohman, L. A., 1978, Loss of central nervous system component of dopaminergic inhibition of prolactin secretion in patients with prolactin-secreting pituitary tumors, *J. Clin. Invest.* **61**:973.

Frohman, L. A., Berelowitz, M., Gonzales, C., Barowsky, H., Rao, R., Lim, V. S., Frohman, M. A., and Lenz Thominet, J., 1981, Studies of dopaminergic mechanisms in hyperprolactinemic states, in: *Endocrinology of Human Infertility: New Aspects* (P. G. Crosignani and B. Rubin, eds.), pp. 39–52, Academic Press, New York.

Healy, D. L., Pepperell, R. J., Stockdale, J., Bremner, W. J., and Burger, H. G., 1977, Pituitary autonomy in hyperprolactinemic secondary amenorrhea: Results of hypothalamic–pituitary testing, *J. Clin. Endocrinol. Metab.* **44**:809.

Hökfelt, T., and Fuxe, K., 1972, Effects of prolactin and ergot alkaloids on the tubero-infundibular dopamine (DA) neurons, *Neuroendocrinology* **9**:100.

Kleinberg, D. L., Noel, G. L., and Frantz, A. G., 1977, Galactorrhea: A study of 235 cases, including 48 with pituitary tumors, *N. Engl. J. Med.* **296**:589.

Lachelin, G. C. L., Leblanc, H., and Yen, S. S. C., 1977, The inhibitory effect of dopamine agonists on LH release in women, *J. Clin. Endocrinol. Metab.* **44**:728.

Lim, V. S., Kathpalia, S. C., and Frohman, L. A., 1979, Hyperprolactinemia and impaired pituitary response to suppression and stimulation in chronic renal failure: Reversal after transplantation, *J. Clin. Endocrinol. Metab.* **48**:101.

Moriondo, P., Travaglini, P., Nissim, M., and Faglia, G., 1980, Evaluation of two inhibitory tests (nomifensine and L-dopa + carbidopa) for the diagnosis of hyperprolactinemic states, *Clin. Endocrinol.* **13**:525.

Müller, E. E., Genazzani, A. R., and Murru, S., 1978, Nomifensine: Diagnostic test in hyperprolactinemic states, *J. Clin. Endocrinol. Metab.* **47**:1352.

Porter, C. C., 1971, Aromatic amino acid decarboxylase inhibitors, *Fed. Proc.* **30**:871.

Quigley, M. E., and Yen, S. S. C., 1980, Evidence for increased dopaminergic inhibition of secretion of thyroid-stimulating hormone in hyperprolactinemic patients with pituitary microadenoma, *Am. J. Obstet. Gynecol.* **137**:653.

Quigley, M. E., Judd, S. J., Gilliland, G. B., and Yen, S. S. C., 1979, Effects of a dopamine antagonist on the release of gonadotropin and prolactin in normal women and women with hyperprolactinemic anovulation, *J. Clin. Endocrinol. Metab.* **48**:718.

Reschini, E., Ferrari, C., Peracchi, M., Fadini, R., Meschia, M., and Crosignani, P. G., 1980, Effect of dopamine infusion on serum prolactin concentration in normal and hyperprolactinaemic subjects, *Clin. Endocrinol.* **13**:519.

Scanlon, M. F., Rodriguez-Arnao, M. D., McGregor, A. M., Weightman, D., Lewis, M., Cook, D. B., Gomez-Pan, A., and Hall, R., 1981, Altered dopaminergic regulation of thyrotrophin release in patients with prolactinomas: Comparison with other tests of hypothalamic–pituitary function, *Clin. Endocrinol.* **14**:133.

Weiner, R. I., and Ganong, W. F., 1978, Role of brain monoamines and histamine in regulation of anterior pituitary secretion, *Physiol. Rev.* **58**:905.

# Index

Acetylcholine
  and dexamethasone suppression test, 408
  and enkephalins, 309
  and TRH, 73–77
  and vasoactive intestinal peptide, 518–519
ACTH
  and adrenergic pathway, 57
  and anorexia nervosa, 131–136
  and behavior, 33–35
  in cerebrospinal fluid, 31–32
  and cholinergic pathway, 57
  and dopaminergic pathway, 57
  inhibiting agents of, 58–59
  in primary affective disorders, 320–323
  in schizophrenia, 317, 323
  secretion, regulation of, 22, 56–57
    stimulating agents of, 57–58
  and sexual behavior in mammals, 145
  and vasopressin, 266, 272–273
$ACTH_{4-10}$
  in attention and arousal, 244–246
  and consolidation of memory, 246–248
  in learning, 232
  and Pavlovian conditioning, 232–241
  and retrieval of memory, 248–250
Affective disorders
  and circadian rhythm of hormones, 466
  dexamethasone suppression test in, 384–388
  methylphenidate stimulation test, 394–395
  TRH stimulation test in, 389–392, 413–425
  urinary MHPG in, 392–394
D-Amphetamine, hormonal alterations due to
  in depression, 361, 368
  in schizophrenia, 549–563
Androgens, in impotency, 149–150

Angiotensin II, and ACTH, 57
Anorexia nervosa
  ACTH secretion in, 131–136
  and affective disorders, 132
  calcitonin serum levels in, 93
  catecholamines in, 168
  clomiphene effect of, 116, 122
  and depression, 358
  dexamethasone suppression test in, 129–134
  endocrine dysfunction in, 118, 358
  and estrogens, 111, 115–116
  etiology of, 129
  gonadal dysfunction in, 111–124
  gonadotropins in, 112–122
  hypothalamic–pituitary–ovarian axis in, 111–118, 360
  neurotransmitter alterations in, 118, 119
  norepinephrine in, 130–132, 136–137
  temperature regulation, 118
  thyroid functions in, 134
Antidiuretic hormone, and GH secretion, 57
Apomorphine
  and dopamine receptors, 144
  and GH secretion, 485–491, 606, 607

Bombesin, 35–36

Calcitonin
  and analgesia, 84
  in anorexia nervosa, 93
  binding sites, 84
  blood–brain barrier permeability to, 103
  and body weight in man, 92–93
  in brain, 84–85
  and conditioned aversion, 99–100

Calcitonin (cont.)
　diuretic effect of, 93–95
　and eating, 91–93, 101–102
　and feeding behavior in rats, 87, 101–102
　and food and water intake, 87, 89–91, 93, 97–98
　and gastric acid secretion, 85
　mode of action, 102
　physiology of, 83, 101
　secretion, and gastrointestinal peptides, 85–86, 101
Calcium, 102
　entry blockers, central effects of, 9
Catecholamines
　assay, methods of, 158–161
　in depression, 340, 347
　plasma levels of, 161–168
　　in anorexia nervosa, 168
　　in essential hypertension, 168
　　in psychosomatic disorders, 168
Circadian rhythm, 465–470
　of cortisol, 130–131
　　in depression, 467
　of gonadotropins
　　in anorexia nervosa, 112–114
　of melatonin, 467–470
　of norepinephrine, 136
　of prolactin, 467–470
　of TSH, in depression, 467–470
Corticotropin-releasing factor
　and acetylcholine, 131
　and ACTH secretion, 22, 56–57
　anterior pituitary, action on, 175
　and dexamethasone suppression test, 130, 137
　and GABA, 131
　and norepinephrine, 131
　and serotonin, 131
Cortisol
　and D-amphetamine, 136
　in anorexia nervosa, 131, 136–137
　circadian rhythm in depression, 474–482
　in depression, 344, 357–376, 445, 453–458
　and lithium, 350
　and schizophrenia, 560, 563

Depression
　α-adrenergic sensitivity in, 361
　amphetamine-induced hormonal alterations in, 381, 386
　animal model for, 332–333, 341
　　brain monoamines changes in, 336–341
　　imipramine, effect on, 335–336

Depression (cont.)
　biogenic amines in, 332
　cholinergic dysfunction in, 363
　circadian rhythm of hormones in, 474–482
　cortisol levels in subtypes of, 357–376
　and Cushing's disease, 360–361, 376
　dexamethasone suppression test in, 358–360, 362
　and dopamine, 386
　endocrine dysfunction in subtypes of, 357–358
　endogenous
　　chronobiology of, 364–366
　methylphenidate, effect of, 386
　MHPG levels in, 348–363, 387, 391
　and norepinephrine, 386
　platelet 5-HT uptake, 473–482
　　and tricyclic antidepressants, 481–482
　psychoneuroendocrine aspects of, 343–344
　and sex hormones, 436
　stress hypothesis of, 361–364
　subtypes of, 346–348
　TRH, therapeutic use in, 433–435
　tricyclic antidepressants in therapy of, 435, 437–438
　TSH response to TRH in, 431–433, 436
　urinary free cortisol in, 364–366, 376
Dexamethasone suppression test
　and acetylcholine, 408
　and affective disorders, 384–388
　and anorexia nervosa, 130–131, 136
　and barbiturates, 402
　and carbamazepine, 402
　in children, 407
　and cholinergic activity, 374–375
　clinical use of, 406–407
　in depression, 358, 362, 376, 405
　and diphenylhydantoin, 402
　in geriatric population, 407
　interpretation of, 401
　method of, 401
　nonsuppression in various psychiatric disorders, 371–374
　rationale of, 400
Dopamine
　agonists of
　　and GH release, 210, 486, 495, 506, 584
　　and prolactin secretion, 211
　　in schizophrenia, 504
　　and TSH secretion, 52
　antagonists of
　　and GH release, 486–495, 506
　　in schizophrenia, 504

Dopamine (cont.)
  antagonists of (cont.)
    and TSH secretion, 53
    and cholecystokinins, 518
    and depression, 336–341, 457
    and FSH, 54
    and gonadotropins, 54
    and lithium, 352
    and luteinizing hormone, 54–55
    and prolactin secretion, 42–43, 349, 613–624
    and schizophrenia, 503–505, 511–513
    and sexual behavior in male rats, 141, 144
Dopamine receptor
  and apomorphine, 485–488, 584, 589
  and clonidine-induced GH secretion, 494–495
  and β-endorphin, 310
  and GH secretion, 485–495
  and hyperprolactinemia, 620–624, 627
  neuroleptic therapy, effect of, 605
  in schizophrenia, 511–513

β-endorphin
  and affective disorders, 294, 310, 314–315
  and analgesia, 33
  assay of, 316–323
  behavioral effects of, 309–310
  CSF levels in schizophrenia, 298
  and dopamine receptors, 310
  functions of, 293–294
  and GH secretion, 48
  and neuroleptics, 299–300
  as neurotransmitter, 309
  and schizophrenia, 297–298, 316–323, 518, 560, 563
des-Tyr-γ-endorphin, in schizophrenia, 300–301, 312
Endorphins, in schizophrenia, 536–537
Epinephrine, plasma levels of, 162–165
Estrogens, in anorexia nervosa, 111

FSH
  in affective disorders, 181–184
  in anorexia nervosa, 122
  and dopamine, 54
  secretion of, 54
  and TRH, 180, 184–186
FSH-releasing factor, 54, 173–175

GABA
  and GH release, 211–212, 215–216
  gonadotropins, effect on, 54
  and prolactin release, 211–212

Glucagon, 47
Gonadotropin-releasing hormone, 41, 53–55
  and dopamine, 54
  and prolactin, 54
Gonadotropins
  acetylcholine, effect of, 54
  in anorexia nervosa, 112–114
  calcitonin, effect of, 55
  and dopamine, 54–55
  and glucocorticoids, 54–55
  and histamine, 54
  hypothalamic control of, 53–55
  and α-MSH, 55
  and prolactin, 54
  and prostaglandins, 55
  and serotonin, 54
  in simple weight loss, 120
  and stress, 120–121
Growth hormone (GH)
  acromegaly, 49, 192–195
  in affective disorders, 181–184
  and anorexia nervosa, 118, 184–186
  and Cushing's syndrome, 360
  in depression, 345–346, 360, 362, 446, 458
  and dopamine agonists, 9, 47, 486, 495, 506, 584
  and dopamine antagonists, 9, 47, 486–495, 506
  and 5-HT, 187, 188
  in Huntington's disease, 210–211, 213, 224
  and hypothyroidism, 190–191
  and lithium, 350
  and neuroleptic therapy, 605–607
  plasma levels of, 9–10
  release of, 9, 46–48, 505
    and somatostatin, 173
    and tardive dyskinesia, 608–609
  and TRH, 173–195
Growth hormone-releasing factor, 46, 173–174
  in Huntington's disease, 223
Growth hormone-release-inhibiting factor, 41, 175

Huntington's disease
  dopaminergic neural activity in, 210–211
  GH release in, 210–211, 213, 224
    dopamine agonist induced, 210
    GABA induced, 215–216
  LH release in, 223–224
  neuroendocrine studies in, 209–226
  prolactin secretion in, 211, 219–222, 224
    dopamine agonist induced, 211
5-Hydroxytryptamine (5-HT)
  antagonists of, in impotency, 143, 148

5-Hydroxytryptamine (cont.)
  in brain
    depletion of, 142
    drug-induced increase of, 143–144
  circadian rhythm of, 348–352
  receptors of
    blockade of, by metergoline, 187–188
  and sexual behavior in male rats, 141–143
Hypothalamic–pituitary–adrenal axis
  and affective disorders, 384–388
  in anorexia nervosa, 130–131, 136
  in depression, 358, 362, 376
  and dexamethasone suppression test, 6
Hypothalamic–pituitary–thyroid axis,
  and affective disorders, 389, 415
  and depression, 431–433
    and therapy of, 433–435
  and neurotensin, 36
  and TRH stimulation test, 6, 422
Hypothalamus
  and ACTH secretion, 56–57
  and anorexia nervosa, 111–124
  and anterior pituitary, 18–19
  calcitonin, effect of, 84–85, 102–103
  and feeding behavior, 35
  and gastric secretion, 85, 103
  and gonadotropin secretion, 53–55
  and growth hormone secretion, 46–47
  and MSH secretion, 58
  regulatory hormones of, 173–197
  in simple weight loss, 119
  and TSH secretion, 49–51
  and vasopressin, 255–257

Leu-enkephalin
  in learning and memory, 280–289
  and psychotropic drugs, 314
β-lipotropin
  behavioral effects of, 518
  secretion of, 22
Lithium, neuroendocrine effects of, 349–352
Luteinizing hormone (LH)
  and affective disorders, 180
  and anorexia nervosa, 112
  and anticonvulsants, 55
  and chlorpromazine, 54
  in depression, 345, 347, 446–447, 455–456, 459
  and dopamine, 54–55
  in FSH-secreting tumors, 196
  5-HT antagonists, effect of, 54
  in Huntington's disease, 223–224
  naloxone, effect of, 56

Luteinizing hormone (LH) (cont.)
  and neuroleptic therapy, 54
  and TRH, 184–186
Luteinizing hormone-releasing hormone, 173–174, 177
  in acromegaly, 192, 195
  in anorexia nervosa, 7, 116, 118, 185
  and cortisol secretion, 195
  in depression, 347
  FSH response to, 6–7
  GH release, effect of, 191–195
    in acromegaly, 192, 195
    in Klinefelter's syndrome, 191
    in schizophrenia, 186
  in impotency, 149
  and LH release, 6–7
  and lithium, 350
  and PRL secretion, in anorexia nervosa, 185
  and sexual behavior in male rats, 141, 146

Median eminence, 20, 44
  and calcitonin, 85
  and gonadotropin secretion, 54
Melanocyte-stimulating hormone (MSH)
  and GH secretion, 47
  regulation of, 59, 174
  releasing factor, 59, 174, 177
  sexual behavior, effect of, 145
Melatonin, 54, 349
  circadian rhythm of, in depression, 467–470
Met-enkephalin, 44, 48
  in learning and memory, 280–289
  and psychotropic drugs, 314
Met-enkephalinamide, 145–146
MK-777, 74, 76, 77
Morphine, in mental disorders, 310

Naloxone
  and affective disorders, 295, 314
  in impotency, 149
  and learning and memory, 280–289, 293
  in schizophrenia, 297–298, 310–311
  in shock states, 301–302
Neuroleptics
  bioavailability of, 579
  extrapyramidal effects of, 570
  and GH secretion, 584–594
    apomorphine induced, 606–607
    clonidine induced, 605–606
  and LH secretion, 603
  potency and dose response, 571
  and prolactin secretion, 8, 569–580, 583, 585–589

Neuroleptics (cont.)
  serum levels of, 574–575
  tolerance to, 573
  and TSH levels, 602–603
Neurosecretion, 15–17
Neurotensin
  in feeding behavior, 35
  and temperature regulation, 36
Norepinephrine
  in depression, 336, 341
  and LH secretion, 54
  and long-term neuroleptic therapy, 603–604
  plasma levels of, 161–165
  and tardive dyskinesia, 609

Opiate peptides
  and affective disorders, 294–296
  and heroin addiction, 301
  in hippocampus, 285
  in learning and memory, 280–284
  receptors, 5, 285
  and reward mediation, 286
  in senile dementia, 287
  and shock state, 301
Oxytocin, 15, 23
  and MSH, 59
  synthesis, of, 17, 255, 257

Parathyroid hormone, and food intake, 101
Peptides
  and behavior, 33–35
  in CNS, 30–37
  in CSF, 31
  CCK, 23, 35–36
    in schizophrenia, 517–519, 538–541
  electrophysiological effects of, 29–30
  endorphin, 518
  enkephalins, 518
  exogenous, 519–536
    access to brain, 519–521
  extrahypothalamic, 2
  gastrointestinal, 23
  MSH-like, 5, 517–518
  and neurons, 27–28, 30
  receptors of, 28–29
  somatostatin, 23
  substance P, 5, 518
  and temperature regulation, 36
  vasoactive intestinal peptide, 518
Pineal gland, and gonadotropin release, 54
Prolactin
  and alytensin, 43
  in anorexia nervosa, 118

Prolactin (cont.)
  and calcitonin, 85
  circadian rhythm of, 467–470
  in cirrhosis of liver, 187
  in depression, 346, 362, 445, 457
  and dopamine, 42–43, 349, 613–624
  GABA, effect of, 43
  hyperprolactinemia
    CNS dopaminergic defects in, 613–627
    LH release in, 624
    and pituitary dopamine, 615–620
    TRH effect of, 614
  hypothalamic regulation of, 42–43, 46
  and lithium, 350
  and neuroleptics, 569–580, 600–602
  and opioid peptides, 21
  releasing factor, 43, 174, 178
  in renal disease, 187
  in schizophrenia, 549–563
  secretion of, 20–21
  and serotonin, 43, 349
  serum levels of
    and neuroleptic therapy, 575–580
  and sexual behavior, 145
  somatostatin, effect of, 173, 179
  and tardive dyskinesia, 609
  and TRH, 21
  and vasoactive intestinal peptide, 43

Schizophrenia
  and acetylcholine, 518–519
  and D-amphetamine
    hormonal alterations due to, 549–564
  and CCK-like peptides, 518, 538–541
  and celiac disease, 521–523, 541
  and cereal grain proteins, 521–536
  cortisol response, 560
  dopamine receptors in, 511–512
  dopaminergic pathways in, 583
  and endorphins, 518, 536–537, 560
  GH release in, 505, 507, 560, 584–595
  neurohormonal changes in, 186
  neuroleptic therapy, 583–594
  and vasoactive intestinal peptide, 518–519
Serotonin, in depression, 336–341; see also 5-hydroxytryptamine
Somatostatin
  in acromegaly, 193
  and ACTH secretion, 59, 195
  and cyclic AMP, 195
  effects of, 4
  and extrapituitary hormones, 47
  functions of, 5

Somatostatin (cont.)
  gastrointestinal hormones, 173
  and GH secretion, 46–47, 173
  pituitary hormones, effect on, 47
  prolactin release effect on, 173, 179
Substance P, 33

Tardive dyskinesia
  and apomorphine-induced hormonal alterations, 608–609
  and GH secretion, 608–609
  norepinephrine levels in, 609
  pathogenesis of, 608
  prolactin levels in, 609
Thyroid-stimulating hormone
  and affective disorder, 389–391
  and anorexia nervosa, 118
  and calcitonin, 52
  circadian rhythm of, 467–470
  and Cushing's syndrome, 360
  and depression, 345–346, 360, 431–433, 437, 446, 467–470
  and dopaminergic drugs, 51, 53
  and ergot derivatives, 52
  and estrogens, 51
  and naloxone, 52
  and neuroleptics, 602–603
  secretion of, 49–53
Thyrotropin-releasing hormone (TRH)
  and acromegaly, 192–195
  ACTH-secreting tumors, effect on, 196
  and affective disorders, 181–184
  and Alzheimer's disease, 78
  and anorexia nervosa, 184–186
  and antimuscarinic agents, 77
  CNS effects of, 3
  and cholinergic transmission, 73–78
  and cyclic AMP, 50

Thyrotropin-releasing hormone (TRH) (cont.)
  and depression, 73, 360, 431–433, 436–37
  in diabetes, 189
  and dopamine, 188
  extrapituitary effects of, 50
  feeding behavior, effect on, 35
  FSH, effect on, 181–186
  gastrointestinal effects of, 76–77
  and GH secretion, 47, 50, 173–195
  and gonadotropin, 50
  and prolactin secretion, 43, 50, 184
  and reticular activating system, 75
  and septohippocampal system, 75
  and temperature regulation, 36–37
  test in depression, 4, 389–391
  and TSH secretion, 6, 49, 51
Tuberoinfundibular DA pathway
  and blood–brain barrier, 8
  and depression, 336–340, 457
  and GH secretion, 583
  and prolactin secretion, 42–43, 570, 583, 592

Vasopressin
  and ACTH secretion, 266, 272–273
  and behavior, 33, 231–257
  blood–brain barrier, 274
  CNS effects of, 258–259
  DDAVP in amnesiac patients, 260–261
  and depression, 346
  electrophysiological studies, 258–261
  and GH secretion, 49
  mechanism of action, 264–275
  memory and learning, 258, 260
  and opioid peptides, 272
  release of, 256
  synthesis of, 17, 255–257

MIX
Papier aus verantwortungsvollen Quellen
Paper from responsible sources
FSC® C105338

If you have any concerns about our products,
you can contact us on
**ProductSafety@springernature.com**

In case Publisher is established outside the EU,
the EU authorized representative is:
**Springer Nature Customer Service Center GmbH
Europaplatz 3, 69115 Heidelberg, Germany**

Printed by Libri Plureos GmbH
in Hamburg, Germany